Studies in Surface Science and Catalysis 29

METAL CLUSTERS IN CATALYSIS

Studies in Surface Science and Catalysis
Advisory Editors: B. Delmon and J.T. Yates

Vol. 29

METAL CLUSTERS IN CATALYSIS

Editors

B.C. Gates
Center for Catalytic Science and Technology, Department of Chemical Engineering, University of Delaware, Newark, DE 19716, U.S.A.

L. Guczi
Institute of Isotopes, Hungarian Academy of Sciences, P.O. Box 77, H-1525 Budapest, Hungary

and

H. Knözinger
Institute of Physical Chemistry, University of Munich, Sophienstrasse 11, 8000 Munich 2, F.R.G.

ELSEVIER Amsterdam — Oxford — New York — Tokyo 1986

ELSEVIER SCIENCE PUBLISHERS B.V.
Sara Burgerhartstraat 25
P.O. Box 211, 1000 AE Amsterdam, The Netherlands

Distributors for the United States and Canada:

ELSEVIER SCIENCE PUBLISHING COMPANY INC.
52, Vanderbilt Avenue
New York, NY 10017, U.S.A.

Library of Congress Cataloging-in-Publication Data

Metal clusters in catalysis.

(Studies in surface science and catalysis ; vol. 29)
Includes bibliographies and index.
1. Metal crystals. 2. Catalysis. 3. Catalysts.
I. Gates, Bruce C. II. Guczi, L., 1932- .
III. Knözinger, H., 1935- . IV. Series: Studies in
surface science and catalysis ; 29.
QD921.M467 1986 546'.3 86-19890
ISBN 0-444-42708-2

ISBN 0-444-42708-2 (Vol. 29)
ISBN 0-444-41801-6 (Series)

Printed in The Netherlands

CONTENTS

STUDIES IN SURFACE SCIENCE AND CATALYSIS

Volume 1 **Preparation of Catalysts I.** Scientific Bases for the Preparation of Heterogeneous Catalysts. Proceedings of the First International Symposium held at the Solvay Research Centre, Brussels, October 14–17, 1975
edited by **B. Delmon, P.A. Jacobs and G. Poncelet**

Volume 2 **The Control of the Reactivity of Solids.** A Critical Survey of the Factors that Influence the Reactivity of Solids, with Special Emphasis on the Control of the Chemical Processes in Relation to Practical Applications
by **V.V. Boldyrev, M. Bulens and B. Delmon**

Volume 3 **Preparation of Catalysts II.** Scientific Bases for the Preparation of Heterogeneous Catalysts. Proceedings of the Second International Symposium, Louvain-la-Neuve, September 4–7, 1978
edited by **B. Delmon, P. Grange, P. Jacobs and G. Poncelet**

Volume 4 **Growth and Properties of Metal Clusters.** Applications to Catalysis and the Photographic Process. Proceedings of the 32nd International Meeting of the Société de Chimie Physique, Villeurbanne, September 24–28, 1979
edited by **J. Bourdon**

Volume 5 **Catalysis by Zeolites.** Proceedings of an International Symposium organized by the Institut de Recherches sur la Catalyse – CNRS – Villeurbanne and sponsored by the Centre National de la Recherche Scientifique, Ecully (Lyon), September 9–11, 1980
edited by **B. Imelik, C. Naccache, Y. Ben Taarit, J.C. Vedrine, G. Coudurier and H. Praliaud**

Volume 6 **Catalyst Deactivation.** Proceedings of the International Symposium, Antwerp, October 13–15, 1980
edited by **B. Delmon and G.F. Froment**

Volume 7 **New Horizons in Catalysis.** Proceedings of the 7th International Congress on Catalysis, Tokyo, June 30–July 4, 1980. Parts A and B
edited by **T. Seiyama and K. Tanabe**

Volume 8 **Catalysis by Supported Complexes**
by **Yu.I. Yermakov, B.N. Kuznetsov and V.A. Zakharov**

Volume 9 **Physics of Solid Surfaces.** Proceedings of the Symposium held in Bechyňe, September 29–October 3, 1980
edited by **M. Láznička**

Volume 10 **Adsorption at the Gas–Solid and Liquid–Solid Interface.** Proceedings of an International Symposium held in Aix-en-Provence, September 21–23, 1981
edited by **J. Rouquerol and K.S.W. Sing**

Volume 11 **Metal-Support and Metal-Additive Effects in Catalysis.** Proceedings of an International Symposium organized by the Institut de Recherches sur la Catalyse – CNRS – Villeurbanne and sponsored by the Centre National de la Recherche Scientifique, Ecully (Lyon), September 14–16, 1982
edited by **B. Imelik, C. Naccache, G. Coudurier, H. Praliaud, P. Meriaudeau, P. Gallezot, G.A. Martin and J.C. Vedrine**

Volume 12 **Metal Microstructures in Zeolites.** Preparation – Properties – Applications. Proceedings of a Workshop, Bremen, September 22–24, 1982
edited by **P.A. Jacobs, N.I. Jaeger, P. Jirů and G. Schulz-Ekloff**

Volume 13 **Adsorption on Metal Surfaces.** An Integrated Approach
edited by **J. Bénard**

Volume 14 **Vibrations at Surfaces.** Proceedings of the Third International Conference, Asilomar, CA, September 1–4, 1982
edited by **C.R. Brundle and H. Morawitz**

In memory of

Yu.I. Yermakov

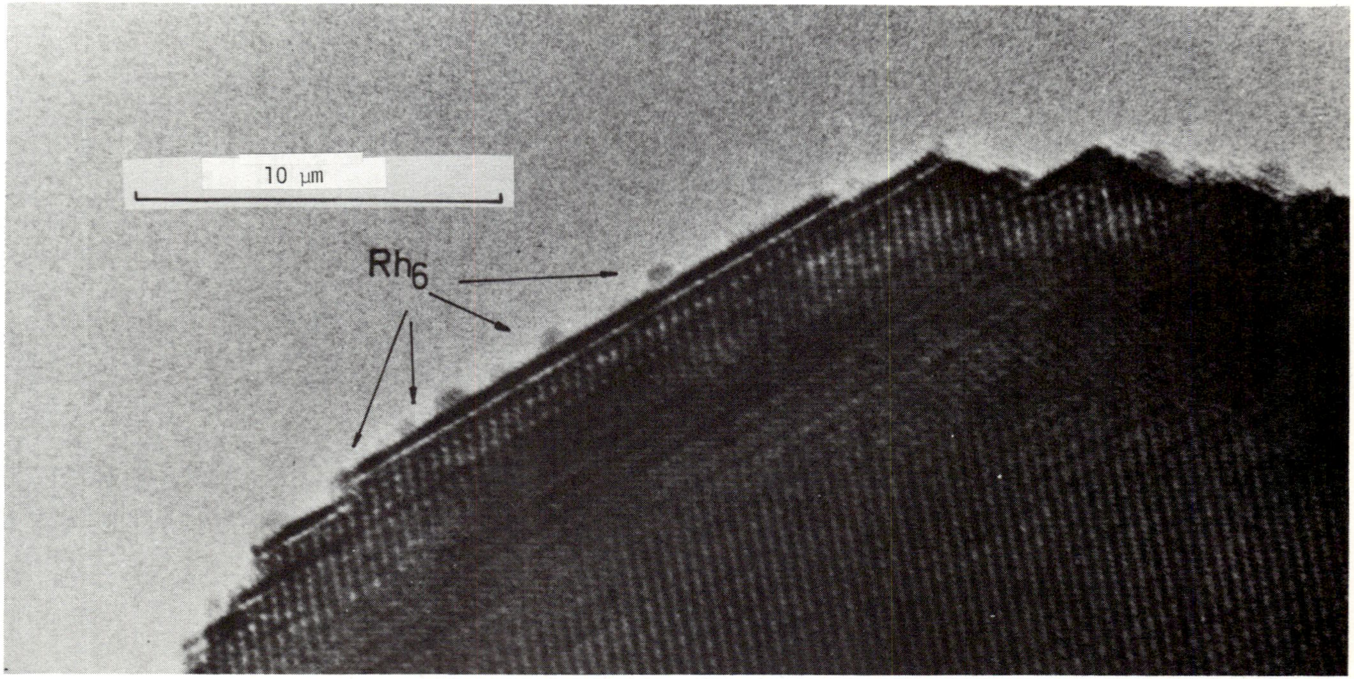

High-resolution electron micrograph showing Rh_6 clusters on the (111) surface of γ-Al_2O_3 (courtesy of Professor M. Ichikawa and Dr. S. Iijima).

PREFACE

Metal cluster chemistry has entered a period of exponential growth in the preceding decade, and much of the driving force, at least nominally, is the prospect of application of metal clusters as catalysts. The goal of this monograph is a thorough and critical evaluation of the literature and prospects of metal clusters in catalysis.

Part I of the book is concerned with molecular chemistry and structures of metal clusters: synthesis, crystal and molecular structure, thermochemistry and bond energies, reactivity, and homogeneous catalysis.

Part II, the largest section of the book, is concerned with metal clusters in and on solid supports, ranging from polymers and oxides functionalized with groups analogous to the ligands in molecular metal clusters to metal oxides such as silica, alumina, and magnesia, and to zeolites, with their molecular-scale cages. The emphasis is on supported metal species with structures analogous to those of molecular species. We restrict the term "cluster" to these "molecular" species; to avoid confusion, the terms "aggregate", "crystallite", and "particle" are used to denote the typical, usually nonuniform, supported metals not having molecular structures (although these are commonly referred to as clusters in the catalysis literature).

Even though they are not the primary focus of the book, the metal aggregates are too interesting and too closely related to the clusters to ignore. The metal aggregates are important subjects of several chapters, including Chapter 8, concerned with metals in zeolites, and parts of Chapter 9 (9.4 and 9.5), concerned with metal-oxide-supported catalysts derived from molecular metal clusters.

Characterization of supported metal clusters is developing rapidly, made possible by the emergence and rapid development of instrumental techniques for surface spectroscopy. A detailed introduction (Chapter 6) summarizing classical and modern methods is included; it is our intent that this section be broadly applicable and of value to researchers characterizing a range of supported structures in addition to metal clusters.

Part III is a single chapter (11) addressing the relations between metal clusters and metal surfaces. Compelling new ideas have arisen that, perhaps no less than the prospects of catalysts made from metal clusters, are driving organometallic chemistry and surface science into common territory.

The subject of metal clusters in catalysis is just emerging from its infancy; the growth is dramatic and exciting, and we believe that understanding of the clusters will propel the advance of fundamental understanding of catalysis, organometallic chemistry, and surface science. These ideas are elaborated in a short summary chapter concluding the book.

We thank the authors for their excellent efforts and their willingness to mold their contributions to our conception of the book. We are grateful to Jeanne Grill, Dale Meyer, and Cecilia Viering, who entered the manuscripts into the word processor at the University of Delaware, and to Venerando Sciuto of the University of Munich, who prepared many of the drawings in final form. We gratefully acknowledge the financial support for our continuing research in this field, provided by the Deutsche Forschungsgemeinschaft (Knözinger), the Hungarian Academy of Science (Guczi), and the National Science Foundation (Gates). Much of the editorial work was done at the University of Munich; B. C. Gates was supported there by grants from the Deutsche Forschungsgemeinschaft, the Fulbright Commission in Bonn, and the National Science Foundation.

B. C. Gates
L. Guczi
H. Knözinger

LIST OF CONTRIBUTORS

M. P. ANDREWS	Bell Laboratories 600 Mountain Avenue Murray Hill, New Jersey, 07974 USA
J. A. CONNOR	University Chemical Laboratory University of Kent at Canterbury Canterbury, Kent, CT2 7NH UK
G. ERTL	Institut für Physikalische Chemie Universität München Sophienstrasse 11 8000 München 2 West Germany
B. C. GATES	Center for Catalytic Science and Technology Department of Chemical Engineering University of Delaware Newark, Delaware 19716 USA
G. L. GEOFFROY	Department of Chemistry Pennsylvania State University University College, Pennsylvania 16802 USA
L. GUCZI	Institute of Isotopes Hungarian Academy of Sciences P.O. Box 77 1525 - Budapest Hungary

P. A. JACOBS

Laboratorium voor Oppervlaktescheikunde
Katholieke Universiteit Leuven
Kardinal Mercierlaan 92
3030 - Leuven
Belgium

H. D. KAESZ

Department of Chemistry
University of California
Los Angeles, California 90024
USA

H. KNÖZINGER

Institut für Physikalische Chemie
Universität München
Sophienstrasse 11
8000 München 2
West Germany

E. LAVIGNE

Department of Chemistry
University of California
Los Angeles, California 90024
USA

G. MAIRE

Institut de Chimie
Université Louis Pasteur de Strasbourg
4, rue Blaise Pascal
67008 Strasbourg Cedex
France

L. MARKÓ

Institute of Organic Chemistry
University of Veszprem
and Research Group for Petrochemistry
Hungarian Academy of Sciences
Schönherz Z. u. 8
8200 Veszprem
Hungary

G. A. OZIN

Department of Chemistry
Lash Miller Chemical Laboratories
University of Toronto
80 St. George Street
Toronto, Ontario M5S 1A1
Canada

R. F. PETTIFER

EMBL Hamburg Outstation
c/o DESY
Notkestrasse 85
2000 Hamburg
West Germany

and Department of Physics
University of Warwick
Coventry, CV4 7AL
UK

R. PSARO

Dipartimento di Chimica Inorganica
e Metallorganica
Universitá degli Studi
Via C. Venezian, 21
20133 Milano
Italy

R. UGO

Dipartimento di Chimica Inorganica
e Metallorganica
Universitá degli Studi
Via C. Venezian, 21
20133 Milano
Italy

A. VIZI-OROSZ

Institute of Organic Chemistry
University of Veszprem
and Research Group of Petrochemistry
Hungarian Academy of Sciences
Schönherz Z. u. 8
8200 Veszprem
Hungary

PART I

MOLECULAR METAL CLUSTERS

Chapter 1

SYNTHESIS OF MONO- AND MULTIMETALLIC METAL CLUSTERS

G. L. GEOFFROY

1.1 INTRODUCTION

Hundreds of di- and polynuclear complexes have been prepared in which transition metal atoms are bonded to each other [1-10]. Synthesis of these complexes, particularly the cluster complexes, has often been considered to be something of a black art. Many of the preparations have been accomplished by placing together mono- or polynuclear transition metal reagents and then allowing them to react under thermal or photochemical conditions, but mixtures of products generally result. However, as a result of a large recent effort to prepare new metal-metal bonded compounds, some general procedures have emerged as rational synthetic methods. Other approaches, while not obvious from a mechanistic standpoint, are at least predictable as to the types of products formed. This chapter briefly summarizes the existing synthetic methodology, beginning with a discussion of synthetic strategies and following with a discussion of the different reaction classes. Examples of reactions are given mainly for mixed-metal compounds, since they better illustrate the chemistry; similar reactions are appropriate for the synthesis of homometallic compounds. Many of the syntheses are presented schematically, with less than complete statements of reaction stoichiometry. Side products are often not identified.

1.2 SYNTHETIC STRATEGIES

The most useful strategy for directing the synthesis of a polymetallic complex is to use monomeric organometallic complexes as building blocks by adding one metal at a time, as illustrated in SCHEME 1 [11, 12]. Thus, to prepare a tetranuclear cluster, one would begin by reacting two mononuclear complexes to form a dinuclear species. This complex could in turn be reacted with another mononuclear complex to yield a closed metal trimer, and the reaction of this trimer with another mononuclear complex could produce the desired tetranuclear species. In many cases it is not necessary to begin with mononuclear precursors if suitable di- or trinuclear complexes are available. With control of reaction conditions and reagents, it should also prove possible to prepare oligomeric compounds via this strategy, as shown in SCHEME 1. Although such an approach seems straightforward, difficulties such as the tendency of oligomeric compounds

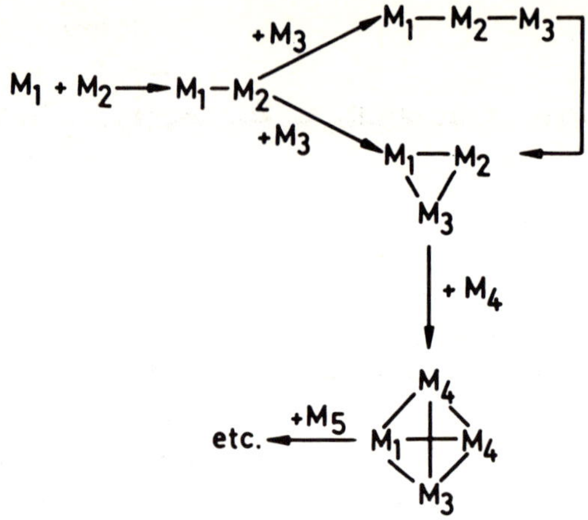

SCHEME 1

to close upon themselves and yield clusters rather than to form extended metal chains must be controlled.

A less attractive synthetic strategy involves the combination of two di- or polynuclear compounds to form a new compound, the metal composition of which is the sum of the compositions of the two original compounds. For example, one could prepare a tetranuclear cluster by reacting two dinuclear complexes as outlined in SCHEME 2.

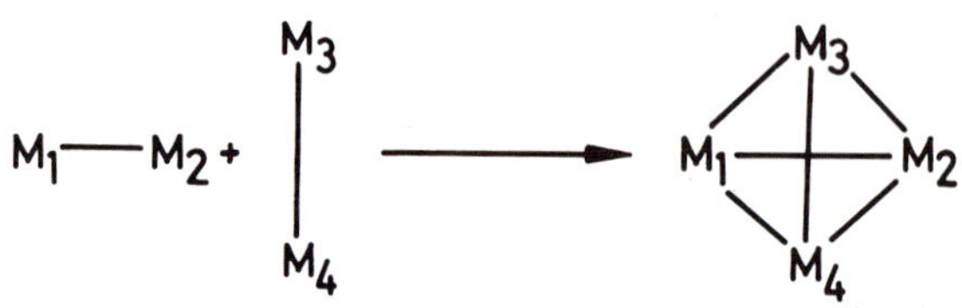

SCHEME 2

This approach has been of limited utility in designed syntheses because of the poor control of nuclearity of the products. The reaction typically requires ligand loss for metal-metal bond formation to occur, and in many polymetallic reagents the metal-ligand bonds are stronger than the metal-metal bonds. Thus a metallic fragment would more likely be eliminated. This problem can be minimized by use of multiply metal-metal bonded or otherwise coordinatively unsaturated compounds which can condense without the necessity of ligand loss. A successful example of such a reaction is shown in Eq. (1) [13].

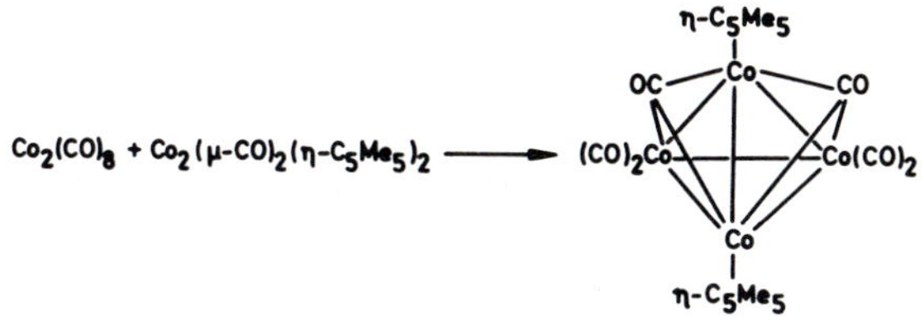

(1)

1.2.1 Synthetic methods

The methods that appear to be most useful for directing the synthesis of metal-metal bonded compounds are discussed in detail in the following paragraphs. These methods have been organized from a phenomenological viewpoint in that the reaction classes are based upon the relationships of the starting materials to the products, minimizing mechanistic considerations.

A large and growing number of synthetic reactions proceed to form complexes which contain bridging ligands such as $-PPh_2$, $-CR_2$, $-CR$, and -SMe. The bridging ligand often plays a key role in these syntheses, and when it is clearly involved, the reaction is referred to as *bridge-assisted*.

These syntheses all involve the other types of reactions described below, but because the bridging ligand plays such an important role in facilitating the synthesis, these reactions have been separately categorized.

1.2.2 Ligand substitution reactions

The replacement of one or more ligands on a transition metal center by another metal fragment is categorized as a ligand substitution reaction. The substitution of an anionic ligand, generally a halide, by an anionic metal complex,

$$[ML_nX] + [M'L_m]^{s-} \qquad X^- + [MM'L_{n+m}]^{(s-1)^-} \qquad (2)$$

has proved to be extremely useful for the synthesis of metal-metal bonded complexes and has provided examples of nearly every dinuclear metal combination possible for groups VIA through IB. Selected examples of the use of this method for the preparation of dinuclear, polynuclear, and oligomeric compounds, respectively, are shown in Eqs. (3) [14], (4) [15], and (5) [16].

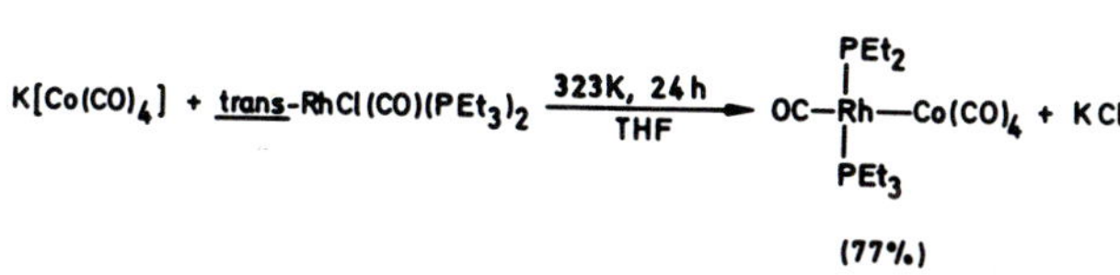

(3)

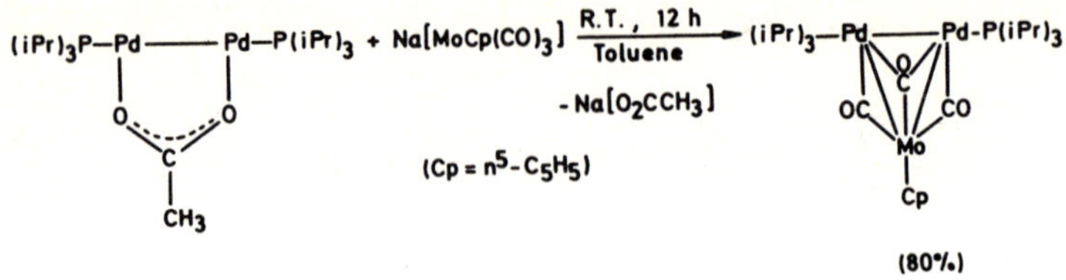

(4)

$$trans\text{-}PtCl_2(py)_2 + 2[Co(CO)_3PPh_3]^- \xrightarrow[THF]{\Delta,\,3h} (PPh_3)(CO)_3Co\text{-}Pt(py)_2\text{-}Co(CO)_3(PPh_3) \qquad (5)$$

$py = C_5H_5N$ $\qquad -2Cl^-$

An anionic metal complex can also displace a neutral ligand:

$$[ML_n] + ML_m{}^{x-} \longrightarrow L + [MM'L_{n+m-1}]^{x-} \qquad (6)$$

In most instances the neutral ligand displaced is carbon monoxide, and the usual reaction is between a carbonylmetallate and a neutral metal carbonyl complex. This type of reaction has often been referred to as a "redox condensation" [17], although a variety of other types of reactions have also been collected under that general heading. Examples of the successful application of this reaction type are shown in Eqs. (7) [18], (8) [19], and (9) [20].

$$[Mn(CO)_5]^- + Cr(CO)_6 \xrightarrow[(CH_3OCH_2CH_2)_2O]{443K,\ 2h} [(CO)_5Cr\text{-}Mn(CO)_5]^- + CO \qquad (7)$$

(71%)

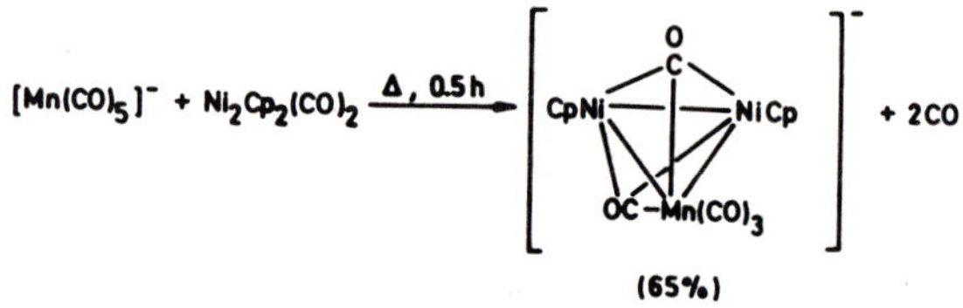

(8)

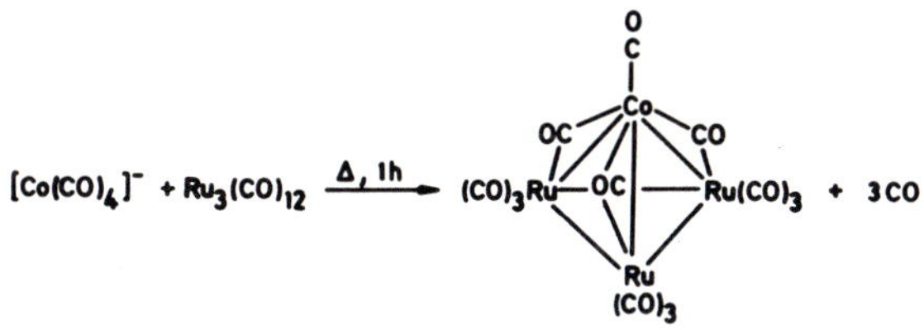

(9)

Although reactions of this type are most useful for the preparation of dinuclear complexes, the substitution of a metal carbonyl anion for CO ligands on a neutral trinuclear cluster as in Eq. (9) has proved to be a convenient synthetic method for the directed synthesis of many tetranuclear clusters [21-24]. Neutral hydride-containing complexes can often be obtained following such reactions by protonation of the anionic products; for example, addition of H_3PO_4 to the product of Eq. (9) gives $HCoRu_3(CO)_{13}$ in high yield [20].

There are some examples of substitution reactions which do not involve a transition metal anion, as illustrated in Eqs. (10) [25] and (11) [26].

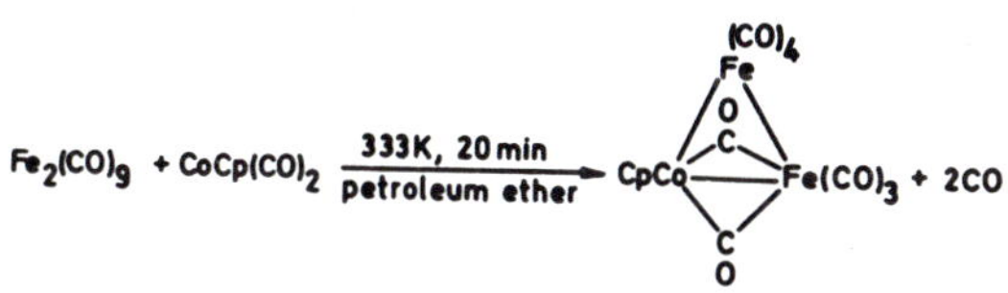

(10)

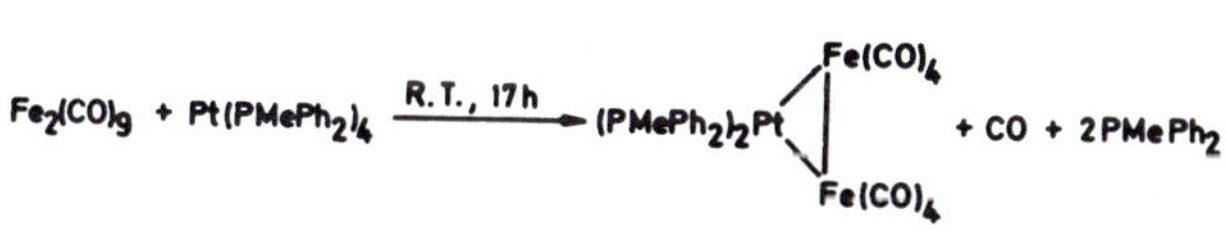

(11)

However, such reactions are generally not as predictable with regard to the nature of the products formed as are reactions which employ anionic metal complexes.

The most serious limitation of this synthetic approach is the limited availability of anionic metal complexes other than carbonylmetallates. The latter are ideally suited for such synthesis, but they of course lead to products containing predominantly carbonyl ligands. These products generally show low reactivity toward extensive ligand substitution or functionalization. Thus the ligand set available in a product from this type of reaction is severely limited. As to predictability of products, the displacement of an anionic ligand, especially a halide, by an anionic metal complex is generally better than the displacement of a neutral ligand. In the latter case, more than one neutral ligand can often be lost and multiple substitution reactions can occur to give a complex product, as illustrated below [27]:

$$[M_2(CO)_{10}]^{2-} + 3Ni(CO)_4 \xrightarrow[THF]{1.5\ h} [M_2Ni_3(CO)_{16}]^{2-} + 6CO \qquad (12)$$

(M = Cr, Mo, W) (30-70%)

1.2.3 Addition reactions

A synthetic approach closely related to that described above is the addition of one metal complex to another by a process *which does not involve ligand loss from one of the metal reagents.* Such reactions are classified as *addition reactions.* Although substitution and addition reactions may bear mechanistic similarities, they do differ in the relationship of the reactants to products.

The simplest type of addition reactions is one involving formation of a Lewis acid-base, donor-acceptor complex. This type of adduct is formed when an electron-rich metal center is added to an electron-deficient metal, and the localization of electrons is more or less retained in the final product. Such complexes must thus possess a polar metal-metal bond. An example is given below:

$$[Fe(mnt)_2]^- + [Mn(CO)_5]^- \xrightarrow[THF]{} [(mnt)_2Fe{\cdot}Mn(CO)_5]^{2-} \qquad (13)$$

(mnt = $C_2S_2(CN)_2{}^{2-}$)

Both $[Mn(CO)_5]^-$ and $[Fe(mnt)_2]^{2-}$ are stable compounds in solution, but when placed together they associate to yield the adduct complex shown in Eq. (13). Such association is not unlike the coordination of a more typical Lewis base such as pyridine to a coordinatively unsaturated metal complex. In this particular case, the 18-valence-electron complex $[Mn(CO)_5]^-$ is the Lewis base and the 16-valence-electron square-planar $[Fe(mnt)_2]^-$ complex is the Lewis acid. Not surprisingly, this interaction is weak; addition of other Lewis bases to solutions of the FeMn complex disrupts the association and separates the metal centers [28].

The reaction shown in Eq. (14) is classified as an addition reaction because there is no ligand loss during the synthesis [29]:

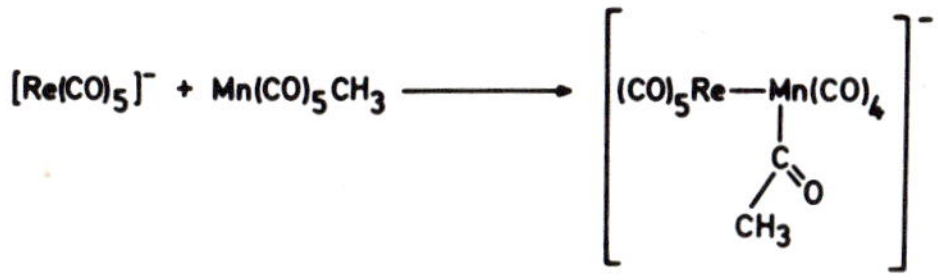

(14)

This reaction probably proceeds through addition of $[Re(CO)_5]^-$ to the coordinatively unsaturated Mn-acyl complex which forms via Eq. (15):

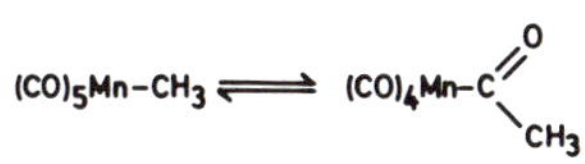

(15)

The addition of a metal complex across a multiple metal-metal bond has emerged as an extremely useful and rational synthetic method. The technique is based on an analogy first described by Stone [11, 12, 30] between the addition of metal nucleophiles to olefins, metal carbenes, and metal-metal double bonds and, similarly, the analogy between the addition of nucleophiles to alkynes, metal carbynes, and metal-metal triple bonds (SCHEME 3).

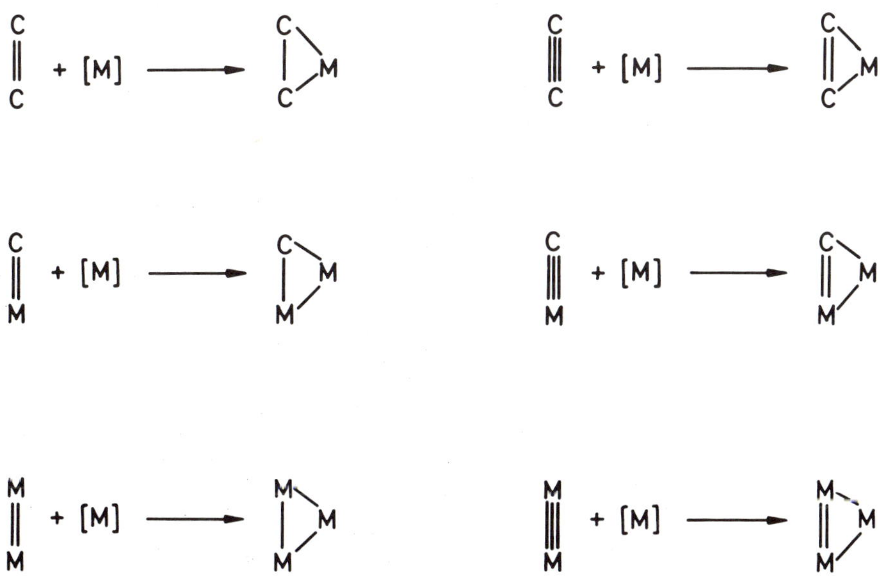

SCHEME 3

The reactions with metal carbene and carbyne complexes will be discussed in the section on bridge-assisted reactions, since the organic ligand ends up as a bridging ligand. The method discussed here involves the addition of a low-valent transition metal complex such as $[Pt(PR_3)(C_2H_4)_2]$ or $[Rh(acac)(C_2H_4)_2]$ across a metal-metal multiple bond. The metal complex added to the multiple metal-metal bond or metal-carbon bond generally has at least one ligand which is readily lost (e.g., C_2H_4), such that the actual adding reagent is a coordinatively unsaturated or solvated metal complex [e.g., $Pt(PPh_3)_2$]. One of the recent examples from Stone's work is given in Eq. (16) [31].

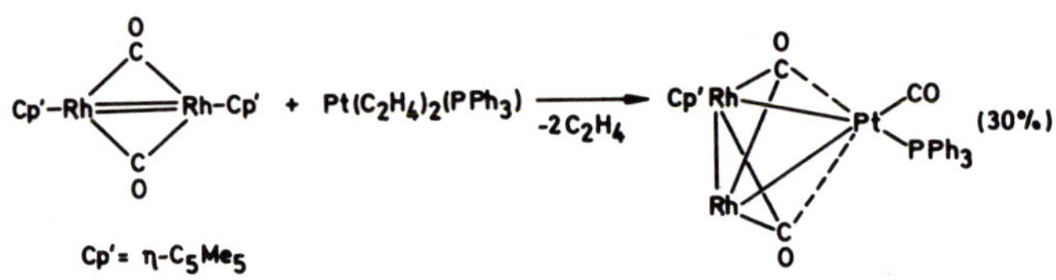

(16)

These reactions generally proceed in very good yield under rather mild reaction conditions and are highly predictable as to the type of product formed. The major disadvantage of this technique is its dependence on the availability of reactive multiply metal-metal bonded complexes. Since there are relatively few of these available, the technique is limited. By its very nature, this reaction can only be used for the preparation of cluster complexes.

Addition of a metal fragment across a metal-ligand bond can also occur:

$$ML_n + M'L_mX \longrightarrow X\text{-}ML_n\text{-}M'L_m \qquad (17)$$

The reaction is analogous to an oxidative addition of an X-Y molecule to an unsaturated metal complex, but it has found limited application in the synthesis of metal-metal bonded compounds. However, the technique is quite appealing for complexes in which X is a halide, since the product would then have a useful functional group for further metal-metal bond-forming reactions. One of the few examples of the successful application of this method is shown below [32]:

$$Pt(PPh_3)_4 + NiI_2(PPh_3)_2 \longrightarrow (PPh_3)_2\overset{I}{\overset{|}{Pt}}\!-\!\underset{I}{\underset{|}{Ni}}(PPh_3)_2 + 2PPh_3 \qquad (18)$$

A series of triosmium-based clusters has been prepared by addition of metal-hydride complexes to the "lightly stabilized" solvated complexes $[Os_3(CO)_{11}(L)]$ and $[Os_3(CO)_{10}(L)_2]$ [where L = CH_3CN, cyclooctane (COT)] as in Eq. (19) [33].

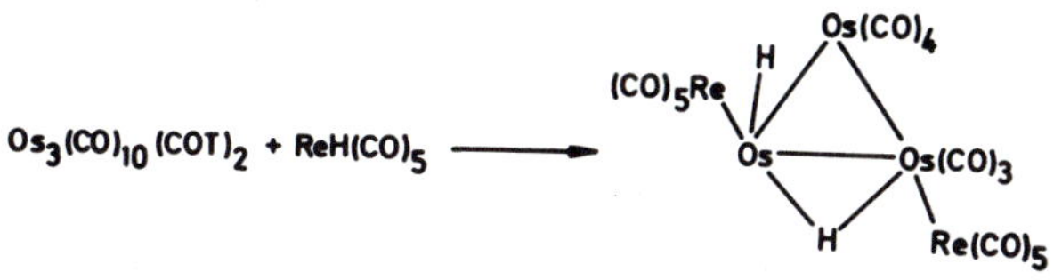

(19)

1.2.4 Condensation reactions

In a condensation reaction a second, generally small, molecule is formed concomitant with formation of the metal-metal bond, and this new molecule is composed of parts provided by each of the initial metal centers. Examples of different types of condensation reactions are given in Eqs. (20-25) [34-39].

Condensation with elimination of H_2 [34]:

$$2HCo(CO)_4 \longrightarrow H_2 + Co_2(CO)_8 \quad (20)$$

Condensation with elimination of an alkane [35]:

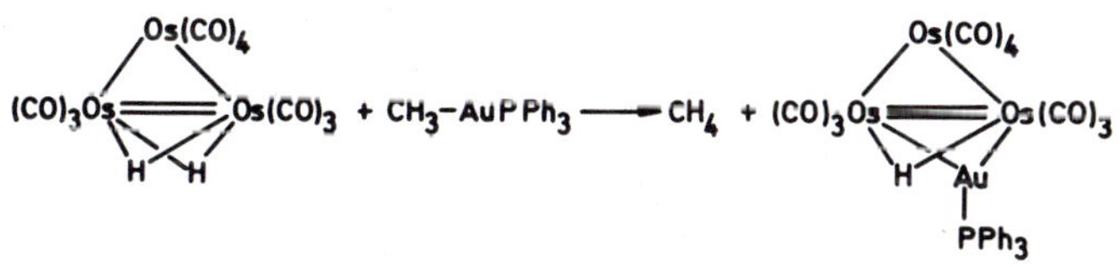

(21)

Condensation with elimination of an alkene [36]:

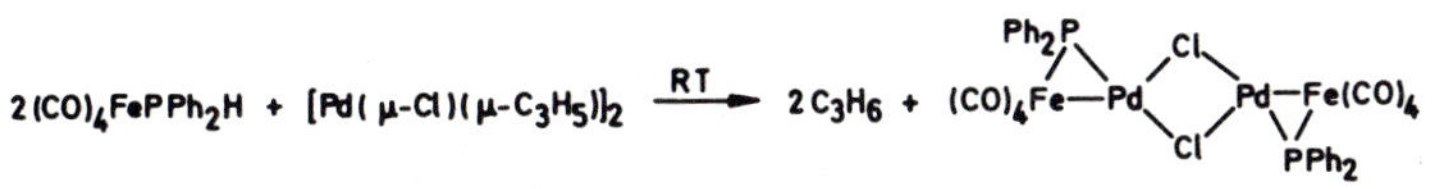

(22)

Condensation with elimination of other molecules [37-39]:

$$(CO)_4Co\text{-}SnMe_3 + ClAu(PPh_3) \longrightarrow Me_3SnCl + PPh_3Au\text{-}Co(CO)_4 \qquad (23)$$

$$Cp(CO)_3Mo\text{-}H + Me_2N\text{-}Ti(OPr^i)_3 \longrightarrow HNMe + Cp(CO)_3Mo\text{-}Ti(OPr^i)_3 \qquad (24)$$

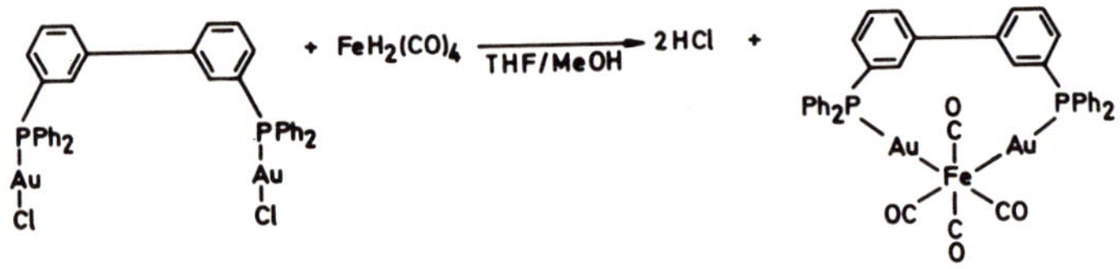

(25)

There are presently no examples of metal-metal bond formation by condensation of a metal hydride complex with a metal allyl complex. However, the similar reaction shown in Eq. (22) occurs between a coordinated PR_2H ligand on one metal and a metal-allyl complex to give an alkene and a μ-PR_2-bridged metal-metal bonded derivative. The elimination of an alkane, especially from CH_3AuPPH_3, is receiving a great deal of current attention and is potentially a very useful synthetic method. Such a reaction has a strong driving force in forming the stable C-H bond of the alkane and is irreversible since the alkane will not undergo subsequent addition and C-H bond scission. The reaction conditions are generally quite mild, and yields are moderate to high. Reactions (23) and (24) illustrate the formation of Me_3SnCl and $HNMe_2$, respectively, both volatile products which are readily removed, providing a driving force and minimizing back reaction. The presence ot a base such as Et_2NH usually facilitates condensation reactions which involve elimination of hydrohalic acids, although in general this is not a good method because of the tendency of the reverse metal-metal bond cleavage reaction to occur.

The primary advantages of the condensation technique are the minimization of back reactions when volatile condensates are formed and the mild conditions, which can limit unwanted side products. Yields are generally high. In some cases the condensation approach can circumvent problems associated with other methods; for example, unwanted redox reactions often occur with the use of strongly reducing carbonylmetallates. This advantage is important in reactions involving the earlier transition metals. The difficulty of this technique is that many of the desired starting materials are not especially stable, particularly so for some of the first-row transition metal alkyl and hydride complexes.

1.2.5 Metal exchange reactions

This method involves mixing together two polymetallic compounds and then applying radiant or thermal energy sufficient for homolytic cleavage of the metal-metal bonds. The resultant fragments can reassemble to give an equilibrium mixture of products:

$$M_2L_m + M'_2L_m \rightleftharpoons 2\ MM'L_{\frac{n+m}{2}} \tag{26}$$

The utility of this method is illustrated by the reactions shown in Eq. (27) [40] and (28) [41].

$$Re_2(CO)_{10} + Mn_2(CO)_{10} \xrightleftharpoons{493\ K,\ 60\ h} 2(CO)_5Re\text{-}Mn(CO)_5 \tag{27}$$

$$Ru_3(CO)_{12} + Os_3(CO)_{12} \rightleftharpoons RuOs_2(CO)_{12} + Ru_2Os(CO)_{12} \tag{28}$$

This method has proved most useful for the preparation of dinuclear complexes, although, as illustrated by Eq. (28), it can be applied to cluster synthesis. The highest yields will generally be obtained when the thermal energy required for metal-metal bond cleavage is about the same for both starting complexes. Low yields would be expected with combinations for which the conditions necessary to cause cleavage of metal-metal bonds for one species are severe enough to cause decomposition of the other. In most cases this synthetic technique will require separation of an equilibrium mixture of products following the reaction. Although one might be skeptical of the utility of this method, in practice it has proved extremely successful for the preparation of heteronuclear complexes.

1.2.6 Bridge-assisted reactions

Bridge-assisted reactions are those which employ a ligand to bring two metal centers together with the ligand functioning as a bridge between the metals in the final product. This synthetic method is receiving increasing attention because of the current interest in ligand-bridged compounds to retard cluster fragmentation reactions. Carbonyl bridges are excluded from the bridge-assisted category even though they may play an integral role in assisting metal-metal bond formation and can end up in bridging positions in some products. As noted earlier, the basic reaction classes are similar to those discussed above; bridged-assisted substitution, addition, and condensation reactions are the most important.

The types of ligand systems which most often participate in *bridge-assisted substitution reactions* are derivatives of the Group V and VI elements, particularly P, As, and S. Examples of such reactions are given in Eqs. (29-32) [42-45].

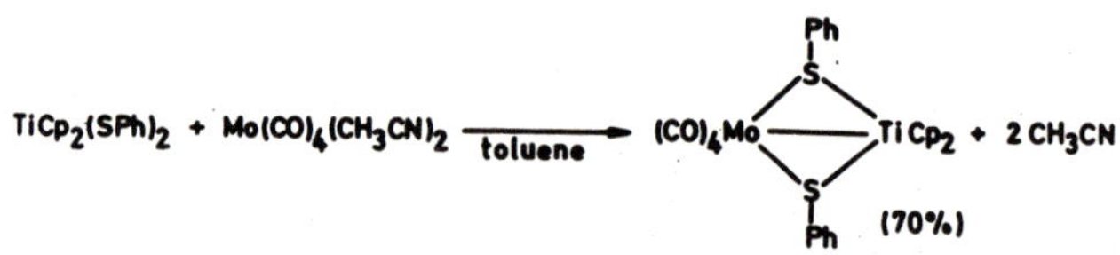

(29)

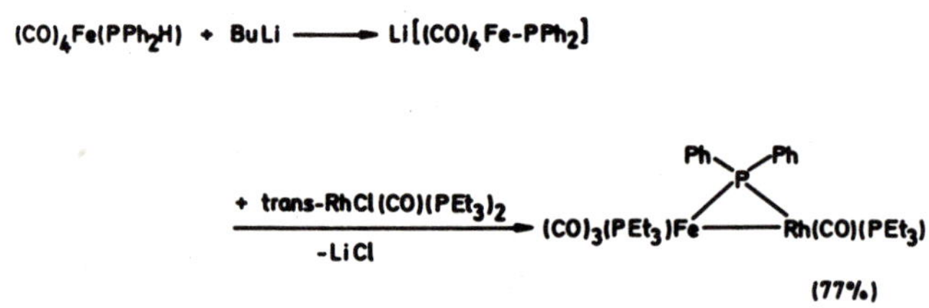

(30)

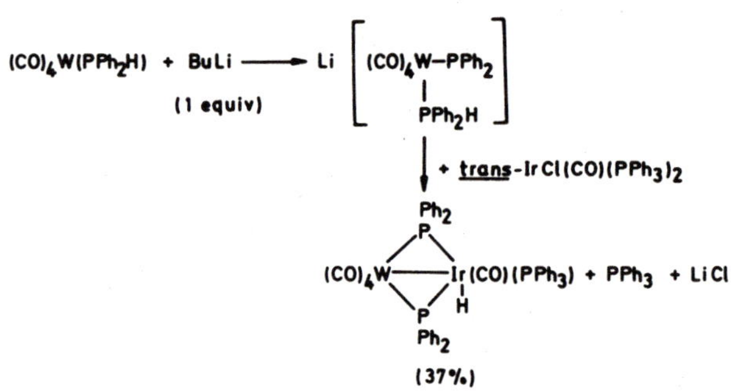

(31)

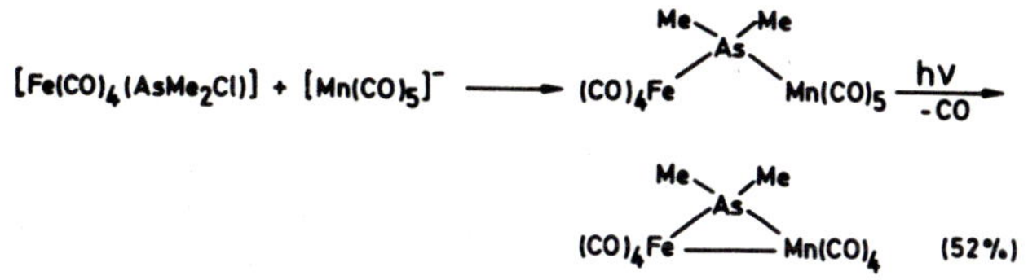

(32)

A ligand with a chalcogen atom bound to a single metal center (e.g., -OR, -SR) will always possess a lone pair of electrons which potentially can be used for donation to a second metal. These metallo-ligands can thus displace a ligand on another metal center, as illustrated by Eq. (29). Ligands with Group V elements (PR_3, AsR_3) generally have only one lone pair of electrons which is used to bind the ligand to the first metal center. A second pair of electrons for metal-ligand bonding can be made available by abstraction of H^+ from a coordinated PR_2H or AsR_2H ligand, as in Eqs. (30) and (31), or by metallation of the P or As, as in Eq. (32). Several reasonably stable neutral $-PR_2$ and $-AsR_2$ reagents for such syntheses have been reported [46], but in generally a large number of potential reagents are available via *in situ* deprotonation of suitable PR_2H and AsR_2H substituted complexes. The initial di- or polynuclear product formed in these reactions does not necessarily have a metal-metal bond as in Eq. (32). However, if not present, a metal-metal bond can often be generated by heating or irradiating the product to induce ligand loss followed by metal-metal bond formation.

Metal-metal bond formation can also occur via *bridge-assisted addition reactions*. The addition of coordinatively unsaturated metal centers across metal-carbene double bonds and metal-carbyne triple bonds, Eqs. (33) and (34), to yield carbene and carbyne bridged products has been successfully exploited by the Stone group [11, 12, 30] for the preparation of a large variety of new heteronuclear compounds.

Addition to Alkylidene Complexes:

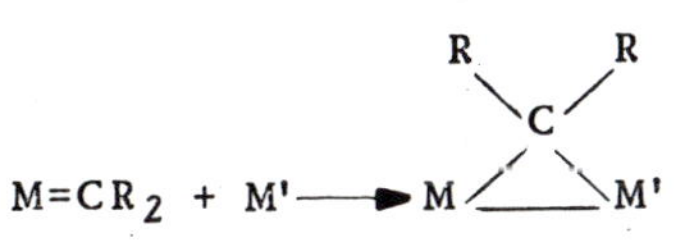

(33)

Addition to Alkylidyne Complexes:

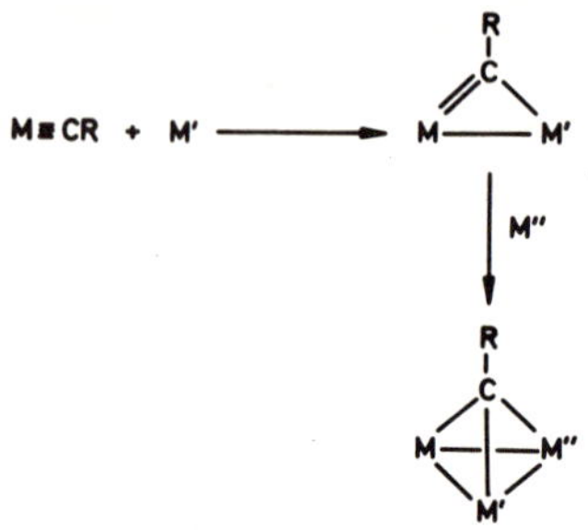

(34)

In general, these reactions proceed in high yield and allow one to direct the synthesis of specific carbyne and carbene derivatives. By carrying out the reaction sequence shown in Eq. (34) one can prepare heteronuclear μ_3-carbyne bridged trinuclear clusters of nearly any composition desired. Two illustrative examples of this approach are given in Eqs. (35) [47] and (36) [48].

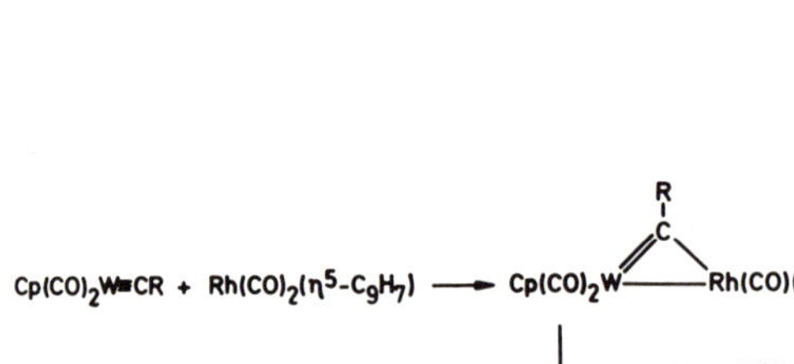

(35)

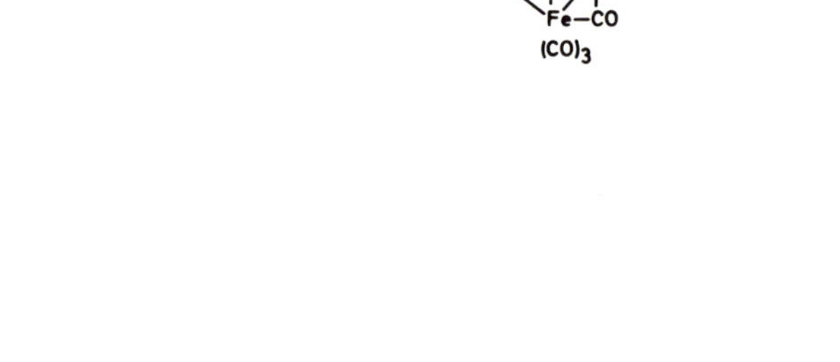

(36)

Additions of metal centers to carbon-carbon bonds of coordinated alkynes have also been demonstrated, e.g., Eq. (37) [49], and this method promises to have considerable utility for directed syntheses.

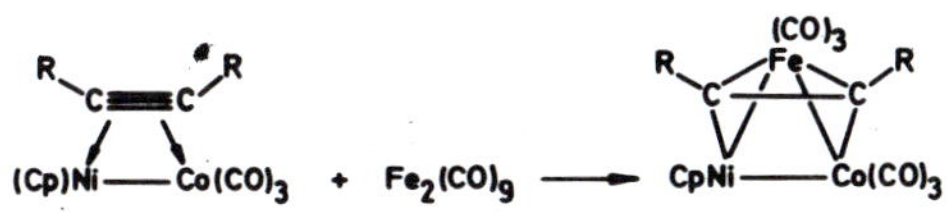

(37)

1.2.7 Bridge-assisted condensation reactions

Bridge-assisted condensation reactions have also been used to prepare metal-metal bonded products. This method has been used almost exclusively for the preparation of phosphido-bridged complexes. Condensation reactions involving elimination of HCl and propene are shown, in Eqs. (38) [47] and (39) [51], respectively.

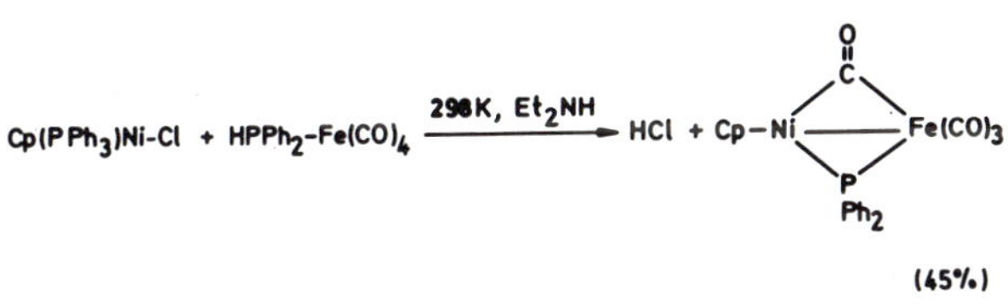

(38)

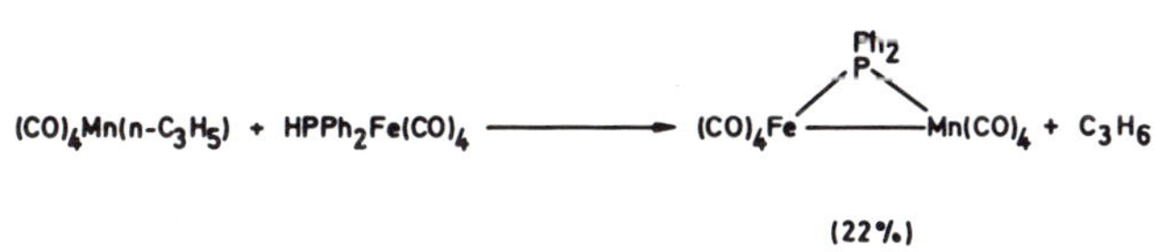

(39)

One potential problem that exists in all of the bridge-assisted reactions is that even after a bridging ligand is fixed between two metals, there is no guarantee that a metal-metal bond will form. For example, the product of Eq. (40) does not yield a metal-metal bonded compound even upon prolonged heating [51].

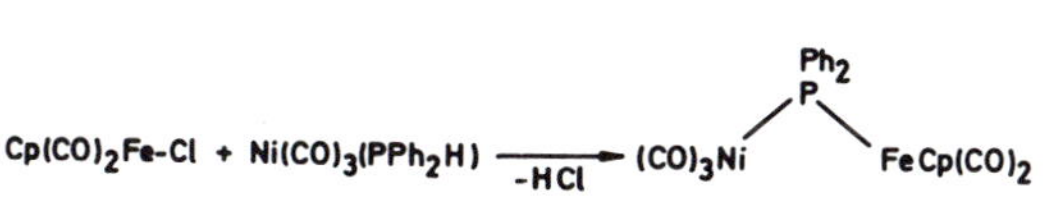

(40)

1.2.8 Serendipitous reactions

A very large number of polynuclear complexes have been prepared by reactions which were not designed to yield the product isolated. Many of these serendipitous syntheses simply involve placing reagents together, applying radiant or thermal energy, and then examining the reaction mixture to determine what kinds of compounds have resulted. Selected examples are given in Eqs. (41) [52], (42) [53], and (43) [54].

$$Ru_3(CO)_{12} + Fe(CO)_5 \xrightarrow[24\ h]{383\ K} FeRu(CO)_{12} + FeRu_2(CO)_{12} + H_2FeRu(CO)_{13} \qquad (41)$$

$$[Co(CO)_4]^- + [IrI_5(CO)]^{2-} \xrightarrow[H_2O]{298\ K} \underset{(70\text{-}80\%)}{Co_2Ir_2(CO)_{12}} \qquad (42)$$

$$Fe(CO)_5 + Co_2(CO)_8 \longrightarrow [FeCo_3(CO)_{12}]^- \qquad (43)$$

This approach does not lead to the designed synthesis of particular clusters, but it is not to be dismissed out-of-hand since so many very interesting compounds have been prepared in this manner. For example, nearly all of the homometallic high-nuclearity (M_n, n > 4) clusters have been prepared by this approach, and Eqs. (44) [55] and (45) [56] illustrate two recent examples:

$$Os_3(CO)_{11}(NC_5H_5) \xrightarrow[solid]{523\ K,\ 24\ h} \cdots \xrightarrow{workup} [Bu_4N]_2[Os_{10}(CO)_{24}C] \qquad (44)$$

$$[Ni_6(CO)_{12}]^{2-} + K_2PtCl_4 \longrightarrow \underset{(\sim 80\%)}{[Ni_{38}Pt_6(CO)_{48}H_{6-n}]^{n-}} + \text{other products} \qquad (45)$$

$$n = 3\text{-}6$$

1.2.9 Synthesis of binary metal carbonyls, $M_x(CO)_y$

Although all of the binary metal carbonyl clusters are available from commercial sources, it is occasionally less expensive to prepare them directly; the original literature syntheses should be consulted for details.

Most of these clusters are prepared by carbonylation of metal salts or coordination complexes in the presence of suitable reducing agents. Selected examples are given in Eqs. (46-49) [57-60].

$$RuCl_3 \cdot 3H_2O \xrightarrow[\Delta]{+CO,\ 1\ bar} "Ru(CO)_nCl_m" \xrightarrow[+CO]{+Zn} Ru_3(CO)_{12} + ZnCl_2 \quad (46)$$

(50-60%)

$$3OsO_4 + 24CO \xrightarrow[CH_3OH]{125^\circ C} Os_3(CO)_{12} + 12\ CO_2 \quad (47)$$

1 bar
12 h (~85%)

$$3Rh_2(O_2CCH_3)_4 + 22CO + 6H_2O \longrightarrow Rh_6(CO)_{16} + 6CO_2 + 12CH_3CO_2H \quad (48)$$

(87%)

$$Na_3[IrCl_6] + CO + NaI \xrightarrow[1\ bar]{} Na[Ir(CO)_2I_2] \xrightarrow[+\ K_2CO_3]{+\ CO} Ir_4(CO)_{12} \quad (49)$$

(~75%)

It should be noted that the yields of these reactions can be dramatically dependent upon subtle experimental variables. For example, the best yields of $Ru_3(CO)_{12}$ are obtained with vigorous CO purging and the use of low-surface-area Zn metal. Apparently, it is essential to have enough CO in solution to trap the reduced Ru species as they are generated and before they can coalesce to Ru metal or to an amorphous $Ru_x(CO)_y$ polymer.

1.3 SUMMARY

It is emphasized that the synthesis of binuclear and polynuclear metal-metal bonded compounds is no longer a black-box exercise and that there are a variety of exceedingly useful methods for directing the synthesis of particularly desired compounds [7]. While it is true that a large number of new and interesting compounds will surely be made by serendipitous reactions, the intelligent synthetic chemist should now be able to design and direct the synthesis of nearly any low-nuclearity metal-metal bonded compound, as long as the product is thermodynamically stable. The synthetic challenge increases as the nuclearity of the desired product increases, but in principle the strategy outlined in SCHEME 1 should still apply.

For those workers who are not interested in synthesizing new compounds, but in using known compounds, e.g., for the preparation of supported cluster catalysts, consultation of the original literature recipe is strongly advised. Several comprehensive tabulations of reported polynuclear compounds facilitate the location of the original literature references [6, 7, 10].

REFERENCES

1. J. Lewis and R. S. Nyholm, Sci. Progr., 52 (1964) 557.
2. H. Vahrenkamp, Angew Chem., Intl. Ed. Engl., 17 (1978) 379.
3. M. Baird, Prog. Inorg. Chem., 9 (1968) 1.
4. J. D. Cotton, Chemical Society Specialist Periodical Repts., Organometallic Chemistry, 1 (1972) 194; N.G. Connelly, ibid., 2 (1973) 234, 3 (1975) 180, 4, (1975) 171, 5 (1976) 176, 6 (1977) 79, 7 (1978) 172, 8 (1979) 193.
5. N. S. Vyazankin, G. A. Razuvaev and O. A. Kruglaya, Organomet. Chem. Rev., 3A (1968) 323.
6. W. L. Gladfelter and G. L. Geoffroy, Advan. Organomet. Chem., 18 (1980) 207.
7. D. A. Roberts and G. L. Geoffroy, in G. Wilkinson, F. G. A. Stone and E. W. Abel (Eds.), Comprehensive Organometallic Chemistry, Pergamon Press, London, 1982, Ch. 40.
8. F. A. Cotton and R. A. Walton, Multiple Bonds Between Metal Atoms, Wiley, New York, 1982.
9. M. H. Chisholm and I. P. Rothwell, Prog. Inorg. Chem., 29 (1982) 1.
10. M. I. Bruce, J. Organomet. Chem., 242 (1982) 147; M. I. Bruce, ibid., in press.
11. T. V. Ashworth, M. J. Chetcuti, L. J. Farrugia, J. A. K. Howard, J. C. Jeffrey, R. Mills, G. N. Pain, F. G. A. Stone and P. Woodward, ACS Symp. Series, 155 (1981) 299.
12. F. G. A. Stone, Phil. Trans. R. Soc. London A, 308 (1982) 87.
13. L. M. Cirjak, R. E. Ginsburg and L. F. Dahl, J. Chem. Soc. Chem. Commun., (1979) 470.
14. D. A. Roberts, W. C. Mercer, S. M. Zahurak, G. L. Geoffroy, C. W. DeBrosse, M. E. Cass and C. G. Pierpont, J. Am. Chem. Soc., 104 (1982) 910.
15. H. Werner, H. J. Kraus and P. Thometzek, Chem. Ber., 115 (1982) 2914.
16. S. Cenini, B. Ratcliff and R. Ugo, Gazz. Chim. Ital., 104 (1974) 1161.
17. P. Chini, Inorg. Chim. Acta Rev., 2 (1968) 31.
18. U. Anders and W. A. G. Graham, J. Am. Chem. Soc., 89 (1967) 539.
19. A. T. T. Hsieh and J. Knight, J. Organomet. Chem., 26 (1971) 125.
20. P. C. Steinhardt, W. L. Gladfelter, A. D. Harley, J. R. Fox and G. L. Geoffroy, Inorg. Chem., 19 (1980) 332.
21. G. L. Geoffroy, Accts. Chem. Res., 13 (1980) 469.
22. J. Knight and M. J. Mays, J. Chem. Soc. Chem. Commun. (1971) 62.
23. J. Knight and M. J. Mays, J. Chem. Soc. Dalton Trans. (1972) 1022.
24. G. L. Geoffroy and W. L. Gladfelter, J. Am. Chem. Soc., 99 (1977) 7565.
25. J. Knight and M. J. Mays, J. Chem. Soc. (A) (1970) 654.
26. M. I. Bruce, G. Shaw and F. G. A. Stone, J. Chem. Soc. Dalton Trans., (1972), 1082.
27. J. K. Ruff, R. P. White, Jr. and L. F. Dahl., J. Am. Chem. Soc., 93 (1971) 2159.

28. R. W. Johnson, W. R. Muir and D. A. Dweigart, J. Chem. Soc. Chem. Commun. (1970) 643.
29. C. P. Casey, C. R. Cyr, R. L. Anderson and D. F. Marten, J. Am. Chem. Soc., 97 (1975) 3053.
30. T. V. Ashworth, J. A. K. Howard and F. G. A. Stone, J. Chem. Soc., Chem. Commun., (1979) 42.
31. M. Green, R. M. Mills, G. R. Pain, F. G. A. Stone and P. Woodward, J. Chem. Soc. Dalton Trans. (1982) 1309.
32. M. Akhtar and H. C. Clark, J. Organomet. Chem., 22 (1970) 233.
33. J. R. Shapley, G. A. Pearson, M. Tachikawa, G. E. Schmidt, M. R. Churchill and F. J. Hollander, J. Am. Chem. Soc., 99 (1977) 8064.
34. H. W. Sternberg, I. Wender, R. A. Friedel and M. Orchin, J. Am. Chem. Soc., 75, (1953) 2717.
35. L. J. Farrugia, J. A. F. Howard, P. Mitrprachachon, J. L. Spender, F. G. A. Stone and P. Woodward, J. Chem. Soc. Chem. Commun. (1978) 260.
36. B. C. Censon, R. Jackson, K. K. Joshi and D. T. Thompson, J. Chem. Soc. Chem. Commun. (1968) 1506.
37. E. W. Abel and G. V. Hutson, J. Inorg. Nucl. Chem., 30 (1968) 2339.
38. O. J. Cardin and M. F. Lappert, J. Chem. Soc. Chem. Commun. (1966) 506.
39. B. Chiswell and L. M. Venanzi, J. Chem. Soc. (A) (1966) 901.
40. E. W. Abel, R. A. N. McLean and S. Moorhouse, Inorg. Nuc. Chem. Lett., 7 (1971) 587.
41. B. F. G. Johnson, J. Lewis and T. W. Matheson, J. Chem. Soc. Chem. Commun. (1974) 441.
42. H. Kopf and K. H. Rathleinm, Angew. Chem. Intl. Ed. Engl., 8 (1969) 980, and P. S. Braterman, V. A. Wilson and K. K. Joshi, J. Chem. Soc. A. (1971) 191.
43. M. J. Breen, M. R. Duttera, G. L. Geoffroy, G. C. Novotnak, D. A. Roberts, P. M. Shulman and G. R. Steinmetz, Organometallics, 1 (1982) 1008.
44. M. J. Breen, G. L. Geoffroy, A. L. Rheingold and W. C. Fultz, J. Am. Chem. Soc., 105 (1983) 1069.
45. W. Ehrl and H. Vahrenkamp, Chem. Ber., 106 (1973) 2563.
46. See Table 3 in ref 7.
47. T. V. Ashworth, J. A. K. Howard, M. Laguna and F. G. A. Stone, J. Chem. Soc. Dalton Trans., (1980) 1593.
48. (a) M. J. Chetcuti, M. Green, J. C. Jeffrey, F. G. A. Stone and A. A. Wilson, J. Chem. Soc. Chem. Commun. (1980) 948.
    (b) M. Green, J. C. Jeffrey, S. J. Porter, H. Razay and F. G. A. Stone J. Chem. Soc. Dalton Trans. (1982) 2475.
49. G. Jaouen, A. Marinetti, B. Mentzen, R. Mutin, J.-Y. Saillard, B. G. Sayer and M. J. McGlinchey, Organometallics, 1 (1982) 753.
50. K. Yasufuku and H. Yamazaki, Bull. Chem. Soc. Jap., 43 (1970) 1588.
51. K. Yasufuku and H. Yamazaki, J. Organomet. Chem., 28 (1971) 415.
52. D. B. W. Yawney and F. G. A. Stone, J. Chem. Soc. (A) (1969) 502.
53. S. Martinengo, P. Chini, V. G. Albano, F. Cariati and T. Salvatori, J. Organomet. Chem., 59 (1973) 379.
54. P. Chini, L. Colli and M. Peraldo, Gazz. Chim. Ital., 90 (1960) 1005.
55. P. F. Jackson, B. F. G. Johnson, J. Lewis, M. McPartlin and W. J. H. Nelson, J. Chem. Soc. Chem. Commun. (1980) 224.
56. A. Ceriotti, F. Demartin, G. Longoni, M. Manassero, M. Marchionva, G. Piva and M. Sansoni, Angew. Chem. Int. Ed. Engl., in press.
57. A. Mantovani and S. Cenini, Inorg. Synth., 16 (1974) 47.
58. B. F. G. Johnson and J. Lewis, Inorg. Synth., 13 (1972) 93.
59. B. R. James, G. L. Renpel and W. K. Teo, Inorg. Synth., 16 (1974) 49.
60. L. Malatesta, G. Caglio and M. Angoletta, Inorg. Synth., 13 (1974) 95.

Chapter 2

STRUCTURES OF METAL CLUSTERS

G. L. GEOFFROY

2.1 INTRODUCTION

The structural diversity ot metal clusters is truly enormous, particularly when the different ligand arrangements on the metallic frameworks are taken into account. Several illustrative examples of cluster structures are shown in the equations in Chapter 1. Here we give a brief summary of the more important metal frameworks for each degree of metal nuclearity, illustrating these with several examples. More comprehensive listings of cluster structures are available in recent reviews [2-6]. Ref. [2] also gives a good discussion of the various electron counting rules for predicting and/or rationalizing the structures of cluster compounds.

2.2 CLUSTER STRUCTURES

2.2.1 Binuclear structures

Binuclear compounds are the simplest metal clusters, but even among these there is considerable structural diversity. Such compounds can exist either with or without metal-metal bonding, although in the latter case bridging ligands are necessary to keep the metals together in one compound. Two interconvertible compounds with and without direct metal-metal bonding are shown in Eq. (1) [1]:

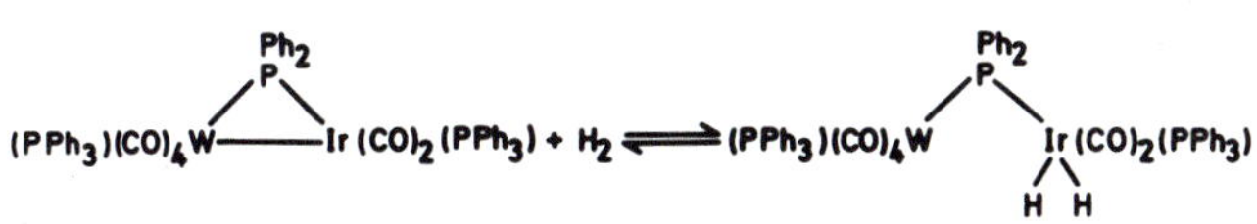

(1)

The bond order between adjacent metals in binuclear compounds is often greater than one, and examples of doubly, triply, and quadruply bonded complexes are illustrated below:

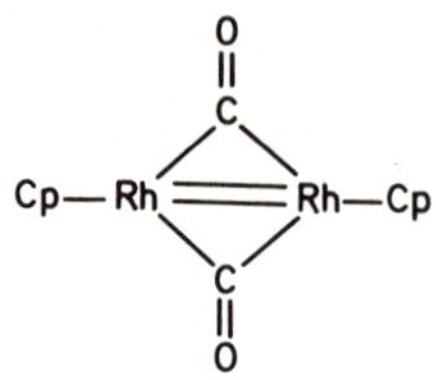

$Rh_2Cp_2(CO)_2$

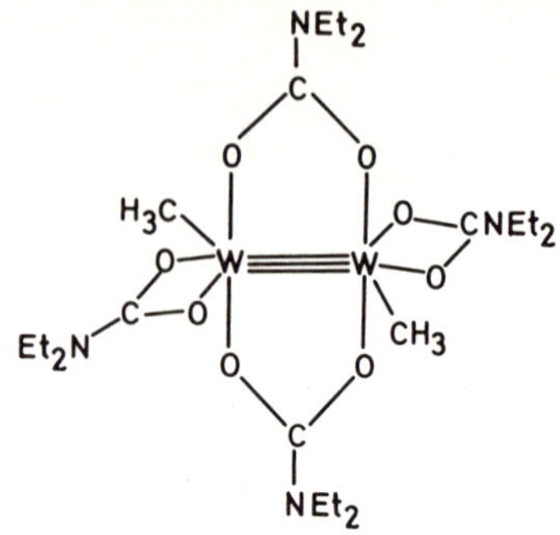

$W_2(CH_3)_2(O_2CNEt_2)_4$

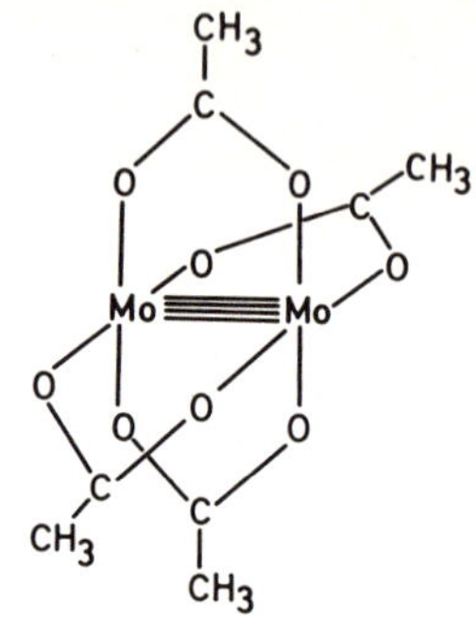

$Mo_2(acetate)_4$

These compounds are not as common as those with single metal-metal bonds, but they are synthetically useful because of their inherent coordinative unsaturation.

2.2.2 Trinuclear structures

Trinuclear compounds can have either open or closed metal frameworks

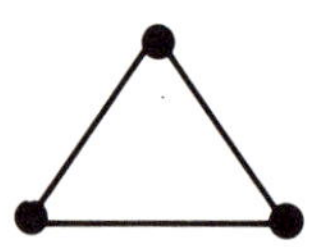

or

with varying degrees of metal-metal bonding. Open structures can be either linear or bent. Closed structures are much more common, partly because of the tendency of open compounds to lose a ligand and close upon themselves. Representative compounds illustrating these structures, particularly for the more important metal carbonyls, are shown below:

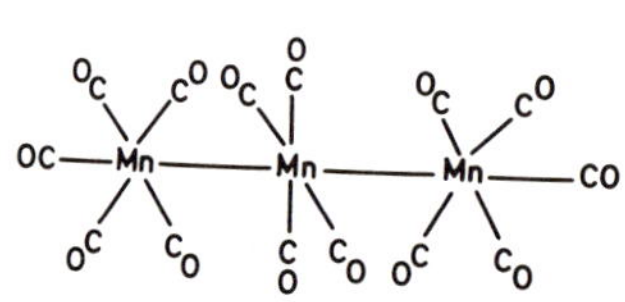

$Mn_3(CO)_{14}$

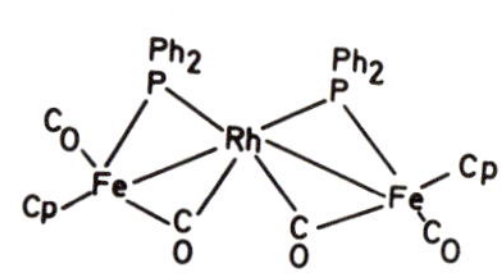

$Fe_2Rh(\mu\text{-}PPh_2)_2(CO)_4Cp_2$

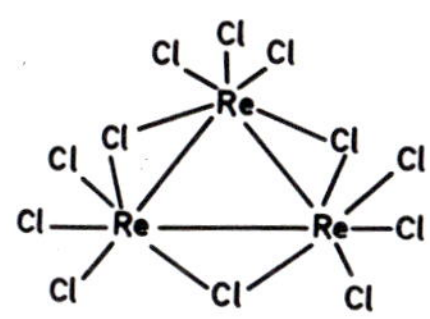

$[Re_3(Cl)_{12}]^{3-}$

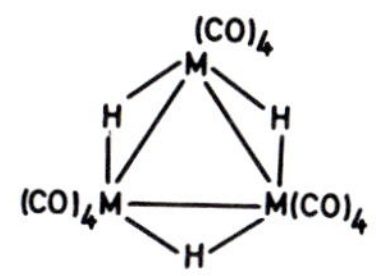

$M_3H_3(CO)_{12}$ (M=Mn, Re)

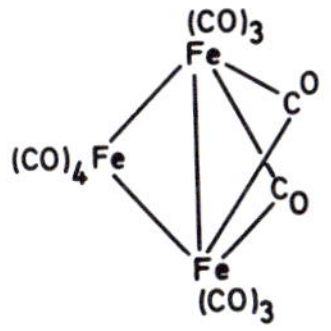

$Fe_3(CO)_{12}$

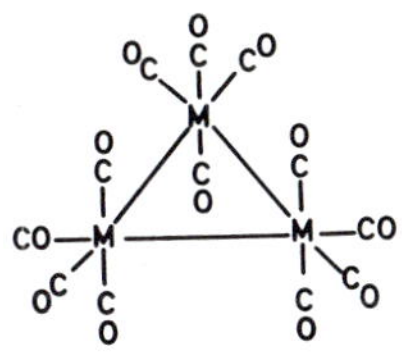

$M_3(CO)_{12}$ (M= Ru, Os)

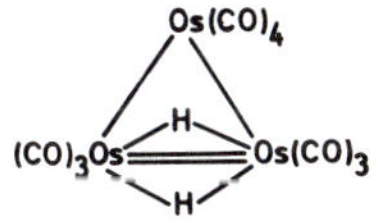

$H_2Os_3(CO)_{10}$

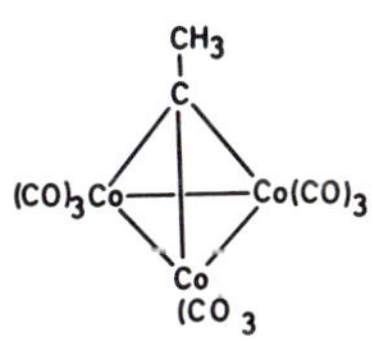

$Co_3(CO)_9(\mu_3\text{-}CCH_3)$

2.2.3 Tetranuclear structures

A series of different structural types exists for clusters having four metal atoms. These can be conceptually related to one another by the addition or removal of electron pairs (SCHEME 1), since a metal-metal bond can be formally replaced by an added pair of framework electrons.

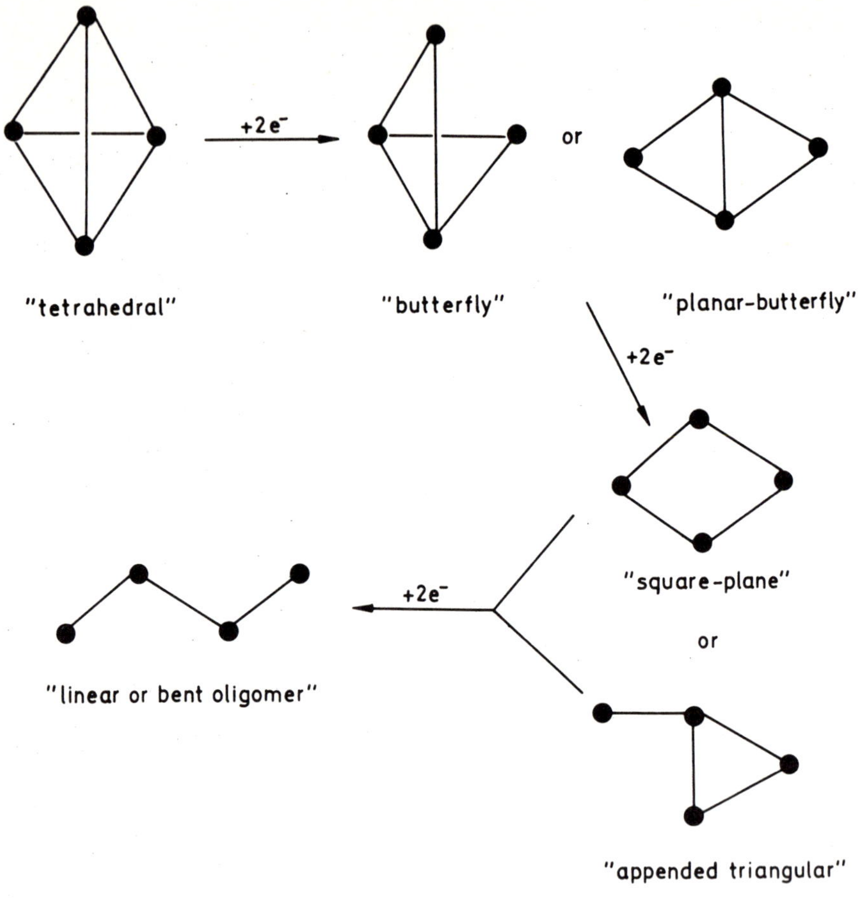

SCHEME 1

In practice it has not proved possible to interconvert these structural forms readily by two-electron oxidation/reduction of clusters or by addition/removal of two-electron ligands, although there is no reason why this should not be the case. Tetrahedral and butterfly clusters are by far the most common tetrametallic frameworks, with the former clearly predominating. Examples illustrating these structures are shown below:

"Tetrahedral"

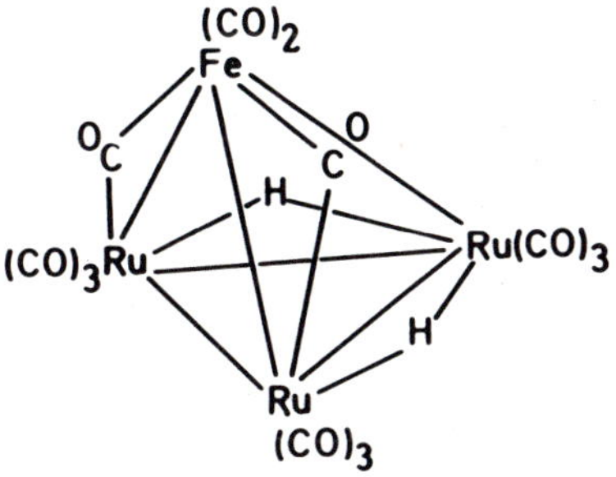

$H_2FeRu_3(CO)_{13}$

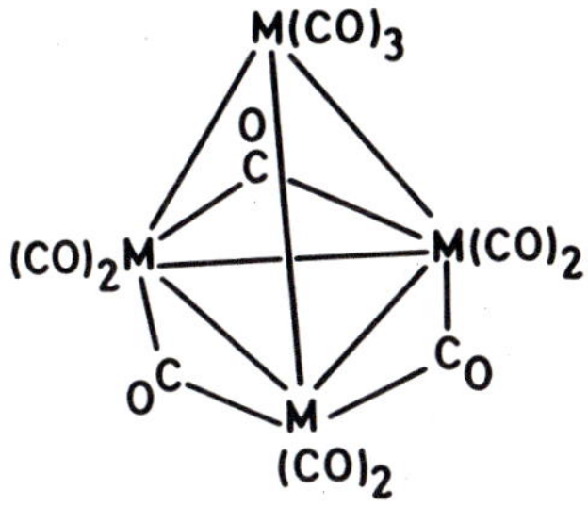

$M_4(CO)_{12}$ (M=Co, Rh)

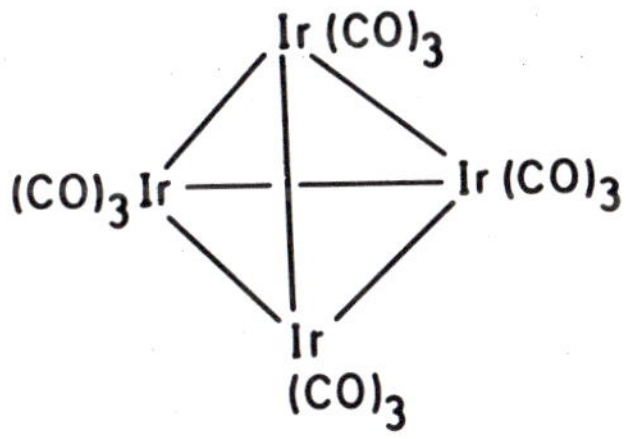

$Ir_4(CO)_{12}$

"Butterfly"

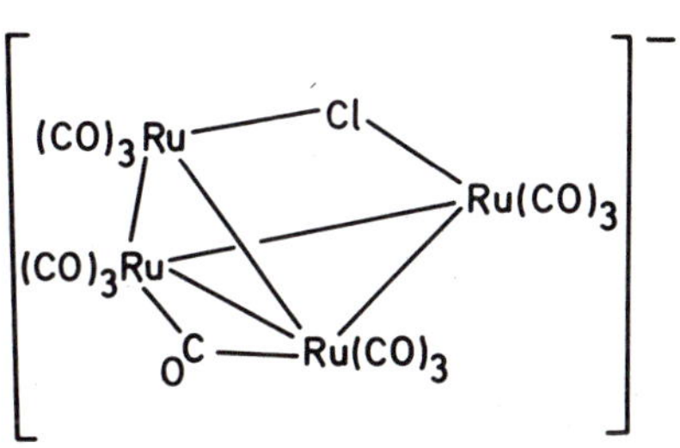

$[Ru_4(CO)_{13}Cl]^-$

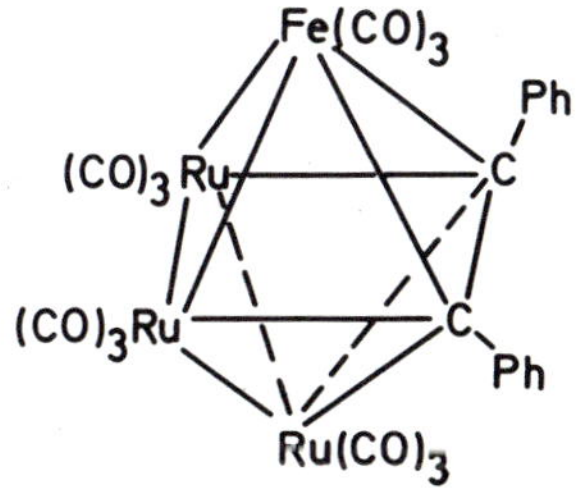

$FeRu_3(CO)_{12}(PhC{\equiv}CPh)$

"Planar-Butterfly"

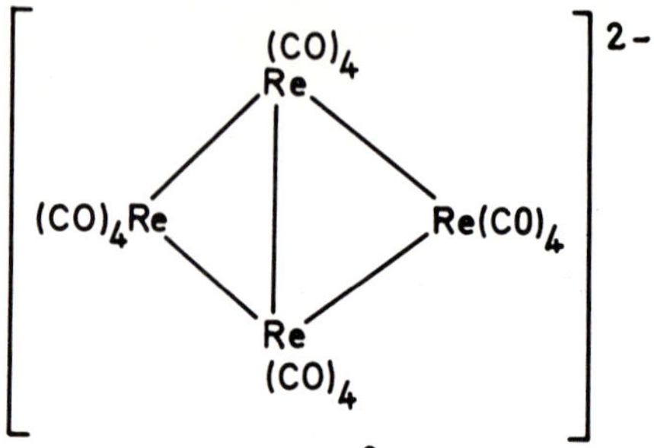

$[Re_4(CO)_{16}]^{2-}$

"Square-Plane"

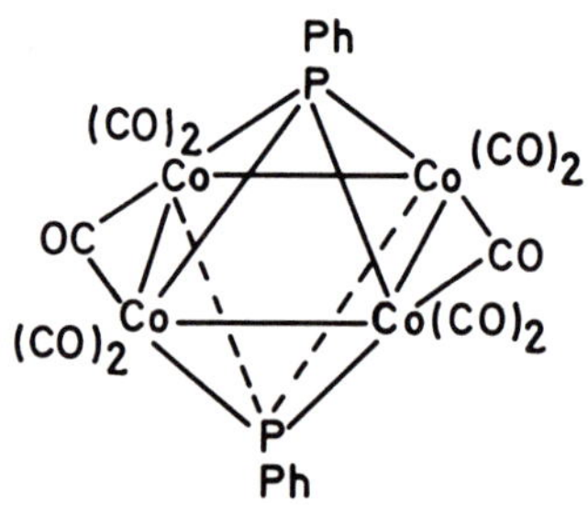

$Co_4(\mu_4\text{-}PPh)_2(CO)_{10}$

"Appended-Triangular"

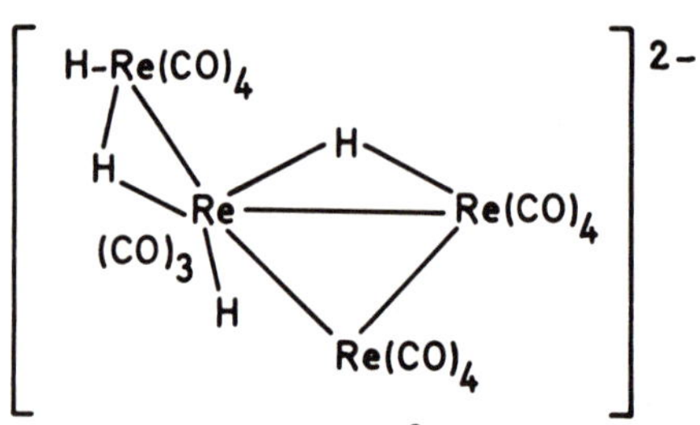

$[H_4Re_4(CO)_{15}]^{2-}$

"Oligomer"

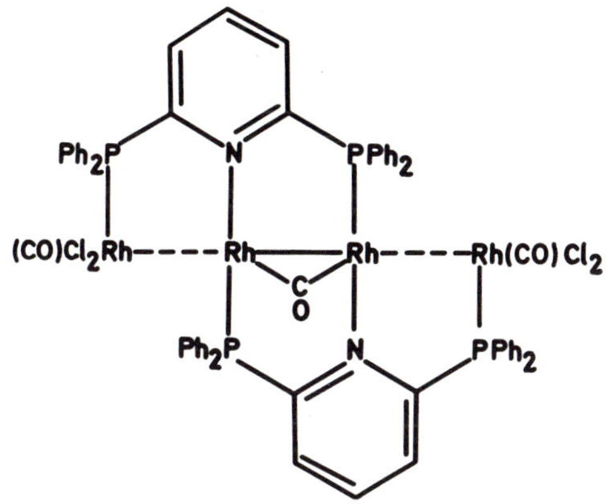

2.2.4 Pentanuclear structures

Pentametallic clusters are not common; the two most typical geometries in this class are trigonal-bypyramidal and square-pyramidal, illustrated below for $Os_5(CO)_{16}$ and $Fe_5(C)(CO)_{15}$, respectively:

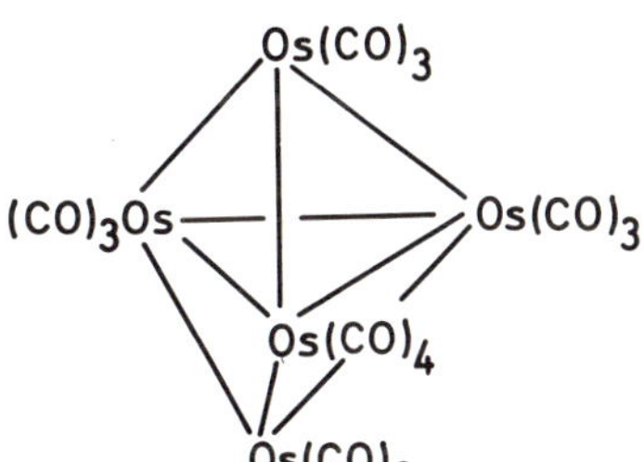

$Os_5(CO)_{16}$

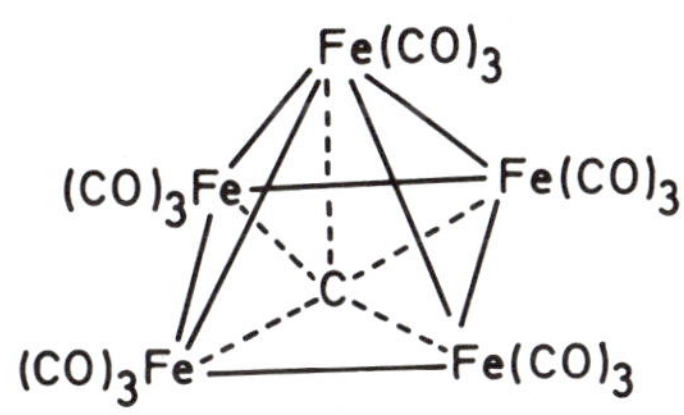

$Fe_5(C)(CO)_{15}$

Another documented structural type is the edge-bridged tetrahedral structure of $H_2Os_5(CO)_{16}$:

"Edge-Bridged Tetrahedron"

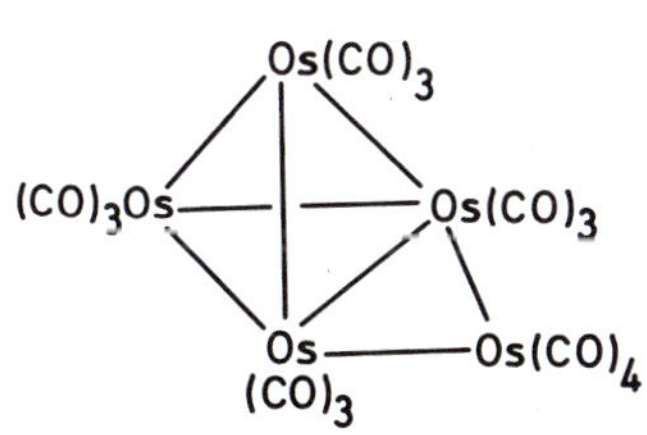

$H_2Os_5(CO)_{16}$

This structure is formally derived from the typical trigonal-bipyramidal structure by replacement of one metal-metal bond with two added framework electrons--$H_2Os_5(CO)_{16}$ is formally derived from $Os_5(CO)_{16}$ by addition of H_2, a net two-electron donor. Other pentametallic structural types are known which are derived from these three by successive formal electron-pair additions with correspondingly decreased metal-metal bonding.

2.2.5 Hexanuclear structures

The most common structures of hexametallic clusters are octahedral, trigonal prismatic, and capped square-pyrimidal, with the former clearly predominating. The octahedral structure is illustrated below for $[HRu_6(CO)_{18}]^-$, in which the hydride ligand is located within the cluster framework in an "interstitial" position.

"Octahedral"

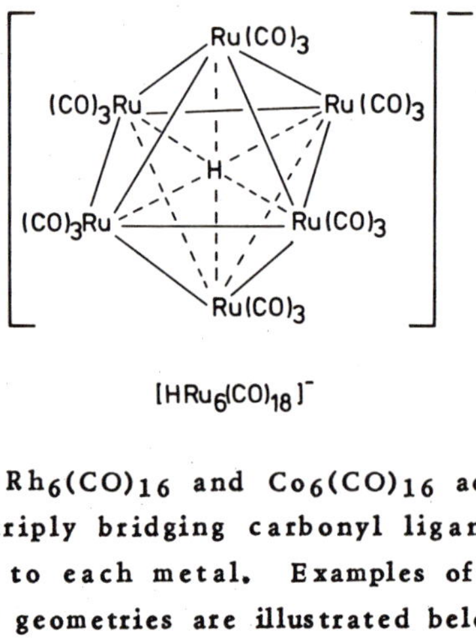

$[HRu_6(CO)_{18}]^-$

the carbonyl clusters $Rh_6(CO)_{16}$ and $Co_6(CO)_{16}$ adopt similar octahedral structures with four triply bridging carbonyl ligands and two terminal carbonyls coordinated to each metal. Examples of trigonal-prismatic and capped square-pyramidal geometries are illustrated below:

"Trigonal-Prismatic"

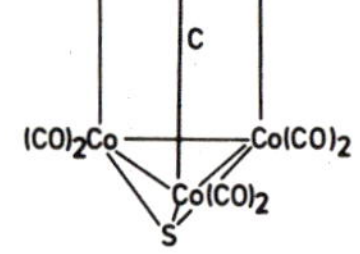

$Co_6(\mu_3\text{-}S)_2(CO)_{12}$

"Capped-Square Pyramidal"

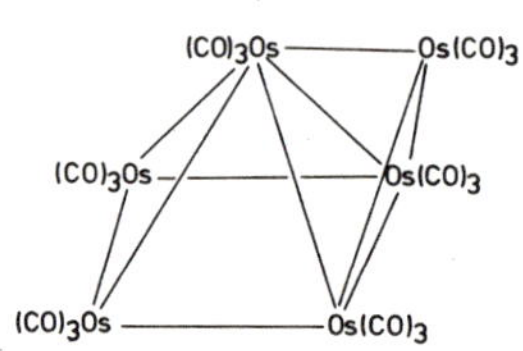

$H_2Os_6(CO)_{18}$

As before, more open structures are known with decreased degrees of metal-metal bonding.

2.2.6 High-nuclearity clusters

A bewildering array of framework structures exists for clusters with nuclearity greater than six. Many of these high-nuclearity clusters incorporate interstitial H, C, S, P, and N atoms within their cores, and several have frameworks which closely resemble the packing of metal atoms in solids. At this stage it becomes difficult to discuss such structures systematically, and we simply illustrate the diversity with the framework structures of representative examples. The surface ligands have been omitted for clarity, but interstitial atoms are indicated. The Rh_x examples shown below illustrate the tremendous variety of cluster frameworks. The $[H_2Rh_{13}(CO)_{24}]^{3-}$ structure closely models a hexagonal close-packed solid lattice.

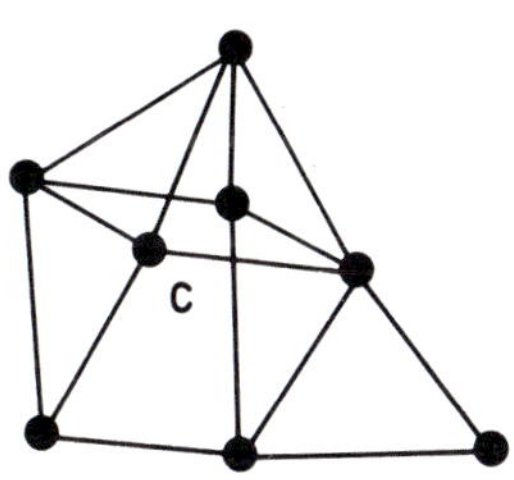

$Rh_8(CO)_{19}C$

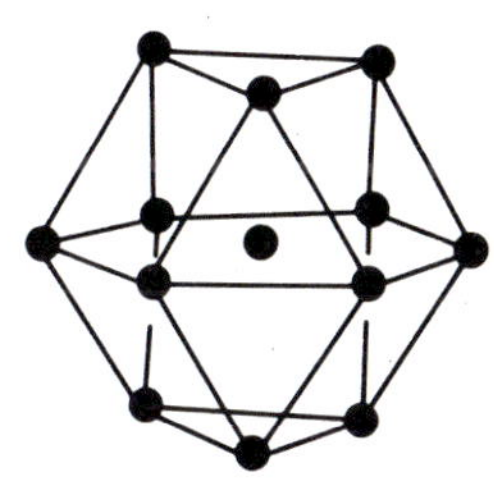

$[H_2Rh_{13}(CO)_{24}]^{3-}$

$[Rh_{12}(CO)_{30}]^{2-}$

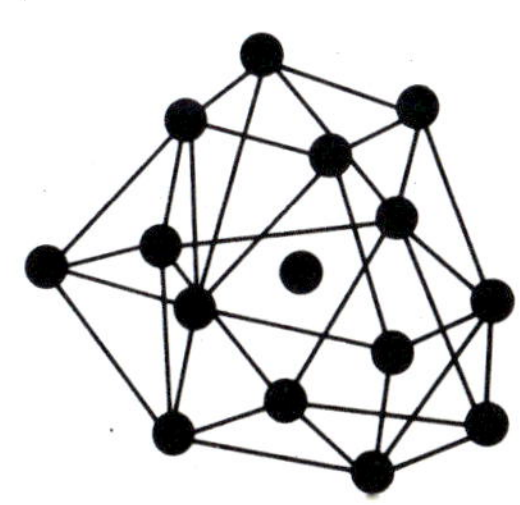

$[Rh_{15}(CO)_{27}]^{3-}$

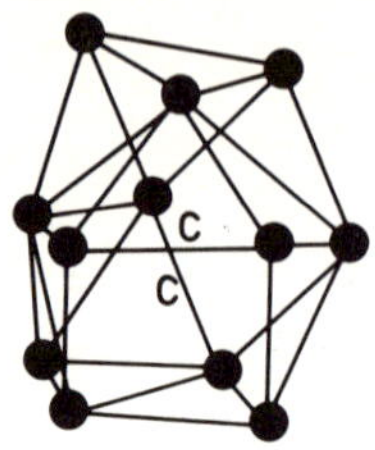

$Rh_{12}(CO)_{25}C_2$

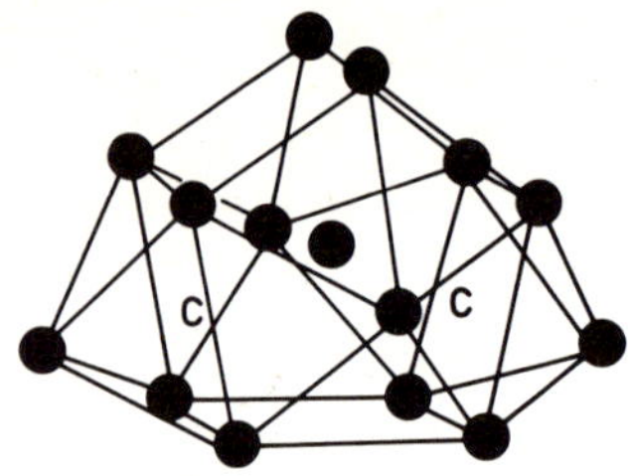

$[Rh_{15}(CO)_{28}C_2]^-$

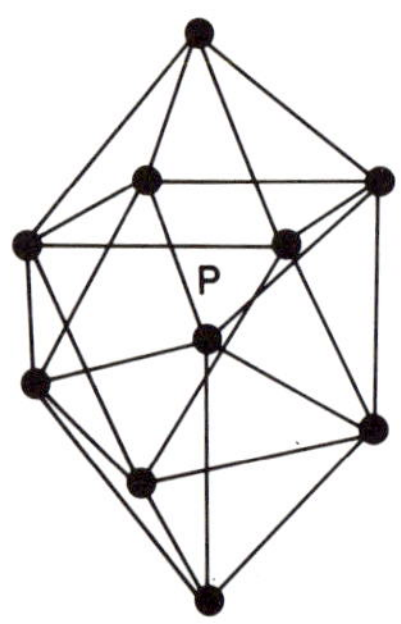

$[Rh_{10}P(CO)_{22}]^{3-}$

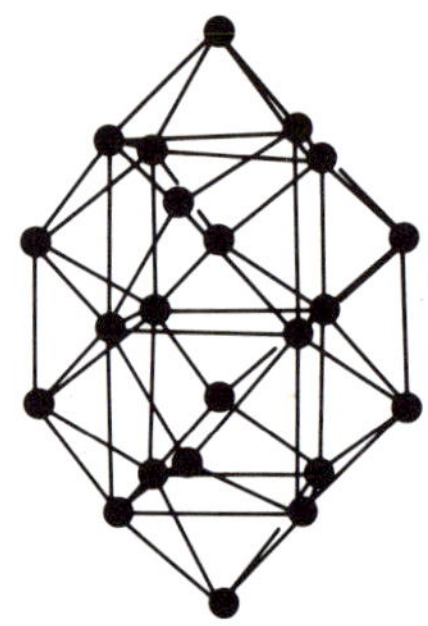

$[Rh_{22}(CO)_{35}H_{X+n}]^{(5-n)-}$

The decaosmium cluster mentioned in Chapter 1, $[Os_{10}(CO)_{24}C]^{2-}$, contains a completely encapsulated carbon atom, and the metal skeleton is a fragment of a cubic close packed array [7]. This cluster is mentioned again in Section 9.5 of this book; it has been identified on the surface of a supported osmium catalyst used for CO hydrogenation.

The remarkable $Ni_{38}Pt_6$ cluster with n = 5 (mentioned in Section 1.2.8) is the largest molecular cluster characterized by X-ray diffraction [8]. Its structure (Fig. 1) consists of an octahedral Pt_6 core surrounded by the 38 nickel atoms in an overall cubic close packed array. It is a beautiful model of a bimetallic particle that is surface-enriched in one of the metal components.

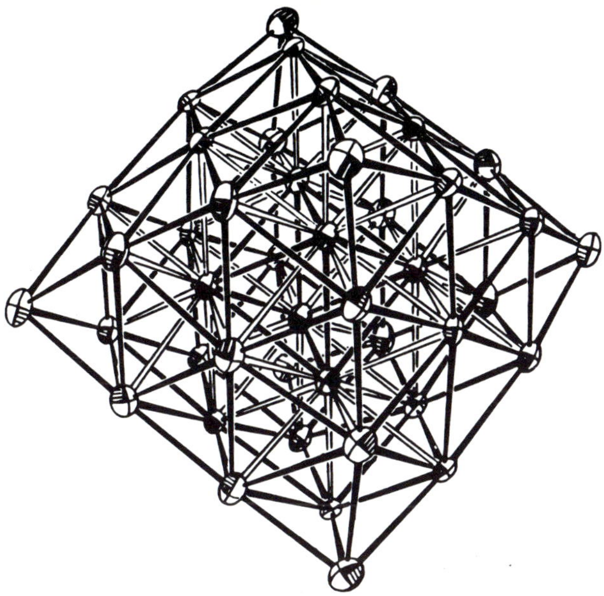

$[Ni_{38}Pt_6(CO)_{48}H]^{5-}$

Fig. 1. ORTEP drawing of the bimetallic skeleton of $[Ni_{38}Pt_6(CO)_{48}H]^{5-}$ [8].

A comprehensive 1976 review [3] lists all examples of high-nuclearity clusters known at that time; this area of cluster chemistry has grown enormously since that writing, and the current literature must be consulted for more recent examples.

REFERENCES

1. M. J. Breen, P. M. Shulman, G. L . Geoffroy, A. L. Rheingold and W. C. Fultz, Organometallics, submitted for publication.
2. B. F. G. Johnson, ed., Transition Metal Clusters, Wiley, New York, 1980.
3. P. Chini, G. Longoni and V. G. Albano, Advan. Organomet. Chem., 14 (1976) 285.
4. D. A. Roberts and G. L. Geoffroy, in Comprehensive Organometallic Chemistry, G. Wilkinson, F. G. A. Stone, and E. W. Abel (eds.), Pergamon, London, 1982, Chapter 40.
5. M. I. Bruce, J. Organomet. Chem., 242 (1982) 147.
6. M. I. Bruce, J. Organomet. Chem., in press.
7. P. F. Jackson, B. F. G. Johnson, J. Lewis, M. McPartlin and W. J. H. Nelson, J. Chem. Soc. Chem. Commun. (1980) 224.
8. A. Ceriotti, F. Demartin, G. Longoni, M. Manassero, M. Marchionva, G. Piva and M. Sansoni, Angew. Chem. Int. Ed. Engl., in press.

Chapter 3

THERMOCHEMICAL PROPERTIES AND BOND ENERGIES OF TRANSITION METAL CLUSTERS

J. A. CONNOR

3.1 INTRODUCTION

The scope of this topic is broad. Several reviews have appeared recently which include relevant material [1-4]. Although there is a steady, if unspectacular, increase in knowledge of the thermochemical properties of metal clusters, very much remains to be done to achieve a proper understanding of the subject. For example, experimental information about the thermochemical properties of diatomic transition metal molecules [5] has increased significantly recently, but there is still relatively little information about triatomic and higher polyatomic molecules containing transition metals alone. The preparation and characterization of complexes containing transition metal clusters with CO and other donor ligands, a field which is now highly developed, offers the possibilty for investigating naked clusters in the future.

3.2 DIATOMIC TRANSITION METAL COMPOUNDS

3.2.1 Homonuclear compounds

Values of D_0^o, the dissociation energy determined from the lowest energy level of the molecule, are known for 18 homonuclear compounds containing d-transition metals and 15 containing f-transition metals. These values are shown in Table 1, which also includes the enthalpy of atomization of the metal and information about the structure of the metal at standard temperature and pressure. The nature of the bonding in these diatomic molecules has attracted much interest, and various approaches to the calculation of D_0^o have been described [6, 7]. The description of, for example, dichromium, Cr_2, obtained from these calculations is closely related to the type of method used, whether *ab initio*, $X\alpha$-SW, or semi-empirical; whether configuration interaction is taken into account and, if so, to what extent; and so on. A further test of the validity of the various computational approaches to dichromium is provided by their ability to match simultaneously the bond length measured [8] (0.168 ± 0.001 nm) in the $^1\Sigma_g^+$ ground state of the molecule, and the experimental value of D_0^o. The low values of D_0^o for Mn_2, Eu_2, and Yb_2 may perhaps indicate that these are van der Waals molecules, although this is disputed [6a]. For the 3d-metals, the pattern of variation of D_0^o is similar to that of the

TABLE 1. Enthalpies of Formation of Gaseous Transition Metal Atoms in their Ground States at 298.15 K, $\Delta H_f^o(g)$; Coordination Number Z, of the Bulk (solid) Metal at S.T.P. and Geometry; Dissociation Energies of Homonuclear Diatomic Molecules $D_o^o(M_2)$.

Metal	ΔH_f^o(M,g), kJ/mol	Z	$D_o^o(M_2)$, kJ/mol
Sc	381.6 ± 1.3	12h	159 ± 21
Ti	470.7	12h	126 ± 17
V	541.6	8	328 ± 21
Cr	396.6 ± 4.2	8	151 ± 21
Mn	279.1	12c	42 ± 29
Fe	416.3 ± 4.2	8	75 ± 17[a]
Co	425.1	12c	167 ± 25
Ni	430.1	12c	230 ± 21
Cu	337.6 ± 1.2	12c	190.2 ± 5.4
Y	424.7 ± 0.8	12h	156 ± 21
Zr	608.4 ± 1.7	12h	
Nb	724 ± 8	8	503 ± 10
Mo	658.1 ± 2.1	8	404 ± 20
Ru	640 ± 8	12h	
Rh	556.5 ± 4.2	12c	281 ± 20.9
Pd	380.7 ± 4.2	12c	99 ± 15[b]
Ag	284.9 ± 0.8	12c	159.0 ± 6.3
La	431.0 ± 0.4	12c	243 ± 21
Ce	423	12c	238 ± 21
Pr	372.8 ± 1.3	12h	126 ± 29
Nd	328	12h	80 ± 29
Sm	207	12h	50 ± 21
Eu	175	8	29.3 ± 16.7
Gd	397.5	12h	172 ± 33
Tb	389	12h	127.6 ± 25.1
Dy	290	12h	67 ± 29
Ho	301	12h	79.5 ± 17
Er	317	12h	71 ± 29
Tm	247.3 ± 0.8	12h	50 ± 17
Yb	151.9 ± 0.4	12c	16.7 ± 17
Lu	428	12h	138 ± 33
Hf	619.2	8	
Ta	786.6 ± 4.0	8	
W	859.9 ± 4.6	8	
Re	783 ± 8	12h	
Os	783 ± 8	12h	
Ir	665 ± 8	12c	
Pt	565.7 ± 4.2	12c	358 ± 7
Au	369.4 ± 3.8	12c	221.3 ± 2.1
Th	598 ± 6	12c	285 ± 21
U	481 ± 13	8	218 ± 21

Values of D_o^o from ref. [9] except

[a] I. Shim and K. A. Gingerich, J. Chem. Phys., 77 (1983) 2490.

[b] I. Shim and K. A. Gingerich, J. Chem. Phys., 80 (1984) 5107.

[h] hexagonal close packed

[c] cubic close packed

enthalpy of atomization $\Delta H_f^o(M,g)$ [or, ΔH_{vap}^o (M,c)]. This quantity is also used in the most recent of the methods proposed for the prediction of D_o^o, the cell model of Miedema [6b], which proposes a relation between ΔH_{vap}^o (M,c), the surface energy of the metal, γ^o, the atomic size parametrized by the molar volume, V_m, and the dissociation energy:

$$D_o^o = 2\Delta H_{vap}^o - \gamma^o V_m^{2/3} \text{ constant.} \quad (1)$$

Another approach emphasizes the contribution of valence state promotion energies [6a]. Recently the bond energy $D^o(Mn_2^+) = (82 \pm 20)$ kJ/mol has been measured [10].

3.2.2 Heteronuclear compounds

Experimental values of D_o^o are available for 31 d/d and 2 d/f heteronuclear diatomic molecules (Table 2). Estimates have been made [9] of the maximum bond energy between two metal atoms; for homonuclear diatomic transition metal molecules, this is (600 ± 40) kJ/mol, and for heteronuclear diatomic molecules, it is expected to be slightly higher, at (640 ± 40) kJ/mol. Possible candidates for these are Ta_2 and PtZr, respectively.

3.3 POLYATOMIC TRANSITION METAL COMPOUNDS

The atomization energies of only very few compounds have been measured. The only consistent series is presented by the triatomic coinage metal molecules, for which the values of D_o^o are [11] Cu_3 (294 ± 13); Ag_3 (253 ± 13); and Au_3 (367 ± 13) kJ/mol. Comparison of these values with those for the diatomic analogues (Table 1) suggests that the binding in Au_3 may be stronger than expected. Other examples are [9] Au_2Eu (549.4 ± 16.7); Au_2Ho (533 ± 42); Au_2Lu (602.1 ± 33.5); Au_2Tb (582 ± 42); and $RhTi_2$ (996 ± 42) kJ/mol. More information is urgently needed here.

3.4 POLYNUCLEAR TRANSITION METAL COMPOUNDS

3.4.1 Compounds containing formal single bonds between metal atoms

The enthalpy of disruption, ΔH_D, of a neutral binuclear metal compound M_2L_{2n} (L is a neutral ligand)

$$M_2L_{2n}\ (g) \longrightarrow 2\ M(g) \ + \ 2\,n\ L(g) \quad (2)$$

can be evaluated and allocated to the M-L and M-M bonds according to the relation

TABLE 2. Dissociation Energies, D_0^o, of Heteronuclear Diatomic Transition Metal Compounds of the d- and f-series.

Series	Metal pair	D_0^o, kJ/mol	Series	Metal pair	D_0^o, kJ/mol
3d/3d	CrCu	155 ± 25	4d/4d	MoNb	448 ± 25
	CoCu	163 ± 21		RhY	441.8 ± 10.5
	CuNi	201 ± 21		PdY	237 ± 15[a]
3d/4d	RuV	410 ± 29	4d/5d	LaY	197 ± 17
	RhSc	440.3 ± 10.5		LaRh	524.7 ± 16.7
	RhTi	387.0 ± 14.6		IrY	452.8 ± 16
	RhV	360 ± 29		PtY	470 ± 8
	AgMn	96 ± 21		AuY	304.1 ± 8.2
	AgCu	169.5 ± 10.5		AuRh	228.9 ± 29
3d/5d	PtTi	394 ± 11		AuPd	151 ± 21
	AuSc	276.6 ± 17		AuAg	200.8 ± 10.5
	AuV	238 ± 12	5d/5d	IrLa	573 ± 12
	AuCr	209 ± 17		LaPt	500 ± 8.1
	AuMn	188 ± 13		AuLa	335 ± 21
	AuFe	188 ± 21	4d/4f	CeOs	503 ± 33
	AuCo	218 ± 17		CeIr	580.7 ± 25
	AuNi	251 ± 21		CePt	551.0 ± 25
	AuCu	224.3 ± 5.1		CeAu	335 ± 21
3d/4f	CuTb	187 ± 19		AuPr	405 ± 21
	CuDy	140 ± 19		AuNd	297 ± 21
	CuHo	139 ± 19		AuEu	238.9 ± 10.5
4d/4f	CeRu	527 ± 25		AuTb	289.5 ± 33
	CeRh	545.6 ± 25		AuDy	254 ± 20
	EuRh	231.8 ± 34		AuHo	236.6 ± 33
	CePd	318.4 ± 17		AuLu	328.4 ± 17
	AgEu	123 ± 12.5		LuPt	397.5 ± 34
	AgDy	124 ± 19	5d/5f	IrTh	570.7 ± 42
	AgHo	119 ± 17		PtTh	546.6 ± 42
4d/4f	RuTh	587.9 ± 42		AuU	318 ± 29
	RhTh	510 ± 21			
	RhU	516 ± 17			

Values taken from ref. 9 except
[a]I. Shim and K. A. Gingerich, Chem. Phys. Lett., 101 (1983) 528.

$$\Delta H_D = 2n\bar{D}(\text{M-L}) + E(\text{M-M}) \quad (3)$$

It is assumed that the value of $\bar{D}$(M-L) (the dissociation energy of the M-L bond), determined from a mononuclear compound ML_{n+1} can be transferred to M_2L_{2n} and to higher homologues, just as is done for p-block molecules. The values of E(M-M), the enthalpy contribution of the M-M bond, obtained in this way can be compared with the value of E(M-M)* derived from the enthalpy of atomization of the crystalline metal by dividing it by half the coordination number, Z, of the metal atom in the lattice (Table 1):

$$E(M\text{-}M)^* = 2\Delta H_f^o(M,g)/Z \tag{4}$$

This procedure implies a simple two-center electron pair bond description of the molecules. Another approach [12] relates E(M-M)* to the length, r, of the metal-metal bond, with the bulk metal taken as the standard. This bond enthalpy/bond length relation,

$$E(M\text{-}M)^* = Ar^{-4.6}, \tag{5}$$

indicates a logarithmic dependence and is based on simple geometric principles. The electron pair bond description can be criticized as naive when bridging ligands are present and the length/strength relation can be criticized as implying that a long bond is necessarily a weak bond--such may be true for bonds involving s and p orbitals, but the more diffuse character of d-orbital wave functions may invalidate the presumption to some extent. The results of the two approaches are indicated by Table 3.

The essence of the problem of interpretation is provided by [$Mn_2(CO)_{10}$]. In this molecule, the metal-metal bond is much longer (0.292 nm) than in the metal (0.274 nm), and manganese is the most volatile of the d-transition elements (see Table 1)--so both the length/strength model [12] (35 kJ/mol) and the electron pair bond model [13] (67 kJ/mol) produce values for E(Mn-Mn) in $Mn_2(CO)_{10}$ which are much lower than those derived from mass spectrometry [14] (104 kJ/mol) and from kinetics [15] (157 kJ/mol).

The data available for heterometallic diatomic molecules (Table 2) indicate that their dissociation enthalpies are not additive with respect to their homonuclear components and that deviations from additivity may be either positive or negative in an apparently random manner. The only information referring to polynuclear heterometallic complexes shows [16] that E(Fe-Ru) in [$FeRu_2(CO)_{12}$] is greater than in [$Fe_2Ru(CO)_{12}$] and that the average value in both compounds (95 kJ/mol) is greater than the mean of the two homonuclear bond enthalpy contributions in the bond length/bond enthalpy scheme.

The presence of ligands which bridge two or more metal atoms in a polynuclear metal complex provides for further (philosophical) problems. Historically, the presence of metal-metal bonds in molecules such as [$Fe_2(CO)_9$] and [$Fe_2(CO)_4(\eta\text{-}C_5H_5)_2$] was proposed on the evidence of their diamagnetism and the observed interatomic distances. More recently, experiments [17] and calculations [18] characterizing these and other

TABLE 3. Metal-metal Bond Enthalpy Contributions, E(M-M), in Polynuclear Metal Carbonyl Complexes, Derived from (i) two-center Electron Pair Bond Description and (ii) Bond Enthalpy/Bond Length Relation.

Metal Complex	ΔH_D, kJ/mol	E(M-M)(i), kJ/mol	E(M-M)(ii), kJ/mol
$Fe_2(CO)_9$	1173	80	70
$Fe_3(CO)_{12}$	1676	80	52, 65
$Ru_3(CO)_{12}$	2414	115	78
$Os_3(CO)_{12}$	2690	128	94
$Co_2(CO)_8$	1160	88	70
$Co_4(CO)_{12}$	2121	88	74
$Rh_4(CO)_{12}$	2648	110	86
$Rh_6(CO)_{16}$	3874	110	80
$Ir_4(CO)_{12}$	3051	127	117

molecules have led to the conclusion that there is no evidence for a direct metal-metal bond and suggested that electron spin correlation occurs as a result of a strong interaction between the metal d-orbitals and the π^* molecular orbitals of the bridging CO ligands in these compounds. It is too early to say whether these results imply that *all* bridged metal-metal axes which might be thought to involve a direct bond in fact achieve electron spin correlation through the bridge in the manner of copper(II) acetate. The consequence of such a change in description for the derivation of E(M-M) values is to increase E(M-M) in the electron pair bond model when it is established that the metal-metal axis is not bridged, but to leave it unchanged in the length/strength model.

3.4.2 Compounds containing bonds of formal multiple order between metal atoms

The interpretation of such thermochemical data as are available is the crucial problem in these systems, whether the metal-metal axis is bridged, as in $[Mo_2(O_2CR)_4]$, or not, as in $[Mo_2(NMe_2)_6]$.

The values of ΔH_f^o $[Mo(NMe_2)_4,g]$ and ΔH_f^o $[W(NMe_2)_6,g]$ have been determined [20], and from them it is possible to calculate $\bar{D}(Mo^{IV}-NMe_2)$ = (255.4 ± 5) kJ/mol and $\bar{D}(W^{VI}-NMe_2)$ = (222.1 ± 5) kJ/mol. Values of $\bar{D}(M-NMe_2)$ (M = Mo, W) in other oxidation states have to be estimated by comparison with the variation in the values of $\bar{D}$(M-X) in MX_n (M = Ti, Zr, Hf, X = Cl, Br, NMe_2, n = 4; M = Ta, n = 5, M = W, n = 6, X = Cl, Br, OMe). This interpolation may be uncertain, but it does enable the evaluation of

$$\Delta H_D = 6D(\text{M-NMe}_2) + D(\text{M}\overset{3}{\text{—}}\text{M}) \qquad (6)$$

in $[M_2(NMe_2)]$ (M = Mo, W). (The superscript 3 in $D(\text{M}\overset{3}{\text{—}}\text{M})$ indicates the formal bond order.) The detailed argument concerning whether these compounds should be regarded in terms of the formal oxidation state of the metal, M(III), or in terms of the hexavalence of M, or in terms of the coordination number (four) of M in the compound have been set out in detail [1, 2, 20]. It is concluded that, until more experimental information becomes available, it is preferable to recognize the metal atom as four-coordinate in these systems. Following this assumption, the values $D(\text{Mo}\overset{3}{\text{—}}\text{Mo})$ = (396 ± 18) and $D(\text{W}\overset{3}{\text{—}}\text{W})$ = (558 ± 20) kJ/mol are obtained. An alternative approach which relates metal-metal bond length to the bond enthalpy contribution in binuclear molybdenum complexes has been proposed [21]. The empirical relation takes the form

$$E(\text{Mo-Mo}) = 4.42\text{x}10^{12}[r(\text{Mo-Mo})]^{-4.29} \qquad (7)$$

and is based on data for molybdenum metal. A value $D(\text{Mo}\overset{3}{\text{—}}\text{Mo})$ = 384 kJ/mol in $Mo_2(NMe_2)_6$ can be calculated using this equation.

The estimation of the metal-metal bond enthalpy contribution in bridged complexes presents additional problems. These have been discussed in detail for the compounds containing a formal quadruple bond between two chromium or two molybdenum atoms in various carboxylate, 1,3-diketonate, and 2-hydroxypyridine complexes [22, 23]. One approach [22], which relies crucially on the assumption that $\bar{D}$(Mo-O) for pentane-2,4-dionato (acac) groups are the same in $[Mo_2(O_2CMe)_2(acac)_2]$ and in $[Mo(acac)_3]$, gives $E(\text{Mo}\overset{4}{\text{—}}\text{Mo})$ = 321 kJ/mol in $[Mo_2(O_2CMe)_4]$. The alternative treatment, which takes account of the bond length r(Mo-Mo) = 0.2093 nm, gives [21, 23] $E(\text{Mo}\overset{4}{\text{—}}\text{Mo})$ = 480 kJ/mol.

In spite of these difficult problems of interpretation, there is a reasonable consistency (Table 4) between measured bond length, bond strength, and formal bond order of the metal-metal bond in molybdenum compounds.

Other methods have been applied to the evaluation of metal-metal bond enthalpies. Birge-Sponer extrapolation of the vibrational progression

TABLE 4. Metal-metal Bond Length, *r*, Bond Enthalpy Contribution, E, and Formal Bond Order in Binuclear Molybdenum Compounds

Metal complex	Bond order	Bond length, *r*, nm	Bond enthalpy E(Mo-Mo), kJ/mol
$Mo_2(CO)_6(\eta\text{-}C_5H_5)_2$	1	0.3235	75
$Mo_2(OPr^i)_8$	2	0.2523	220
$Mo_2(OPr^i)_6$	3	0.2222	378
$Mo_2(NMe_2)_6$	3	0.2214	384
$Mo_2(O_2CMe)_4$	4	0.2093	480
$Mo_2(mhp)_4$	4	0.2065	505

[a] Hmhp = 6-methyl-2-hydroxypyridine

in (Re-Re) observed in the resonance Raman spectrum of the $[Re_2Br_8]^{2-}$ ion in the solid state gives $D(Re\overset{4}{—}Re)$ *ca.* 500 kJ/mol. This value can be compared with the results of measurements of the heat of oxidation of $Cs_2[Re_2Br_8]$ in acidic bromate solution [25], which lead to $E(Re\overset{4}{—}Re)$ = 408 ± 50 kJ/mol. Resonance Raman spectra of $[Mo_2X_8]^{4-}$ (X = Cl, Br) ions lead to estimates [26] of $E(Mo^4\text{-}Mo)$ in the range 480-550 kJ/mol.

3.5 IMPLICATIONS FOR CATALYSIS

The importance to catalysis of the thermochemical properties of metal clusters is clear. Often metal clusters are not stable under conditions of catalytic conversion because the metal-metal bonds are relatively weak, and the conditions required to create sites of coordinative unsaturation for bonding and activation of reactants are sufficient to destroy the clusters. There are still far too few thermochemical data to allow accurate predictions. The availability of such data would facilitate the prediction of whether intact clusters of a given structure would be stable under catalytic conditions; such data would also provide a basis for judging the plausibility of proposed reaction mechanisms.

REFERENCES

1. J. A. Connor, in B. F. G. Johnson (Ed.), Transition Metal Clusters, Wiley, Chichester, 1980, 345.
2. J. A. Connor and H. A. Skinner, ACS Symp. Ser., 155 (1981) 197.
3. G. Pilcher and H. A. Skinner, in F. R. Hartley and S. Patai (Eds.), The Chemistry of the Metal-Carbon Bond, Wiley,

Chichester, 1982, 43.
4. M. N. Jones and H. A. Skinner, Ann. Rep. Chem. Soc., (1982) 3.
5. Faraday Symp. Chem. Soc., 14 (1980).
6. a) L. Brewer and J. S. Winn, Ref. [5], p. 126;
b) A. R. Miedema, Ref. [5], p. 136.
7. N. A. Baykara, B. N. McMaster, and D. R. Salahub, Mol. Phys., 52 (1984) 891.
8. D. L. Michalopoulos, M. E. Geusic, S. G. Hansen, D. E. Powers, and R. E. Smalley, J. Phys. Chem., 86 (1982) 3914; M. D. Morse, S. G. Hansen, P. R. R. Langridge-Smith, Z. D. Zheng, M. E. Geusic, D. L. Michalopoulos, and R. E. Smalley, J. Chem. Phys., 80 (1984) 5400.
9. K. A. Gingerich, Ref. [5], p. 109.
10. K. Arvin, S. K. Loh, N. Aristov, and P. B. Armentrout, J. Phys. Chem., 87 (1983) 3593.
11. K. Hilpert and K. A. Gingerich, Ber. Bunsenges. Phys. Chem., 84 (1980) 739.
12. C. E. Housecroft, K. Wade, and B. C. Smith, J. Chem. Soc. Chem. Commun., (1978) 765.
13. J. A. Connor, M. T. Zafarani-Moattar, J. Bickerton, N. I. El Saied, S. Suradi, R. Carson, G. Al Takhin, and H. A. Skinner, Organometallics, 1 (1982) 1166.
14. D. R. Bidinosti and N. S. McIntyre, Canad. J. Chem., 48 (1970) 593.
15. A. J. Poë, ACS Symp. Ser., 155 (1981) 135.
16. A. K. Baev, J. A. Connor, N. I. El Saied, and H. A. Skinner, J. Organometal. Chem., 213 (1981) 151.
17. A. Mitschler, B. Rees, and M. S. Lehmann, J. Amer. Chem. Soc., 100 (1978) 3390.
18. W. J. Heijser, E. J. Baerends, and B. Ros, Faraday Symp. Chem. Soc., 14 (1980) 211.
19. F. A. Cotton and R. A. Walton, Multiple Bonds between Metal Atoms, Wiley, Chichester, 1982.
20. F. A. Adedeji, K. J. Cavell, S. Cavell, J. A. Connor, G. Pilcher, H. A. Skinner, and M. T. Zafarani-Moattar, J. Chem. Soc. Faraday I, 75 (1979) 603.
21. K. J. Cavell, J. A. Connor, G. Pilcher, M. A. V. Ribeiro da Silva, M. D. M. C. Ribeiro da Silva, H. A. Skinner, Y. Virmani, and M. T. Zafarani-Moattar, J. Chem. Soc. Faraday I, 77 (1981) 1585.
22. K. J. Cavell, C. D. Garner, G. Pilcher, and S. Parkes, J. Chem. Soc. Dalton, (1979) 1714.
23. K. J. Cavell, C. D. Garner, J. A. Martinho-Simoes, G. Pilcher, H. Al Samman, H. A. Skinner, G. Al-Takhin, I. B. Walton, and M. T. Zafarani-Moattar, J. Chem. Soc. Faraday I, 77 (1981) 2927.
24. R. J. H. Clark and N. R. D'Urso, J. Amer. Chem. Soc., 100 (1978) 3088.
25. L. R. Mross, R. J. Porcja, J. W. Nicoletti, J. San Filippo, and H. D. B. Jenkins, J. Amer. Chem. Soc., 102 (1980) 1923.
26. W. C. Trogler, C. D. Cowman, H. B. Gray, and F. A. Cotton, J. Amer. Chem. Soc., 99 (1977) 2993.

Chapter 4

REACTIVITIES OF METAL CLUSTERS

G. LAVIGNE and H. D. KAESZ

4.1 INTRODUCTION

Historically, studies of metal cluster complexes have dealt primarily with their synthesis, structure determination, and their possible relationship to metal surfaces. Recently, efforts have turned towards a systematic understanding of the chemical reactivity of this new class of compounds. Developments in this field are well represented by current review articles [1-6] and a comprehensive treatise which includes the chemistry of cluster complexes [7]; specific citations within this work as well as citations to more specialized reviews are made in the following pages.

In the present chapter, we emphasize aspects of cluster reactivity that have relevance to catalysis. First, we examine the pathways available to cluster complexes for coordination of substrates (reactants). We find cluster complexes possessing low-energy activation pathways leading to reactions with high selectivity in the products. Next, we present recently discovered means to enhance the reactivity of cluster complexes which are otherwise inert under mild conditions. Finally, to illustrate possibilities and limits to the use of cluster complexes as synthetic tools for organic synthesis and catalysis, we analyze cluster-assisted transformations of organic substrates.

4.2 PATHWAYS FOR COORDINATION OF SUBSTRATE (SUBSTRATE ACTIVATION)

4.2.1 Introduction

Except for the few cases of cluster complexes which are electronically unsaturated, reactivity of a cluster requires activation by some means. This is usually achieved by loss of ligand or by opening of a metal-metal bond.

Since metal-ligand and metal-metal bond energies in many cases are of comparable magnitude [6, 8], the activation pathways may involve a combination of these two transformations. When breaking of M-M bonds is prevalent, opening of the polyhedral structure is observed in the presence of nucleophiles. This opening may also lead to cluster fragmentation. In attempts to prevent such fragmentation, cluster complexes have been placed on solid organic or inorganic supports (see Chapter 9). The goal of such cluster immobilization is to keep the nuclearity of the cluster intact while permitting processes involving reversible metal-metal bond opening.

By contrast, selective ligand dissociation is achieved through "lightly stabilized" clusters. Historically [9], this term describes clusters containing labile mono*hapto* ligands whose dissociation is preferred over the breaking of metal-metal bonds. We extend the use of this term to include cluster complexes containing labile poly*hapto* ligands or labilizing groups, as explained in Section 4.2.3.

4.2.2 Additions to metal clusters

4.2.2.a Additions to multiple bonds in unsaturated clusters. A cluster is considered electronically unsaturated if it contains less than a closed shell of electrons. This is determined by the filling of cluster valence molecular orbitals (CVMO's) [10], or derived from electron counting rules such as the Polyhedral Skeletal Electron Pair theory (PSEPT) [11, 12], or, more recently, determined by a Topological Electron Counting (TEC) theory [13, 14].

It should be noted that a cluster will also exhibit a deficiency of 2 e^- for each d^{10} metal fragment [e.g., $Pt(PPh_3)_2$] it contains. Such clusters do not generally exhibit the reactivity associated with electronic unsaturation, except for one remarkable case mentioned below.

Unsaturated clusters are indeed few in number. These are summarized in Table 1 [15-31]. The prototype for this class of compound is $Os_3\{\mu\text{-}H\}_2(CO)_{10}$, **1**, which contains 46 e^-, as compared with 48 e^- required for a saturated trinuclear cluster. A localized H-bridged double bond is seen in its structure [32]. This complex is obtained in high yield by reaction of hydrogen with the parent complex $Os_3(CO)_{12}$ [33].

Complex 7 in Table 1 deserves special comment. The unsaturated metal-metal bond in this cluster is bridged on one side by a hydrogen atom but on the second side by the *ortho*-carbon atom of a phenyl group on phosphorus [21].

Addition of electron pair donors. The unsaturation of $Os_3\{\mu\text{-}H\}_2(CO)_{10}$, 1, is reflected in its reaction with electron donors under mild conditions (*ca.* 25°C) (see SCHEME 1) [35, 36].

Through such reactivity, the trinuclear unit has been attached to polymers (Chapter 9.1) and to inorganic supports such as silica and alumina [37, 38] (Chapter 9.2). Supported clusters may also be prepared through the use of lightly stabilized cluster complexes (see Section 3.2.3.a).

Hydrogenation of alkenes. The low-activation-energy pathway to add electron pair donors enables 1 to effect the hydrogenation of alkenes [39]; prior coordination of the olefin is required. Hydrogenation occurs in competition with alkene isomerization which takes place in the absence of

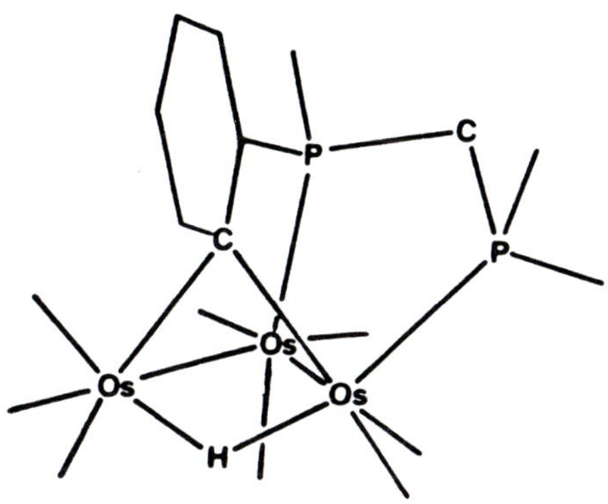

7

STRUCTURE 7*

*In the structures for complexes 7 and for 1 and 13 in SCHEME 1 below, we have omitted the line joining the two metal atoms bridged by one or two hydrogen atoms. This representation conveys three-center bonding in which the metal-metal interaction is part of the bridged M-X-M system [34].

This convention is not commonly practiced. More often, the line or lines are taken to represent bond order between the metal atoms: a double bond (two lines) in $Os_3\{\mu\text{-}H\}_2(CO)_{10}$, with 4 e^- in the bridging system [32]; a single bond in $M_3\{\mu\text{-}H,\mu\text{-}X\}(CO)_{10}$ (X = halogen or OR) with 6 e^- in the bridging system; no line in $M_3\{\mu\text{-}X\}_2(CO)_{10}$ (with 8 e^- in the bridging system the M-M bond order = 0) [34c, d]. In polyhydrido complexes, lines between metal atoms help to indicate the underlying connectivity in the cluster polyhedron. Thus, we have included such lines between bridged metal atoms where the bond order is 1 and have omitted such lines where the metal-metal bond order is 0 in structures following SCHEME 1.

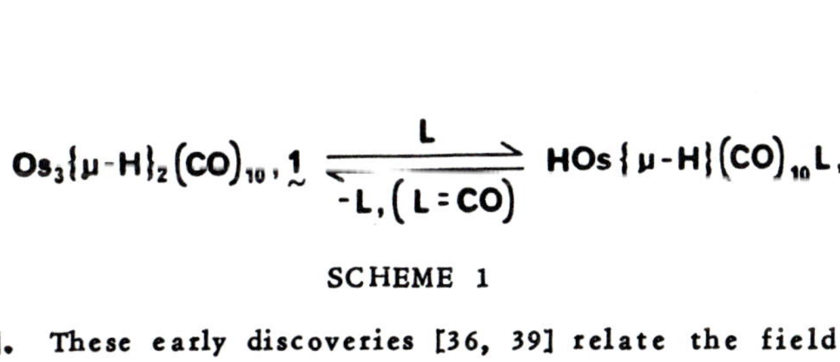

$$Os_3\{\mu\text{-}H\}_2(CO)_{10},\ \mathbf{1} \underset{-L,\ (L = CO)}{\overset{L}{\rightleftharpoons}} HOs\{\mu\text{-}H\}(CO)_{10}L,\ \mathbf{13}$$

SCHEME 1

hydrogen [36]. These early discoveries [36, 39] relate the field of cluster complexes to homogeneous catalysis (also see Chapter 5). Subsequent

TABLE 1.

Electronically Unsaturated Metal Clusters.

Cluster complex	Chemical reactivity	Reference
1 $Os_3\{\mu\text{-}H\}_2(CO)_{10}$	many reactions; see text	[15]
2 $Os_3\{\mu\text{-}H,\mu\text{-}AuPR_3\}(CO)_{10}$	reversible CO uptake; yields **4a** upon reaction with PPNCl	[16]
3 $Os_3\{\mu\text{-}AuPR_3\}_2(CO)_{10}$	unreactive	[17]
4a [PPN] $[Au\{Os_3\{\mu\text{-}H\}(CO)_{10}\}_2]^a$		[18a]
4b [PPN] $[Ag\{Os_3\{\mu\text{-}H\}(CO)_{10}\}_2]$	the anion accepts 2 e^- (electrochemically)	[18b]
5a $Fe_3\{\mu,\eta^2\text{-}Ph_2C_2\}(CO)_9$		[19a]
5b $Os_3\{\mu,\eta^2\text{-}Ph_2C_2\}(CO)_9{}^b$	CO; PPh_3; CH_2N_2**	[19b]
6 $Os_3\{\mu\text{-}H\}_3(\eta^1\text{-}SiPh_3)(CO)_9$		[20]
7 $Os_3\{\mu\text{-}H,\mu\text{-}C_6H_4\}\text{-}(P(Ph)CH_2P(Ph)_2)(CO)_8$	reversible CO uptake	[21]
8 $Re_4\{\mu\text{-}H\}_4(CO)_{12}$	decomposes in presence of L	[22, 23]
9 $[PPN][Re_3\{\mu\text{-}H\}_4(CO)_{10}]^c$	reacts mainly with electrophiles (e.g., halogens, HX, carboxylic acids)	[24]
10 $Fe_4\{\mu_4\text{-}PR\}_2(\mu\text{-}CO)(CO)_{10}{}^d$	adds CO (reversibly); adds phosphorus ligands	[26]
11 $Rh_4\{\mu_4\text{-}PR\}_2(cod)_4$	may be labile in presence of H_2	[28]
12 $[RhHL_2]_x$;		
a, L = $P(OCH_3)_3$; x = 3	catalyst precursors for olefin and acetylene hydrogenations	[29]
b, $L_2 = R_2PCH_2CH_2PR_2$; x = 4		[30]
c, L_2 = cod; x = 4	reacts with H_2 and ethylene	[31]

[a]$PPN^+ = (PPh_3)_2N^+$; cod = cyclo-octa-1,5-diene.

[b]See Section 3.4.7, below.

[c]Other examples are known in rhenium chemistry [25].

[d]Equivalent complex known in ruthenium chemistry [27].

TABLE 2.

"Lightly stabilized" clusters[a].

Cluster complex		Reacts with:	References
36	$Os_3(CO)_{11}CH_3CN$	CO, PPh_3, C_2H_4, Py, H_2, HX; $HRe(CO)_5$; $OsH_2(CO)_4$	[64, 65, 66]
37	$Os_3(CO)_{10}(CH_3CN)_2$	PR_3, $P(OR)_3$, CNR, dienes, acetylenes, py, Benzene; thioamides; $HRe(CO)_5$;	[9, 67, 68]
		$OsH_2(CO)_4$, $CpWH(CO)_3$,	[65, 69]
		$Os_3(CO)_9(\mu_3S)_2$;	[70]
		silica or alumina	[71]
38	$Os_3(CO)_{10}(\eta^4\text{-}C_6H_8)$	ROH; RSH; amines; X_2; RCO_2H; vinyl benzene	[72]
39	$Os_3(CO)_{10}(\eta^2\text{-}C_8H_{14})_2$	$HRe(CO)_5$	[65]
40	$[Os_4\{\mu\text{-}H\}_3(CO)_{12}(CH_3CN)_2][BF_4]$	reaction with PPN Cl or PPN NO_2	[73]
41	$Fe_3\{\mu_3\text{-}PPh\}_2(CO)_8CH_3CN$		
42	$Fe_3\{\mu\text{-}_3PPh\}_2(CO)_7(CH_3CN)_2$	CO, RCN, PR_3, $P(OR)_3$, AsR_3 , CO site exchange localized on individual atoms	[74]
43	$Co_4\{\mu_4PPh\}_2(CO)_9CH_3CN$		
44	$Re_3\{\mu\text{-}H\}_3(CO)_{10}(CH_3CN)_2$	I^-, RCO_2^-, MeO^-, Py, PPh_3; (i) OH^-, (ii) H^+ gives benzamide	[75, 76]

[a] See also $Rh_4[\mu_4\text{-}PR\}_2(cod)_4$ in Table 1.

studies have provided many other examples of transformations of unsaturated alkenes [40]. The unsaturated cluster 1 has proved to be an ideal complex on which either organic or isolobal inorganic fragments may become activated, as discussed below.

Reaction with diazomethane. Studies of the reaction of CH_2N_2 with 1 are shown in SCHEME 2 [41]. These have given species in which hydrocarbon fragments are bound in several ways to two or three metal centers, much as might be envisaged for the reactions of hydrocarbons with metal surfaces [42]. Other reactions of diazomethane leading to the coupling of CH_2 with hydrocarbon groups on cluster complexes are presented in Section 3.4.7.

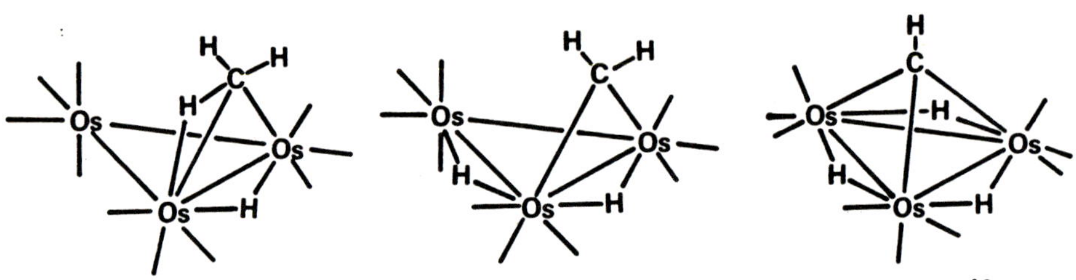

14 15 16

SCHEME 2

Building up of clusters and synthesis of mixed metal clusters. A generally useful synthetic route to larger metal cluster complexes and to mixed-metal complexes is provided by the addition of ligated metal fragments to the unsaturated clusters. The first demonstration of this principle was the synthesis or the mixed metal derivative $Os_3Pt\{\mu\text{-}H\}_2(CO)_{10}PR_3$, **17**, obtained from the reaction of $Os_3\{\mu\text{-}H\}_2(CO)_{10}$, 1 with $Pt(C_2H_4)_2PR_3$ [43]. This 58-e^- complex reacts irreversibly with phosphines or reversibly with CO or H_2 to give 60-e^- complexes, as shown in SCHEME 3 [43]. The butterfly structure for **18** continues to be electronically unsaturated, the closed valence shell for this structure requiring 62 e^-. It should be noted here that complexes **14** and **18** illustrate the isolobal principle developed by Hoffmann [44] and discussed more recently by Stone [45]. This principle can be used for logical syntheses of many mixed metal clusters, further indicated in SCHEME 4 and in SCHEME 26 in Section 4.4.6. A more detailed discussion of cluster synthesis is given in Chapter 1.

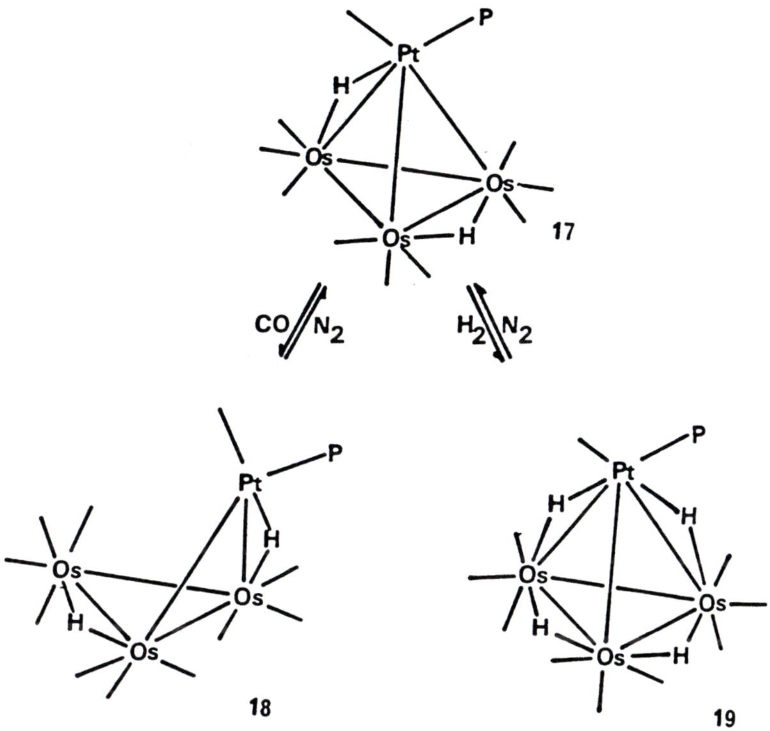

SCHEME 3

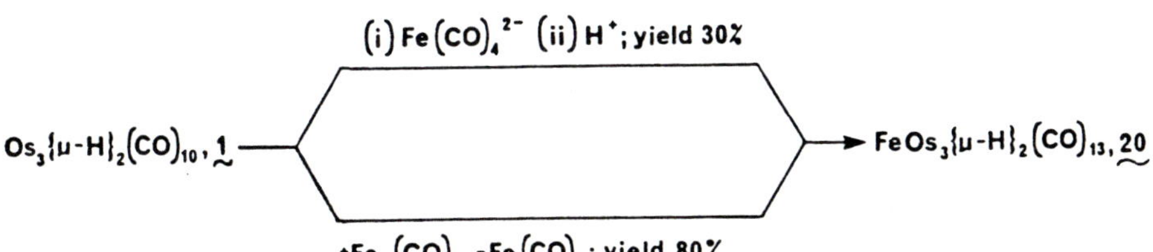

SCHEME 4.

Two further examples of the building up of clusters are seen in reactions of $Os_3\{\mu\text{-}H\}_2(CO)_{10}$, **1** (SCHEME 4) [46]. In these we can identify amphoteric properties for **1**. This complex is serving as an electron acceptor (Lewis acid) by adding $Fe(CO)_4^{2-}$ (top path) or as an electron donor (Lewis base) by displacing $Fe(CO)_5$ from $Fe_2(CO)_9$ (lower path).

4.2.2.b Disruption of metal-metal bonds. *Cluster degradation (in the presence of nucleophiles).* The kinetics of fragmentation of 60-e$^-$ tetranuclear clusters in the presence of CO has been investigated [47], the results indicating that fragmentation of $H_2Ru_4(CO)_{13}$, **21**, and related mixed metal derivatives proceeds by a second-order [CO]-dependent path. This conclusion is consistent with a mechanism involving intermediate association of CO somewhere on the cluster to give a 62-e$^-$ species, which is believed to contain a butterfly arrangement of metals [47]. Two other cluster fragmentations (not necessarily connected with nucleophilic attack) are discussed below in the reactions leading to **68** and **96** in SCHEMES 14 and 26, respectively.

Cluster nuclearity intact (opening of polyhedral structures). Examples of cluster opening (and unfolding) have been realized for the extensive series of transformations starting with the polynuclear complex $Os_6(CO)_{18}$, **22** (SCHEME 5) [48-50]. The structures of the starting complex and the tetrakis substituted raft complex $Os_6(CO)_{17}(P(OMe)_3)_4$, **24**, have been established [49].

The raft complex is one of the first such structures to be observed for a metal complex and bears relation to the types of structures observed for metal aggregates on solid supports [51]. As shown in SCHEME 5, a change in synthesis conditions may yield products of different nuclearity.

To provide more control of the nuclearity of the clusters, investigators have introduced a variety of non-fluxional polydentate phosphines or arsines [52-54] or non-fluxional bridging atoms [55]. Another means of cluster stabilization may be achieved by encapsulated atoms, exemplified by the phosphorus in $Rh_9P(CO)_{21}$, **27** [56].

Using the phosphinidene ligand as a bridge between three metal atoms, Huttner *et al.* [57] observed the reversible opening of a metal-metal bond concomitant with reversible CO addition (SCHEME 6).

Further examples of such reversible metal-metal bond cleavage have been observed for tellurium-bridged complexes [58] and are also believed to play a role in hydroformylation reactions catalyzed by $Co_3\{\mu_3\text{-}CPh\}(CO)_9$, **30**, or $Co_4\{\mu_4\text{-}PPh\}_2\{\mu\text{-}CO\}_2(CO)_8$, **31** [55, 59].

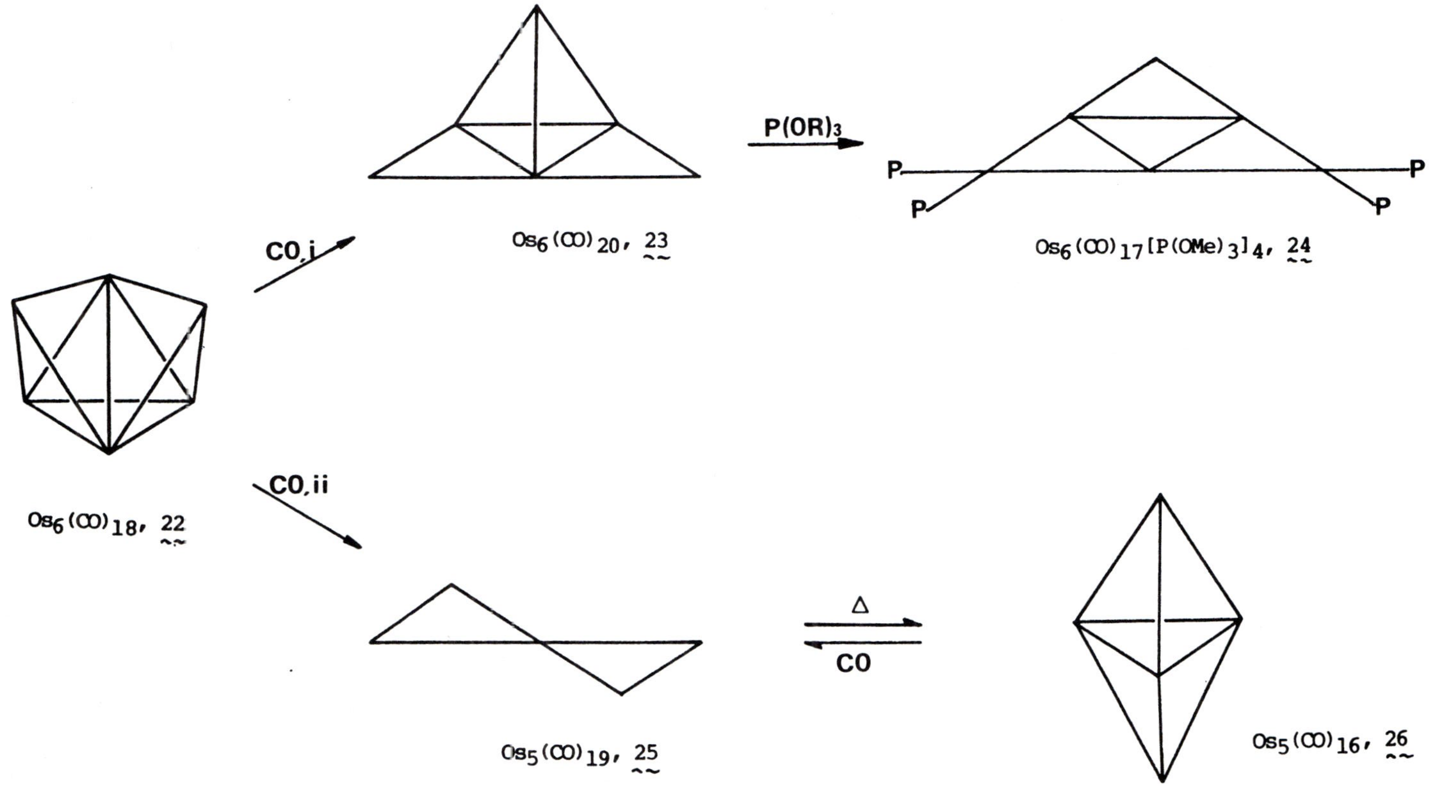

SCHEME 5
i) solid state
ii) heptane solution

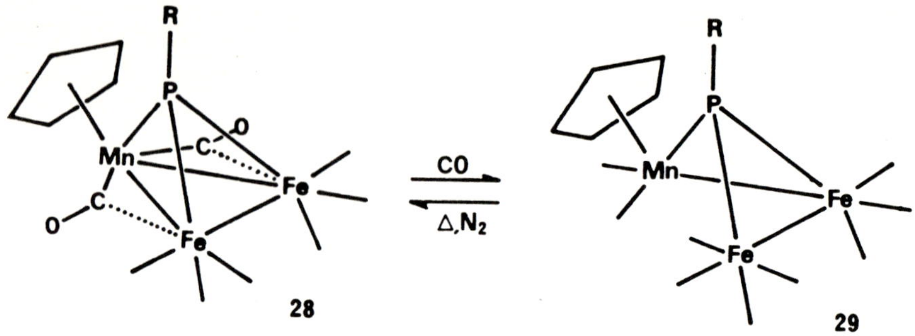

28 29

SCHEME 6

In electron-precise cluster complexes, metal-metal bond cleavage is observed by the addition of electrons through donor groups or by simple reduction of the cluster. Where the LUMO is localized, one metal-metal bond may be seen to be broken. Where the LUMO is delocalized over several metal-metal bonds, addition of electrons causes weakening of several of these, as first observed for the tetrameric cyclopentadienyliron sulfide series, $[Fe_4(\eta^5\text{-}C_5H_5)_4)\mu_3\text{-}S)_4]^{n+}$, n = 2, **32a** or n = 0, **32b** (SCHEME 7) [60]. This and other cubane cluster systems showing similar behavior have been reviewed by Garner [61].

Further examples of this phenomenon may be seen in comparison of the 90-valence-electron cluster, $Os_6(CO)_{17}\{P(OMe)_3]_4$, **24** [49], containing a fully bonded internal triangle of metal atoms with that of the 92-valence-electron cluster anion, $[Ru_6H(\mu\text{-}O{=}CNMe_2)_2(\mu\text{-}CO)_4(CO)_{14}]^-$, **33**, [62] (see Fig. 1); all three of the internal metal-metal interactions in 33 are seen to be weakened equally. This is believed to result from occupation of an anti-bonding orbital symmetrically disposed over all three metal atoms of the internal triangle in **33**.

In some cases such electron-rich species may exhibit reactivity associated with complete breaking of the weakest bonds. This is seen in the 64-e$^-$ cluster $Os_4(CO)_{12}(\mu_3\text{-}S)_2$, **34** (SCHEME 8) [63a]. Addition of carbon monoxide to this cluster results in complete dissolution of the formerly weakened metal-metal interactions, a reaction also observable for the closely related mixed metal (Os_3W) species [63b].

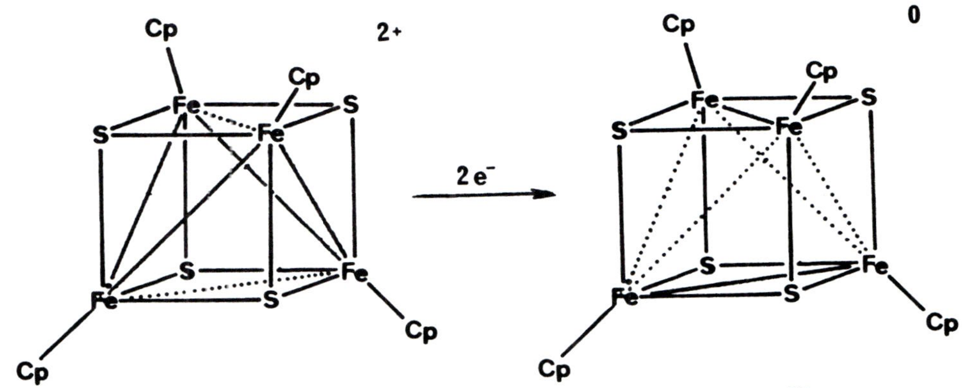

32 a

32 b

SCHEME 7

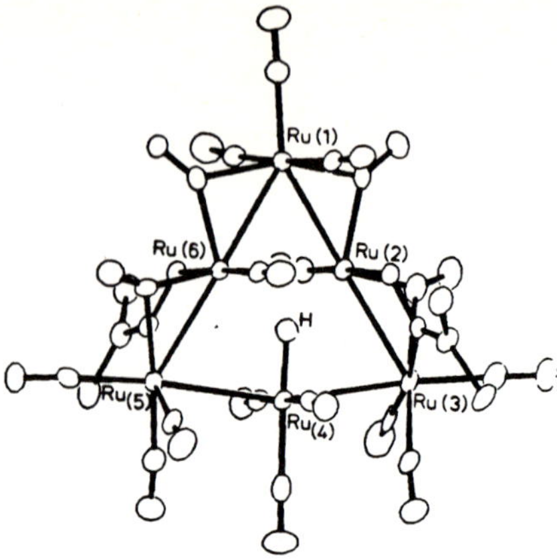

Fig. 1. ORTEP projection of the hexa-nuclear anion, $[Ru_6H\{\mu\text{-}O{=}CNMe_2\}_2\{\mu\text{-}CO\}_4(CO)_{14}]^-$. Copyright, Gesellschaft Deutscher Chemiker, 1983; reproduced with permission.

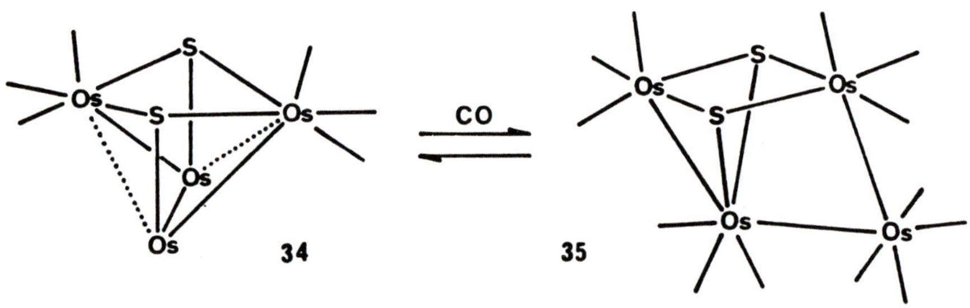

34 35

SCHEME 8

4.2.3 Dissociation of metal-ligand bonds or reductive elimination of substrates

A variety of reactive complexes may be obtained by replacement of one or more CO groups with less tightly held ligands or by using bridging ligands that may change their degree of attachment to bring about coordinative unsaturation. In some of these latter complexes, we see labilization of the cluster, especially of carbonyl groups on the bridged metal atoms, without apparent overall change in the coordination of the bridging atom or ligand. Such cases are treated at the end of this section.

4.2.3.a Lightly stabilized clusters. *Labile Mono-hapto Ligands.* Once again the prototypes for this class are found in osmium chemistry (Table 2) [9, 64-76]. Complexes $Os_3(CO)_{12-x}(CH_3CN)_x$ (x = 1), **36**, and (x = 2), **37** are prepared from $Os_3(CO)_{12}$ by oxidation of carbonyl groups by $(CH_3)_3NO$ in acetonitrile. In this method, CO_2 is formed *in situ* and is eliminated as

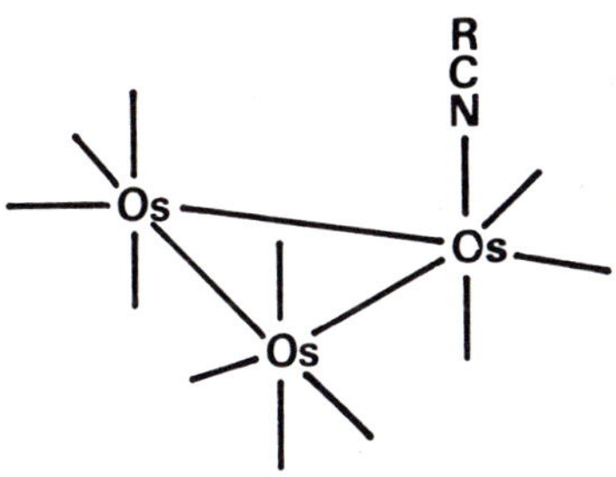

36

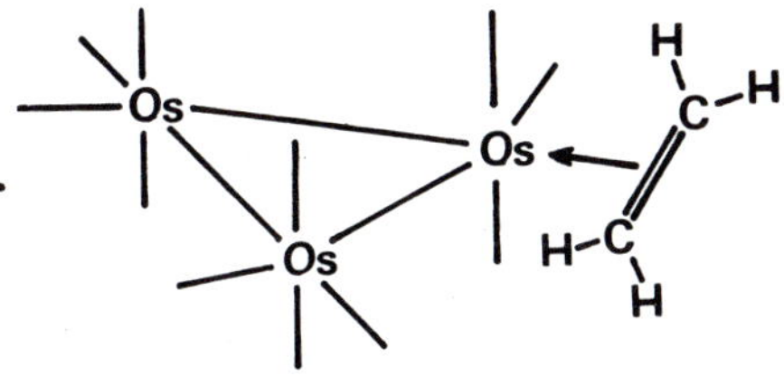

45

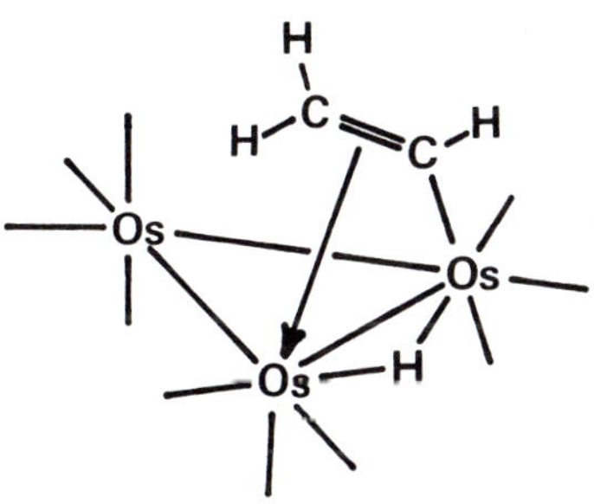

46

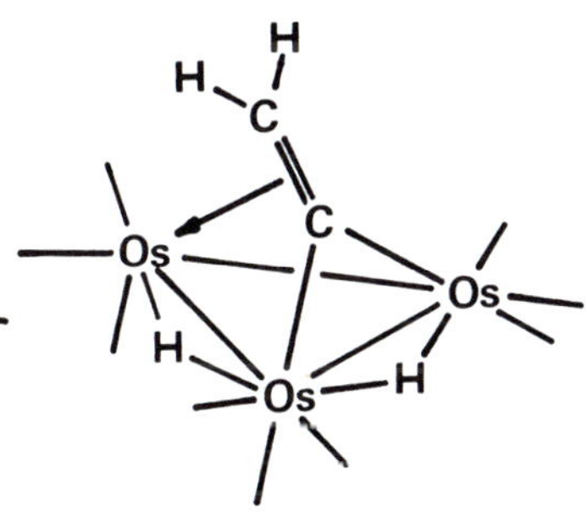

47

SCHEME 9

a small molecule (see also discussion in Section 4.2.3.b). Both are useful starting materials for permitting a wide range of organic or inorganic substrates to become coordinated to the clusters under mild conditions, also given in Table 2. For instance, these permit isolation of ethylene complexes such as those shown in SCHEME 9, which are not stable under the thermal reactions required to displace CO from the parent metal clusters [77].

Another example of the use of labile ligands is in attachment of metal clusters to catalyst supports (see Chapter 9). For example, reactive surface hydroxyl groups on the support become involved in bonding to the cluster, as shown in the structure of **48** [71]:

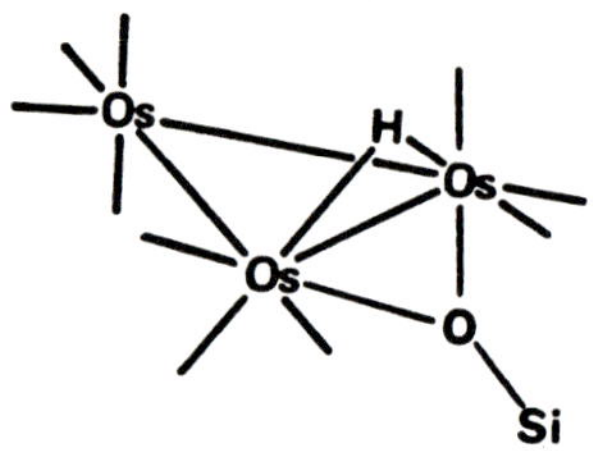

48

Labile polyhapto ligands. In this category we find that the heavier electron donor atoms exhibit lability between several degrees of coordination to a cluster, as shown in SCHEMES 10 [64, 78] and 11 [79]. Whereas in the first case [64, 78] the transformation is spontaneous, the second [79] requires prior alkylation of the sulfur atom. This reduces the availability of electrons from 5 to 3, releasing an orbital on the cluster for reversible activation of ethylene or irreversible addition of CO, CH_3CN, or PPh_3.

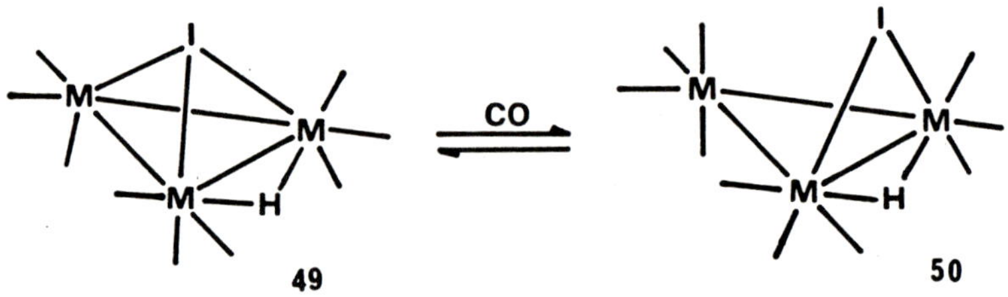

SCHEME 10

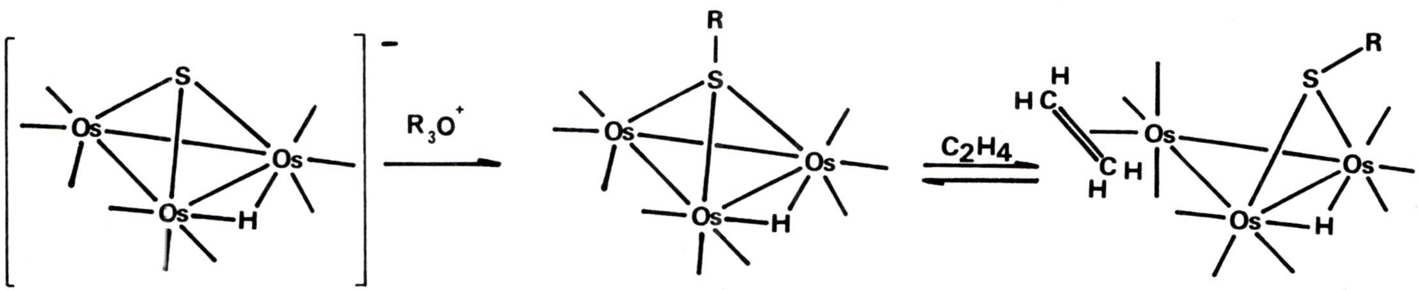

51 52 53

SCHEME 11

A comparable process of variable coordination of a bridging group has been postulated to provide the catalytic activity in the hydrogenation of ethylene by the silica-grafted cluster $Os_3\{\mu\text{-H},\mu\text{-OSi}\equiv\}(CO)_{10}$, **48** [80] (Chapter 9.2).

In this connection, we should also mention the formally unsaturated complex $Ru_3\{\mu\text{-H},\mu\text{-PPh}_2\}(CO)_9$, **54** (SCHEME 12) [81]. In this complex, the PPh_2 unit is seen to be blocking an axial coordination site on the unique ruthenium atom. Complex **54** reacts with CO to give $Ru_3\{\mu\text{-H},\mu\text{-PPh}_2(CO)_{10}\}$ **55**, or, with acetylene (without CO loss), yielding the 50-e^- open cluster $Ru_3\{\mu\text{-H},\mu\text{-PPh}_2,\mu\text{-}\eta^2\text{-C}_2Ph_2\}(CO)_9$, 56, (SCHEME 12) [81].

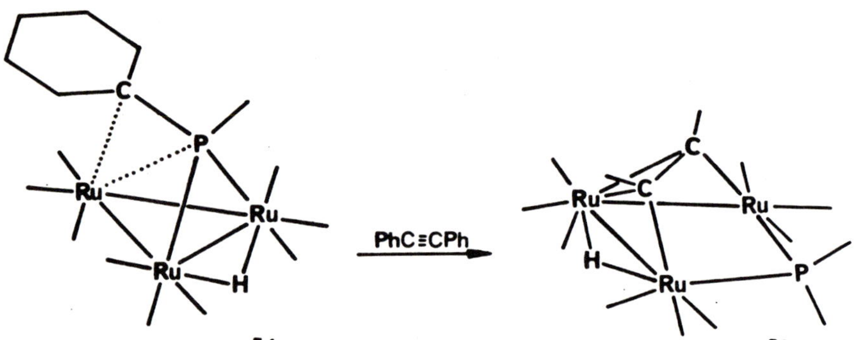

54 **56**

SCHEME 12

A similar unsaturation due to steric factors is seen in the acetylene complex $Fe_3\{\mu,\eta^2\text{-C}_2Ph_2\}(CO)_9$, **5a** (Table 1); due to the extremely low yields in its synthesis, the chemistry of this complex has not been examined. Reactions of the osmium analogue **5b** are discussed in Section 3.4.7.

Although bridging phosphido groups are generally associated with a role of maintaining cluster integrity as nonfluxional ligands (see previous section), bridging phosphido groups at binuclear centers have recently been observed to engage in a variety of insertion reactions with alkynes [82, 83], alkenes [84], hydrides [85], or the formyl group [86]. In the latter connection, we cite the doubly bridged binuclear anion $[(CO)_4W\{\mu\text{-PPh}_2\}\{\mu\text{-P(Ph}_2)C(O)H)\}Os(CO)_2PMePh_2]^-$, arising from hydride attack on the doubly bridged precursor, $(CO)_4W\{\mu\text{-PPh}_2\}_2Os(CO)_2PMePh_2$, 57, to give one example of the type of insertion (*i.e.*, migration) that is being observed.

A final example of bridge opening reactions is that for the orthometallated complex $Ru_3\{(C_6H_4\text{-P(Ph)CH}_2P(Ph)\}(CO)_9$, **58** (SCHEME 13) [87]. Addition or CO to **58** causes migration of the o-metallated phenyl

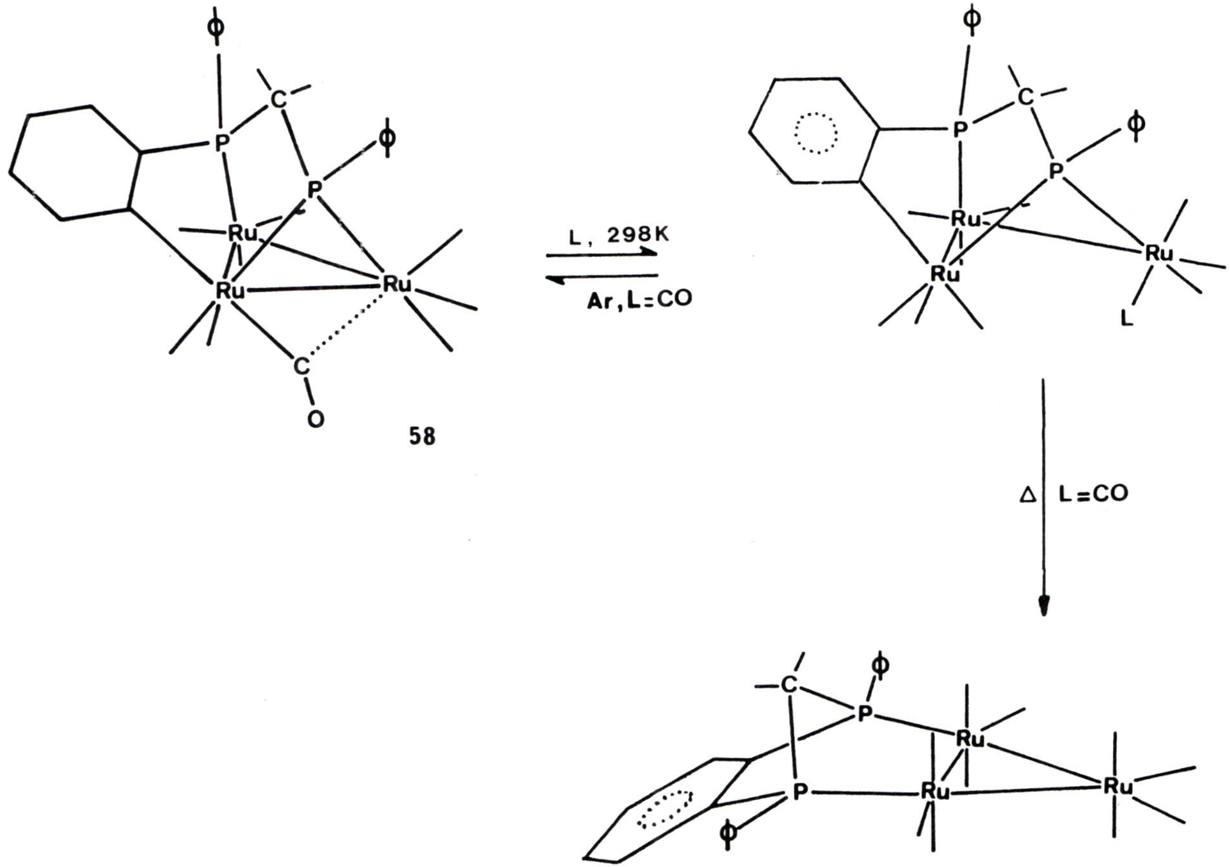

58

59

SCHEME 13

group to the bridging phosphide (reductive elimination), converting the coordinated group to a new chelating ligand in complex **59**.

Labilizing groups. Until recently it was tantalizing that no ruthenium analogues of the two reactive triosmium complexes, $Os_3\{\mu\text{-}H\}_2(CO)_{10}$, **1**, or $Os_3(CO)_{11}CH_3CN$, **36**, were known. In 1983, Knox and coworkers announced the synthesis of $[\eta^5\text{-}C_5H_5Ru(CO)H]_3$, **60** [88]. This trimer reacts readily with unsaturated hydrocarbons to give the triply bridged alkylidene complexes, $[\eta^5\text{-}C_5H_5Ru(CO)H]_3\{\mu_3C\text{-}R\}$, **61**. It is noteworthy that its reactivity is derived neither from unsaturation nor from possession of labile ligands *per se*.

High reactivity is also observed in the series of edge double-bridged complexes $Ru_3\{\mu\text{-}NO\}_2(CO)_{10}$, **62** [89], $Ru_3\{\mu\text{-}H,\mu\text{-}X\}(CO)_{10}$, where X = carboxamido (O=CNR$_2$), **63** [90-92], acyl (O=CR), **64** [93], halide (Cl, Br, or I) **60a, b, c** [78, 93] or in their conjugate base edge-bridged anions [91-93]. In the latter group, we also count $[Ru_3\{\mu\text{-}OC(R)O\}(CO)_{10}]^-$, where OC(R)O = acetate or formate, **65a, b** [94]. These complexes exchange CO readily under ambient temperature and pressure and lead readily to phosphine-substituted products, specifically on the metal atoms along the bridged or double-bridged edge [95, 96]. Hydrido-bridged complexes react with ethylene at ambient temperature and 3 atm to give acyl bridged products (see SCHEME 14) [97]. The second reaction in SCHEME 14 illustrates one step of a hydroformylation cycle that the hydrido-bridged anion is known to catalyze (SCHEME 15) [98].

A comparison of the ease of substitution of the edge double-bridged series, including the binuclear complex obtained from the reaction with ethylene, is shown in SCHEME 16 [96].

4.2.3.b Elimination of small molecules. Reductive elimination of small molecules creates coordinative unsaturation at one or more metal centers. This is of course an important component of any catalytic cycle, but in the case of cluster-coordinated substrates, desorption of the product is usually the point at which many cluster complexes fail to complete their catalytic cycles. It is, for instance, the reason that trinuclear cluster complexes do not catalyze the hydrogenation of nitriles [99]; the reductive elimination of an amine is extremely unfavorable from the triply bridged hydrido-nitrene complexes, $M_3\{\mu\text{-}H\}_2\{\mu_3\text{-}N(R)\}$, where M = Fe, **73a** [100]; M = Ru, **73b** [101]; or M = Os **73c** [102].

In contrast, the elimination of RH is favored, as observed in the cycle for hydrogenation of olefin complexes by $Os_3\{\mu\text{-}H\}_2(CO)_{10}$ [34, 35] (see also Section 3.2.2).

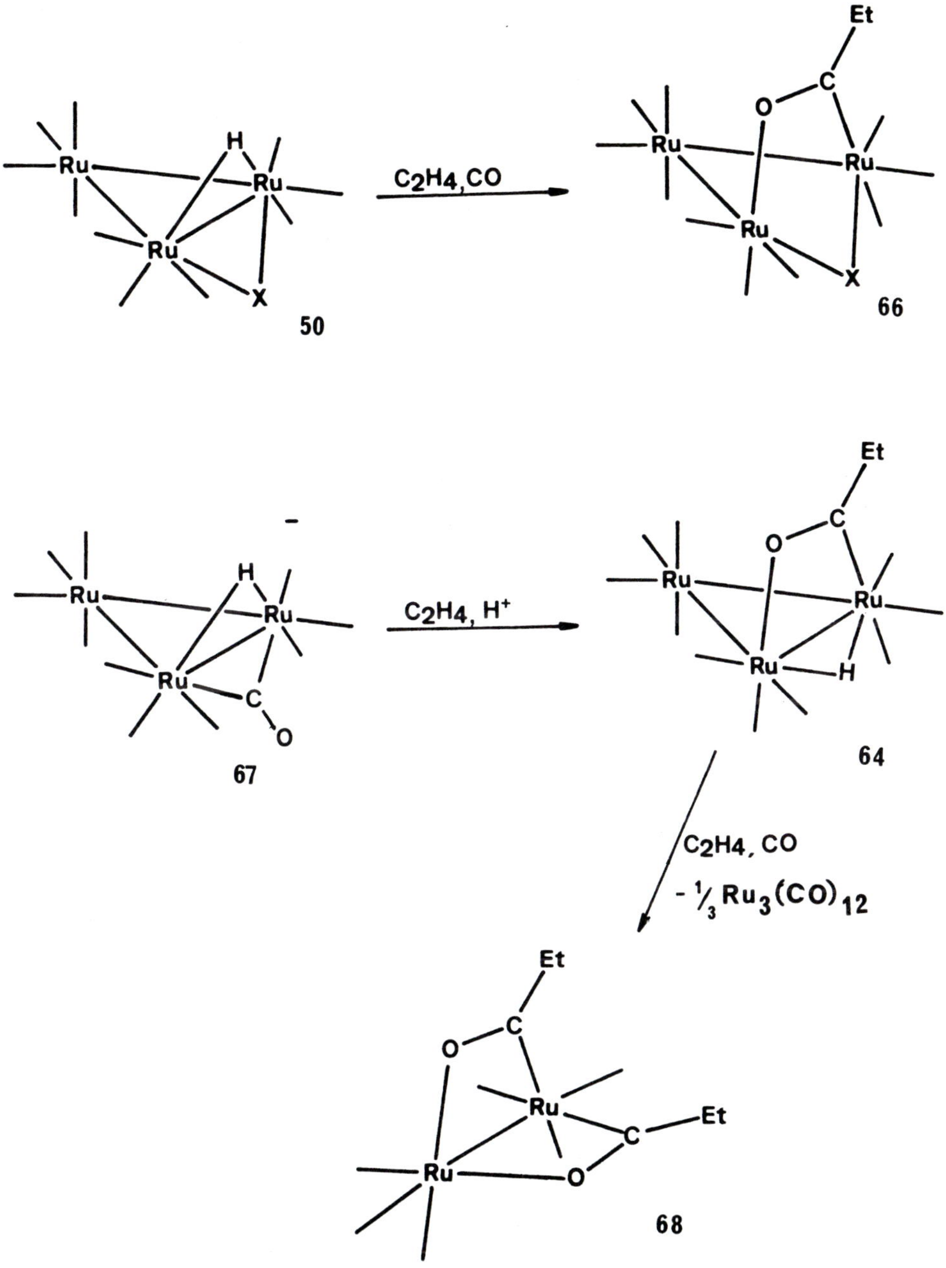

SCHEME 14

$$CO/C_2H_4/H_2 \xrightarrow[\substack{C_2H_4,\ (13\ bar) \\ H_2,\ (13\ bar) \\ CO,\ (26\ bar) \\ 373\ K,\ 5h}]{[HRu_3(CO)_{11}]^-} CH_3CH_2CHO$$

74% Conversion of Ethylene
TN = 315

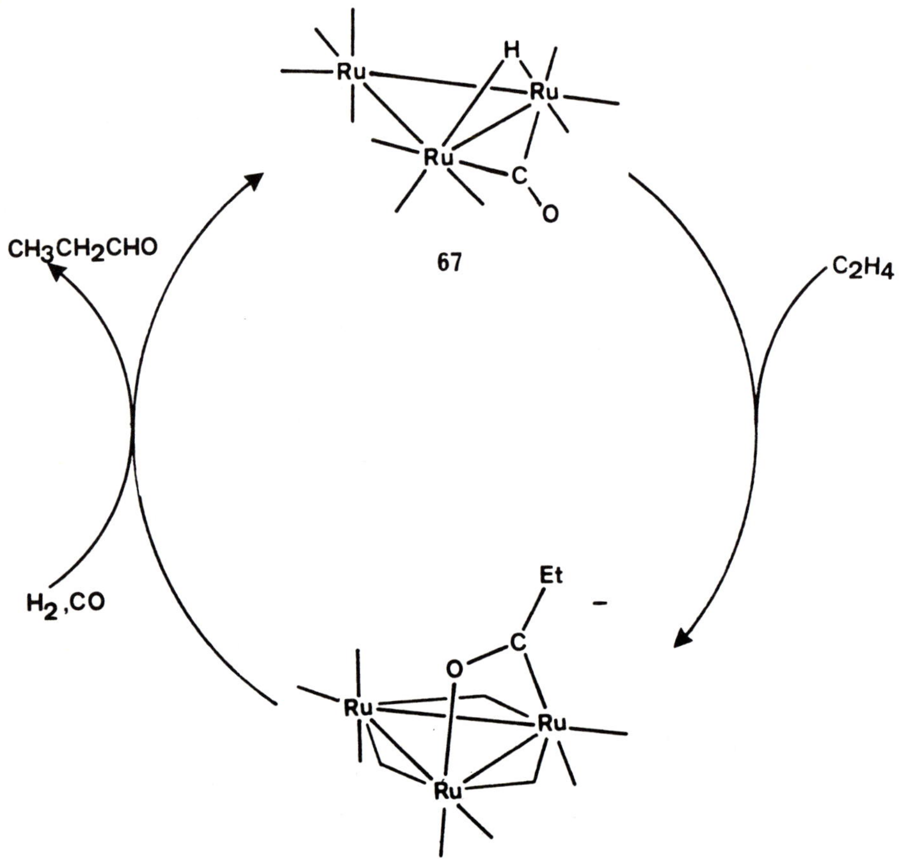

67

64⁻

SCHEME 15

SUBSTITUTION BY PHOSPHORUS LIGANDS

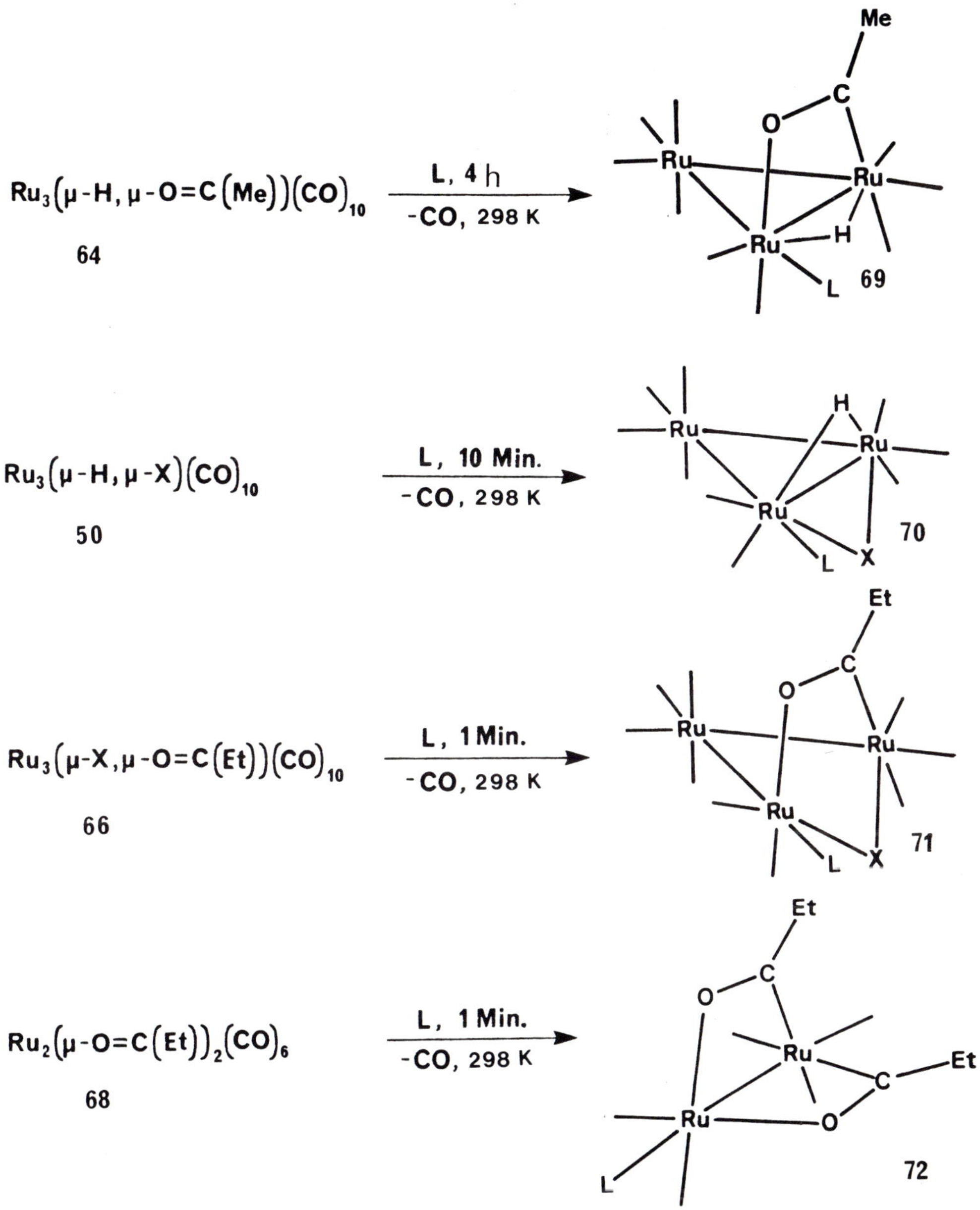

L = PPh_3 $P(OMe)_3$ or $P(OPh)_3$

SCHEME 16

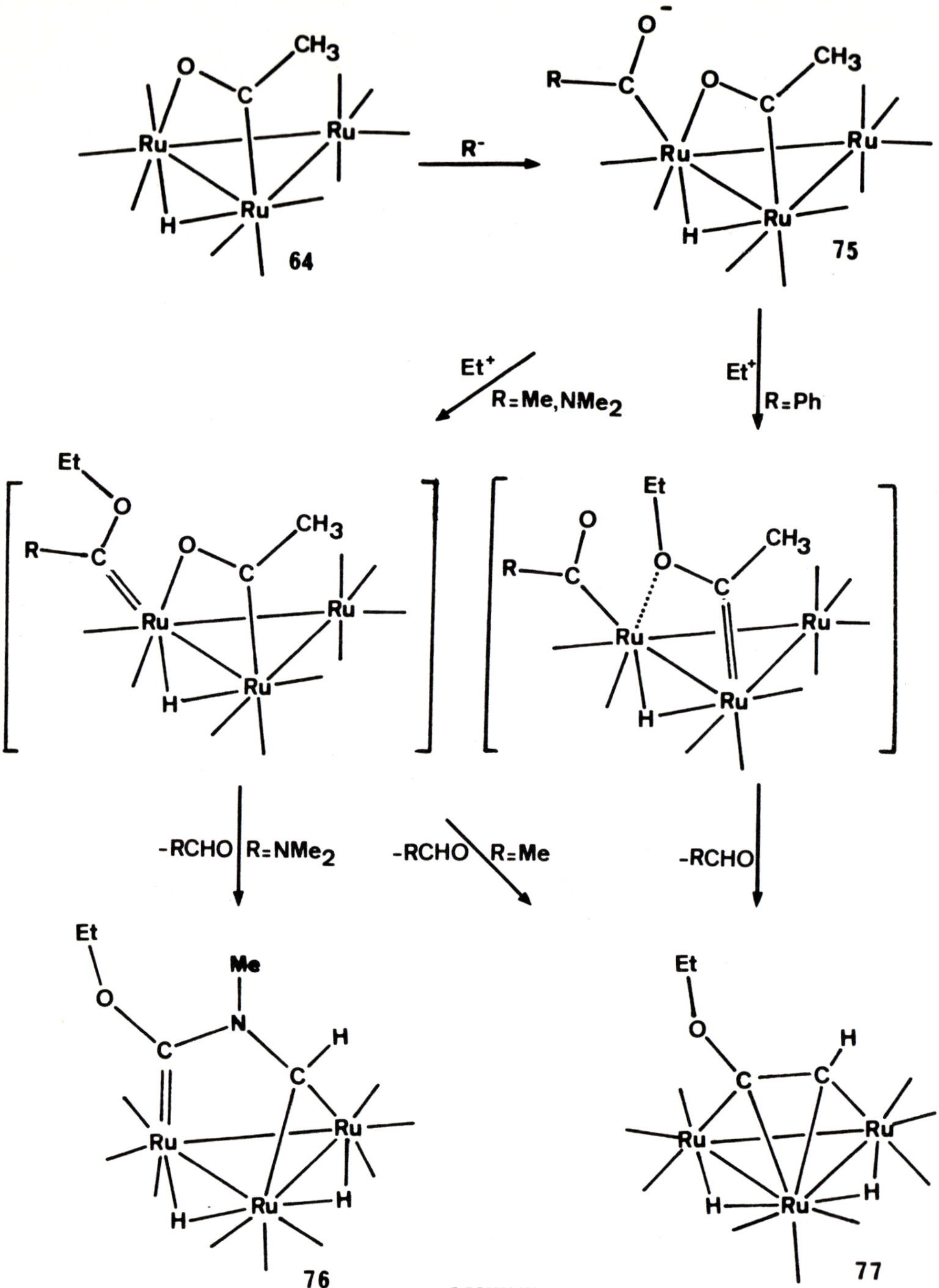

SCHEME 17

The well-known use of $(CH_3)_3NO$ [103], cited above in the synthesis of "lightly stabilized" cluster complexes, also operates by this pathway, namely, the elimination of CO_2 formed *in situ* by nucleophillic attack on a coordinated CO group.

Another example of this phenomenon is the hydrogenation of the methoxylcarbyne complex, $Ru_3\{\mu\text{-}H,\mu\text{-}C(OCH_3)\}(CO)_{10}$, **74** [104]. At 130°C, dimethyl ether is eliminated to generate $Ru_3(CO)_{12}$ under a CO atmosphere. More recently, elimination of aldehyde at room temperature following the alkylation of η^1-μ-*bis*-acyl anions, $[Ru_3\{\eta^1\text{-}C(O)R\}\{\mu\text{-}O{=}OCH_3\}(CO)_9]^-$, **75**, has been observed to lead to double oxidative addition of an organic group already attached to the trinuclear complex (SCHEME 17) [105].

What is remarkable about these transformations is the mild temperatures at which C-H oxidative addition is seen to occur, a fact which was masked in other derivatives in which such cyclometalations have taken place because of the high temperatures required to produce coordinative unsaturation on the cluster [1]. It should also be noted that product **77** in SCHEME 17 is identical to the types of derivatives seen in low-temperature reactions of alkynes with triruthenium cluster complexes [40].

4.3 ACTIVATION OF METAL CLUSTER COMPLEXES

4.3.1 Introduction

Catalytic activation of saturated cluster complexes which are otherwise either unreactive or poorly reactive under mild conditions has recently been discovered.

4.3.2 Radical ion catalysis

Bruce and coworkers found that sodium benzophenone-ketyl is a catalyst for the substitution of CO by phosphines, arsines, and isonitriles for a variety of cluster complexes including $M_3(CO)_{12}$ (M = Fe, Ru, Os), $Ru_4\{\mu\text{-}H\}_4(CO)_{12}$, $Co_3(\mu_3\text{-}CCl)(CO)_9$, and $Rh_6(CO)_{16}$ [106, 107]. The mechanism proposed for the activation of the cluster complex towards the substitution reactions is shown in SCHEME 18. Such reactions are closely related to the general class of electron-induced nucleophilic substitutions, which are proving extremely useful in cluster chemistry [108].

$$\begin{array}{llllll}
Ru_3(CO)_{12} & + [Ph_2CO\cdot]^- & = & [Ru_3(CO)_{12}\cdot]^- & + & Ph_2CO \\
[Ru_3(CO)_{12}\cdot]^- & + L & = & [Ru_3(CO)_{11}L\cdot]^- & + & CO \\
[Ru_3(CO)_{11}L\cdot]^- & + Ru_3(CO)_{12} & = & Ru_3(CO)_{11}L & + & [Ru_3(CO)_{12}\cdot]^-
\end{array}$$

SCHEME 18

4.3.3 Catalytic activation by nucleophiles. The [PPN^+] salts of the anions CN^-, $CH_3CO_2^-$, F^- Cl^-, I^-, and NO_2^- have been shown to be effective catalysts for substitution of CO on $Ru_3(CO)_{12}$ by a variety of phosphine ligands [109]. Substitutions are also promoted by catalytic amounts of $C(NMe_2)_4$ [110], which dissociates into [$C(N(CH_3)_2)_3^+$] and [$N(CH_3)_2^-$] in polar solvents such as THF. The labilization may be understood in terms of the enhanced activity towards substitution by the methoxycarbonyl adduct $Ru_3\{\eta^1C(O)OMe\}(CO)_{11}$ [111]. Mechanistic pathways for this activation and related transformations of cluster complexes are discussed in the following section.

4.4 TRANSFORMATION OF COORDINATED SUBSTRATES

4.4.1 Introduction

Examples of the activation of single bonds are discussed in this section followed by a discussion of activation of CO and related molecules. The conversion of CO into various organic products, notably, ethylene glycol or (as in the Fischer-Tropsch process) hydrocarbons, is of interest both from the practical and the fundamental point of view. Cluster complexes have been identified as being present or serving as catalyst prescursors in the conversion of synthesis gas ($CO + H_2$) into ethylene glycol under homogeneous conditions [112-114] (see Chapter 3). The precise mechanism of this transformation is still unknown [115], but a number of fundamental reaction steps in the transformation of CO on metal cluster complexes which may be related to this important reaction are beginning to emerge.

The surface-catalyzed Fischer-Tropsch reaction to give hydrocarbons is now thought to proceed via dissociative chemisorption of CO to give surface-coordinated carbon [116]. A parallel reaction is now observed in metal cluster complexes in which decomposition of CO is seen to give carbido-carbonyl complexes [117]. This and related transformations are presented in the following sections.

4.4.2 Activation of single bonds (H-H, C-H and P-C)

The activation of molecular hydrogen is one of the most important reactions in catalysis. Its activation on metal cluster complexes leading to hydrido-metal cluster complexes has been reviewed in detail [1]. Recently, the first study of the kinetics and mechanism of oxidative addition and reductive elimination of H_2 at a metal cluster has been reported. The results are summarized in equation (1a) [118]; the results of the kinetics study are consistent with the mechanism represented by equations (1b) and (1c):

$$Ru_3\{\mu\text{-}H,\mu\text{-}COMe\}(CO)_{10} + H_2 \rightleftharpoons Ru_3\{\mu\text{-}H\}_3\{\mu_3\text{-}COMe\}(CO)_9 + CO \quad (1a)$$

$$Ru_3\{\mu\text{-}H,\mu\text{-}COMe\}(CO)_{10} \rightleftharpoons Ru_3\{\mu\text{-}H,\mu\text{-}COMe\}(CO)_9 + CO \quad (1b)$$

$$H_2 + Ru_3\{\mu\text{-}H,\mu\text{-}COMe\}(CO)_9 \rightleftharpoons Ru_3\{\mu\text{-}H\}_3\{\mu_3\text{-}COMe\}(CO)_9 \quad (1c)$$

In parallel studies of the closely related iron system, $Fe_3\{\mu\text{-}H\}_3\{\mu_3\text{-}CMe\}(CO)_9$, the Fe-H-Fe bond energy was shown to be between 339 and 356 kJ mol^{-1} [119]. Theoretical questions dealing with reductive elimination or oxidative addition of H_2 at dinuclear centers have recently been examined [120].

Oxidative addition of C-H bonds of attached ligands (cyclometalation) and the breaking of C-P bonds in such ligands has similarly been reviewed in some detail [1]. Examples of C-H activation are shown in structure 7, and the transformations illustrated in SCHEMES 12, 13, and 17. Evidence of C-H activation in coordinated ligands is obtained through the observations of H-D exchange during catalytic processes [121]. An extreme case of C-P bond cleavage can be seen in the encapsulation of the donor atom in formation of the anion $[Rh_9P(CO)_{12}]^{2-}$ [122].

4.4.3 Reaction of nucleophiles with $M_3(CO)_{12}$ (M = Fe, Ru, Os) and formation of acyl derivatives. The first interaction of a nucleophile with a coordinatively saturated metal carbonyl derivative is formation of a donor-acceptor complex by attachment of the nucleophile to the carbon atom of a coordinated CO group [123]. This requires nucleophiles which are sufficiently reactive, in general found only for anionic nucleophiles, some only under highly dissociating conditions [109]. The order of reactivity for trinuclear clusters is $Fe_3(CO)_{12}$ > $Os_3(CO)_{12}$; in contrast, the sequence observed for the corresponding mononuclear carbonyls is $Os(CO)_5$ > $Ru(CO)_5$ >> $Fe(CO)_5$ [124].

With $M_3(CO)_{12}$, the first adduct has the formula $[M_3\{\eta^1C(O)Nu\}(CO)_{11}^-]$ (see A^- (M = Ru), SCHEME 19). Such intermediates have been observed spectroscopically, when Nu^- = NMe_2^- [91], CH_3^- [125], and H^- [126, 127], and they have been isolated when Nu^- = OR^- [111]. With OH^- [128] or NO_2^- [129], initial attack is followed by elimination of CO_2. In the former, this is accompanied by a hydride shift, whereas in the latter it is accompanied by conversion to other coordinated anions (*i.e.*, NO_2^- to NO^- [129]; N_3^- reacts with CO to give NCO^- [130].

Labilization of CO is seen in the adducts A^- (SCHEME 19), leading to anions of the type $[M_3\{\mu^x C(O)Nu\}(CO)_{10}^-]$, B^-. The labilization of CO parallels what has been observed for mononuclear derivatives [131, 132], with the additional possibility in a metal cluster complex that the labilization may be *vicinal* rather than *geminal*. This labilization may provide the basis for catalysis of substitution reactions for highly dissociated salts, as discussed in Section 3.3.3 [109].

In the absence of an incoming ligand, the vacant coordination site in B^- may become occupied by two alternate processes. These are controlled by the relative affinity of the nucleophile for carbon or for metal. In the event of high carbon affinity, the nucleophile remains on the carbonyl group which is converted from an η^1- to a μ-acyl group (proceeding to the right from B^- to B''^- in SCHEME 19). Such behavior is observed for $Nu^- = NMe_2^-$ [91] or CH_3^- [125]. An example involving similar nucleophilic attack on $Ru_5C(CO)_{15}$ has been reported recently [133].

Alternatively, the nucleophile can disengage at carbon and migrate to the metal (process B^- to B'^- in SCHEME 19), whereby the nucleophile is either terminally bonded ($Nu^- = NCO^-$ [130]) or bridging ($Nu^- = H^-$ [127]), with several possible CO configurations alternative to that represented for B'^- in SCHEME 19).

Where the nucleophile can act as a three-electron donor, further loss of CO is observed to form species of the type C^- (SCHEME 19) for $Nu^- = CH_3CO_2^-$ [94], Cl^- [109, 110], NO^- [129], or NCO^- [130]. With heating, further transformations take place, such as the reaction of [PPN]Cl or $[PPN]N_3$ with $Ru_3(CO)_{12}$ in refluxing tetrahydrofuran, which give principally $[Ru_4\{\mu\text{-}Nu\}(CO)_{13}]^-$ ($Nu^- = Cl^-$ [134] or NCO^- [130]).

Under a CO atmosphere, formation of C^- is prevented, and species such as B^- are observed ($Nu^- = C(O)OCH_3^-$ [110]). Species B^- may also be intercepted by other external nucleophiles such as L [$L = PPh_3$ or $P(OR)_3$]. The triruthenium salt A^-, with $Nu^- = CH_3O^-$, has recently been isolated under a CO atmosphere and shown to undergo rapid substitution by phosphites [111], represented by the transformation A^- to D^- to $Ru_3(CO)_{11}L$ in SCHEME 19.

The sequence of CO loss and migration of nucleophile to the metal must also take place in treatment of $Os_3(CO)_{12}$ with H^-; a triosmium analogue of A^- in SCHEME 19 ($Nu^- = H^-$) is observed spectroscopically, leading to the isolatable salt $[Os_3\{\mu\text{-}H,\mu\text{-}(CO)\}(CO)_{10}^-]$, the analogue of C^- in SCHEME 19 [126, 127].

Alternate preparations of $[M_3\{\mu\text{-}H,\mu\text{-}(CO)\}(CO)_{10}]^-$ anions involve reduction of the starting carbonyl with $[Mn(CO)_5]^-$ [135] or reaction with

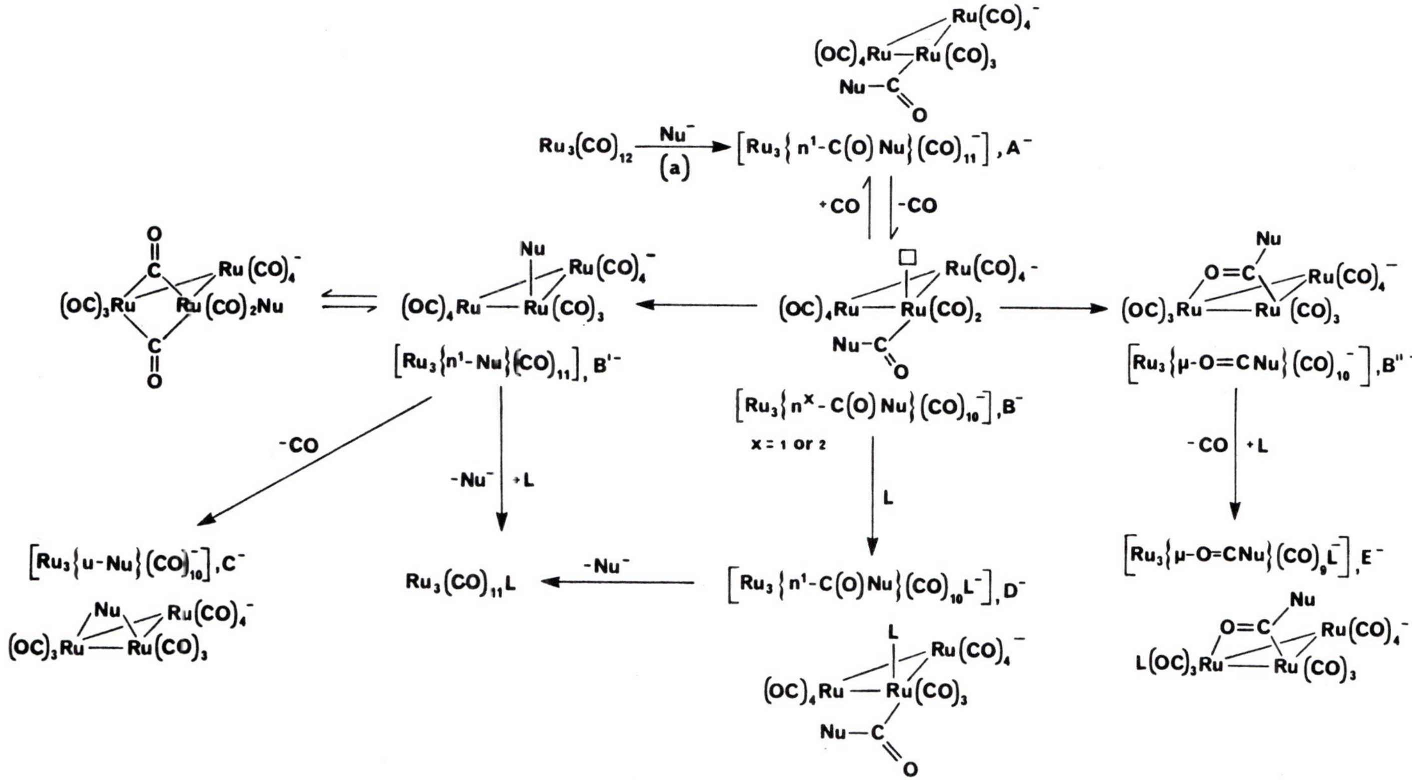
Ru3(CO)12
Nu⁻
(a)
[Ru3{n¹-C(O)Nu}(CO)11⁻], A⁻
+CO
-CO
[Ru3{nˣ-C(O)Nu}(CO)10⁻], B⁻
x = 1 or 2
[Ru3{n¹-Nu}(CO)11], B'⁻
[Ru3{μ-O=CNu}(CO)10⁻], B''⁻
-CO
-Nu⁻ +L
L
-CO +L
[Ru3{u-Nu}(CO)10⁻], C⁻
Ru3(CO)11L
-Nu⁻
[Ru3{n¹-C(O)Nu}(CO)10L⁻], D⁻
[Ru3{μ-O=CNu}(CO)9L⁻], E⁻

SCHEME 19

various bases such as KOH/CH_3OH, [136], $NaBH_4$ [128], $(C_2H_5)_3N/H_2O$ [137], or KH (Reaction (2) [138]):

$$Ru_3(CO)_{12} + KH \xrightarrow{\text{tetrahydrofuran}} K[Ru_3\{\mu\text{-}H,\mu(CO)\}(CO)_{10}] + CO \quad (2)$$

$$K[Ru_3\{\mu\text{-}D,\mu(CO)\}(CO)_{10}] + CO + H_2O \xrightarrow{H_2O} HD + Ru_3(CO)_{12} + KOH \quad (3)$$

The reversibility of reaction (2) indicates that $[Ru_3\{\mu\text{-}H, (CO)\}(CO)_{10}]^-$ can function as a *hydride donor* in the presence of CO, which was subsequently demonstrated by Reaction (3). From these results, Bricker and coworkers [138] suggested a cycle for the water gas shift reaction in which the trinuclear species in Reactions (4) and (5) play a major role irrespective of whether the reaction is initiated by tri- or tetranuclear species:

$$Ru_3(CO)_{12} + OH^- = [HRu_3(CO)_{11}]^- + CO_2 \quad (4)$$

$$[HRu_3(CO)_{11}]^- + CO = Ru_3(CO)_{12} + H^- \quad (5)$$

$$H^- + H_2O = H_2 + OH^- \quad (6)$$

Although not directly observed, a formyl intermediate of the type A^- in SCHEME 19 ($Nu^- = H^-$) may be the actual hydride transfer agent.

4.4.4 Transformation of coordinated CO to carbide and to hydrocarbon fragments

4.4.4.a Reduction of CO initiated by nucleophilic attack of hydride; the formyl intermediate. We enter this sequence through the anion **78**, analogous to intermediate A^- in SCHEME 19, with M = Os and $Nu^- = H^-$ [127]. Acidification of this formyl intermediate leads through a sequence of steps to the methylene complex, **79**, as shown in SCHEME 20; the overall reaction is represented as follows:

$$\underset{\mathbf{78}}{[Os_3\{\eta^1C(O)H\}(CO)_{11}]^-} \xrightarrow{H^+} \underset{\mathbf{79}}{Os_3\{\mu\text{-}CH_2,\mu\text{-}CO\}(CO)_{10}} \quad (7)$$

Results of labelling experiments are consistent with intermediatesa traversed on mononuclar complexes [139, 140] containing the {=C(H)OH} group after the first protonation, which is converted to the anionic $\{-CH_2OH\}^-$ group by hydride transfer from formyl complexes in solution. The OH^- group is removed through a second protonation that is followed by

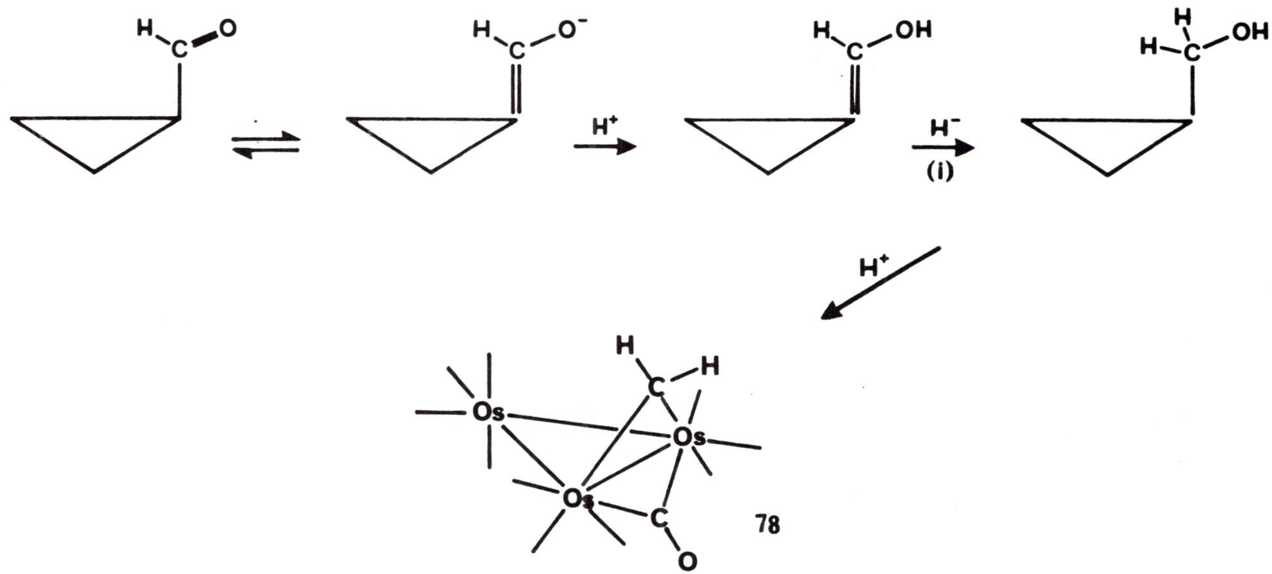

78

SCHEME 20

i) hydride transfer from formyl intermediate

evolution of H_2O. The {=CH_2} group is thus formed through electrophilic attack (see the following section) in a process that is initiated by nucleophilic attack. Complex **79** evolves CH_4 in treatment with H_2.

4.4.4.b **C-O bond cleavage initiated by electrophilic attack.** This is illustrated by the alkylation of $[Os_3\{\mu\text{-}H,\mu\text{-}CO\}(CO)_{10}]^-$, **80**, which occurs at oxygen of the bridging CO group, giving the neutral alkoxyalkylidyne derivative **81** (SCHEME 21) [141]. Treatment of **81** with $LiBEt_3H$ followed by protonation leads to complex **83**, containing an unusual triply bridging methylidyne. An internal donor-acceptor interaction is seen between the electronically saturated metal atom of the $Os(CO)_4$ group and the electron-deficient carbon of the methylidyne group. The electron deficiency of the methylidene carbon leads to attachment of H^- in reaction of **83** with $LiBEt_3H$:

$$\underset{\mathbf{83}}{HOs_3\{\mu_3\text{-}CH\}(CO)_{10}} \xrightarrow{LiBEt_3H} [HOs_3\{\mu\text{-}CH_2\}(CO)_{10}]^- \xrightarrow{H^+} HOs_3\{\mu\text{-}CH_3\}(CO)_{10} \qquad (8)$$

It is noteworthy that the overall reduction of CH to CH_3 cannot be effected with H_2 due to the thermal rearrangement of **83** shown in Section 3.4.7 (equation 10).

Another case of C-O cleavage is observed in the acyl derivative **84** (SCHEME 22) [143]. Alkylation of this derivative occurs on the oxygen atom of the triply bridged acyl group, leading to the alkoxy-carbyne complex **86**. In this transformation, the cluster acts as an electron sink by cleavage of one of the metal-metal bonds in the original triangular array. Isolation of **86** provides a molecular parallel for the proposed formation of surface-bound alkylidenes.

Further such examples are seen in the thermal cleavage of the triply bridged acyl group in complex **87** (SCHEME 23) [144]. In this reaction, the oxo-alkyne complex **88** is produced, the transformation being accompanied by folding at the Os-W hinge; the metal atoms are rearranged into a pseudo-tetrahedral array. This fundamental transformation may contribute to understanding of the role of acyl intermediates postulated in the Fischer-Tropsch process.

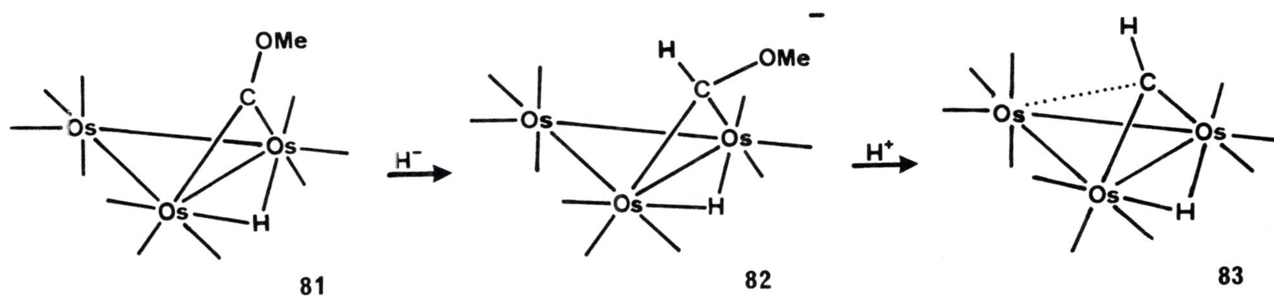

SCHEME 21

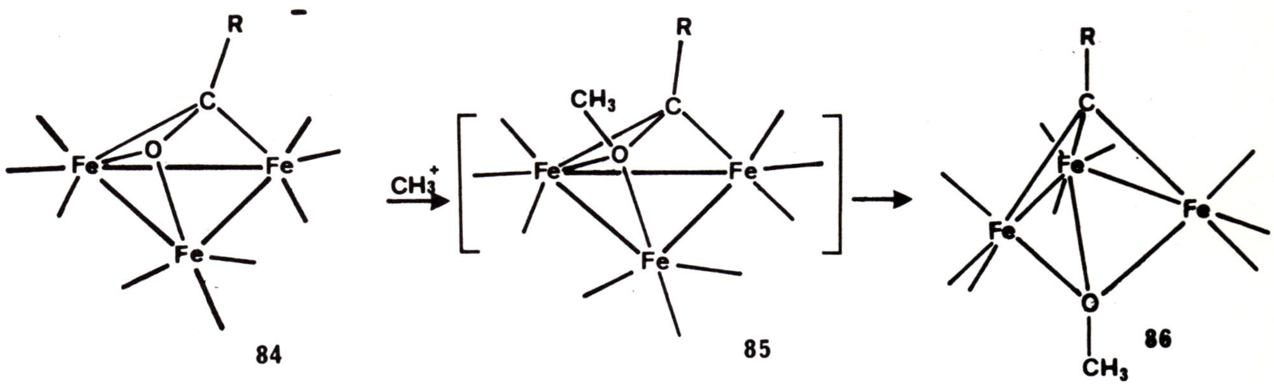

SCHEME 22

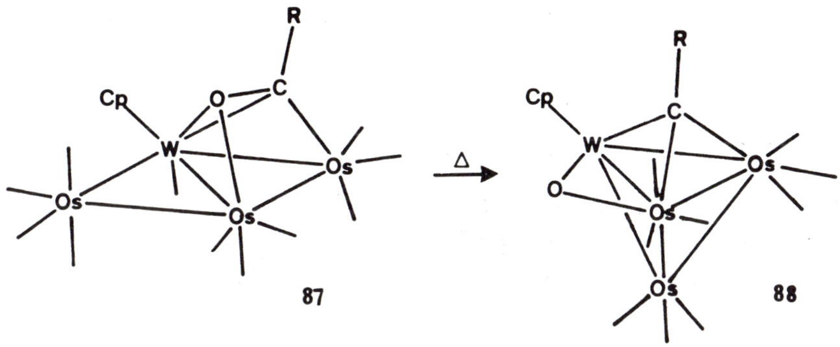

SCHEME 23

C-O cleavage induced under protonation is yet another example of the transformation of coordinated CO, in this case to isolatable carbido cluster complexes (SCHEME 24) [145-147]. The transformation begins with the dianion $[Fe_4(CO)_{13}]^{2-}$ (**89** in SCHEME 24). This anion contains one face-bridging in CO in a 60-e^- cluster of the required tetrahedral geometry. First protonation of this cluster occurs on the metal framework, leading to the butterfly complex **90**. In this anion, a CO group is found coordinated to all four of the metal atoms (and thus acting as a four-electron donor). The butterfly configuration requires a total electron count of 62 e^-, which indeed is achieved through the μ_4-(CO). A recent study of the CO-fluxional processes in this butterfly complex indicates that the active intermediate (or transition state) for exchange may be a closed (protonated tetrahedral) metal cluster. In any case, it is the open cluster (in which the oxygen as well as the carbon of a CO group is coordinated to the metal) that is the important form for the chemical steps which follow.

Acidification of **90** leads to protonation at oxygen of μ_4-(CO). The overall reaction leads to production of methane through loss of H_2O by steps that are not yet fully understood.

4.4.5 Cleavage of alkynes to dialkylidynes

Parallel to observations of the cleavage of such high-bond-energy species as CO (bond dissociation energy 1076 kJ mol^{-1}) is the conversion of a complexed acetylene into two coordinated alkylidyne groups (SCHEME 25) [148].

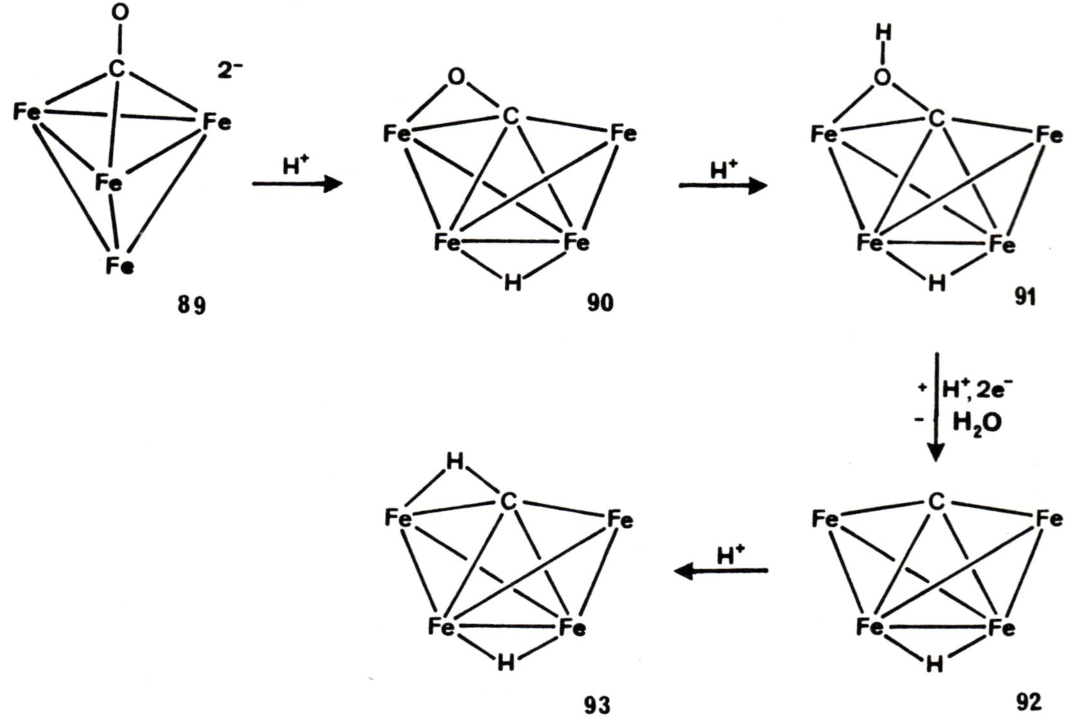

SCHEME 24

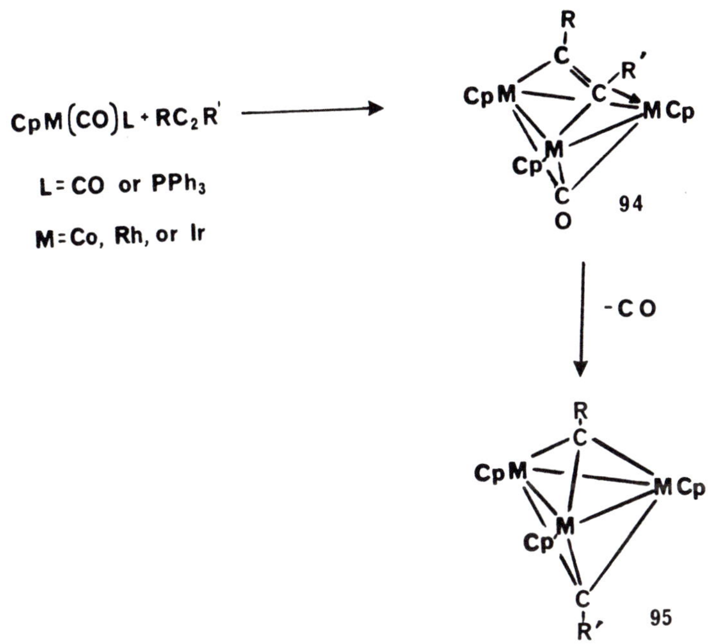

SCHEME 25

The coordinated alkyne in complex **94** serves as a four-electron donor on the cluster. Under conditions of flash vacuum pyrolysis, **94** (M = Rh or Ir [148]) loses a CO group. The resulting electronic unsaturation on the cluster causes the scission of the acetylene, giving two 3-e^- donors in **95**. In this transformation, electrons removed by the departing CO are restored to the trinuclear unit. Alkyne scission has also been observed on a tetranuclear cluster complex [149], and that occurring on the trinuclear cluster complexes has been examined theoretically [148]. Electron deficiency by itself, however, is not enough to induce alkyne scission; note the electronically unsaturated alkyne complexes $M_3\{\mu,\eta 2\text{-}Ph_2C_2\}(CO)_9$, M = Fe, **5a**, or M = Os, **5b** (Table 1).

4.4.6 Cleavage of NO and nonmetal oxo-anions

Heating of the nitrosyl mixed metal anionic cluster complex $[FeRu_3(CO)_{12}(NO)]^-$ at 65°C in tetrahydrofuran solution smoothly converts it into the nitrido complex $[FeRu_3\{\mu_4\text{-}N\}(CO)_{12}]^-$ with evolution of CO_2 [150]. This transformation is sufficiently clean to have been studied kinetically; the reaction is first order in cluster concentration. A review of nitrido-cluster complexes has been prepared [151].

In this connection, we also note the deoxygenations seen in the syntheses of the cluster complexes $[Ru_3\{\mu_3\text{-}X\}(CO)_9]^-$ (X = S, Se, or Te) starting from $Ru_3(CO)_{12}$ and XO_3^- [152]. No intermediate oxygenate complexes were observed in these reactions, however.

4.4.7 Coupling reactions of carbon fragments and of CO with carbon fragments on metal cluster complexes

The reverse of alkyne scission, namely, the formation of a coordinated alkyne from two alkylidynes, is observed in the reaction depicted in SCHEME 26 [153, 154]. In the alkyne complex **96** a rather unusual cleavage

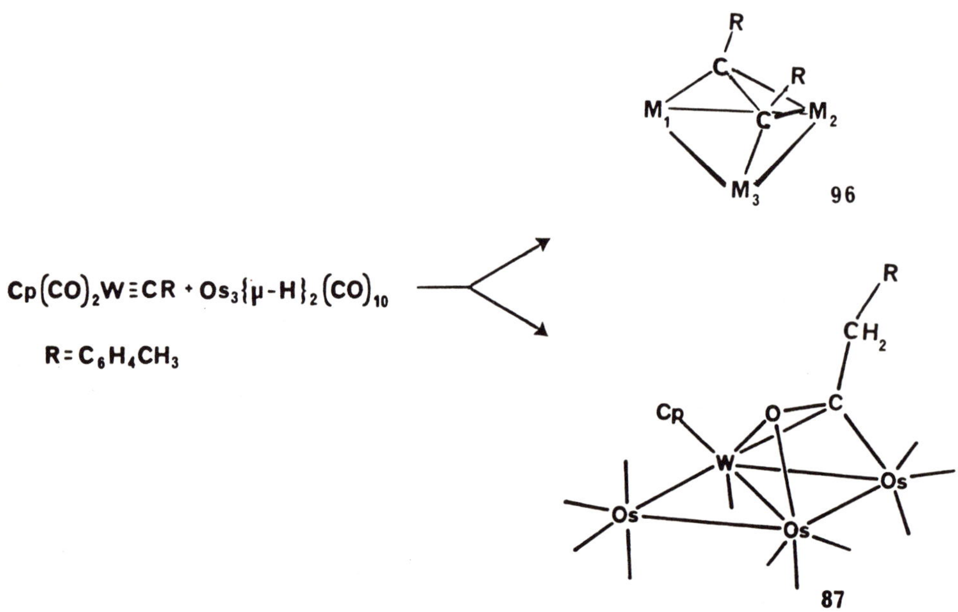

96a, $M_1 = M_3 = C_p(CO)W$; $M_2 = Os(CO)_3$

96b, $M_1 = M_2 = C_p(CO)W$; $M_3 = Os(CO)_3$

SCHEME 26

of one $Os(CO)_4$ unit has occurred, resulting in a mixed metal cluster; two isomers **96a** and **96b** were observed [153a, b]. In complex **87** (mentioned above for its subsequent C-O cleavage reaction), the Os_3 triangle is retained, with the tungsten unit adding on to give a triangulated (butterfly) arrangement of metal atoms. In **87** we witness the coupling of the alkylidene unit, originally attached to tungsten, with a CO group leading to the triply bridged acyl unit [153a, b].

The coupling of a triply bridged alkylidyne to an acetylene is seen in the transformation shown in SCHEME 27 [155].

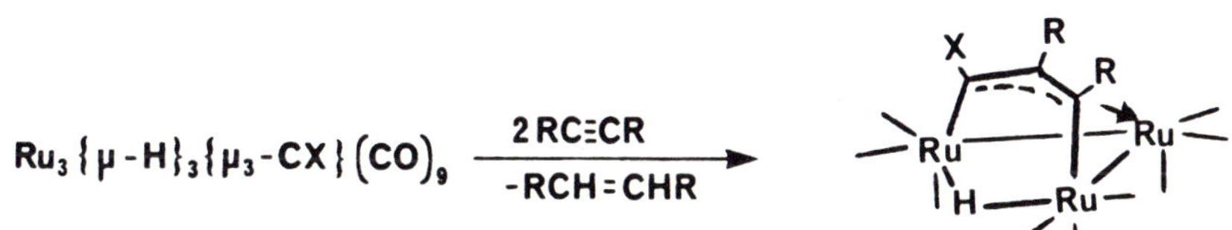

97 **98**

X = OMe, Me, or Ph

R = Me or Ph

SCHEME 27

Bridged allyl complexes of structures similar to that of **98** are obtained in the coupling of a methylene unit with a coordinated acetylene, shown in the following reaction [156]:

$$Os_3\{\mu\text{-}CH_2,\mu_3,\eta^2\text{-}C_2Ph_2\}(CO)_9 \xrightarrow{135^oC} Os_3\{\mu\text{-}H,\mu_3,\eta^3\text{-}CHC_2Ph_2\}(CO)_9 \qquad (9)$$

99 **100**

Complex **99** is obtained by room-temperature reaction of diazomethane with the electronically unsaturated $Os_3\{\mu,\eta^2Ph_2C_2\}(CO)_9$, **5b** (Table 1); a diazomethane adduct, $Os_3\{\mu,\eta^2\text{-}Ph_2C_2\}\{CH_2N_2\}(CO)_9$, **101**, is also isolated, which is converted into **99** photochemically [156]. Complex **5b** is obtained from the coordinatively saturated precursor $Os_3\{\mu,\eta^2\text{-}Ph_2C_2\}(CO)_{10}$, **102**, by heating at 250°C [19b, 156].

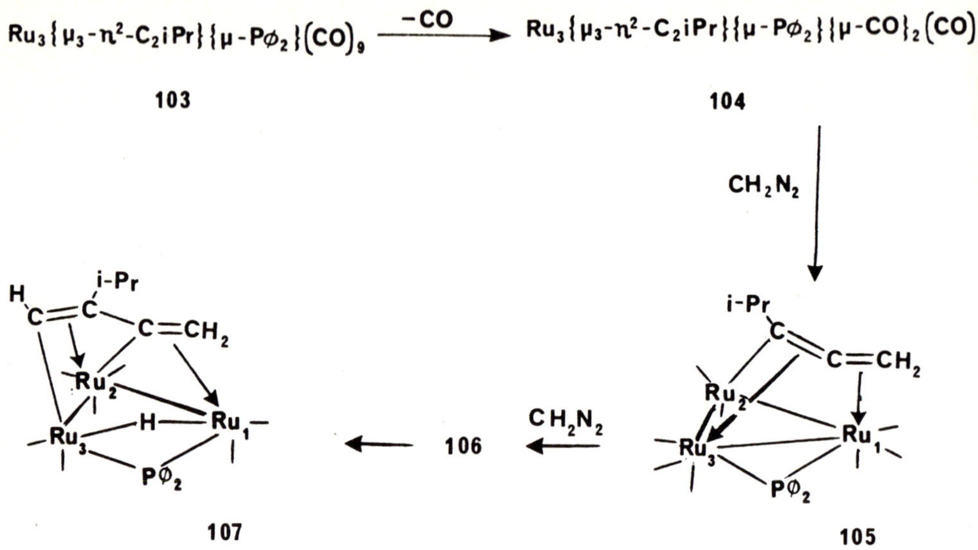

SCHEME 28

Further examples of hydrocarbon chain growth by treatment of metal cluster complexes with diazomethane are shown in SCHEME 28 [157]. Here, too, as in the reactions with triosmium clusters mentioned above, the starting acetylide **103** must be heated (for 3h at 60°C) to give electronically unsaturated **104**. This reacts with diazomethane at 0°C to give the allenyl complex **105**. Further treatment of **105** with diazomethane (at 20°C) gives, first, a methylene complex, **106**, in which the μ-CH_2 group is seen bridging between Ru atoms (1) and (2) in a structure similar to **105**. This μ-methylene, μ-allenyl complex isomerizes to the butadienylidene complex **107** at 20°C over a period of 3 days [157].

A further example of the coupling of a μ-CH_2 group, in this case with CO on a metal cluster complex, is shown in SCHEME 29 [158]. The methylene complex **78** has been mentioned above in connection with the reduction of coordinated CO to hydrocarbon groups. The conversion shown in SCHEME 29 requires 4 h to go to completion. In contrast, a halogen-bridged derivative of **78**, $Os_3\{\mu$-X, -$CH_2\}(CO)_{10}$ (X = Cl, Br, I, or NCO), obtained by the treatment of **78** with PPN^+ salts of X^- [159], also undergoes CO coupling with μ-CH_2. The X-bridged complexes, however, show an acceleration of *ca.* one hundred fold for this reaction compared with **78** [159].

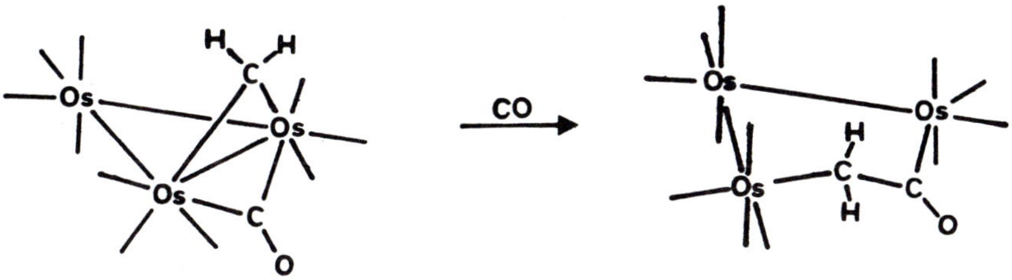

78 108

SCHEME 29

The coupling of an alkylidyne group with CO (accompanied by hydrogen transfer to the metal atoms in the cluster) is seen in the thermal rearrangement of **83** [142]:

$$Os_3\{\mu\text{-}H,\mu_3\text{-}CH\}(CO)_{10} \xrightarrow[\Delta]{} Os_3\{\mu\text{-}H\}_2\{\mu_3,\eta^2\text{-}C{=}C\text{-}O\}(CO)_9 \qquad (10)$$

83 **i-83**

Complex **83** could not be hydrogenated, owing to its thermal isomerization to **i-83**. The latter, however, does react with H_2 to give $Os_3\{\mu\text{-}H\}_3\{\mu_3\text{-}CH\}(CO)_9$.

The migration of CO to a coordinated carbide is observed to occur in the presence of a nucleophilic reagent, as shown in SCHEME 30 [160]. In the product anion $[Fe_4\{\mu_4\text{-}C\text{-}C(O)OMe\}(CO)_{12}]^-$, **110**, the wingtip iron atoms are folded back from what they are in the starting material $Fe_4\{\mu_4\text{-}C\}(CO)_{13}$, **109**; the wingtip Fe-C-Fe distance in the latter is 0.180 nm, whereas it is 0.202 nm in the product anion.

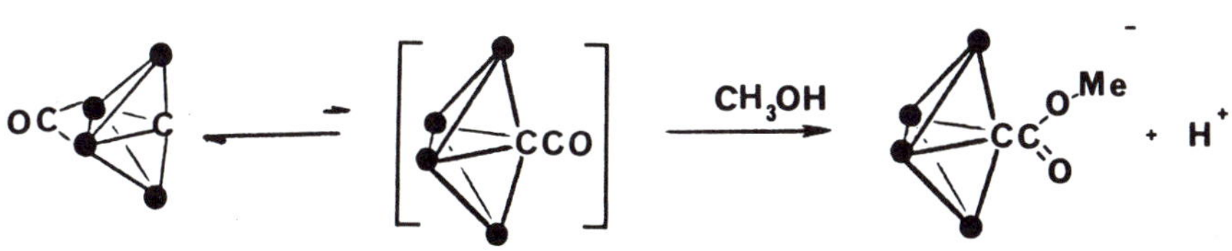

109 i-109 110

● = $Fe(CO)_3$

SCHEME 30

The transformation shown in SCHEME 30 has been analyzed in molecular orbital studies [161, 162]. Both on steric and electronic grounds, the carbide carbon is inferred not to be available for nucleophilic attack in the starting material **109** *as long as its geometry is maintained.* A ketenylidene complex, **i-109**, is believed to be the important intermediate, the formation of which must be preceded by a deformation from the skeleton of **109** to that in the product **110**. To date, no spectroscopic or other evidence has been obtained for the existence of intermediate **i-109**.

We conclude our discussion with the report of an elegant series of reactions tracing in some detail the pathway leading to coupling between an alkylidyne group and CO on a cluster; these are depicted in SCHEME 31 [163, 164].

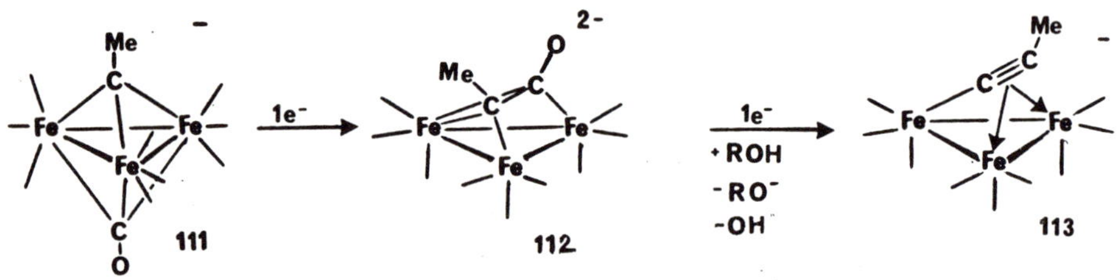

SCHEME 31

An overall 2-e⁻ reduction of **111** gives the μ_3,η^2-acetylide complex **113** [163]. This reduction demonstrates the *reverse* of scission reactions which occur when electrons are *lost* from a metal cluster (Section 3.4.5). The paramagnetic intermediate **112** has recently been isolated; it has been demonstrated by a structure study that the coupling of the triply bridging alkylidyne with CO occurs after reduction of **111** with one electron [164]. These observations may also apply to the coupling (with CH_3O^- loss) observed upon reduction of $[Fe_3\{\mu\text{-}C(OMe)\}(CO)_{10}]^-$ giving $[Fe_3\{\mu_3,\eta^2\text{-}C{=}C{=}O\}(CO)_9]^{2-}$ [165].

4.5 CONCLUDING REMARKS

In addition to the reviews cited in this section [1-7, 40, 116, 117, 151], we wish to draw attention to additional sources which may also prove useful to readers interested in further details of the transformations discussed here: reviews of alkylidene complexes [166], vinylidene and propadienylidene (allenylidene) complexes [167], and complexes containing C- and O-metal bonded carbonyl groups [168].

REFERENCES

1. A. P. Humphries and H. D. Kaesz, Prog. Inorg. Chem., 25 (1979) 145.
2. A. J. Deeming in B. F. G. Johnson (Ed.), Transition Metal Clusters, Wiley, Chichester, 1980, pp. 391-469.
3. H. Vahrenkamp, Advan. Organometal. Chem., 22 (1983) 169.
4. R. D. Adams and I. Horvath, Prog. Inorg. Chem., 33 (1985), in press.
5. E. L. Muetterties, R. R. Burch and A. M. Stolzenberg, Ann. Rev. Phys. Chem., (1982) 33, 89.
6. H. Vahrenkamp, Angew. Chem., Int. Ed. Engl., 17 (1978) 379.
7. G. Wilkinson, F. G. A. Stone and E. W. Abel (Eds.), Comprehensive Organometallic Chemistry, Peramon, Oxford, 1982.
8. J. A. Connor in B. F. G. Johnson (Ed.), Transition Metal Clusters, Wiley, Chichester, 1980, Chapter V.
9. M. Tachikawa and J. R. Shapley, J. Organomet. Chem., 124 (1977) C19.
10. J. W. Lauher, J. Am. Chem. Soc., 100 (1978) 5305.
11. a. K. Wade, J. Chem. Soc. Chem. Commun. (1971) 792.;
    b. *idem.*, Advan. Inorg. Chem. Radiochem. 18 (1976) 1.
12. a. D. M. P. Mingos, Nature (London) Phys. Sci., 236 (1972) 99.;
    b. *idem.*, Acc. Chem. Res., 17 (1984) 311.
13. B. K. Teo, Inorg. Chem., 23 (1984) 1251, 1257.
14. B. K. Teo, G. Longoni and F. R. K. Chung, Inorg. Chem., 23 (1984) 1257.
15. R. D. Adams and J. P. Selegue, in G. Wilkinson, F. G. A. Stone and E. W. Abel (Eds.), Comprehensive Organometallic Chemistry, Vol. 4, Pergamon, Oxford, 982, pp. 1023 ff.
16. B. F. G. Johnson, D. A. Kaner, J. Lewis and P. R. Raithby, J. Organomet. Chem., 215 (1981) C33.
17. K. Burgess, B. F. G. Johnson, D. A. Kaner, J. Lewis, P. R. Raithby and S. N. A. B. Syed-Mustaffa, J. Chem. Soc. Chem. Commun. (1983) 455.
18. a. B. F. G. Johnson, D. A. Kaner, J. Lewis and P. R. Raithby, J. Chem. Soc. Chem. Commun. (1981) 752;
    b. M. Fajardo, M. P. Gomez-sal, H. D. Holden, B. F. G. Johnson, J. Lewis, R. C. S. McQueen and P. R. Raithby, J. Organomet. Chem., 267 (1984) C25.
19. a. J. F. Blount, L. F. Dahl, C. Hoogzand and W. Hubel, J. Am. Chem. Soc., 88 (1966) 292;
    b. M. Tachikawa, J. R. Shapley and C. G. Pierpont, J. Am. Chem. Soc., 97 (1975) 7172.
20. A. C. Willis, F. W. B. Einstein, R. M. Ramadan, R. C. Pomeroy, Organometallics, 2 (1983) 935.
21. J. A. Clucas, D. F. Foster, M. M. Harding and A. K. Smith, J. Chem. Soc. Chem. Commun. (1984) 949.
22. R. Saillant, G. Barcelo and H. D. Kaesz, J. Am. Chem. Soc., 92 (1970) 5739.
23. R. D. Wilson and R. Bau, J. Am. Chem. Soc., 98 (1976) 4687.
24. G. Ciani, G. D'Alfonso, P. Romiti, A. Sironi and M. Freni, Inorg. Chem., 22 (1983) 3115 and references cited therein.
25. M. J. Mays, D. W. Prest and P. R. Raithby, J. Chem. Soc. Chem. Commun. (1980) 171 and references cited therein.
26. H. Vahrenkamp and D. Wolters, Organometallics, 1 (1982) 874.
27. H. Vahrenkamp, E. J. Wucherer and D. Wolters, Chem. Ber., 116 (1983) 1219.
28. E. W. Burkhart, W. C. Mercer, G. L. Geoffroy, A. L. Rheingold and W. C. Fultz, J. Chem. Soc. Chem. Commun. (1983) 1251.
29. A. J. Sivak and E. L. Muetterties, J. Am. Chem. Soc., 101 (1979) 4878.
30. M. D. Fryzuk, Organometallics, 1 (1982) 408.

31. M. Kulzick, R. T. Price, E. L. Muetterties and V. W. Day, Organometallics, 1 (1982) 1256.
32. R. W. Broach and J. M. Williams, Inorg. Chem., 18 (1979) 314 and references cited therein.
33. S. A. R. Knox, J. W. Koepke, M. A. Andrews and H. D. Kaesz, J. Am. Chem. Soc., 97 (1975) 3942.
34. a. P. T. Chesky and M. B. Hall, Inorg. Chem., 22 (1983) 3327;
 b. M. Berry, N. J. Cooper, M. L. H. Green and S. J. Simpson, J. Chem. Soc. Dalton Trans. (1980) 37;
 c. B. K. Teo, M. B. Hall, R. F. Fenske and L. F. Dahl, J. Organomet. Chem., 70 (1974) 413.
 d. R. Mason and D. M. P. Mingos, J. Organomet. Chem., 50 (1973) 53.
35. a. J. R. Shapley, J. B. Keister, M. R. Churchill and B. G. de Boer, J. Am. Chem. Soc., 97 (1975) 4145.
 b. M. R. Churchill and B. G. de Boer, Inorg. Chem., 16 (1977) 878; *ibid.*, 16 (1977) 2397.
36. A. J. Deeming and S. Hasso, J. Organomet. Chem., 114 (1976) 313.
37. R. Psaro, R. Ugo, G. M. Zanderighi, B. Besson, A. K. Smith and J.-M. Basset, J. Organomet. Chem., 213 (1981) 215.
38. S. C. Brown and J. Evans, J. Mol. Catal., 11 (1981) 143.
39. J. B. Keister, J. R. Shapley, J. Am. Chem. Soc., 98 (1976) 1056.
40. E. Sappa, A. Tiripicchio and P. Braunstein, Chem. Rev., 83 (1983) 203.
41. R. B. Calvert and J. R. Shapley, J. Am. Chem. Soc., 99 (1977) 5525; *ibid.*, 100 (1978) 7726.
42. R. C. Brady III and R. Pettit, J. Am. Chem. Soc., 103 (1981) 1287.
43. L. J. Farrugia, J. A. K. Howard, P. Mitrprachachon, J. L. Spencer, F. G. A. Stone and P. Woodward, J. Chem. Soc. Chem. Commun. (1978) 260.
44. R. Hoffmann, Angew. Chem. Int. Ed. Engl., 21 (1982) 711.
45. F. G. A. Stone, Angew. Chem. Int. Ed. Engl., 23 (1984) 89.
46. J. S. Plotkin, D. G. Alway, C. R. Weisenberger and S. G. Shore, J. Am. Chem. Soc., 102 (1980) 6156.
47. J. R. Fox, W. L. Gladfelter and G. L. Geoffroy, Inorg. Chem., 19 (1980) 2574.
48. D. H. Farrar, B. F. G. Johnson, J. Lewis, J. Nicola Nicholls, P. R. Raithby and M. J. Rosales, J. Chem. Soc. Chem. Commun. (1981) 273.
49. R. J. Goudsmith, B. F. G. Johnson, J. Lewis, P. R. Raithby and K. H. Whitmire, J. Chem. Soc. Chem. Commun. (1982) 640.
50. B. F. G. Johnson and J. Lewis in Huang Yaozeng, A. Uamamoto and B.-K. Teo (Eds.) New Frontiers in Organometallic and Inorganic Chemistry, Science Press, Beijing, 1984, pp. 359-377.
51. E. B. Prestridge, G. H. Via and J. H. Sinfelt, J. Catal., 50 (1977) 115.
52. J. J. De Boer, J. A. Van Doorn, C. Masters, J. Chem. Soc. Chem. Commun. (1978) 1005.
53. A. A. Bahsoun, J. A. Osborn, C. Voelker, J.-J. Bonnet and G. Lavigne, Organometallics, 1 (1982) 1114.
54. G. Lavigne, N. Lugan and J.-J. Bonnet, Nouv. J. Chim., 5 (1981) 423 and references cited therein.
55. R. C. Ryan, C. U. Pittman Jr. and J. P. O'Connor, J. Am. Chem. Soc., 99 (1977) 1986.
56. J. L. Vidal, W. E. Walker, R. L. Pruett and R. C. Schoening, Inorg. Chem. 18 (1979) 129.
57. G. H. Huttner, J. Schneider, H.-D. Müller, G. Mohr, J. von Seyerl and L. Wohlfahrt, Angew. Chem. Int. Ed. Engl., 18 (1979) 76.
58. D. A. Lesch and T. B. Rauchfuss, Organometallics, 1 (1982) 499.
59. C. U. Pittman Jr., G. M. Wilemont, W. D. Wilson and R. C. Ryan, Angew. Chem. Int. Ed. Engl., 19 (1980) 478.

60. Trinh-Toan, B. K. Teo, J. A. Ferguson, T. J. Meyer and L. F. Dahl, J. Am. Chem. Soc., 99 (1977) 408.
61. C. D. Garner, B. F. G. Johnson (Ed.), Transition Metal Clusters, Wiley, Chichester, 1980, Chapter IV.
62. N. M. Boag, C. B. Knobler and H. D. Kaesz, Angew. Chem. Int. Ed. Engl., 22 (1983) 249.
63. a. R. D. Adams and Li-Wu Yang, J. Am. Chem. Soc., 105 (1983) 235.
b. R. D. Adams, I. T. Horvath and P. Mathur, J. Am. Chem. Soc., 106 (1984) 6269.
64. B. F. G. Johnson, J. Lewis and D. A. Pippard, J. Chem. Soc. Dalton Trans. (1981) 407.
65. J. R. Shapley, G. A. Pearson, M. Tachikawa, G. E. Schmidt, M. R. Churchill and F. J. Hollander, J. Am. Chem. Soc., 99 (1977) 8064.
66. E. J. Ditzel, H. D. Holden, B. F. G. Johnson, J. Lewis, A. Saunders and M. J. Taylor, J. Chem. Soc. Chem. Commun. (1982) 1373.
67. J. R. Goudsmith, B. F. G. Johnson, J. Lewis, P. R. Raithby and M. J. Rosales, J. Chem. Soc. Dalton Trans. (1983) 2257.
68. A. M. Brodie, H. D. Holden, J. Lewis and M. J. Taylor, J. Organomet. Chem., 253 (1983) Cl.
69. M. R. Churchill, F. J. Hollander, J. R. Shapley and D. S. Foose, J. Chem. Soc. Chem. Commun. (1978) 534.
70. R. D. Adams, I. Horvath and L. W. Yang, J. Am. Chem. Soc., 105 (1983) 1533.
71. B. Besson, B. Moraweck, A. K. Smith, J.-M. Basset, R. Psaro, A. Fusi and R. Ugo, J. Chem. Soc. Chem. Commun. (1980) 569.
72. E. G. Bryan, B. F. G. Johnson and J. Lewis, J. Chem. Soc. Dalton Trans. (1977) 1328.
73. B. F. Johnson, J. Lewis, W. J. H. Nelson, J. Puga, P. R. Raithby and K. H. Whitmire, J. Chem. Soc. Dalton Trans. (1983) 1339.
74. J. K. Kouba, E. L. Muetterties, M. R. Thompson and V. W. Day, Organometallics, 2 (1983) 1065.
75. G. Ciani, G. D'Alfonso, M. Freni, P. Romiti, A. Sironi, Congr. Naz. Chim. Inorg. [Atti] 16th (1983) 367; C.A. 100 (1984) 28882w.
76. T. Beringhelli, E. Cesarotti, G. Ciani, G. D'Alfonso and A. Sironi, J. Organomet. Chem., 268 (1984) C18.
77. A. J. Deeming and M. Underhill, J. Chem. Soc. Dalton Trans. (1974) 1415.
78. C. E. Kampe, N. M. Boag, C. B. Knobler and H. D. Kaesz, Inorg. Chem., 23 (1984) 1390.
79. B. F. G. Johnson, J. Lewis and D. A. Pippard, J. Organomet. Chem., 160 (1978) 263; *ibid.*, 213 (1981) 249.
80. B. Besson, A. Choplin, L. D'Ornelas and J. M. Basset, J. Chem. Soc. Chem. Commun. (1982) 843.
81. a. S. A. MacLaughin, A. J. Carty and N. J. Taylor, Can. J. Chem., 60 (1982) 87;
b. S. A. MacLaughin, N. J. Taylor and A. J. Carty, Organometallics, 3 (1984) 392.
82. W. F. Smith, N. J. Taylor and A. J. Carty, J. Chem. Soc. Chem. Commun. (1976) 896.
83. R. Regragui, P. H. Dixneuf, N. J. Taylor and A. J. Carty, Organometallics, 3 (1984) 814.
84. A. D. Harley, G. J. Guskey and G. L. Geoffroy, Organometallics, 2 (1983) 53.
85. S-G. Shyu and A. Wojcicki, Organometallics, 3 (1984) 809 and references cited therein.
86. S. Rosenberg, R. R. Whittle and G. L. Geoffroy, J. Am. Chem. Soc., 106 (1984) 5934.
87. N. Lugan, J.-J. Bonnet and J. A. Ibers, J. Am. Chem. Soc., 107 (1985), in press.

88. N. J. Forrow, S. A. R. Knox, M. J. Morris and A. G. Orpen, J. Chem. Soc. Chem. Commun. (1983) 234.
89. J. R. Norton and J. P. Collman, Inorg. Chem., 12 (1973) 476.
90. R. Szostak, C. E. Strouse and H. D. Kaesz, J. Organomet. Chem., 191 (1980) 2432.
91. A. Mayr, Y. C. Lin, N. M. Boag and H. D. Kaesz, Inorg. Chem. 21 (1982) 1704.
92. H. D. Kaesz, A. Mayr, Y. C. Lin, N. M. Boag and C. E. Kampe, in Huang Yaozeng, A. Yamamoto and B.-K. Teo (Eds.), New Frontiers in Organometallic and Inorganic Chemistry, Science Press, Beijing, 1984, pp. 301-308.
93. N. M. Boag, C. E. Kampe, Y. C. Lin and H. D. Kaesz, Inorg. Chem., 21 (1982) 1706.
94. D. J. Darensbourg, M. Pala and J. Waller, Organometallics, 2 (1983) 1285.
95. A. Mayr, Y.-C. Lin, N. M. Boag, C. E. Kampe, C. B. Knobler and H. D. Kaesz, Inorg. Chem., 23 (1984) 4640.
96. C. E. Kampe and H. D. Kaesz, Inorg. Chem., 23 (1984) 4646.
97. C. E. Kampe, N. M. Boag and H. D. Kaesz, J. Mol. Catal., 21 (1983) 297.
98. G. Süss-Fink and J. Reiner, J. Mol. Catal., 16 (1982) 231.
99. H. D. Kaesz, C. B. Knobler, M. A. Andrews, G. Van Buskirk, R. Szostak, C. E. Strouse, Y.-C. Lin and A. Mayr, Pure Appl. Chem., 54 (1982) 131.
100. M. A. Andrews and H. D. Kaesz, J. Am. Chem. Soc., 101 (1979) 7255.
101. W. Bernhardt and H. Vahrenkamp, Angew. Chem., 96 (1984) 362.
102. a. J. Banford, X. Dawoodi, J. Henrick and M. J. Mays, J. Chem. Soc. Chem. Commun. (1982) 554;
b. Z. Dawoodi, M. J. Mays and K. Henrick, J. Chem. Soc. Dalton Trans. (1984) 433.
103. Y. Shvo and E. Hazum, J. Chem. Soc. Chem. Commun. (1974) 336.
104. J. B. Keister, J. Chem. Soc. Chem. Commun. (1979) 214.
105. C. M. Jensen and H. D. Kaesz, J. Am. Chem. Soc., 105 (1983) 6969.
106. M. I. Bruce, D. C. Kehoe, J. G. Matisons, B. K. Nicholson, P. H. Rieger and M. L. Williams, J. Chem. Soc. Chem. Commun. (1982) 442.
107. M. I. Bruce, J. G. Matisons and B. K. Nicholson, J. Organomet. Chem., 247 (1983) 321.
108. M. Arewgoda, B. H. Robinson and J. Simpson, J. Am. Chem. Soc., 105 (1983) 1893.
109. G. Lavigne and H. D. Kaesz, J. Am. Chem. Soc., 106 (1984) 4647.
110. C. E. Kampe, G. Lavigne and H. D. Kaesz, in preparation.
111. M. Amstock, D. Taube, D. C. Gross and P. C. Ford, J. Am. Chem. Soc., 106 (1984) 3696.
112. R. L. Pruett, Ann. N. Y. Acad. Sci., 295 (1977) 239.
113. J. F. Knifton, J. Am. Chem. Soc., 103 (1981) 3959.
114. a. B. D. Dombek, J. Am. Chem. Soc., 103 (1981) 6508;
b. B. D. Dombek, J. Organomet. Chem., 250 (1983) 467.
115. a. E. L. Muetterties and J. Stein, Chem. Rev., 79 (1979) 479;
b. E. L. Muetterties and M. J. Krause, Angew. Chem. Int. Ed. Engl., 22 (1983) 135.
116. W. A. Herrmann, Angew. Chem. Int. Ed. Engl., 21 (1982) 117 and references cited therein.
117. J. S. Bradley, Advan. Organomet. Chem., 22 (1983) 1 and references cited therein.
118. L. M. Bavaro, P. Montangero and J. B. Keister, J. Am. Chem. Soc., 105 (1983) 4877.
119. J. Vites and T. P. Fehlner, Organometallics, 3 (1984) 491.
120. G. Trinquier and R. Hoffmann, Organometallics, 3 (1984) 370.

121. R. M. Laine, D. W. Thomas, L. W. Cary and S. E. Buttrill, J. Am. Chem. Soc., 100 (1978) 6527.
122. J. L. Vidal, W. E. Walker, R. L. Pruett and R. C. Schoening, Inorg. Chem., 18 (1979) 129.
123. R. J. Angelici, Acc. Chem. Res., 5 (1972) 335.
124. D. C. Gross and P. C. Ford, Inorg. Chem., 21 (1982) 1702.
125. C. M. Jensen, C. B. Knobler and H. D. Kaesz, J. Am. Chem. Soc., 106 (1984) 5926.
126. R. C. Schoening, J. L. Vidal and R. A. Fiato, J. Organomet. Chem., 206 (1981) C43.
127. a. G. R. Steinmetz and G. L. Geoffroy, J. Am. Chem. Soc., 103 (1981) 1278;
  b. G. R. Steinmetz, E. D. Morrison and G. L. Geoffroy, J. Am. Chem. Soc., 106 (1984) 2559.
128. C. R. Eady, J. J. Guy, B. F. G. Johnson, J. Lewis, M. C. Malatesta and G. M. Sheldrick, J. Chem. Soc. Dalton Trans. (1978) 1358.
129. R. E. Stevens and W. L. Gladfelter, Inorg. Chem., 22 (1983) 2034.
130. D. E. Fjare, J. A. Jensen and W. L. Gladfelter, Inorg. Chem., 22 (1983) 1774.
131. F. Basolo and D. E. Morris, J. Am. Chem. Soc., 90, (1968) 2536.
132. T. L. Brown and P. A. Bellus, Inorg. Chem., 17 (1978) 3726.
133. A. G. Cowie, B. F. G. Johnson, J. Lewis, J. Nicola Nicholls, P. R. Raithby and A. G. Swanson, J. Chem. Soc. Chem. Commun. (1984) 637.
134. G. R. Steinmetz, A. D. Harley and G. L. Geoffroy, Inorg. Chem., 19 (1980) 2895.
135. J. Knight and M. J. Mays, J. Chem. Soc. Dalton Trans. (1972) 1022.
136. W. McFarlane and G. Wilkinson, Inorg. Synth., 8 (1966) 178.
137. J. B. Keister, M. W. Payne and M. J. Muscatella, Organometallics, 2 (1983) 219.
138. J. C. Bricker, C. C. Nagel and S. G. Shore, J. Am. Chem. Soc., 104 (1982) 1444; cf. *ibid.* 107 (1985) 377.
139. C. P. Casey and S. M. Neumann, J. Am. Chem. Soc., 100 (1978) 2544.
140. S. Tam, G. Lin, W. Wong, W. A. Kiel, V. K. Wong and J. A. Gladysz, J. Am. Chem. Soc., 104 (1982) 141.
141. J. R. Shapley, M. E. Cree-Uchiyama, G. M. St. George, M. R. Churchill and C. Bueno, J. Am. Chem. Soc., 105 (1983) 140.
142. J. R. Shapley, D. S. Strickland, G. M. St. George, M. R. Churchill and C. Bueno, Organometallics, 2 (1983) 185.
143. a. W.-K. Wong, G. Wilkinson, A. M. Galas, M. B. Hursthouse and M. Thornton-Pett, J. Chem. Soc. Dalton Trans. (1981) 2496;
  b. W.-K. Wong, K. W. Chiu, G. Wilkinson, A. M. R. Galas, M. Thornton-Pett and M. B. Hursthouse, J. Chem. Soc. Dalton Trans. (1983) 1557.
144. J. R. Shapley, J. T. Park, M. R. Churchill, J. W. Ziller and L. R. Beanan, J. Am. Chem. Soc., 106 (1984) 1144.
145. M. A. Drezdon, K. H. Whitmire, A. A. Battacharya, W.-L. Hsu, C. C. Nagel, S. G. Shore and D. F. Shriver, J. Am. Chem. Soc., 104 (1982) 5630.
146. M. A. Drezdon and D. F. Shriver, J. Mol. Catal., 21 (1983) 81.
147. C. P. Horwitz and D. F. Shriver, Organometallics, 3 (1984) 756.
148. A. D. Clauss, J. R. Shapley, C. N. Wilker and R. Hoffmann, Organometallics, 3 (1984) 619.
149. J. T. Park, J. R. Shapley, M. R. Churchill and C. Bueno, J. Am. Chem. Soc., 105 (1983) 6182.
150. D. E. Fjare and W. L. Gladfelter, J. Am. Chem. Soc., 106 (1984) 4799.
151. W. L. Gladfelter, Advan. Organomet. Chem., 24 (1985), in press.
152. E. Sappa, O. Gambino and G. Cetini, J. Organomet. Chem., 35 (1972) 375.

153. a. J. R. Shapley, J. T. Park, M. R. Churchill, C. Bueno and H. J. Wasserman, J. Am. Chem. Soc., 103 (1981) 7385;
b. M. R. Churchill, C. Bueno and H. J. Wasserman, Inorg. Chem., 21 (1982) 640;
c. J. T. Park, J. R. Shapley, M. R. Churchill and C. Bueno, Inorg. Chem., 22 (1983) 1579.
154. L. Busetto, M. Green, J. A. K. Howard, B. Hessner, J. C. Jeffrey, R. M. Mills, F. G. A. Stone and P. Woodward, J. Chem. Soc. Chem. Commun. (1981) 1101.
155. L. R. Beanan, Z. A. Rahman and J. B. Keister, Organometallics, 2 (1983) 1062.
156. A. D. Clauss, J. R. Shapley and S. R. Wilson, J. Am. Chem. Soc., 103 (1983) 7387.
157. D. Nucciarone, N. J. Taylor and A. J. Carty, Organometallics, 3 (1984) 177.
158. E. D. Morrison, G. R. Steinmetz, G. L. Geoffroy, W. C. Fultz and A. L. Rheingold, J. Am. Chem. Soc., 105 (1983) 4104; 106 (1984) 2559.
159. E. D. Morrison, F. L. Geoffroy and A. L. Rheingold, J. Am. Chem. Soc., 107 (1985) 254.
160. a. J. S. Bradley, G. B. Ansell, M. E. Leonowicz and E. W. Hill, J. Am. Chem. Soc., 103 (1981) 4968;
b. J. S. Bradley, E. W. Hill, G. B. Ansell and M. A. Modrick, Organometallics, 1 (1982) 1634.
161. S. D. Wijeyesekera, R. Hoffmann and C. N. Wilker, Organometallics, 3 (1984) 962.
162. S. Harris and J. S. Bradley, Organometallics, 3 (1984) 1086.
163. D. de Montauzon and R. Mathieu, J. Organomet. Chem., 252 (1983) C83.
164. F. Dahan and R. Mathieu, J. Chem. Soc. Chem. Commun. (1984) 432.
165. J. W. Kolis, E. M. Holt, M. Drezdzon, K. H. Whitmire and D. F. Shriver, J. Am. Chem. Soc., 104 (1982) 6134.
166. J. E. Hahn, Prog. Inorg. Chem., 31 (1984) 205.
167. M. I. Bruce and A. G. Swincer, Advan. Organomet. Chem., 22 (1983) 60.
168. C. P. Horwitz and D. F. Shriver, Advan. Organomet. Chem. 23 (1984) 219.

Chapter 5

HOMOGENEOUS CATALYSIS BY METAL CLUSTERS

LASZLO MARKÓ AND ANNA VIZI-OROSZ

5.1 INTRODUCTION

There has been a surge in interest in the homogeneous catalytic activity of metal clusters [1-3]. This development may be regarded as a natural consequence of the rapidly increasing number of well-characterized transition metal clusters and the improved understanding of homogeneous catalytic reactions involving mononuclear transition metal complexes. Research in this field has been stimulated not only by academic interest, but also by hopes of finding new and selective catalysts for technologically important reactions.

The aim of this chapter is to give an overall view of the subject, based on the literature available at the end of 1983. This review is not intended to be exhaustive, and the selection of references is to some degree arbitrary because of the ambiguity in defining homogeneous catalysis by metal clusters.

Several definitions of metal clusters have been proposed [1, 3, 4-6], but for present purposes a very simple and flexible definition is needed, since otherwise some of the many types of possible catalytic cycles might be excluded. Therefore, we regard metal clusters as compounds containing more than one transition metal atom. This broad interpretation allows us to consider a few interesting examples for which the two or more metal atoms necessary for catalysis are not directly bonded to each other. There are also many borderline cases involving fractional metal-metal bonds.

When is a reaction regarded as being catalyzed by such di- (or multi-) nuclear complexes? Consider the catalytic cycle of SCHEME 1 for hydroformylation [7, 8]. Clearly, $Co_2(CO)_8$ is essential for the reaction and $Co_2(CO)_8$ is definitely a cluster compound, but is hydroformylation indeed a cluster-catalyzed reaction? Although many scientists working in this field might answer no, we include a few examples, attempting to compile results for catalytic reactions in solution in which clusters play an important role.

Frequently the catalytically active metal has been applied in the form of a cluster compound as the catalyst precursor, since most of these complexes are relatively stable and easy to handle. When nothing is known about the species formed from the precursor cluster under reaction conditions, the work has usually been omitted from this compilation. In at least one case [hydroformylation of olefins with $Rh_4(CO)_{12}$ or $Rh_6(CO)_{16}$ as

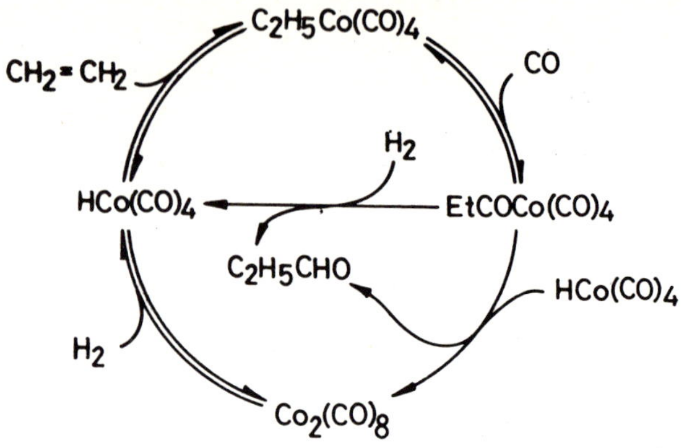

SCHEME 1

catalyst precursors] it has been shown by kinetics measurements that the clusters were fragmented to mononuclear complexes under reaction conditions [9].

In addition to catalytic reactions, a few examples illustrating the use of cluster compounds to effect stoichiometric organic transformations are also included. The only examples included in this category involve reasonably well understood mechanisms which can serve as models for the catalytic transformations of organic molecules on clusters.

In the following pages, the examples of metal-cluster-catalyzed reactions are presented for the transition metals in turn; a few mechanistic conclusions are summarized in the final section.

5.2 MOLYBDENUM COMPLEXES

The dinuclear complexes $Cp_2Mo_2(CO)_4(RCCR)$ containing a bridging acetylene unit (Fig. 1) react with H_2 at temperatures >373 K to form *cis*-RCH=CHR and $Cp_2Mo_2(CO)_4$.

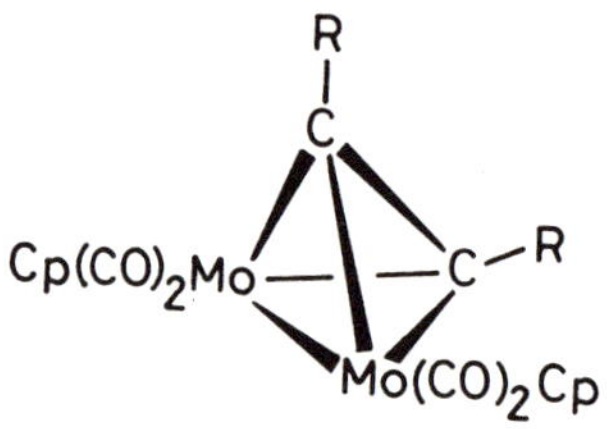

Fig. 1. Structure of $Cp_2Mo_2(CO)_4(RCCR)$ complexes.

In the presence of excess alkyne, the reaction is catalytic with turnover rates of 5×10^{-4} s^{-1} at 423 K. Experiments performed with a mixture of $(C_5H_5)_2Mo_2$ and $(CH_3C_5H_4)_2Mo_2$ showed that no ligand interchange occurs and thus that fragmentation of the dinuclear complex into mononuclear species is not an integral part of the catalytic cycle. CO inhibited the catalytic reaction, and therefore the dissociation of a CO ligand is proposed to be the first step [10]:

$$Cp_2Mo_2(CO)_4(RCCR) \rightleftharpoons Cp_2Mo_2(CO)_3(RCCR) + CO \qquad (1)$$

The protonated derivatives of the above complexes formed by reaction with CF_3COOH also react with H_2 in a stoichiometric manner to yield alkenes, but they are not suitable as catalysts for hydrogenation. In the presence of excess alkyne, polymerization becomes the main reaction [11].

The stoichiometric hydrogenation of alkynes to give *cis*-alkenes has been realized with the dinuclear molybdenum complexes, as follows:

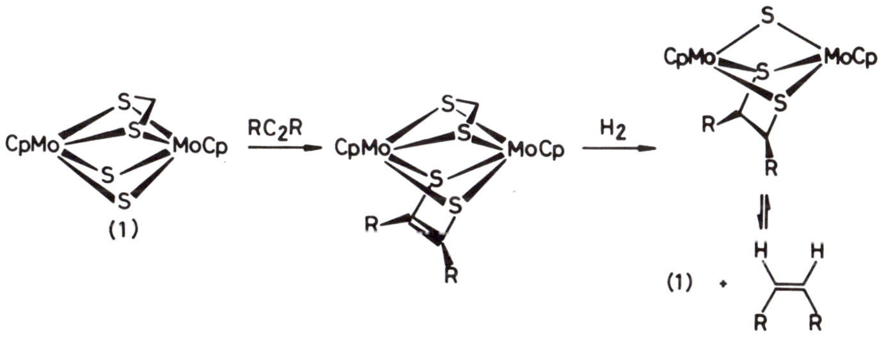

SCHEME 2

Despite regeneration of the starting complex, however, a catalytic cycle could not be closed [12].

5.3 IRON COMPLEXES

Some years ago it was proposed that the Reppe synthesis of alcohols

$$CH_2\text{-}CH_2 + 3CO + 2H_2O \longrightarrow CH_3CH_2CH_2OH + 2CO_2 \qquad (2)$$

is catalyzed by the $HFe_3(CO)_{11}^-$ anion [13]. This conclusion was based on the observation that R_3NH^+ salts of this cluster could be isolated from the product and that these salts may be used as catalyst precursors for the reaction [13-16]. More recent work, however, suggests that under the reaction conditions (373-433 K, 50-200 bar of CO) all iron carbonyl clusters are transformed into mononuclear complexes, and among these $H_2Fe(CO)_4$ and $HFe(CO)_4^-$ are the catalytically effective species [17-20]. The same anionic hydride cluster has been used in the form of $(Et_4N)[HFe_3(CO)_{11}]$ for the stoichiometric reduction of nitroarenes, α,β-unsaturated carbonyl compounds, and organic halides [21].

The neutral trimeric cluster $Fe_3(CO)_{12}$ is also a useful stoichiometric reducing agent if applied in the presence of a base. Under such conditions, $HFe_3(CO)_{11}^-$ is formed *in situ* and reduces nitroarenes to aniline derivatives. Two variants of this system have been described. In the earlier method, one applies $Fe_3(CO)_{12}$ in the presence of methanol in benzene solution (as a Lewis base) at reflux temperature [22]; in the other method, one uses a benzene-aqueous NaOH two-phase system in the presence of benzyltriethylammonium chloride as a phase transfer catalyst at room temperature [23]. Yields are usually high (80-90%). Only little is known about the mechanisms of these reductions, and even the stoichiometry is unclear. Part of the oxygen contained in the nitro groups appears as CO_2 [22].

The iron cluster $[CpFe(CO)]_4$ (Fig. 2) catalyzes the hydrogenation of alkynes to give *cis*-alkenes and the hydrogenation of aryl nitro groups to give aniline derivatives at 373-403 K and 7-70 bar [24].

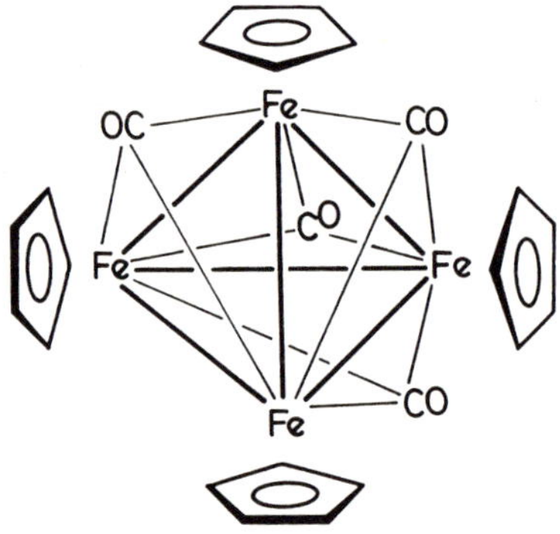

Fig. 2. Structure of $[CpFe(CO)]_4$.

Terminal activated double bonds (e.g., in methyl acrylate) are also hydrogenated with moderate yields. The cluster concentration, monitored by high-pressure liquid chromatography, was unchanged after more than 1000 turnovers, and the complex could be isolated in 95-97% yield from the reaction product. No other iron-containing species were detected. Fragmentation of the tetrameric complex to give $CpFe(CO)_2H$ is unlikely, since the dimeric $[CpFe(CO)_2]_2$, which gives this mononuclear hydride under H_2 pressure, was not active for the hydrogenation of pent-1-yne under the reaction conditions. These results support the suggestion of cluster catalysis, but no other information is available about the mechanism of the reaction.

The cluster $Fe_4S_4Cl_4^{2-}$ catalyzes the hydrogenation of octenes in the presence of phenyllithium [25].

Collman and coworkers have thoroughly studied the stoichiometric reduction of α,β-unsaturated carbonyl compounds by the dinuclear $NaHFe_2(CO)_8$ [26,27]. In the presence of acetic acid, the stoichiometry of the reaction is the following:

$$2NaHFe_2(CO)_8 + RCH{=}CHCOR' + HOAc \longrightarrow RCH_2CH_2COR' + NaHFe_3(CO)_{11} + Fe(CO)_5 + NaOAc \quad (3)$$

The mechanism proposed for these reactions is shown in SCHEME 3. The key pieces of evidence supporting this mechanism are the following:

(1) A first order of each reaction on the concentration of metal hydride and of activated alkene.

(2) Isotopic labeling experiments which demonstrate a reversible, regioselective hydride addition step, and a large inverse isotope effect (k_D/k_H = 3.5).

According to the simplest interpretation of the experimental results, the active reagent in these reductions is binuclear $NaHFe_2(CO)_8$. The addition of $HFe_2(CO)_8^-$ to activated alkenes is regarded as a concerted reaction.

5.4 RUTHENIUM COMPLEXES

The water gas shift reaction

$$CO + H_2O \longrightarrow CO_2 + H_2 \quad (4)$$

is catalyzed by $Ru_3(CO)_{12}$ in aqueous ethoxyethanol solutions in the presence of KOH at 363-393 K and 0.5-2.0 bar [28-30]. The catalyst

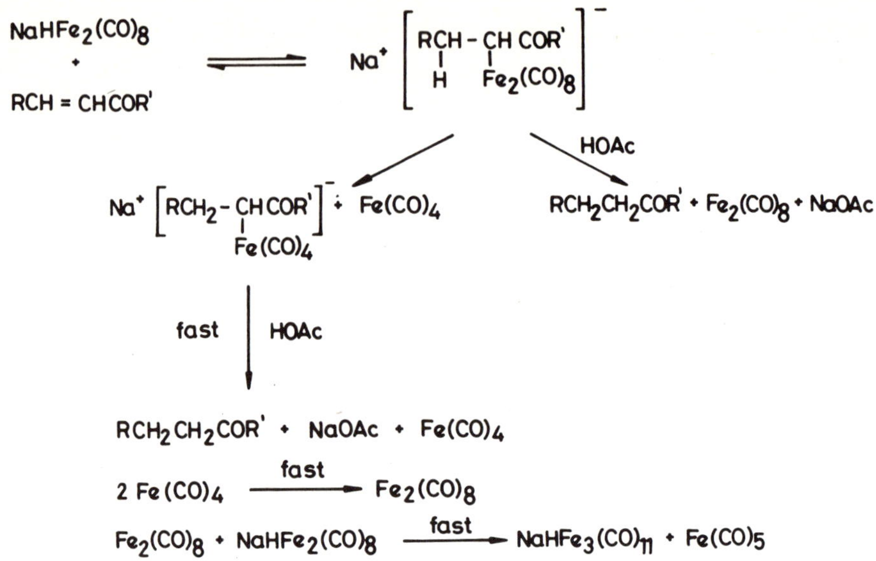

SCHEME 3

system reaches equilibrium after a few hours and then maintains constant activity [~3 turnovers/day per mol of $Ru_3(CO)_{12}$]. This maturation process is characterized by the formation of formate and bicarbonate (which buffer the solution to pH ~9.5) and the transformation of $Ru_3(CO)_{12}$ into a mixture of trinuclear and tetranuclear ruthenium clusters. Ruthenium is present mainly in the form of $HRu_3(CO)_{11}^-$ and $H_3Ru_4(CO)_{12}^-$, but as H_2 accumulates in the product gases the tetranuclear anion becomes predominant. This shift in composition has been shown to be due to the following equilibrium [31]:

$$3H_3Ru_4(CO)_{12}^- + 9CO \rightleftharpoons 3HRu_3(CO)_{11}^- + Ru_3(CO)_{12} + 3H_2 \qquad (5)$$

The catalytic reaction shows a first-order dependence on the CO partial pressure and on the total ruthenium concentration. Two catalytic cycles have been proposed to operate under these reaction conditions, one involving tetranuclear species (SCHEME 4) and the other trinuclear species (SCHEME 5) [28-30]:

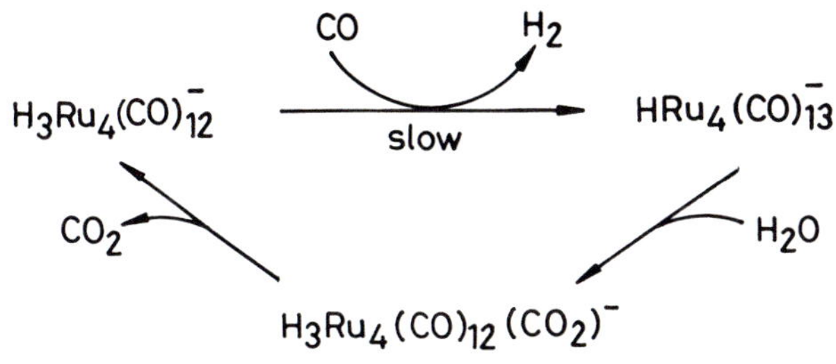

SCHEME 4

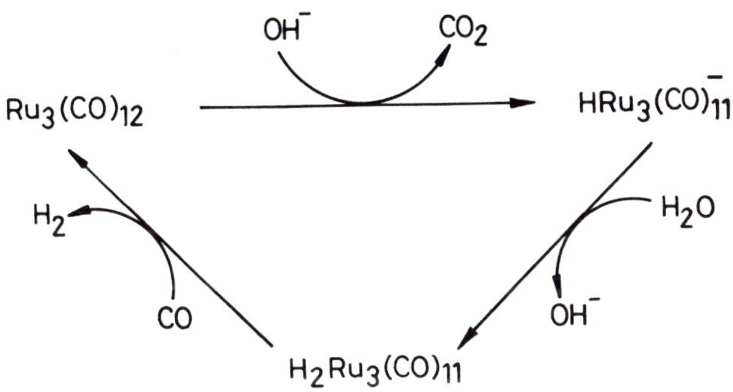

SCHEME 5

It is conceivable that catalysis occurs by both mechanisms and that the relative importance of each depends on the CO and H_2 partial pressures which shift equilibrium (5). In accordance with this suggestion, when H_2 was continuously removed from the reaction mixture, only $HRu_3(CO)_{11}^-$ was observed in the reaction mixture and the turnover was approximately four times that observed when H_2 was allowed to accumulate [29]. This result suggests that the cycle involving the trinuclear complex is the more effective one.

Homogeneous catalysis of the water gas shift reaction has been demonstrated with $Ru_3(CO)_{12}$ as a catalyst precursor in acidic solutions [32]. Under such conditions, the predominant ruthenium species is $[HRu_2(CO)_{9-x}(OR_2)_x]^+$ (OR_2 = water or diglyme solvent), and a catalytic cycle based on dinuclear Ru carbonyl derivatives has been proposed.

Soluble ruthenium catalysts have been found to hydrogenate CO to give (1) methanol and methyl formate [33], (2) methyl, ethyl, and glycol esters (with carboxylic acids as solvents [34, 35]), or (3) methanol, ethylene glycol, and ethanol {with halides (especially iodide) used as promoters

[36-39]}. Spectroscopic evidence suggests that the anionic ruthenium cluster $HRu_3(CO)_{11}^-$ may play an important role in these latter catalyst systems which are formed *in situ* under reaction conditions from $Ru_3(CO)_{12}$ and I^- in polar solvents like sulfolane or N-methylpyrrolidone.

The activity of such catalyst systems depends on the I^-/Ru ratio; best results are obtained when the ratio is about 0.5. This result is in accordance with the observation that $HRu_3(CO)_{11}^-$ and $Ru(CO)_3I_3^-$ are the principal species present in the reaction mixture and that their ratio corresponds to that demanded by the following reaction:

$$7/3Ru_3(CO)_{12} + 3I^- + H_2 \longrightarrow 2HRu_3(CO)_{11}^- + Ru(CO)_3I_3 + 3CO \quad (6)$$

Neither $Ru(CO)_3I_3^-$ nor $HRu_3(CO)_{11}^-$ alone has catalytic activity for glycol formation, and the mononuclear complex is inactive even in the presence of added iodide. The cluster $HRu_3(CO)_{11}^-$, however, produces glycol at a low rate when iodide is present. On the basis of this information, the following cycle has been proposed to explain the chemistry of the catalytic process [38]. (SCHEME 6 does not provide any details about the reactions leading to C_1 and C_2 alcohols.)

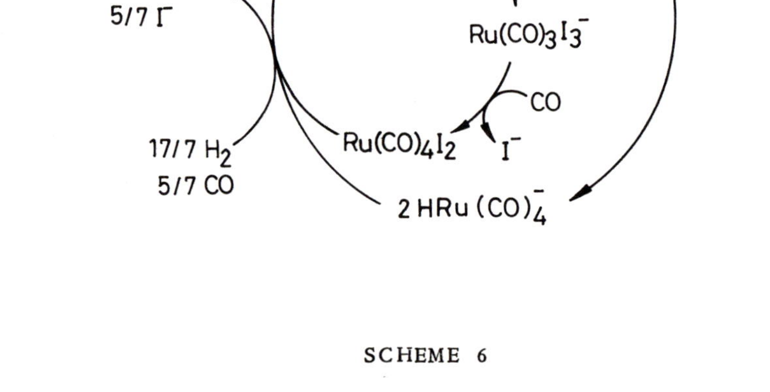

SCHEME 6

It has been found that $HRu_3(CO)_{11}^-$ is an effective catalyst precursor for the production of methyl formate from CO_2 at 398 K and 30 bar:

$$CO_2 + H_2 + CH_3OH \longrightarrow HCOOCH_3 + H_2O \quad (7)$$

It is probable that the tetranuclear species $H_3Ru_4(CO)_{12}^-$ is the catalytically active one. The results suggest that the reaction is not preceded by the water gas shift reaction [40].

The isomerization of pent-1-ene is catalyzed by $Ru_3(CO)_{12}$ under thermal (343 K) and photochemical conditions. The thermal reaction is inhibited by CO, which suggests that the first step of the process is the dissociation of a CO ligand [41]. Acetic acid strongly promotes the reaction, but experiments with CH_3COOD show that D is not incorporated into the product and that the mechanism therefore involves not an addition-elimination of a metal hydride but rather the formation of a π-allyl type intermediate. Photochemical experiments also support a mechanism starting with CO loss without fragmentation of the Ru_3 cluster [42].

The tetranuclear hydrido clusters $H_4Ru_4(CO)_{12}$ [43-46] and $H_2Ru_4(CO)_{13}$ [47] are also active catalysts for the isomerization of pent-1-ene. The reaction is inhibited by CO and acetic acid [43-47], and the effect of H_2 is complex [45]. Experiments with deuterated alkene suggest a metal hydride addition-elimination mechanism [43, 47]. Accordingly, the loss of a CO ligand with the formation of an alkene complex is regarded as the first step. Such a mechanism is supported by results of photochemical investigations [46]. Derivatives of $H_4Ru_4(CO)_{12}$ substituted by phosphorus donor ligands show similar catalytic behavior [44, 45].

In the presence of H_2, $H_4Ru_4(CO)_{12}$ is the principal species present in solution, and homogeneous hydrogenation of alkenes [46, 48-50] or alkynes [51, 52] occurs. On the basis of kinetics, infrared, and mass spectrometric measurements, the following mechanism has been proposed for the hydrogenation of ethylene [48, 50]. The rate of hydrogenation increases upon irradiation of the solution, and this effect is reversible [49]. $Na[HRu_3(CO)_{11}]$ is also a catalyst for the hydrogenation of alkenes [53].

The hydrogenation of polynuclear heteroaromatic nitrogen compounds like quinoline, acridine, or different benzoquinolines is catalyzed by $H_4Ru_4(CO)_{12}$ at 423 K and 25 bar of H_2. Reduction proceeds regioselectively at the N-containing ring [54].

Asymmetric hydrogenation of prochiral alkenes has been studied with the chiral ruthenium cluster $H_4Ru_4(CO)_8L_2$ as catalyst, where L is (-)-DIOP, a chiral ditertiary phosphine [55]. Reaction conditions were 353-393 K and

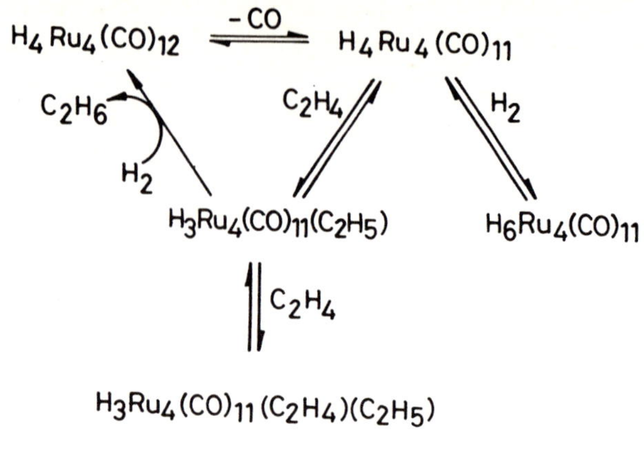

SCHEME 7

100 bar of H_2 partial pressure. In the case of α,β-unsaturated carboxylic acids as substrates, optical yields up to 68% have been achieved [56], but phenylbutenes could be hydrogenated with only a much lower enantioselectivity [57]. Infrared spectroscopic investigations suggest, that the structure of the complex is not much affected during the catalytic process [56].

Under 100 bar of H_2 pressure, $H_4Ru_4(CO)_{12}$ catalyzes the hydrogenation of ketones at 373 K [58]. The reaction is first order in cluster, in substrate, and in H_2, and the cluster can be recovered unchanged after the reaction. These findings support the inference of cluster catalysis. With the derivative $H_4Ru_4(CO)_8L_2$ as catalyst [L = (-)-DIOP] in the hydrogenation of prochiral ketones, secondary alcohols were obtained with low (0.1-5.4%) optical yields [59]. When isopropyl alcohol was used as a hydrogen source instead of H_2, transfer hydrogenation with the same catalyst resulted in somewhat higher enantioselectivities [59]. Saturated aliphatic monocarboxylic and dicarboxylic acids were hydrogenated to give the corresponding alcohols (present in the reaction product as esters) with $H_4Ru_4(CO)_8(PBu_3)_4$ catalyst at 373-473 K and 100-200 bar of H_2. Anhydrides yielded esters or lactones, but esters were not reduced further. It is probable that ruthenium carbonyl carboxylates are the catalytically active intermediates, since the dinuclear complex $[Ru(CO)_2(AcO)(PBu_3)]_2$ has been isolated from the reaction product obtained by hydrogenating acetic acid. This complex is also an active catalyst for the hydrogenation of acetic acid and itself gives rise to ethyl acetate and ethyl alcohol if heated under H_2 pressure [60].

The hydroformylation of alkenes takes place in the presence of $Ru_3(CO)_{12}$ with CO + H_2 [61] or CO + H_2O (eq. 2) [62, 63]. In both cases the rate of reaction shows a maxium depending on the partial pressure of CO, which strongly suggests cluster fragmentation under reaction conditions. This suggestion is supported also by infrared studies [63]. It is probable that the same is true when $H_4Ru_4(CO)_{12}$ [62] or $(Et_4N)[HRu_3(CO)_{11}]$ [53] is used as a catalyst. Under hydroformylation conditions (373 K, 13 bar of H_2, and 26 bar of CO), the two cluster anions $H_3Ru_4(CO)_{12}^-$ and $HRu_3(CO)_{11}^-$ were found to be readily interconverted [53], and this process presumably proceeds through mononuclear species. Cluster fragmentation is obviously the favored process under the relatively high CO partial pressures required for hydroformylation.

The $HRu_3(CO)_{11}^-$ anion also catalyzes hydrosilylation

$$CH_2{=}CH_2 + Et_3SiH \longrightarrow C_2H_5\text{-}SiEt_3 \qquad (8)$$

and silacarbonylation of alkenes at 373 K [53].

$$CH_2{=}CH_2 + CO + Et_3SiH \longrightarrow CH_3CH{=}CHOSiEt_3 \qquad (9)$$

(*cis* and *trans*)

In the presence of CO and Et_3SiH, a new cluster anion is formed in an equilibrium reaction:

$$HRu_3(CO)_4^- + 2Et_3SiH \rightleftharpoons HRu_3(CO)_{10}(SiEt_3)_2^- + CO + H_2 \qquad (10)$$

The two anionic clusters showed the same catalytic activity in both reactions (eqs. 8 and 9) but were stable only in the presence of CO. When used as catalysts for hydrosilylation, they were transformed into an unidentified mixture of ruthenium carbonyl anions [53].

Solutions of $[Ru_3O(OAc)_6(H_2O)_3]OAc$ in dimethyl formamide (DMF) were found to hydrogenate alkenes and alkynes at 353 K under 1 bar of H_2 [64]. The starting complex is transformed into the complex $[Ru_3O(OAc)_4(DMF))_n]OAc$, which serves as the active catalyst. The mechanism shown in SCHEME 8 has been suggested on the basis of a detailed kinetics investigation. The hydride path proceeding through $H_2Ru_3O(OAc)_4^+$ is preferred. Experimental results suggest that only one

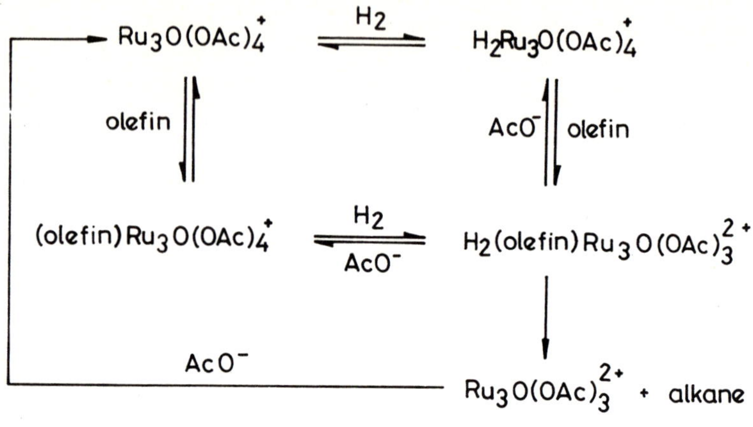

SCHEME 8

ruthenium center of the $H_2(alkene)Ru_3O(OAc)_3^{2+}$ cluster intermediate is active in coordinating both the hydrogen and the alkene molecule (Fig. 3). The other two ruthenium centers of the cluster function merely as metal entities present in a polydenate ligand.

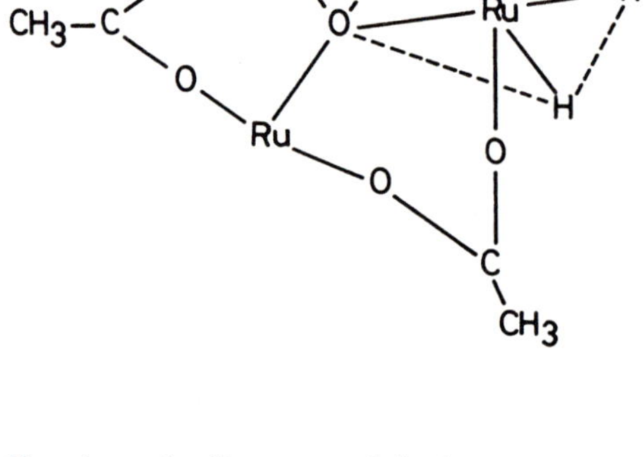

Fig. 3. Coordination of alkenes and hydrogen to Ru_3O-acetate clusters.

5.5 OSMIUM COMPLEXES

Hydrogenation of hex-1-ene is catalyzed by $H_2Os_3(CO)_{10}$ at 323 K and 3 bar [65], and the reaction is rather slow. The low catalytic activity of this cluster makes possible the isolation of intermediate species; good evidence has been provided for intermediates usually assumed in catalytic cycles for alkene hydrogenation. Upon reaction of $H_2Os_3(CO)_{10}$ with diethyl fumarate or diethyl maleate at 298 K, a hydridoalkyl intermediate cluster is formed which may be isolated and reacted separately with H_2 (at 323 K and 3 bar) to yield diethyl succinate and the starting complex [65]:

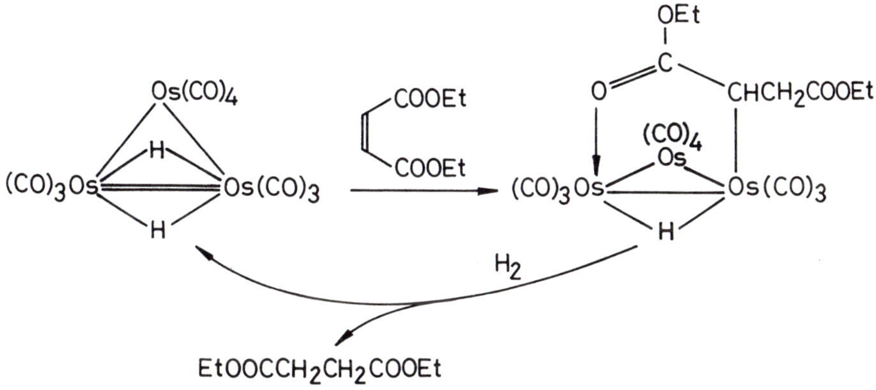

SCHEME 9

An even more complete series of intermediates has been presented for the hydrogenation of alkynes to give alkenes in the presence of $H_2Os_3(CO)_{10}$; $CF_3C_2CF_3$ was used as the model alkyne [66]. The sequence of stoichiometric reactions realized separately is shown in SCHEME 10.

5.6 COBALT COMPLEXES

Hydroformylation of hex-1-ene is catalyzed by the polymeric phosphido-bridged complex $[Co(\mu\text{-}PPh_2)(CO_3)_x]$ at 383 K and 90 bar of (CO + H_2). The activity of this complex is, however, much less than that of $Co_2(CO)_8$, and it is transformed into the dinuclear $Co_2(CO)_6[PPh_2(hexyl)]_2$ under reaction conditions [67]. This observation points to cluster fragmentation. Clusters of the type $Co_3(CO)_9(S)(LX)$ prepared from $Co_2(CO)_8$ and thioamides (e.g., LX = $CH_3C{=}NC_6H_{11}$) were also tested as hydroformylation catalysts, but evidence again suggests that only complexes which can be fragmented under reaction conditions are active [68].

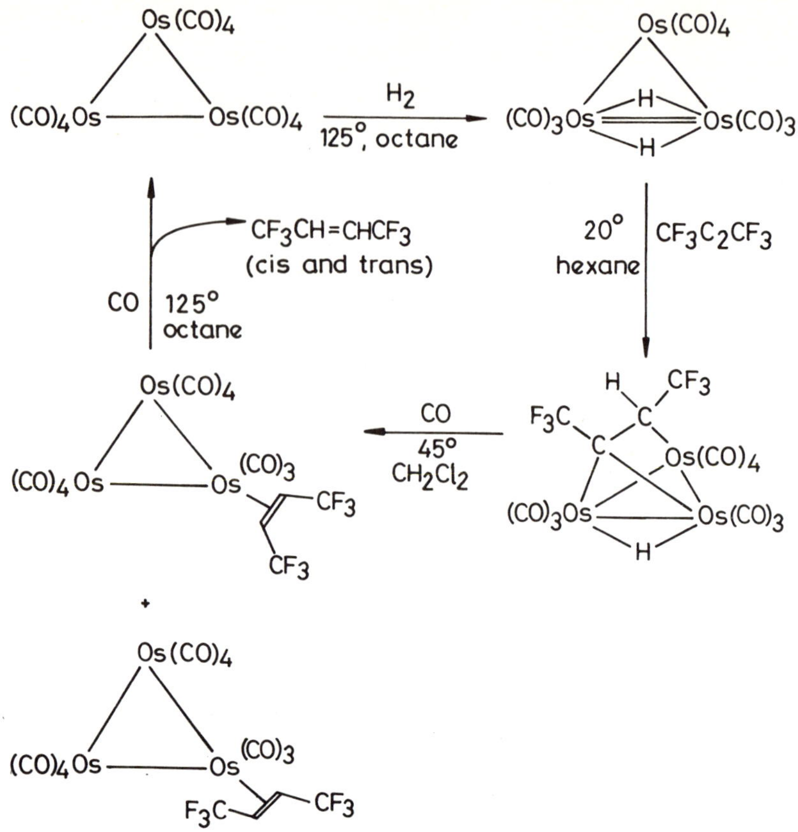

SCHEME 10

Catalytic activity of intact clusters has been repeatedly claimed for the complexes $Co_4(CO)_{10}(PPh)_2$ and $Co_4(CO)_8(PPh_3)_2(PPh)_2$ in hydroformylation of alkenes [69-71] and in homologation of methanol [72]:

$$CH_3OH + CO + 2\ H_2 \longrightarrow CH_3CH_2OH + H_2O \qquad (11)$$

These clusters have an especially stable framework (Fig. 4) and could be recovered with about 99.5% yield from the reaction products. Furthermore, no other complexes could be detected in the products by high-pressure liquid chromatography. Unfortunately, however, no kinetics measurements have been performed and no intermediates have been isolated. The rather low reaction rates observed suggest the formation of minute amounts of decomposition products which may be responsible for the catalytic activity. As already noted in connection with Ru complexes in hydroformylation, the

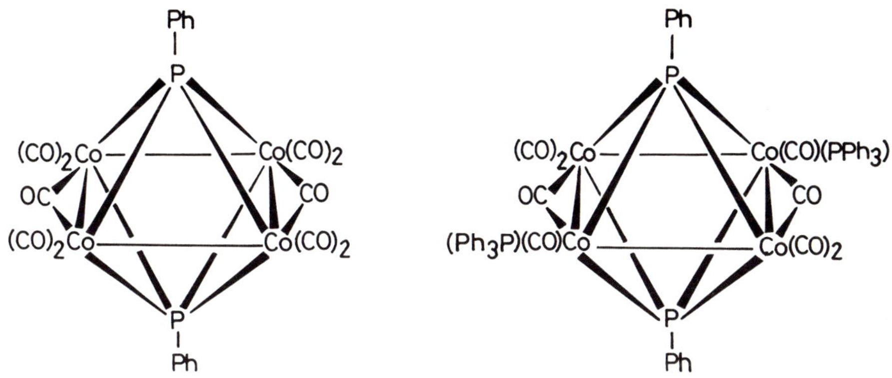

Fig. 4. Structures of $Co_4(CO)_{10}(PPh)_2$ and $Co_4(CO)_8(PPh_3)_2$ $(PPh)_2$ complexes.

formation of mononuclear complexes is possible. The tendency to form such complexes should be even stronger in the case of cobalt, a 3d transition metal.

The trimerization of alkynes to give aromatic compounds is catalyzed by many transition metal complexes, among them cobalt carbonyls, including $Co_2(CO)_8$ [73]. Under mild conditions, several cobalt carbonyl alkyne complexes have been isolated from these systems, and they may be regarded as intermediates of the catalytic cyclization process. Among these, $Co_2(CO)_4(Bu^tC_2H)_2(C_2H_2)$ deserves special attention; its structure is shown in Fig. 5. Upon treatment with Br_2, this complex decomposes, yielding o-di-*tert*-butylbenzene [75]. These stoichiometric transformations suggest dinuclear cobalt carbonyls as the species mainly responsible for trimerization, a remarkable result achieved many years before the concept of cluster catalysis was born.

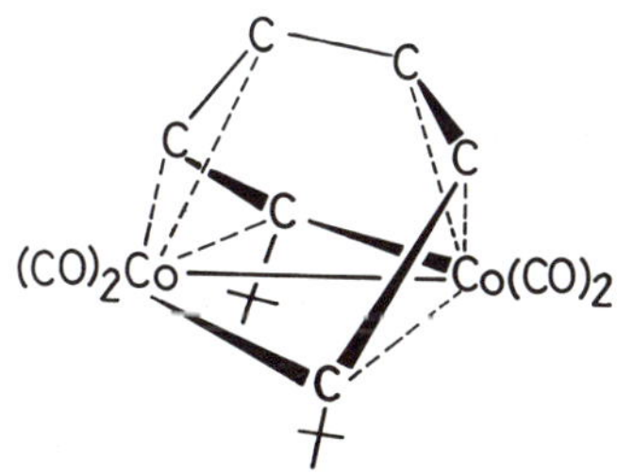

Fig. 5. Structure of $Co_2(CO)_4(Bu^tC_2H)_2(C_2H_2)$ [74].

5.7 RHODIUM COMPLEXES

Rhodium, particularly in combination with alkali metal salts, catalyzes the formation of ethylene glycol from CO and H_2 at 483-523 K and 1200-2000 bar [76]. The catalyst may be added in any convenient form, such as $Rh(acac)(CO)_2$ or $Rh_6(CO)_{16}$, and it is transformed under reaction conditions into a mixture of anionic rhodium carbonyl clusters. High-pressure infrared [77, 78] and NMR [79] studies showed that $[Rh_5(CO)_{15}]^-$ is the principal species present.

The above observations support the inference of cluster catalysis, although kinetics measurements are lacking, and no intermediates of glycol formation have been observed. Obviously, a rather complicated sequence of reactions takes place under such extreme conditions. This complexity is also shown by the formation of several byproducts such as methanol and glycerol. Additional support for the catalytic role of Rh clusters has been provided by experiments with rhodium carbonyl clusters containing encapsulated heteroatoms such as C, P, or S. It was found that $[Rh_6(CO)_{15}C]^{2-}$, $[Rh_9P(CO)_{21}]^{2-}$, and $[Rh_{17}S_2(CO)_{32}]^{3-}$ catalyze the formation of polyols under conditions hindering their fragmentation into other species [77].

The stoichiometric hydroformylation of olefins with $Rh_4(CO)_{12}$ [80]

$$4\ CH_2{=}CH_2 + 3\ Rh_4(CO)_{12} + 4\ H_2 \longrightarrow 4\ C_2H_5CHO + 2\ Rh_6(CO)_{16} \qquad (12)$$

is a related reaction which gives some insight into the role of anionic rhodium carbonyl clusters in catalytic reactions of CO. If the reaction is performed in a more polar solvent like acetone, and if water is present, an acyl derivative is formed:

$$2\ CH_2{=}CH_2 + 3\ Rh_4(CO)_{12} \longrightarrow 2\ H^+ + 2\ [Rh_6(CO)_{15}(COEt)]^- + 2\ CO_2 + 2\ CO \qquad (13)$$

In the presence of triphenylphosphine, hydroformylation becomes catalytic, but in this case it is likely that mononuclear $HRh(CO)_n(PPh_3)_{4-n}$ complexes may be formed, which are active catalysts even at room temperature and atmospheric pressure [81].

Catalytic hydroformylation with dinuclear rhodium carbonyls containing strong bridging ligands has been observed recently. Thiolato bridged complexes, particularly $Rh_2(\mu\text{-}SBu^t)_2(CO)_2[P(OMe)_3]_2$, were found to be active at 5 bar 358 K and for hydroformylation of hept-1-ene [82]. The structurally similar $Rh_2(\mu\text{-}PBu^t_2)(CO)_4$ catalyzes the hydroformylation of hex-1-ene at room temperature and 2 bar [83]. In both cases the complexes applied as catalyst precursors remained unchanged under reaction conditions, and no apparent fragmentation could be observed. No information is available concerning possible intermediates.

The dinuclear and trinuclear rhodium hydrides $[HRhL_2]_2$ [L = $P(OPr^i)_3$] and $[HRhL_2]_3$ [L = $P(OMe)_3$ or $P(OEt)_3$] are highly active catalysts for the hydrogenation of alkenes [84] and alkynes [85, 86]. Analogous complexes containing the bidentate ligands $R_2PCH_2CH_2PR_2$ (R = OMe or OEt) are also catalysts for alkene hydrogenation [87]. In all these clusters, the metal atoms are held together by bridging hydrogen ligands. Some details of the catalytic cycles with dinuclear complexes are known; a mechanism suggested for ethylene hydrogenation is shown in SCHEME 11 [84].

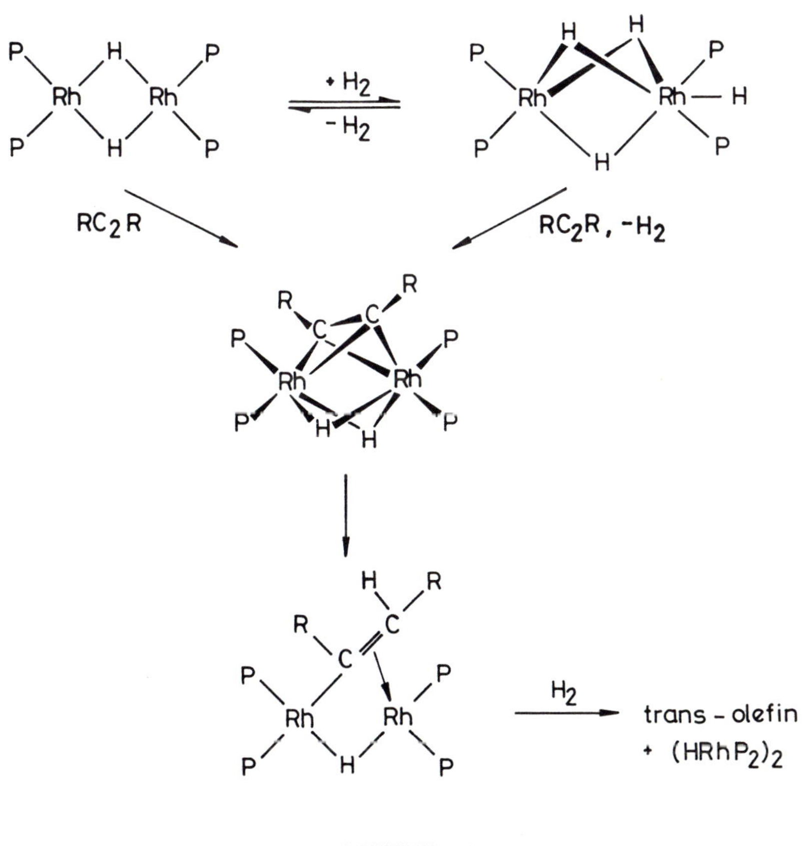

SCHEME 11

An interesting aspect of this proposal is that hydrogen adds oxidatively to one rhodium atom and the alkene to the other rhodium atom--this feature formally models metal-surface-catalyzed alkene hydrogenations on adjacent metal atom sites.

Alkyne hydrogenation by the same dinuclear rhodium hydrides is unique in the respect that it furnishes *trans* olefins in the first few minutes of the catalytic reaction. This selectivity is ascribed to the formation of an intermediate bridged vinyl complex, which was isolated (Fig. 6) and which has a trans arrangement of alkyl or aryl groups [86]. The rate-determining step of the catalytic cycle appears to be H_2 addition to this vinyl derivative. Competitive with hydrogen addition is alkyne addition, which leads to cluster fragmentation and to mononuclear complexes catalyzing the hydrogenation of the alkyne to the corresponding *cis*-alkyne. Accordingly, stereoselectivity of alkyne hydrogenation is rapidly reversed from *trans*- to *cis*-alkene formation. This observation underscores an often-described flaw in cluster catalysis, namely, the generation of mononuclear species.

Benzaldehyde is hydrogenated at 383 K and 70 bar with a (4 H_2 + CO) gas mixture in the presence of $Rh_6(CO)_{16}$ as a catalyst precursor. The observation of a reaction order of 0.92 in $Rh_6(CO)_{16}$ supports the suggestion that clusters are the active species [88].

Several reactions are catalyzed by $Rh_6(CO)_{16}$ under basic conditions when CO + H_2O is used as a hydrogen source. We emphasize that formation of mononuclear species is probably not important under such

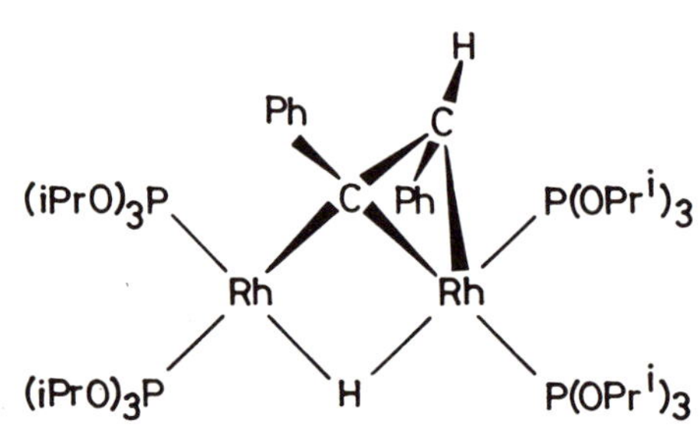

Fig. 6. Isolated intermediates of diphenyl acetylene hydrogenation catalyzed by $H_2Rh_2[P(OPr^i)_3]_4$.

conditions (even at high pressures) because in the presence of bases anionic carbonyl clusters are formed which are more stable to fragmentation by CO than their neutral analogues. Results have been published on the hydrogenation of α,β-unsaturated carbonyl and nitrile compounds [89, 90], the hydrogenation of pyridine [91], the hydrogenolysis of trialkyl amines [92], and the H-D exchange between triethylamine and D_2 [93]. Little is known about the active species in these catalyst systems, but the isolation of an anionic rhodium carbonyl cluster derivative and the unusual deuterium distribution in the product support a cluster-catalyzed H-D exchange mechanism for the last reaction [93]:

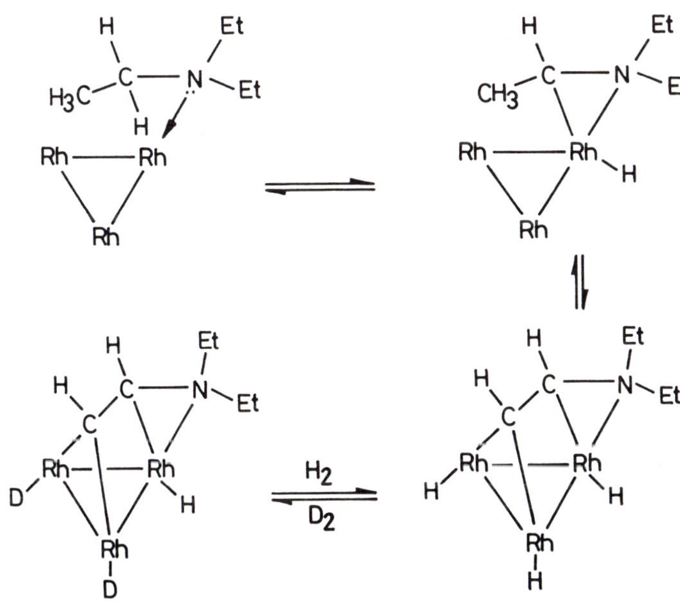

SCHEME 12

The oxidation of cyclohexanone by O_2 to adipic acid is catalyzed by $Rh_6(CO)_{16}$ in the presence of CO (373 K, 34 bar, CO:O_2 = 2:1) [94]. CO is concomitantly oxidized to CO_2. Attempts to oxidize the ketone under pure O_2 were unsuccessful as a result of decomposition of the cluster, but catalyst could be regenerated by treating the reaction mixture with CO + O_2 at 373 K. Rhodium carbonyl carboxylate complexes such as $Rh_2(OAc)_2(CO)_4$ are probable intermediates in the reaction.

5.8 NICKEL COMPLEXES

Several reactions are catalyzed by the tetranuclear Ni isocyanide complexes $Ni_4(CNR)_7$. These include the trimerization of alkyne derivatives,

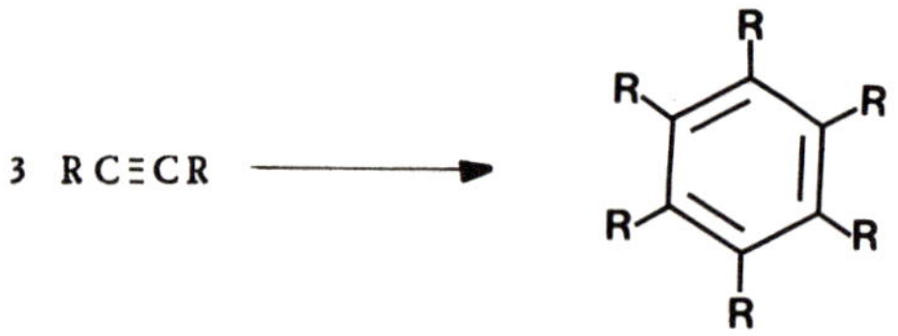

(14)

the dimerization of butadiene to 1,5-cyclooctadiene,

(15)

and the hydrogenation of disubstituted acetylenes to give *cis*-alkenes [95, 96].

The hydrogenation of alkynes catalyzed by the complexes $Ni_4(CNBu^t)_7$ and $Ni_4(CNBu^t)_4(PhC_2Ph)_3$ has been studied in most detail. The structures of these clusters are shown in Figs. 7 and 8 [96].

The cluster $Ni_4(CNBu^t)_7$ is insoluble in pentane but dissolves readily in the presence of diphenyl acetylene, which suggests the formation of alkyne complexes similar to $Ni_4(CNBu^t)_4(PhC_2Ph)_3$ as a first step which is followed by the reaction with H_2 [95].

$Ni_4(CNBu^t)_7$ itself undergoes no reaction with H_2, but the alkyne cluster reacts with H_2 at room temperature to form *cis*-stilbene. In the presence of excess diphenyl acetylene, catalytic hydrogenation starts, and the alkyne cluster later can be recovered unchanged from the reaction mixture. The rate-determining step in the catalytic cycle appears to be the addition of hydrogen to this cluster.

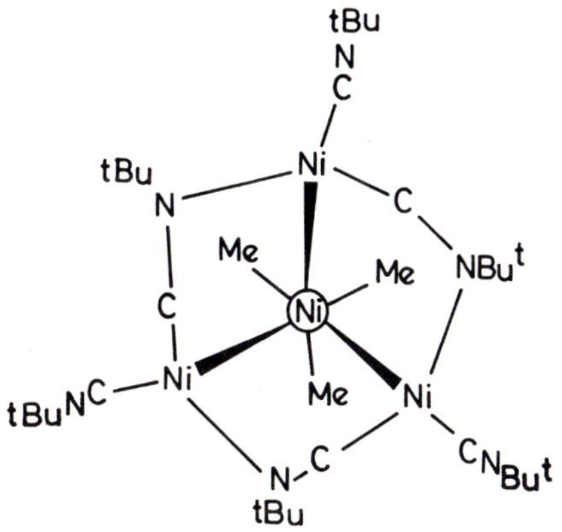

Fig. 7. Structure of $Ni_4(CNBu^t)_7$.

A rather unusual catalytic isocyanide hydrogenation was achieved in toluene solution by using the mononuclear $Ni(CNBu^t)_4$ as an isocyanide source and the tetranuclear $Ni_4(CNBu^t)$ as a catalyst precursor [95]:

$$4\ Ni(CNBu^t)_4 + 18\ H_2 \xrightarrow{Ni_4(CNBu^t)_7} Ni_4(CNBu^t)_7 + 9\ Bu^tNHCH_3 \quad (16)$$

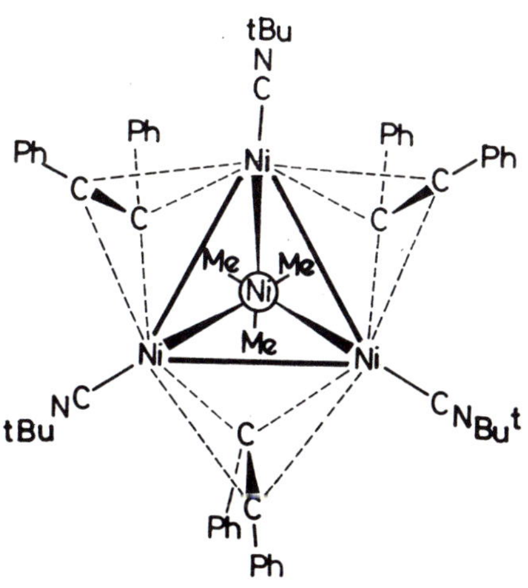

Fig. 8. Structure of $Ni_4(CNBu^t)_4(PhC_2Ph)_3$.

In an attempt to generalize this reaction, acetonitrile was hydrogenated to ethylamine with $Ni_4(CNBu^t)_7$ as a catalyst, but the reaction rate was very low even at 363 K.

The dinuclear Ni complexes $Ni_2(COD)_2(RC_2R)$ (where COD = 1,5-cyclooctadiene: R = Ph or Et) catalyze the hydrogenation of the corresponding alkynes to give *cis*-alkenes at low rates (2-3 turnovers/h). Trimerization to give benzene derivatives (eq. 14) is a competing reaction [97]. The structurally analogous iron complex $Fe_2(CO)_6(tBuC_2Bu^t)$ reacts rapidly with hydrogen, but the reaction is not catalytic.

The complex $Ni_2(COT)_2$ (COT = cyclooctatetraene) catalyzes the cyclotetramerization of acetylene:

4 HC≡CH ⟶ 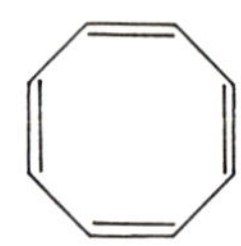 (17)

The mechanism depicted in SCHEME 13 has been suggested for this reaction [98]:

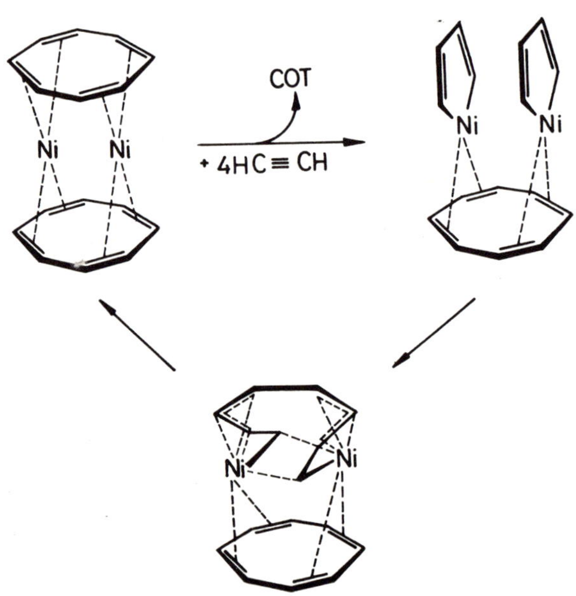

SCHEME 13

5.9 PALLADIUM COMPLEXES

Hydrogenation of palladium carboxylates $[Pd(OOCR)_2]_3$ (R = Me, tBu or Ph) in the presence of tertiary phosphines yields Pd-phosphine clusters of different compositions, such as $Pd_5(PPh)_2(PPh_3)_2$ or $[Pd_5(PPhPr^i)_3]_n$, which catalyze the hydrogenation of alkynes, dienes, and monoenes [99]. Only traces of alkanes could be detected as long as alkynes or dienes were present, and internal alkynes gave *cis*-alkenes selectively. The analogous cluster $Pd_5(PPh)_2$ was found to be an effective catalyst for the hydrogenation of alkenes, dienes, alkynes, aldehydes, quinones, Schiff bases, and nitro, nitroso, or azo compounds [100]. The structures of these clusters with unusually low ligand:metal ratios are unknown, and no detailed information is available about the mechanisms of these reactions.

5.10 PLATINUM COMPLEXES

Platinum cluster carbonyl anions of the general formula $[Pt_3(CO)_9]_n^{2-}$ (n = 1-6) catalyze the formation of water from H_2 and O_2 [101]. The stoichiometries of the reactions have been formulated as follows:

$$3\ [Pt_3(CO)_9]_4^{2-} + H_2 \longrightarrow 4\ [Pt_3(CO)_9]_3^{2-} + 2\ H^+ \qquad (18)$$

$$4\ [Pt_3(CO)_9]_3^{2-} + 2\ H^+ + 1/2\ O_2 \rightleftharpoons 3\ [Pt_3(CO)_9]_4^{2-} + H_2O \qquad (19)$$

Since the trimer is instantaneously oxidized by *p*-benzoquinone to give the tetramer, a catalytic cycle can be developed which results in the reduction of *p*-benzoquinone to hydroquinone by water [102]:

$$H_2O + 3\ [Pt_3(CO)_9]_4^{2-} \rightleftharpoons 4\ [Pt_3(CO)_9]_3^{2-} + 2\ H^+ + 1/2\ O_2 \qquad (20)$$

$$4\ [Pt_3(CO)_9]_3^{2-} + C_6H_4O_2 + 2\ H^+ \longrightarrow 3\ [Pt_3(CO)_9]_4^{2-} + C_6H_4(OH)_2 \qquad (21)$$

Eq. (20) is the reverse of Eq. (19). Upon purging of the reaction mixture with nitrogen, O_2 is removed, and the reaction becomes catalytic in platinum.

The homogeneous hydrogenation of aldehydes, ketones, alkenes, and nitriles has been studied with $(Bu_4N)_2[Pt_3(CO)_6]_5$ as a catalyst precursor [103]. Before reaction starts, the cluster is transformed into another species, which has been characterized only by infrared spectroscopy. The homologues $[Pt_3(CO)_6]_3^{2-}$ and $[Pt_3(CO)_6]_4^{2-}$ yield the some species and are, therefore, equally suitable as catalyst precursors. Benzaldehyde is readily hydrogenated at 355 K, and the products depend on the H_2 partial pressure;

at lower pressures (<2 bar) benzyl alcohol is formed, whereas at higher pressures methanol and benzene are the main products. The reaction is first order in catalyst precursor, which supports the suggestion of cluster catalysis.

Under more drastic conditions (393 K and 100 bar of H_2 + CO), $[Pt_3(CO)_6]_4^{2-}$ catalyzes the hydroformylation of pent-1-ene [104]. An almost equimolar mixture of normal and iso aldehydes is formed. At the end of the reaction, platinum can be recovered from the product as a mixture of $[Pt_3(CO)_6]_3^{2-}$ and $[Pt_6(CO)_6]_4^{2-}$.

5.11 MIXED-METAL CATALYSTS

Mixed-metal homogeneous catalysis can be carried out with a mixture of two different metal complexes or with a mixed-metal cluster as a catalyst precursor.

Several ruthenium carbonyls, discussed above, are active catalysts for the water gas shift reaction under basic conditions [28-31]. Interestingly, it has been found that $Fe(CO)_5$ increases the activity of the ruthenium complexes several fold, and in accordance with this observation, the mixed complex $H_2FeRu_3(CO)_{13}$ proved to be the most active of the metal carbonyls tested (Table 1) [105].

The rhodium-catalyzed synthesis of ethylene glycol from CO + H_2 [76-78] is promoted by ruthenium carbonyl complexes [106, 107]. Optimum results were achieved with Rh:Ru ratios of 1:1 when quaternary phosphonium salts were used as solvents [106], and in this case a mixed Rh-Ru carbonyl cluster has been isolated from the reaction product. With tetraglyme as a solvent, the highest selectivities and reaction rates were observed at surprisingly low Rh concentrations (Rh:Ru = 1:10) [107], and no mixed-metal clusters have been found in the reaction mixtures. Evidently this synthesis is a complex reaction, and the most probably explanation for these findings is that some of the individual steps are catalyzed by rhodium and others more effectively by ruthenium complexes.

The hydrogenation of alkenes, dialkenes, and alkynes was investigated with $Pt_2Co_2(CO)_8(PPh_3)_2$ and $Pt(C_6H_{11}NC)_2[Mo(CO)_3Cp]_2$ as catalysts [108]. These heteronuclear clusters showed poor activity for alkene hydrogenation but were more active for the hydrogenation of alkynes. The $PtMo_2$ cluster could be recovered unchanged from the products. The Pt_2Co_2 cluster was much less stable and was transformed during olefin hydrogenation into $Pt_5(CO)_6(PPh_3)_4$, which itself is a hydrogenation catalyst. In the presence of alkynes, other complexes were formed from the Pt_2Co_2 cluster, which, however, could not be identified.

TABLE 1. Activities of Metal Carbonyl Catalysts for the Water Gas Shift Reaction in Alkaline Aqueous Ethoxyethanol Solution[a]

Catalyst Precursor[b]	Activity[c]
$Ir_4(CO)_{12}$	5.3
$H_2Ru_4(CO)_{13}$	4.4
$H_4Ru_4(CO)_{12}$	3.7
$Ru_3(CO)_{12}$	2.8
$Fe(CO)_5$	1.0
$Rh_6(CO)_{16}$	3.4[d]
$Ru_6C(CO)_{17}$	1.5
$H_3Re_3(CO)_{12}$	0.15
$Re_2(CO)_{10}$	0.1
$H_2FeRu_3(CO)_{13}$	10.3

[a]T = 373 K; P_{CO} = 0.9 bar.
[b]Solution prepared from 0.04 mmol of complex, 2 mmol of KOH, 0.02 mol of H_2O, and 3 mL of ethoxyethanol.
[c]Mol of H_2 per mol of complex per day.
[d]May be heterogeneous.

Ruthenium compounds have recently been employed as cocatalysts with $Co_2(CO)_8$ to improve the selectivity of methanol homologation [109] (Eq. 11). The influence of the Ru:Co ratio on the yields of ethanol and acetaldehyde is shown in Fig. 9. In accordance with these results, the mixed clusters $M[RuCo_3(CO)_{12}]$ (M = Na, Cs, Et_4N) [110] and $[Et_4N][Ru_3Co(CO)_{13}]$ [109] have significantly better activity and selectivity than $Co_2(CO)_8$. Since iodine is added as a promoter in the form of CH_3I, it is assumed that Ru-Co mixed-carbonyl clusters containing iodide ligands are formed, but no details of the mechanism have been elucidated.

An analogous synergistic effect of $Mn_2(CO)_{10}$ has been noted with another type of methanol homologation catalyzed by $Fe(CO)_5$ [111]:

$$CH_3OH + H_2 + 2\ CO \longrightarrow CH_3CH_2OH + CO_2 \qquad (22)$$

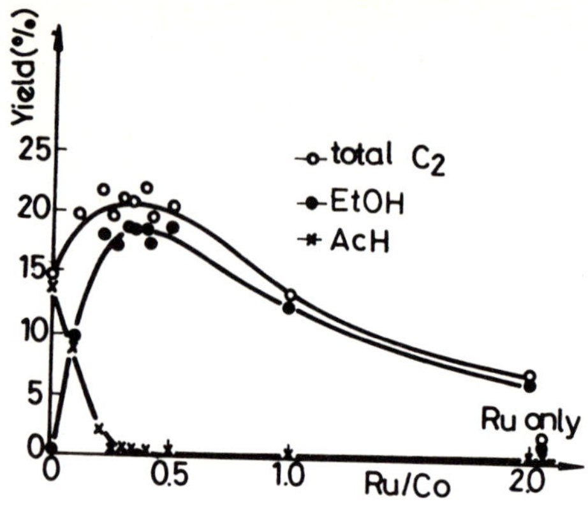

Fig. 9. Influence of the Ru/Co molar ratio on the yields of C_2 compounds in methanol homologation.

No mixed-metal clusters are assumed to be formed in this system; the effectiveness of the Mn-Fe catalyst solutions is ascribed to a favorable combination of the individual steps in the catalytic cycle, each being catalyzed by one metal complex or the other.

The aminomethylation reaction is catalyzed by Fe, Ru, Rh, and Ir carbonyls:

$$RCH{=}CH_2 + 3\ CO + H_2O + HNR'_2 \longrightarrow R(CH_2)_3NR'_2 + 2\ CO_2 \qquad (23)$$

The highest rates and selectivities were obtained with mixtures of $Fe_3(CO)_{12}$ and $Ru_3(CO)_{12}$, but no details of the mechanism are known [112]. Infrared spectroscopy revealed the formation of mixed-metal clusters.

γ-Keto acids are obtained by the phase-transfer-catalyzed reaction of alkynes with CO and methyl iodide under basic conditions in the presence of catalytic amounts of $Co_2(CO)_8$ and $Ru_3(CO)_{12}$ (1:1) [113]:

$$RC{\equiv}CH + MeI + CO \longrightarrow R\underset{\displaystyle COOH}{\underset{|}{C}}HCH_2COMe \qquad (24)$$

RC≡CH + MeO⁻ + CO ⟶ 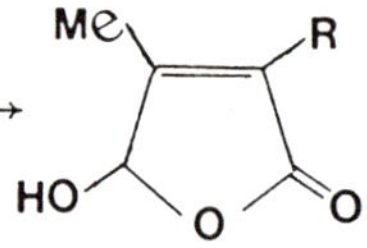 (25)

With $Co_2(CO)_8$ alone, but-2-enolides are formed [114] (Eq. 24); $Ru_3(CO)_{12}$ alone is inactive. The surprising effect of the Ru carbonyl is ascribed not to the formation of a mixed-metal cluster catalyst, but rather to the nucleophilic addition to the intermediate acylcobalt carbonyl of $HRu_3(CO)_{11}^-$ formed from $Ru_3(CO)_{12}$ under the influence of NaOH (SCHEME 14):

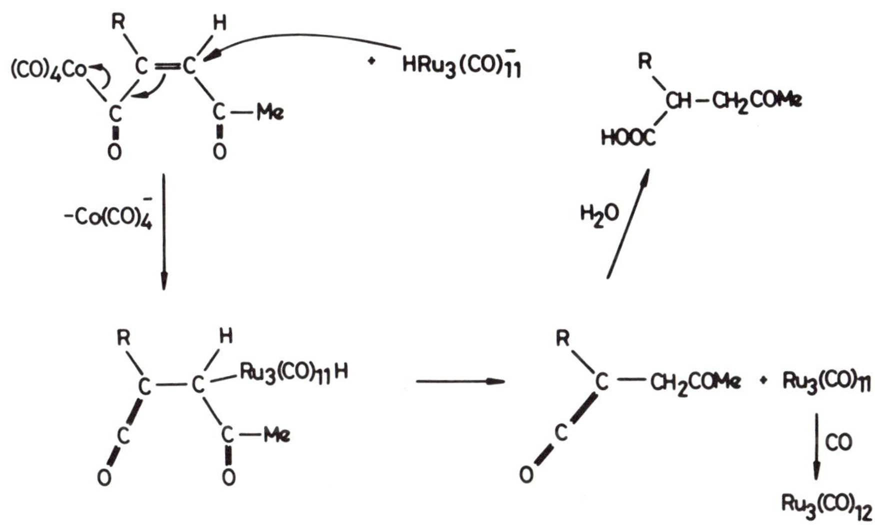

SCHEME 14

The use of mixed-metal clusters in homogeneous catalysis is in its very early beginnings; almost nothing is known about the actual role of such species in catalytic transformations. This is no wonder, however, since the difficulties and ambiguities of homogeneous cluster catalysis are even further enhanced when more than one metal is present in the system.

5.12 CONCLUSIONS

It is evident from the preceding sections that our knowledge of homogeneous catalysis by metal clusters is still rather vague; very much experimental work has yet to be done to provide better insights into this exciting area.

Aside from the difficulties in defining the subject, the principal problem is the reliable identification of the catalytically active species. In homogeneous hydrogenation with mononuclear Rh-phosphine complexes, it has been shown that the catalytic reaction cycle includes intermediates which cannot be observed spectroscopically because of their high reactivities and low concentrations in the reaction medium [115]. Species which can be detected by spectroscopic means are actually only reservoirs of the metal which furnish the active species in fast equilibrium reactions and thus participate in the catalytic cycle only indirectly. Therefore, the presence of the really important complexes can be established only by kinetics measurements, preferably not just of the overall reaction (which is usually much too complex) but also of the individual steps of the catalytic cycle.

This conclusion is probably valid for most fast homogeneous catalytic reactions, and therefore many statements concerning the involvement of clusters in homogeneous reactions must be treated with caution. In our opinion, only a few reactions catalyzed by triruthenium and/or tetraruthenium carbonyl clusters may be regarded as being reasonably well understood and mechanistically clear. This is especially true for the water-gas shift reaction catalyzed by these clusters; at present, we may regard this as the clearest example of homogeneous cluster catalysis.

The prospects in this area are obviously very bright, however, since clearly there is no major obstacle to the discovery of other cluster systems showing catalytic properties. The main problem will always be the tendency of clusters to fragment, but several means to minimize this tendency can be visualized:

1. Use of 4d or 4d metals instead of 3d metals, because of the stronger metal-metal bonds they form.

2. Use of ligands which can function as bridges between two or three metal atoms, such as R_2P, RS, RP, the halides, etc. (concerning the lability of such bridges, however, see ref. [116]).

3. Application of relatively high temperatures and/or small concentrations of ligands to favor ligand dissociation and metal association.

The most promising potential advantage of homogeneous cluster catalysis is the possiblity of multisite activation of a substrate leading to an unusual product (cf. SCHEME 11). Such processes may fill the gap that still exists between homogeneous and heterogeneous catalysis by transition metals and open areas for homogeneous catalysis which, so far, are restricted to heterogeneous catalysis.

REFERENCES

1. R. M. Laine, J. Mol. Catal., 14 (1982) 137.
2. R. Ugo and P. Psaro, J. Mol. Catal., 20 (1983) 53.
3. E. L. Muetterties and M. J. Krause, Angew. Chem., 95 (1983) 135.
4. F. A. Cotton, Quart. Rev., (1976) 389.
5. P. Chini, Inorg. Chim. Acta. Rev. (1968) 31.
6. R. B. King, Progr. Inorg. Chem. 15 (1972) 287; and J. Am. Chem. Soc., 94 (1972) 95.
7. R. F. Heck and D. S. Breslow, J. Am. Chem. Soc., 83 (1961) 4023.
8. L. Markó, Fundamental Res. Homogeneous Catal. 4 (1984) 1.
9. G. Csontos, B. Heil and L. Markó, Ann. N. Y. Acad. Sci., 239 (1974) 47.
10. S. Slater and E. L. Muetterties, Inorg. Chem., 19 (1980) 3337.
11. R. F. Gerlach, D. N. Duffy and M. D. Curtis, Organometallics, 2 (1983) 1172.
12. M. McKenna, L. L. Wright, D. J. Miller, L. Tanner, R. C. Haltiwanger and M. Rakowski DuBois, J. Am. Chem. Soc., 105 (1983) 5329.
13. N. von Kutepov and H. Kindler, Angew. Chem., 72 (1960) 82.
14. A. L. Lapidus, E. Z. Gildenberg and Ja. T. Eidus, Kinet. Katal., 16 (1975) 252.
15. E. Z. Gildenberg and A. L. Lapidus, Kinet. Katal., 18 (1977) 124.
16. E. Z. Gildenberg, V. A. Svec, A. L. Lapidus and Ja. G. Eidus, Kinet. Katal., 16 (1975) 1077.
17. R. Pettit, C. H. Mauldin, T. Cole and H. Kang, Ann. N. Y. Acad. Sci., 295 (1977) 151.
18. H. Kang, C. H. Mauldin, T. Cole, W. Slegeir, K. Cann and R. Pettit, J. Am. Chem. Soc., 99 (1977) 8323.
19. L. Markó, M. A. Radhi and I. Ötvös, J. Organometal. Chem. 218 (1981) 369.
20. J. Palágyi and L. Markó, J. Organometal. Chem. 236 (1982) 343.
21. G. P. Boldrini, A. Umani-Ronchi and M. Panunzio, J. Organometal. Chem. 171 (1979) 85.
22. J. M. Landesberg, L. Katz, and C. Olsen, J. Org. Chem., 37 (1972) 930.
23. H. des Abbayes and H. Alper, J. Am. Chem. Soc., 99 (1977) 98.
24. C. U. Pittman, R. C. Ryan, and J. McGee, J. Organometal. Chem. 178 (1979) C43.
25. H. Inone and M. Sato, J. Chem. Soc. Chem. Commun., (1983) 983.
26. J. P. Collman, R. G. Finke, P. L. Matlock, R. Wahren and J. I. Brauman, J. Am. Chem. Soc., 98 (1976) 4685.
27. J. P. Collman, R. G. Finke, P. L. Matlock, R. Wahren, R. G. Komoto, J. I. Brauman, J. Am. Chem. Soc., 100 (1978) 119.
28. R. M. Laine, R. G. Rinker and P. C. Ford, J. Am. Chem. Soc., 99 (1977) 252.
29. C. Ungermann, V. Landis, S. A. Moya, H. Cohen, H. Walker, R. G. Pearson, R. G. Rinker and P. C. Ford, J. Am. Chem. Soc., 101 (1979) 5922.
30. P. C. Ford, Acc. Chem. Res., 14 (1981) 31.
31. J. C. Bricker, C. C. Nagel, and S. G. Shore, J. Am. Chem. Soc., 104 (1982) 1444.
32. P. Yarrow, H. Cohen, C. Ungermann, D. Vandenberg, P. C. Ford and R. G. Rinker, J. Mol. Catal., 22 (1983) 231.
33. J. S. Bradley, J. Am. Chem. Soc., 101 (1979) 7419.
34. B. D. Dombek, J. Am. Chem. Soc., 102 (1980) 6855.
35. J. F. Knifton, J. Chem. Soc. Chem. Comm., (1981) 188.
36. B. D. Dombek, J. Am. Chem. Soc., 103 (1981) 6508.
37. J. F. Knifton, J. Am. Chem. Soc., 103 (1981) 3959.

38. B. D. Dombek, J. Organometal. Chem., 250 (1983) 467.
39. B. K. Warren and B. D. Dombek, J. Catal., 79 (1983) 334.
40. D. J. Darensbourg, C. Oralles and M. Pala, J. Am. Chem. Soc., 105 (1983) 5937.
41. M. Castiglioni, L. Mildne, D. Osella, G. A. Vaglio and M. Valle, Inorg. Chem., 15 (1976) 394.
42. J. L. Graff, R. D. Sanner and M. S. Wrighton, J. Am. Chem. Soc., 101 (1979) 273.
43. M. Valle, D. Osella and G. A. Vaglio, Inorg. Chim. Acta., 20 (1976) 213.
44. G. A. Vaglio and M. Valle, Inorg. Chim. Acta., 30 (1978) 161.
45. P. M. Lausarot, G. A. Vaglio and M. Valle, Transition Met. Chem. 4 (1979) 39.
46. J. L. Graff and M. S. Wrighton, J. Am. Chem. Soc., 102 (1980) 2123.
47. G. A. Vaglio, D. Osella and M. Valle, Transition Met. Chem. 2 (1977) 94.
48. Y. Doi, K. Koshizuka and T. Keii, Inorg. Chem., 21 (1982) 2732.
49. Y. Doi, S. Tamura and K. Koshizuka, Inorg. Chim. Acta., 65 (1982) L63.
50. Y. Doi, S. Tamura and K. Koshizuka, J. Mol. Catal., 19 (1983) 213.
51. P. M. Lausarot, G. A. Vaglio and M. Valle, Inorg. Chim. Acta, 36 (1979) 213.
52. P. M. Lausarot, G. A. Vaglio and M. Valle, Inorg. Chim. Acta, 25 (1977) L107.
53. G. Süss-Fink and J. Reiner, J. Mol. Catal., 16 (1982) 213.
54. R. H. Fish, A. D. Thormodsen and G. A. Cremer, J. Am. Chem. Soc., 104 (1982) 5234.
55. T. P. Dang, J. C. Poulin and H. B. Kagan, J. Organomet. Chem., 91 (1975) 105.
56. C. Botteghi, S. Gladiali, M. Bianchi, U. Matteoli, P. Frediani, P. G. Vergamini and E. Benedetti, J. Organomet. Chem., 140 (1977) 221.
57. M. Bianchi, U. Matteoli, P. Frediani, G. Menchi, F. Piacenti, C. Botteghi and M. Marchetti, J. Organomet. Chem., 252 (1983) 317.
58. P. Frediani, U. Matteoli, M. Bianchi, F. Piacenti and G. Menchi, J. Organomet. Chem., 150 (1978) 273.
59. M. Bianchi, U. Matteoli, G. Menchi, P. Frediani, S Pratesi and F. Piacenti, J. Organomet. Chem., 198 (1980) 73.
60. M. Bianchi, G. Menchi, F. Francalanci, F. Piacenti, U. Matteoli, P. Frediani and C. Botteghi, J. Organomet. Chem. 188 (1980) 109.
61. G. Braca, G. Sbrana, F. Piacenti and P. Pino, Chim. Ind., (Milan), 52 (1970) 1091.
62. R. M. Laine, J. Am. Chem. Soc.,100 (1978) 64.
63. R. M. Laine, Ann. N. Y. Acad. Sci., 333 (1980) 124.
64. S. A. Fouda and G. L. Rempel, Inorg. Chem., 18 (1979) 1.
65. J. B. Keister and J. R. Shapley, J. Am. Chem. Soc., 98 (1976) 1056.
66. Z. Dawoodi, K. Henrick and M. J. Mays, J. Chem. Soc. Chem. Commun., (1982) 696.
67. A. D. Harley, G. J. Guskey, and G. L. Geoffroy, Organometallics, 2 (1983) 53.
68. C. Make, H. Patin, J.-Y. LeMarouille and A. Benoit, Organometallics, 2 (1983) 1051.
69. R. C. Ryan, C. U. Pittman and J. P. Connor, J. Am. Chem. Soc., 99 (1977) 1986.
70. C. U. Pittman, G. M. Wilemon, Jr., W. D. Wilson and R. C. Ryan, Angew. Chem., 92 (1980) 494.
71. C. U. Pittman, R. H. Ryan, D. W. Wilson, G. Wilemon and M. Absi-Halbi, Prepr. Am. Chem. Soc. Div. Pet. Chem., 25 (1980) 714.

72. B. H. Chang, Inorg. Chim. Acta., 65 (1982) L189.
73. C. Hoogzand and W. Hübel, Organic Syntheses via Metal Carbonyls (Eds. I. Wender and P. Pino), Vol. I., Wiley, New York, 1968.
74. O. S. Mills and G. Robinson, Proc. Chem. Soc., (1964) 187.
75. C. Hoogzand and W. Hübel, personal communication quoted by ref. [74]. See also U. Kruerke, C. Hoogzand and W. Hubel, Chem. Ber., 94 (1961) 2817.
76. R. L. Pruett, Ann. N. Y. Acad. Sci. 295 (1977) 239.
77. J. L. Vidal and W. E. Walker, Inorg. Chem., (1980) 896.
78. R. Whyman, Phil. Trans. R. Soc. Lond. A 308 (1982) 131.
79. B. T. Heaton, J. Jonas, T. Eguchi and G. A. Hoffman, J. Chem. Soc. Chem. Commun. (981) 331.
80. P. Chini, S. Martinengo, and G. Garlaschelli, J. Chem. Soc. Chem. Commun., (1972) 709.
81. G. H. Brown and G. Wilkinson, J. Chem. Soc. A., (1970) 2753.
82. P. Kalck, J.-K. Frances, P.-M. Pfister, T. G. Southern and A. Thorez, J. Chem. Soc. Chem. Commun., (1983) 510.
83. R. A. Jones, T. C. Wright, J. L. Atwood and W. E. Hunter, Organometallics, 2 (1983) 470.
84. A. J. Sivak and E. L. Muetterties, J. Am. Chem. Soc., 101 (1979) 4878.
85. R. R. Burch, E. L. Muetterties, R. G. Teller, and J. M. Williams, J. Am. Chem. Soc., 104 (1982) 4257.
86. R. R. Burch, A. J. Shusterman, E. L. Muetterties, R. G. Teller, and J. M. Williams, J. Am. Chem. Soc., 105 (1983) 3546.
87. M. D. Fryzuk, Can. J. Chem., 61 (1983) 1347.
88. B. R. Cho and R. M. Laine, J. Mol. Catal., 15 (1982) 383.
89. T. Kitamura, M. Sakamoto and T. Joh, Chem. Lett., (1972) 379.
90. T. Kitamura, T. Joh and N. Hagihara, Chem. Lett., (1975) 203.
91. R. M. Laine, D. W. Thomas and L. W. Cary, J. Org. Chem. 44 (1979) 4964.
92. R. M. Laine, D. W. Thomas and L. W. Cary, J. Am. Chem. Soc. (1982) 1763.
93. R. M. Laine, D. W. Thomas, L. W. Cary and S. E. Buttrill, J. Am. Chem. Soc., 100 (1978) 6527.
94. G. D. Mercer, J. Shing Shu, T. B. Brauchfuss, D. M. Roundhill, J. Am. Chem. Soc., 971 (1975) 1967.
95. M. G. Thomas, W. R. Pretzer, B. F. Beier, F. J. Hirsekorn and E. L. Muctterties, J. Am. Chem Soc., 99 (1977) 743.
96. E. L. Muetterties, E. Band, A. Kokorin, W. R. Pretzer and M. G. Thomas, Inorg. Chem., (1980) 1552.
97. E. L. Muetterties, W. R. Pretzer, M. G. Thomas, B. F. Beier, D. L. Thorn, V. W. Day and A. B. Anderson, J. Am. Chem. Soc., 100 (1978) 2090.
98. G. Wilke, Pure App. Chem., 50 (1978) 677.
99. A. Sisak, F. Ungváry and K. Kiss, J. Mol. Catal., 18 (1983) 223.
100. A. S. Berenblyum, A. G. Knizhnik, S. L. Mond, and I. I. Moiseev, J. Organometal. Chem., 234 (1982) 219.
101. G. Longoni and P. Chini, J. Am. Chem. Soc. 98 (1976) 7225.
102. S. Bhaduri and K. R. Sharma, J. Chem. Soc. Chem. Commun. (1983) 1412.
103. S. Bhaduri and K. R. Sharma, J. Chem. Soc. Dalton Trans. (1982) 727.
104. G. Longoni and P. Chini, J. Am. Chem. Soc., 98 (1976) 7725.
105. P. C. Ford, R.G. Rinker, C. Ungermann, R. M. Laine, V. Landis, and S. A. Moya, J. Am. Chem. Soc., 100 (1978) 4595.
106. J. F. Knifton, J. Chem. Soc. Chem. Commun. (1983) 729.
107. R. Whyman, J. Chem. Soc. Chem Commun. (1983) 1439.

108. A. Fusi, R. Ugo, R. Psaro, P. B. Braunstein and J. Dehand, J. Mol.Catal., 16 (1982) 217.
109. M. Hidai, M. Orisako, M. Ue, Y. Koyasu, T. Kodama and Y. Uchida, Organometallics, 2 (1983) 292.
110. M. Hidai, M. Orisako, M. Ue, Y. Uchida, K. Yasufuku and H. Yamazaki, Chem. Lett., (1981) 143.
111. M. J. Chen, H. M. Feder, and J. W. Rathke, J. Mol. Catal., 17 (1982) 331; J. Am. Chem. Soc., 104 (1982) 7346.
112. M. Laine, J. Org. Chem., 45 (1980) 3370.
113. H. Alper and J.-F. Petrignani, J. Chem. Soc. Chem. Commun., (1983) 1154.
114. H. Alper, J. K. Currie and H. des Abbayes, J. Chem. Soc. Chem. Commun., (1978) 311.
115. J. Halpern, Inorg. Chim. Acta, 50 (1981) 11.
116. G. L. Geoffroy, S. Rosenberg, P. M. Shulman, and R. R. Whittle, J. Am. Chem. Soc., 106 (1984) 1519.

PART II

METAL CLUSTERS IN AND ON SUPPORTS

Chapter 6

CHARACTERIZATION BY PHYSICAL METHODS

Section 6.1

INTRODUCTION

H. KNÖZINGER

The identification and characterization of supported metal clusters and related species are indispensible to determination of relationships between catalytic properties and the nature and structure of the active species. When clusters are brought in contact with supports or subjected to catalytic reaction conditions, chemical and structural changes of the original molecular cluster compound are to be anticipated. These may involve (1) modification of the ligand sphere with the metal-metal bonded cluster frame intact; (2) rearrangement of the cluster frame with the nuclearity of the cluster; (4) reaction of the cluster with surface groups leading to changes of the metal oxidation state and degradation of the cluster frame; and (5) metal aggregation when the clusters are stripped of their ligands. In many cases, mixtures of several surface species coexist. Characterization techniques should therefore provide opportunities for discriminating between the surface entities by probing the ligand sphere and/or the metal species.

A wide variety of spectroscopic methods are available which provide sufficient sensitivity and surface specificity for a meaningful qualitative and quantitative analysis of supported clusters, complexes, and metal aggregates [1-5]. It is emphasized, however, that the detectable species need not be related simply to the catalytic properties of the material under study. As an example, supported catalysts derived from molecular carbonyl cluster compounds are most frequently characterized by their infrared spectra in the carbonyl region. For catalysis to occur, however, the cluster may be stripped of at least some of its ligands, and hence the active species may remain undetected by this approach. The need for the use of suitable catalytic test reactions as complementary tools for the characterization of active species is obvious. Such studies should include *in situ* spectroscopic experiments with catalysts under working conditions whenever possible.

Vibrational spectroscopies are among the most promising and widely used methods for the characterization of cluster-derived catalysts. The value of the methods derives from the very detailed molecular structure data they provide. Photons, electrons, and neutrons are used principally as probes. A wide variety of special techniques have been developed for the photon vibrational spectroscopies, which have to be selected according to the surface and optical properties of the sample. Because of its relative simplicity and wide applicability, infrared transmission absorption spectroscopy is most frequently used today, while laser Raman, diffuse reflectance, internal reflection, and optoacoustic spectroscopies have found less application (or none) for vibrational studies of cluster-derived materials.

Two electron vibrational spectroscopies have been developed. Electron energy loss spectroscopy (EELS) can best be applied to the study of model systems, rather than polydisperse, high-surface-area materials. Inelastic electron tunneling spectroscopy (IETS) is an interesting, sophisticated technique permitting high-resolution spectra within the entire vibrational frequency range. The experiments, however, require liquid helium temperature and special specimen preparation.

Inelastic neutron scattering (INS) is an attractive spectroscopy for vibrational analysis. It provides particular advantages for the characterization of hydride species such as hydride cluster compounds or hydrogen chemisorbed on small metal aggregates, but it requires a neutron source with sufficiently high flux.

The diffuse reflectance technique is almost exclusively applied when *optical spectra* in the visible and/or ultraviolet region of the electromagnetic spectrum are studied. The method is experimentally simple and *in situ* sample treatment is possible. Interpretation of the spectra, which can provide information about bonding characteristics of metal clusters and of cluster-derived complexes, is difficult in most cases, since the theoretical background is not generally available. Nevertheless, optical spectroscopy (mostly in conjunction with other spectroscopic methods such as infrared spectroscopy) is finding increasing application for the characterization of supported cluster materials.

Electronic configurations and bonding characteristics of cluster compounds can also be studied by *electron spin resonance* (ESR). Although most molecular carbonyl clusters are diamagnetic, paramagnetic derivatives are frequently obtained by oxoreductive treatments, and species (poly- or mononuclear) having unpaired electrons may be generated on cluster-derived supported catalysts.

Nuclear magnetic resonance (NMR) techniques are extremely valuable for molecular structure determination. ^{1}H- and ^{13}C-NMR can be used most directly for characterization of cluster ligands, and other nuclei which frequently occur in cluster ligands (such as ^{31}P) have become accessible by modern instrument developments. In addition, NMR spectra of many metal nuclei can now be measured, and the modern instruments provide the opportunity for probing structure and bonding in metal clusters. Of great significance for the application of NMR to the study of supported samples is the use of magic angle spinning and cross-polarization techniques, which serve to reduce the dipolar line broadening typical of the solid state. Pulsed NMR techniques also allow the measurement of chemical shifts and Knight shifts for metal nuclei such as ^{103}Rh and ^{195}Pt present in small metal aggregates and particles. For nuclei having a quadrupole moment, NMR line shapes contain information about the local environment of the nucleus. Quadrupole interaction parameters such as quadrupole coupling constants and electric field gradient asymmetry parameters can be determined more directly by *nuclear quadropole resonance* (NQR) [6]. Although these data would provide significant information about the bonding, NQR has been applied only infrequently to the characterization of molecular metal clusters. To our knowledge, there has been no NQR investigation of supported clusters, probably an indication of the low sensitivity of the technique.

The same holds true for another nuclear spectroscopy, namely *time differential perturbed angular correlation* (TDPAC) [7-9]. This technique would also provide information concerning the local environment of a given element and bonding characteristics via hyperfine interactions. The limitations of the technique lie in the fact that only selected isotopes can be investigated. Among the metals that can be characterized are some of great relevance for catalysis, such as ruthenium (^{99}Ru) and rhodium (^{100}Rh).

In contrast, *Mössbauer* or *nuclear γ-resonance spectroscopy*, although again limited to selected so-called Mössbauer nuclei (e.g., ^{57}Fe), has been applied frequently to the characterization of supported clusters, namely, $Fe_3(CO)_{12}$ or bimetallic clusters containing iron. The principal experimental parameters, the isomer shift and the quadrupole splitting, are extremely sensitive to the valence and spin states of the Mössbauer atom and hence provide detailed information about bonding.

The advantage of the nuclear resonance spectroscopies is that they probe the nuclear environment in molecules (mononuclear or polynuclear complexes) as well as in "infinite" lattices (metal aggregates or particles). Moreover, those techniques using γ-quanta as probes (TDPAC, Mössbauer spectroscopy) can be applied with relative ease for characterization of working catalysts.

The electron spectroscopies of interest for the characterization of clusters and cluster-derived materials are *X-ray* and *ultraviolet photoelectron spectroscopies* [XPS (or ESCA)] and UPS, respectively. XPS, which measures the energy distribution of inner shell photoelectrons, permits an elemental analysis within the topmost atomic layers of a solid, the only element of chemical interest excluded from detection being hydrogen. In favorable cases, chemical shifts observed in XP spectra also allow conclusions regarding the chemical environment and/or oxidation state of an element. Due to the far lower photon energy, UPS measures the energy distribution of photoelectrons emitted from molecular orbitals and thus provides direct insight into the chemical bonding. Both techniques have been applied successfully in studies of molecular carbonyl clusters and chemisorbed carbon monoxide, and XPS has found application for the characterization of supported cluster-derived materials. A disadvantage of the electron spectroscopies is that they require ultrahigh vacuum conditions.

Among the spectroscopies which use ion probes, *secondary ion mass spectroscopy* (SIMS) is the most promising in the present context. It has been shown in recent years that SIMS permits determination of structural details of surface groups and molecular species from molecular fragments which are observed in the secondary ion mass spectra when the specimen is bombarded with primary ion probes. This is possible at least under favorable so-called static conditions, when sputter rates are maintained low. The SIMS spectra detect ion characteristic of molecular weight and also structurally diagnostic fragment ions, and they may therefore provide information about the nuclearity of supported cluster-derived species and the ligand environment of the metal atoms of the cluster frame.

Electron microscopy, in particular *conventional transmission electron microscopy* (CTEM) has been developed in recent years to yield resolutions in the 0.2 to 0.3 nm range, provided that suitable prepared, extremely thin (<20 nm) specimens are available. Electron micrographs can therefore establish the size ranges of metal aggregates or clusters, and in exception cases individual metal atoms may be visible. It is emphasized, however, that particularly in the size range of interest for molecular clusters ($\leqslant 0.5$ nm) supported on polydisperse high-surface-area supports, contrast interpretation is critical and may be subject to error. In such situations, the use of model systems has proved to be advantageous. These may consist of very thin (<10 nm), nonporous, and self-supporting oxide films as supports for the metal clusters. In some cases, particle shapes have been discerned by secondary image treatments.

Since crystalline long range order does not exist in highly dispersed supported cluster-derived materials or supported metal aggregates, conventional diffraction techniques cannot be applied for structural characterization. The most promising modern technique for this purpose is undoubtedly EXAFS (*extended X-ray adsorption fine structure*), which measures oscillatory variations of X-ray absorption cross-sections near absorption edges. The principal information that can be deduced from this technique includes interatomic distances for the first and subsequent coordination shells, coordination numbers, and Debye-Waller factors. These data can be obtained for several elements of interest in the same material. The method therefore can be used to probe the local environment of a variety of elements in a given catalytic material, even if it is amorphous. A significant improvement of the sensitivity can be obtained by measuring the X-ray fluorescence due to core-hole filling instead of absorption cross-sections. This modification of the EXAFS technique becomes highly relevant when elements of low abundance in the catalytic material are to be detected, a situation which the experimentalist frequently faces when dealing with supported molecular clusters. EXAFS measurements can be performed under controlled atmospheres or under working conditions. Although conventional X-ray sources can be used for EXAFS (at the expense of data acquisition time due to their comparably low X-ray intensity), most experiments today are carried out with high-intensity synchrotron radiation. It is important to realize that the theoretical treatment of experimental EXAFS results still seems to be a matter of discussion. In most cases of catalyst characterization, structurally well-defined reference compounds are therefore being used. Even with its limitations, EXAFS probably offers the best opportunities for physical structure elucidation of supported clusters or cluster-derived materials, and it can be expected to provide some of the most relevant information for establishing structure-reactivity relationships.

Thermoanalytical methods have found wide application to characterization of surface species. The methods of most interest for characterizing supported cluster-derived materials are *temperature-programmed deomposition (TPDE) and temperature-programmed reduction (TPR)*. In both techniques the sample is subjected to heating, usually with a linear temperature program. A gas flows over the sample, usually H_2 for TPR and an inert gas (He, N_2) for TPDE. TPDE measures the evolution of gases during the thermal decomposition of anchored clusters as catalyst precursors, and TPR characterizes the reduction of precursor species via H_2 consumption. In connection with analytical methods to determine the metal content in a given material, TPDE provides information about the stoichiometries of transformations of anchored clusters and cluster derivatives, and TPR provides information about the oxidation states of surface metal species. These two relatively simple and inexpensive techniques are extremely valuable for the quantitative characterization of supported cluster-derived materials, especially when they are combined with other techniques, e.g., infrared spectroscopy.

The following sections of this chapter include brief descriptions of the physical background of the techniques mentioned above. Typical clusters and cluster-derived catalysts are described and discussed.

REFERENCES

1. R. B. Anderson and P. T. Dawson (Eds.), Experimental Methods in Catalytic Research, vols. 2 and 3, Academic Press, New York, San Francisco, London, 1976.
2. W. N. Delgass, G. L. Haller, R. Kellerman, and J. H. Lunsford, Spectroscopy in Heterogeneous Catalysis, Academic Press, New York, San Francisco, London, 1979.
3. J. M. Thomas and R. M. Lambert (Eds.), Characterization of Catalysts, Wiley, New York, 1980.
4. H.-P. Boehm and H. Knözinger, in J. R. Anderson and M. Boudart, Eds., Catalysis--Science and Technology, Vol. 4, Springer, Berlin, 1983, p. 39.
5. F. Delannay (Ed.), Characterization of Heterogeneous Catalysis, Dekker, Basel, 1984.
6. E. A. C. Lucken, Nuclear Quadrupole Coupling Constants, Academic Press, London and New York, 1969.
7. H. Frauenfelder and R. M. Steffen, in K. Siegbahn (Ed.), Alpha-, Beta-, and Gamma-Ray Spectroscopy, vol. 2, North Holland, Amsterdam, 1965, p. 997.
8. D. A. Shirley and H. Haas, Ann. Rev. Phys. Chem., 23 (1972) 385.
9. H. H. Rinneberg, Atomic Energy Rev., 172 (1979) 477.

Section 6.2

VIBRATIONAL SPECTROSCOPIES

H. KNÖZINGER

Beyond doubt, vibrational spectroscopy--particularly in the conventional transmission mode--is the preferred technique for structural characterization of supported clusters and their derivatives. This preference is a consequence of the relative simplicity of transmission infrared spectroscopy and its wide applicability. In most cases, vibrational spectra have been recorded in the limited spectral range of the carbonyl stretching region, the results being compared with the carbonyl spectral fingerprints of the authentic reference compounds. Only in few cases have low-frequency Raman spectra been reported.

Since unsupported molecular compounds are used for reference purposes, the first part of this section is devoted to a brief description and interpretation of vibrational spectra of selected compounds. Typical examples of the vibrational spectroscopic characterization of supported clusters and cluster-derived surface species (including mononuclear complexes and metal aggregates) are discussed in Section 6.2.3.

6.2.1 VIBRATIONAL SPECTRA OF MOLECULAR CLUSTERS AND COMPLEXES

6.2.1.a Carbonyl clusters

The vibrational frequencies of carbonyl clusters are expected in three typical regions, namely, 2150-1800, 650-300, and <200 cm^{-1}. Bands in the higher-frequency region arise from carbonyl stretching, with the 2150-1950 cm^{-1} region characteristic of terminal carbonyl ligands and the 1950-1800 cm^{-1} region characteristic of bridging carbonyls. Bands in the 650-300 cm^{-1} region arise from metal-carbon stretching and carbonyl bending vibrations, and those in the low-frequency region <200 cm^{-1} arise from vibrations involving metal-metal stretching and metal-carbon deformation.

Only a few examples of carbonyl clusters of particular interest for catalyst preparation are discussed here. These examples include $Os_3(CO)_{12}$, $Ru_3(CO)_{12}$, $Ir_4(CO)_{12}$, $Co_4(CO)_{12}$, $Rh_4(CO)_{12}$, $Co_6(CO)_{16}$, and $Rh_6(CO)_{16}$. Various cluster frameworks (triangular, tetrahedral, and octahedral) are represented by these clusters; some of these structures incorporate only terminal carbonyl ligands (Os_3, Ru_3, and Ir_4), and some incorporate both terminal and bridging carbonyl ligands (Co_4, Rh_4, Co_6, Rh_6). For more detailed and comprehensive summaries of vibrational spectra of carbonyl clusters, various review articles [1,2] and books [3,4] are recommended.

$Os_3(CO)_{12}$ and $Ru_3(CO)_{12}$ are isostructural, having the following idealized structure:

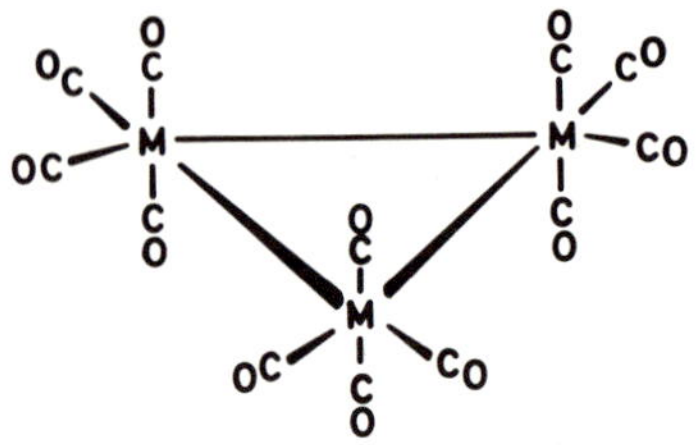

The metal atoms form a triangular frame, with each metal atom bearing four CO ligands, two in equatorial and two in axial positions. The $M(CO)_4$ units in this structure have local C_{2v} symmetry, whereas the $M_3(CO)_{12}$ cluster has D_{3h} symmetry. As shown in Fig. 1, $Os_3(CO)_{12}$ has four infrared bands in the carbonyl stretching region at 2069, 2037, 2016, and 2005 cm^{-1}. The first band assignments for $Os_3(CO)_{12}$ were made on the basis of the local C_{2v} symmetry by Huggins *et al.* [5]. This treatment necessarily neglects all interactions between CO oscillators attached to different osmium atoms. However, Quicksall and Spiro [6] have shown that the carbonyl stretch-stretch interaction across the cluster is of a magnitude comparable to that of the interaction between CO oscillators attached to the same metal atom. Consequently, additional band splittings must be expected for the D_{3h} symmetry of the cluster. These splittings, shown schematically in Fig. 2, lead to eight vibrational modes in the carbonyl stretching region, four of which are infrared active, of symmetry species $1A_2'' + 3E'$. This predicted number of bands is in agreement with the experimental carbonyl infrared spectrum shown in Fig. 1.

The most detailed spectral analysis has been published by Battiston *et al.* [7]; their band assignments for the carbonyl stretching region of $Os_3(CO)_{12}$ and $Ru_3(CO)_{12}$ are summarized in Table 1. The same authors [7] also reported the infrared carbonyl spectra of the two mixed metal clusters $Ru_2Os(CO)_{12}$ and $RuOs_2(CO)_{12}$. In the medium frequency region (650-300 cm^{-1}), there are bands originating from metal-carbon stretching and carbonyl bending vibrations. Kettle and Stanghellini [8] have shown that this spectral region can be assigned by an isolated C_{2v} $M(CO)_4$ unit model to a good first approximation, i.e., intramolecular coupling is effectively very weak and can be neglected.

A detailed vibrational analysis of the low-frequency region (<200 cm^{-1}) has been given by Quicksall and Spiro [6]; their proposed model, based on the assumption of mechanical coupling only, was later supported by Kettle

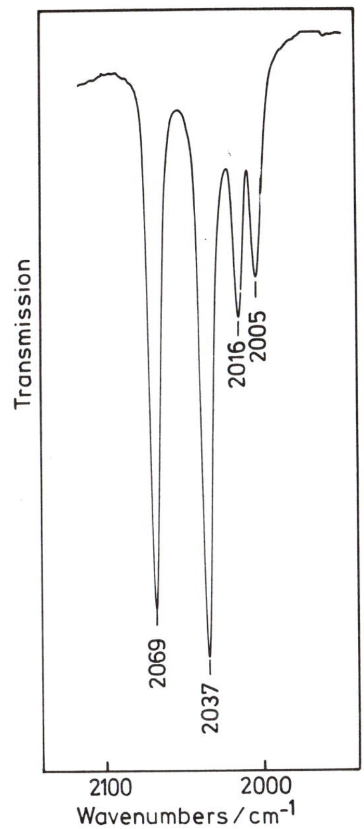

Fig. 1. Infrared carbonyl spectrum of $Os_3(CO)_{12}$ (in octane).

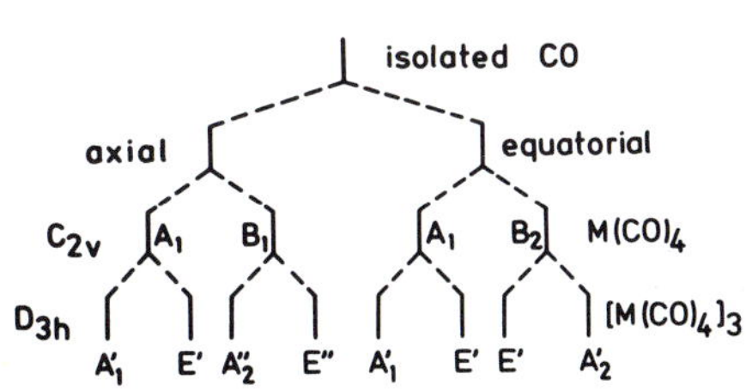

Fig. 2. Splitting of carbonyl stretching modes in $M_3(CO)_{12}$ clusters.

and Stanghellini [8]. This model describes the cluster as a very plastic unit, in which a change in one metal-metal bond length has only a small effect on the energy required to change another.

TABLE 1. Carbonyl Stretching Frequencies and Assignments for $Os_3(CO)_{12}$ and $Ru_3(CO)_{12}$ [7].

Assignment	Activity	ν_{CO}/cm^{-1}	
		$Ru_3(CO)_{12}$	$Os_3(CO)_{12}$
$\nu_1(A_1')$	Raman	2120	2130
$\nu_5(E')$	IR, Raman	2061	2069
$\nu_2(A_1')$	Raman	2036	2038
$\nu_4(A_2'')$	IR	2031	2037
$\nu_6(E')$	IR, Raman	2019	2016
$\nu_7(E')$	IR, Raman	2013	2005
$\nu_3(A_2')$		2000	1995
$\nu_8(E'')$	Raman	1995	1987

Raman spectra of $Os_3(CO)_{12}$ and $Ru_3(CO)_{12}$ at frequencies <200 cm^{-1} are shown in Fig. 3. The most prominent features in these spectra are 158 and 117 cm^{-1} and at 186 and (less intense) 151 cm^{-1}, respectively. The bands at higher frequencies are assigned as A_1' symmetry species and have predominantly metal-metal stretching character; the E' mode at the lower frequency is closer to the metal-carbon deformation modes, which mix in to a greater extent, but metal-metal stretching is still predominant [6, 8]. These two bands are therefore very important for diagnostic purposes when the existence of metal-metal bonds is to be examined.

It is emphasized that extensive intermolecular coupling occurs in the solid state [7, 9, 145], which leads to further band splitting, as shown by the carbonyl spectra of $Os_3(CO)_{12}$ and $Ru_3(CO)_{12}$ in Fig. 4. These correlation-field splitting effects are significant, particularly in the carbonyl stretching region [7, 9], whereas to a first approximation they can be neglected in the medium- and low-frequency regions [8, 9].

The tetrairidium cluster $Ir_4(CO)_{12}$ has T_d symmetry, with the four iridium atoms forming a tetrahedron. Each of these bears three carbonyl ligands in local C_{3v} symmetry. The solution infrared spectra of this cluster show intense bands at 2071 and 2032 cm^{-1} in the terminal carbonyl stretching region [10]. According to group theoretical considerations, these two bands are assigned as the A_1- and E-derived T_2 modes as shown by the band splitting diagram in Fig. 5. On the basis of T_d symmetry, four

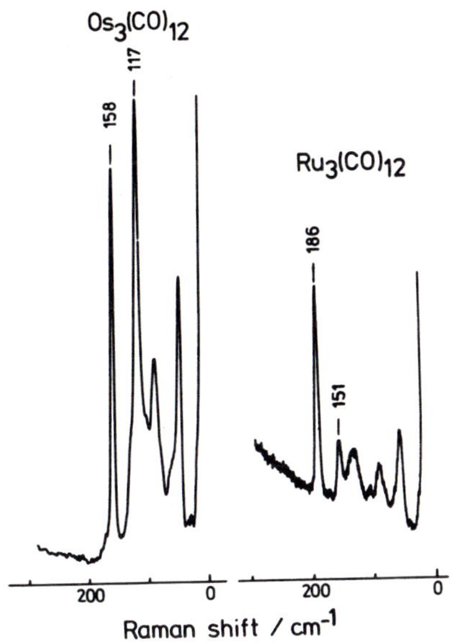

Fig. 3. Raman spectra in the frequency region below 200 cm^{-1} of $Os_3(CO)_{12}$ and $Ru_3(CO)_{12}$ (in KBr matrix).

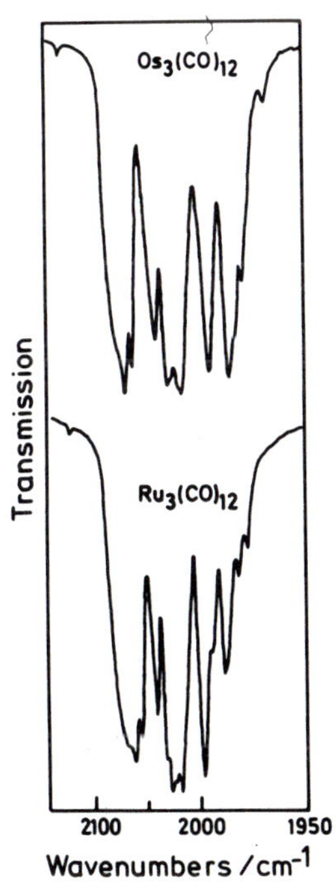

Fig. 4. Infrared carbonyl spectra of $Os_3(CO)_{12}$ and $Ru_3(CO)_{12}$ in the solid state at 100 K in CsI matrix (from ref. [9]).

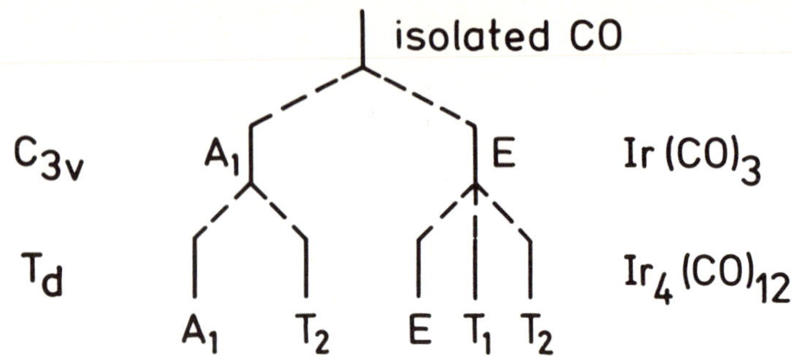

Fig. 5. Splitting of carbonyl stretching modes in $Ir_4(CO)_{12}$.

Raman-active species--namely (A_1 + E + $2T_2$)--are expected, which have indeed been observed [10] in solid-state spectra. These spectra, however, contain additional weak features which are indicative of intermolecular interactions in the solid state, as becomes evident from the rather more complex infrared spectra observed at low temperatures [11].

The middle frequency range from 600-350 cm^{-1} provides the (IrCO) and (IrC) modes [10, 11]. The number of observed infrared and Raman bands in this range clearly suggests that there is significant coupling between the four sets of $Ir(CO)_3$ groups; intermolecular interactions do not seem to affect the spectrum [11]. Three strong bands in the low-frequency region appear at 207, 161, and 131 cm^{-1} in the Raman [12], and only one strong band at 163 cm^{-1} was observed in the far infrared spectrum [11]. The Raman bands have been assigned as the A_1, T_2, and E Ir_4 cluster modes [12], a view which was supported later by Adams and Taylor [11]. Again, intermolecular interactions do not seem to play a significant role in this spectral region in the solid state.

$Co_4(CO)_{12}$ and $Rh_4(CO)_{12}$ both belong to point group C_{3v} and have very similar spectra in the carbonyl stretching region [13]. As shown by their molecular configuration [14, 15]

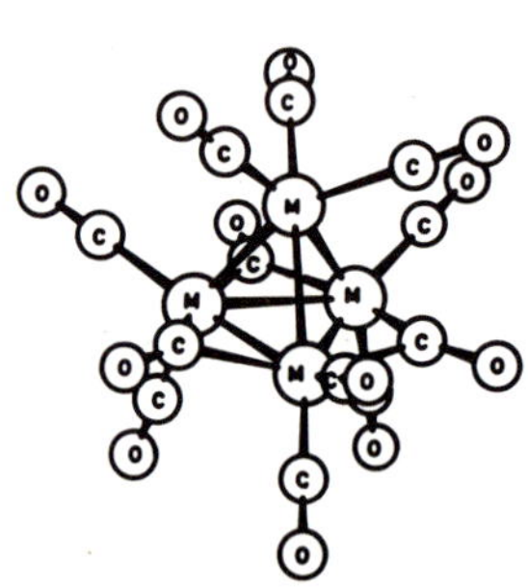

these molecules consist of four tetrahedrally arranged metal atoms in which an apical $M(CO)_3$ group is symmetrically coordinated to a basal $M_3(CO)_9$ fragment by only M-M bonds. The latter fragment contains three equivalent $M(CO)_2$ units, each located at a vertex of a nearly equilateral triangle. Each pair of $M(CO)_2$ groups is linked by both a M-M bond and a bridging carbonyl group. Consequently, these cluster carbonyls have infrared carbonyl bands characteristic of terminal ($\nu_{CO} > 2000\ cm^{-1}$) and bridging ($\nu_{CO} < 1900\ cm^{-1}$) CO ligands.

For $Co_4(CO)_{12}$, for example, six terminal stretching modes were observed [at 2104(vvw), 2063(vs), 2055(vs), 2048(vvw), 2038(w), and 2028(w) cm^{-1}] and two bridging carbonyl stretching modes were observed [at 1898(vvw) and 1867(m) cm^{-1}] [16]. This number of infrared-active C-O stretching bands is consistent with the C_{3v} symmetry of the tetracobalt cluster.

For $Rh_4(CO)_{12}$ (solid), Raman bands at 2109(w), 2066(w), 2026(s), 1828(w), and 1817(w) cm^{-1} and infrared bands at 2105(w), 2077(vs), 2055(w), 2028(s), and 1848(s) cm^{-1} were reported by Griffith and Wickham [17]. Evidently, some of the weak bands escaped detection in these experiments. The same authors [17] reported two Raman bands at 221 and 173 cm^{-1} for solid $Rh_4(CO)_{12}$ which were assigned as the symmetric and asymmetric metal-metal stretching modes of the tetrahedral Rh frame.

$Co_6(CO)_{16}$ and $Rh_6(CO)_{16}$ have overall T_d symmetry:

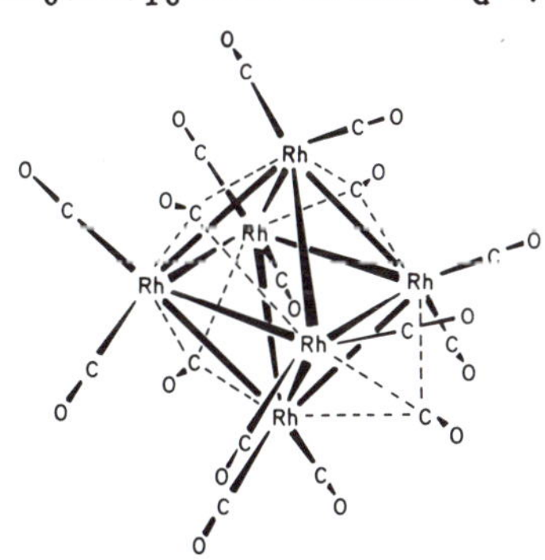

This structure is based on an octahedron of six metal atoms with two terminal CO groups per metal atom and four face-bridging CO ligands. Group theory, therefore, predicts five terminal carbonyl stretching modes ($A_1 + E + T_1 + 2T_2$), all of which should be Raman-active; only the two T_2 species are infrared-active. The bridging carbonyl ligands give rise to two ($A_1 + T_2$) stretching modes of which only the T_2 mode is infrared-active.

Table 2 summarizes the observed carbonyl stretching frequencies for $Co_6(CO)_{16}$ and $Rh_6(CO)_{16}$. The predicted number of bands was observed in the solution infrared spectra, but the Raman spectra seem to be incomplete, probably because these clusters are rather weak Raman scatterers. Griffith and Wickham [17] reported two Raman bands for $Rh_6(CO)_{16}$ in the low-frequency region at 199 and 172 cm^{-1}, which were assigned as the symmetric and asymmetric stretching modes (A_1 + E) of the Rh atom octahedron.

A variety of carbonyl cluster compounds have been structurally characterized which have carbonyl ligands bonded through both the carbon and the oxygen atoms. These cluster structures have been suggested as molecular analogues of an intermediate state of the dissociation of a CO molecule chemisorbed on a metal surface [19-21]. This type of CO

TABLE 2. Carbonyl Stretching Frequencies and Assignments for $Co_6(CO)_{16}$ and $Rh_6(CO)_{16}$ [18].

		ν_{CO}/cm^{-1}	
Assignment	Activity	$Co_6(CO)_{16}$	$Rh_6(CO)_{16}$
A_1	Raman	2115^d	2111^b
T_2	IR, Raman	2063^c	2076^a
T_2	IR, Raman	2022^c	2043^a
A_1	Raman	1808^d	1834^b
T_2	IR, Raman	1791^c	1805^a

^{a}in CCl_4; ^{b}in KBr; ^{c}in *n*-heptane; ^{d}in Nujol.

coordination leads to a significant reduction in bond order (bond distance is 0.126 nm for the μ-η^2-CO ligand in $[HFe_4(\mu\text{-}\eta^2\text{-}CO)(CO)_{12}]^-$ as compared to typical bond distances of 0.11-0.114 nm for CO ligands bonded through the carbon atom only). The most extreme case of reduction of CO bond order was found with $Cp_3Nb_3(CO)_7$, in which one CO ligand acts as a six-electron donor. The unique CO ligand donates electrons to one niobium atom through the carbon lone pair and to the remaining two niobium atoms using the two orthogonal carbon-oxygen bonds. This bonding leads to the longest C-O bond yet observed (0.13 nm) in a carbonyl ligand, and the exceptional reduction in bond order is reflected in the low C-O stretching frequency of 1330 cm^{-1} [22].

6.2.1.b **Hydrido carbonyl clusters**

Although many hydrido carbonyl clusters are known, the M-H vibrational modes have been reported and assigned in but a few cases. This lack of data is attributed to the low infrared extinction coefficients and Raman scattering cross sections. No inelastic neutron scattering spectra have been published, to our knowledge. It has been shown, however, that the M-H band intensities in the infrared spectra increase significantly on cooling to liquid-nitrogen temperature [23]. Simultaneously, the bands show a marked sharpening so that better resolution can be obtained.

The cluster $H_4Ru_4(CO)_{12}$ has been characterized by Kaesz and coworkers [24, 25]. They reported two weak, broad Raman bands at 1585 and 1290 cm^{-1} [24, 25]. The corresponding infrared bands were observed at 1605 and 1272 cm^{-1} [25]. These band pairs shifted to 1153 and 909 cm^{-1} and to 1095 and 895 cm^{-1}, respectively, in the Raman and infrared spectra, for the deuterated compound $D_4Ru_4(CO)_{12}$.

The cluster $H_2Os_3(CO)_{10}$ contains two bridging hydride ligands. The temperature effect on the intensity of the M-H-M modes of this cluster was investigated by Andrews *et al.* [26]; spectra are shown in Fig. 6. The observed frequencies are summarized in Table 3 together with their assignments. The table also contains the frequencies of stretching modes of

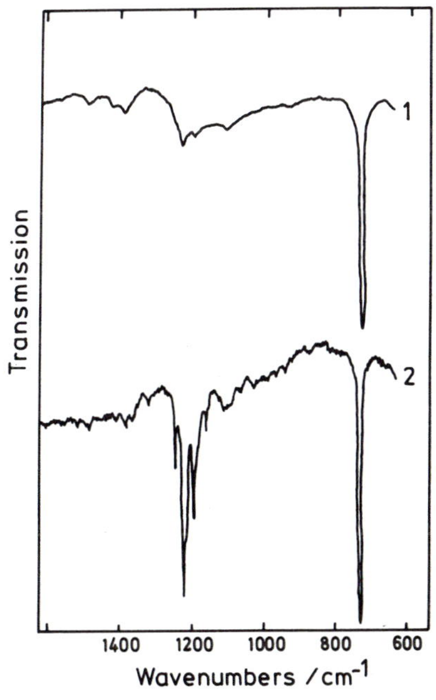

Fig. 6. Os-H-Os infrared spectra of $H_2Os_3(CO)_{10}$ at room temperature (1) and at 77 K (2) (from ref. [26]).

TABLE 3. Frequencies Associated with Metal-Hydrogen Vibrations of Various Hydridotriosmium clusters (data from Ref. [26]).

Assignment	Frequencies/cm^{-1}					
	$HOs_3(CH=CH_2)(CO)_{10}$		$H_2Os_3(C=CH_2)(CO)_9$		$H_2Os_3(CO)_{10}$	$D_2Os_3(CO)_{10}$
	IR	Raman	IR	Raman	IR	IR
Os-H-Os (asym)	1392w	1390w	1398m	1401w		
			1375vvw	1373vvw	1245w	905w
			1349w	1352w	1225s	890s
					1216sh	887sh
Os-H-Os (sym)	1286w	1285w	1302w	1308w	1194m	880m
			1288vw	1292vw	1160w	875w

Os-H-Os vibrations for $HOs_3(CH=CH_2)(CO)_{10}$ and $H_2Os_3(C=CH_2)(CO)_9$. The multiplicity of bands for the vinylidene derivative clearly shows that the two hydrogen bridges are nonequivalent.

These triosmium clusters have three metal-metal stretching modes which are infrared and Raman active. The increased number of bands as compared with $Os_3(CO)_{12}$ (section 6.2.1.a) is a consequence of the lowered symmetry, which leads to a splitting of the E' mode of $Os_3(CO)_{12}$. The $(2A_1 + B_2)$ modes of $(\mu\text{-H})_2Os_3(CO)_{10}$ have been reported to occur at 187, 142, and 97 cm^{-1} [27]. On the basis of a detailed normal coordinate analysis and force constant calculation, Oxton [27] concluded that the vibrational spectrum of this cluster may be described adequately in terms of the μ_2-bridging hydride system without the need to invoke a direct Os-Os bond. The low-frequency modes in the metal-metal stretching region may thus be used as a structural diagnostic, and this prospect is particularly attractive for clusters anchored to surfaces.

6.2.1.c **Substituted carbonyl cluster derivatives**

This class of cluster compounds is of special interest for the characterization of supported clusters, since chemical anchoring onto surfaces often occurs via substitution reactions involving surface functional groups. These may be phosphine, amine, or thiol groups in the cases of functionalized polymers or oxides (also see Chapter 9.1), or they may be surface hydroxyl groups on unmodified oxide surfaces. Moreover, the identification of, for example, hydrocarbon intermediates involved in catalytic transformations requires analysis of vibrational spectra of the respective ligands.

Substitution on the cluster frame with retention of its structural characteristics will influence the carbonyl stretching region by changes of the overall molecular symmetry; this symmetry is normally reduced on substitution, which generally leads to a larger number of carbonyl stretching bands. Further, the carbonyl bands may be shifted to higher or lower frequencies, depending on whether the substituted ligand has weaker or stronger electron-donor properties than a carbonyl ligand. For example, the carbonyl infrared spectra of a series of phosphine-substituted tetrairidium clusters are compared in Table 4.

Symmetry reduction on substitution will also lead to spectral changes in the low-frequency metal-metal stretching region, as was already mentioned in the preceding paragraph for the hydrido cluster $(\mu\text{-H})_2Os_3(CO)_{10}$. $(\mu\text{-O})$-bridged triosmium clusters are interesting reference compounds for the identification of triosmium clusters anchored to oxide surfaces (see section

6.2.3.b.2 and Chapter 9.2). The lowering of the symmetry in comparison with that of $Os_3(CO)_{12}$ must lead to a splitting of the E' mode, so that three Raman bands are expected in the low-frequency region. The frequencies of these bands are summarized in Table 5 for the derivatives $(\mu\text{-H})(\mu\text{-SCH}_3)Os_3(CO)_{10}$, $(\mu\text{-H})(\mu\text{-OPh})Os_3(CO)_{10}$, $(\mu\text{-H})(\mu\text{-OC}_2H_5)Os_3(CO)_{10}$, and $(\mu\text{-OC}_2H_5)_2Os_3(CO)_{10}$.

Infrared band intensities of organic ligands are typically approximately one order of magnitude smaller than those of carbonyl ligands [26]. Therefore, accumulation of spectra is necessary for acceptable signal-to-noise ratios for characterization of not only anchored clusters but also molecular clusters. Recently, ligand spectra of various clusters incorporating hydrocarbon fragments have been reported and assigned, including CH $\{H_3Ru_3(\mu_3\text{-CH})(CO)_9\}$ [32], CH_2 $\{(\mu_2\text{-H})_2Os_3(CO)_{10}(\mu_2\text{-CH}_2)$ and $Os_3(CO)_{10}(\mu_2\text{-CO})(\mu_2\text{-CH}_2)\}$ [33], C_2H_2 $\{Os_3(CO)_9(\mu_2\text{-CO})(\mu_3\text{-}\eta^2\text{-C}_2H_2)\}$ [34], $HC{=}CH_2$ $\{HOs_3(CH{=}CH_2)(CO)_{10}\}$ [26], and $C{=}CH_2$ $\{H_2Os_3(C{=}CH_2)(CO)_9\}$ [26]. These clusters may serve as reference compounds of corresponding anchored clusters and they provide models for the bonding of hydrocarbon fragments chemisorbed on metal surfaces (Chapter 11).

TABLE 4. Carbonyl Infrared Frequencies of Phosphine-Substituted Tetrairidium Clusters.

Compound	ν_{CO}/cm^{-1}	Ref.
$Ir_4(CO)_{12}$	2072, 2032;	[10]
$Ir_4(CO)_{11}(PPh_3)$	2087(m), 2069(w), 2054(vs), 2033(m), 2023(vs), 2015(m), 1888(vw), 1854(m), 1831(m);	[28]
$Ir_4(CO)_{10}(PPh_3)_2$	2070(s), 2043(s), 2010(s), 1835(m), 1795(m);	[28]
$Ir_4(CO)_9(PPh_3)_3$	2042(m), 2015(sh), 1982(vs), 1962(sh), 1772(s);	[29]

TABLE 5. Raman Frequencies Associated with Metal-Metal Stretching Modes of Triosmium Clusters.

Compound	$\nu_{Os\text{-}Os}/cm^{-1}$	Ref.
$Os_3(CO)_{12}$	158(A_1'); 117(E')	[6, 8]
$(\mu\text{-H})(\mu\text{-SCH}_3)Os_3(CO)_{10}$	164(A_1); 125(A_1); 104(B_2)	[26]
$(\mu\text{-H})(\mu\text{-OC}_2H_5)Os_3(CO)_{10}$[a]	152; 114; ~100(sh)	[30]
$(\mu\text{-H})(\mu\text{-OPh})Os_3(CO)_{10}$[a]	154; 116; ~100(sh)	[30]
$(\mu\text{-OC}_2H_5)_2Os_3(CO)_{10}$	172(A_1); 136(A_1); 119(B_2)	[26]
$H_2(\mu_3\text{-S})Os_3(CO)_9$	172; 144	[31]

[a]Compounds were characterized by their infrared carbonyl spectra.

To elucidate the bonding of a cluster onto a functional group by which it is anchored onto a support, the positions of bands associated with vibrations between metal atoms in a cluster and the bonding partner in the functional group are of prime interest. Metal-sulfur stretching frequencies in the region 250-350 cm^{-1} have been reported for the clusters $Ni_3(\pi\text{-}C_5H_5)_3(\mu_3\text{-S})_2$ [35] and $Co_6C(CO)_{12}(\mu_3\text{-S})_2$ [36], and the trinuclear clusters of the type $H_2M_3(CO)_9(\mu_3\text{-S})$ with M = Fe, Ru, and Os [31].

6.2.1.d **Interactions of Lewis acids with carbonyl ligands**

Coordination of Lewis acids to the oxygen of metal-bound CO is well known [37-40]. This type of adduct formation may occur as a primary step during the interaction of metal carbonyl complexes or clusters with oxide surfaces [42, 43] and it may play a major role in CO activation during catalytic hydrogenation [42, 44-47]. The adduct formation with a Lewis acid via the carbonyl oxygen leads to marked shifts of the carbonyl stretching bands to lower frequencies as exemplified by data of Fig. 7 for the adduct formed from *cis*-$[(\pi\text{-}C_5H_5)Fe(CO)_2]_2$ and $Al(C_2H_5)_3$. This frequency shift depends strongly on the strength of the Lewis acid. For the adduct formed from $[(\pi\text{-}C_5H_5)Fe(CO)]_4$ and BCl_3, a band at 1292 cm^{-1} was observed [40]. Some data for polynuclear carbonyl clusters forming

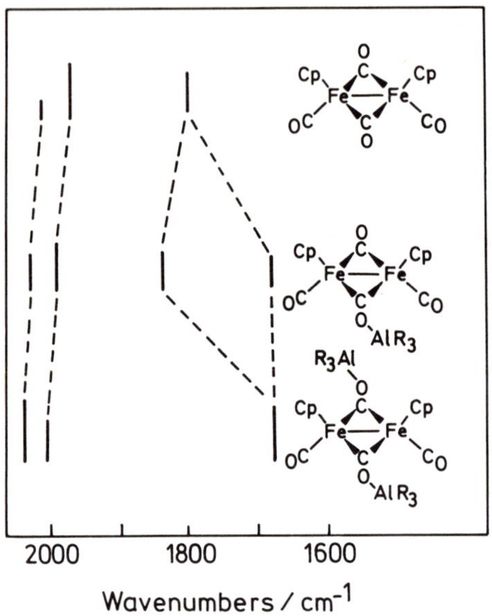

Fig. 7. Infrared carbonyl stretching frequencies for $[(C_5H_5)Fe(CO)_2]_2$, $[(C_5H_5)Fe(CO)_2]_2 \cdot AlR_3$, and $[(C_5H_5)Fe(CO)_2]_2 \cdot 2AlR_3$ (from ref. [41]).

adducts with $AlBr_3$ are summarized in Table 6. The frequency shifts are indicative of a significant reduction in C-O bond order. A relation between bond order and CO stretching frequency, determined by Shriver [46] and Stimson and Shriver [48], is shown in Fig 8.

TABLE 6. Carbonyl Stretching Frequencies for Adducts of Metal Clusters with $AlBr_3$ (from ref. [40]).

Carbonyl Cluster	ν_{CO}/cm^{-1} (M-CO-Al)
$Co_2(CO)_8$	1600
$Fe_3(CO)_{12}$	1548
$Fe_2(CO)_9$	1523(w), 1557(s)
$Ru_3(CO)_{12}$	1535

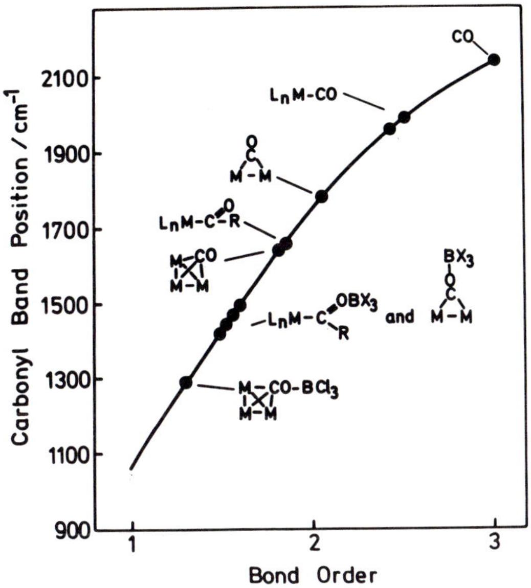

Fig. 8. Relationship between carbonyl stretching frequency and bond order (from ref. [46]).

6.2.1.e Mononuclear carbonyl complexes and CO chemisorption on metal surfaces

A discussion of vibrational spectra of mononuclear carbonyl complexes and of chemisorbed CO is beyond the scope of this book. There are excellent reviews concerned with these subjects: vibrational spectra of carbonyl complexes were reviewed by Haines and Stiddard [2], Kettle [49], Nakamoto [3], and Braterman [4]. The spectroscopy of chemisorbed CO was reviewed by Sheppard and Nguyen [50].

6.2.2 VIBRATIONAL SPECTROSCOPIC TECHNIQUES FOR SURFACE CHARACTERIZATION

6.2.2.a Infrared transmission-absorption spectroscopy

Transmission infrared spectroscopy and its application to surface chemistry have been reviewed extensively [51-59]. The principles are well known from conventional transmission infrared spectroscopy of solids; thin, pressed self-supporting wafers are usually used for surface characterization [53, 59]. The applicability of the transmission technique is determined by the properties of the solid powder sample. Samples exhibiting only weak bulk absorption and having an average particle size less than the wavelength of the incident infrared radiation are optimal. The particle size condition which determines the wavelength range of suitably low scattering losses is usually met in the mid and far infrared region, and scattering losses become strong in the near infrared region. On the other hand, most samples show strong bulk absorption in the low-wavenumber region (roughly <1000 cm^{-1}). As a result, the accessible wavenumber range for transmission infrared spectroscopy of surfaces is generally limited to the mid-infrared region ($4000 > \nu > 1000$ cm^{-1}). However, when less than optimal spectroscopic conditions can be accepted, spectra may also be obtained in the near infrared.

When polymer-supported organometallics are to be characterized, thin (~10-μm) polymer films are best, and these require special preparation methods [60].

The sensitivity of the infrared technique is dependent on the extinction coefficients of the surface groups, which may vary from 5×10^{-18} cm^2 $molecule^{-1}$ for the carbonyl stretching mode in CO ligands to between 10^{-20} and 10^{-19} cm^2 $molecule^{-1}$ for C-H stretching modes in saturated hydrocarbon chains. The sensitivity of transmission infrared spectroscopy also depends on the physical properties of the solid. Assuming typical values of 100 m^2 g^{-1} for the adsorbent surface area, 20 mg cm^{-2} for the weight of the irradiated geometric area, 10^{-19} cm^2 $molecule^{-1}$ for the extinction coefficient, and 5% for a desirable absorption in order to obtain good quality spectra with standard infrared spectrometers, one estimates a lower limit for the surface coverage of 0.02. This would correspond to approximately 10^{17} oscillators per m^2. Assuming, for supported clusters, that a 1:1 stoichiometry exists between metal atoms and oscillators, the detection limit would correspond to a loading of ~0.1 wt% for Fe and ~0.4 wt% for Os.

With the application of modern data acquisition techniques, the sensitivity of the technique can be increased further, even with conventional dispersive spectrometers. Quantitative measurements of surface group densities are possible, provided the Lambert-Beer law is applicable. It must be kept in mind, however, that this law is valid only for optically homogeneous materials and deviations may occur for dispersed samples.

A number of different transmission cells have been described in the literature. Typical designs of cells for *in-situ* work at low and high temperatures are discussed in the reviews mentioned above [51-54, 59].

Transmission infrared measurements have been carried out largely with conventional dispersive instruments. Long recording times are necessary to obtain sufficiently high signal-to-noise ratios, because of the generally low energy level (low transmittance of samples). Moreover, the scattering and absorption background of the solid adsorbent are superimposed in the spectra. This can be compensated for by the use of an optically equivalent wafer in the reference beam or by computer subtraction of the signals. The advantages of Fourier transform (FT) infrared spectrometers over dispersive spectrometers are twofold. First, the energy input through the relatively large circular entrance aperture is rather high. This, however, cannot always be utilized fully, since it may be difficult to prepare sample wafers of equivalent size. The second and more important advantage of FT spectroscopy is the so-called Felgett or multiplex advantage, which allows a spectrum to be measured with a Fourier transform spectrometer in the same time but with considerably better signal-to-noise ratios than with a conventional spectrometer, or in much shorter time with an equivalent signal-to-noise ratio. Critical evaluations of the opportunities provided by Fourier transform spectrometers were published by Peri [59] and Bell [61].

6.2.2.b **Infrared reflection spectroscopies**

The transmission technique fails when opaque and/or strongly scattering materials are to be investigated. A variety of infrared reflection methods have been developed which make use of the specular or diffuse reflection, depending on the nature of the solid adsorbent and its optical properties [62]. Most important for the detection of surface groups among the reflection techniques are *internal reflection spectroscopy* (IRS, a modification of the well-known attenuated total reflection method) [63] and *diffuse reflectance spectroscopy* (DRS) [55, 64, 65]. Although they are potentially useful, these two techniques do not seem to have found application to vibrational spectroscopy of supported clusters. DRS is discussed in more detail as the major technique for UV-VIS spectroscopy (Section 6.3).

6.2.2.c **Raman spectroscopy**

The application of Raman spectroscopy to surface characterization has been reviewed by several authors [55, 58, 66-70]. Raman spectroscopy provides several unique opportunities:

(1) The interference of gas-phase spectra is negligibly weak. *In-situ* Raman spectra of working catalysts at elevated temperatures and pressures can therefore be obtained.

(2) Simple glass or quartz cells can be used.

(3) The Raman scattering probabilities of the solid adsorbent or support (especially of most oxides) are frequently very low, so that Raman spectra of surface species can be recorded in the low-frequency

region down to approximately 50 cm^{-1}. Thus, the metal-metal stretching region becomes accessible for oxide-anchored clusters, whereas this region cannot be investigated in the far infrared because of the strong lattice absorptions (Section 6.2.2.a).

The basic experimental requirements for conventional laser Raman spectroscopy, including data manipulation and signal averaging, are described in the review articles mentioned above [66, 67, 69]. Suitable *in-situ* Raman cells for surface investigations have been described in the literature [71-73].

Some major problems encountered in laser Raman spectroscopy are the following:

(1) Samples are sensitive to heating by the laser beam; cluster degradation is a likely consequence. This phenomenon may become severe when colored samples are used.

(2) The sensitivity of the technique is low.

(3) Background fluorescence is a problem which is sometimes so severe with oxide surfaces that weak Raman signals remain undetectable.

The heating effects of the laser beam can be reduced by simply applying low laser power levels ($\leq$100 mW). Alternatively, cylindrical lens focus techniques have been applied to spread the beam into a line image on the sample. This approach may additionally provide increased signal intensity [74]. Rotation of the sample also decreases the incident energy flux.

As a consequence of the inherent low sensitivity of Raman spectroscopy, the few published investigations of anchored clusters have been restricted to the metal-metal stretching region, since the metal-metal vibrations fortunately have rather high Raman scattering cross sections. Still, detection limits in terms of metal loading are frequently too high to allow recording of reliable Raman spectra of anchored clusters at typical loadings <1-2 wt% metal.

An optimization of spectrometer sensitivities is therefore urgently needed. The intensity of light reaching the detector depends on a variety of parameters, of which only the spectral slit width, the scanning rate, and the laser power can be changed when commercial spectrometers are used. All other parameters are normally invariably preselected and determined by the instrument's design. However, the illumination and viewing of the sample (lens aperture, focal length and demagnification) strongly influence the instrument's sensitivity, and a given commercial instrument does not necessarily provide the optimum for surface studies [74]. One may therefore wish to change illumination and viewing of the sample in critical cases.

Most Raman spectrometers work in the scanning mode using photomultipliers or photon counting detector systems. The signal-to-noise ratio in the final spectrum is proportional to the time during which data are accumulated within the band pass of the monochromator. The use of "multiplex" (position sensitive) detection, whereby the spectral information of the entire spectral range is available simultaneously at the optical multichannel detector (OMA), improves the instrument sensitivity considerably (>x1000) over that of conventional scanning instruments [74]. Multiplex detection therefore yields significantly higher signal-to-noise ratios with considerably lower data acquisition times. This detection method will, therefore, permit Raman spectroscopy of weak Raman scatterers even at low surface concentrations. Moreover, the reduced measuring times required permit a more efficient use of the laser light source within the limited lifetime of the expensive plasma tubes.

The fluorescence background often encountered in Raman spectra of oxide surfaces [55, 66, 67, 70] must be attributed partly to contamination by hydrocarbons. These can be burned off by treatment of the oxide in oxygen at elevated temperatures, but with air- and temperature-sensitive anchored clusters, this method is obviously not applicable. Moreover, considerable contributions to the fluorescence background from hydroxylated or partially hydroxylated oxide surfaces are due to laser-induced electronic excitations of surface hydroxide ions [75]. Thus, fluorescence seems to be an intrinsic property of oxide surfaces which bear basic hydroxide ions.

In some cases, a simple change of the exciting frequency may reduce the fluoresence background in the spectral range of interest [76]. Alternatively, time-resolved Raman techniques may be applied. These methods, however, are experimentally complicated and depend on the fluorescence decay times of the samples; they can hardly be developed as routine techniques for surface studies.

Frequency modulation Raman spectroscopy [77,78] is a relatively simple technique, which produces the final spectrum in the form of the first derivative of the Raman intensity. The background is therefore effectively eliminated and the signal-to-noise ratio is improved over that of a conventional Raman spectrum recorded under comparable conditions. Bruckner *et al.* [79] described an experimental design.

Difference Raman spectroscopy [80] may be of interest when background spectra are to be eliminated.

6.2.2.d **Electron vibrational spectroscopies**

High-resolution electron energy loss spectroscopy (EELS) [55, 57, 58] is a powerful technique for the detection of vibrations of atoms and molecules chemisorbed on metal surfaces. EELS provides a very high sensitivity; depending on the surface dipole moment, as few as 10^{10} molecules may be detected on an area of only 1 mm^2. The spectral resolution, however, is restricted to the order of about 60 cm^{-1}, whereas infrared and Raman spectroscopies easily achieve resolutions of 2 cm^{-1} and better. Although EELS has the great advantage of opening up the entire vibrational region from <200 to 4000 cm^{-1}, its application remains restricted to ultrahigh vacuum conditions and to geometrically flat surfaces, since the method uses the specular reflection of electrons.

Attempts have been reported [81] with model supported metal catalysts which were fabricated by evaporating a small quantity of rhodium onto an oxidized aluminum support. The positive results of this work lend some support to the possibility of detecting vibrational modes associated with anchored cluster-derived species by EELS on model systems consisting of thin insulating oxide layers grown or deposited on metal supports.

The second electron vibrational spectroscopy is inelastic electron tunneling spectroscopy (IETS). Excellent reviews dealing with theoretical and experimental aspects of the technique have been published [55, 57, 82-86]. The IETS sample is a metal-insulator-metal tunnel junction. The current produced by electrons tunneling through the insulating barrier between the metal electrodes is measured when a bias voltage is applied. Since additional channels open up for the tunneling electrons which traverse the insulating barrier inelastically by exciting vibrational modes within the barrier, the electron current across the barrier exhibits a small, abrupt increase at bias voltages which correspond to the vibrational excitation energy of the respective oscillator. The second derivative of the current as a function of the applied voltage corresponds to infrared absorption or Raman spectra [87]. The most attractive features of the IETS technique are the following:

(1) The entire vibrational spectral region from 200-4000 cm^{-1} is accessible.

(2) The sensitivity is comparable to that of EELS.

(3) Resolutions better than 10 cm^{-1}, and in some cases near 1 cm^{-1}, can be obtained.

(4) For molecular adsorbates and surface groups, infrared-active, Raman-active, and optically forbidden transitions are all allowed and observed in IETS with intensities of nearly the same magnitude [87]. IETS therefore permits the detection of a complete vibrational spectrum in a single experiment.

The technique also suffers from some disadvantages which restrict its routine application:

(1) The measurements must usually be made at temperatures ≤ 4.2 K to allow the high resolutions mentioned above.

(2) The metal-insulator-metal tunnel junction requires a sample preparation which usually involves growing a thin (typically 2-3 nm thick) oxide layer on the surface of the corresponding metal film. The oxide layer is then covered by the second metal electrode. Normally, chemisorption on and treatment of the oxide layer must be carried out before the second metal is deposited. However, Jaklevic and Gaerttner [88] recently reported an infusion technique permitting the external doping of a complete tunnel junction with organic molecules.

The tunnel junction which has been investigated most frequently is the $Al-Al_2O_3-Pb$ junction, where the Al_2O_3 layer is used as an adsorbent. Another reported tunnel junction is Mg-MgO-Pb [89]. Investigations by IETS of physically adsorbed clusters have been reviewed by Weinberg [90].

6.2.2.e **Miscellaneous vibrational spectroscopic techniques**

Inelastic neutron scattering spectroscopy (INS) is a promising technique for investigation of adsorbed molecules and surface groups, although it has the disadvantage of the need of a central neutron source. The origin of neutron scattering is the interaction between the neutron and the nucleus of the scattering atom. Dipole or polarizability selection rules are not obeyed in INS, and therefore all normal modes of vibration are active. The intensity of scattering is weighted by the nuclear incoherent inelastic cross section of the scattering nucleus and the square of the amplitude of vibration of the nucleus in the particular mode [91]. Consequently, INS is particularly sensitive to H-containing species, since both these parameters are greater for protons than for any other nucleus. Due to the high penetration power of neutrons, INS can be applied to any kind of solid material, including metals, but for the same reason it is obviously not surface sensitive, and possible bulk contributions to neutron scattering spectra must be eliminated if surface phenomena are to be resolved. The recent literature of INS, as applied to investigations of adsorbed molecules and surfaces, has been reviewed by Wright [92]. Results have been obtained for hydrogen chemisorbed on metals [93-95], but studies of anchored clusters or cluster-derived catalysts by this technique are lacking.

In *photoacoustic spectroscopy* [55, 57, 96-98], PAS, a sample is subjected to illumination with chopped radiation. This may produce temperature pulsations in the surface layers of the specimen, the amplitudes of which can be made to depend on the absorption spectrum of the sample. The temperature fluctuations can be measured photoacoustically through the sound waves which will be generated in a gas (usually He) atmosphere in contact with the sample or in a piezoelectric substrate. Any sample having an optical absorption length similar to or larger than the thermal diffusion length is well suited to characterization by PAS. This requirement is easily fulfilled for infrared chromophores. The most typical characteristics of PAS are as follows:

(1) The samples need not be thin, flat, smooth, or powdered; minimum sample preparation is therefore needed. This advantage is especially attractive when preparation such as grinding, pressing, etc. may alter the sample properties.

(2) Optically opaque samples can be investigated.

(3) The spectral information is simply determined by the absorption of the sample. It does not depend on the totally or diffusely reflected components of the radiation (as in IRS and DRS, respectively), which may be weak and difficult to interpret.

(4) The thermal diffusion length can be varied by using different light modulation frequencies. This provides the possibility to deduce depth profiles and to discriminate between surface and bulk contributions to the experimental PA spectrum.

(5) Low acoustical background may be achieved, which is independent of the light irradiation intensity. This enables one to improve the signal-to-noise level linearly with the incident light intensity.

Low and Parodi [99] modified a conventional dispersive infrared spectrometer to allow its use for computer-controlled IR-PAS. They described photoacoustic cells for investigations of chemisorption phenomena on dispersed high-area adsorbents. From their results, it can be concluded that the technique has submonolayer sensitivity, and high-quality spectra of surface groups in the mid-infrared region can be obtained; these are equivalent to transmission-absorption spectra. When one must deal with samples having low transmission (because of strong absorption and/or scattering losses), and must avoid subjecting the sample to severe treatment, PAS might be superior to transmission spectroscopy.

Since the sensitivity of the technique is not very high, dispersive spectrometers need unfavorably long measurement times. Fourier transform infrared photoacoustic spectroscopy (FTIR-PAS) has therefore been advanced recently [96, 100]. This is a technique for surface groups which can well compete with IRS and DRS. It will be especially advantageous for opaque samples having large particle sizes, which would cause problems due to poor optical contact in IRS and of extremely low reflectivity in DRS. The contact-free spectroscopy by photothermal radiometry (PTR — pulsations in the thermal reradiation produced by modulated sample irradiation are measured) might be further developed as useful techniques for certain applications in surface group spectroscopy [100-102].

Woo and Hill [146] reported Fourier transform PA spectra (mid-infrared region) of a sample which was prepared by grinding $Co_2(CO)_6(PPh_3)_2$ and γ-Al_2O_3 together.

6.2.3 APPLICATION OF VIBRATIONAL SPECTROSCOPIES TO THE CHARACTERIZATION OF SUPPORTED CLUSTERS AND OF CLUSTER-DERIVED SPECIES

This section is an evaluation of the application of vibrational spectroscopies to the characterization of supported clusters and derivatives thereof. Typical examples have been selected, with emphasis placed on the vibrational spectroscopic information. The surface organometallic chemistry and catalysis are discussed in Chapter 9.

6.2.3.a Physically adsorbed clusters

When a cluster is adsorbed physically in molecular form, only minor perturbations of its vibrational frequencies are expected. Physical adsorption may particularly occur when the adsorption of the cluster is performed from dilute solution with the solvent being subsequently removed at sufficiently low temperature. Choplin *et al.* [103] have impregnated the bimetallic cluster $H_2FeOs_3(CO)_{13}$ onto a silica surface and observed infrared carbonyl stretching bands at 2087(s), 2075(s), 2041(vs), and 2031(br,sh) cm^{-1}, which match very well those reported for the same cluster in *n*-heptane solution [104] 2086(s), 2072(s), 2040(vs), 2032(m), 2025(m), 2015(m), 1994(w), 1875(w), and 1848(m) cm^{-1}. The weak bands of the molecular cluster in solution appearing at <2000 cm^{-1} could probably not be observed for the physisorbed species since they would interfere with relatively strong absorptions of the SiO_2 support. In addition, a Raman band was reported for the physisorbed cluster at 192 cm^{-1}, which was assigned as a metal-metal stretching mode indicating the retention of the metal frame. It was therefore concluded that individual molecular cluster species were adsorbed physically on the SiO_2 surface.

If the same impregnation procedure is applied to the preparation of samples having higher loadings of clusters, the formation of microcrystals of the cluster is possible. This must consequently lead to a more complex

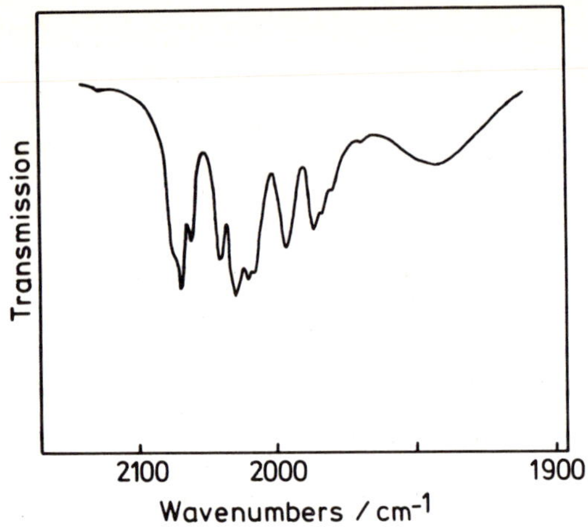

Fig. 9. Infrared carbonyl spectrum of $Os_3(CO)_{12}$ crystallites on Al_2O_3 (Nujol mull); (from ref. [105]).

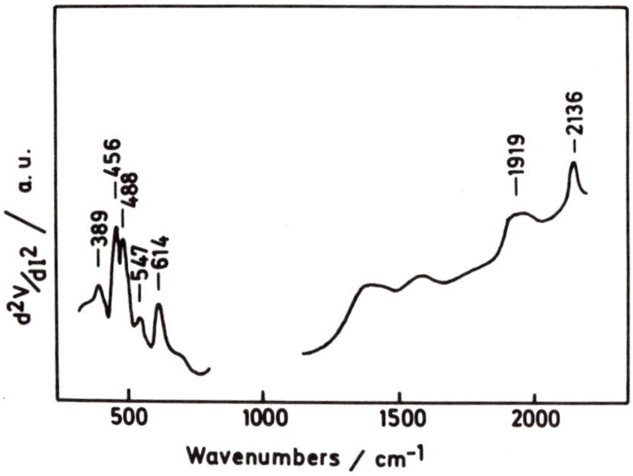

Fig. 10. Inelastic electron tunneling spectrum of $Ru_3(CO)_{12}$ on Al_2O_3 (from ref. [106].

carbonyl stretching spectrum due to intermolecular interactions between the individual cluster molecules (see Section 6.2.1.a). A typical example is shown in Fig. 9 for $Os_3(CO)_{12}$ on Al_2O_3. The infrared carbonyl spectrum, shown in Fig. 9, strongly resembles that shown in Fig. 4 for $Os_3(CO)_{12}$ in the solid state, thus indicating the formation of crystallites on the support surface having a metal loading of 2 wt%.

Physisorbed clusters can also be produced by sublimation of the cluster onto a support which is kept at low temperature. This route is a preferred procedure for the preparation of tunnel junctions for IET spectroscopy. An IET spectrum of $Ru_3(CO)_{12}$ on Al_2O_3 is shown in Fig. 10. The low-frequency bands at 389, 456, 488, 547, and 614 cm^{-1} fall into the range of vibrations which are assigned as predominantly Ru-C stretching and predominantly Ru-C-O deformation modes (see Section 6.2.1.a). They coincide well with band positions reported for the molecular $Ru_3(CO)_{12}$ cluster in the solid state [6, 8, 9]. Two bands are observed in the IET spectrum in the carbonyl stretching region at 2136 and 1919 cm^{-1}. The band at 2136 cm^{-1} was assigned as the Raman-active A_1' mode [106]. The band at 1919 cm^{-1} is broad; it was related to a number of closely spaced bands appearing around 2020 cm^{-1} in the Raman spectrum of $Ru_3(CO)_{12}$ (resolution of the IET spectrum was approximately 15 cm^{-1}).. Hence, the IETS bands are shifted to lower frequency by approximately 100 cm^{-1}. This downshift was attributed by Bowser and Weinberg [106] to an effect of the top Pb electrode of the tunnel junction, and the shift could be estimated quantitatively on the basis of a theory outlined by Kirtley and Hansma [107]. Bowser and Weinberg [106] concluded that the $Ru_3(CO)_{12}$ cluster was indeed adsorbed physically on the Al_2O_3 of the tunnel junction. Their results, however, also demonstrate that the interpretation of IET spectra, particularly in the CO stretching region, may be complicated by the effects of the top Pb electrode. Since Raman- and infrared-active modes should be seen in IETS, it may be surprising that the infrared- and Raman-active E' mode at 2061 cm^{-1} (see Table 1) of molecular $Ru_3(CO)_{12}$ is not observed in the IET spectrum of the physisorbed cluster (Fig. 10). Bowser and Weinberg [106] did not comment on this observation.

6.2.3.b. **Chemically anchored clusters**

Functionalized supports. When clusters are anchored onto functionalized polymers or oxide surfaces, ligand-substituted cluster derivatives are formed (see Chapter 9.1). Their vibrational spectroscopic characterization can be performed most easily by direct comparison of the spectra of the anchored cluster species with those of homogeneous analogues. This approach has been applied almost exclusively for the carbonyl stretching region. As an example, Fig. 11 shows the comparison of infrared carbonyl spectra of the cluster $(\mu\text{-H})_2RhOs_3(CO)_{10}(acac)PPh_3$ in solution and anchored onto a phosphine-functionalized polymer [108]. The coincidence of the observed band positions for the two samples clearly indicates the presence of the phosphine-substitued polymer-bound cluster $(\mu\text{-H})_2RhOs_3(CO)_{10}(acac)PPh_2$-Ⓟ.

Obviously, the same procedure must be applicable to the characterization of clusters anchored to functionalized oxide surfaces. Table 7 is a comparison of the carbonyl infrared spectra of several thiol-derivatives of $Ru_3(CO)_{12}$ and $Os_3(CO)_{12}$ and their surface-bound analogues formed by anchoring the clusters to thiol-modified silica surfaces.

There may be cases in which molecular analogues of surface-anchored cluster-derivatives are not available and consequently their vibrational

TABLE 7. Infrared Carbonyl Bands of Molecular and Surface-bound Thiol Derivates of Triruthenium and Triosmium Clusters.

Compound	ν_{CO}/cm^{-1}	Ref.
$HRu_3(CO)_{10}SC_2H_5$ [a]	2105(m), 2064(s), 2056(s), 2026(vs), 2012(m), 2008(s), 1994(m)	[110]
$HRu_3(CO)_{10}SC_4H_9$ [a]	2105(m), 2064(s), 2056(s), 2025(vs), 2012(m), 2007(s), 1994(m)	[110]
$HRu_3(CO)_{10}S(CH_2)_3Si(OCH_3)(CH_3)_2$ [a]	2103(m), 2063(s), 2054(s), 2022(s), 2006(s), 1933(m)	[109]
$HRu_3(CO)_{10}S(CH_2)_3Si(CH_3)_2O$〰 (anchored)	2106(m), 2065(s), 2058(s), 2026(s), 2010(s,sh), 1996(m,sh)	[109]
$HOs_3(CO)_{10}SH$ [b]	2108(m), 2066(vs), 2057(s), 2022(vs), 2018(sh), 1998(m), 1987(m), 1976(w)	[111]
$HOs_3(CO)_{10}SPr^i$ [b]	2108(m), 2065(vs), 2057(s), 2021(vs), 2017(sh), 1995(m), 1986(m), 1979(w)	[111]
$HOs_3(CO)_{10}S(C_6H_6)$ [b]	2106(m), 2067(vs), 2057(s), 2024(vs), 2018(m), 2000(s), 1980(m)	[110]
$HOs_3(CO)_{10}S(CH_2)_3Si(OCH_3)(CH_3)_2$ [b]	2107(m), 2065(vs), 2058(s), 2020(vs), 1996(m), 1987(w)	[109]
$HOs_3(CO)_{10}S(CH_2)_3Si(CH_3)_2O$〰 (anchored)	2110(m), 2070(vs), 2061(s), 2027(vs), 2003(m)	[109]

[a] in CCl_4 solution; [b] in cyclohexane solution.

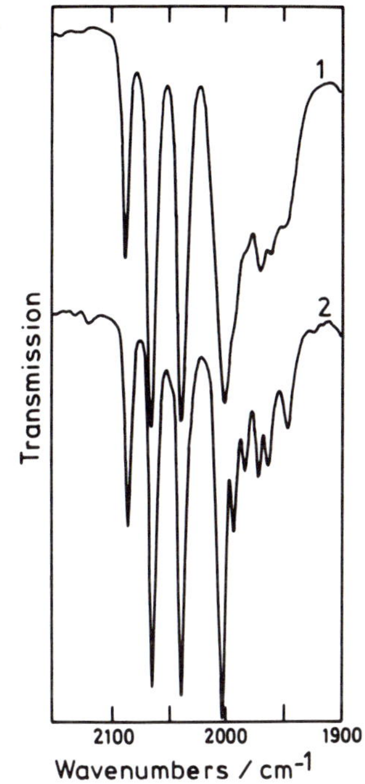

Fig. 11. Carbonyl infrared spectra of $H_2RhOs_3(CO)_{10}(acac)PPh_3$ and of its polymer-supported analog $H_2RhOs_3(CO)_{10}(acac)PPh_2$-Ⓟ (from ref. [108]).

spectra remain unknown. For reasons of site isolation of phosphine-functionalized silica having a low phosphorus content, a monosubstituted cluster $Os_3(CO)_9(\mu\text{-}Cl)_2PPh_2(CH_2)_2$-SIL (SIL indicates the silica surface) could be anchored to the surface [112], although the ligand exchange reaction in solution leads to rapid formation of the disubstituted derivative, and the monosubstitued compound could not be isolated. This reactivity is associated with the presence of two labile CO ligands in the (μ-Cl)-bridged cluster $(\mu\text{-}Cl)_2Os_3(CO)_{10}$. The formation of the monosubstituted surface-bound cluster was inferred by consideration of substituent effects on the carbonyl frequencies, since substitution of CO by phosphine ligands usually shifts carbonyl bands to lower frequencies (see 6.2.1.c.). The greater the number of phosphine ligands, the lower the frequency of the CO bands (Table 4). Consistent with this pattern, the carbonyl bands of the anchored cluster inferred to be monosubstituted did fall between those of the unsubstituted and disubstituted clusters.

Unfunctionalized oxide surfaces. Triosmium and triruthenium carbonyl clusters $M_3(CO)_{12}$ react with surface hydroxyl groups of SiO_2 and Al_2O_3 to form a (μ-H)(μ-O)-bridged trinuclear cluster (Chapter 9.2):

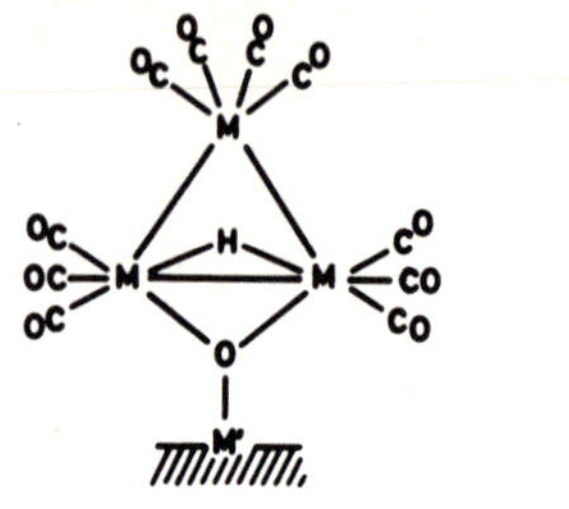

M = Os or Ru ; M' = Si or Al

For $Os_3(CO)_{12}$, the formation of this surface species has been inferred by comparison of its infrared carbonyl spectrum with that of the molecular compound $HOs_3(CO)_{10}(OSiPh_3)$ [113-116]. Table 8 is a summary of the reported carbonyl frequencies for a few molecular (μ-O)-bridged analogues and surface-bound clusters. These spectra may also be compared with those of the corresponding (μ-S)-bridged clusters which are summarized in Table 7. Additional support for this surface-bound cluster species comes from its Raman spectrum in the frequency region <200 cm^{-1} (metal-metal stretching). Two bands were observed at 160 and 119 cm^{-1} (Fig. 12) which were

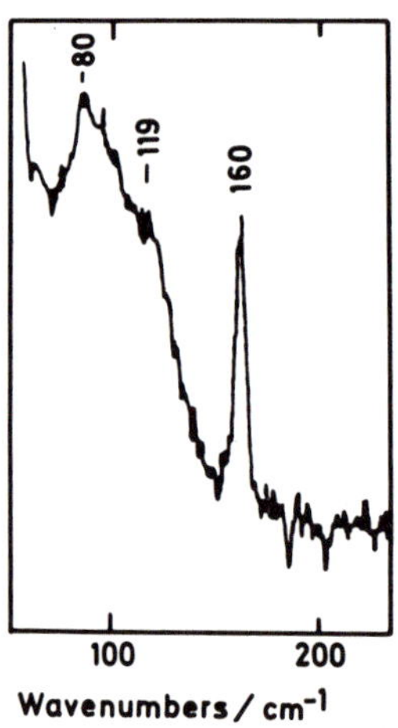

Fig. 12. Raman spectrum of a powder of Al_2O_3-supported triosmium carbonyl cluster $(\mu\text{-H})Os_3(CO)_{10}(\mu\text{-O-Al}\leqslant)$; (from ref. [119]).

assigned as metal-metal stretching modes of the (μ-O)-bridged cluster by comparison with the low-frequency Raman spectrum of $Os_3(CO)_{12}$ [119]. As discussed in Section 6.2.1.c, however, the symmetry reduction in the proposed surface-bound cluster as compared to $Os_3(CO)_{12}$ should lead to three metal-metal stretching modes, of which the band at the lowest frequency (B_2 mode near 100 cm^{-1}) may not be resolved from the A_1 mode

TABLE 8. Infrared Carbonyl Frequencies of (μ-O)-bridged Triosmium Clusters.

Cluster	ν_{CO}/cm^{-1}	Ref.
$[HOs_3(CO)_{10}OCH(CH_3)_2]$	2110w, 2072s, 2059ms, 2023vs, 1998m, 1988	[117]
$[HOs_3(CO)_{10}OH]$	2110w, 2072s, 2060m, 2027vs, 2025sh, 2006m	[118]
$[HOs_3(CO)_{10}$-O-Si(Ph)$_3]$	2107w, 2069s, 2055s, 2017s, 1997m, 1980m, 1907w, 1887w	[113, 114]
Surface-bound		
$HOs_3(CO)_{10}$-O-Al$\leqslant$	2107w, 2068s, 2056s, 2023vs, 2005m	[116]
$HOs_3(CO)_{10}$-O-Al$\leqslant$	2109w, 2070m(sh), 2058s, 2030vs, 2018s, 2012s, 2000ms, 1990m(sh)	[114]
$HOs_3(CO)_{10}$-O-Ti$\leqslant$	2107mw, 2067s, 2055s, 2026s,b, 2010m, 2000sh	[116]
$HOs_3(CO)_{10}$-O-Zn$\leqslant$	2107m, 2067s, 2050s, 2030vs, 2006m, 2000sh	[116]
$HOs_3(CO)_{10}$-O-Si$\leqslant$	2115m, 2080s, 2067s, 2031s, 2013sh, 1995sh	[116]
$HOs_3(CO)_{10}$-O-Si$\leqslant$	2122w, 2088s, 2075s, 2037vs(br), 1955(sh)	[113, 114]

near 115 cm^{-1}. Therefore, a reassignment on the basis of the spectra of reference compounds (Table 5) of the Raman bands reported by Deeba *et al.* [119] must attribute the bands at 160 and 119 cm^{-1} to the two A_1 modes while the B_2 mode (see Table 5) remains unresolved and superimposed on the broad band centered at 80 cm^{-1} in Fig. 12. Beyond doubt, however, this Raman spectrum supports the conclusion of the formation of the proposed (μ-O)-bridged triosmium cluster.

A first attempt has also been made to detect the Os-H-Os vibrations of the (μ-H)-bridging hydrogen in the surface-bound cluster. Hilliard and Gold [120] reported an IET spectrum of $Os_3(CO)_{12}$ on Al_2O_3, showing a band near 1500 cm^{-1} which was attributed to a Os-H-Os vibration in the proposed (μ-H)(μ-O)-bridged triosmium cluster bound to the Al_2O_3 surface. This frequency appears to be relatively high if compared with band positions of molecular hydrido-triosmium clusters in Table 3, which fall in the range <1400 cm^{-1}. There is still no report of the IET spectrum of the deuterated surface-bound cluster.

Theolier *et al.* [121] suggested that an analogous triruthenium cluster forms on SiO_2 surfaces. Bands at 2111(w), 2077(s), 2066(s), 2033(s,br), and 1995(sh) cm^{-1} were attributed to the surface-bound cluster (μ-H)$Ru_3(CO)_{10}$(μ-O-Si$\lessgtr$) by comparison with the molecular compound (μ-H)$Ru_3(CO)_{10}$(μ-SC_2H_5) (Table 7). In addition, the authors reported two low-frequency Raman bands assigned as Ru-Ru stretching modes. Evans and McNulty [122] later supported the formation of the (μ-O)-bridged triruthenium cluster on SiO_2. The latter authors also successfully demonstrated the usefulness of ^{13}CO substitution experiments and of spectrum simulation techniques for the elucidation of supported cluster species [122, 123].

A few cases have been reported in which an adduct formation between a carbonyl ligand of a cluster with a Lewis acid site of the oxide support surface had been inferred from infrared spectra. Ballivet-Tkatchenko and Coudurier [124] reported band positions of 1795 and 1760 cm^{-1} for bridging carbonyls when $Fe_3(CO)_{12}$ was brought in contact with a dehydrated HY zeolite. They assumed that these bands arose from adduct formation involving a Lewis acid site with $Fe_3(CO)_{12}$ via the oxygen of the bridging carbonyl ligand. Evans [125], on the other hand, considered this interaction to be weak and possibly to involve hydroxyl protons in the supercage. Lazar *et al.* [126] observed a bridging carbonyl band near 1800 cm^{-1} (corresponding to a frequency shift of 30-40 cm^{-1}) when they adsorbed $Fe_3(CO)_{12}$ on silica. They inferred that three bridging carbonyl ligands were interacting with silanol groups, thus keeping the three Fe atoms in a plane parallel to the surface. Hugues *et al.* [127] showed spectroscopically that $Fe_3(CO)_{12}$, when adsorbed on alumina, forms the anionic hydride $HFe_3(CO)_{11}^-$. This species gives bands at 2082, 2020, and 2008 cm^{-1} for the linear carbonyl ligands and a broad band of lower intensity in the bridging region at 1598 cm^{-1}. The corresponding band of the molecular analogue $[N(C_2H_5)_4][HFe_3(CO)_{11}]$ is located at 1709 cm^{-1}. The frequency shift of 111 cm^{-1} was considered characteristic of adduct formation and the following structure was therefore suggested for the surface-bound triiron cluster:

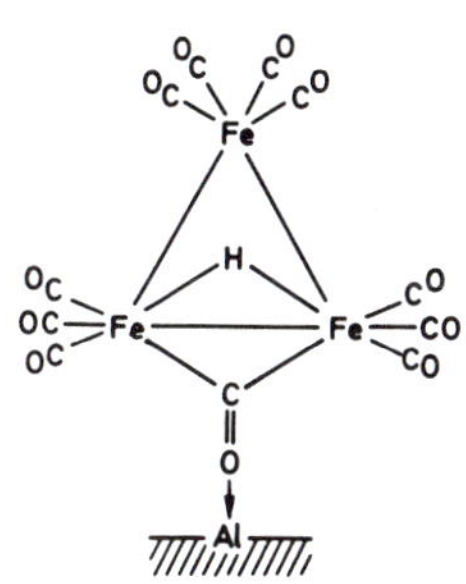

Still larger red shifts were reported by Tessier-Youngs *et al.* [43] for the interaction of $(\pi\text{-}C_5H_5)_4Fe_4(CO)_4$, $(\pi\text{-}C_5H_5)Ni_3(CO)_2$, and $(\pi\text{-}C_5H_5)_2Ni_2(CO)_2$ with dehydroxylated $\gamma\text{-}Al_2O_3$. The observed frequency shifts fell between those measured for the corresponding 1:1 adducts with $Al(C_2H_5)_3$ and $AlBr_3$. It was therefore concluded that the clusters formed the surface adducts

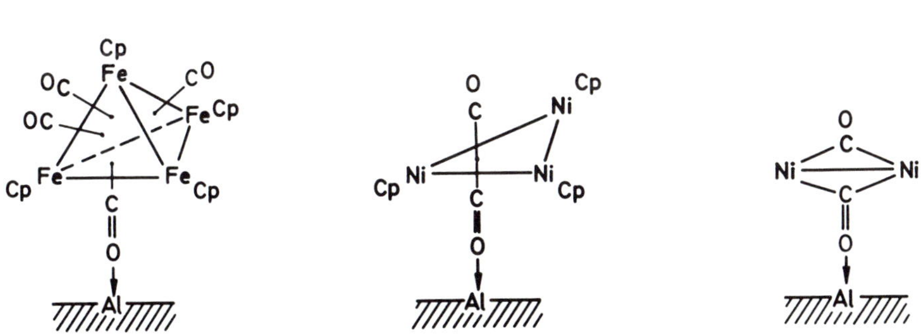

where Al denotes a coordinatively unsaturated surface Al^{3+} cation. In fact, the band positions reported by Hugues *et al.* [127] and by Tessier-Youngs *et al.* [43] are very close to those summarized in Table 6 for molecular 1:1 adducts between carbonyl clusters and $AlBr_3$.

6.2.3.c Infrared spectroscopy of supported carbonyl clusters under reaction conditions

There are but a few reports in which supported intact clusters could be inferred to be the catalytically active species. One approach to determining the catalytic role of supported clusters involves infrared spectroscopy of catalysts under reaction conditions. A reactor/infrared cell combination for these purposes has been described [128]. Besson *et al.* [129] suggested for the catalytic hydrogenation of ethylene in the presence of $(\mu\text{-H})Os_3(CO)_{10}(\mu\text{-O-Si}\leqslant)$ that the crucial step was a reversible opening of one oxygen-osmium bond. This step would provide free coordination sites for the activation of H_2 and C_2H_4. This mechanistic model was based on relatively minor but reversible changes in the infrared carbonyl spectra of the surface-bound cluster when it was exposed to the reactant gases. Another example [112] is alkene isomerization catalyzed by $[Os_3(CO)_9(\mu\text{-Cl})_2PPh_2(CH_2)_2\text{-SIL}]$. Fig. 13 shows the spectral changes which

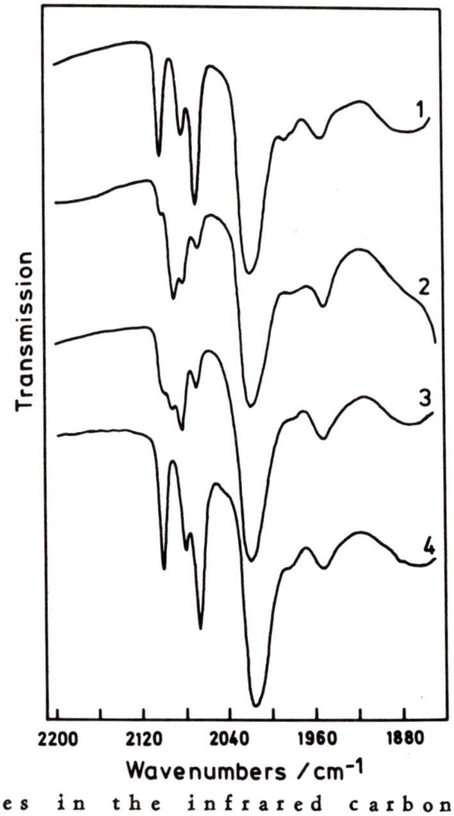

Fig. 13. Changes in the infrared carbonyl spectra of $Os_3(CO)_9(\mu\text{-Cl})_2PPh_2(CH_2)_2SIL$: (1) initial spectrum; (2) after exposure to flowing but-1-ene at 318 K for 2 h; (3) after exposure to flowing He or H_2 at 318 K for 30 minutes; (4) after carbonylation at 318 K for 30 minutes (from ref. [112]).

occur in the carbonyl region when the catalyst is exposed to but-1-ene, He, H_2, and CO. These changes were inferred to originate from dissociation of the one labile CO ligand in the original surface-bound cluster and its substitution by but-1-ene. But-1-ene is a weak ligand itself and can easily be dissociated in He or H_2, while the original cluster is reversibly obtained after final exposure to CO. Interpretation of spectra in these cases is indirect, since reference spectra of molecular analogues of the proposed species are usually not available. Therefore, the spectral assignments may remain uncertain. However, the reversibility of the spectra at least indicates the possible interaction/coordination of reactant molecules with the surface-bound cluster without detectable cluster degradation.

6.2.3.d Products of cluster degradation

Mononuclear carbonyl species are frequently produced during cluster fragmentation on support surfaces. These have been identified by comparison of their infrared carbonyl spectra with those of well-characterized mononuclear carbonyl complexes. Table 9 is a summary of some spectroscopic data of the most extensively studied systems. The assignments for rhodium and osmium species given by various research groups agree quite well. For ruthenium, on the other hand, there are still significant discrepancies, as can be seen from Table 9. In general, the comparison of carbonyl band positions of surface complexes with those of molecular reference compounds may be misleading. Additional spectroscopic information is therefore required. ^{13}CO isotopic exchange experiments have been performed by several research groups for rhodium [123, 132, 133] and ruthenium [122, 136]. The spectral features can be predicted for the particular symmetry [e.g., C_{2v} for $M(CO)_2$ or C_{3v} for $M(CO)_3$ species] when isotopically mixed species are formed. Thus, a dicarbonyl species with local C_{2v} symmetry would give three band pairs in the carbonyl stretching region for $M(^{12}CO)_2$, $M(^{12}CO)(^{13}CO)$, and $M(^{13}CO)_2$ species, respectively. In contrast, four band pairs would be expected for a tricarbonyl of local C_{3v} symmetry for the four possible species $M(^{12}CO)_3$, $M(^{12}CO)_2(^{13}CO)$, $M(^{12}CO)(^{13}CO)_2$, and $M(^{13}CO)_3$. Moreover, frequency shifts can be calculated quantitatively [4] for the predicted structures and compared with the experimental results. An example is given in Table 10 for $Rh^{1+}(CO)_2$ species with local C_{2v} symmetry which were formed during fragmentation of $Rh_6(CO)_{16}$ on a diamine-functionalized SiO_2-surface [133]. The agreement between observed and calculated frequencies (within 2 cm^{-1}) strongly supports the assignment of the observed band pair to the $Rh(CO)_2(C_{2v})$ species.

TABLE 9. Infrared Carbonyl Frequencies of Supported Mononuclear Carbonyl Species Produced by Cluster Fragmentation.

Cluster Precursor	Support	Carbonyl Frequencies cm^{-1}	Assignment	Ref.
$Rh_6(CO)_{16}$	η-Al_2O_3	2105; 2040	$Rh^+(CO)_2$	[130]
$Rh_6(CO)_{16}$	γ-Al_2O_3	2090; 2020	$Rh(CO)_2$	[131]
$Rh_6(CO)_{16}$	γ-Al_2O_3	2082; 2002	$Rh(CO)_2$	[123]
$Rh_4(CO)_{12}$, $Rh_6(CO)_{16}$	γ-Al_2O_3	2085; 2012	$Rh^+(CO)_2$	[132]
$Rh_6(CO)_{16}$	MgO	2085; 2007	$Rh^+(CO)_2$	[130]
$Rh_6(CO)_{16}$	SiO_2, amine-modified	2092; 2024	$Rh^+(CO)_2$	[133]
		1970	$Rh(CO)$	[133]
$Ru_3(CO)_{12}$	γ-Al_2O_3	2045-2050; 1965-1970	$[Ru(CO)_2X_2]_n$[a]	[134]
		2130-2140; 2070-2080; 1990-2000	$[Ru(CO)_3X_2]_n$ and/or $[Ru(CO)_4X_2]_n$	[134]
	γ-Al_2O_3	2138; 2075	$Ru^{3+}(CO)_2$	[135,136]
		2072; 2005	$Ru^{2+}(CO)_2$	[135,136]
		2054; 1977	$[Ru^o(CO)_2]_n$	[135,136]
$Ru_3(CO)_{12}$	γ-Al_2O_3	2063; 1986	$Ru^{2+}(CO)_2$	[122]
	SiO_2	2078; 2014	$Ru^{2+}(CO)_2$	[122]
		2144	$Ru^{4+}CO$	[122]
		2076	$Ru^{3+}CO$	[122]
$Os_3(CO)_{12}$	γ-Al_2O_3	2118-2135;2034-2052; 1955-1970	$Os^{2+}(CO)_2$ and $Os^{2+}(CO)_3$	[114-116]
	γ-Al_2O_3	2128; 2037-2045; 1970	$Os^{2+}(CO)_2$ and $Os^{2+}(CO)_3$	[137]
	MgO	2117; 2033; 1941	$Os^{2+}(CO)_2$ and $Os^{2+}(CO)_3$	[138]

[a] X denotes a surface oxygen ion.

TABLE 10. Observed and Calculated Carbonyl Stretching Frequencies of $Rh(CO)_2$ and Rh(CO) Species Bonded to Diamine Modified SiO_2 (data from ref. [133]).

Species	ν_{sym}/cm^{-1}			ν_{asym}/cm^{-1}		
	obs.	calc.	Δ	obs.	calc.	Δ
$Rh(^{12}CO)_2$	2092	--	--	2024	--	--
$Rh(^{12}CO)(^{13}CO)$	2075	2076	-1	1996	1994	-2
$Rh(^{13}CO)_2$	2043	2045	-2	1977	1979	-2

Evans and McNulty, in two careful studies, have further improved the vibrational analysis of infrared carbonyl spectra of mononuclear rhodium [123] and ruthenium [122] complexes on oxide surfaces by force field calculations and spectrum simulation techniques. From intensity ratios of carbonyl bands, bond angles can be calculated [4]. For the proposed $Rh(CO)_2$ species (C_{2v}), the C-Rh-C angle was 94° as determined by Evans and McNulty [123], whereas Knözinger *et al.* [133] found a value of 100° for the $Rh(CO)_2$ unit bonded to diamine-modified SiO_2. Bond angles were also determined [137] for the $Os(CO)_3$ unit produced on Al_2O_3 by $Os_3(CO)_{12}$ fragmentation. Local C_{3v} symmetry with the C_3 axis being oriented perpendicular to the surface was assumed:

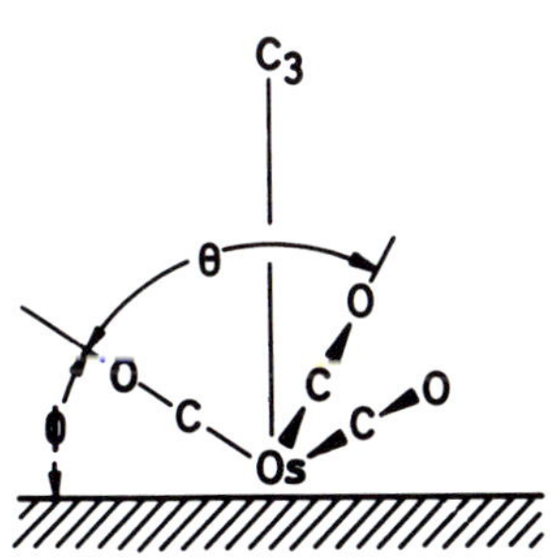

The angles ϕ and Θ are related to the intensities of the A_1 and E mode by the following equation [4]:

$$I(A_1)/I(E) = \tan^2\phi = \frac{1}{4}\,[3\,\cot^2(\frac{\Theta}{2}) - 1] \quad (1)$$

The two angles were estimated to be $\Theta = 97°$ and $\phi = 30°$. Together with the spatial requirements and bond lengths of this carbonyl complex, a model for the nearest-neighbor approach of $Os(CO)_3$ units was suggested; ensembles of three osmium atoms are evidently formed when $Os_3(CO)_{12}$ is fragmented on the Al_2O_3 surface. The osmium-osmium distance in these ensembles was estimated to be >0.59 nm.

6.2.3.e Metal aggregation

Oxide-supported carbonyl clusters may react to form metal aggregates when they are decarbonylated, particularly when they are treated in H_2 at elevated temperatures. After adsorption of CO on these metal aggregates, carbonyl spectra are expected which are typical of CO adsorbed on metallic particles. Infrared spectra of chemisorbed CO on metal surfaces have been extensively reviewed by Sheppard and Nguyen [50] and are not discussed in detail here. It is, however, important to note that positions of carbonyl bands of chemisorbed CO may depend strongly on the surface coverage (blue shift with increasing coverage). This phenomenon is a consequence of dipole-dipole coupling of neighboring CO molecules [50, 139]. The carbonyl band positions depend on the particle size. When CO is adsorbed on rough surfaces or small particles of group VIII metals, carbonyl bands are found at frequencies lower than those observed for plane crystal faces (presuming that the surface coverages are the same) [140]. It must also be noted that due to the metal-surface selection rule some absorption bands may be absent from the spectra of chemisorbed molecules on metal particles [50, 141]. These selection rules come into play when metal particles larger than about 2 nm are present; for smaller particles the rule is weakened [141].

6.2.3.f Disintegration of metal aggregates

When highly dispersed metal aggregates are exposed to a CO atmosphere they may be disrupted, at least at elevated temperatures. For example, Fig. 14 shows carbonyl infrared spectra of an $Os(CO)_5$-derived catalyst supported on alumina. The catalyst was used for CO hydrogenation, and metal aggregates in the 1- to 2-nm diameter range were detected by transmission electron microscopy [47]. When this catalyst was exposed to CO at room temperature, spectrum 1 of Fig. 14 was obtained, having two

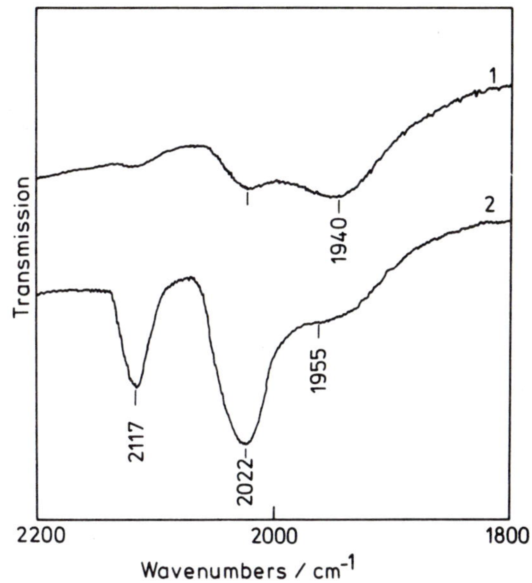

Fig. 14. Carbonyl infrared spectra of an $Os(CO)_5$-derived Al_2O_3-supported osmium catalyst after adsorption of CO at room temperature (1) and after exposure to CO at 605 K for 5 h (2) [47].

broad bands centered around 2020 and 1940 cm^{-1}. These bands are assigned as carbonyl stretching bands of linear and bridge-bonded CO, respectively, on the metal aggregates. Exposure of the sample to CO at 605 K resulted in spectrum 2, which now shows two bands at 2117 and 2022 cm^{-1} and a broad shoulder near 1955 cm^{-1}. The two sharper bands closely resemble those of mononuclear tricarbonyl species (Table 9). The increased temperature required to produce these species may suggest a disintegration of small aggregates (or dissociation of edge atoms from larger aggregates) and oxidation of the isolated atoms with the assistance of CO. In this case the interpretation is still speculative.

Metal particle disruption in a CO environment was demonstrated very convincingly for ultra-dispersed Rh aggregates supported on alumina [142, 143]. The Rh aggregates were characterized by EXAFS; they were extremely small, the Rh atoms having a low average coordination number of 3.7. Temperataure-programmed reduction experiments showed that rhodium was not dispersed to a significant degree as isolated Rh^+ ions. When this sample was exposed to CO at room temperature, only two infrared carbonyl bands (at 2095 and 2023 cm^{-1}) were observed; these are assigned as the symmetric and asymmetric stretching modes of a $Rh^+(CO)_2$ unit (Table 9). The oxidation state of rhodium was confirmed by XPS, and the absence of isolated Rh^0 atoms was demonstrated by ESR. EXAFS results indicated that after CO admission the Rh-Rh distances had increased somewhat, with the appearance of a large spread; in other words, the small metal aggregates expanded under the influence of CO. Since this process is reversible after desorption of CO at elevated temperatures, the metal aggregates were described as "breathing particles" [143, 144]. The oxidation of rhodium to form $Rh^+(CO)_2$ units was described by two possible alternative routes [142]:

(1) Rhodium is oxidized by hydroxyl groups of the alumina support with evolution of H_2:

$$2\ Rh + 4CO + 2Al\text{-}OH \longrightarrow H_2 + 2Al\text{-}O\text{-}Rh^+(CO)_2 \tag{2}$$

(2) Alternatively, rhodium is oxidized by a dissociatively adsorbed CO molecule:

$$CO + 3Rh \longrightarrow RhC + 2Rh^+ + O^{2-} \tag{3}$$

$$2Rh^+ + 4CO \longrightarrow 2Rh^+(CO)_2 \tag{4}$$

If graphitic carbon would be formed ($n\ RhC \longrightarrow nRh + C_n$), a large portion of the rhodium present in the catalyst could be oxidized.

Disintegration of highly dispersed rhodium particles on alumina in the presence of CO was also demonstrated by low-temperature infrared spectroscopy [147]. When the freshly reduced catalyst was exposed to CO at 80 K, a single broad band at 2055 cm^{-1} was observed. This band is characteristic of CO chemisorbed terminally on metallic rhodium particles or aggregates [50]. After warming of the sample to 295K, the two bands characteristic of the Rh^+ dicarbonyl complex at 2085 and 2015 cm^{-1} appeared (Table 9); the results indicate the existence of small metal particles after reduction and their disintegration in a CO atmosphere near room temperature.

6.2.3.g **Metal Clusters in Zeolites**

Vibrations of zeolite frameworks occur between 250-400 cm^{-1} [148], while those of metal atoms, metal ions, and metal clusters immobilized within the framework are expected between 250 and 30 cm^{-1} [149-151].

Far infrared Fourier transform spectroscopy (FIR-FTS) is therefore extremely informative regarding metal atom/ion locations and cluster formation in zeolitic frameworks. This has recently been demonstrated by Ozin and coworkers [152]. As an example, results reported for Ag-exchanged zeolites will be briefly summarized. The tri-silver redox cycle:

$$Ag_3^{2+} \underset{O_2}{\overset{H_2}{\rightleftharpoons}} Ag_3^{o}$$

is well known (see Chapter 8). This redox couple could be observed directly by FIR-FTS, and two band pairs at 149 and 110 cm^{-1} and at 183 and 91 cm^{-1} could be assigned as the infrared-active stretching and deformation modes of linear Ag_3^{2+} and Ag_3^{o} clusters, respectively, which are located within the β-cage of the partially exchanged Ag_6Na_6A-zeolite [152, 153]. For the fully exchanged red form of $Ag_{12}A$-zeolite, bands at 150 and 105 cm^{-1} were observed and attributed to a Ag_3^{x+} cluster in the β-cage. Weakly interacting Ag_3^{x+} clusters in the β-cage (maximum four clusters) were produced by intrazeolitic autoreduction of α-cage Ag^+ ions and migration into the β-cage. These coupled tri-silver clusters gave rise to four bands in the far infrared spectrum at 168, 153, 110, and 63 cm^{-1} [154]. Group theoretical considerations suggested an assignment in terms of either D_{2d} or D_{2h} $(Ag_3^{x+})_2$ clusters, although this assignment was not entirely conclusive. Therefore, the authors regard the observed spectra as supportive of the formation of "some kind of $(Ag_3^{x+})_n$ silver clusters" [152].

Bands at 89, 80, and 77 cm^{-1} observed in Ag-exchanged Y and A zeolites were associated with the cage modes of site I Ag^o atoms, Ag_2^+, and Ag_3^{2+} species, respectively [152, 155]. The small red shift observed for Ag_2^+ and Ag_3^{2+} is remarkable and suggests a model involving only minor perturbation of the site I Ag^o cage mode through weak coupling with one or two Ag^+ ions in site I'. A band was observed at 133 cm^{-1} for $Ag_{55}Y$-zeolite [154]. This band was associated with Ag^{q+} β-cage clusters. Since the band position fell between that of small silver clusters (typically 200-150 cm^{-1}) and that of larger microcrystals (65 cm^{-1}), a cluster of intermediate nuclearity in the range of 5 to 13 atoms was suggested for the Ag^{q+} species. The formation of β-cage Ag_3^+, Ag_6^{y+} and Ag_6^{z+} cluster cations could be observed during H_2 reduction and Ag_m^o microcrystals of around 5 nm diameter (located on the external zeolite Y surface) could also be detected (broad absorption centered at 65 cm^{-1}). The Ag-Ag stretching frequencies tend to decrease with increasing nuclearity of the zeolite entrapped clusters [152, 154].

These most recent results demonstrate the fascinating potential of FIR-FTS for *in-situ* observation of cluster growth and degradation of catalytic materials.

REFERENCES

1. K. L. Watters and W. M. Risen, Jr., Inorg. Chim. Acta Rev., 3 (1969) 129.
2. L. M. Haines and M. H. B. Stiddard, Advan. Inorg. Chem. Radiochem., 12 (1969) 53.
3. K. Nakamoto, Infrared Spectra of Inorganic and Coordination Compounds, Wiley Interscience, New York, London, Sydney, Toronto, 1970.
4. P. S. Braterman, Metal Carbonyl Spectra, Academic Press, London, New York, San Francisco, 1975.
5. D. K. Huggins, N. Flitcroft and H. D. Kaesz, Inorg. Chem., 4 (1965) 166.
6. C. O. Quicksall and T. G. Spiro, Inorg. Chem., 7 (1968) 2365.
7. G. A. Battiston, G. Bor, U. K. Dietler, S. F. A. Kettle, R. Rossetti, G. Sbrignadello and P. L. Stanghellini, Inorg. Chem., 19 (1980) 1961.
8. S. F. A. Kettle and P. L. Stanghellini, Inorg. Chem., 18 (1979) 2749.
9. D. M. Adams and I. D. Taylor, J. Chem. Soc. Faraday II, 78 (1982) 1561.
10. E. W. Abel, P. J. Hendra, R. A. N. McLean and M. M. Quarishi, Inorg. Chim. Acta, 3 (1969) 77.
11. D. M. Adams and I. D. Taylor, J. Chem. Soc. Faraday II, 78 (1982) 1573.
12. C. O. Quicksall and T. G. Spiro, Inorg. Chem., 8 (1969) 2011.
13. W. Beck and K. Lottes, Chem. Ber., 94 (1961) 2578.
14. P. Corradini, J. Chem. Phys., 31 (1959) 1676.
15. H. S. Wei, Inorg. Chem., 8 (1969) 2384.
16. G. Bor, Spectrochim. Acta, 19A (1963) 1209.
17. W. P. Griffith and A. J. Wickham, J. Chem. Soc. A (1969) 834 .
18. F. Cariati, V. Valenti and P. Barone, Gazz. Chim. Ital., 99 (1969) 1327.
19. E. L. Muetterties and J. Stein, Chem. Rev., 79 (1979) 479.
20. E. L. Muetterties, Catal. Rev.-Sci. Eng., 23 (1981) 69.
21. E. L. Muetterties and M. J. Krause, Angew. Chem., 95 (1983) 135; Int. Ed. Engl., 22 (1983) 135.
22. W. A. Herrmann, M. L. Ziegler, K. Weidenhammer and H. Biersack, Angew. Chem., 91 (1979) 1026; Int. Ed. Engl., 18 (1979) 960.
23. J. R. Andrews, K. A. Jayasooriya, I. A. Oxton, D. B. Powell, and N. Sheppard, Inorg. Chem., 19 (1980) 3033.
24. H. D. Kaesz and R. B. Saillant, Chem. Rev., 72 (1972) 231.
25. S. A. R. Knox J. W. Koepke, M. A. Andrews and H. D. Kaesz, J. Am. Chem. Soc., 97 (1975) 3942.
26. J. R. Andrews, S. F. A. Kettle, D. B. Powell and N. Sheppard, Inorg. Chem., 21 (1982) 2874
27. I. O. Oxton, J. Mol. Struct., 79 (1982) 243.
28. G. F. Stuntz and J. R. Shapley, Inorg. Chem., 15 (1976) 1994.
29. A. J. Drakesmith and R. Whyman, J. Chem. Soc. Dalton Trans. (1973) 362.
30. B. Vielhaber and H. Knözinger, unpublished results.
31. A. J. Hempelman, I. A. Oxton, D. B. Powell, P. Skinner, A. J. Deeming and L. Markó, J. Chem. Soc. Faraday II, 77 (1981) 1669.
32. I. A. Oxton, Spectrochim. Acta, 38A (1982) 181.

33. I. A. Oxton, D. B. Powell, N. Sheppard, K. Burgess, B. F. G. Johnson and J. Lewis, J. Chem. Soc. Chem. Commun. (1982) 719.
34. C. A. Anson, B. T. Keiller, I. A. Oxton, D. B. Powell and N. Sheppard, J. Chem. Soc. Chem. Commun., (1983) 470.
35. I. A. Oxton, D. B. Powell, P. Skinner, L. Markó, and H. Werner, Inorg. Chim. Acta, 47 (1981) 177.
36. G. Bor and P. L. Stanghellini, J. Chem. Soc. Chem. Commun. (1979) 886.
37. S. W. Ulmer, P. M. Skarstadt, J. M. Burlitch and R. E. Hughes, J. Am. Chem. Soc., 95 (1973) 4469.
38. M. Y. Darensbourg, D. J. Darensbourg, D. Burns and D. A. Drew, J. Am. Chem. Soc., 98 (1976) 3127.
39. A. Wong, M. Harris and J. D. Atwood, J. Am. Chem. Soc., 102 (1980) 4529.
40. J. S. Kristoff and D. F. Shriver, Inorg. Chem., 13 (1974) 499.
41. D. F. Shriver, J. Organomet. Chem., 94 (1975) 259.
42. T. L. Brown, J. Mol. Catal., 12 (1981) 41.
43. C. Tessier-Youngs, F. Correa, D. Pioch, R. L. Burwell, Jr., and D. F. Shriver, Organometallics, 2 (1983) 898.
44. Hsu-Kun Wang, H. W. Choi and E. L. Muetterties, Inorg. Chem., 20 (1981) 2661.
45. H. W. Choi and E. L. Muetterties, Inorg. Chem., 20 (1981) 2664.
46. D. F. Shriver, in P. C. Ford (Ed.), Catalytic Activation of Carbon Monoxide, ACS Symp. Ser., 152 (1981) 1.
47. E. O. Odebunmi, Y. Zhao, H. Knözinger, B. Tesche, W. H. Manogue, B. C. Gates and J. Hulse, J. Catal., 86 (1984) 95.
48. R. E. Stimson and D. F. Shriver, Inorg. Chem., 19 (1980) 1141.
49. S. F. A. Kettle, Topics Current Chem., 71 (1977) 111.
50. N. Sheppard and T. T. Nguyen, Advan. Infrared Raman Spectrosc., 5 (1978) 67.
51. L. H. Little, Infrared Spectroscopy of Adsorbed Species, Academic Press, London, New York, San Francisco, 1966.
52. M. L. Hair, Infrared Spectroscopy in Surface Chemistry, Dekker, New York, 1967.
53. H. Knözinger, Acta Cient. Venez. 24, Supl. 2 (1973) 76.
54. A. V. Kiselev and V. I. Lygin, Infrared Spectra of Surface Compounds, J. Wiley, New York, Toronto, 1975.
55. W. N. Delgass, G. L. Haller, R. Kellerman and J. H. Lunsford, Spectroscopy in Heterogeneous Catalysis, Academic Press, New York, San Francisco, London, 1979.
56. N. Sheppard, in R. F. Willis, Ed., Vibrational Spectroscopy of Adsorbates, Springer Verlag, Berlin, Heidelberg, New York, 1980.
57. G. L. Haller, Catal. Rev.-Sci. Eng., 23 (1981) 477.
58. N. Sheppard, J. Mol. Struct., 80 (1982) 163.
59. J. B. Peri, in J. R. Anderson and M. Boudart, Eds., Catalysis-Science and Technology, Vol. 5, Springer Verlag, Berlin, Heidelberg, New York, Toronto, 1984, p. 171.
60. G. Zundel, Hydration and Intermolecular Interaction, Academic Press, New York, London, 1969.
61. A. T. Bell, ACS Symp. Ser., 137 (1980) 13.
62. C. E. Jones and K. Klier, Ann. Rev. Materials Sci., 2 (1972) 1.
63. N. J. Harrick, Internal Reflection Spectroscopy, Interscience Publ., New York, 1967.
64. K. Klier, Catal. Rev.-Sci. Eng., 1 (1967) 207
65. K. Klier, ref. [61], p. 141.
66. T. A. Egerton and A. H. Hardin, Catal. Rev.-Sci. Eng. 11 (1975) 1.
67. R. P. Cooney, G. Curthoys and T. T. Nguyen, Advan. Catal., 24 (1975) 293.
68. T. Takenaka, Advan. Coll. Interface Sci., 11 (1979) 291.

69. P. J. Hendra and M. Fleischmann, in E. Kay and P. S. Bagus (Eds.), Topics in Surface Chemistry, Plenum Press, New York, 1978.
70. B. A. Morrow, ACS Symp. Ser., 137 (1980) 119.
71. F. R. Brown, L. E. Makovsky and K. H. Rhee, Appl. Spectrosc., 31 (1977) 563.
72. C. P. Cheng, J. D. Ludowise and G. L. Schrader, Appl. Spectrosc., 34 (1980) 146.
73. H. Jeziorowski, B. Vielhaber and H. Knözinger, unpublished.
74. J. J. Freeman, J. Heaviside, P. J. Hendra, J. Prior and E. W. Reid, Appl. Spectrosc., 35 (1981) 196.
75. H. Jeziorowski and H. Knözinger, Chem. Phys. Letters, 51 (1977) 519.
76. H. Jeziorowski and H. Knözinger, Chem. Phys. Letters, 42 (1976) 162.
77. M. Cardona, in F. Seitz, D. Turnbull and H. Ehrenreich (Eds.), Solid State Physics, Suppl. 2, Academic Press, New York, 1969.
78. F. L. Galeener, Chem. Phys. Letters, 48 (1977) 7.
79. S. Bruckner, H. Jeziorowski and H. Knözinger, Chem. Phys. Letters, 105 (1984) 218.
80. W. Kiefer, Advan. Infrared Raman Spectrosc., 3 (1977) 1.
81. L. H. Dubois, P. K. Hansma and G. A. Somorjai, Appl. Surface Sci., 6, (1980) 173.
82. W. H. Weinberg, Ann. Rev. Phys. Chem., 29 (1978) 115.
83. D. G. Walmsley, ref. [56], p. 67.
84. J. Kirtley, ACS Symp. Ser., 137 (1980) 217.
85. R. M. Kroeker and P. K. Hansma, Catal. Rev.-Sci. Eng., 23 (1981) 553.
86. P. K. Hansma, J. Electron Spectrosc. Rel. Phenom., 30 (1983) 163.
87. J. Lambé and R. C. Jaklevic, Phys. Rev., 165 (1968) 821.
88. R. C. Jaklevic and M. R. Gaerttner, Appl. Surface Sci., 1 (1978) 479.
89. D. G. Walmsley, W. J. Nelson, N. M. D. Brown and R. B. Floyd, Appl. Surface Sci., 5 (1980) 107.
90. W. H. Weinberg, in J. R. Durig (Ed), Vibrational Spectra and Structure, 11 (1982) 1.
91. J. Howard and T. C. Waddinton, Appl. Surface Sci., 2 (1978) 102.
92. C. J. Wright, ref. [56], p. 111.
93. A. J. Renouprez, P. Fouilloux, G. Coudurier, D. Tochetti and R. Stockmeyer, J. Chem. Soc. Faraday, I 73 (1977) 1.
94. D. Graham, J. Howard and T. C. Waddington, J. Chem. Soc. Faraday I, 79 (1983) 1281.
95. R. D. Kelley, R. R. Cavanagh and J. J. Rush, J. Catal., 83 (1983) 464.
96. J. F. McClelland, Anal. Chem., 55 (1983) 89A.
97. T. J. Chuang, H. Coufal and F. Trager, J. Vac. Sci. Technol., A1 (1983) 1236.
98. J. B. Kinney and R. H. Staley, Anal. Chem., 55 (1983) 343.
99. M. J. D. Low and G. A. Parodi, Appl. Spectrosc., 34 (1980) 76.
100. M. J. D. Low and M. Lacroix, Infrared Phys., 22 (1982) 139.
101. S. O. Kanstad and P. E. Nordal, Appl. Surface Sci., 6 (1980) 372.
102. M. J. D. Low, C. Morterra and A. G. Severdia, J. Mol. Catal., 20 (1983) 311.
103. A. Choplin, M. Leconte and J. M. Basset, J. Mol. Catal., 21 (1983) 389.
104. J. R. Moss and A. G. Graham, J. Organomet. Chem., 23 (1970) C 23.
105. G. Collier, D. J. Hunt, S. D. Jackson, R. B. Moyes, I. A. Pickering, P. B. Wells, A. F. Simpson and R. Whyman, J. Catal., 80 (1983) 154.
106. W. M. Bowser and W. H. Weinberg, J. Am. Chem. Soc., 102 (1980) 4720.
107. J. R. Kirtley and P. K. Hansma, Phys. Rev. B, 13 (1976) 2910.
108. J. Lieto, M. Wolf, B. A. Matrana, M. Prochazka, B. Tesche, H. Knözinger and B. C. Gates, J. Phys. Chem., 89 (1985) 991.

109. T. Catrillo, H. Knözinger and M. Wolf, Inorg. Chim. Acta, 45 (1980) L 235.
110. G. R. Crooks, B. F. G. Johnson, J. Lewis and I. G. Williams, J. Chem. Soc., A, (1969) 797 .
111. E. G. Bryan, B. F. G. Johnson and J. Lewis, J. Chem. Soc. Dalton Trans. (1977) 1328.
112. M. Wolf, J. Lieto, B. A. Matrana, D. B. Arnold, B. C. Gates and H. Knözinger, J. Catal., 89 (1984) 100.
113. R. Ugo, R. Psaro, G. M. Zanderighi, J. Basset, A. Theolier and A. K. Smith, in M. Tsutsui, Ed., Fundamental Research in Homogeneous Catalysis, Vol. 3, Plenum Press, New York, London, 1979, p. 579.
114. A. K. Smith, B. Besson, J. M. Basset, R. Psaro, A. Fusi and R. Ugo, J. Organomet. Chem., 192 (1980) C 31.
115. B. Besson, B. Moraweck, A. K. Smith, J. M. Basset, R. Psaro, A. Fusi and R. Ugo, J. Chem. Soc. Chem. Commun. (1980) 569.
116. M. Deeba and B. C. Gates, J. Catal., 67 (1981) 303.
117. K. A. Azam, A. J. Deeming, R. F. Kimber and P. R. Shukla, J. Chem. Soc. Dalton Trans. (1976) 1853.
118. B. F. G. Johnson and J. Lewis, J. Chem. Soc. A (1968) 2859.
119. M. Deeba, B. J. Streusand, G. L. Schrader and B. C. Gates, J. Catal., 69 (1981) 218.
120. L. J. Hilliard and H. S. Gold, Appl. Spectrosc., 39 (1985) 124.
121. A. Theolier, A. Choplin, L. d'Ornelas, J. M. Basset, G. Zanderrighi, R. Ugo and C. Sourisseau, Polyhedron, 2 (1983) 119.
122. J. Evans and G. S. McNulty, J. Chem. Soc. Dalton Trans. (1984) 1123.
123. J. Evans and G. S. McNulty, J. Chem. Soc. Dalton Trans. (1984) 587.
124. D. Ballivet-Tkatchenko and G. Coudurier, Inorg. Chem., 18 (1979) 558.
125. J. Evans, Chem. Soc. Rev., 10 (1981) 159.
126. K. Lázár, K. Matusek, J. Mink, S. Dobos, L. Guczi, A. Vizi-Orosz, L. Markó and W. M. Reiff, J. Catal., 87 (1984) 163.
127. F. Hugues, J. M. Basset, Y. Ben Taarit, A. Choplin, M. Primet, D. Rojas and A. K. Smith, J. Am. Chem. Soc., 104 (1982) 7020.
128. R. Barth, B. C. Gates, Y. Zhao, H. Knözinger and J. Hulse, J. Catal., 82 (1983) 147.
129. B. Besson, A. Choplin, L. d'Ornelas and J. M. Basset, J. Chem. Soc. Chem Commun. (1982) 843; J. M. Basset, B. Besson, A. Choplin and A, Theolier, Phil. Trans. R. Soc. London A, 308 (1982) 115.
130. A. K. Smith, F. Hugues, A. Theolier, J. M. Basset, R. Ugo, G. M. Zanderighi, J. L. Bilhou, V. Bilhou-Bougnol and W. F. Graydon, Inorg. Chem., 18 (1979) 3104.
131. K. L. Watters, R. F. Howe, T. P. Chojnacki, C. M. Fu, R. L. Schneider and N. B. Wong, J. Catal., 66 (1980) 424.
132. V. L. Kuznetsov, A. V. Romanenko, A. P. Shepelin, P. A. Zhdan and Yu. I. Yermakov, React. Kinet. Catal. Letters, 22 (1983) 119.
133. H. Knözinger, E. W. Thornton, and M. Wolf, J. Chem. Soc. Faraday I, 75 (1979) 1888.
134. V. L. Kuznetsov, A. T. Bell and Yu. I. Yermakov, J. Catal., 65 (1980) 374.
135. A. Zecchina, E. Guglielminotti, A. Bossi and M. Camia, J. Catal., 74 (1982) 225.
136. E. Guglielminotti, A. Zecchina, A. Bossi and M. Camia, J. Catal., 74 (1982) 240.
137. H. Knözinger and Y. Zhao, J. Catal., 71 (1981) 337.
138. R. Psaro, C. Dossi and R. Ugo, J. Mol. Catal., 21 (1983) 331.
139. B. N. J. Persson and R. Ryberg, Phys. Rev. B, 24 (1981) 6954.
140. F. J. C. M. Toolenaar G. J. van der Poort, F. Stoop and V. Ponec, J. Chim. Phys., 78 (1981) 927.
141. R. G. Greenler D. R. Snider, D. Witt and R. S. Sorbello, Surface Sci., 118 (1982) 415.

142. a. H. F. J. van't Blik, PhD Thesis, Technical University of Eindhoven, 1984.
b. H. F. J. van't Blik, J. B. A. D. van Zon, T. Huizinga, J. C. Vis, D. C. Koningsberger and R. Prins, J. Phys. Chem., 87 (1983) 2264.
143. H. F. J. van't Blik, J. B. A. D. van Zon, D. C. Koningsberger and R. Prins, J. Mol. Catal., 25 (1984) 379.
144. D. J. C. Yates, L. L. Murrell and E. B. Prestridge, in J. Bourdon, Ed., Growth and Properties of Metal Clusters, Elsevier, Amsterdam (1980) p. 137.
145. T. R. Gilson and J. Evans, J. Chem. Soc. Dalton Trans. (1984) 155.
146. S. I. Woo and C. G. Hill, Jr., J. Mol. Catal., 24 (1984) 165.
147. M. Zaki, G. Kunzmann, B. C. Gates and H. Knözinger, to be published.
148. E. M. Flanigen, H. A. Szymanksi and H. Khatami, Advan. Chem. Ser., 101 (1971) 201.
149. C. Penker, K. Moeller and D. Kunath, J. Mol. Struct., 114 (1984) 215.
150. T. Stock, D. Dombrowski, J. Fruwert and H. Ratajczak, J. Chem. Soc. Faraday, I 79 (1983) 2773.
151. C. A. Ozin, M. D. Baker and J. M. Parnis, Angew. Chem., 95 (1983) 813.
152. M. D. Baker, G. A. Ozin and J. Godber, Catal. Rev.-Sci. Eng., in press.
153. M. D. Baker, G. A. Ozin and J. Godber, J. Phys. Chem., submitted.
154. M. D. Baker, J. Godber and G. A. Ozin, J. Phys. Chem., submitted.
155. M. D. Baker, G. A. Ozin and J. Godber, J. Phys. Chem., submitted.

Section 6.3

OPTICAL SPECTROSCOPY

H. KNÖZINGER

6.3.1 UV-VISIBLE SPECTRA OF MOLECULAR CLUSTERS

Most metal clusters are highly colored [1] and show rich UV-visible spectra. As pointed out by Chini [2], the electronic spectra can be reasonably assumed to reflect the separation of frontier orbitals. Generally the absorption bands of cluster compounds are shifted to longer wavelengths (lower energy) on increasing the cluster nuclearity. This trend is clearly demonstrated by the data for iron and rhodium carbonyl cluster anions in Table 1. The limiting case can be considered [2] to be that of the continuous absorption of bulk metals associated with the high delocalization of bonding electrons.

As the observed absorption bands are likely to be related to metal-metal bonds, their positions may reflect the metal-metal bond strength. The metal-metal orbital interaction should increase as one passes from first-row to second- and third-row elements, which would cause increasing bonding-antibonding energy splitting, and consequently a blue shift of the respective electronic transitions. This trend is clearly shown by the data in Table 2. The positions of the electronic absorption bands of monometallic carbonyl clusters having comparable structures [e.g., $Ru_3(CO)_{12}$ and $Os_3(CO)_{12}$ or $Co_4(CO)_{12}$ and $Rh_4(CO)_{12}$] shift to shorter wavelength with descending position in the same transition metal group in the periodic table. Similarly, λ_{max} decreases as increasing numbers of second-row atoms are incorporated from first-row parent clusters [compare, for example, the $Fe_{3-x}Ru_x(CO)_{12}$ or $Co_{4-x}Rh_x(CO)_{12}$ series].

Theoretical treatments and detailed assignments of electronic transitions in cluster compounds are scarce. Levenson and Gray [7] reported a semiempirical molecular orbital calculation for dinuclear $M_2(CO)_{10}$ clusters (M = Mn, Tc, Re). The calculation was consistent with the assignment of electronic absorption bands in the range 300-340 nm to transitions between σ and σ^* orbitals associated with the metal-metal bond. The position of the band was sensitive to the metal-metal bond strength. Intense absorptions in the 200-300-nm range could be attributed to $\sigma \longrightarrow \pi^*(CO)$ and $d\pi \longrightarrow \pi^*(CO)$ metal-to-ligand charge transfer (MLCT) transitions.

Abrahamson *et al.* [8] carried out an analogous study of dinuclear cobalt carbonyl complexes, particularly $Co_2(CO)_8$, which is known [9] to exist in solution as a mixture of bridged and nonbridged isomers. Absorption bands

TABLE 1. Colors and Absorption Band Maxima of Carbonyl Cluster Anions: Dependence on Cluster Nuclearity.

Cluster	Color	λ max/nm	Ref.
$[Fe(CO)_4]^{2-}$	Colorless	300	[3]
$[Fe_2(CO)_8]^{2-}$	Orange	347	[3]
$[Fe_3(CO)_{11}]^{2-}$	Red	485	[3]
$[Fe_4(CO)_{13}]^{2-}$	Brown	500	[3]
$[Rh(CO)_4]^-$	Colorless	—	[4]
$[Rh_3(CO)_{10}]^-$	Yellow	—	[4]
$[Rh_6(CO)_{14}]^{4-}$	Red-brown	—	[4]
$[Rh_7(CO)_{16}]^{3-}$	Dark green	—	[4]
$[Rh_{12}(CO)_{30}]^{2-}$	Violet	—	[4]

at 280 and 350 nm could be assigned as $\sigma \longrightarrow \sigma^*$ transitions in these two isomers, respectively.

Phosphine substitution led to a decrease in energy for the $\sigma \longrightarrow \sigma^*$ transitions for the $Mn_2(CO)_{10}$ [7] and $Co_2(CO)_8$ dimers [8].

An extended Hückel molecular orbital calculation for D_{3h} $Ru_3(CO)_{12}$ was performed by Tyler *et al.* [6]. Molecular orbital energy levels for $Ru(CO)_4$ were calculated; the $Ru(CO)_4$ fragments were combined to form $Ru_3(CO)_{12}$. The energy levels of the metal orbitals in the cluster as they develop on combination of the $Ru(CO)_4$ fragments are shown in Fig. 1. The assignments given in Table 2 are based on these calculations; with the following notation being used [6]: the e'(xz) bonding orbital is σ, whereas the $a_2'(xz)$ antibonding orbital is σ^* (this orbital being the LUMO in $Ru_3(CO)_{12}$); the $a_1'(z^2)$ bonding orbital is σ' and the $e'(z^2)$ antibonding orbital is $\sigma^{*\prime}$. The band at 390 nm in the spectrum of $Ru_3(CO)_{12}$ could be assigned to the $^1A_1' \longrightarrow {}^1E'$ (xz [bonding] $\longrightarrow$ xz [antibonding]), or

TABLE 2. Electronic Absorption Band Maxima of Tri- and Tetra-nuclear Carbonyl Clusters: Dependence on Metal-Metal Bond Strength.[a]

Cluster	λ_{max}/nm	Assignment	Ref.
$Fe_3(CO)_{12}$	597		[5]
	602	$\sigma^{*\prime} \longrightarrow \sigma^*$	
	437 sh	$\sigma \longrightarrow \sigma^*$	[6]
$Ru_3(CO)_{12}$	392		[5]
	390	$\sigma \longrightarrow \sigma^*$	
	320 sh	$\sigma^{*\prime} \longrightarrow \sigma^*$	[6]
$Os_3(CO)_{12}$	385 sh	$\sigma^{*\prime} \longrightarrow \sigma^*$	[6]
	330	$\sigma^* \longrightarrow \sigma^*$	[6]
$Fe_2Ru(CO)_{12}$	548		[5]
$FeRu_3(CO)_{12}$	476		[5]
	470 sh	$\sigma^{*\prime} \longrightarrow \sigma^*$	
	390	$\sigma \longrightarrow \sigma^*$	[6]
$Mn_2Fe(CO)_{14}$	432	$\sigma \longrightarrow \sigma^*$	[6]
$MnReFe(CO)_{14}$	403	$\sigma \longrightarrow \sigma^*$	[6]
$Re_2Fe(CO)_{14}$	380	$\sigma \longrightarrow \sigma^*$	[6]
$Co_4(CO)_{12}$	375	—	[2]
$Co_3Rh(CO)_{12}$	365	—	[2]
$Co_2Rh_2(CO)_{12}$	348	—	[2]
$Rh_4(CO)_{12}$	300	—	[2]
$Rh_3Ir(CO)_{12}$	300	—	[2]
$Ir_4(CO)_{12}$	319	—	[2]

[a]Spectra recorded at room temperature

$\sigma \longrightarrow \sigma^*$ transition. The weaker band at 320 nm was assigned to the $^1A'_1 \longrightarrow {}^1E'$ (z^2 [antibonding] $\longrightarrow$ xz[antibonding]), or $\sigma^{*\prime} \longrightarrow \sigma^*$ transition.

Schilling and Hoffmann [10] calculated the MO scheme of $Fe_3(CO)_{12}$, which was analogously built up from three $Fe(CO)_4$ fragments.

MLCT transitions were observed below 300 nm for the trinuclear clusters [6]: the intense bands at 263 nm in $Fe_3(CO)_{12}$ and at 238 nm

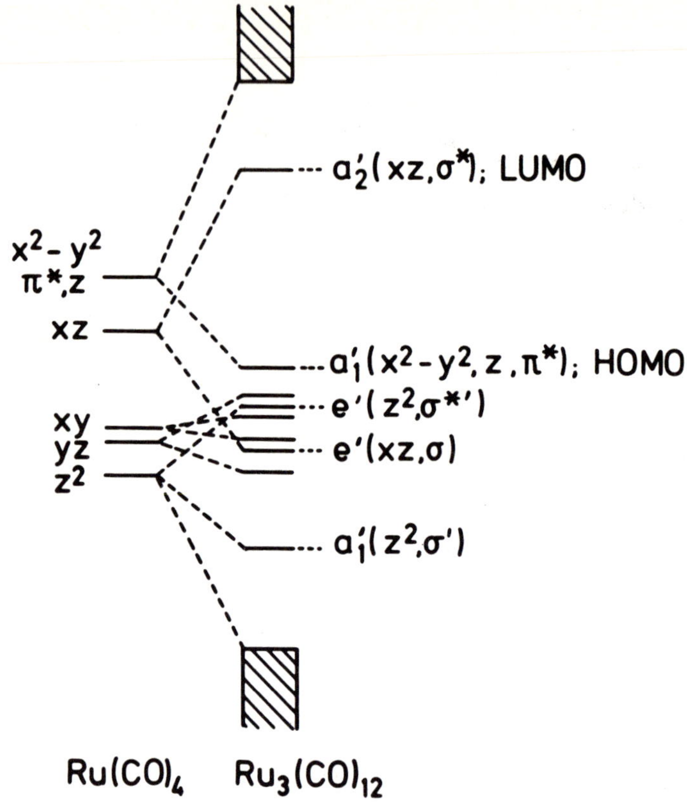

Fig. 1. Molecular orbital scheme for $Ru_3(CO)_{12}$, constructed by combination of $Ru(CO)_4$ molecular orbitals [6].

in $Ru_3(CO)_{12}$ were assigned to MLCT transitions by comparison with the MLCT transitions in $Fe(CO)_5$ and $Ru(CO)_5$, respectively. By analogy, the band at 240 nm in the spectrum of $Os_3(CO)_{12}$ was attributed to a similar MLCT transition.

Phosphine substitution on the triruthenium cluster led to the appearance of an absorption band near 500 nm for $Ru_3(CO)_9(PR_3)_3$ [6]. This band was assigned as the $\sigma^{*\prime} \longrightarrow \sigma^*$ transition, which is observed at 320 nm for the unsubstituted cluster. The shift to lower energy of this transition was explained by the fact that phosphine substitution occurs at a site that is in the Ru_3 plane and, hence, primarily affects the energy of the xz orbital. Replacement of CO by phosphine makes this orbital less antibonding and, consequently, the energy difference between xz and z^2 orbitals will be lower in the $Ru(CO)_3(PR_3)$ than in the $Ru(CO)_4$ fragment. This then results in a z^2 antibonding orbital $\sigma^{*\prime}$ which is higher in energy than the xz bonding orbital. Therefore, the $\sigma^{*\prime} \longrightarrow \sigma^*$ transition will be red shifted on going from $Ru_3(CO)_{12}$ to $Ru_3(CO)_9(PR_3)_3$.

Tetranuclear hydrido clusters of the iron triad absorb in the 300-400 nm wavelength range [11]; λ_{max} values for mono- and multimetallic clusters are summarized in Table 3. This series of data again demonstrates the shift of the absorption band to higher energy as the metal-metal bond

TABLE 3. Colors and Absorption Band Maxima of Tetrahedral Hydrido Clusters of the Iron Triad [11].

Cluster	Color	λ_{max}/nm
$H_2FeRu_3(CO)_{13}$	Red	344, 383 sh
$H_2FeRu_2Os(CO)_{13}$	Orange-red	312, 374
$H_2FeRuOs_2(CO)_{13}$	Orange	305 sh, 364
$H_2Fe_2Ru_2(CO)_{13}$	Red	-
$H_2FeOs_3(CO)_{13}$	Yellow-orange	-
$H_2Ru_4(CO)_{13}$	Red	-
$H_4Ru_4(CO)_{12}$	Yellow	360
$H_4Os_4(CO)_{12}$	Pale yellow	314

strength increases with incorporation of more third-row character. Hoffman *et al.* [12] analyzed the electronic structure of tetranuclear hydrido clusters $M_4(CO)_{12}H_n$ and $M_4Cp_4H_n$ (n = 3, 4, and 6) with the aid of symmetry arguments and molecular orbital calculations.

Schilling and Hoffmann [10] have shown by molecular orbital calculations that $M_3(CO)_9$ fragments of C_{3v} symmetry provide three axial acceptor orbitals of the cyclopropenium type which are ideally set up to interact with a capping ligand. The electronic and geometrical structures of various clusters of this type, such as M_3L_9X [where X is CO, S, $(CO)_2$, S_2, CR, CCH_2^+, ethylene, acetylene, or acyclic and cyclic polyenes] were discussed by these authors.

Geoffroy and Epstein [13] analyzed the electronic spectra of a series of $YCCo_3(CO)_9$ clusters which are all very similar in color regardless of the nature of the Y group; λ_{max} values for these compounds (measured at 300 K) are summarized in Table 4, which, for comparison, also contains spectral data for phosphine-substituted compounds, $HFeCo_3(CO)_{12}$, and $Na[Co_3(CO)_{10}]$. These data clearly indicate that the nature of the apical group perturbs only slightly the low-lying electronic transitions in these clusters. Therefore, the orbitals involved in these transitions should be localized primarily within the Co_3 framework. Molecular orbital considerations [10, 14] support this conclusion.

A summary of the λ_{max} values for a few triosmium-derived clusters, which are of interest as reference compounds for supported clusters, is given in Table 5.

Little is known about electronic transitions in carbonyl cluster compounds in the vacuum ultraviolet region, since this spectral region is not easily accessible experimentally. Netzer *et al.* [17] investigated $Ir_4(CO)_{14}$ by means of *electron-energy loss spectroscopy* (EELS) and observed loss peaks at 7.5 and 14.2 eV, corresponding to 60,495 and 114,537 cm^{-1}, respectively. The lower energy loss at 7.5 eV could be unambiguously assigned as a d $\longrightarrow$ $2\pi^*$(CO) transition; the higher energy loss was thought to be rather more intrinsic to the CO ligand; it was attributed (in one-electron terminology) to excitation of 1π or 5σ electrons to excited $2\pi^*$ states. Loss features at similar energies were observed for CO adsorbed on the Ir(111) face. A close correspondence between the electronic properties of polynuclear metal carbonyl clusters and strongly adsorbed molecular CO on metal surfaces was established (see Chapter 11), and significant distinctions between polynuclear and mononuclear carbonyls were emphasized. Relations between these electronic excitations and XPS shake-up satellites (see section 6.5) were discussed.

6.3.2 DIFFUSE REFLECTANCE SPECTROSCOPY

Electronic spectra of supported clusters cannot be obtained by the transmission technique since scattering losses are too severe with disperse high-surface area supports in the visible and ultra-violet spectral region. Diffuse reflectance spectroscopy can be applied to this type of material. The technique has been reviewed by Kortüm [18] and by Wendlandt and Hecht [19] and is discussed in several other articles [20-22].

TABLE 4. Absorption Band Maxima (300 K) of Capped Tricobalt Clusters [13].

Cluster	λ_{max}/nm
$HCCo_3(CO)_9$	505, 370sh, 290sh
$CH_3CCo_3(CO)_9$	510, 370sh, 295sh
$CH_3CCo_3(CO)_8PPh_3$	502, 400sh, 310sh
$HFeCo_3(CO)_{12}$	528, 380sh, 324sh, 280sh
$Na[Co_3(CO)_{10}]$	505sh, 371sh

TABLE 5. Absorption Band Maxima of Clusters Derived from Triosmium Clusters.

Cluster	λ_{max}/nm	Ref.
$Os_3(CO)_{12}$	330, 385sh	[6]
$H_2Os_3(CO)_{10}PPh_3$	328, 385	[15]
$HOs_3(CO)_{10}(OC_6H_5)$	312, 390	[15]
$Os_3(CO)_{10}Cl_2$	348, 410vw	[16]
$HAuOs_3(CO)_{10}PPh_3$	348, 590vw	[15]
$ClAuOs_3(CO)_{10}PPh_3$	337sh, 480	[15]

Diffuse reflectance spectroscopy can be used with samples in the form of powders or pellets, which are usually placed between two parallel surfaces such as the quartz windows of an optical cell. The front surface is illuminated by diffuse light or by a parallel beam entering at a fixed angle of incidence [18].

Assuming isotropic scattering, the reflectance R, defined as the ratio of backward and forward fluxes at the illuminated front surface, can be expressed in terms of phenomenologically defined absorption and scattering coefficients K and S, respectively, according to the Schuster-Kubelka-Munk theory [18, 19, 21, 22]:

$$R = \frac{1 + R_g[b \cdot \coth Y - a]}{a + b \cdot \coth Y - R_g} \tag{1}$$

In this equation, Y = Sbd; R_g is the reflectance of the backing; d is the thickness of the sample layer; a = 1 + K/S; and b = $\sqrt{(a^2-1)}$.

For an infinitely thick sample, $Y \longrightarrow \infty$. In the UV-visible region, d must be of the order of magnitude of 0.1 cm [18]. With these simplifications, the equation reduces to the so-called Kubelka-Munk-function:

$$F(R_\infty) = \frac{(1 - R_\infty)^2}{2R_\infty} = \frac{K}{S} \tag{2}$$

where R_∞ is the reflectance of a sample with infinite thickness. The ratio K/S is determined as a function of the wavelength, and this ratio is equivalent to the true absorption spectrum, provided that the scattering coefficient is independent of the wavelength. This condition is usually met in the UV-visible region.

The diffuse scattered radiation from the sample is collected in an integrating sphere to reach the detector. Large detector areas (about 1 cm^2) can be allowed in the UV-visible range, and highly sensitive detectors are available.

As an alternative to diffuse reflectance spectroscopy, photoacoustic spectroscopy [20, 23, 24] may be applied to supported clusters in the visible region.

6.3.3 DIFFUSE REFLECTANCE SPECTROSCOPY OF SUPPORTED CLUSTERS

Absorption band maxima λ_{max} for tri- and tetraruthenium clusters supported on alumina, silica, and functionalized silicas are shown in Table 6. $Ru_3(CO)_{12}$ physically adsorbed on a hydroxylated Al_2O_3 surface gave a band at 400 nm [25], which is close to the $\sigma \longrightarrow \sigma^*$ transition observed for this cluster in solution (Tables 2 and 6). Decarbonylation of the physisorbed cluster led to a decrease in intensity of the 400-nm band and the appearance of bands near 320 and 500 nm. This process was reversed on reexposure of the sample to CO at temperatures up to 320 K. These results were attributed [25] to a ligand displacement reaction by surface OH groups:

$$Ru_3(CO)_{12} + x(OH) \rightleftharpoons Ru_3(CO)_{12-x}(OH)_x + xCO \qquad (3)$$

This assignment was based on the comparison of the spectra of the surface species with those of the substituted cluster $Ru_3(CO)_9[P(C_2H_5)Ph_2]$ (see Section 6.3.1 and ref. 6).

When $Ru_3(CO)_{12}$ was chemically anchored onto a silica surface (by refluxing in *n*-hexane solution), bands were observed at 270sh, 325, and 438 nm [26] (Table 6). By comparison with results from vibrational spectroscopy [26, 28, 29] (Section 6.2), these electronic transitions were attributed to the surface species $(\mu_2\text{-}H)Ru_3(CO)_{10}(\mu_2\text{-}O_s)$ which is formed in an oxidative addition involving a surface silanol group. The cluster $(\mu_2\text{-}H)Ru_3(CO)_{10}\text{-}[\mu_2\text{-}S(CH_2)_2SIL]$, which is formed by reaction of $Ru_3(CO)_{12}$ with a thiol-modified SiO_2 surface (Section 9.1), gave a very similar spectrum (Table 6), with electronic transitions appearing at 288sh, 350, and 413 nm [26].

A band was observed near 360 nm when $H_4Ru_4(CO)_{12}$ was adsorbed on SiO_2 [26, 27], this being characteristic of electronic transitions associated with the metal-metal bonds in this hydrido cluster (Tables 3 and 6). As shown in Fig. 2 and Table 6, a red shift occurs when the carbonyl ligands in this cluster are replaced by phosphine ligands [26]. Doi *et al.* [27]

TABLE 6. Electronic Spectra of Supported Tri- and Tetraruthenium Clusters and a Molecular Analogue.

Cluster Precursor	Support	λ_{max}/nm	Surface Species	Ref.
Supported Clusters				
$Ru_3(CO)_{12}$	-	390, 320sh	-	[6]
	Al_2O_3	400	$Ru_3(CO)_{12}$	[25]
	SiO_2	270sh, 325, 438	$(\mu_2\text{-H})Ru_3(CO)_{10}(\mu_2\text{-}O_s)^a$	[26]
	SiO_2, thiol-modified	288sh, 350, 413	$(\mu_2\text{-H})Ru_3(CO)_{10}[\mu_2\text{-S}(CH_2)_3SIL]^b$	[26]
$H_4Ru_4(CO)_{12}$	-	350	-	[11]
	SiO_2	360	$H_4Ru_4(CO)_{12}$	[27]
	SiO_2	363, 313sh	$H_4Ru_4(CO)_{12}$	[26]
	SiO_2, phosphine modified	383, 320sh	$H_4Ru_4(CO)_{11}[PPh_2(CH_2)_2SIL]^b$	[26]
	"	400, 325sh	$H_4Ru_4(CO)_{10}(PPh_3)[PPh_2(CH_2)_2SIL]^b$	[26]
	"	413, 325	$H_4Ru_4(CO)_9(PPh_3)_2[PPh_2(CH_2)_2SIL]^b$	[26]
	"	405	$H_4Ru_4(CO)_8[PPh_2(CH_2)_2SIL]_4$	[27]
Molecular Cluster				
$H_4Ru_4(CO)_8[PPh_2(CH_2)_2Si(OC_2H_5)_3]_4$		407	-	[27]

[a] O_s indicates surface oxygen; [b] SIL indicates silica, the ligand being anchored by Si-O-Si links.

reported a transition at 405 nm for the tetra-phosphine substituted cluster. Decarbonylation by evacuation at increasing temperature of the hydrido tetraruthenium clusters anchored onto the phosphine-modified SiO_2 surface led to a gradual decrease in intensity of the characteristic electronic transitions (Fig. 3). In agreement with infrared results, this observation is reasonably explained by cluster fragmentation.

Iwasawa *et al.* [30] investigated the transformations of $Co_2(CO)_8$ supported on SiO_2. One CO molecule per Co atom was liberated, an electronic transition at 375 nm being observed, which could be interpreted as indicative of the formation of the tetranuclear cluster $Co_4(CO)_{12}$ (Table 2). This interpretation was further supported by results of a variety of other techniques.

Diffuse reflectance spectra of supported osmium clusters have been reported by two groups. The product of the reaction of $Os_3(CO)_{12}$ with a silica surface (see Section 9.2.4.4) gave bands at 309, 360, and 410 nm [31]. By comparison with the electronic spectrum of $(\mu_2\text{-H})Os_3(CO)_{10}(\mu_2\text{-OPh})$ (Table 5), the bands at 309 and 410 nm can be attributed to the surface analogue $(\mu_2\text{-H})Os_3(CO)_{10}(\mu_2\text{-O}_s)$. The relatively weak band at 360 nm remains unassigned and probably indicates the presence of a second species. The sample was lemon yellow. On

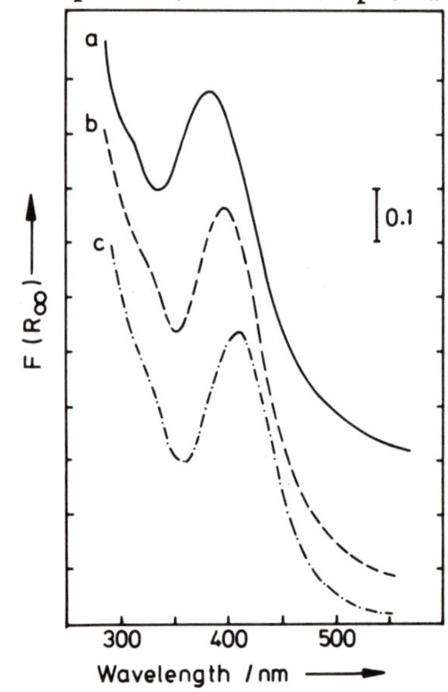

Fig. 2. Electronic spectra of phosphine-substituted $H_4Ru_4(CO)_{12}$ anchored to phosphine-modified SiO_2: a, $H_4Ru_4(CO)_{11}[PPh_2(CH_2)_2SIL]$; b, $H_4Ru_4(CO)_{10}(PPh_3)[PPh_2(CH_2)_2SIL]$; and c, $H_4Ru_4(CO)_9(PPh_3)_2$-$[PPh_2(CH_2)_2SIL]$ [26].

evacuation at 420 K for >90 h, the sample turned red; the intensity of the three bands decreased and additional weak bands appeared at 270, 390, and 470 nm. These bands were attributed to a partially decarbonylated cluster with an intact metal framework. Recarbonylation led to a new species giving rise to electronic transitions at 305 and 355 nm. This brownish-yellow species was suggested to be a cluster different from the original one, its structure remaining undetermined.

Collier *et al.* [32] and Hunt *et al.* [33] reported diffuse reflectance spectra of Al_2O_3- and SiO_2-supported catalysts prepared from $Os_3(CO)_{12}$ and $Os_6(CO)_{18}$ by impregnation from solution. The spectra of the freshly prepared samples clearly gave the most intense bands in the range characteristic of the $\sigma \longrightarrow \sigma^*$ transitions (320-340 nm, Tables 2 and 5). These authors also observed a large number of weaker bands, particularly in the region >400 nm; these bands remain unassigned. Thermal treatment of the sample led to spectral changes, namely, a gradual erosion of the features characteristic of the $\sigma \longrightarrow \sigma^*$ transitions in the parent clusters; numerous weak bands were observed in the entire wavelength range 250-900 nm. Since both $Os_6(CO)_{18}$ and $Os_{10}C(CO)_{24}^{2-}$ in solution give bands at >400 nm, these low-energy transitions in the supported species were considered as being "mostly characteristic of a clustered state of considerable nuclearity" [32]. The spectral properties of Al_2O_3- and SiO_2-supported

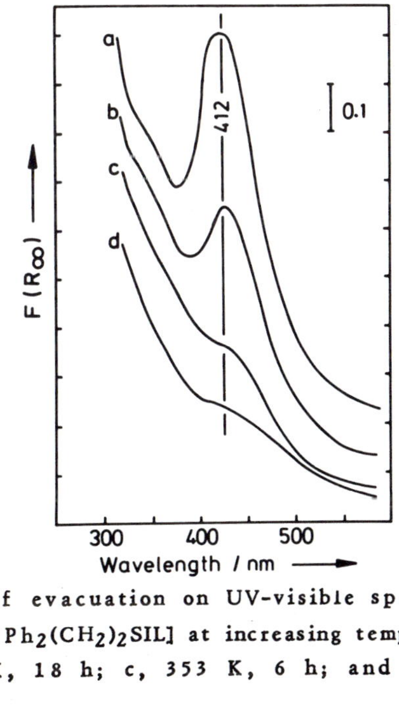

Fig. 3. Effect of evacuation on UV-visible spectrum of $H_4Ru_4(CO)_9$-$(PPh_3)_2[PPh_2(CH_2)_2SIL]$ at increasing temperatures: a, 300K, 20 h; b, 323 K, 18 h; c, 353 K, 6 h; and d, 393 K, 4 h [26].

catalysts prepared from $H_4Os_4(CO)_{12}$ are similar to those of the $Os_3(CO)_{12}$-derived materials [34].

Diffuse reflectance spectra of $Os_3(CO)_9(\mu_2\text{-}Cl)_2PPh_2(CH_2)_2SIL$ were reported by Wolf *et al.* [16]. Gold-osmium bimetallic clusters anchored onto phosphine-modified silica were also investigated [35]. Cluster fragmentation with aggregation of the gold could be followed in this case. The aggregated gold particles gave rise to an absorption band at 480 nm.

Diffuse reflectance spectroscopy in the UV-visible region has been used to characterize magnesia-supported CO hydrogenation catalysts prepared from a mononuclear precursor, $H_2Os(CO)_4$ [36]. When the catalyst, after use in a flow reactor at 548 K and 10 bar, was brought in contact with a solution of $(CH_3)_4NCl$ in isopropanol, $H_3Os_4(CO)_{12}^-$ was extracted from the surface. The remaining solid was characterized by UV-visible reflectance spectroscopy. The results, including a comparison of the solution spectrum of $Os_{10}C(CO)_{24}^{2-}$ and of its salt supported on magnesia (Fig. 4), provide a good indication (confirming the results of infrared spectroscopy) that $Os_{10}C(CO)_{24}^{2-}$ was the predominant organometallic species on the surface. These results are elaborated in Section 9.5.

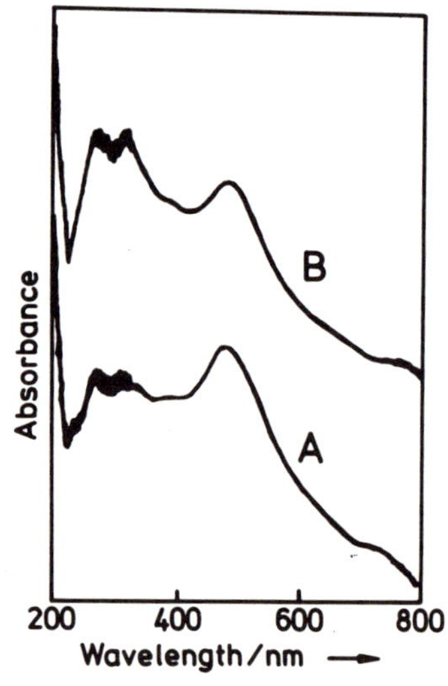

Fig. 4. Electronic spectra of A, $[(CH_3)_4NCl]_2[Os_{10}C(CO)_{24}]$ adsorbed on magnesia and B, a magnesia-supported catalyst prepared from $H_2Os(CO)_4$ after use for CO hydrogenation at 548 K and 10 bar followed by extraction with $(CH_3)_4NCl$ in isopropanol [36].

6.3.4 SUMMARY

In conclusion, although optical diffuse reflectance spectroscopy has not been applied very frequently for the characterization of supported clusters, the technique is promising, at least as a tool complementing other spectroscopies. The methods are easy to apply, and cells for *in situ* sample treatment can be constructed simply and economically. The lack of a sufficient theoretical background and of sufficiently broad collections of reference spectra imply that optical spectroscopy cannot yet be used for diagnostic purposes. However, the detection of metal-metal bonded species is certainly far easier with this method than by Raman spectroscopy (Section 6.2) or EXAFS (Section 6.7). The sensitivity of λ_{max} to cluster size, metal-metal bond strength, and substitution on the cluster framework can aid in characterizing supported clusters and cluster-derived species.

REFERENCES

1. R. B. King, Progr. Inorg. Chem., 15 (1972) 237.
2. P. Chini, Pure Appl. Chem., 23 (1970) 489.
3. W. Hieber and E. H. Schubert, Z. Anorg. Allg. Chem., 338 (1965) 32.
4. P. Chini and S. Martinengo, J. Chem. Soc. Chem. Commun. (1969) 1092.
5. D. B. W. Yawney and F. G. A. Stone, J. Chem. Soc. (A) (1969) 502.
6. D. R. Tyler, R. A. Levenson and H. B. Gray, J. Am. Chem. Soc., 100 (1978) 7888.
7. R. A. Levenson and H. B. Gray, J. Am. Chem. Soc., 97 (1975) 6042.
8. H. B. Abrahamson, C. C. Frazier, D. S. Ginley, H. B. Gray, J. Lilienthal, D. R. Tyler and M. S. Wrighton, Inorg. Chem., 16 (1977) 1554.
9. S. Onaka and D. F. Shriver, Inorg. Chem., 15 (1976) 915.
10. B. E. R. Schilling and R. Hoffmann, J. Am. Chem. Soc., 101 (1979) 3456.
11. G. L. Geoffroy and W. L. Gladfelter, J. Am. Chem. Soc., 99 (1977) 7565.
12. R. Hoffmann, B. E. R. Schilling, R. Bau, H. D. Kaesz and D. M. P. Mingos, J. Am. Chem. Soc., 100 (1978) 6088.
13. G. L. Geoffroy and R. A. Epstein, Inorg. Chem., 16 (1977) 2795.
14. C. E. Strouse and L. F. Dahl, J. Am. Chem. Soc., 93 (1971) 6032.
15. M. Wolf and H. Knözinger, unpublished results.
16. M. Wolf, J. Lieto, B. A. Matrana, D. B. Arnold, B. C. Gates and H. Knözinger, J. Catal., 89 (1984) 100.
17. F. P. Netzer, E. Bertel and J. A. D. Matthew, J. Electron Spectrosc. Rel. Phen., 18 (1980) 199.
18. G. Kortüm, Reflexionsspektroskopie, Springer, Berlin, 1969.
19. W. W. Wendlandt and H. G. Hecht, Reflectance Spectroscopy, Interscience, New York, 1966.
20. W. N. Delgass, G. L. Haller, R. Kellerman and J. H. Lunsford, Spectroscopy in Heterogenous Catalysis, Academic Press, New York, 1979.
21. W. W. Wendlandt, Ed., Modern Aspects of Reflectance Spectroscopy, Plenum, New York, 1968.
22. K. Klier, Catal. Rev., 1 (1967) 207.

23. Y. H. Pao, Optoacoustic Spectroscopy and Detection, Academic Press, New York, 1977.
24. A. Rosencwaig, Photoacoustics and Photoacoustic Spectroscopy, Wiley, Chichester, 1980.
25. A. Zecchina, E. Guglielminotti, A. Bossi and M. Camia, J. Catal., 74 (1982) 225.
26. T. Castrillo and H. Knözinger, unpublished results.
27. Y. Doi, K. Soga, K. Yano and Y. Ono, J. Mol. Catal., 19 (1983) 359.
28. A. Theolier, A. Choplin, L. d'Ornelas, J.-M. Basset, G. Zanderighi, R. Ugo and C. Sourisseau, Polyhedron, 2 (1983) 119.
29. J. Evans and G. S. McNulty, J. Chem. Soc. Dalton Trans. (1984) 1123.
30. Y. Iwasawa, M. Yamada, Y. Sato and H. Kuroda, J. Mol. Catal., 23 (1984) 95.
31. R. Barth, B. C. Gates, Y. Zhao, H. Knözinger and J. Hulse, J. Catal., 82 (1983) 147.
32. G. Collier, D. J. Hunt, S. D. Jackson, R. B. Moyes, I. A. Pickering, P. B. Wells, A. F. Simpson and R. Whyman, J. Catal., 80 (1983) 154.
33. D. J. Hunt, S. D. Jackson, R. B. Moyes, P. B. Wells, P. Worthington and R. Whyman, J. Mol. Catal., 20 (1983) 289.
34. D. J. Hunt, R. B. Moyes, P. B. Wells, S. D. Jackson and R. Whyman, Proc. 8th Int. Congr. Catal., Berlin, 1984, Vol. V, p. 27.
35. M. Wolf, H. Knözinger and B. Tesche, J. Mol. Catal., 25 (1984) 273.
36. H. H. Lamb and B. C. Gates, J. Am. Chem. Soc., 108 (1986) 81.

Section 6.4

MAGNETIC RESONANCE

H. KNÖZINGER

6.4.1 INTRODUCTION

Magnetic resonance techniques, namely electron paramagnetic (or spin) resonance (EPR or ESR) and nuclear magnetic resonance (NMR), are well known and widely used in analytical chemistry. It is therefore beyond the scope of this volume to discuss the basic principles of these techniques in detail. The application of magnetic resonance techniques to surface chemical and physical problems is more restricted, and high-resolution NMR has found significant application only in recent years with the advent of modern Fourier transform (FT) spectrometers and superconducting magnets. Several reviews are available describing the application of EPR [1-6] and NMR [3, 6, 7-11] to the analysis of surface species.

6.4.2 ELECTRON PARAMAGNETIC RESONANCE (EPR)

EPR is restricted to the observation of species having unpaired electrons. Since most molecular clusters are diamagnetic, the application of EPR characterization of this class of compounds is naturally limited. However, paramagnetic species may be formed under certain conditions when clusters are supported on surfaces. Due to the high sensitivity of the EPR technique (as few as 10^{11} spins can be detected in favorable cases), very small percentages of the total metal loading become measurable in these systems and information about the structure, environment, and electronic configuration can be gained.

If we neglect direct interactions of nuclear spins with the external magnetic field, interactions of the electron spin with nuclear quadrupoles, and zero field splitting, then the interaction of a paramagnetic species containing a single unpaired electron with the external magnetic field H can be described by the approximate spin Hamiltonian:

$$\hat{H} = \beta \cdot H \cdot g \cdot S + \Sigma\ S \cdot A \cdot I \qquad (1)$$

In eq. (1), β is the Bohr magneton, g the g-tensor, S the electron spin, A the nuclear hyperfine tensor, and I a nuclear spin. The first term in eq. (1) describes the interaction of the electron spin with the external field, and the second term introduces the hyperfine interactions between the electron spin and all nuclear spins in its vicinity. Hence, the hyperfine splitting analysis can permit discrimination between mononuclear and polynuclear paramagnetic species, provided that the unpaired electron is located in orbitals of primarily metal character. This approach may be used in studying cluster degradation of supported species; however, resolution may be limited in these materials.

Single-crystal samples would be required for full determination of the anisotropic spin Hamiltonian parameters g and A. This may be possible in favorable cases for molecular clusters. In studies of supported clusters and their derivatives, only polycrystalline powder samples can be used. The resulting random powder patterns are correspondingly more difficult to interpret.

6.4.2.a **EPR of molecular clusters**

As mentioned above, most molecular clusters are diamagnetic. Examples of the few existing paramagnetic clusters are $Co_3(CO)_9S$ [12] and $Ni_3(\pi\text{-}Cp)_3(CO)_2$ [13]. In addition, paramagnetic clusters can be produced in various cases by chemical, electrochemical, or photochemical one-electron reduction reactions [14, 15].

Longuet-Higgins and Stone [16] measured the EPR spectrum of $Ni_3(\pi\text{-}Cp)_3(CO)_2$ and reported $g_{\parallel}$ and $g_{\perp}$ values of 2.11 and 2.02. Strouse and Dahl [17] observed $g_{\parallel}$ = 2.04 and $g_{\perp}$ = 2.02 for $Co_3(CO)_9S$. These authors also showed that the molecular orbital containing the unpaired electron is an antibonding a_2 orbital for both these clusters, which consists primarily of Co (Ni) d_{xz} atomic orbitals localized in the tricobalt(nickel) plane.

The $YCCo_3(CO)_9$ tetrahedral clusters can be easily reduced chemically and electrochemically to form stable radical anions [14, 18]. The hyperfine interaction of the unpaired electron with three equivalent cobalt nuclei (I = 7/2 for ^{59}Co) gave rise to a well-resolved hyperfine structure of 22 lines [14]. This feature could be interpreted successfully by assuming a single hyperfine coupling constant and g-value. The value of the hyperfine coupling constant did not vary significantly with the nature of the apical substituent Y. It was suggested, as in the previous cases, that the unpaired electron occupies a nondegenerate level of largely metal character.

The paramagnetic mixed-metal cluster anion $[Fe_3Pt_3(CO)_{15}]^-$ can be obtained by mild oxidation of $[Fe_3Pt_3(CO)_{15}]^{2-}$, and the EPR parameters of the former have been reported [19]. Beringhelli *et al.* [20] have shown that the g-values for this cluster could be compared with those of the mononuclear $[PtCl_3]^{2-}$ complex. In this complex, Pt has d^9 electron configuration, and the unpaired electron is located in the in plane $d_{x^2-y^2}$ orbital [21]. In generalizing this observation, Beringhelli *et al.* [20]

concluded that agreement between experimental g-values for the cluster and a mononuclear model compound of the same metal in the same d electron configuration suggests that the HOMO in the cluster is mainly composed of atomic orbitals of the kind of the ground state in the model mononuclear compound.

This principle was successfully applied to the interpretation of the EPR parameters of $Ni_3(\pi\text{-}Cp)_3(CO)_2$. Beringhelli *et al.* also studied by EPR the paramagnetic high-nuclearity cluster anions $[Rh_{12}(CO)_{13}(\mu_2\text{-}CO)_{10}(C)_2]^{3-}$, $[Co_{13}(CO)_{12}(\mu_2\text{-}CO)_{12}(C)_2]^{4-}$, and $[Co_6(CO)_8(\mu_2\text{-}CO)_6C]^-$. The assignment of the HOMO was based on a mixed valence model in which again the g-tensor components of the clusters were related to those of an appropriate conventional paramagnetic center. On this basis, the HOMO's of the Rh_{12} and Co_{13} cluster anions could be described as mainly metal d_{z^2} atomic orbitals. In contrast, a large overlap between d atomic orbitals and ligand orbitals was suggested for the Co_6 cluster anion.

6.4.2.b EPR of supported cluster-derived species

EPR investigations of supported clusters or cluster-derived species are scarce. The few studies which have been reported involve samples prepared from $Rh_4(CO)_{12}$, $Rh_6(CO)_{16}$, $Os_3(CO)_{12}$, and $Fe_3(CO)_{12}$.

Watters *et al.* [22], in a study of $Rh_6(CO)_{16}$ supported on $\gamma\text{-}Al_2O_3$, reported the observation of two very weak EPR signals in the decarbonylated state, which corresponded to $g_{\parallel} = 2.10$ and 2.05 and $g_{\perp} \approx 1.99$. These signals could be enhanced by heating the sample in O_2. They were thought to probably originate from Rh^{2+} species or, alternatively, from O_2^- formed on rhodium sites. The concentration of these paramagnetic species was <5% of the total Rh loading; a detailed analysis of the EPR spectra was not given. Kuznetsov *et al.* [23] reported the appearance of a symmetrical signal with $g = 2.21$ when $Rh_4(CO)_{12}/\gamma\text{-}Al_2O_3$ was heated in a $CO + O_2$ mixture at 423 K. This signal corresponded to 10-15% of the total rhodium in the sample. In addition, a weak signal was also observed with $g = 1.995$. At increasing treatment temperatures, the signal with $g = 2.21$ vanished and the intensity of the signal with $g = 1.995$ increased slightly. These two signals were attributed to Rh^{2+} and Rh^{4+} species, respectively, and the possible transformations

$$Rh^{+} \rightleftharpoons Rh^{2+} \rightleftharpoons Rh^{3+}$$

were suggested, with the more stable oxidation states Rh^+ and Rh^{3+} being the most abundant.

Beringhelli *et al.* [24, 25] showed that a resonance signal corresponding to 0.02% of the total number of clusters in the sample was produced in a sample prepared from $Rh_4(CO)_{12}$ on γ-Al_2O_3 in the decarbonylated state. The strongly asymmetrical lines ($g_{||}$ = 2.21, $g_{\perp}$ = 2.11) suggested location of the unpaired electron on Rh centers. The lines were strongly temperature dependent, and their intensities decreased and their line widths increased with temperature, as expected for 4d transition metal ion resonances. The presence of hyperfine structure in the perpendicular region of the signal indicated interaction with more than one Rh nucleus (I = 1/2). Hence, the multiline features seemed to arise from contributions of several polynuclear paramagnetic species with very similar g components rather than from the hyperfine coupling of a single species.

Analysis of the resonance signal observed by adsorption of O_2 suggested the formation of a Rh^{3+}-(O_2^-) species. The paramagnetic rhodium species originally present was therefore inferred to be most likely a Rh^{2+} (d^7) species. Due to the width of the signal, the hyperfine structure could not be resolved completely, and hence the nuclearity of the polynuclear paramagnetic center could not be determined.

Although the EPR signals of a catalyst formed from $RhCl_3$ are significantly broader than those of the material prepared from $Rh_4(CO)_{12}$, it was shown that the g-tensor components were the same in the two samples [25]. Beringhelli *et al.* [24] suggested that the paramagnetic species were formed from $Rh_4(CO)_{12}$ via electronic interactions with Lewis acid sites of the γ-Al_2O_3 support. This inference was supported by the fact that the paramagnetic Rh species was not formed on the less acidic ZrO_2 support [24, 25].

The formation of paramagnetic species during thermal decomposition and oxidation of samples prepared from $Os_3(CO)_{12}$ supported on γ-Al_2O_3 [26], CeO_2, and ThO_2 [27] have been reported by Shvets *et al.* With γ-Al_2O_3, EPR signals provided evidence for the intermediate formation of Os^+ and Os^{3+} ions during reduction-oxidation cycles [26]. These low-valent osmium ions formed paramagnetic surface carbonyl complexes, the majority of the surface osmium species (>98% of the total osmium atoms), however, being in a diamagnetic state. In contrast, high-valent osmium ions (Os^{5+} or Os^{7+}) were formed after oxidation on the surfaces of the basic oxides CeO_2 and ThO_2 [27].

When $Fe_3(CO)_{12}$ is thermally decomposed on Al_2O_3 or on MgO, decarbonylation with aggregation to small metallic particles occurs (see Section 9.2). Using ferromagnetic resonance (FMR) techniques, Hugues *et al.* [28] have shown that small superparamagnetic particles were formed on dehydrated supports (thermal pretreatment at 573 K), and the presence of Fe^{3+} could be excluded. It was concluded that the diameters of the metal particles on MgO were well below 3.5 nm, whereas the large linewidth of the Al_2O_3-supported sample suggested the formation of still smaller homodispersed particles. Hydrated MgO and Al_2O_3 supports led to the formation of larger particles and also of iron oxides.

6.4.3 NUCLEAR MAGNETIC RESONANCE (NMR)

Although less sensitive than EPR, NMR is far more widely applicable, and is being applied all the more as modern Fourier transform spectrometers and high-field magnets open the way to more and more nuclei. Besides ^{1}H and ^{13}C, many elements of catalytic relevance immediately come to mind, e.g., ^{15}N and ^{31}P as ligand atoms, ^{27}Al and ^{29}Si as constituents of oxide matrices, and a number of metal nuclei, including ^{51}V, ^{59}Co, ^{95}Mo, ^{103}Rh, ^{183}W, and ^{195}Pt.

The approximate spin Hamiltonian describing the interaction of a nuclear spin I with an external magnetic field is

$$\hat{H} = g_n\beta_n I(1 - \sigma)H + \sum_{i<k} I_i J_{ik} I_k + \sum_{i<k} \hat{H}_{ik}^{dip} \tag{2}$$

Here, g_n is the nuclear g-value, β_n the nuclear magneton, σ the chemical shift tensor, and J_{ik} the indirect nuclear coupling tensor. Nuclear quadrupole interactions are not considered in eq. (2). The first term in eq. (2) is the result of the nuclear Zeeman interaction with the magnetic field $(1 - \sigma)H$ at the nucleus, which deviates from the external field due to screening from surrounding electrons by the factor $(1 - \sigma)$; σ is known as the chemical shift. Screening effects are generally anisotropic, and hence σ is a tensor. The second term describes the indirect magnetic interaction between different nuclei mediated by bonding electrons. The third term describes the direct dipolar interaction between nuclear spins and is hence given by the sum of the dipolar Hamiltonians $\hat{H}_{ik}^{dip}$ for the interaction between pairs of nuclear spins i and k.

It is this third term which dominates in NMR of solids and which renders resolution of the hyperfine interaction and of chemical shifts impossible in conventional NMR spectroscopy of solids and adsorbed systems unless physically adsorbed molecules have sufficient mobility. In the latter case, rapid motion of the adsorbed molecule on the NMR time scale of approximately 10^{-6} s will average dipolar interactions and the chemical shift anisotropy to zero, a situation which is characteristic of liquids. Average chemical shifts

$$\langle\sigma\rangle = \frac{1}{3}(\sigma_{xx} + \sigma_{yy} + \sigma_{zz}) \quad (3)$$

can be measured for the adsorbed molecules in favorable cases. In general, however, all anisotropic interactions will be present in solid samples and the resulting NMR spectrum of a powder sample will be a broad envelope of superimposed individual spectra from all possible orientations in the magnetic field.

Specialized techniques have been developed for high-resolution NMR of solids. These are mentioned only briefly, and the reader is referred to the relevant literature [3, 6, 7-11, 29-32]. Dipolar interactions between dissimilar nuclei (e.g., ^{1}H and ^{13}C) can be readily removed by the *decoupling technique*, in which intense RF radiation at the resonance frequency of the second nucleus (^{1}H in ^{13}C-NMR) is applied. This effectively decouples it from the first. Simultaneously, the decoupling technique can produce a significant sensitivity enhancement by *cross-polarization* for the first nucleus in cases where this is magnetically more dilute than the second. Since this is particularly the case at natural ^{13}C abundance, the decoupling technique combined with cross-polarization is successfully applied in ^{13}C NMR. The enhancement factor is given by the ratio of the gyromagnetic ratios γ; for ^{13}C NMR this ratio is $\gamma_H/\gamma_C \approx 4$.

An effective technique to achieve line narrowing in solid samples is based on motional averaging by *spinning* the sample at the *magic angle* [3, 6, 11, 29]. In MAS-NMR, dipolar interactions and anisotropy in the chemical shift tensor are effectively averaged. The angular dependence of the dipolar interactions is given by the term $(1 - 3\cos^2\theta)$, where θ is the angle between the external magnetic field and the radius vector connecting the dipoles. If the nuclear spin system is rotated about an axis inclined at an angle α to the applied field at an angular velocity ω, the $\cos\theta$ term can be written as

$$\cos\theta = \cos\alpha \cdot \cos\psi + \sin\alpha \cdot \sin\psi \cdot \cos(\omega t) \quad (4)$$

where ψ represents the angle between the radius vector of the two nuclei and the rotation axis. The time-dependent part of the $(1 - 3\cos^2\theta)$ term then becomes

$$\frac{1}{2}(3\cos^2\alpha - 1)(1 - 3\cos^2\psi).$$

Hence, when $3\cos^2\theta = 1$ (i.e., when $\alpha = 54.7^o$, the magic angle), the dipolar interaction term is equal to zero and only isotropic interactions will be detected in the spectrum. Therefore, only the isotropic chemical shift

$$\langle\sigma\rangle = \frac{1}{3}(\sigma_{\parallel} + 2\sigma_{\perp}) \tag{5}$$

will be measured, where $\sigma_{\parallel}$ and $\sigma_{\perp}$ are the principal values of an axially symmetric chemical shift tensor parallel and perpendicular to the applied magnetic field, respectively. Sidebands in the spectra at integral multiples of ω on either side of the central spectrum are produced by the time-dependent term in eq. (4). For rotational frequencies greater than the static linewidth, the satellites do not interfere with the main spectrum, and their intensity becomes weaker with increasing rotational frequency. For ^{13}C with a shift range of approximately 300 ppm, a spinning frequency of about 2.0 - 2.2 kHz suffices at a carbon resonance frequency of 15 MHz.

High resolution NMR of solids has also become possible by means of *multiple pulse sequences* [3, 6, 30-32]. By applying intense multiple-pulse cycles of electromagnetic radiation, static magnetic field inhomogeneities as well as homonuclear and heteronuclear dipolar interactions can be essentially averaged. The line width of absorption bands in solid samples can thus be reduced by several orders of magnitude, so that chemical shifts of nonequivalent nuclei can be resolved, and typical powder patterns can be detected, which then permit the chemical shift tensor to be determined.

The *spin echo technique* is frequently applied in solid state NMR of metals [29, 33, 34]. In metals large local fields are produced by the interaction between the metal nuclei and the polarized conduction electron spins. This interaction shifts the metal resonance to higher fields. This shift is rather large compared with chemical shifts observed in nonmetallic diamagnetic compounds of the same metal and is called the *Knight shift*.

6.4.3.a **NMR spectroscopy of molecular clusters**

^1H and ^{13}C NMR have been used extensively for both structural characterization of molecular carbonyl and hydrido carbonyl clusters and their fluxional behavior. A compilation of ^1H and ^{13}C NMR data was published by Evans [35], and the fluxional properties of these clusters was discussed by Johnson and Benfield [36].

Proton chemical shifts generally lie in the τ 15-45 range, this large shift to high fields being considered as indicative of bridging hydride ligands [37, 38]. Proton chemical shifts in polycrystalline samples of $H_2Os_3(CO)_{10}$ have been reported by Nicol and Vaughan [39, 40]. A multiple

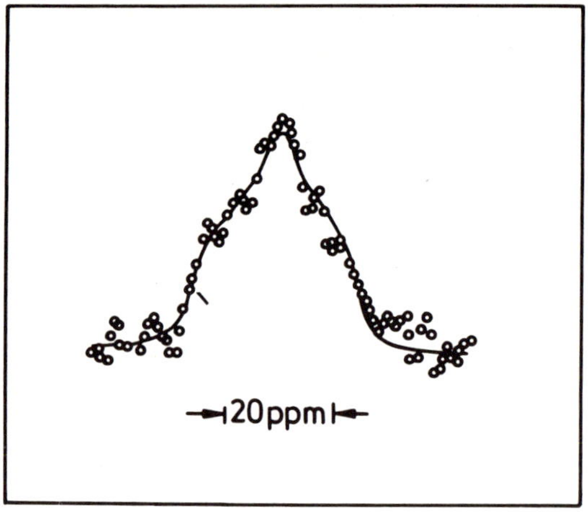

Fig. 1. Multiple-pulse ^1H NMR spectra of polycrystalline $H_2Os_3(CO)_{10}$ at 300 K [39, 40].

pulse spectrum of $H_2Os_3(CO)_{10}$ (eight-pulse sequence) is shown in Fig. 1. The isotropic chemical shifts for the solids $H_2Os_3(CO)_{10}$ and $H_4Ru_4(CO)_{12}$ were found to be close to the shifts observed in solution, this result providing evidence of the integrity of the molecular unit in the solid state. It was demonstrated that the major portion of the line width of the spectrum shown in Fig. 1 came from inhomogeneity broadening due to chemical shift tensor and heteronuclear ^{189}Os-^1H dipolar interactions (16.1% of osmium ^{189}Os with nuclear spin 3/2 and nonzero magnetic moment). From the fit of the data points, principal values of the chemical shift tensor of -2.5, 9.8, and 21.5 $\pm$ 1 ppm at 230 K were obtained. In contrast, the 8-pulse spectrum of $H_4Ru_4(CO)_{12}$ was characteristic of a powder pattern averaged by restricted motion of the protons. This conclusion was also substantiated from dipolar line shape measurements. An axially symmetric chemical shift tensor with an anisotropy of about 26 ppm at 300 K was observed for $H_4Os_4(CO)_{12}$ [39].

^{13}C NMR of solutions has been applied extensively to characterize the fluxional properties of metal clusters [36]; structural information has also been obtained. Chemical shift data for a variety of cluster compounds are compiled in the review by Evans [35]. It has been shown that resonances due to bridging carbonyls lie at much lower fields than those of terminally bound carbonyls in the same cluster. Moreover, terminal carbonyls in bimetallic clusters are often grouped together in regions typical of carbonyl groups connected with a given metal. For example, in $H_2FeRuOs_2(CO)_{13}$, chemical shifts of carbonyls bound to osmium lie between 168 and 177 ppm; those of carbonyls associated with ruthenium occur at 184 to 189 ppm; and those of carbonyl groups bound terminally to iron occur at 204 to 211 ppm [41].

^{13}C solid-state NMR of pure metal carbonyls is difficult since cross polarization is not possible for signal enhancement and circumventing long carbon T_1 relaxation times. However, solid-state spectra provide inherently a greater amount of information since chemical shift interactions are not averaged by molecular motions. Typical examples of ^{13}C solid-state spectra [42] are shown in Fig. 2 for $Mo(CO)_6$, $Ru_3(CO)_{12}$, and $Rh_6(CO)_{16}$, and the principal components of chemical shift tensors and average chemical shifts as obtained for polycrystalline samples of a variety of binary carbonyls are compared with the corresponding solution spectra in Table 1. The solid lines in Fig. 2 are fits of theoretical chemical shift powder patterns. In the case of $Rh_6(CO)_{16}$, the spectrum was deconvoluted into contributions from terminal and bridging carbonyl groups. From an analysis of the data presented in Table 1, Gleeson and Vaughan [42] came to the following conclusions:

(1) The chemical shift tensors of terminal carbonyl ligands are axially symmetric and highly anisotropic ($\Delta\sigma = |\sigma_{\perp} - \sigma_{\parallel}| = 360 \pm 60$ ppm). They vary only slightly with the transition metal. The electron orbitals are symmetric about the carbon-oxygen axis.

(2) In contrast, chemical shift tensors of bridging carbonyl ligands are much less anisotropic, since electron orbitals about the carbon-oxygen axis are significantly asymmetric, suggesting greater perturbation of the electronic structure of CO in bridged than in terminal bonding to a metal. It is important to note that the bonding in metal-bridging CO is distinct from that in organic carbonyls, this being apparent from isotropic chemical shifts or vibrational frequencies.

In conclusion, on the basis of chemical shift powder patterns and/or isotropic chemical shifts, terminal and bridged carbonyl ligands can be easily distinguished from each other and from CO that has reacted with

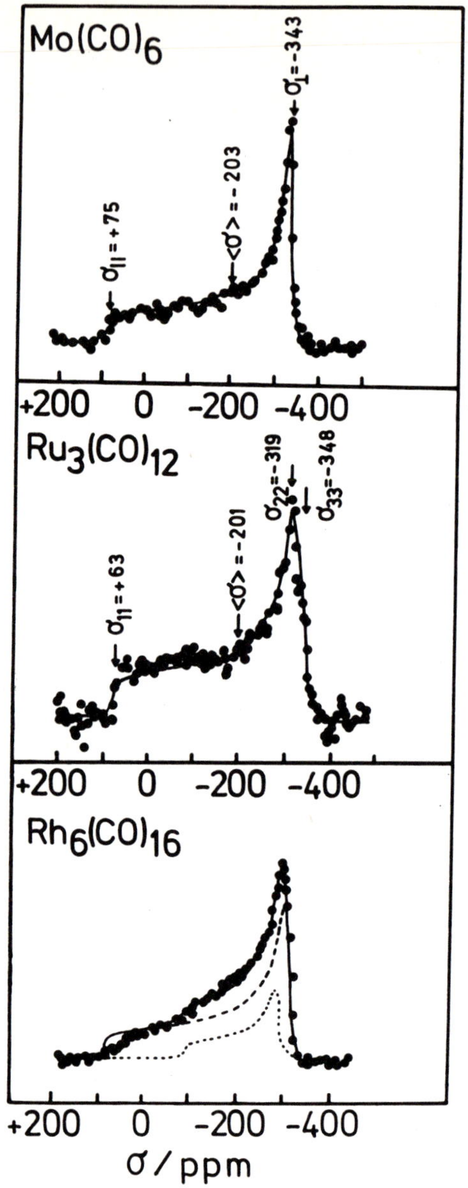

Fig. 2. ^{13}C NMR spectra of polycrystalline $Mo(CO)_6$, $Ru_3(CO)_{12}$, and $Rh_6(CO)_{16}$ at 300 K. Solid lines are theoretical fits for chemical shift powder patterns; for $Rh_6(CO)_{16}$, the spectrum was deconvoluted into contributions from terminal (---) and bridging (····) CO ligands [42].

TABLE 1. Principal Components, Isotropic Values, and Anisotropies of ^{13}C NMR Chemical Shift Tensors of Metal Carbonyls.[a]

Compound	σ_{11}	σ_{22}	σ_{33}	$\langle\sigma\rangle$	$\Delta\sigma$	Ref.	$\sigma_{solution}$	Ref.
$Cr(CO)_6$	70	-353	-353		423	[42]	-212	[45]
$Mo(CO)_6$	75	-343	-343	-203	417	[42]	-202	[45]
$W(CO)_6$	71	-324	-324		395	[42]	-192	[45]
$Ni(CO)_4$	69	-326	-326	-194	395	[43]	-191	[46]
$Fe(CO)_5$	70	-355	-355	-213	425	[43]	-212	[47]
$Rh_2(CO)_4Cl_2$	99	-299	-306	-186	402	[44]	-180	[48]
$Ru_3(CO)_{12}$	63	-319	-348	-201	397	[42]	-200	[49]
$Os_3(CO)_{12}$	55	-292	-292		347	[42]	-176	[49]
$Ir_4(CO)_{12}$	53	-277	-277		322	[42]	-167	
$Rh_6(CO)_{16}$								
terminal	80	-305	-315		390	[42]	-181	[50]
bridging	-102	-296	-296		194		-231	
$Fe_3(CO)_{12}$	6	-316	-325		327	[42]	-212	[49, 51]

[a]Values in ppm relative to TMS

hydrogen or other organic groups. Therefore, the results described above can serve as valuable references for the interpretation of ^{13}C NMR spectra of chemisorbed CO or surface groups derived from CO.

^{31}P NMR data have been reported for a few cluster compounds with phosphine ligands [35, 52].

Metal NMR data characterizing cluster compounds are scarce. ^{59}Co chemical shift data have been reported for $Co_2(CO)_8$ and $Co_4(CO)_{12}$ [53, 54]. ^{195}Pt resonances were observed for the cluster anions $[Pt_3(CO)_6]^{2-}$, $[Pt_6(CO)_9]^{2-}$, and $[Pt_9(CO)_{18}]^{2-}$ by Brown *et al.* [55] and for $[Pt_{38}(CO)_{44}]^{2-}$ by Stokes *et al.* [56]. The ^{195}Pt NMR data are discussed in more detail below in connection with ^{195}Pt NMR of supported Pt metal particles.

6.4.3.b **NMR spectroscopy of supported clusters and complexes**

Analogues of substituted carbonyl clusters are frequently synthesized on chemically modified oxide surfaces, the surface ligands (e.g., phosphines or amines) being produced from silane coupling agents (see Section 9.1). Cross polarization magic angle spinning NMR (CP-MAS-NMR) with ^{29}Si, ^{13}C, ^{31}P, and ^{15}N can provide useful information about the structures of these surface ligands to which a metal cluster or complex can be anchored.

^{13}C chemical shifts of a variety of organosilanes bonded to silica surfaces have been reported by several groups [11, 57-59]. Sindorf and Maciel [60, 61] demonstrated by combined ^{13}C and ^{29}Si CP-MAS-NMR that the bonding of the ligand to the surface can easily be monitored when ethoxy silanes are used, and details on the nature of the secondary surface structures that are formed can be obtained.

Bemi *et al.* [62] reported ^{31}P proton decoupling CP-MAS-NMR spectra of $Ph_2PCH_2CH_2Si(OEt)_3$ and its Ni^{2+}, Pd^{2+}, and Pt^{2+} complexes anchored to silica surfaces. The results demonstrated that the geometry of surface-bound complexes may be determined and the surface reactions may be monitored. Lui *et al.* [63] followed the silylation of silica by $Ru_3(CO)_9[PPh_2CH_2CH_2Si(OEt)_3]$ using ^{31}P and ^{29}Si CP-MAS-NMR. The closely similar ^{31}P chemical shift values for the compound in solution (35.0 ppm) and for the derivatized silica (34.5 ppm) were taken as evidence for their structural analogy.

^{13}C solid-state NMR can be used for the characterization of metal carbonyls on unmodified oxide surfaces. Since these surfaces bear hydroxyl groups, sensitivity enhancements by cross-polarization can be achieved. Simple physisorption of $Ni(CO)_4$ [64], $Fe(CO)_5$ [64], $Cr(CO)_6$ [65], and $Mo(CO)_6$ [65] on various oxide surfaces has been characterized without magic angle spinning. Solution-like ^{13}C NMR spectra have been obtained

TABLE 2. ^{13}C Chemical Shifts of Adsorbed Metal Carbonyls.

Metal Carbonyl	Support	Chemical Shift/ppm	Ref.
$Ni(CO)_4$	γ-Al_2O_3	190.4	[64]
	NaY zeolite	193.8	
	HY zeolite	184.7	
$Fe(CO)_5$	HY zeolite	211.5	[64]
$Cr(CO)_6$	NaY zeolite	210	[65]
$Mo(CO)_6$	NaY zeolite	210	[65]
	γ-Al_2O_3	201	[66, 67]

due to rapid molecular tumbling in these systems. The chemical shifts of these physically adsorbed metal carbonyls were typically very close to those observed in solution (Table 2). However, $Ni(CO)_4$ in HY zeolite showed an upfield shift relative to that observed in NaY, which was attributed to interaction with an extra-framework proton [64].

Shirley *et al.* [66] attempted to follow the decarbonylation of $Mo(CO)_6$ adsorbed on alumina using conventional ^{13}C Fourier transform NMR and ^{13}C-enriched samples. A ^{13}C resonance could be observed for $Mo(CO)_6$ only; subcarbonyls could not be detected. It was therefore concluded that strong interactions of the subcarbonyls with the support surface reduced their mobility so as to erode the signal. However, CP-MAS ^{13}C NMR spectroscopy has recently been demonstrated to be capable of detecting molybdenum subcarbonyls on aluminas, these results indicating the potential of the technique for identifying supported organometallics. For example, Fig. 3 shows the development of CP-MAS ^{13}C NMR spectra [67] with time during decomposition of $Mo(CO)_6$ on γ-Al_2O_3 at 373 K, which was shown to proceed through the following steps [68]:

$$Mo(CO)_{6,phys} \underset{+CO}{\overset{-CO}{\rightleftharpoons}} Mo(CO)_{5,ads} \underset{+CO}{\overset{-CO}{\rightleftharpoons}} Mo(CO)_{3,ads} \qquad (6)$$

Physisorbed $Mo(CO)_6$ has a chemical shift of 201 ppm and was present in the original sample treated at 303 K, but was not observed by the cross polarization experiment. The resonance occurring at 203 ppm in Fig. 3 was attributed to the species $Mo(CO)_{5,ads}$, and that developing at 223 ppm with

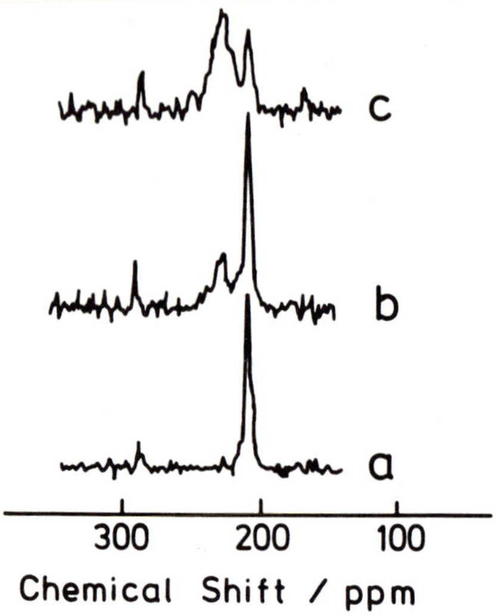

Fig. 3. ^{13}C CP-MAS-NMR spectra of $Mo(CO)_6$ on γ-Al_2O_3; (a) after activation in a He flow at 303 K for 1 h; (b) same at 373 K for 1 h; (c) same at 373 K for 3 h [67].

time at 373 K was assigned to the species $Mo(CO)_{3,ads}$. The results of variable-temperature experiments suggested that these subcarbonyls are fluxional [67].

The anion $[HFe_3(CO)_{11}]^-$ can easily be generated on hydroxylated alumina surfaces (see Sections 6.2 and 9.2). The ^{13}C CP-MAS-NMR spectrum of this surface species gave a broad resonance at 221 ppm and a sharp peak at 210 ppm [69]. These chemical shifts correspond closely to solution chemical shifts for $[HFe_3(CO)_{11}]^-$ and $Fe_3(CO)_{12}$, respectively. The $Fe_3(CO)_{12}$ cluster was physisorbed and was present due to high loading and incomplete reaction. The anionic cluster $[HFe_3(CO)_{11}]^-$ was also suggested to be physically adsorbed, a conclusion based on its chemical shift, although its charge should indicate strong bonding to the surface.

The broadness of the corresponding band was assumed to be due either to a dynamic process or the possibility that physisorption occurs on inequivalent sites, thereby producing slightly different shifts.

^{13}C CP-MAS NMR spectra of $Rh_4(CO)_{12}$ in faujasites were reported by Gelin *et al.* [70], who observed chemical shifts almost identical to those of the molecular cluster in a Nujol mull. Chemical transformations of the zeolite-entrapped cluster were also followed (see Chapter 8).

Kuznetsov *et al.* [71] investigated the formation of surface species during catalytic CO hydrogenation on SiO_2- and La_2O_3-supported catalysts prepared from $Rh_4(CO)_{12}$. This study clearly demonstrated the capability of ^{13}C NMR to detect intermediate surface species such as $-C{\lesssim}^{O}_{R}$, HCO, $-OCH_2-$, $-OCH_3$, $-CH_2-$, and $-CH_3$ as well as adsorbed final products.

^{13}C NMR spectroscopy has also been used to characterize chemisorbed CO, e.g., CO chemisorbed on alumina-supported rhodium catalysts [44, 72]. The spectra reported were broadened due to chemical shift anisotropy, since magic angle spinning was not applied. Homonuclear broadening was significantly less than the chemical shift anisotropy at the ^{13}CO concentrations used, and heteronuclear dipolar broadening from ^1H, ^{27}Al, or ^{103}Rh was negligible. The ^{13}C spectra contained contributions from two overlapping signals which could be distinguished by their different spin-lattice relaxation times T_1. The two signals had average chemical shifts of -177 and -199 ppm and T_1 values of 5.6 and 64 ms, respectively. The rapidly relaxing signal ($\langle\sigma\rangle$ = -177 ppm; T_1 = 5.6 ms) was assigned to isolated $Rh(CO)_2$ surface complexes. On metal surfaces, terminal and bridging chemisorbed CO is expected to be formed, and these species should rapidly exchange on the NMR time scale, thus having the same ^{13}C spin-lattice relaxation times. The chemical shift of the second signal ($\langle\sigma\rangle$ = -199 ppm) falls between that for terminal CO groups ($\langle\sigma\rangle \approx$ -185 ppm) and that for bridging CO groups ($\langle\sigma\rangle \approx$ -230 ppm) (see Section 6.4.3.a and Table 1). Therefore, this signal was attributed to an average of terminal and bridging CO ligands chemisorbed on small rhodium metal particles.

Metal NMR studies of supported molecular metal clusters have not yet been reported. However, there are several publications reporting ^{195}Pt NMR of supported platinum catalysts [33, 56, 73-77]. Also, in a preliminary study, ^{63}Cu resonances of a silica-supported copper catalyst were observed [78]. Spin-echo or multiple pulse techniques have been used in these experiments.

Yu and Halperin [74] measured ^{195}Pt NMR line widths, lineshapes, Knight shifts, and nuclear spin relaxation times for silica-supported platinum catalysts. The metal particle sizes ranged from 3 to 20 nm. The data provided evidence for size effects in the electronic structure, which were interpreted in terms of surface electron spin density oscillations. When alumina-supported platinum catalysts having varying dispersions were investigated [33, 56, 73, 75], resonance lines with widths of about 4 kG were observed. This large line width was assumed to be due to variations of the Knight shift for Pt atoms near a surface as compared with those atoms in the bulk. The perturbation of the state of conduction electrons near a surface tends to reduce the Knight shift. A broad particle size distribution would then lead to a broad smear of the Knight shift and hence to extremely broad resonance lines.

Fig. 4 shows typical ^{195}Pt NMR absorption lines for two catalysts having 15 and 46% dispersion, respectively. These samples were exposed to

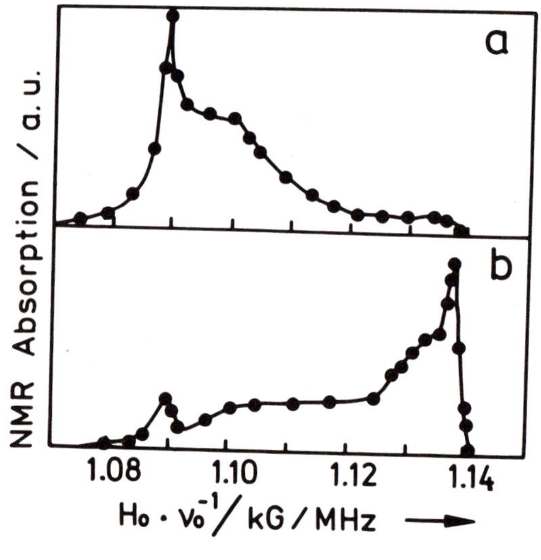

Fig. 4. ^{195}Pt NMR absorption vs. H_o/ν_o for Pt/Al_2O_3 catalysts: (a) 46% dispersion; (b) 15% dispersion; samples were exposed to the atmosphere before the NMR experiment [33, 56, 73].

the atmosphere before the measurements were carried out. Significant structure can be clearly seen in these spectra. The sharp peak at H_0/ν_0 = 1.136 kG/MHz is very close to the metal peak of bulk platinum (Knight shift) and dominates the feature observed for the catalyst having low dispersion. In contrast, the catalyst with high dispersion (46%) produces a sharp peak at H_0/ν_0 = 1.088 kG/MHz, the position of which almost exactly coincides with that of the complex $H_2Pt(OH)_6$. It was therefore suggested that the surfaces of small platinum particles were covered by $Pt(OH)_6^{2-}$ species after exposure of the sample to the atmosphere [33, 56, 73]. Support for this assignment was provided by the result that the peak at 1.088 kG/MHz disappeared after cleaning of the surface by exposure to O_2/H_2 cycles at 582 K [73]. The important inference from these results is that surface atoms can apparently be detected by ^{195}Pt NMR.

The presence of chemisorbed hydrogen gave a surface peak with still another position and width [73]. Experiments of this type were complemented by ^1H magnetic resonance of the chemisorbed hydrogen [79-81], which showed significant particle size effects. The resonance line of hydrogen on platinum was observed at 60 ppm upfield of tetramethylsilane, its position being a function of hydrogen coverage and metal dispersion [79]. The data were interpreted to indicate spin pairing in the Pt crystallites, due to a covalent component in the adsorption bond.

Of particular interest in this context are ^{195}Pt NMR studies performed with supported platinum catalysts after chemisorption of CO [33, 56, 82]. The ^{195}Pt NMR line shapes of CO-covered particles were clearly different from those of bare Pt particles, indicating that the electronic environment of the Pt nuclei near or at the surface of the particles is significantly altered by the presence of CO. It was possible by a spin echo double resonance (SEDOR) experiment to separate the resonance of surface Pt from the rest of the lineshape. The results indicated that the lineshapes due to surface Pt on catalysts having different dispersions (26 and 76%) were strikingly similar. It was therefore concluded that the NMR characteristics of the surface Pt are largely determined by the nature of the chemical bonding to the nearby CO molecules and not by the size of the particle itself. A comparison of the ^{195}Pt NMR peak positions of surface Pt in supported catalysts with those of carbonyl cluster anions is given in Table 3. Most interestingly, the positions of ^{195}Pt NMR peaks for surface Pt are found to be very close to those of the peaks characterizing platinum carbonyl clusters, particularly when typical variations among Pt compounds are considered.

TABLE 3. Comparison of ^{195}Pt NMR Peak Positions in Alumina-supported Pt Catalysts and Various Platinum Carbonyl Cluster Anions.

	$H_o \cdot \nu_o^{-1}$/kG·MHz^{-1}	Refs.
Surface Pt in supported catalysts	1.096	[33, 56]
$[Pt_{38}(CO)_{44}]^{2-}$	1.096 (77K)	[33, 56]
$[Pt_9(CO)_{18}]^{2-}$	1.0975, 1.0980	[55]
$[Pt_6(CO)_9]^{2-}$	1.0975	[55]
$[Pt_3(CO)_6]^{2-}$	1.0975	[55]

6.4.4 CONCLUSIONS

The application of the EPR technique is restricted to a few materials, but already the relatively few examples of the application of NMR techniques to the characterization of supported clusters, cluster-derived species, and small metal aggregates demonstrate the great potential of the method. Particularly attractive is the possibility of simultaneously characterizing a given sample by means of several nuclear probes including, for example, the metal nuclei, ^{13}C, and ^{1}H. Valuable structural information regarding the ligand sphere and the metal framework, the electronic structure of the metal framework, the electronic structure of the metal components (and related particle size effects) and the ligand bonding (and the chemisorption bond) can be gained. Moreover, intermediates in catalytic reactions can be traced, at least under favorable circumstances. In summary, the future of NMR techniques for surface analysis in general and for the characterization of cluster-derived catalysts in particular looks very promising.

REFERENCES

1. J. H. Lunsford, in Chemistry and Physics of Solid Surfaces, Vol. I, R. Vanselow and S. Y. Tong (Eds.), CRC Press, Boca Raton, Florida, 1976, p. 255.
2. P. H. Kasai and R. J. Bishop, ACS Monograph Ser., 171 (1976) 350.
3. J. H. Lunsford, in Spectroscopy in Heterogeneous Catalysis, W. N. Delgass, G. Haller, R. Kellerman and J. H. Lunsford (Eds.), Academic Press, New York, San Francisco, London, 1979, p. 183.
4. M. Che, in Magnetic Resonance in Colloid and Interface Science, J. P. Fraissard and H. A. Resing (Eds.), Reidel, Dordrecht, London, 1980, p. 79.
5. R. F. Howe, Advan. Colloid Interf. Sci., 18 (1982) 1.

6. R. F. Howe, in Chemistry and Physics of Solid Surfaces, Vol. V, R. Vanselow and R. Howe (Eds.), Springer, Berlin, Heidelberg, New York, Tokyo, 1984, p. 39.
7. H. Pfeiffer, Phys. Rep., 26 (1976) 293.
8. J. Tabony, Progr. Nucl. Magn. Spectr. Reson., 14 (1980) 1.
9. T. M. Duncan and C. Dybowski, Surface Sci. Rep., 1 (1981) 157.
10. W. Derbyshire, Roy. Soc. Chem., Specialist Periodical Rep., Nucl. Magn. Res., 11 (1981) 264.
11. G. E. Maciel, in Heterogeneous Catalysis, B. L. Shapiro (Ed.), Texas A & M University Press, College Station, Texas, 1984, p. 349.
12. E. Klumpp, L. Markó and G. Bor, Ber., 97 (1964) 926.
13. E. O. Fischer and C. Palm, Ber., 91 (1958) 1725.
14. B. M. Peake, B. H. Robinson, J. Simpson and D. J. Watson, Inorg. Chem., 16 (1977) 405.
15. B. M. Peake, P. H. Rieger, B. H. Robinson and J. Simpson, Inorg. Chem., 20 (1981) 2540.
16. H. C. Longuet-Higgins and A. Stone, Mol. Phys., 5 (1962) 417.
17. C. E. Strouse and L. F. Dahl, Disc. Faraday Soc., 47 (1969) 93.
18. J. Kotz, J. V. Peterson and R. C. Reed, J. Organomet. Chem., 120 (1976) 433.
19. G. Longoni and F. Morazzoni, J. Chem. Soc. Dalton Trans. (1981) 1735.
20. T. Beringhelli, F. Morazzoni and D. Strumolo, J. Organomet. Chem., 236 (1982) 109.
21. T. Krigas and M. T. Rogers, J. Chem. Phys., 55 (1971) 3035.
22. K. L. Watters, R. F. Howe, T. P. Chojnacki, C.-M. Fu, R. L. Schneider and N.-B. Wong, J. Catal., 66 (1980) 424.
23. V. L. Kuznetsov, A. V. Romanenko, A. P. Shepelin, P. A. Zhdan and Yu. I. Yermakov, React. Kinet. Catal. Lett., 22 (1983) 119.
24. T. Beringhelli, A. Gervasini, S. Martinengo, F. Morazzoni, D. Strumolo and L. Zanderighi, J. Chem. Soc. Faraday I, 80 (1984) 1479.
25. T. Beringhelli, A. Gervasini, F. Morazzoni, D. Strumolo, S. Martinengo, L. Zanderighi, F. Pinna and G. Strukul, Proc. 8th Int. Congr. Catal., Dechema, Frankfurt, 1984, Vol. V, p. 63.
26. A. V. Shvets, A. L. Tarasov, V. B. Kazansky and H. Knözinger, J. Catal., 86 (1984) 223.
27. A. V. Shvets, A. L. Tarasov, V. B. Kazansky and H. Knözinger, Chem. Phys. Lett., 115 (1985) 515.
28. F. Hugues, P. Bussière, J. M. Basset, D. Commereuc, Y. Chauvin, L. Bonneviot and D. Olivier, Proc. 7th Int. Congr. Catal., Kodansha, Tokyo, and Elsevier, Amsterdam, Oxford, New York, 1981, Part A, p. 418.
29. C. P. Slichter, Principles of Magnetic Resonance, Springer, New York, 1980.
30. U. Haeberlen, High Resolution NMR Spectroscopy in Solids: Selective Averaging, Academic Press, New York, 1976.
31. M. Mehring, High Resolution NMR Spectroscopy in Solids, Springer, Berlin, Heidelberg, New York, 1976.
32. B. C. Gerstein, in Magnetic Resonance in Colloid and Interface Science, J. P. Fraissard and H. A. Resing (Eds.), Reidel, Dordrecht, London, 1980, p. 177.
33. H. T. Stokes, in Catalytic Materials-Relationships between Structure and Reactivity, T. E. Whyte, Jr., R. A. Dalla Betta, E. G. Derouane, and R. T. K. Baker (Eds.), ACS Symp. Ser. 248 (1984) 385.
34. L. H. Bennett, S. M. Bhagat, J. H. Schelleng, L. J. Swartzendruber and L. D. Weisman, in Measurement of Physical Properties, Part 2: Magnetic Properties and Mössbauer Effect, E. Passaglia (Ed.), Wiley, New York, London, Sydney, Toronto, 1973, p. 165.
35. J. Evans, Advan. Organomet. Chem., 16 (1977) 319.

36. B. F. G. Johnson and R. E. Benfield, in Transition Metal Clusters, B. F. G. Johnson (Ed.), Wiley, Chichester, New York, Brisbane, Toronto, 1980, p. 471.
37. S. Aime and L. Milone, Progr. Nucl. Magn. Reson. Spectrosc., 11 (1977) 183.
38. A. P. Humphries and H. D. Kaesz, Progr. Inorg. Chem., 25 (1979) 145.
39. A. T. Nicol and R. W. Vaughan, J. Am. Chem. Soc., 101 (1979) 583.
40. A. T. Nicol, in Magnetic Resonance in Colloid and Interface Science, J. P. Fraissard and H. A. Resing (Eds.), Reidel, Dordrecht, London, 1980, p. 207.
41. G. L. Geoffroy and W. L. Gladfelter, J. Am. Chem. Soc., 99 (1977) 6775.
42. J. W. Gleeson and R. W. Vaughan, J. Chem. Phys., 78 (1983) 5384.
43. H. W. Spiess, R. Grosescu and U. Haeberlen, Chem. Phys., 6 (1974) 226.
44. T. M. Duncan, J. T. Yates, Jr. and R. W. Vaughan, J. Chem. Phys., 73 (1980) 975.
45. O. A. Gansow, B. Y. Kimura, G. R. Dobson and R. A. Brown, J. Am. Chem. Soc., 93 (1971) 5922.
46. B. E. Mann, J. Chem. Soc. Chem. Commun. (1971) 1173.
47. P. C. Lauterbur and R. B. King, J. Am. Chem. Soc., 87 (1965) 3266.
48. P. Chini, S. Martinengo, D. J. A. McCaffrey and B. T. Heaton, J. Chem. Soc. Chem. Commun. (1974) 310.
49. A. Forster, B. F. G. Johnson, J. Lewis, T. W. Matheson, B. H. Robinson and W. G. Jackson, J. Chem. Soc. Chem. Commun. (1974) 1042.
50. B. T. Heaton, A. D. C. Towl, P. Chini, A. Fumgilli, D. J. A. McCaffrey and S. Martinengo, J. Chem. Soc. Chem. Commun. (1975) 523.
51. O. A. Gansow, A. R. Burke and W. D. Vernon, J. Am. Chem. Soc., 94 (1972) 1550.
52. S. R. Bunkhall, H. D. Holden, B. F. G. Johnson, J. Lewis, G. N. Pain, P. R. Raithby and M. J. Taylor, J. Chem. Soc. Chem. Commun. (1984) 25.
53. E. A. C. Lucken, K. Noack and D. F. Williams, J. Chem. Soc. A (1967) 148.
54. H. Haas and R. K. Sheline, J. Inorg. Nucl. Chem., 29 (1967) 693.
55. C. Brown, B. T. Heaton, P. Chini, A. Fumagalli, and G. Longoni, J. Chem. Soc. Chem. Commun. (1977) 309.
56. H. T. Stokes, C. D. Makowka, P.-K. Wang, S. L. Rudaz, C. P. Slichter and J. H. Sinfelt, J. Mol. Catal., 20 (1983) 321.
57. H. A. Resing, D. Slotfeld-Ellingsen, A. N. Garroway, D. C. Weber, T. J. Pinnavaia and K. Unger, in Magnetic Resonance in Colloid and Interface Science, J. P. Fraissard and H. A. Resing (Eds.), Reidel, Dordrecht, London, 1980, p. 239.
58. D. E. Leyden, D. S. Kendall, and T. G. Waddell, Anal. Chim. Acta, 126 (1981) 207.
59. C.-H. Chiang, Nan-I Liu and J. L. Koenig, J. Colloid Interface Sci., 86 (1982) 26.
60. D. W. Sindorf and G. E. Maciel, J. Am. Chem. Soc., 103 (1981) 4263.
61. D. W. Sindorf and G. E. Maciel, J. Am. Chem. Soc., 105 (1983) 3767.
62. L. Berni, H. C. Clark, J. A. Davies, C. A. Fyfe and R. E. Wasylishen, J. Am. Chem. Soc., 104 (1982) 438.
63. D. K. Lui, M. S. Wrighton, D. R. McKay and G. E. Maciel, Inorg. Chem., 23 (1984) 212.
64. J. B. Nagy, M. van Eenoo, E. G. Derouane and J. C. Vedrine, in Magnetic Resonance in Colloid and Interface Science, J. P. Fraissard and H. A. Resing (Eds.), Reidel, Dordrecht, London, 1980, p. 591.
65. Y. Ben Taarit, G. Wicker and C. Naccache, in Magnetic Resonance in Colloid and Interface Science, J. P. Fraissard and H. A. Resing (Eds.), Reidel, Dordrecht, London, 1980, p. 497.

66. W. M. Shirley, B. R. McGarvey, B. Maiti, A. Brenner and A. Cichowlas, J. Mol. Catal., 29 (1985) 259.
67. B. E. Hanson, G. W. Wagner, R. J. Davis and E. Motell, Inorg. Chem., 23 (1984) 1635.
68. A. Brenner and R. L. Burwell, Jr., J. Am. Chem. Soc., 97 (1975) 2565.
69. B. E. Hanson and J. Petty, Inorg. Chem., submitted.
70. P. Gelin, F. Lefebvre, B. Elleuch, C. Naccache and Y. Ben Taarit, in Intrazeolite Chemistry, G. D. Stucky and F. G. Dwyer (Eds.), ACS Symp. Ser. 28, 1983, p. 455.
71. V. L. Kuznetsov, A. V. Romanenko, I. L. Mudrakovskii, V. M. Mastikhin, Yu. I. Yermakov and V. A. Shmachkov, Proc. 8th Int. Congr. Catal., Dechema, Frankfurt, 1984, Vol. V, p. 3.
72. T. M. Duncan, J. T. Yates and R. W. Vaughan, J. Chem. Phys., 71 (1979) 3129.
73. C. P. Slichter, Surface Sci., 106 (1981) 382.
74. I. Yu and W. P. Halperin, J. Chem. Phys., 78 (1981) 901.
75. H. E. Rhodes, P.-K. Wang, H. T. Stokes, C. P. Slichter and J. H. Sinfelt, Phys. Rev. B, 26 (1982) 3559.
76. H. E. Rhodes, P.-K. Wang, C. D. Makowka, S. L. Rudaz, H. T. Stokes, C. P. Slichter and J. H. Sinfelt, Phys. Rev. B, 26 (1982) 3569.
77. H. T. Stokes, H. E. Rhodes, P.-K. Wang, C. P. Slichter and J. H. Sinfelt, Phys. Rev. B, 26 (1982) 3575.
78. H. Ebert, L. Renbiao, J. Voitländer and H. Knözinger, unpublished.
79. T.-C. Sheng and I. D. Gay, J. Catal., 71 (1981) 119.
80. T.-C. Sheng and I. D. Gay, J. Catal., 77 (1982) 53.
81. L. C. de Menorval and J. P. Fraissard, Chem. Phys. Lett., 77 (1981) 309.
82. C. D. Makowka, C. P. Slichter and J. H. Sinfelt, Phys. Rev. Lett., 49 (1982) 379.

Section 6.5

PHOTOELECTRON SPECTROSCOPIES

L. GUCZI

6.5.1 PHYSICAL BACKGROUND

The physical background and principles of electron spectroscopy have been reviewed frequently [1, 2]. Therefore, only brief descriptions of X-ray photoelectron spectroscopy (XPS) and UV-photoelectron spectroscopy (UPS) are given here.

A solid sample is bombarded by photons, causing photoelectrons to be ejected. The kinetic energy E_{kin} of these photoelectrons is related to the photon energy $h\nu$ and the binding energy E_B of the electron, as follows:

$$h\nu = E_B + E_{kin} + \phi \qquad (1)$$

where ϕ is the spectrometer work function. In XPS, the AlK_α or MgK_α lines with photon energies of 1487 and 1254 eV, respectively, are normally used for excitation (since nonmonochromatized radiation is normally used, satellite peaks of lower intensity occur in XP spectra due to the $K_{\alpha 3-6}$ and K_β lines), while He I and He II lines with photon energies of 21.2 and 40.8 eV, respectively, are applied for UPS. The escape depth of the photoelectrons depends on their kinetic energy. The spectra typically contain information from several atomic layers (depth range 0.3-3 nm); the contributions of these layers decrease exponentially with increasing depth.

A serious problem is the positive charging of insulating samples, which leads to an apparent shift of the binding energy E_B to higher values. This charging effect can be accounted for by using the C 1s signal (285 eV) as an internal standard, since this signal appears from carbon on almost any dispersed sample. However, the origin of carbon contamination is usually unknown and may vary for different samples, so that the C 1s binding energy may be subject to uncertainty. Therefore, gold is frequently evaporated onto the sample, the Au $4f_{7/2}$ peak being used as a reference. Alternatively, the binding energies of constituents of the support, such as Si 2p (103 eV) for SiO_2 and Al 2s (119.5 eV) or Al 2p (73.5 eV) for Al_2O_3, may be used as internal standards.

Photon energies of X-rays are in the energy region of ionization energies of core levels, and hence the photoelectrons stem from core level ionizations. For the heavier elements, p, d, f, etc. levels are split due to spin orbit coupling and hence lead to doublets in the XP spectra. The theoretical intensity ratios are 1:2, 2:3, and 3:4 for $p_{1/2}$ to $p_{3/2}$, $d_{3/2}$ to $d_{5/2}$, and $f_{5/2}$ to $f_{7/2}$ peaks, respectively [2].

Core level energies are affected by the electron density around the nucleus and therefore depend on the valence state and the chemical environment (charge transfer with surrounding atoms or ligands) of the element under consideration. These effects are reflected in the *chemical shifts* which therefore may, in principle, be interpreted in terms of chemical bonding. However, binding energy shifts can also result from relaxation phenomena (final state effects). The electronic relaxation energy associated with the core level ionization of an atom consists of two contributions: one which is associated with the contraction of the local charge distribution around the ionized atom and a second which is associated with the flow of electron density towards the core hole generated. Therefore, XP binding energies are not generally typical of the ground state of the atom. For example, due to an unusually low relaxation energy, the C 1s binding energy of CO is much higher than that observed for other compounds of carbon. In contrast, the O 1s binding energy in carbonyls is lower than the predicted value. This difference is ascribed to a large relaxation owing to a transfer of π-electron density towards the carbonyl oxygen atom from ligands bonded to the carbonyl group ("π-donor relaxation") [3-5]. A more detailed discussion of these phenomena is given in ref. [6].

Satellite peaks and peak tailing may be observed in XP spectra. Satellite peaks at binding energies higher than that of the principal peak are due to electronically excited final states. These "shake-up" transitions are to be considered as combinations of core ionization and molecular electronic excitations. At very high energies, a broad "tailing" may occur, which is due to the simultaneous loss of a valence electron and a core electron ("shake-off") [7]. Shake-up peaks are generally the more intense the higher the electronic relaxation energy. For the free CO molecule, for example, the shake-up peak is relatively weak [8], whereas it becomes very intense for metal carbonyls [9, 10]. Intense satellite peaks were observed for $Cr(CO)_6$ at about 5.5 eV from the principal peak in the C 1s and O 1s spectra. These were attributed to a metal$\longrightarrow$ligand charge transfer transition which was observed at 5.5 eV in the UP spectra of $Cr(CO)_6$ vapor [11, 12].

Due to the far lower photon energies used in UPS, valence ionization spectra are measured, and these provide direct insight into the chemical bonding.

6.5.2 PHOTOELECTRON SPECTROSCOPY OF MOLECULAR CLUSTERS

In a number of studies, valence and core photoionization spectra of the carbonyl ligand of polynuclear transition metal clusters have been investigated [13, 14]. The goal of these experiments was an understanding of the bonding characteristics of CO chemisorbed on transition metal surfaces in comparison with polynuclear carbonyl compounds. This aspect is discussed by Ertl in Chapter 11 of this book.

On coordination, the C 1s and O 1s core level binding energies as well as ionization energies of the molecular orbitals are shifted to lower values. The satellite peaks already present in the spectra of the free CO molecule follow the same trend, and additional shake-up satellite peaks with sometimes considerable intensity appear in the spectra of coordinated CO. An example is shown in Fig. 1 for $Co_4(CO)_{12}$. These phenomena have been discussed by Freund *et al.* [15]. Binding energy shifts in the core level spectra that occur upon coordination are explained as being due to differences in relaxation energies in the ion rather than a chemical shift in the initial state. The pronounced intensity of shake-up satellites in XP spectra can be interpreted by intensity borrowing from the principal line.

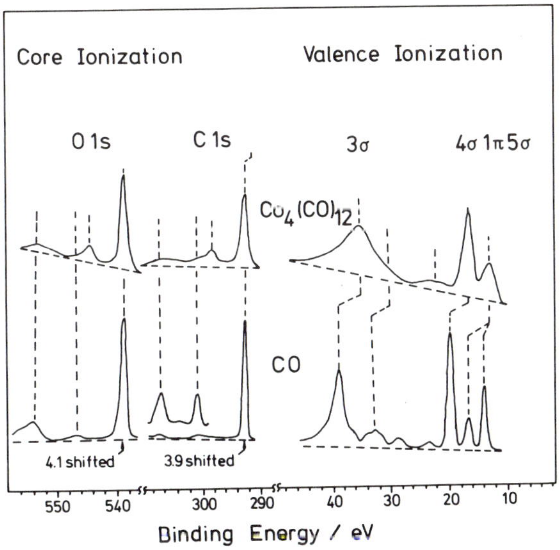

Fig. 1. Valence and core ionization spectra of free carbon monoxide and of tetracobaltdodecacarbonyl $Co_4(CO)_{12}$ (binding energies are referenced to the vacuum level) (from ref. [15c]).

In the UP spectra of transition metal clusters in the gas phase, the d-states lie between 7 and 11 eV and are thus clearly separated from the bands characteristic of emissions from the carbonyl ligands ($5\sigma + 1\pi$ at 13-16 eV; 4σ at 18-21 eV). Table 1 is a summary of ionization energies reported for several carbonyl clusters. These spectra can be interpreted on the basis of a simplified molecular orbital scheme. For trinuclear clusters of the type $M_3(CO)_{12}$, this may be constructed by combination of the frontier molecular orbitals of individual $M(CO)_4$ fragments with local C_{2v} symmetry [18]. These fragments have three filled orbitals, which correspond approximately to the t_{2g} orbital set of an octahedral $M(CO)_6$ molecule, and at higher energies two single occupied orbitals of a_1 and b_2 symmetry. The radial characteristics of the latter two orbitals are such that they permit effective bonding with the corresponding orbitals of the other $M(CO)_4$ unit in the triangle. A molecular orbital diagram according to Schilling and Hoffmann [18] is shown in Fig. 2. The 2e' and 2a' molecular orbitals are responsible for the primary bonding. There are significant interactions between the t_{2g}-like orbitals of the $M(CO)_4$ fragments, and mixing between these orbitals and the higher lying 2e' and 2a' orbitals is possible. Band A (in Table 1) is assigned to the $2a_1'$ orbital and bands B and C to the 2e' orbital, and bands D, E, and F are assigned to the t_{2g}-derived set of orbitals. It has been noted [17] that the first ionization energy is lowered as the cluster size increases, although the decrease is only 0.3 eV for $Os_3(CO)_{12}$ and $Os_6(CO)_{16}$ (Table 1).

TABLE 1. Ionization Energies of Metal d-states of Transition Metal Carbonyl Clusters as Determined from Gas Phase UP Spectra (data from refs. 16 and 17).

Band	Ionization energy, eV				Assignment
	$Ru_3(CO)_{12}$	$Os_3(CO)_{12}$	$Os_6(CO)_{18}$	$Co_4(CO)_{12}$	
A	7.5	7.83	7.52	8.9	M-M
B	7.8	8.28	7.94		
C	9.1	8.48	8.09		
D	9.2	9.24	9.48		t_{2g}
E	9.8	9.60	10.55		
F		10.44			

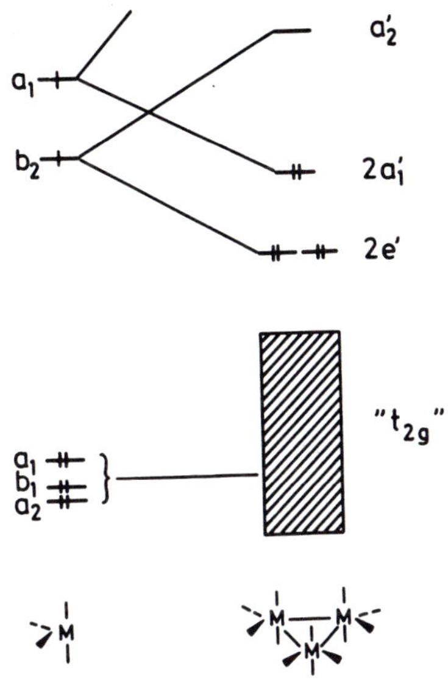

Fig. 2. Simplified molecular orbital scheme of $M_3(CO)_{12}$ (D_{3h}) constructed from $M(CO)_4$ fragments.

For hydrido carbonyl clusters, additional peaks are observed near 12 eV in gas phase UP spectra [17, 19]. Among the clusters investigated were $H_3Re_3(CO)_{12}$, $H_2Os_3(CO)_{10}$, and $H_4Os_4(CO)_{12}$, and the additional photoemission was interpreted essentially in terms of a localized three-center two-electron M-H-M bond model. Brucker *et al.* [20] investigated the effect of ligands bonded through one or two metal atoms on the photoionization spectra of condensed hydridotriosmium clusters. As a result of cluster substitution, the t_{2g} bands were modified, the frontier orbital set was strongly perturbed, and additional lower-lying orbitals were observed in most cases which were considered most likely to be localized on the unique ligand.

The metal core level binding energies are dependent on cluster size. Apai *et al.* [21] measured the binding energies of 4f and 4d core levels of a series of platinum cluster anions $[Pt_3(CO)_6]_n^{2-}$ with n = 2, 3, 4, 5, 6, and ~10. All core-level binding energies increased asymptotically with increasing cluster size and closely approached the asymptotic values at about n = 5, as shown in Fig. 3. The absolute values of the Pt 4f and 4d core levels were appreciably larger than the corresponding values for bulk Pt metal (Fig. 3). The asymptotic value of the Pt $4f_{7/2}$ binding energy for the cluster anions was found to be much closer to that of PtO (73.4 eV) than to that of Pt metal (71.0 eV). The increased binding energies relative to Pt

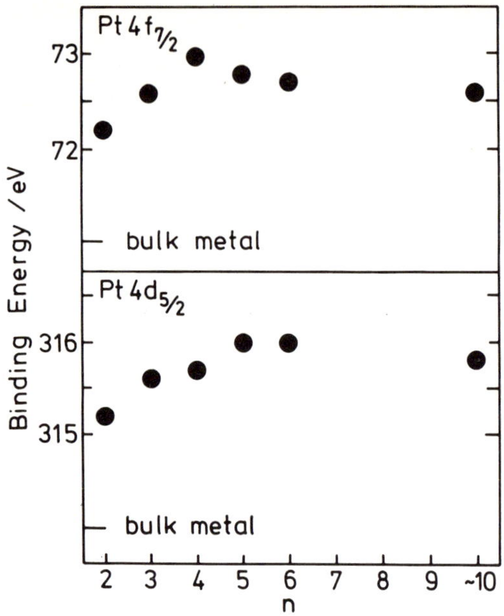

Fig. 3. Pt $4f_{7/2}$ and Pt $4d_{5/2}$ binding energies of $[Pt_3(CO)_6]_n^{2-}$: dependence on cluster size n (from ref. 21).

metal are explained as follows: First, the CO ligands withdraw electrons from the Pt frame. This has been shown by photoemission [22] and work-function [23] measurements to occur also when CO is chemisorbed on Pt metal. Moreover, metal carbonyls generally show core level binding energies higher than those of the bulk metal [24]. Second, there is an incomplete extraatomic relaxation. As shown by Apai *et al.* [21], there is very little interaction between the individual $[Pt_3(CO)_6]^{2-}$ units, and therefore the extraatomic relaxation must be reduced as compared with the bulk metal, and the core level binding energies are correspondingly higher.

Incomplete extraatomic relaxation is also an important factor when bare metal aggregates or clusters are considered. It has been established that the metal core level binding energies for small naked metal aggregates are always higher than those of the bulk metal. For the Pd $3d_{5/2}$ level, for example, a shift of the binding energy of 0.7-2.3 eV towards higher energy has been reported, depending on the nature of the support and its pretreatment [25-27]. In addition to the effects of cluster or aggregate size on the core level binding energies, the d-band structure is also significantly modified. The electronic properties of small metal aggregates and clusters have recently been reviewed by Baetzold and Hamilton [28].

6.5.3 PHOTOELECTRON SPECTROSCOPY OF SUPPORTED CLUSTERS AND CLUSTER-DERIVED SPECIES

Although potentially very useful for the characterization of supported clusters and surface complexes or metal aggregates derived from supported molecular clusters, photoelectron spectroscopic studies of these systems are rather scarce. Only a few XPS studies have been published, while UPS does not appear to have been applied yet. The lack of application of UPS is associated with the interference of support emissions. These are particularly strong in the O 1s region for oxide supports and in the C 1s region for polymer supports. Valence band photoemission from the support may obscure the emission from the d-states. Therefore, experiments have been limited to core level photoionization of the transition metal atoms of the supported clusters or species derived from them.

Anderson *et al.* [29] investigated the behavior of $Rh_6(CO)_{16}$ deposited on Al_2O_3. The Rh $3d_{5/2}$ binding energy of the parent cluster was found to be significantly higher than that of the bulk metal, in agreement with the observations for $[Pt_3(CO)_6]_n^{2-}$, as discussed in the previous section. When the hexarhodium cluster was brought in contact with Al_2O_3, an X-ray-induced sample decomposition was observed. For the fully decarbonylated cluster, the observed Rh $3d_{5/2}$ binding energy was 1.3 eV higher than that of the parent cluster. This was considered to be indicative of an oxidation of each Rh atom relative to $Rh_6(CO)_{16}$, and the oxidized species was tentatively described as $(Rh^+)_6(O_2^-)_6$. On exposure of the decarbonylated species to water-saturated CO, the binding energy shifted back to its initial value, suggesting the occurrence of reversible decarbonylation/carbonylation.

Knözinger [30] reported Rh 3d binding energies for $Rh_6(CO)_{16}$ anchored onto ligand-modified silicas, the values falling into the range of Rh^{1+} complexes. This result is in agreement with the conclusions from infrared carbonyl spectra which had indicated the formation $Rh^+(CO)_2$ and $Rh^+(CO)$ complexes.

In a study of $Os_3(CO)_{12}$ supported on Al_2O_3, Os 4f binding energies provided support for the suggestion that Os^{2+} surface complexes formed after cluster degradation [31]. A shift in binding energies towards lower values after H_2 treatment indicated reduction of these complexes.

XPS data are available for the bimetallic cluster $[H_2RhOs_3(acac)(CO)_{10}]$ anchored to phosphine-functionalized supports [32]. The binding energies of both Rh 3d and Os 4f levels were higher than those for the bulk metals. An interpretation of this result follows the same reasoning as for hexarhodium clusters and triplatinum cluster anions (see above). Also in this case, possible X-ray-induced cluster decomposition could not be completely excluded. Data interpretation was possible only with the aid of additional spectroscopic information.

XP spectra provided evidence for the formation of aggregated Ru metal and cationic complexes of Ru and Os in alumina-supported catalysts prepared from $H_2RuOs_3(CO)_{13}$ or from the combination of $H_4Ru_4(CO)_{12}$ and $H_2Os_3(CO)_{10}$ after their use for CO hydrogenation [33]. Catalysts prepared from $Rh_4(CO)_{12}$ and from $H_2RhOs_3(CO)_{10}(acac)$ supported on alumina gave a Rh $3d_{5/2}$ peak at 307 eV after their use for CO hydrogenation. This binding energy is indicative of zero-valent rhodium; the high binding energies found for the Os 4f levels in the bimetallic catalysts implied that osmium was present in an oxidized state [34].

In connection with bimetallic catalysts, it is worthwhile to mention briefly some investigations performed with highly dispersed materials prepared from precursors other than clusters. The elemental distribution was measured by XPS for the Ru-Cu bimetallic system [35]. Shifts in the binding energies relative to monometallic materials were observed neither for copper nor for ruthenium core levels. This result is suggestive of negligible electronic interaction between the immiscible metals. In contrast, Ioffe *et al.* [36] reported that Pt $4f_{7/2}$ binding energies shifted by 0.5 eV towards higher energy when associated with Re, Mo, or W on a silica support. This increased binding energy was attributed to a decrease in electron density at the Pt atoms by interaction with the modifying ions. As an alternative interpretation based on a comparison of valence band emission from Pt/SiO_2 and $PtMo/SiO_2$, one could argue that an increased dispersion (smaller aggregate size) was responsible for the increased binding energy rather than a direct electronic interaction.

It is concluded that photoelectron spectroscopies, although potentially very useful, may suffer from some sample-dependent drawbacks, e.g., sample sensitivity towards X-ray excitation and masking of emission spectra from clusters or cluster-derived species by photoemission from support materials. However, the very limited experimental results available to date do not permit final conclusions, and more work is required to test the potential of photoelectron spectroscopies for the characterization of supported clusters. Model systems such as clusters anchored onto thin oxide layers which are grown on metal supports may play a key role in this respect. Such model catalysts could be particularly advantageous for UPS studies, the more so since they do not suffer from charging effects [37].

6.5.4 ADDENDUM (RECENT RESULTS)

Kawai *et al.* [38] reported the effect of the nature of the support on the XP binding energies of supported rhodium derived from $Rh_4(CO)_{12}$ and $Rh_6(CO)_{16}$. They observed binding energies of 307.0 and 308.4 eV for the Rh $3d_{5/2}$ level for rhodium supported on SiO_2 and ZnO, respectively. Hence, Rh metal was inferred to be present on SiO_2 and Rh^+ species on ZnO, and relations between the rhodium oxidation state and selectivities in the catalytic hydrogenation of CO were suggested (Section 9.5).

Zanoni *et al.* [39] reported XP spectra of various molecular osmium clusters and complexes. The observed binding energies are summarized in Table 2 together with those of several osmium species supported on SiO_2 [40]. Effects of thermal treatment of the supported materials were also characterized. These experiments confirmed the transformation of physically adsorbed $Os_3(CO)_{12}$ into $HOs_3(CO)_{10}(OSi\equiv)$ at 470-570 K (Section 9.2). Evacuation of the latter sample at 623K gave an additional species with E_b = 50.5 eV. It was therefore inferred that metallic osmium had formed.

TABLE 2. $Os4f_{7/12}$ Binding Energies of Molecular and SiO_2-Supported Osmium Clusters and Complexes (data from refs. [39] and [40]).

Compound	Os Content/ wt%	Q^a	E_b/eV^b
Os, powder	---	0	50.7
$Os_3(CO)_{12}$	---	0	51.6
$Os_3(CO)_{12}/SiO_2$	1.0	0	51.2
(physisorbed)	5.0	0	51.2
$HOs_3(CO)_{10}(OSi\equiv)$	1.0	+0.67	52.5
	3.0	+0.67	52.3
$[Os(CO)_3Cl_2]_2$	---	+2.0	53.1
$[Os(CO)_3Cl_2]_2/SiO_2$	1.0	+2.0	53.4
$Os(CO)_4Cl_2$	---	+2.0	53.4
$Os(CO)_4Cl_2/SiO_2$ (physisorbed)	1.0	+2.0	53.1

[a] Formal oxidation state.

[b] E_b values referenced to C 1s at 285 eV for pure compounds and to Si 2p at 103.5 eV for supported compounds.

REFERENCES

1. W. N. Delgass, G. Haller, R. Kellerman and J. H. Lunsford, Spectroscopy in Heterogeneous Catalysis, Academic Press, New York, San Francisco, London, 1979.
2. C. D. Wagner, W. Riggs, L. E. Davis, J. F. Moulder, and G. E. Muilenberg (Eds.), Handbook of X-ray Photoelectron Spectroscopy, Perkin-Elmer Corp., Eden Prairie, MN (1979).
3. D. W. Davis and D. A. Shirley, Chem. Phys. Lett., 15 (1972) 185.
4. W. L. Jolly and W. B. Perry, J. Am. Chem. Soc., 95 (1973) 5442.
5. W. L. Jolly and T. F. Schaaf, J. Am. Chem. Soc., 98 (1976) 3178.
6. W. L. Jolly, Topics Current Chem., 71 (1977) 149.
7. K. Siegbahn *et al.*, ESCA Applied to Free Molecules, North Holland, Amsterdam, 1979.
8. V. Gelius, Electron Spectr. Rel. Phenom., 5 (1974) 985.
9. M. Barber, J. A. Connor and I. H. Hillier, Chem. Phys. Lett., 9 (1971) 570.
10. I. H. Hillier and J. Kendrich, J. Chem. Soc. Faraday II, 71 (1975) 1369.
11. S. Pignataro, A. Foffani, and G. DiStefano, Chem. Phys. Lett., 20 (1973) 350.
12. N. A. Beach and H. B. Gray, J. Am. Chem. Soc., 90 (1978) 5713.
13. E. W. Plummer, W. R. Salaneck and J. S. Miller, Phys. Rev. B, 18 (1978) 1673.
14. H. Conrad, G. Ertl, H. Knözinger, J. Küppers and E. E. Latta, Chem. Phys. Lett., 42 (1976) 115.
15. a. H.-J. Freund and E. W. Plummer, Phys. Rev. B, 23 (1981) 4859.
    b. H.-J. Freund, E. W. Plummer, W. R. Salaneck and R. W. Bigelow, J. Chem. Phys., 75 (1981) 4275.
    c. H.-J. Freund, F. Greuter, D. Heskett and E. W. Plummer, Phys. Rev. B, 28 (1983) 1727.
16. J. C. Green, E. A. Seddon and D. M. P. Mingos, J. Chem. Soc. Chem. Commun. (1979) 94.
17. J. C. Green, D. M. P. Mingos, and E. A. Seddon, Inorg. Chem., 20 (1981) 2595.
18. B. E. R. Schilling and R. Hoffmann, J. Am. Chem. Soc., 101 (1979) 3456.
19. J. C. Green, D. M. P. Mingos and E. A. Seddon, J. Organomet. Chem., 185 (1980) C20.
20. C. F. Brucker, T. N. Rhodin, S. Wijeyesekera, G. St. George and J. R. Shapley, J. Chim. Phys. Phys.-Chim. Biol., 78 (1981) 897.
21. G. Apai, S.-T. Lee, M. G. Mason, L. J. Gerenser and S. A. Gardner, J. Am. Chem. Soc., 101 (1979) 6880.
22. G. Apai, P. S. Wehner, J. Stohr, R. S. Williams and D. A. Shirley, Solid State Commun., 20 (1976) 1141.
23. A. E. Morgan and G. A. Somorjai, J. Chem. Phys., 51 (1969) 3309.
24. M. Barber, J. A. Connor, M. F. Guest, M. B. Hall, I. H. Hillier and W. N. E. Meredith, Faraday Disc. Chem. Soc., 72 (1972) 219.
25. E. J. Robbins, R. E. Leckenby and P. Willis, Advan. Phys., 16 (1967) 739.
26. M. G. Mason, L. G. Gerenser and S. T. Lee, Phys. Rev. Lett., 39 (1977) 288.
27. Y. Takasu, R. Unwin, B. Tesche and A. M. Bradshaw, Surface Sci., 77 (1978) 219.
28. R. C. Baetzold and J. F. Hamilton, Progr. Solid State Chem., 15 (1983) 1.
29. S. L. T. Andersson, K. L. Watters and R. F. Howe, J. Catal., 69 (1981) 212.
30. H. Knözinger, Inorg. Chim. Acta, 37 (1979) L537.

31. H. Knözinger, Y. Zhao, B. Tesche, R. Barth, R. Epstein, B. C. Gates and J. P. Scott, Faraday Disc. Chem. Soc., 72 (1981) 53.
32. J. Lieto, M. Wolf, B. A. Matrana, M. Prochazka, B. C. Gates, H. Knözinger and B. Tesche, J. Phys. Chem., 89 (1985) 991.
33. J. R. Budge, B. F. Lücke, J. P. Scott and B. C. Gates, Proc. 8th Int. Congr. Catal., Berlin, Dechema, Frankfurt, 1984, Vol. 5, p. 89.
34. J. R. Budge, B. F. Lücke, B. C. Gates, and J. Toran, J. Catal., 91 (1985) 272.
35. C. R. Helms and J. H. Sinfelt, Surface Sci., 72 (1978) 229.
36. M. S. Ioffe, Yu. M. Shulga, Yu. A. Ryndin, B. N. Kuznetsov, A. N. Startsev, Yu. G. Boronko and Yu. I. Yermakov, React. Kinet. Catal. Lett., 4 (1976) 229.
37. K. Wandelt, personal communication.
38. M. Kawai, M. Uda and M. Ichikawa, J. Phys. Chem., 89 (1985) 1654.
39. R. Zanoni, V. Carinci, Raja H. Abu Samn, R. Psaro and C. Dossi, J. Mol. Struct., in press.
40. C. Furlani, R. Zanoni, R. Psaro and C. Dossi, paper presented at Sixteenth Italian Congr. Inorg. Chem., 1985; to be published.

Section 6.6

MÖSSBAUER SPECTROSCOPY

L. GUCZI

6.6.1 INTRODUCTION

Mössbauer spectroscopy is a relatively sensitive and sophisticated technique providing detailed information about the bonding of atoms of a few particular elements. The chief disadvantage is that but a few elements (e.g., Fe) have Mössbauer nuclei with low enough γ-transition energies (<100 keV) to provide a sufficiently high recoil-free fraction (see below) at room temperature. For other nuclei (e.g., W, Pt, Ir, Au, etc.) low-temperature measurements are required. Since the method is primarily feasible for characterizing bulk properties of solid samples, its applicability to highly dispersed systems such as oxide-supported metal clusters or aggregates sometimes requires highly developed and advanced techniques. The experiment involves the application of low-temperature and an external magnetic field. Despite the difficulties caused by the highly dispersed state, a great advantage of Mössbauer spectroscopy is that it provides the opportunity to characterize catalysts under reaction conditions.

In this section a brief description of the physical principle is given and the applicability of Mössbauer spectroscopy for characterization of highly dispersed supported metals is illustrated.

6.6.2 PHYSICAL BACKGROUND

Mössbauer spectroscopy or nuclear gamma resonance [NGR] spectroscopy is a sensitive tool for investigation of the solid state. Small changes of nuclear transition energies can be detected. These energy shifts can be related to changes of the chemical state and/or of the magnetic and electronic environments of an atom. Details of the Mössbauer effect as well as its theory have been discussed in several reviews [1-4].

The basis of the Mössbauer technique is the fact that resonant absorption and emission of γ-quanta can occur when the emitting and absorbing nuclei are incorporated in a solid matrix. Due to the relatively long lifetime of the excited nuclear state (typically 10^{-8} s, corresponding to a natural linewidth of 10^{-8} eV), the emitting photon has a very narrow line width. Hence, small variations in the nuclear energy levels induced by changes in the chemical state and/or the environment of the Mössbauer nucleus can be probed. Experimentally, changes in the nuclear energy levels are measured by modifying the photon energy of the probing γ-quanta by applying a Doppler shift (*transmission Mössbauer spectroscopy*). The

γ-source is attached to a transducer, the velocity of which is controlled by a ramp wave. This is simultaneously fed into a multichannel analyzer. The transmitted γ-quanta are detected by a scintillation crystal or a gas-filled proportional counter. The spectrum is obtained by plotting the count rate in each channel vs. the corresponding velocity of the source.

Unlike transmission Mössbauer spectroscopy, *conversion electron Mössbauer spectroscopy* can be applied as a surface-sensitive technique. Once a Mössbauer nucleus is excited, deexcitation produces a high yield of emitted conversion electrons, which for K-shell emission typically have initial kinetic energies of the order of several keV. Therefore these are surface-sensitive probes for the topmost atomic layers of the sample (probing depth *ca.* 200 nm).

The Mössbauer effect can be observed provided that the Mössbauer nucleus is incorporated into a solid matrix. The absorption or emission of γ-quanta will be recoil free when the final state of the solid is identical to the initial state. If the recoil energy E_R is smaller than the phonon energy $\hbar\omega_E$ of an Einstein solid having equally spaced vibrational levels, emission and absorption will be recoil free with the probability

$$f = \exp(-E_R/\hbar\omega_E), \tag{1}$$

or (in a more general form)

$$f = \exp[-\langle x^2\rangle E_0^2/(\hbar c)^2], \tag{2}$$

where f is the recoil-free fraction of the emission or absorption, E_0 is the nuclear transition energy, and $\langle x^2\rangle$ is the mean square displacement of the Mössbauer atom. It can be seen from Eq. (2) that f is nearly unity in a rigid lattice, whereas this value is low for a soft vibrational mode. When, instead of the Einstein model, the Debye model is applied to approximate the phonon spectrum of the solid, the recoil-free fraction f is represented by eq. (3):

$$f = \exp\left[\frac{-6E_R}{k\Theta_D}\left\{\frac{1}{4} + \left(\frac{T}{\Theta_D}\right)^2 \int_0^{\Theta_D/T} \frac{x\,dx}{e^x - 1}\right\}\right], \tag{3}$$

where Θ_D is the Debye temperature and k the Boltzmann constant. Θ_D is normally high for a rigid lattice and low for a lattice having soft vibrational modes; f increases with decreasing temperature, as shown in Fig.

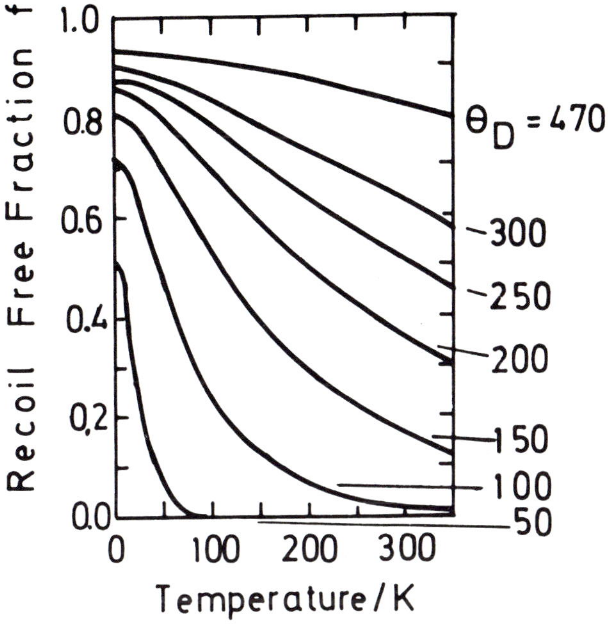

Fig. 1. Temperature dependence of recoil free fraction f for varying Θ_D values.

1 [5]. There is a practical upper limit of approximately 100 keV on E_0. Low temperature is needed for measurable f values when E_0 is high or Θ_D is low. This is what limits the applicability of Mössbauer spectroscopy at room temperature.

As mentioned above, transitions between energy states of a nucleus are sensitive to the environment of the atom studied. The corresponding interactions are governed by a nuclear parameter and by an electronic parameter. Three main parameters characteristic of these nuclear hyperfine interaction parameters are to be considered: (1) the isomer shift, (2) the quadrupole splitting, and (3) the magnetic hyperfine splitting.

The *isomer shift* (or chemical shift) is a measure of the difference of the electron density within the finite nuclear volume of an atom in the absorber relative to that of an atom in the source and is a consequence of the Coulomb interaction between the positively charged nucleus and the electrons [5].

The isomer shift IS can be expressed as follows [1-4]:

$$IS = \frac{4\pi Ze^2R^2c}{5E_0} \cdot \frac{\Delta R}{R} \left[\sum_A |\Psi(0)|^2 - \sum_S |\Psi(0)|^2 \right], \tag{4}$$

where Z is the atomic number, e the charge on an electron, R the average nuclear radius $(R_e + R_g)/2$, (R_e and R_g are the radii of nuclear excited and ground states, respectively), E_0 the transition energy, and $\Delta R = R_e - R_g$. The last term describes the chemical contribution to the isomer shift, $\sum_A |\psi(0)|^2$ and $\sum_S |\psi(0)|^2$ representing the electron densities at the nucleus of the absorber and of the source, respectively. The s and relativistic $p_{1/2}$ electrons have a finite probability of being at the nucleus. Screening effects from orbitals which do not have a finite electron density at the nucleus will modify the IS value. It is therefore possible to monitor chemical characteristics of the Mössbauer atom such as the oxidation state and strength of bonding between the Mössbauer atom and the surrounding atoms or ions.

Quadrupole splitting arises with nuclei having spin quantum numbers greater than $I = \frac{1}{2}$, which therefore have nonspherical charge distributions. The nuclear quadrupole moment Q is a measure of the magnitude of this nonsphericity of the charge distribution. As a result of nonspherical electronic charge of the atom (or noncubic symmetry of surrounding ions) an electric field gradient q will be produced; the interaction between the nuclear quadrupole moment and the electric field gradient lifts the degeneracy of the nuclear spin and hence leads to a multiplicity of transitions which are observed in the Mössbauer spectrum. The number of transitions depends on the spin quantum numbers I_g and I_e of the nuclear ground and excited states. Assuming spin quantum numbers $I_g = \frac{1}{2}$ and $I_e = \frac{3}{2}$ (e.g., ^{57}Fe), the spectrum will consist of two lines. In this case, and assuming axial symmetry (i.e., the electric field gradients V_{xx} and V_{yy} are equal and $|V_{xx}|,|V_{yy}|>|V_{zz}|$, z being the direction of the nuclear spin axis), the magnitude of the quadrupole splitting QS is given as

$$QS = \frac{1}{2}e^2gQ(1 + \frac{1}{3}\eta^2)^{1/2}, \tag{5}$$

where $\eta = \frac{(V_{xx}-V_{yy})}{V_{zz}}$ is the asymmetry parameter. The magnitude of QS is determined by the two parameters eg and eQ, where eQ is a nuclear constant for a given isotope and eg depends on the intra- and interelectronic configuration around the isotope.

Magnetic hyperfine splitting occurs as a result of the interaction of a nuclear magnetic dipole moment with a magnetic field. As a result, the degeneracy of the 2I + 1 magnetic sublevels will be lifted. For iron, for example, this results in a characteristic six-line spectrum. The measurement of the magnetic hyperfine splittings permits determination of magnetic properties of the sample and may also give information concerning particle sizes, since the temperature at which magnetic ordering occurs depends critically on the dimensions of the particle [1-4].

6.6.3 MÖSSBAUER SPECTROSCOPY OF SUPPORTED METAL CLUSTERS AND PARTICLES

The transformations of $Fe(CO)_5$ occurring on the surface of TiO_2 were investigated with Mössbauer spectroscopy at room temperature [6]. A zerovalent species detected after partial decomposition was inferred to be a $Fe(CO)_2$ subcarbonyl associated with surface hydroxyl groups. After decomposition under more severe conditions, high-spin Fe^{2+} was observed.

Investigations of $Fe_3(CO)_{12}$ on Cab-O-Sil and Al_2O_3 clearly revealed the necessity of low-temperature measurements for unambiguous interpretation of the Mössbauer spectra [7, 8]. The freshly prepared material gave a quadrupole doublet at 77 K, as shown in Fig. 2, spectrum b (IS = 0.42 mm s^{-1}; QS = 0.95 mm s^{-1}), which arises from a high-spin Fe^{3+} component. The results of complementary methods such as ^{13}CO exchange experiments and infrared spectroscopy [7] demonstrated the presence of the original triiron cluster in addition to the oxidized species. The conclusions were confirmed by the Mössbauer spectra taken at 1.6 K (Fig. 2, spectrum c), which showed the magnetic hyperfine splitting characteristic of α-Fe_2O_3 and in addition included a three-line feature typical of molecular $Fe_3(CO)_{12}$, as shown in Fig. 2, spectrum a. Consequently, only the low-temperature spectra give unambiguous information regarding the structures of highly dispersed catalysts [9]. Spin-to-spin relaxation is a characteristic of the highly dispersed state [10]. Adsorption of Fe^{3+} ions from aqueous solution at pH = 1 or, alternatively, pH = 3, gives different Mössbauer spectra of the original surface species, and also of the iron catalysts formed by reduction. The Mössbauer spectra of these samples in the initial, calcined, and reduced states are shown in Fig. 3. The hyperfine field of 574 kOe characterizing the sample prepared at pH = 1 represents the spin-to-spin relaxation resulting from the quasi-isolation of the Fe^{3+} ions on the surface. In contrast, the observation of the sample prepared at pH = 3, shows that partial precipitation occurred during drying, resulting in the formation of large iron oxyhydrate particles, a high nucleation rate during the reduction of the sample in hydrogen, and, as a result, to the formation of large α-iron particles (Fig. 3B, curve c). In this series of experiments low-temperature measurements were necessary.

Since in highly dispersed catalysts the proportion of the surface atoms is high, attempts have been made to assign particular peaks in the Mössbauer spectra to surface components. Vannice *et al.* [11] investigated a conventionally prepared Ru-Fe catalyst supported on silica. After

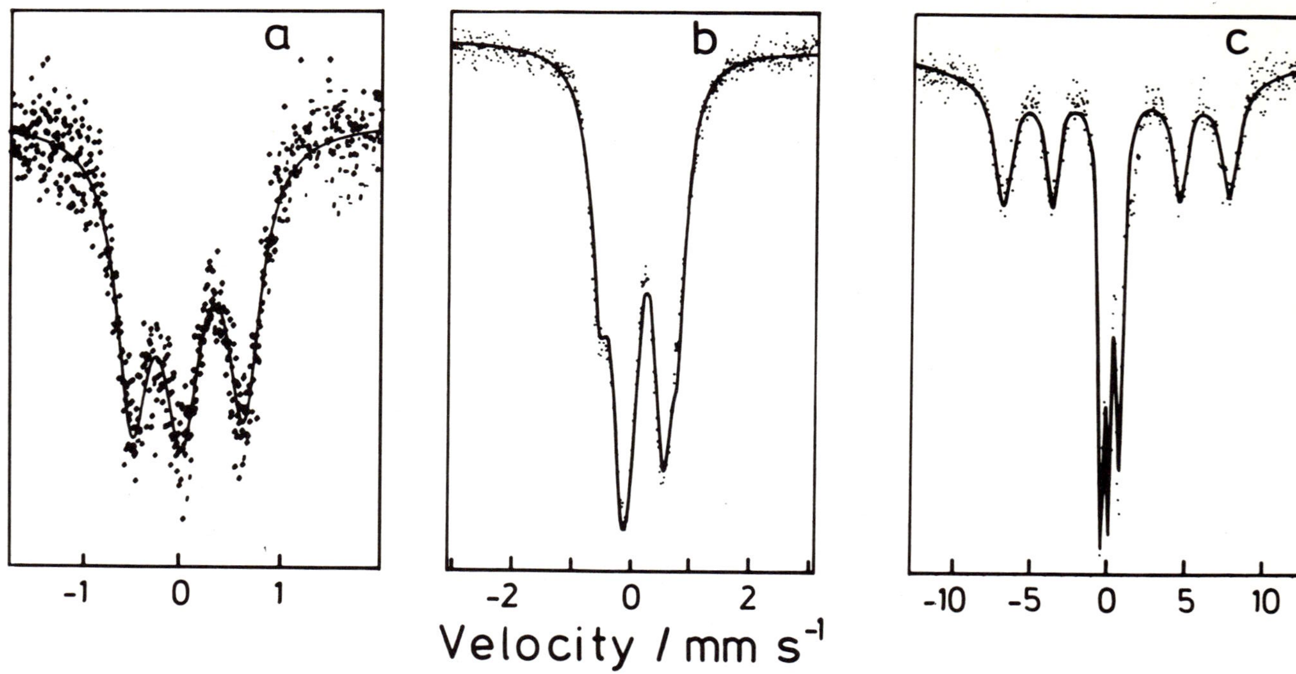

Fig. 2. Mössbauer spectra of $Fe_3CO)_{12}$ (a) in hexane solution at 77 K; (b) deposited on Cab-O-Sil, measured at 77 K and (c) measured at 1.6 K [8, 9].

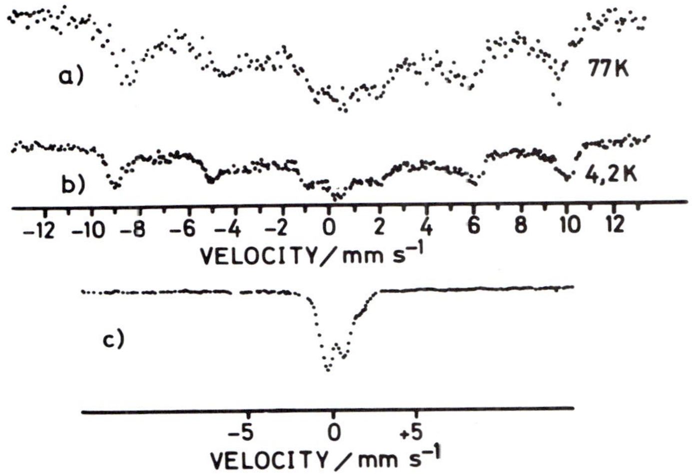

Fig. 3A. Mössbauer spectra of 0.5 wt% Fe on SiO_2 impregnated at pH = 1: after calcination measured at a) 77 K; b) 4.2 K; and c, after reduction in H_2 at 970 K [10].

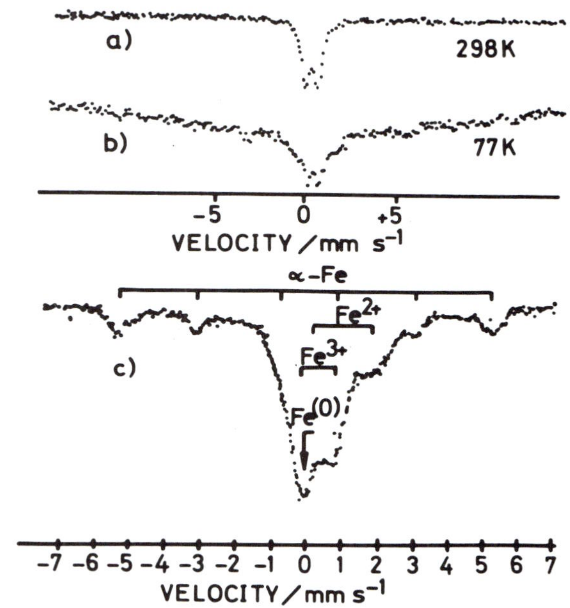

Fig. 3B. Mössbauer spectra of 0.5 wt% Fe on SiO_2 impregnated at pH = 3: after calcination measured at a) 298 K; and b) 77 K; and c, after reduction in H_2 at 970 K [10].

reduction at 773 K, a singlet and a doublet were observed. The authors assigned the singlet to an Fe-Ru particle in the bulk and the doublet to a surface component of the bimetallic aggregate. Others have disagreed with the assignment [12], and Niemantsverdriet *et al.* [13-15] found evidence that the doublet is not due to a surface component of FeRu aggregates, but instead to the presence of unreduced high-spin Fe^{3+} oxide.

Recent studies of reduced, cluster-derived Fe-Ru catalysts supported on Cab-O-Sil at low metal loading give further evidence for this assignment [16]. The Mössbauer spectrum measured at 4.4 K of a reduced bimetallic cluster displayed a singlet (IS = 0.0 mm s^{-1}) and a sextet assigned as α-Fe_2O_3 with a hyperfine field of 431 kOe (IS = 0.52 mm s^{-1}) (Fig. 4). The doublet evidently corresponds to Fe_2O_3, which can be identified only by the hyperfine splitting observed at low temperature. The singlet can be assigned to zerovalent iron of very small particle size since a finite hyperfine field is not detectable even at 4.4 K.

High-field measurements provide evidence of small Fe-Ru particles. The effect of an external magnetic field on the line at IS = 0.0 mm s^{-1} is shown in Fig. 5. The 2T field leads only to a line broadening (the half width) increases from 0.51 mm s^{-1} to 1.9 mm s^{-1}. If small ferromagnetic particles were present, complete splitting of the central line by 2T magnetic field would have been expected. On the other hand, FeRu alloys are Pauli paramagnets in the concentration range of 12-100% Ru, and consequently, the line splitting is expected to be proportional to the magnetic field strength. In fact, the central pair of lines of the 6T spectrum also includes the split IS = 0.0 mm s^{-1} line. The 2.0 mm s^{-1} line separation obtained for the paramagnetic component corresponds to a 62 kOe field. These results suggest the formation of highly dispersed iron-ruthenium particles.

Characterization of small particles can also be carried out by measuring the Mössbauer spectra in the absence and in the presence of H_2 and CO adsorbed on the surface. With iron particles prepared by the impregnation of Cab-O-Sil using an aqueous solution of $Fe(NO_3)_3$, Clausen *et al.* [17] found the first Mössbauer evidence that magnetic and electronic properties of iron atoms at a solid-gas interface are different from those of bulk iron atoms. It was found that the CO and H_2 chemisorption give rise to an increase of the magnetic hyperfine field, the isomer shift, and the relative spectral area of the surface component. These results suggest that the chemisorption-induced changes are localized and only the surface is affected.

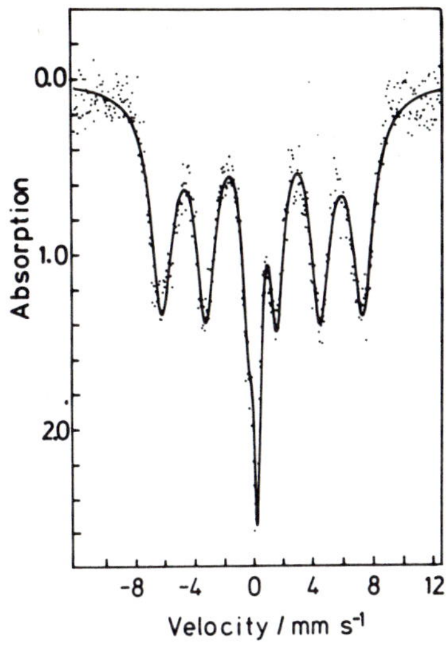

Fig. 4. Mössbauer spectra of FeRu/Cab-O-Sil catalyst measured at 4.4K [16].

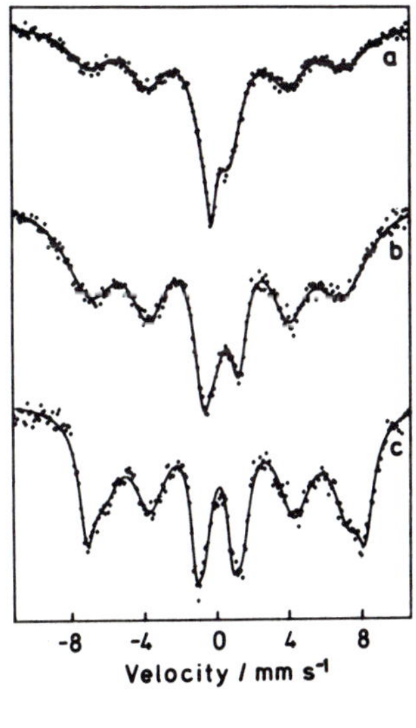

Fig. 5. Mössbauer spectra of FeRu/Cab-O-Sil catalyst in the presence of an external magnetic field at 4.2 K; at magnetic field a) 0 T, b) 2T, c) 6 T.

6.6.4 SUMMARY

The determination of hyperfine parameters by means of Mössbauer spectrscopy may provide detailed information characterizing the physical state of cluster-derived iron species on support surfaces. However, even for iron, low-temperature (liquid He) is frequently required to elucidate fully the nature of the iron species present in the material under study. In conclusion, the application of Mössbauer spectroscopy for characterization of cluster-derived catalysts is undoubtedly far more limited than the application of the other physical techniques discussed in this chapter.

REFERENCES

1. W. N. Delgass, G. L. Haller, R. Kellerman and J.H. Lunsford, Spectroscopy in Heterogeneous Catalysis, Academic Press, New York, 1979.
2. N. N. Greenwood and T. C. Gibbs, Mössbauer Spectroscopy, Chapman and Hall, London, 1971.
3. V. I. Gol'danskii and R. H. Herber, (Eds.), Chemical Application of Mössbauer Spectroscopy, Academic Press, New York, 1968.
4. J. A. Dumesic and H. Topsøe, Advan. Catal., 26 (1976) 122.
5. J. W. Niemantsverdriet, Ph.D. Thesis, Delft, 1983.
6. J. Phillips and J. A. Dumesic, Appl. Catal., 9 (1984) 1.
7. K. Lázár, K. Matusek, J. Mink, S. Dobos, L. Guczi, L. Markó, A. Vizi-Orosz and W. M. Reiff, J. Catal., 87 (1984) 163.
8. I. Böszörmenyi, S. Dobos, L. Guczi, L. Markó, K. Lázár, W. M. Reiff, Z. Schay, L. Takács and A. Vizi-Orosz, Proc. 8th Intern. Congr. Catal., Berlin, 1984, Vol. V, p. 183.
9. K. Burger and Z. Vertes, Mössbauer Spectroscopy, Akadémiai Kiadó, Budapest, 1982.
10. M. Eszterle and L. Guczi, Magyar Kemiai Folyóirat, 36 (1978) 429.
11. M. A. Vannice, Y. L. Lam and R. L. Garten, Prepr. ACS Div. Petrol. Chem., 23 (1978) 495.
12. L. Guczi, Catal. Rev.-Sci. Eng., 23 (1981) 329.
13. J. W. Niemantsverdriet, A. M. van der Kraan and J. J. van Loef, J. Phys. Chem., 87 (1983) 1292.
14. H. F. J. van't Blik and J. W. Niemantsverdriet, Appl. Catal., in press.
15. J. W. Niemantsverdriet, D. P. Aschenbeck, F. A. Fortunato and W. N. Delgass, J. Mol. Catal., in press.
16. K. Lázár, L. Guczi and W. M. Reiff, to be published.
17. B. S. Clausen, S. Mørup and H. Topsøe, Surface Sci., 106 (1981) 438.

Section 6.7

X-RAY ABSORPTION SPECTROSCOPY APPLIED TO METAL CLUSTERS

R. F. PETTIFER

6.7.1 INTRODUCTION

X-ray absorption spectroscopy (XAS) offers exciting potential for probing the local structure surrounding a particular type of atom in a material. This is the salient feature that sets this technique apart fom X-ray or neutron scattering. Despite this tremendous advantage, X-ray absorption spectroscopy is regarded as only a complementary technique to other scattering-based measurements for structure determination. This point is amplified in Section 6.7.3 below.

The X-ray absorption edges of two elements with atomic numbers differing by 1 are illustrated in Fig. 1; the elements (As and Se) are

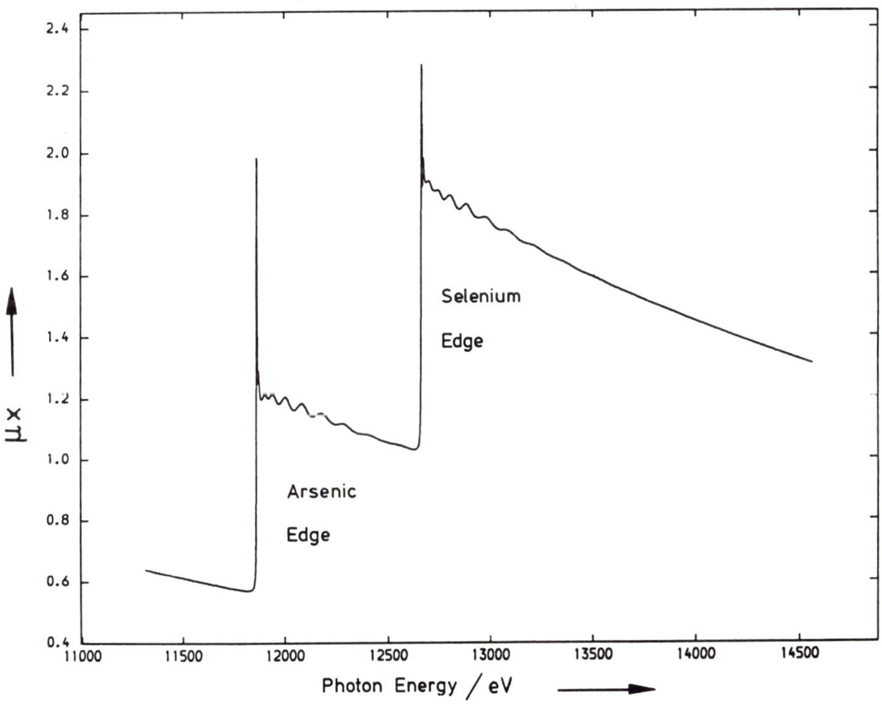

Fig. 1. The absorption thickness product μx for glassy As_2Se_3. The figure illustrates the separation of two absorption edges so that structural information can be obtained surrounding two atomic species. Both edges have a very sharp singularity of absorption at the edge, resulting from essentially atomic-like transitions.

combined in a glassy material As_2Se_3. The spectrum illustrates several important points: Above both edges there are oscillations which can be interpreted to give structural information. Close to the onset of absorption of the arsenic edge (at ~11865 eV) and the selenium edge (at ~12655 eV) there are sharp features called white lines. These were so named because the spectra were recorded on films, and points of high absorption produced no film darkening. These features, together with any other fine structure found within ~50 eV from the absorption edge, are referred to as XANES (X-ray absorption near-edge structure), Kossell structure, or near-edge structure. The oscillations extending beyond ~50 eV from the edge are called EXAFS (extended X-ray absorption fine structure) or Kronig structure. The delimiting energy between XANES and EXAFS is a rather fuzzy parameter depending on the material under investigation.

EXAFS and XANES have a common origin, but a different emphasis is placed in the interpretation of the two regions. Both sets of oscillations arise from an electron scattering process. When the X-ray energy reaches a sufficiently high value to promote an electron from a deep level, energy is absorbed from the incident photon beam. This results in a sudden increase in the absorption coefficient, i.e., the X-ray absorption edge has been reached. Just at the onset of absorption, the electron from the deep core level has been promoted to a higher unoccupied level of the atom but is possibly still trapped in the atomic potential, i.e., it is not free to escape the parent atom.

In solids or polyatomic gases, these atomic levels are of course modified into molecular orbitals or valence bands. This is the basic origin of the white lines shown in Fig. 1. They are transitions to conduction bands originating from empty atomic p states. This type of description is familiar to atomic or molecular spectroscopists; there is an equivalent explanation based on scattering theory which provides a unified insight into the interpretation of the complete spectrum, as follows:

The incident X-ray photon creates a perturbation of the electrons in the system, principally via its electric field. Symmetry arguments show that the excitation of a 1s electron must result in a final state of p symmetry. In fact the usual dipole selection rule $\Delta l = \pm 1$ applies. This is why p states were specified for the origin of the white lines observed at the arsenic and selenium edges in Fig. 1. The scattering theory interpretation states that the electron wave, created by photoabsorption, is strongly scattered to produce a standing wave. If the standing wave overlaps with a perturbed core state then strong absorption results--hence the white lines observed in Fig. 1.

Scattering in this case comes principally from the atomic potential of the atom being excited. When the atom is embedded in a solid or molecular environment the boundary conditions of the atomic potential are perturbed by the presence of the surrounding atoms. This in turn shifts the energy and possibly the width of the white line, depending on the environment. thus structural information is present from the immediate near edge data. As the energy of the incident photon is increased, then the energy of the excited photoelectron is also increased. Eventually the atomic potential is not sufficiently strong to retain the electron, and scattering from surrounding atoms may then build up standing waves which result in changes in the absorption spectrum.

At low energies, atomic potentials may scatter strongly in all directions (Fig. 2), and the paths taken by the electron to form a standing wave may

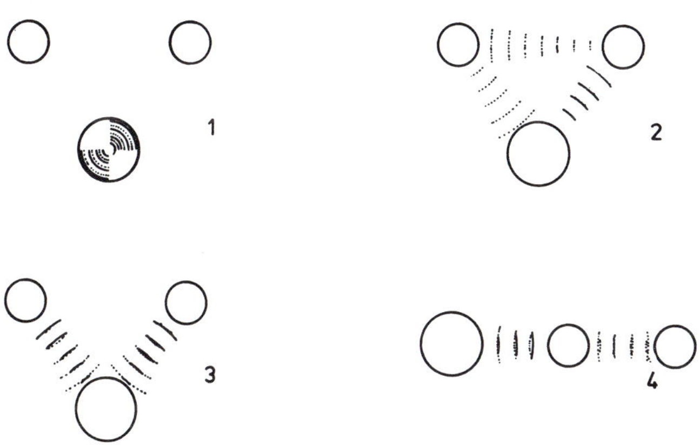

Fig. 2. An illustration of the various scattering processes involved in X-ray absorption. (1) Scattering by the potential of the absorbing atom. The type of scattering is principally responsible for the white line at the edge and is largest for low energy photoelectrons. (2) Multiple scattering, where the electron is scattered a significant amount through large angles. Paths such as that shown are also expected to contribute significantly in the near-edge region. (3) Back-scattering of the photoelectron which is dominant when its energy is large in the EXAFS region. (4) A special type of multiple scattering which persists at high energies and occurs when the emitter and two scatterers are colinear. (The emitted wave is drawn as a solid line, and the scattered wave is dotted.)

be complex and sensitive to the geometry of the surrounding ligands. This relative isotropy of the scattering depends in general on the ligand and on the energy of the photoelectron, usually persisting up to about 50 eV from the edge and, together with the long mean free paths of low-energy photoelectrons, gives rise to multiple scattering in the XANES region.

As the photoelectron energy is increased further beyond 50 eV, the scattering processes are simplified, in that there are basically only two directions of scattering, forward and backward. Thus, unless atoms lie in line with each other, multiple scattering becomes small in comparison with single back-scattering events. This region is called EXAFS, and oscillations in the absorption coefficient may well persist up to 2000 eV from the absorption edge.

The various regions are represented diagrammatically in Fig. 2. The effects illustrated in Fig. 2 do not have sharp transition regions and in fact overlap considerably. With some materials, for example, a white line is not observed in the atomic case, simply because the atomic potential is not sufficiently strong to trap an electron with the appropriate symmetry.

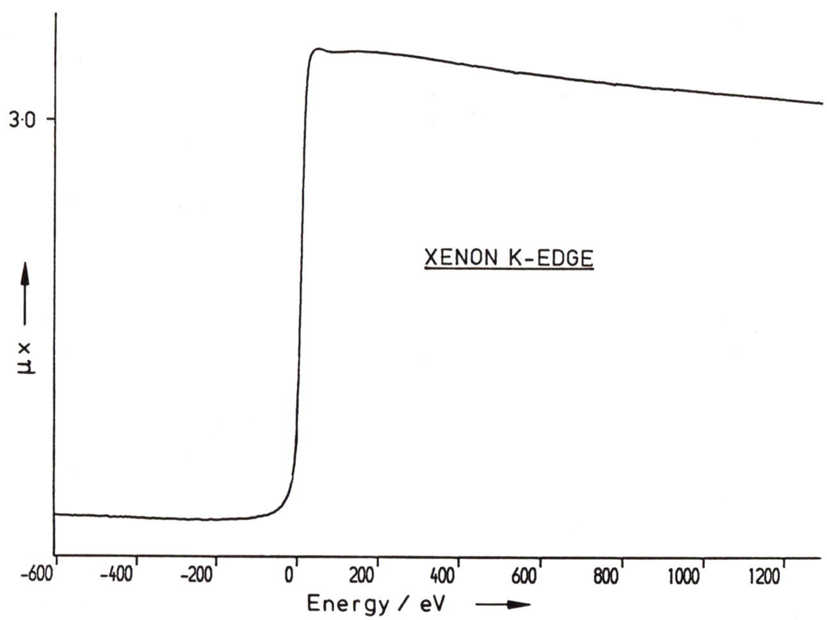

Fig. 3. The X-ray absorption edge of monatomic gas, Xe, which shows essentially featureless behavior appart from atomic scattering. The energy axis is relative to the absorption edge and is practically the same as the photoelectron energy.

An example is given in Fig. 3, where the atomic Xe K absorption edge has been measured. No white line is observed since Xe has no bound excited p state. No EXAFS is observed since the Xe atoms have no near neighbors. The slight fluctuations in the absorption coefficient result from the weak atomic scattering still evident 100 eV above the absorption edge. The extent to which atomic scattering (process 1 in Fig. 2) rather than molecular scattering (process 2 in Fig. 2) dominates the near-edge spectrum is determined by considerations closely analogous to those determining the use of tight binding or pseudo-potential techniques in solid state calculations. Thus, the spectrum depends on the materials; ionic solids have a stronger atomic influence at the edge than metals.

At this point it is worthwhile to emphasize that, despite the name X-ray spectroscopy, we are concerned with *electron* scattering. It is the final state of the electron that controls the absorption coefficient. The essential difference between XAS and other scattering techniques lies in the fact that for XAS the source and detector are the same, i.e., the atoms being excited, and they are embedded inside the material under study. A free atom will produce virtually a structureless absorption edge, apart from atomic scattering, whereas an atom in a molecule or solid will cause the emitted photoelectron to be scattered; the scattered wave interferes with the source and thus changes the absorption coefficient.

A good analogy has been given by Ashley and Doniach [1], who called EXAFS "spherical low energy electron diffraction." Indeed the vast progress in the field of XAS can be directly traced to the enormous theoretical progress in low energy electron diffraction (LEED) theory during the 1970's. The ranges of energies involved in XAS and LEED are the same, so that many of the electron atom scattering problems are identical. We introduce here a further analogy for EXAFS oscillations. We consider the EXAFS process to be similar to what occurs in an interferometer. Scanning the wavelength of light incident on an interferometer is equilavent to scanning the wavelength or energy of the photoelectron. At low energies, scattering may be very strong, and as a result narrow resonances may form--analogous to a narrow band-pass for highly silvered mirrors in the interferometer. At high energies, scattering is weak and the resonances are broad, suggesting the weakly reflecting mirrors in the interferometer.

Each atom scatters electrons differently, and in principle this information can tell us the type of ligand surrounding a particular type of atom. For example, examining an osmium absorption edge, we can in principle distinguish between osmium neighbors and (say) carbon or oxygen.

This qualitative description also explains other qualitative observations. In Fig. 1, we observe that the fine structure in the EXAFS region appears simply as an amplitude-modulated sine wave. The amplitude modulation reflects the variation in scattering with energy of the surrounding ligands; a maximum in the scattering occurs at approximately 200 eV from the edge for both the arsenic and selenium edges because the surrounding ligands are of similar atomic number and back-scatter in a similar manner. As in X-ray scattering, discrimination between atoms of similar atomic number is very difficult.

The simplicity of fine structure seen in Fig. 1 can be contrasted with the fine structure shown in Fig. 4 for a niobium edge observed for the solid with a body-centered cubic (bcc) structure. The niobium edge for the crystalline solid is far more complex than that shown in Fig. 1 for the glass. Indeed the complexity of fine structure seen in Fig. 4 is to be expected because there are many atoms surrounding an absorbing atom in a solid, and these all contribute coherent scattering paths like 3 and 4 in Fig. 2 and give rise to a complex interference pattern. The fact that the

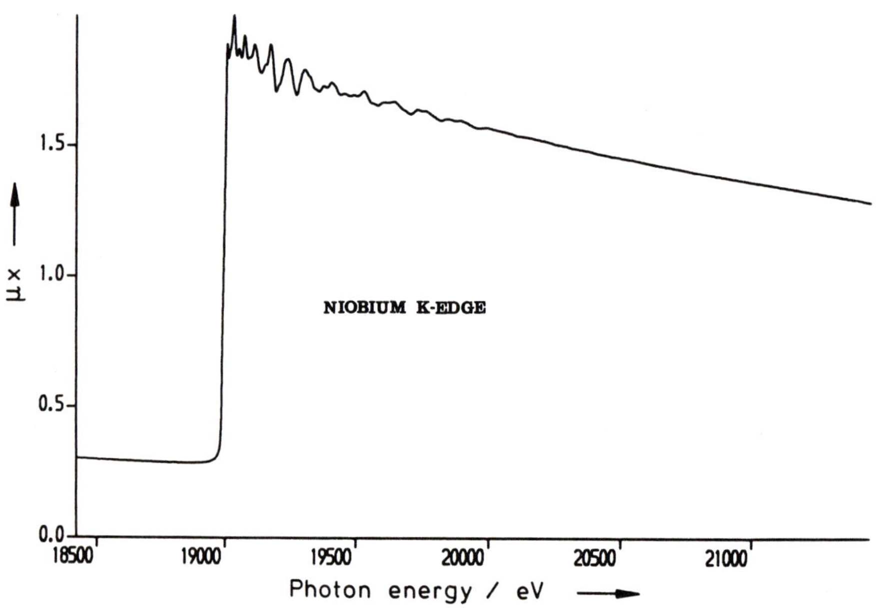

Fig. 4. The X-ray absorption edge of a body-centered cubic material, niobium. The complexity of the fine structure testifies to the many paths which contribute to scattering in this well-ordered material.

spectrum of the glass does not show this complexity relates to the fact that the atoms surrounding As and Se are ordered only in the immediate vicinity of the absorbing atom. Other atoms outside the first coordination shell of either arsenic or selenium obviously are present, but the distances from the absorbing atom are random, which removes phase coherence of the scattered waves when viewed as an ensemble average.

Of course, all macroscopic scattering techniques yield an ensemble average, and this frequently complicates analysis when multiphase materials are examined. More seriously, despite the fact that atoms are present in the glassy material beyond the first coordination shell, they are not detetable with XAS.

Experimentally it is found that the amplitude of the fine structure is temperature sensitive. At some hundreds of electron volts above the edge, the fine structure amplitude always increases in magnitude as the temperature is lowered. The origin of this phenomenon is the same as that of the removal of higher shell scattering in glasses. Simply, thermal vibrations cause a loss of phase coherence in the scattering paths. The higher the temperature, the larger the thermal vibrational amplitudes and the greater the loss of coherence and hence fine structure amplitude. Disorder, both thermal and static, may be evaluated quantitatively; a discussion of this is presented in Section 6.7.3.

In summary, this section is a qualitative, oversimplified explanation of the origins of structure in XAS. The reader is referred to excellent reviews of XAS for more detail [2, 3]. In Section 6.7.2 some experimental aspects are presented, and in Section 6.7.3 the simple quantitative theory is presented together with a statement of its limitations. Section 6.7.4 is a review of data characterizing osmium clusters assessed in light of the foregoing discussion.

6.7.2 EXPERIMENTAL TECHNIQUES

6.7.2.a **Data collection.** Compared with the interpretation of the XAS, the measurements themselves are straightforward. In principle, either radiation from a conventional X-ray tube or from a synchrotron can be used as a source. Both sources have their advantages, and the main difference is in flux for an equivalent bandwidth. Conventional sources provide 10^5 times less flux at the sample than synchrotron sources. Notwithstanding the flux difference, the quality of the spectrum taken with a conventional source is usually only a factor of 10^2 worse for selected wavelengths, as measured in terms of signal to noise ratio, bandwidth, etc.

This statement is true because time, which is highly restricted at synchrotron sources, can be traded for intensity.

The following is a description of the essential properties of a synchrotron radiation spectrometer to perform X-ray absorption measurements. A review of X-ray absorption spectrometers using conventional sources is contained in the monograph edited by Stern [4]. Synchrotron radiation [5, 6] produces a continuous range of wavelengths. The short wavelength limit of the radiation is controlled by the bending radius and energy of the high-energy particles (electrons or positrons) stored or accelerated in the accelerator ring. A schematic diagram of a synchtrotron spectrometer is shown in Fig. 5. This spectrometer is described in detail by Hermes et al. [7] and is installed in Hasylab on the synchrotron storage ring DORIS II at DESY, Hamburg. The essentially white radiation from the ring is incident on a crystal surface of silicon

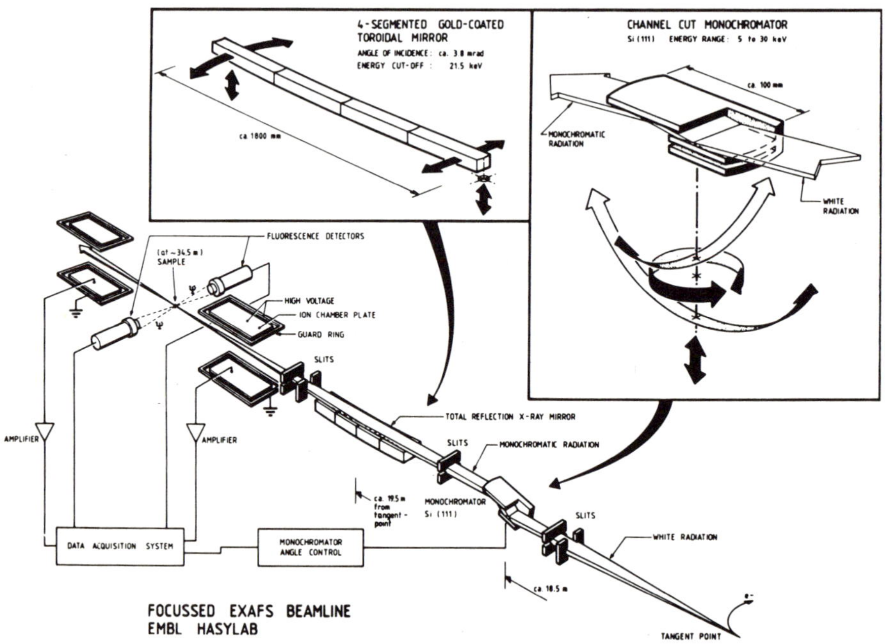

Fig. 5. Schematic diagram of a synchrotron radiation X-ray spectrometer. The insets detail the monochromator crystal and the focussing mirror together with their essential adjustable controls.

which diffracts radiation at specific energies λ (the desired radiation), λ/n (harmonic contamination), and $\alpha\lambda$ (other excited Bragg diffraction resulting from planes not parallel to the surface of the crystal). If a single diffraction were used, then as the crystal were rotated to select different energies, the entire detection system would have to be rotated. Therefore, a second diffraction is used to send the diffracted beam on a course parallel to the incident white radiation.

Unfortunately, if the crystals are perfectly aligned (which will be the case if the two crystal faces are still attached to the same block of single crystal), then the unwanted radiation $\lambda/3$, $\lambda/4$, ..., λ/n, and $\alpha\lambda$ also gets diffracted. Two general approaches have been used to limit this unwanted radiation: (1) the crystals can be shifted from their parallel setting, and since the angular bandwidth of first-order radiation is greater than that from $\lambda/3$, λ/n, and $\alpha\lambda$, then, at the expense of a reduction of primary intensity, the unwanted radiation is suppressed. Unfortunately, the angular offset is ~1 arcsecond, and this places stringent conditions on the stability of the system. This stability is easily disturbed by heating or vibrational effects (2). The radiation $\alpha\lambda$ as it originates from Bragg planes not parallel to the surface is physically displaced from the λ, $\lambda/3$, λ/n components. If the slits of the system are sufficiently narrow, then it is frequently possible to select the main diffracted beam λ and its harmonics λ/n.

Implicit in the previous discussion is the fact that the second-order radiation $\lambda/2$ is absent, which is true if, for example, a silicon (111) monochromator is used. With the radiation removed, the harmonics can be reduced by total external reflection from a mirror. This occurs because there is a cut-off in reflectivity as a function of wavelength:

$$\Theta_c = \frac{e\lambda}{c} \sqrt{\frac{\rho}{\pi \cdot m}} \tag{1}$$

where Θ_c is the cut-off angle for total external reflection, e the electron charge, c velocity of light, λ the wavelength of X-ray radiation, ρ the electronic charge density of the mirror material, and m the mass of the electron. Thus, by choice of Θ the grazing angle can be made smaller than Θ_c for λ (giving reflection) but larger than Θ_c for radiation which is $\lambda/3$ or shorter (the unwanted radiation is suppressed).

By choice of the crystal and slit width, the resolution can easily be adjusted to ~1 eV, which is consistent with the natural energy width associated with spectral features in the X-ray region. The sharpness of features is to a large extent dominated by "lifetime effects", i.e., the duration of the excited state, which is typically ~10^{-16} s.

As well as reducing harmonic content, the mirror also serves to focus the radiation onto the specimen, which is usually placed between the two ion chambers. A typical cross-sectional area for the beam may be 0.1 cm high by 1 cm wide. The specimen may be supported in air, a cryostat, or even a controlled-atmosphere reactor. Thus, *in situ* experiments are possible, provided that the absorption of X-rays by cell windows is not too great [8]. In practical terms, this limits measurements to elements with atomic numbers greater than 19, although special techniques, e.g., those used in surface EXAFS, where photoelectron yield mimics the absorption process, allow extension of the range to the entire periodic table.

A choice of detectors is possible (Fig. 5), depending on the type of measurement to be performed. If high concentrations of the element to be studied (>2 wt%) are embedded in a matrix with a low absorbance, as in the case of say Os dispersed on Al_2O_3, then conventional transmission measurements are possible. Under these circumstances, the incident flux onto the specimen I_o^s is measured with the first ionization chamber, and the transmitted flux I^s is measured with the rear chamber. Both detectors have to be corrected for detector efficiency to enable an absorption thickness product μx to be found. This is easily achieved by measuring a spectrum with no specimen, I_o^n, I^n, and then using the relationship

$$\ln\left(\frac{I_o^s\ I^n}{I_o^n\ I^s}\right) = \mu x \tag{2}$$

In fact, the right-hand side of this equation is the quantity required, since the thickness x can be eliminated when the fine structure is placed on an absolute basis. It should be noted that eq. (2) relies on the familiar law of absorption

$$I = I_o\ \exp(-\mu x) \tag{3}$$

and this is valid only if we consider a uniformly thick slab of material.

Variations in the thickness distort the measured value of μx such that the fine structure cannot be placed on an absolute basis. This distortion may place rather stringent requirements on the samples. A pure material measured with a radiation of ~1 Å will have a thickness of ~30 μm to give μx ~1, an optimal value for the measurements. Thus, the sample should not have thickness variations of more than ± 3 μm; otherwise the amplitude of the fine structure is reduced.

Similar reductions of the fine structure amplitude are also associated with the presence of harmonics in the X-ray beam. Such reductions of fine structure amplitude caused by experimental artifacts directly affect structural information, and in particular a determination of the coordination number. At present, there is little that can be done except to repeat the experiments with different specimens of varying thickness until a consistent amplitude of the fine structure is obtained.

6.7.2.b **Data reduction.** The analysis of the fine structure rests on defining a function which compares the size of the fine structure oscillations to the size of the X-ray absorption edge for an equivalent number of isolated atoms. We are interested in absorption from a particular atomic core level, usually the K or one of the three L levels L_I, L_{II}, L_{III}. Usually (but not always) the absorption below an absorption edge is featureless and over an extended range can be approximated by an empirical expression originally given by Victoreen:

$$\mu = A\lambda^3 + B\lambda^4 \tag{4}$$

Above an absorption edge this formula is no longer valid for an isolated atom, and another approximation has to be adopted. Since the exact form of the atomic absorption is not known, it is usually approximated by a polynominal in E. The inadequacies of this polynomial have to be borne in mind when the subsequent analysis is performed. If μ_1 is the absorption below the edge, μ_2 the isolated atom absorption above the edge, and μ_3 the measured absorption including fine-structure, then

$$\chi(E) = \frac{\mu_3(E) - \mu_1(E)}{\mu_2(E)} \tag{5}$$

and the energy axis is defined at some position close to the absorption edge. For the subsequent analysis, E should represent the kinetic energy of an excited photoelectron when it propagates in the almost constant part of the interatomic potential. Determination of the value of the origin of the kinetic energy (E_0) is a major problem in the interpretation of EXAFS data. Unfortunately, errors in determining E_0 are coupled to the determination of shell radii. An example of the resulting fine-structure is shown in Fig. 6.

6.7.3 THEORY

6.7.3.a **Direct calculation of fine structure.** Many interpretations of the function $\chi(E)$ have been made using a single scattering equation which has evolved from the work of many authors [1, 9, 10, 11, 12]:

$$\chi(k) = -|S_o|^2 \sum_j \frac{N_j}{R_j^2} \frac{|f_j(\pi,k)|}{k} \exp\left[-\frac{2R_j}{\lambda(k)}\right] \exp\left[-2\sigma_j^2 k^2\right] \cdot \sin\left[2k R_j + 2\delta_1(k) + \eta_j(k)\right] \quad (6)$$

It is worth itemizing these parameters and their meanings:

(1) $|S_o|^2$ is a factor having a magnitude of ~0.8 for many materials and

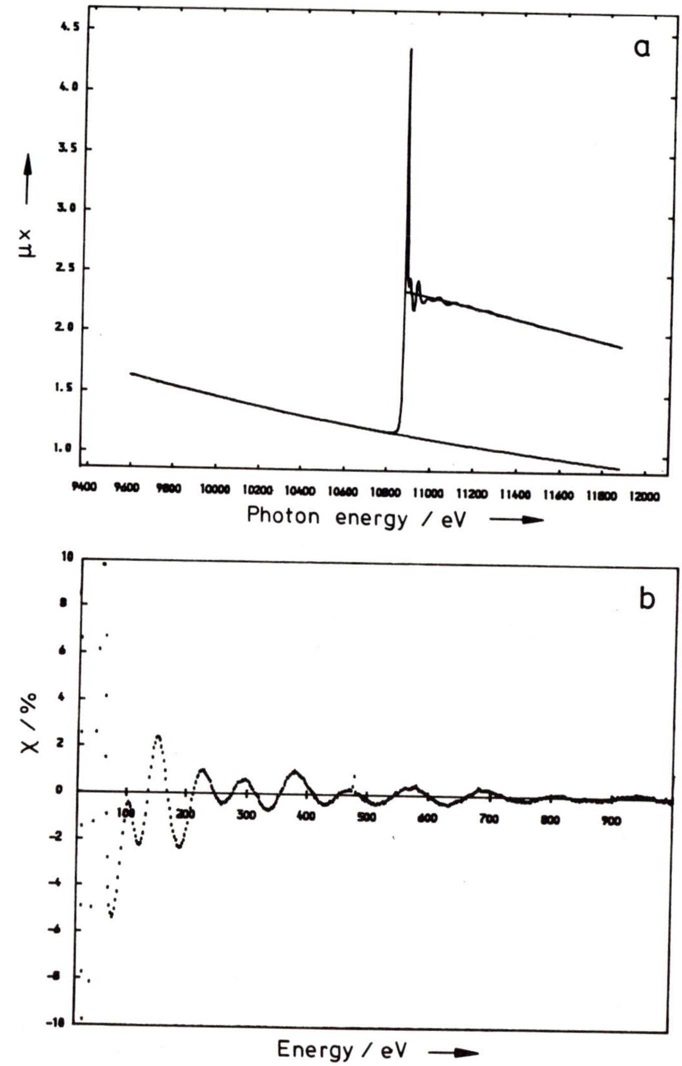

Fig. 6. (a) The L_{III} X-ray absorption edge of $Os_3(CO)_{12}$ together with the extrapolated pre-edge absorption and the polynominal approximation to the atomic absorption above the edge; (b) the resulting fine structure from the absorption edge (a).

representing the effect of multielectron excitation. It is believed that this factor is constant for photoelectron energies greater than 100 eV where 80% of the events are single electron events and the rest are multielectron excitations which do not contribute to the fine structure. However, this asumption has not been tested for a wide range of materials.

(2) The term j is the summation index counting shells which are defined as a collection of identical scatterers at a given distance.

(3) N_j represents the number of scatterers in shell j; this may be integral or fractional, the latter especially if the atom type excited exists in two or more sites.

(4) R_j is the radius of the jth shell.

(5) The term k represents the photoelectron wave vector defined by

$$c\ k^2 = E - E_o \tag{7}$$

with c taking the numerical value 3.81 for k in Angstrom units and E in electron volts. E_o marks the origin of the excited photoelectron's kinetic energy.

(6) The term $|f_j(\pi, k)|$ is the magnitude of the backscattering amplitude of atoms in shell j with a photoelectron wave vector of k. The k dependence of this function can frequently be used to discriminate between atoms of different atomic number Z. In general, high-Z elements scatter more strongly at high k [13, 14].

(7) The $\eta_j(k)$ term is the phase of backscattering corresponding to the amplitude given in (6) above. In principle, this also allows atom type discrimination.

(8) The exponential term $\exp[-2R_j/\lambda(k)]$ represents the fact that the photoelectron loses phase coherence on its passage through the material.

(9) The σ_j^2 term is the mean square relative displacement due to thermal or static variations in R_j. Note that this is relative to the emitter and should not be confused with the thermal ellipsoids given in crystallography, which are relative to a lattice point. The latter depend on all vibrational frequencies present in the solid, whereas for EXAFS it is principally the stretch frequencies that are relevant.

(10) The $\delta_1(k)$ term is the emitter potential phase-shift which is created as the photoelectron emerges from its excited potential; 2kR, $2\delta_1(k)$, and $n_j(k)$ form the argument of the sine term responsible for the EXAFS oscillations. The terms in the argument are listed in order of greatest k dependence.

It is now important to review the approximations used in deriving eq. (6):

(1) The equation implies single scattering only; no multiple scattering between atoms is accounted for.

(2) The electron wave emitted by the absorbing atom is assumed to be planar at the scatterer and also to take on its asymptotic form. This results in a modification to the amplitude term $|f_j(\pi,k)|$ and $\eta_j(k)$.

(3) The atoms surrounding the absorber are assumed to be in a Gaussian distribution about their mean shell radius. This may not be true for either weakly bonded solids which vibrate anharmonically or for solids with a non-Gaussian static disorder. This approximation can lead to very serious errors in the determination of both distances and coordination numbers.

(4) The material is assumed to be isotropic. This criterion is satisfied for powders but not for crystals, except those of cubic materials. The equation can, however, be modified to take into account polarization-dependent properties.

(5) Only one final state having definite angular momentum is assumed. This is true for K-edge spectra, as the selection rules for allowed transitions are $\Delta l = \pm 1$, which implies emission into states of p-symmetry. On the other hand, if an L_{III} edge is measured, this being the optimal choice for spectra of heavy metal spectra such as Os, then the final state is an admixture of s and d states. Fortunately, the d states dominate, and unless very subtle effects are sought, the appropriate choice of $\delta(k)$ is $\delta_2(k)$ rather than $\delta_0(k)$. The fact that $\Delta l = \pm 1$ gives rise to a p-type final state for K-edge spectra with the p lobe pointing along the electric field vector. This is the origin of the polarization dependence, with those atoms lying in the direction of the E vector contributing most to the fine structure for a perfectly polarized X-ray beam.

(6) The equation is valid only for high energies, say greater than 50 eV beyond the absorption edge. This restriction pertains because $|S_0|^2$ may well be energy dependent at lower energies, $|f_j(\pi,k)|$ becomes environmentally sensitive, and the plane wave approximation breaks down completely [point (3) above]; there is little knowledge of the factor $\lambda(k)$ in this region. Further, multiple scattering is expected to become more important owing to the fact that $\lambda(k) \longrightarrow \infty$ at the edge and the electron atom scattering becomes more isotropic.

The limitation of no low-energy (low-k) data is serious for EXAFS in that the signal is usually strongest at the edge, but the problems of interpretation become severe. Nevertheless, the benefits that can result have prompted a considerable effort in interpreting the XANES region.

We should note that eq. (6) has a factor 2kR in the sine term. This term accounts for the fact that the electron suffers a momentum change of 2k on back-scattering. This can be compared with the momentum transfer $Q = \frac{4 \sin\Theta}{\lambda}$, familiar from X-ray or neutron scattering. For scattering with a radiation having 0.154 nm, the range of Q accessible corresponds to $0 \leq Q \leq 8 Å^{-1}$. On the other hand, the range of EXAFS is typically 50-1000 eV, and this yields $8 \leq Q(=2k) \leq 32\ Å^{-1}$. In this sense the techniques are complementary in that the momentum transfer ranges do not overlap. Also it should be noted that it is in the high momentum transfer regime that the data are most sensitive to the details of the pair distribution function, e.g., deviations from a Gaussian form.

The approximations listed above should be assessed in light of presumptions concerning the possible structure under investigation. In particular, for a measurement of the L_{III} edge of Os in $Os_3(CO)_{12}$, we can anticipate that approximation (2) is invalid since Os is a large atom and the photoelectron wavefront is appreciably curved over the spatial extent of the Os atom scattering core. Fortunately, a computer code exists to remove this approximation with the so called "curved wave" theory [15]. Approximation (3) is valid as the Os-C, Os-Os, and C-O bonds are strong, and hence a Gaussian distribution appears to be a reasonable prediction.

The most serious approximation for the materials of specific interest to this review is (1), namely, the neglect of multiple scattering. The effect of multiple scattering has been treated by several authors, and it is well known that these effect are most serious in materials in which two scattering atom centers lie in a straight line with the emitter. This is indeed the case for carbonyl groups attached to a metal center which is being excited. This point can be seen most clearly by plotting $|f(\Theta,k)|$ as a function of energy and angle, as shown in Fig. 7 for carbon. Many points can be made from this diagram. First, the scattering cone at high energies is strongly peaked in the forward direction, and this enhances the photoelectron wave at the second shell atom and consequently the scattering contribution from oxygen. In the back-scattered direction, the amplitude decays quickly with energy, and this means that the contribution of backscattering from carbon is very weak in the energy regime above the edge, which is accessible to interpretation by an equation of the form of eq. (6).

Recently Boland and Baldeschwieler [16] considered the multiple scattering for a simple linear system within certain approximations. It is clear that the forward scattering cone is so narrow that librational modes, which cause a bend in the metal-carbon-oxygen bond angle, should influence

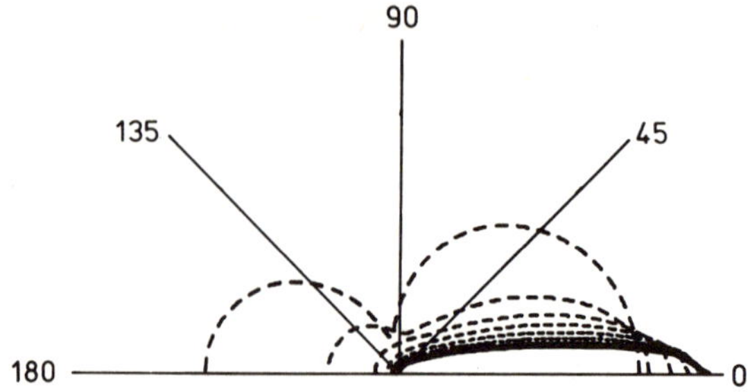

Fig. 7. The magnitude of scattering $|f(\Theta, k)|$ for carbon. The large dashed contours correspond to low energies and the more solid lines to high energies. The lowest k vector is at 19 nm^{-1} and the others are calculated at increasing increments of 25 nm^{-1}. Note that the back-scattering collapses quickly to small values, whereas the forward scattering persists in a narrowing cone as the energy increases.

the forward scattering by averaging over the forward cone. Including these effects, the authors found that double and triple scattering events were at least a factor of four greater than the single scattering contribution for an intervening oxygen atom. Comparable effects are expected for carbon forward scattering in carbonyl groups. The work of Boland *et al.* [16-18] has extended the theory, but it is still within the approximations (2), (5), and (6) above. Also, it contains parameters which are difficult to establish.

6.7.3.b. **The inversion problem.** From the preceding section we see that the fundamental physics needs careful consideration prior to attempts to answer specific structural questions. For some materials the approximations inherent in eq. (6) can be circumvented, and hence this equation can be manipulated to yield structural information for an unknown material.

Two types of analysis are possible, but there are many variations on these two basic themes. The first technique is based on a comparison of the fine structure measured from a standard material with that of an unknown substance. If this type of analysis can be performed, then it should be regarded as the most reliable.

We can rewrite eq. (6) as follows:

$$\chi(k) = \sum_j A_j(k) \frac{N_j}{R_j^2} \exp(-2\sigma_j^2 k^2) \sin[2kR_j + \phi_j(k)] \tag{8}$$

or

$$\chi(k) = \sum_j B_j(k) \frac{N_j}{R_j^2} \sin[2kR_j + \phi_j(k)] \tag{9}$$

We now proceed to establish $B_j(k)$ and $\phi_j(k)$ to fit the fine structure. The problem is still ill-defined, and thus an implicit assumption is made that $B_j(k)$ and $\phi_j(k)$ are smoothly dependent on k and thus can be modelled using a polynomial or other functional form with a small number of fitting parameters.

As an example, it is found that typically four parameters each for $B_j(k)$ and $\phi_j(k)$ are necessary. Thus, there are eight parameters per shell of atoms contributing to the fine structure. In this context, it is worth repeating that a shell of atoms is defined as all atoms lying at the same radius from the emitter and having equivalent bonding. Thus, in $Os_3(CO)_{12}$, for example, each Os atom has three shells, one consisting of four carbon atoms, one consisting of two osmium atoms, and one consisting of four oxygen atoms, and these will dominate the fine structure. Other shells exist, but owing to the term $\exp(-2\sigma^2k^2)$ and the increased path length, their contributions should be small. Thus, to parametrize the amplitude and phase for the interactions Os-Os, Os-C, and Os-O would require 24 parameters.

This type of analysis can be performed successfully if attention is paid to the possible existence of local minima in a least-squares fit and if different forms for the $B_j(k)$ and $\phi_j(k)$ functions are tried. However, the problem is ill-conditioned, and it may well be necessary to combine information from a number of reference compounds to improve conditioning. Once the functions $B_j(k)$ and $\phi_j(k)$ have been established, we fit the unknown spectrum by varying N_j, R_j, and ϕ_j^2. Implicit in this procedure is the assumption of phase $\phi_j(k)$ and amplitude $B_j(k)$ transferability. This assumption is thought to be valid for photoelectron energies >50 eV, although exhaustive tests have not been performed. Further, the origin of the photoelectron kinetic energy may well be different in absolute terms for the standard compounds and for the unknown. Thus, a fit to the unknown spectrum will include three parameters per shell plus an adjustable parameter (E_0) to account for energy shifts of core and interstitial energies.

A somewhat cruder treatment of the above can be adopted by use of the Fourier transform of $\chi(k)$ [11]. The justification for this approach lies in the existence of an almost linear form for $\phi_j(k)$ for some pairs of atoms. An idea of the realism of this approximation can be seen from the tables of calculated $\phi_j(k)$ presented by Teo and Lee [14].

For example, an Os-O pair may have a reasonably linear behavior as a function of k, but Os-Os pairs will not. Thus, if we write

$$\phi_j(k) = -2a_jk + b_j \tag{10}$$

then

$$\chi(k) = \sum_j B_j(k) \frac{N_j}{R_j^2} \sin[2k(R_j - a_j) + b_j] \tag{11}$$

If the b_j's were zero, then the sine transform with respect to 2k would peak at values of $(R_j - a_j)$, and a real space representation of the data would be obtained. Unfortunately, b_j is rarely zero, and hence a variety of shapes in *r*-space can be observed. To combat this, it is usual to plot the modulus of the Fourier transform, which is the square root of the power density, which should always be positive and located around $(R_j - a_j)$.

The peaks in the *r*-space representation have a width which is controlled by (1) truncation effects (this comes from the finite range of data over which the transform can be performed, typically from k = 4 to 16 $\mathring{A}^{-1}$); (2) amplitude modulation caused by $B_j(k)$; and (3) phase modulation due to the nonlinearity of $\phi_j(k)$. The minimization of termination-derived width can be made by use of a window function which pre-emphasises the data in k-space before transformation.

Examples of the Fourier transform of a spectrum are shown in Fig. 8. Two transforms are given for the range of data from 4 to 12 $\mathring{A}^{-1}$ determined on the basis of (1) the unmodified fine structure (no window function) and (2) multiplication of the fine structure by a window function which decreases smoothly from one to zero at the ends of the data range:

$$W(k) = \frac{1}{2}\left[1 - \cos\left(\frac{(k - k_{min})\pi}{\Delta k \cdot FW}\right)\right], \text{ if } \frac{k - k_{min}}{\Delta k \cdot FW} \leq 1 \tag{12}$$

$$= 1, \text{ if } \frac{k - k_{min}}{\Delta k \cdot FW} > 1 \tag{13}$$

where FW is the tapering parameter. If FW = 0.5, the tapering extends over the entire k range. Further, the transform also depends on the range of data. Compensation for the amplitude modulation of the data can also

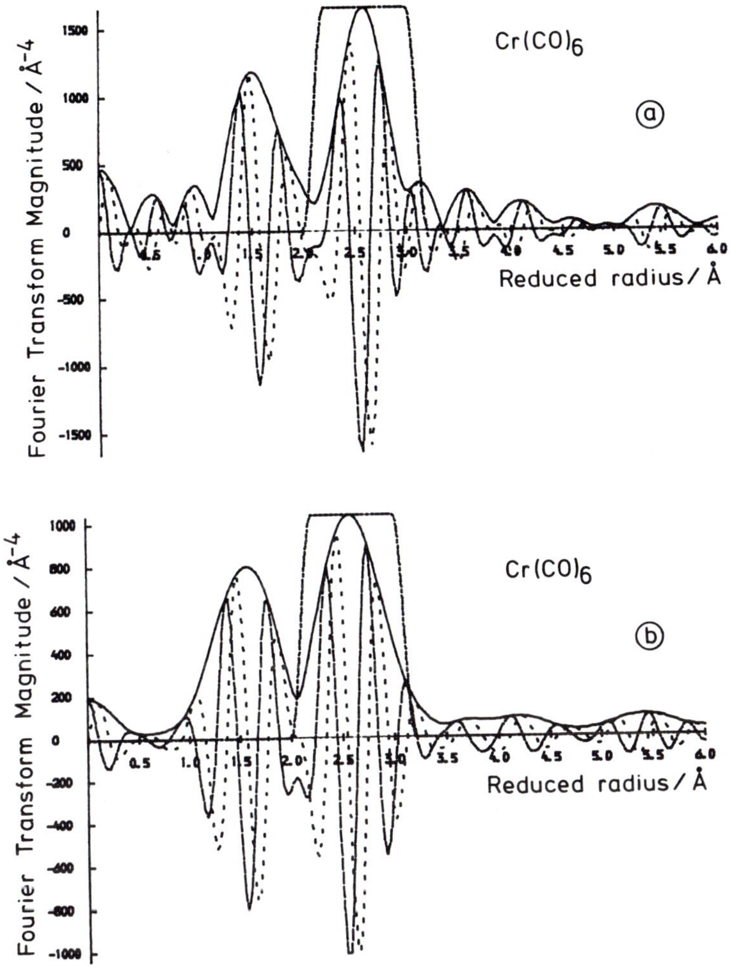

Fig. 8. Curves a and b are Fourier transforms of the k^3 weighted fine structure from $Cr(CO)_6$ calculated using different window functions. The solid line is the square root of the power density, with the long-dashed and short-dashed lines being the real and imaginary parts of the Fourier transform, respectively. The peaks centered at about 1.5 Å and 2.6 Å result from C and O scattering, respectively. The dotted rectangular window is typical of a back transform window which in this case isolates the oxygen scattering signal for subsequent transformation back into wave-vector k-space. The Fourier transform was performed between 40 eV and 580 eV, with a tapering parameter FW = 0.1 for (a) and FW = 0.5 for (b).

be made by multiplying $\chi(k)$ by k^n, where n is an integer between 0 and 3. For light-atom scatterers, k^3 is usually used to compensate for the rapidly decaying back scattering $|f(\pi,k)|$ from the light elements. In the case of heavy-metal scatterers such as Os, k^0 is more appropriate. Multiplication of $\chi(k)$ by $k^nW(k)$ constitutes the pre-emphasis applied, but this varies from one author to another.

It is evident from Fig. 8 that quite different structures are obtained, depending on the treatment of the data. Thus it is important to determine exactly the treatment of the data when comparing Fourier transforms from different sources. Thus, it is always wise for authors to present their results as fine structure plots. It is this writer's opinion that the Fourier-based techniques are highly susceptible to the generation of artifacts in the data, unless exhaustive tests are made of the sensitivity of the final results to the procedures adopted.

A further step in the analysis can be made in that the structure which appears from shells other than that of interest can be filtered out in r-space by multiplying the r-space data by a further window function such as that shown by the fine dotted line in Fig. 8. In this case, the r-space contribution from oxygen is isolated. When this has been done, the transform can be reversed and the scattering contributions from that particular shell alone can be isolated. Also, the amplitude $B_j(k)$ and phase $\phi_j(k)$ for this shell can be isolated. Unfortunately, this isolation procedure is also sensitive to the data treatment, especially the choice of the windowing function. If it is possible to isolate scattering shell contributions, it is possible to simplify the fitting procedure via this technique.

The second major technique for interpreting data relies on the theoretically calculated $A_j(k)$ and $\Phi_j(k)$ functions. This technique is not as simple as it might at first appear. The calculation of these functions requires the solution of Schrödinger's equation in the potential of the material under investigation. This is a fairly difficult problem, and simplifying assumptions are necessary. The validity of these assumptions has not been fully tested. Usually, the atomic potentials of the atom are considered spherically symmetric and a one-electron model is adopted, i.e., the photoelectron is assumed to move in the potential created by the nuclei and the other electrons in the system. The electron also suffers a loss of phase coherence as it propagates and during excitation owing to inelastic electron-electron effects. Parameters are chosen to account for these effects together with estimates of the range of the approximate spherical potentials in the material. These problems are still of interest to physicists and are topics of current research. Some progress in this field has been made, and the reader is referred to the literature [9, 13, 19] for details.

Fitting of spectra is thus accomplished with an incomplete theory. The most serious problem for characterization of metal clusters is the neglect of multiple scattering. As a check on the parameters found from this type of analysis, the reader should note the number of parameters, which is equal to 3NS + 1, where NS is the number of shells. For a realistic interpretation, the values of E_0, R, N, and σ^2 should be consistent with values inferred from other measurements. For example, density (which affects N_j and R_j), infrared and Raman stretch frequencies (which are related to σ^2), and SCF X-alpha or band structure calculations should yield one value of E_0 for similar types of materials. This check is important, since the mere existence of a good fit to the data does not imply that the correct structural information has been found.

6.7.4 EXAFS OF CLUSTER-DERIVED CATALYSTS

6.7.4.a Case study of EXAFS of supported osmium catalysts. Data are presented in this section for osmium catalysts supported on γ-Al_2O_3. The goal is to illustrate some of the points made in the preceding sections. These data come from the unpublished work of Knözinger, Kozlowski, and Pettifer [20]; related unpublished work has been done by another group [21]. $Os_3(CO)_{12}$-derived catalysts were characterized, the Os concentration on the γ-Al_2O_3 being 0.67 wt%; with this low concentration of Os, it is profitable to make fluoresence EXAFS measurements instead of ordinary transmission studies.

The absorption spectrum taken by transmission (Fig. 9) shows that the L_{III} osmium edge is very weak in comparison with the other absorption due to the Al_2O_3 support. In contrast, the fluorescence spectrum is given in Fig. 10. Clearly, the fluorescence signal has less noise than that of the transmission spectrum. It is interesting to note that the ratio of the white line intensity to that of the absorption edge is 4.8 in the case of the fluorescence data, compared with 2.1 for the transmission spectrum.

This result clearly indicates that new physics is involved with the fluorescence efficiency, which is affected by the exciting X-ray energy involved. Thus, a type of resonant fluorescence is present when the photoelectron has a low kinetic energy. Nevertheless, it is believed [20] that the fine structure above the edge is not influenced by this effect. A series of three catalysts was characterized in the fluorescence mode, and the results from four scans were added to improve the signal-to-noise ratio.

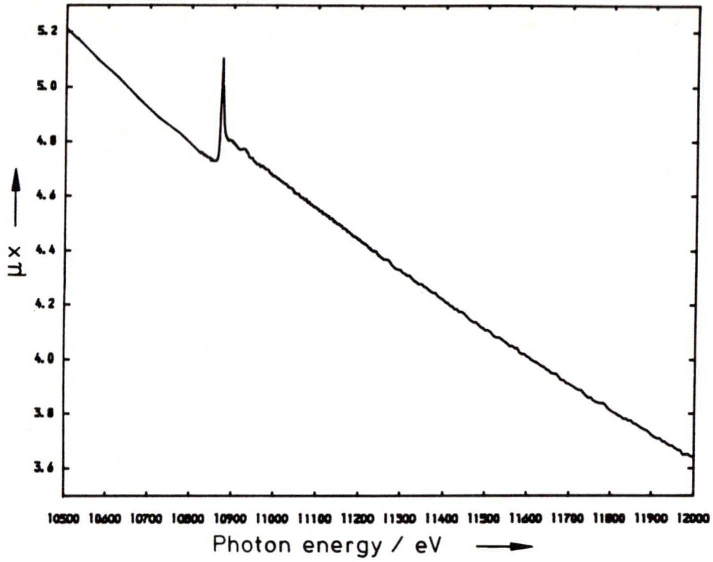

Fig. 9. Transmission spectrum of 0.6 wt% Os-on-Al_2O_3 catalyst.

The results are shown in Fig. 11. Even after the averaging procedure, the curves indicate a large amount of noise.

For comparison, a series of standard compounds was characterized, and their fine structures are shown in Fig. 12. The bottom curve characterizes the simplest compound which closely resembles the structure of the supported catalyst, namely, $Os_3(CO)_{12}$. Oscillations persist to 1000 eV above the absorption edge. The source of these ripples at high energies can be either backscattering from Os (this seems likely since Os is a heavy metal and scatters strongly even for high-kinetic-energy photoelectrons) or backscattering from oxygen. This latter possibility seems unlikely, but we must remember that the fine structure contribution from oxygen is enhanced by the forward scattering of the intervening carbon atom in the carbonyl group. Since multiple scattering calculations have not been performed for this system, we cannot reject this possibility. (See ref. [21] in this regard [Ed.].)

Unfortunately, Fourier techniques do not help here, as the Os-O distance (2.93 Å) is comparable to the Os-Os distance (2.86 Å). For Os metal, however, the fine structure amplitude gives a $\Delta\chi \sim 3\%$ for 12 nearest neighbors in the energy region 700 to 1000 eV. In $Os_3(CO)_{12}$, $\Delta\chi \sim 0.5\%$ over the same region. This result is in good accord, because each Os atom in the cluster is surrounded by two neighbors rather than the 12 found in the metal.

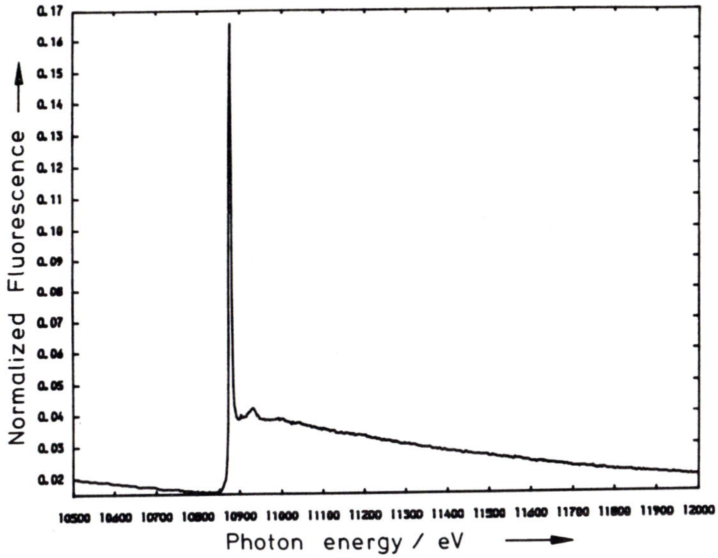

Fig. 10. The flourescence spectrum of the material referred to in Fig. 9.

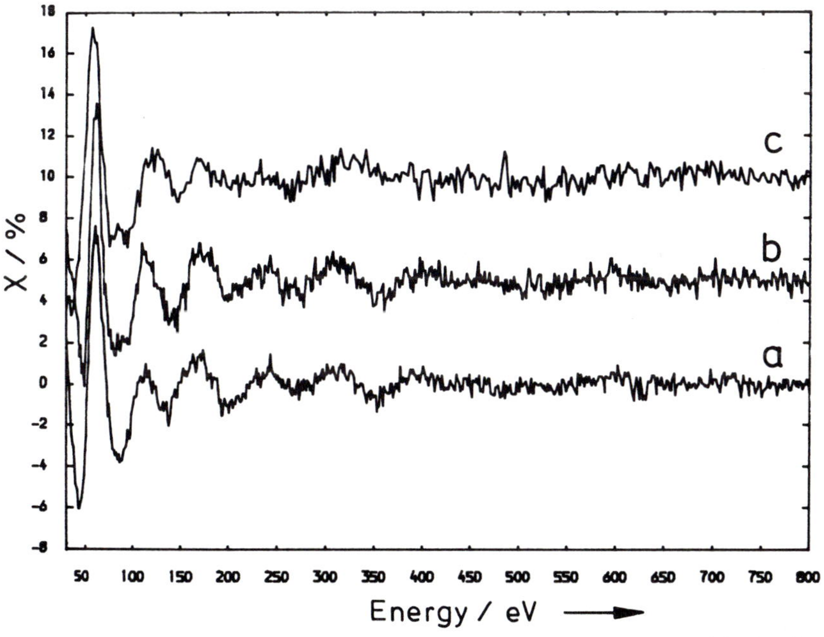

Fig. 11. Fine structure for Os cluster-derived catalysts: (a) $Os_3(CO)_{12}$ deposited on γ-Al_2O_3 (0.67 wt% Os); (b) the same material after thermal treatment under 0.53 MPa of CO at 670 K; (c) same as (a) but after heating under vacuum at 670 K (the curves have been displaced vertically by 5% for clarity).

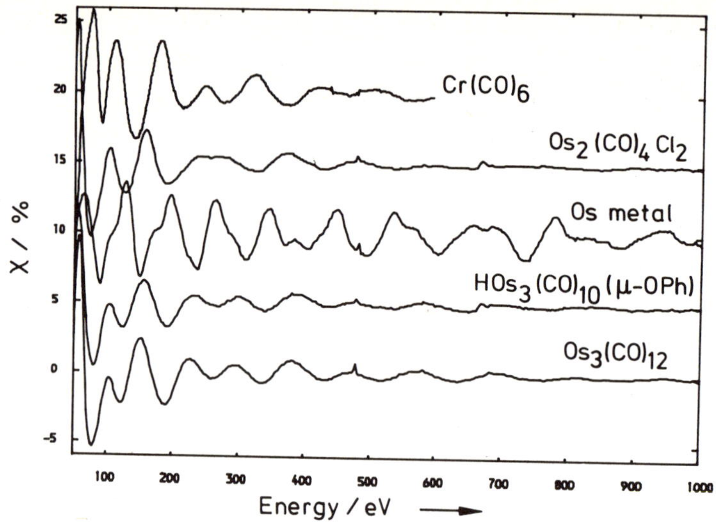

Fig. 12. Pure compound fine structures; all measurements made at room temperature; the curves have been displaced by 5% vertically for clarity.

The (μ-O) bridged Os cluster compound (μ-$HOs_3(CO)_{10}$-μ-O-Ph) also shows evidence of a similar magnitude of scattering in this high-energy region. Support for this idea comes from the $Os_2(CO)_4Cl_2$ spectrum, where direct Os-Os interactions are not present, and indeed the fine structure is certainly weaker than that observed for $Os_3(CO)_{12}$, although each has four carbonyl groups per osmium. Thus, at least in the region of the spectrum E>700 eV, Os-Os interactions dominate.

If we return to Fig. 11, however, we note that variations of $\Delta\chi \sim 0.5\%$ will be buried in the noise. To answer the questions relating to persisting Os-Os interactions with certainty would require somewhat greater statistical accuracy than that obtained in this study. Below 600 eV, the contributions from multiply scattered oxygen and osmium are mixed. The three spectra of $Os_2(CO)_4Cl_2$, μ-$HOs_3(CO)_{10}(\mu$-OPh), and $Os_3(CO)_{12}$ all have fine structures which are dominated by carbonyl scattering below 600 eV. This inference is confirmed by the observation that the periodicities of the fine structures are practically identical to below 100 eV. Subtleties exist in the amplitude, however, which, with care and very accurate modelling of carbonyl scattering could provide evidence of the weaker ligand contributions.

It is clear that further progress cannot be made with this study, for two reasons: First, the signal-to-noise ratio of the catalyst measurements is insufficiently high. This is not an impossible problem, but access to synchrotron radiation for a matter of several days per specimen (rather than hours) is required. The second limitation comes from the lack of availability of calculations including multiple scattering effects. Again, this difficulty is not insuperable as the mathematical formulation of the problem has already been given. We lack, however, the coding of this problem into efficient software. Recently collected transmission data characterizing alumina-supported triosmium clusters with a higher Os loading than that achieved in the above study promise to yield better-quality EXAFS spectra, which may solve the Os-Os interaction problem for the material containing the broken-up cluster [21].

6.7.4.b **Other studies of cluster-derived catalysts.** Via *et al.* [22] characterized various Pt, Ir, and Os catalysts prepared by conventional impregnation techniques, and the quality of their spectra is sufficiently high to demonstrate conclusively the existence of metal-metal interactions on the support surface. The mean square relative displacements were approximately double those found in the bulk parent metal at the same temperature. Metal-metal bond lengths were, however, practically the same as those of the parent metal. However, no information regarding the metal-support interaction could be derived. One might assume that a range of metal-support distances is present; this gives rise to a loss of phase coherence for the photoelectron scattering, and hence a loss of fine structure.

Hunt *et al.* [23] investigated $Os_6(CO)_{18}$/silica-derived catalysts, both in the freshly prepared and in the thermally activated states. Also measured was the fine structure of a conventionally prepared catalyst (2% Os on silica) prepared from osmium trichloride. The results clearly show differences in structure between the conventional catalyst and the cluster-derived material. The data extend to 700 eV above the edge. The authors state that the carbonyl contribution to the fine structure is negligibly small beyond 400 eV from the edge. From measurements with $Cr(CO)_6$, we know that substantial fine structure exists beyond 400 eV from CO groups. The present author assumes that the estimate of carbonyl scattering was based on single scattering theory, which is known to be incorrect.

Cook *et al.* [24] calculated the bond lengths by fitting EXAFS spectra for a set of four osmium cluster compounds (Table 1). The theory used was embodied in a computer package at Daresbury Laboratory, England. This theory is based on single scattering only, with theoretical phase shifts calculated in a non-self-consistent manner. Table 1 shows the resulting bond distances; not included here are the values of coordination numbers or thermal mean-square relative displacements which could be used to check the viability of the fitting procedures used. It is surprising that the bond lengths derived show a wide variation from those established by X-ray crystallography (Table 1).

TABLE 1. EXAFS-Derived Distances for a Number of Model Compounds (Data from Cook *et al.* [24])[a]

Compound	Distance/Å					
	Os-C		Os-Os		Os-O	
$Os_3(CO)_{12}$	1.94	(1.93)	2.79	(2.88)	2.93	(3.07)
$Os_6(CO)_{18}$	1.91	(1.89)	2.78	(2.78)	2.82	(3.03)
$[N(PPh_3)_2]_2Os_6(CO)_{18}$	1.90	(1.83)	2.82	(2.86)	2.87	(2.93)
$[N(PPh_3)_2]_2Os_{10}C(CO)_{24}$	1.94	(1.90)	2.81	(2.84)	2.89	(3.00)

[a]Numbers in parentheses refer to the mean X-ray diffraction results.

Again, it is the present author's opinion that this discrepancy results from an inadequacy of the physics. In particular, multiple scattering and/or sensitivity analysis of the structural parameters and their relation to physical assumptions should be undertaken before investigation of unknown materials.

Binstead *et al.* [25] have improved the analysis described above to obtain better estimates of bond lengths, but they still performed their calculations within the single scattering regime. Nevertheless, these authors also concluded that multiple scattering should be included in a further refinement of the work.

The work of Cook *et al.* [24, 26] and Binstead *et al.* [25] relied on theoretical fitting of the data, whereas in the work of Yokoyama *et al.* [27, 28] the Fourier transform technique was used in comparing standard $Rh_4(CO)_{12}$, $RhCo_3(CO)_{12}$, and $Rh_2Co_2(CO)_{12}$ compounds with the cluster-derived catalysts. Unfortunately, merely the Fourier transforms but no fine structure data were presented. Again, it was stated in these reports that carbonyl contributions to the fine structure were sufficiently small in the 6 $Å^{-1}$ (136 eV) to 12 $Å^{-1}$ (548 eV) region of the spectrum to be neglected. This assumption, however, again conflicts with the measurements of $Cr(CO)_6$ and the other standard compounds (Fig. 12). Nonetheless, the authors stated that the RuCo cluster frameworks were retained from the precursor compound when they were activated on the support. Further, they claimed that gross sintering of the conventionally impregnated catalysts was observed and that this had a structure quite different from that of the catalyst derived from the cluster. At least in the latter conclusion, the work of Via *et al.* [22], Hunt *et al.* [23], and Yokoyama *et al.* [27, 28] is in agreement.

6.7.5 CONCLUSIONS

EXAFS spectroscopy has many attractive features for structural research, in particular for metal clusters and materials derived from them. The number of applications of this technique has grown enormously over the last decade, but unfortunately the developments in experimental technique and theoretical analysis have not moved at the same pace as the applications. The situation is likely to change, however, to bring this spectroscopy to maturity. In particular, the availability of high-intensity, high-sensitivity apparatus, such as that reviewed above and described by Hermes *et al.* [7], together with development of the analysis as in the treatment of multiple scattering by Boland *et al.* [16-18] and Teo [29], and the use of the correct spherical wave expansions as shown by Pettifer and Cox [15] and Binsted *et al.* [25], together with the recent work of Natoli *et al.* [19] which removed the muffin-tin approximation, should enhance reliability of structural conclusions.

For the present, those involved in applications of EXAFS will have to appreciate the limitations imposed by the lack of development in the basic experimental and theoretical physics necessary for this spectroscopy. We should remember that in the early days of X-ray diffraction a similar situation existed, and it took nearly 50 years to make X-ray diffraction the routine technique it is today. EXAFS will develop in a similar manner, but at present the detailed structural conclusions often demanded (e.g., coordination numbers) are still unreliable.

REFERENCES

1. C. A. Ashley and S. Doniach, Phys. Rev. B, 11 (1975) 1279.
2. P. A. Lee, P. Citriu, P. Eisenberger and B. Kincaid, Rev. Mod. Phys., 53 (1981) 769.
3. T. M. Hayes and J. B. Boyce, Solid State Phys., 37 (1982) 173.
4. E. A. Stern (ed.), AIP Conference Proceedings No. 64, (1980).
5. K. Codling, Rep. Progr. Phys., 36 (1973) 541.
6. C. Kunz (ed.), Synchrotron Radiation, Springer Verlag, Heidelberg, Berlin, New York, (1979).

7. C. Hermes, E. Gilberg and M. H. J. Koch, Nucl. Inst. Methods, 222 (1984) 207.
8. M. Boudart, R. Dalla Betta, K. Foger and D. G. Loffler in EXAFS and Near Edge Structure III, K. O. Hodgson, B. Hedman and J. E. Penner-Hahn (Eds.), Springer Verlag, Berlin, Heidelberg, New York, 1984, p. 187.
9. P. A. Lee and J. B. Pendry, Phys. Rev. B, 11 (1975) 2795.
10. D. E. Sayers, E. A. Stern and F. W. Lytle, Phys. Rev. Lett., 35 (1975) 584.
11. E. A. Stern, D. E. Sayers and F. W. Lytle, Phys. Rev. B, 11 (1975) 4836.
12. F. W. Lytle, D. E. Sayers and E. A. Stern, Phys. Rev. B, 11 (1975) 4825.
13. R. F. Pettifer and P. W. McMillan, Phil. Mag., 35 (1977) 871.
14. B. K. Teo and P. A. Lee, J. Am. Chem. Soc., 101 (1979) 2815.
15. R. F. Pettifer and A. D. Cox in EXAFS and Near Edge Structure, A. Bianconi, L. Incoccia and S. Stipcich (Eds.), Springer Verlag, Berlin, Heidelberg, New York, 1982, p. 66.
16. J. J. Boland and J. D. Baldeschwieler, J. Chem. Phys., 81 (1984) 1145.
17. J. J. Boland, S. E. Crane, and J. D. Baldeschwieler, J. Chem. Phys., 77 (1982) 162.
18. J. J. Boland, S. E. Crane and J. D. Baldeschwieler, J. Chem. Phys., 80 (1984) 3005.
19. C. R. Natoli, M. Benfatto and S. Doniach, to be published.
20. H. Knözinger, R. Kozlowski and R. F. Pettifer, unpublished data.
21. F. B. M. Duivenvoorden, D. C. Koningsberger, Y. S. Uh and B. C. Gates, J. Am. Chem. Soc., in press.
22. G. H. Via, J. H. Sinfelt and F. W. Lytle, J. Chem. Phys., 71 (1979) 690.
23. D. J. Hunt, J. D. Jackson, R. B. Moyes, P. B. Wells, P. Worthington and R. Whyman, J. Mol. Catal., 20 (1983) 289.
24. S. L. Cook, J. Evans, G. N. Greaves, B. F. G. Johnson, J. Lewis, P. R. Raithby, P. B. Wells and P. Worthington, J. Chem. Soc. Chem. Commun. (1983) 777.
25. N. Binstead, S. L. Cook, J. Evans and G. N. Greaves in EXAFS and Near Edge Structure III, K. O. Hodgson, B. Hedman and J. E. Penner-Hahn (Eds.), Springer Verlag, Berlin, Heidelberg, New York, 1984, p. 226.
26. S. L. Cook, J. Evans and G. N. Greaves, J. Chem. Soc. Chem. Commun. (1983) 1287.
27. T. Yokoyama, K. Yamazaki, N. Kosagi, H. Kuroda, M. Ichikawa and T. Fukushima, J. Chem. Soc. Chem. Commun. (1984) 962.
28. T. Yokoyama, K. Yamazaki, N. Kosagi, H. Kuroda, M. Ichikawa and T. Fukushima in EXAFS and Near Edge Structure III, K. O. Hodgson, B. Hedman and J. E. Penner-Hahn (Eds.), Springer Verlag, Berlin, Heidelberg, New York, 1984, p. 192.
29. B. K. Teo, J. Am. Chem. Soc., 103 (1981) 3390.

Section 6.8

THERMOANALYTICAL METHODS

H. KNÖZINGER

Thermoanalytical methods as applied in surface chemistry and physics are largely based on *thermal desorption spectroscopy* (TDS), a method for the study of desorption kinetics [1-5]. This technique involves measurement of the desorption rate of surface species into an inert carrier gas (or into vacuum) as a function of the linearly programmed temperature. The evolution of the desorbed gases into the carrier gas can easily be followed quantitatively by conventional gas chromatographic detectors [thermal conductively detectors (TCD) or flame ionization detectors (FID)], which may be combined with a quadrupole mass spectrometer for analysis of the desorbed material.

TDS experiments provide the following information:

(1) The initial surface population by the desorbing species or its adsorbed precursors from the desorption peak area.

(2) The order of the desorption process.

(3) The activation energy E_d of the desorption process from the position of the maximum of the desorption peak on a temperature scale. E_d is a measure of the adsorption bond strength.

(4) The number of energetically or chemically distinct adsorption states from the number of resolved desorption peaks (an energy resolution of approximately 10-20 kJ mol^{-1} can typically be achieved).

(5) The desorption peak shape, determined by the order of desorption for energetically homogeneous surfaces, which may provide information about the energy distribution for energetically heterogeneous surfaces.

Modifications of the TDS technique are *temperature programmed reduction* (TDR) [6, 7] and *temperature programmed decomposition* (TPDE) [8, 9]. In TPR, an unreduced catalytic material is heated at a linear rate in a H_2/inert gas stream, and the H_2 consumption is measured as a function of temperature. Hence, the temperature regimes in which reduction occurs and the degrees of reduction of the catalyst can be determined.

TPDE is essentially the same method as TDS, the term TPDE being used when the material under study is not simply a chemisorption system, but contains molecular species such as mononuclear or polynuclear carbonyls, which may be precursors of catalytically active species. The thermal decomposition can be carried out in pure inert gas streams or inert gas streams containing reactive gases (e.g., H_2); the product evolution as a function of temperature provides information concerning the reactivity of the supported catalyst precursor and the formation of catalytically active species.

Another version of the TDS method is *temperature programmed reaction spectroscopy* (TPRS) [5], a method in which a reactive gas or gas mixture flows over the catalyst during linear temperature programming. The rate of reaction is measured by continuous monitoring of the gas phase concentrations as a function of time. This method may also involve reaction of a gas with chemisorbed species (e.g., hydrogenation of predeposited carbon). Important mechanistic information can be provided by TPRS.

In principle, experiments of this kind can be performed in microbalances, the results indicating weight loss (or gain) of the solid sample as complementary information. This type of experiment relates to conventional thermogravimetric analysis.

The most important of these thermoanalytical methods for the characterization of supported clusters and their surface transformations is undoubtedly TPDE. Brenner *et al.* [8] investigated the surface chemistry of chromium, molybdenum, and tungsten hexacarbonyls on silica. One of the striking results was the evolution of H_2 according to the stoichiometry

$$M(CO)_{6,ads} + n(\sigma\text{-OH}) \xrightarrow{\Delta} (\sigma\text{-O}^-)_nM^{n+} + \frac{n}{2}H_2 + 6\ CO \qquad (1)$$

(σ-OH and σ-O^- refer to surface hydroxyl groups and oxide ions, respectively). This result indicates oxidation of the metal. Obviously quantitative determination of the metal content and of the amount of H_2 evolved permits an estimate of the average metal oxidation state.

Hucul and Brenner [10] extended these studies to a wide range of carbonyl metal cluster/support combinations. Further examples are discussed in Section 9.2 and in the review by Brenner [9]. Basset *et al.* [11] have also compiled TPDE data for a variety of supported cluster compounds. It has been suggested [11] that the TPDE pattern of supported clusters could be related to the properties of cluster-derived materials in CO + H_2 reactions (see Section 9.5).

The essential features of TPDE for the characterization of supported clusters are illustrated by a few examples. Fig. 1 shows the scheme of an apparatus that can be applied for TPDE, TDS, TPR, and TPRS and also for measurements of catalytic reaction rates under stationary or pulse conditions. Product analysis is performed by gas chromatography using both TC and FI detectors. Hence, analysis in this case is not continuous, and

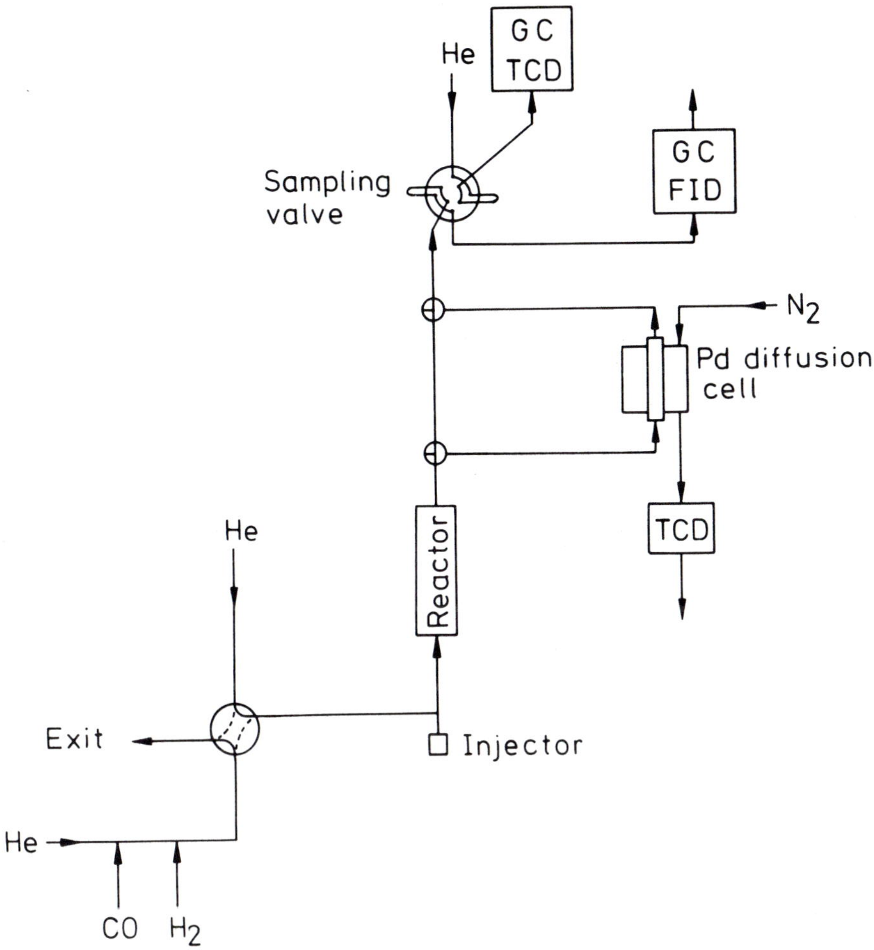

Fig. 1. Schematic representation of a flow system for TDS, TPDE, TPR, and TPRS and for stationary or pulse catalytic reaction experiments [12].

and heating rates must be chosen according to analysis time; in other words, suitable temperature resolution requires sufficiently short analysis times. In the system shown, typical temperature resolutions of 20-30 K could be achieved at heating rates of 2.5 K min^{-1}. Since a TC detector is required for analysis of CO, CO_2, and H_2O, helium is to be used as a carrier gas—which does not permit detection of H_2. Therefore, an essential part of the apparatus is a Pd diffusion cell connected to an additional TC detector which is run with N_2 carrier gas. Continuous mass spectroscopic analysis may improve the temperature resolution; however, when complex hydrocarbon mixtures are evolved, mass spectroscopic analysis can become seriously complicated.

Fig. 2 shows TPDE spectra obtained for $Os_3(CO)_{12}$ supported on SiO_2 and ZrO_2 [12]. The materials initially contained the oxygen-bridged triosmium cluster species $(\mu\text{-}O)(\mu\text{-}H)Os_3(CO)_{10}$ described in Section 9.2. It is obvious from these spectra that product distributions and the temperature regimes of maximum evolution rates for the various decomposition products, namely, CO, CO_2, CH_4, and H_2, depend on the support.

During decomposition of the supported cluster in helium (for both supports), CO_2 evolution begins at temperatures below 500 K, with a

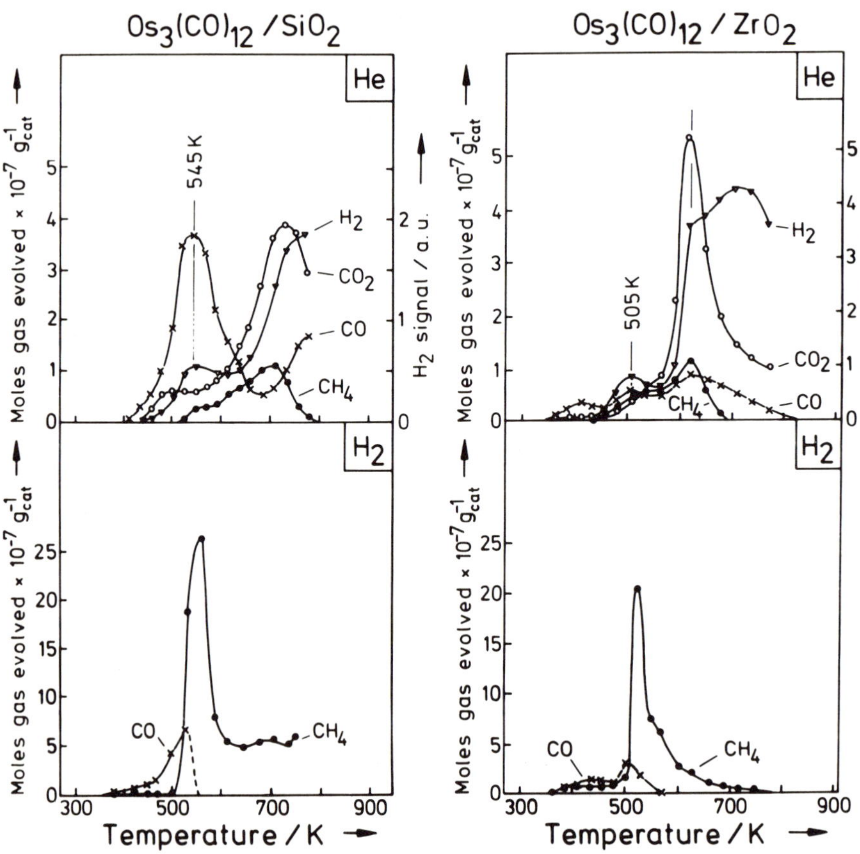

Fig. 2. TPDE spectra of samples prepared from $Os_3(CO)_{12}$ on SiO_2 and ZrO_2 supports; the decomposition was carried out in helium and in hydrogen [12].

maximum being developed at 500 K for the sample supported on SiO_2. This CO_2 evolution must be attributed to CO disproportionation:

$$2\ CO \longrightarrow CO_2 + C \qquad (2)$$

Maxima appear simultaneously at 545 K for the evolution of CO and H_2 from the SiO_2-supported sample and at 505 K for the ZrO_2-supported sample during decomposition in helium. The fact that both evolution curves pass through maxima at the same temperature suggests that both gases were formed in the same decomposition step. The evolution can be identified with the cluster oxidation according to eq. (1). At higher temperatures, CH_4 is produced and CO_2 and H_2 evolutions pass through sharp maxima, although within different temperature regimes for the two supports. The high-temperature CO_2 and H_2 evolution may be due to the water gas shift reaction, which may involve surface hydroxyl groups.

Decomposition in hydrogen preferenitally leads to the evolution of CH_4, with some CO. Only very small amounts of CO_2 and H_2O were detected. Water that should be produced may be retained on the support surface at temperatures <770 K.

In conclusion, the TPDE technique is extremely useful for the quantitative characterization of transformations of supported clusters. The technique becomes particularly informative when combined with other spectroscopic methods (e.g., vibrational spectroscopy) permitting identification of surface species that are responsible for specific gas evolutions in a given temperature regime. A major advantage of TPDE is that it provides quantitative data with relative ease, thus allowing the stoichiometry of surface transformations to be established and average oxidation states of surface species to be estimated.

REFERENCES

1. R. J. Cvetanović and Y. Amenomiya, Advan. Catal., 17 (1967) 103.
2. M. Smutek, S. Černý and F. Buzek, Advan. Catal., 14 (1975) 343.
3. D. A. King, Surface Sci., 47 (1975) 384.
4. P. T. Dawson and P. C. Walker in Experimental Methods in Catalytic Research, R. B. Anderson and P. T. Dawson (Eds.), Academic Press, New York, San Francisco, London, 1976, Vol. 3, p. 211.
5. J. L. Falconer and J. A. Schwarz, Catal. Rev.-Sci. Eng., 25 (1983) 141.
6. S. D. Robertson, B. D. McNicol, J. H. de Baas, S. C. Kloet and J. W. Jenkins, J. Catal., 37 (1975) 424.
7. J. W. Jenkins, Preprints 6th Canad. Symp. Catal., Ottawa, 1979, p. 21.
8. A. Brenner, D. A. Hucul and S. J. Hardwick, Inorg. Chem., 18 (1979) 1478.

9. A. Brenner, in Metal Clusters, M. Moskovits (Ed.), Wiley, New York, in press.
10. D. A. Hucul and A. Brenner, J. Phys. Chem., 85 (1981) 496.
11. J. M. Basset, B. Besson, A. Choplin, F. Hugues, M. Leconte, D. Rojas, A. K. Smith, A. Theolier, Y. Chauvin, C. Commereuc, R. Psaro and R. Ugo, in Fundamental Research in Homogeneous Catalysis, Vol. 4., M. Graziani and M. Giongo (Eds.), Plenum Press, New York, London, 1984, p. 19.
12. E. Alizo Delgado and H. Knözinger, unpublished data.

Chapter 7

DISPERSED METAL CLUSTERS FROM METAL VAPOR CHEMISTRY

GEOFFREY A. OZIN and MARK P. ANDREWS

7.1. INTRODUCTION

It would seem that realization of the potential of metal clusters in catalysis can be achieved by harnessing the unique properties that are anticipated for groupings of metal atoms having low nuclearity. Chini's proposition [1] that large anionic metal-carbonyl clusters might serve as models for small metallic crystallites covered (or poisoned) by ligands is based on the recognition that it is necessary to understand how the electronic, magnetic, thermodynamic, structural, and reactivity properties of the condensed state are altered in ways that modify catalytic behavior as metal atoms are associated to form, first, multimers and then groupings of increased nuclearity, and, finally, the bulk solid. Thus the anion $[Pt_{38}(CO)_{44}]^{2-}$ (one of the largest cluster carbonyls yet identified [1b]), which contains a fragment of ccp platinum metal (Fig. 1A) and has a metal skeleton diameter of 1.16 nm, is larger than the smaller immobilized metal particles in highly dispersed supported metal catalysts.

The structures of high-nuclearity close-packed metal cluster compounds with arrangements resembling fragments of metallic lattices and/or surfaces point to the general importance of 5-fold symmetry growth patterns in a number of cluster systems, such as "ligand-free" Mn_5 [2], Li_7 [3h] and Sc_{13}

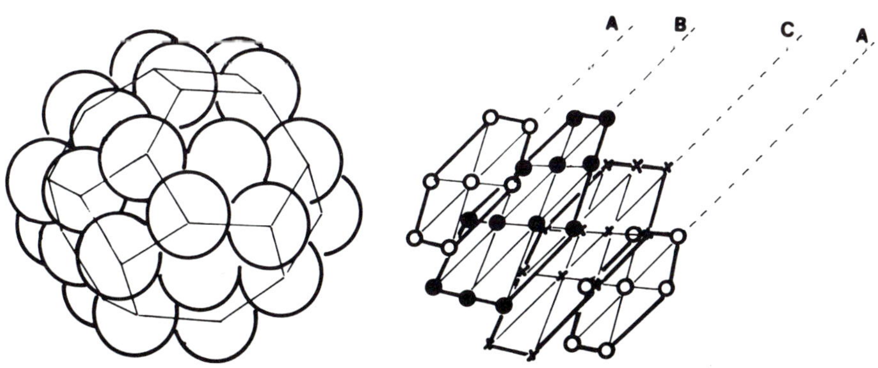

Fig. 1A. The truncated octahedron of platinum atoms (O_h) found in the $[Pt_{38}(CO)_{44}]^{2-}$ cluster and the sequence of ccp layers [1b].

[4] entrapped in rare gas solids, and $[Au_{13}(PMe_2Ph)_{10}Cl_2]^{3+}$, $[Pt_{19}(CO)_{22}]^{4-}$, and $[Au_{13}Ag_{12}(PPh_3)_{12}Cl_6]^{m+}$ containing "ligand encapsulated" metal frameworks of Au_{13} [5], Pt_{19} [6] and $Au_{13}Ag_{12}$ [7], respectively, as illustrated in Fig. 1B. The exceptional stability of pentagonal-bipyramidal alkali metal septemers M_7 (M = Li, Na, K) is in reasonable agreement with the predictions of recent *ab initio* calculations [8]. Moreover, irregularities in the normal intensity distribution of microcluster mass spectra have been observed for several metals, such as Pb_n [9] and Ag_n [10] at so-called "magic-numbers" n = 7, 13, 19, 25, 55, 71, 87 and 147. In this context, we note that energy and sphere packing calculations [11] predict stable configurations at n = 7, 13, 19, 55, and 147 involving full shell icosahedra (n = 13, 55, 147) with one or two added icosahedral caps. Also the

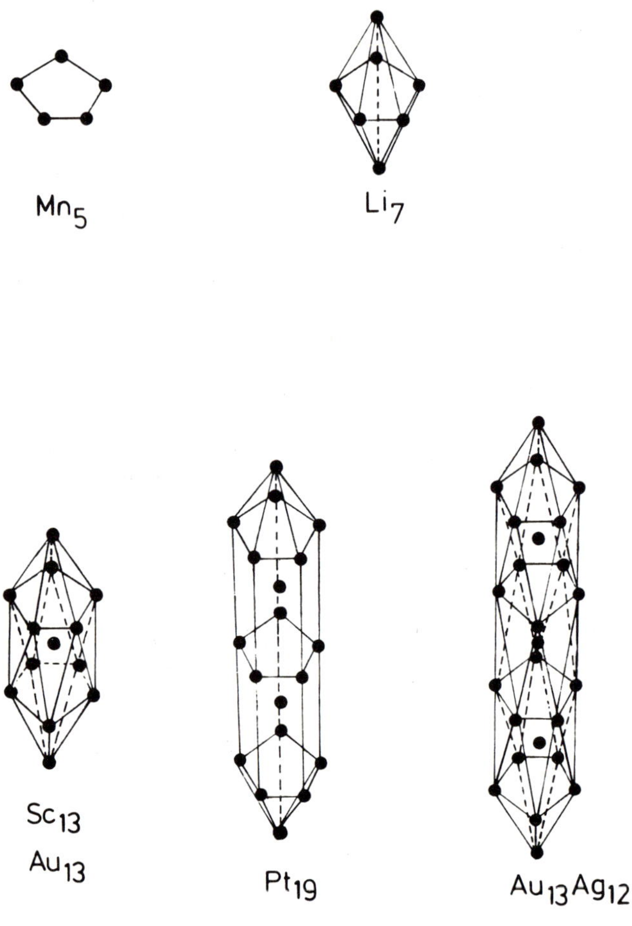

Fig. 1B. Metal frameworks of clusters exhibiting five-fold symmetry.

prevalence of polyicosahedral structures has been observed by electron microscopy of large supported metal aggregates and by electron diffraction of gas phase Ar_n clusters, where n = 20-50 [12].

In this regard we note that the cationic complex of Teo and Keating [7], $[Au_{13}Ag_{12}(PPh_3)_{12}Cl_6]^{m+}$ (where m $\simeq$ 3-5), shown in Fig. 1B, has an unprecedented geometry of three interpenetrating icosahedra and represents the largest metal alloy (mixed-metal) cluster reported to date. For the record, some other known unimetallic high-nuclearity clusters approaching small metal crystallites in size and shape include (1) the previously mentioned $[Pt_{38}(CO)_{44}]^{2-}$ with a truncated octahedral structure with 32 surface and six bulk atoms and a fcc arrangement, (2) $[Rh_{22}(CO)_{37}]^{4-}$, with 21 surface and one bulk atom and a mixed fcc/hcp arrangement [13], (3) $[Rh_{22}(CO)_{35}H_{5-q+x}]^{q-}$, with 20 surface and two bulk atoms and a mixed fcc/bcc arrangement [14], (4) $[Ni_{12}(CO)_{21}H_{4-n}]^{n-}$, with hcp([15], (5) $[Os_{10}C(CO)_{24}]^{2-}$, with a carbide-centered tetracapped octahedral structure [16], and (6) $[Ru_{10}C_2(CO)_{24}]^{2-}$, with edge-sharing bioctahedral geometry and two encapsulated carbides [17].

Size-dependent properties of small metal particles are subjects of intense current research [18]. Not unexpectedly, size-induced changes in structure-property relationships manifest themselves in catalysis. A succinct illustration of the unusual properties of metal microparticles has been given by Takasu *et al.* [19], who found that the phenomenological activation energy for H_2/D_2 exchange on Pd/carbon increased with decreasing Pd catalyst particle size. Size-dependent selectivity for hydrogenolysis and isomerization has been found by Dartigues *et al.* [20], who studied the catalytic behavior of 0.85-17 nm Pt/alumina particles. The fine spin-echo NMR studies by Rhodes and co-workers [21] of ^{195}Pt in small particles of Pt/alumina showed that aggregates as large as 10 nm in diameter do not exhibit properties (Knight shift) of bulk Pt atoms; examination of spin-lattice relaxation times corresponding to metal atoms at the surface of the particles revealed that these atoms are essentially non-metallic in nature, the conduction electron spins being largely tied up by the adsorbate. Pd/mica and Ni/mica model catalysts showed dramatic decreases in the rate of carbon formation as the average particle size was reduced [22]. In our own laboratory, iron and cobalt particles with diameters <1.2 nm deposited using solution phase/metal vapor techniques (described later) [23] onto the surface and into the α-cages of NaY zeolite showed a marked selectivity for low-carbon-number olefins, a finding which was considered to reinforce the idea of Jacobs and Nijs [24] that

Fischer-Tropsch product distributions are directly dependent on particle size.

This interest in small particles in catalysis takes us into the cryogenic matrix isolation area of research, where detailed information is emerging concerning the interactions of inorganic and organic molecules and fragments with low-nuclearity transition metal clusters. An interdisciplinary thrust, particularly from the groups of Ozin and Moskovits [25, 26], focuses on correlating the bonding of "supercoordinatively unsaturated" M_nL (low-temperature, matrix-isolated cluster systems) with that of L chemisorbed on the corresponding bulk metal M. These complexes have been variously advanced [25, 26d, 27] as models for reactive intermediates involved in heterogeneously catalyzed reactions, in particular, as these complexes have been proposed as models for localized bonding theories of catalysis [25a, 28, 29] (also see Chapter 11). Matrix-isolated ligand-free metal clusters have also received a great deal of attention. Although not always concerned explicitly with catalysis, the results of studies here do hold considerable import for the subject, inasmuch as ligand-free metal clusters are thought to represent the most chemically and physically realistic bridge between isolated metal atoms and extended arrays of atoms found in bulk metals [30, 31] and inasmuch as their ligated congeners may represent a chemical and physical bridge between homogeneous and heterogeneous catalytic systems.

To summarize many of the features of metal clusters in matrices and on supports—which are the subject of this chapter—we have devised SCHEME 1, which interrelates various classes of transition metal compositions. In this scheme, L and S respectively represent ligand and support (either liquid or solid). Variations in ligand coordination number, metal nuclearity, and support are built into the illustration. The scheme is intended to represent the hierarchy of unsupported and supported metal atom and aggregate regions, interconnecting (for bare metals) isolated metal atoms, molecular metal atom clusters, and metal microparticles (amorphous and crystalline); supercoordinatively unsaturated metal complexes; coordinatively saturated or unsaturated metal complexes; and supported metals. In the first part of this chapter, we concentrate on metal clusters (ligated and unligated) in matrices. The distinction between a matrix and a support will be made on the basis of conventional usage (a support, for example, may be alumina, silica, titania, magnesia, carbon, etc.). By default, everything else will be considered a matrix, allowing us a fuller range of survey than just the cryogenic experiments, although the latter experiments will dominate this

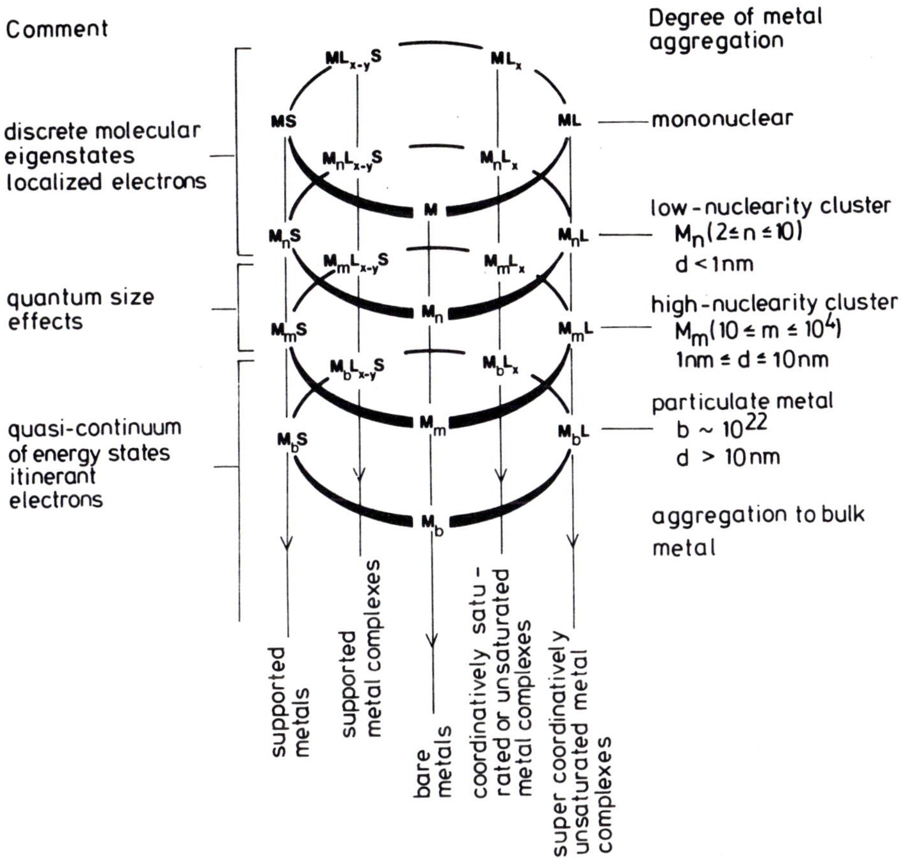

SCHEME 1

presentation. Polymers have been used both as matrices and supports, so that here the distinction is most easily made according to context. Of course, the low-temperature matrix isolation of clusters can be considered a special case of preparation of supported metal clusters. Depending on whether or not the matrix is doped with a reactive guest, one can in principle control and understand the various stages of a complex reaction network such as that illustrated in SCHEME 2. When it can be realized, this control continues to be one of the attractive features of matrix isolation synthesis.

Noteworthy spectroscopic and theoretical developments have emerged from studies of the dynamic and energetic processes of atomic and small

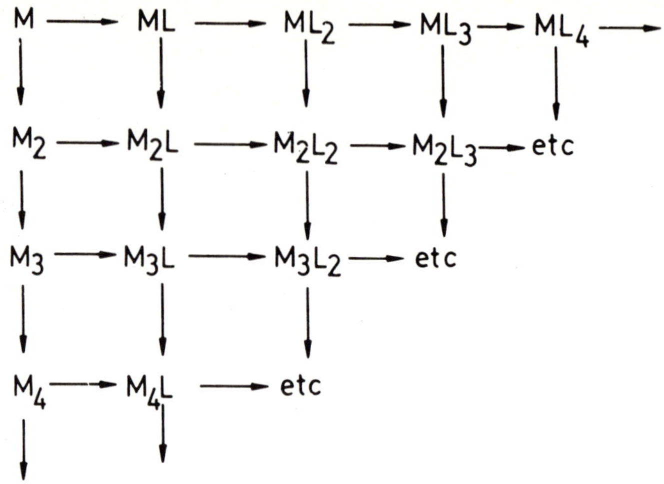

SCHEME 2

unimetallic and bimetallic cluster guests in rare gas hosts [32]. Such studies have been concerned with understanding photoinduced diffusion and agglomeration processes of metal atoms to give small metal clusters of well-defined nuclearity, photofragmentation-recombination channels of metal diatomics, and photo-isomerization and dissociation of metal triatomics and higher cluster species. The discovery of geometrical isomers of Cu_3 and Ag_3 and their ready photo and thermal interconvertibility has raised the question of whether ligand-free clusters have a preferred shape in rare gas solids. Much of the literature devoted to matrix-isolated metal clusters contains explicit reference to some aspect of the research ideology relating unresolved issues in catalysis to the role that small, well-characterized metal clusters can play in illuminating such issues, e.g., inquiry into the question of the number of metal atoms required by a cluster for it to display properties characteristic of the bulk metal (such as cohesive energy, band structure, work function, magnetism). Such inquiry resurrects questions concerning the size and geometric dependence of electronic, magnetic, molecular, and chemical properties.

7.2 METAL CLUSTERS IN MATRICES

7.2.1 Ligand-free clusters in weakly interacting matrices

The literature in this field is too large to summarize in detail here. The relevant references are those of Ozin and Mitchell [32], and prior reviews and references are available [25b, 25d, 33, and 34]. Since it is critical to the success of matrix isolation synthesis to be able to correctly attribute observed spectral features to the appropriate trapped species, it is natural to select from the broad range of products according to a particular spectroscopy. Therefore, the presentation of this section has been organized according to the spectroscopic probe type (but the list is not complete [32]) and by a sampling of metals which have been examined over the broadest range of particle sizes, as in SCHEME 1. For convenience, we have collected in Table 1 a list of the spectroscopic techniques applied to the study of ligand-free metal dimers and trimers in cryogenic matrices, the electronic ground states and geometries established to date, and a self-contained bibliography. In the section that follows, selected examples from this table (as well as other clusters) are discussed.

Procedures for synthesizing small metal clusters in chemically inert matrices have been developed from a patient accumulation and exercise of a combination of physical and chemical methods. The practice of matrix isolation and strategies for relaxing matrix constraints to prescribed low-nuclearity ligand-free clusters are described in reviews [32, 33] and monographs [26a, 34].

As mentioned, the term "ligand free" carries with it an implicit reference to the resolution of the problem of assigning the observed spectral features to the appropriate carrier of them. This is by no means a straightforward problem, requiring for its solution the combined efforts and painstaking analysis of a variety of spectroscopies, and at times some rather subtle understanding and manipulation of the matrix materials. Ozin and Moskovits [26a] gave some prescriptions in their monograph for assigning nuclearities of ligand-free metal clusters.

Despite the small perturbation to cluster properties presumed to occur on stabilizing the cluster in a weakly interacting support, it remains the rule that the properties actually observed reflect the compound state of the guest and host [35]. Such interactions must obviously be understood for proper assignment of intrinsic guest properties. In this context, an appeal to some other complementary area of research can provide verification and supplementary information regarding assignments. For example, the ideal, collision-free environment of a molecular beam of metal clusters can contribute selectively to information bearing on the size-dependent properties of aggregates of metal atoms in their equilibrium configuration [36].

TABLE 1. Matrix Isolation Studies of Ligand-free Metal Dimers and Trimers of the First Transition Series Elements

TECHNIQUE	n	Sc_n	Ti_n	V_n	Cr_n	Mn_n	Fe_n	Co_n	Ni_n	Cu_n	Zn_n
UV-visible absorption	2	1	2	2,3	4	6,7,8	7,9,10	11	7,12	13,14,15	16
	3				5	6			12	13,14	16
Fluorescence	2				17				18	13,19	
	3										
Raman	2	20			22	20		21			
	3	20			22						
Resonance Raman	2	20	21	21	22	20	9				
	3	20			22				23	24	
Far IR	2										
	3				25						
UPS	2									26	
	3									26	
MCD	2					8				27	
	3									27	
ESR	2	28				30				31	
	3	29									
EXAFS	2						32,33				
	3						33				
Mössbauer	2						34,35				
	3						35				
PROPERTIES											
Electronic G.S.	2	$^5\Sigma_u^-$		$^3\Sigma_g^-$	$^1\Sigma_g^+$	$^1\Sigma_g^+$	$^7\Sigma_g^+$			$^1\Sigma_g^+$	$^1\Sigma_g^+$
	3	$^2A'_1$								$^2E'$, $^2B_2$	
Geometry	3	D_{3h}			D_{3h}, C_{2v}				C_{2v}	D_{3h}, C_{2v}	

REFERENCES TO TABLE 1

1. R. Busby, W. Klotzbucher and G. A. Ozin, J. Am. Chem. Soc., 98 (1976) 4013.
2. G. A. Ozin and W. Klotzbucher, Inorg. Chem., 16 (1977) 984.
3. T. A. Ford, H. Huber, W. Klotzbucher, E. P. Kundig, M. Moskovits, and G. A. Ozin, J. Chem. Phys., 66 (1977) 524.

4. W. Klotzbucher, G. A. Ozin, J. Norman and H. Kolari, Inorg. Chem., 16 (1977) 2871; E. P. Kundig, M. Moskovits and G. A. Ozin, Nature, 254 (1975) 503.
5. G. A. Ozin and W. Klotzbucher, J. Am. Chem. Soc., 100 (1978) 2262; Inorg. Chem., 18 (1979) 2101; J. Mol. Catal., 3 (1977) 195.
6. W. E. Klotzbucher and G. A. Ozin, Inorg. Chem., 19 (1980) 3776.
7. T. C. DeVore, A. Ewing, H. F. Franzen and V. Calder, Chem. Phys. Letters, 35 (1975) 78.
8. J. Rivoal, J. Shakhs Emampour, K. J. Zeringue and M. Vala, Chem. Phys. Letters, 92 (1982) 313.
9. M. Moskovits and D. P. DiLella, J. Chem. Phys., 73 (1980) 4917.
10. G. A. Ozin and J. McCaftrey, Inorg. Chem., 22 (1983) 1397.
11. G. A. Ozin and A. J. L. Hanlan, J. Chem. Phys., 18 (1979) 1781.
12. M. Moskovits and J. Hulse, J. Chem. Phys., 66 (1977) 3988.
13. G. A. Ozin, S. A. Mitchell and J. Garcia-Prieto, J. Phys. Chem., 86 (1982) 473; G. A. Ozin, S. A. Mitchell, J. Garcia-Prieto and S. Mattar, J. Phys. Chem., 87 (1983) 4666, 4651; G. A. Ozin, H. Huber, D. McIntosh, S. A. Mitchell, J. G. Norman, Jr., and L. Noodleman, J. Am. Chem. Soc., 101 (1979) 3504; G. A. Ozin, Appl. Spectros., 30 (1976) 573.
14. M. Moskovits and J. Hulse, J. Chem. Phys., 67 (1977) 4271.
15. D. M. Kolb, H. H. Rotermund, W. Schrittenlacher and W. Schroeder, J. Chem. Phys., 80 (1984) 695.
16. B. S. Ault and L. Andrews, J. Mol. Spec., 65 (1977) 102.
17. M. J. Pellin and D. M. Gruen, J. Chem. Phys., 79 (1983) 5887; M. J. Pellin, D. M. Gruen, T. Fischer and J. Foosnaes, J. Chem. Phys., 79 (1983) 5871.
18. F. Ahmed and E. R. Nixon, J. Chem. Phys., 71 (1979) 3547.
19. V. E. Bondybey and J. H. English, J. Phys. Chem., 87 (1983) 4647.
20. M. Moskovits, D. P. DiLella and W. Limm, J. Chem. Phys., 80 (1984) 626.
21. G. Cosse, M. Fouassier, T. Mejean, M. Tranquille, D. P. DiLella and M. Moskovits, J. Chem. Phys., 73 (1980) 6076.
22. D. P. DiLella, W. Limm, R. H. Lipson, M. Moskovits and K. V. Taylor, J. Chem. Phys., 77 (1982) 5263.
23. M. Moskovits and D. P. DiLella, J. Chem. Phys., 72 (1980) 2267.
24. D. P. DiLella, K. V. Taylor and M. Moskovits, J. Phys. Chem., 87 (1983) 524.
25. G. A. Ozin, S. A. Mitchell and M. D. Baker, Angew. Chem. Suppl., 92 (1983).
26. D. Schmeisser, K. Jacobi and D. M. Kolb, J. Phys. Chem., 75 (1981) 5300.
27. R. Grinter, S. Armstrong, U. A. Jayasooriya, J. McCombie, D. Norris and J. P. Springall, Faraday Symp. Chem. Soc., 14 (1980) 94.
28. L. B. Knight, R. J. Van Zee and W. Weltner, Chem. Phys. Letters, 94 (1983) 296.
29. L. B. Knight, R. W. Woodward, R. J. Van Zee and W. Weltner, J. Chem. Phys., 79 (1983) 5820.
30. C. A. Baumann, R. J. Van Zee, S. V. Bhat, and W. Weltner, J. Chem. Phys., 78 (1983) 190.
31. J. A. Howard, K. F. Preston, R. Sutcliffe and B. Mile, J. Phys. Chem., 87 (1983) 536.
32. P. A. Montano and G. K. Shenoy, Solid State Comm., 35 (1980) 53.
33. H. Purdum, P. A. Montano, G. K. Shenoy and T. Morrison, Phys. Rev. B., 25 (1982) 4412.
34. H. M. Nagarathna, P. A. Montano and V. M. Naik, J. Am. Chem. Soc., 105 (1983) 2938.
35. W. Dyson and P. A. Montano, Solid State Comm., 33 (1980) 191.

7.2.2 Spectroscopy of ligand-free metal clusters in weakly interacting matrices

Low-nuclearity metal atom clusters represent a strong perturbation of the bulk-phase metal. It is interesting that where analogies have been drawn between metal clusters and fragments of a metal surface or bulk solid [1, 16, 26d, 27], one of the descriptors, "metal", contentiously resists definition [37-41]. Interest in what constitutes a metal perhaps underscores the significance of the size-dependent catalytic behavior referred to in the opening remarks.

As Baetzold and Hamilton pointed out [41], the transition between the atomic state and the bulk is not necessarily the simple monotonic one classically intuited. The complexity of the changes is indicated both by theoretical calculations and by experimental evidence. It is now quite apparent that metallic status can be conferred by the particular property one choses to describe [18a, 37]. In terms of transport properties, the appellation "metallic" might, for example, connote itinerant electrons in a (nearly) free electron metal at absolute zero temperature. In sampling the range of naked metal clusters symbolized in SCHEME 1, ESR measurements have been found to be rewarding, as they are sensitive not only to complete delocalization of the wavefunction (free ionization of the electron from its atomic site), but also to partial (spin) delocalization, possibly involving extension of a primarily localized electron wavefunction over several sites [42]. The latter process need only be sensitive to those changes in metal cluster nuclearity that modify inter- and intra-site exchange energies of the order of the magnetic Zeeman or hyperfine energies (~10^{-4} eV).

7.2.2.a Electron spin resonance (ESR) spectroscopy of low-nuclearity metal clusters. Although they are used primarily as promoters in catalysis [43], the alkali metals have so far revealed a rich ESR spectroscopy for clusters in the size range $3 \leq n \leq 7$. Here there are obvious and distinguishable molecular properties. Lindsay and coworkers have been instrumental in preparing and characterizing the small alkali metal clusters, Li_3, Na_3, K_3, Li_7, Na_7, K_7 [3]. The successful isolation of these molecules depends critically on the metal:rare gas dilution ratio, the atomic number of the rare gas, and the temperature of the cold substrate during the deposition process. For example, in Lindsay's experimental configuration [3b], Na_3 was isolated in argon during cocondensation only when the 100-kHz magnetic field modulation was used to cause Joule heating in a localized region of the sapphire target. The researchers found that clustering to give the sodium trimer in argon:alkali gas mole ratios of 100:1 occurred during deposition and not by subsequent annealing. With a different experimental configuration [3a], K_3 could be isolated with no magnetic field modulation, a result which emphasizes the subtle differences associated with the two experimental stations. $^6Li_3$ and $^7Li_3$ molecules were observed [3g], following photoaggregation at 30 K of ^6Li and ^7Li atoms, respectively, which were isolated initially in argon matrices (75:1, Ar:Li) at 4.2 K.

Analysis of the ESR spectra reported for Na_3 and K_3 indicated linear ($^2\Sigma_u^+$ ground state) or slightly bent ($^2B_2$ ground state) structures. Moreover, these results showed that the trimer (neutral) is chemically bound, with predominantly s rather than p character. The spectra showed an isotropic hyperfine structure arising from the Fermi contact interaction of a single unpaired electron and three nuclei (Fig. 2). Two nuclei were found to be magnetically equivalent, having hyperfine constants approximately half of that of the free atom. The hyperfine coupling to the third nucleus was very small. For both Na_3 and K_3, evidence was found for a second form of each trimer exhibiting a dynamic structure which fluctuated between three equivalent bent forms of the molecule, the three atoms alternately occupying the apical position. Distortion of the alkali metal trimer away from the equilateral triangle (D_{3h}) can be rationalized within the Jahn-Teller theorem (Fig. 3). The orbitally degenerate $^2E'$ ground electronic state (equal spin populations on all atoms) splits into $^2A_1$ and $^2B_2$ states as the triangle is distorted towards C_{2v} symmetry. The $^2B_2$ state lies lowest in energy [3f]. In this way the unequal spin density distribution is explained. However, in the distortion of the D_{3h} trimer, no preference can be shown for angle bending from any particular atom; therefore, there must exist three equivalent potential wells, each with a $^2B_2$ electronic symmetry. This has some important structural implicationsfor clusters trapped in matrices. On the ESR time scale, a trimer may appear to be isolated in the three stationary $^2B_2$ configurations, or it may exchange by tunnelling between adjacent potential wells. The latter process results in three equivalent alkali atoms on the average. Na_3 and K_3 show both stationary and pseudo-rotating spectra in argon [3a, b, f]. As expected, interconversion between the two forms is temperature dependent. In contrast, Li_3 shows no stationary spectra between 4.2 and 40 K. Thus the observation (Fig. 2) of seven equally spaced hyperfine transitions with measured intensities close to 1:3:6:7:6:3:1 is consistent with the coupling of the unpaired electron with three equivalent I = 1 nuclei, hence with the assignment to a pseudorotating $^6Li_3$. This result was

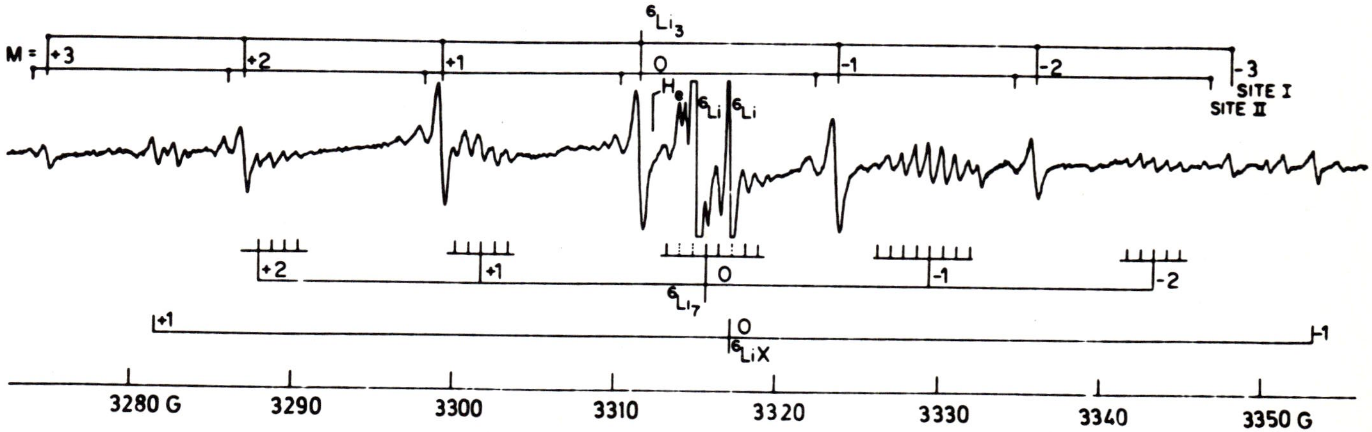

Fig. 2. ESR spectrum of pseudorotationg $^6Li_3$ in an argon matrix at 28.5 K. Roman numerals designate matrix sites. Also shown are atomic lithium, $^6Li_7$, and a radical designated ^6LiX. H_o = 3312.27 G is the resonance field of a free electron at a cavity frequency of 9.8255 GHz [3g].

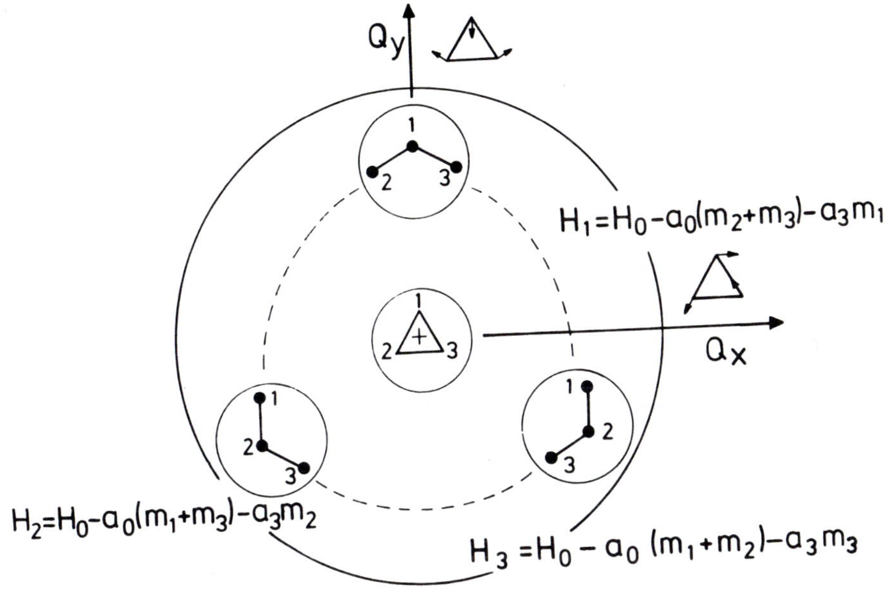

Fig. 3. Normal coordinate and real space description of the lowest $^2E'$ surface of pseudorotating M_3. Circles represent energy contours for the three minima of the trimer potential surface, and the dashed lines represent three barriers to pseudorotation [3g].

anticipated by the *ab initio* calculations of Gerber *et al.* [45], whose explorations indicated a $^2B_2$ symmetry potential too shallow to support a vibrational level for Li_3.

Lithium, sodium, and potassium heptamers have been reported by Lindsay [3d, h] (see, for example, Fig. 4). The single unpaired electron in the sodium heptamer shows an isotropic hyperfine interaction arising from an approximate 37% spin population on each of two equivalent atoms. There is a residual 2% spin population on each of (at least) five atoms. Through a series of exploratory simple Hückel calculations, the authors recognized the plausibility of the pentagonal bipyramidal geometry for Li_7, Na_7, and K_7 [3d]. *Ab initio* calculations by Fantucci *et al.* [46] have indicated a high stability for the pentagonal bipyramidal alkali metal clusters. Their calculations, confirm that such a structure exhibits a group of two equivalent and another group of five equivalent centers.

There has been some attempt [3g, h] to relate the "Fermi electron" s-character in these clusters to the degree of sp hybridization in the bulk metals, and therefore to assess qualitatively the onset of metalization in

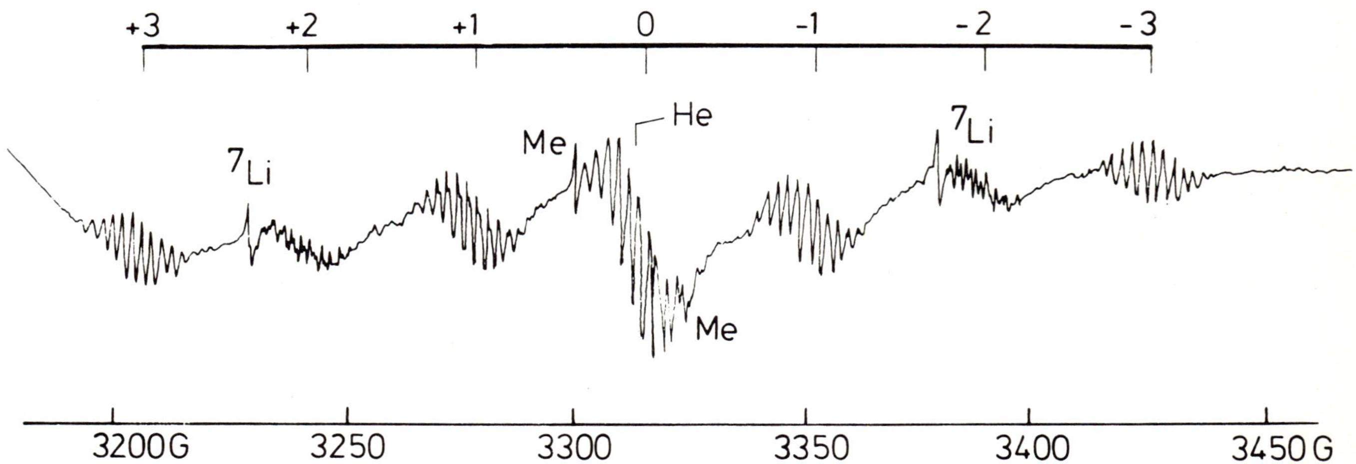

Fig. 4. ESR spectrum of $^{7}Li_{7}$ in Ar at 30 K showing hfs into seven groups from a large isotropic spin density on two equivalent I = 3/2 ^{7}Li atoms. The shfs within each group corresponds to a much smaller spin population (~2%) on each of at least five equivalent atoms [3h].

small alkali metal clusters. The difference $1 - \Sigma\rho$, where $\Sigma\rho$ is the total isotropic spin population, gives an approximate measure of the average p character of the unpaired electron. There is a general trend towards increasing p character with increasing cluster size. For sodium and potassium heptamers, the mean hybridization approaches 30-38% of the value of the bulk metal [3d, g]. Bulk lithium has the smallest Fermi electron s character, and in the series Li_{2-7} shows an sp hybridization consistently larger than that of sodium or potassium molecules or bulk metal [3d, 46].

Howard *et al.* [47, 48] described the ESR spectrum of Ag_3 molecules isolated in solid perdeuteriobenzene. Found within the complex spectrum were essentially isotropic features consisting of four ~40 G doublets with a_{107} (2) = 295 G, a_{107} (1) = 38.5 G, and g_{iso} = 1.9622. The separation of the central pair of doublets is as expected for a second-order splitting associated with equal (~295 G) isotropic hyperfine interactions from two nuclear spins of magnitude 1/2. Spin populations of 44% were found for each terminal Ag(5s) and 6% for the central Ag(5s) atomic orbital. The composition of the singly occupied molecular orbital of Ag_3 is thus quite similar to those of Na_3 and K_3. On the basis of an analysis of the spectrum of Ag_3 in terms of an orthorhombic g tensor having all three principal values less than the free-spin value, the authors favored a bent C_{2v} molecule having a $^2B_2$ ground state. Similarly, an obtuse triangular structure with a $^2B_2$ ground state was attributed to Cu_3 isolated in an adamantane matrix at 77 K [49].

Sc_3, Y_3, and La_3 have been isolated in rare gas matrices [4] and studied by ESR spectroscopy. The Sc_3 molecule has also been observed in solid argon by resonance and preresonance Raman spectroscopy, and in well-annealed matrices it has been assigned an equilateral geometry [50]. In accord with this proposal, the ESR spectrum of Sc_3 has been interpreted in terms of an equilateral triangular geometry at 4-30 K with a $^2A'_1$ ground state. Thus one finds that the ESR spectrum of Sc_3 (^{45}Sc, 100% natural abundance, I = 7/2) is centered near the free-electron value and shows a well-resolved 22-line hyperfine pattern (Fig. 5) expected from one unpaired electron interacting with three equivalent scandium nuclei. The spectrum is essentially invariant over the temperature range 2-32 K. The data therefore point toward a D_{3h} equilateral triangular Sc_3 molecule in which the unpaired electron is orbitally nondegenerate, although the possibility is recognized that Sc_3 is a fluxional bent Jahn-Teller molecule with a very low barrier to pseudo rotation even at 2 K. Recall that Li_3 undergoes pseudorotation under similar conditions [3g]. The observed ^{45}Sc hfs provides clues concerning the most likely configurations adopted by the nine

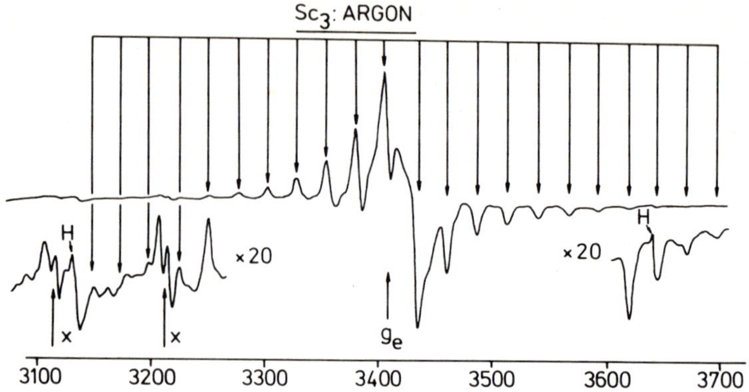

Fig. 5. ESR spectrum of Sc_3 molecules isolated in solid argon at 4 K with 22 hyperfine lines of ^{45}Sc (I = 7/2) assigned (cavity frequency = 9.538 GHz). The magnetic field is in G. H indicates hydrogen atom signals, and X designates a few of the lines previously assigned to Sc_2 [4].

valence electrons of Sc_3 [4]. The choice narrows down to

$$(a_1'\text{s}\sigma)^2\ (e'\text{s}\sigma)^4\ (a_1'\text{d}\sigma)^2\ (a_1'\text{d}\delta)^1$$

or

$$(a_1'\text{s}\sigma)^2\ (e'\text{s}\sigma)^4\ (a_1'\text{d}\delta)^2\ (a_1'\text{d}\sigma)^1$$

based on the observation that $A_{iso}(^{45}Sc)$ is quite small in comparison with the hfs expected for a single 4s electron on the atom (indicating that only about 3% of the total unpaired spin density is due to s-electron character). This is reasonable as the s-orbitals are expected to be largely involved in bonding. In support of this proposal, the observed anisotropic ^{45}Sc hfs is in accord with the unpaired spin residing in a molecular orbital comprising mainly 3d atomic orbitals. Whereas the pentamers of the alkali metals have so far eluded detection, Cu_5 [49], Ag_5 [49], and Mn_5 [2] have been isolated and probed by ESR. Calculations of Cu and Ag cluster geometries by the CNDO method, with parameter sets improved to allow greater spd hybridization, have shown that there exists a delicate energetic balance between linear and three-dimensional geometries [41]. For Cu_5, for example, the bipyramid structure is favored over the linear chain. The ESR spectra of both Cu_5 and Ag_5 can be explained in terms of trigonal bipyramidal geometries, Jahn-Teller distorted to C_{2v} "opened" structures. These clusters were synthesized by simultaneous photolysis ($\lambda > 320$ nm) of the metal atoms codeposited at 77 K with perdeuterocyclohexane within a rotating cryostat. The ESR spectrum of the Cu_5 molecule displayed a

16-line pattern, each component of which was further split into a 13-line multiplet. The unequally spaced 16 lines were assigned to a large hyperfine splitting from two equivalent Cu atoms and were fitted with the Breit-Rabi equation [51]. This fitting yielded a hyperfine value of a(2) = 608G and an average g value of 2.055. Examination of the highest field hyperfine line showed it to be split into the (1:2:4:6:7:8:8:8:7:6:4:2:1) 13-line pattern with an intensity distribution close to that expected for two equivalent Cu atoms with a hyperfine interaction of 15G, and a third Cu atom having twice this interaction. Exploratory INDO/2 calculations for Cu_5 suggested that the unpaired electron occupies an e' orbital formed by a linear combination of equatorial Cu orbitals, in agreement with *ab initio* calculations [52]. Jahn-Teller distortion of the $^2E'$ electronic ground state toward an open C_{2v} geometry places the unpaired electron density in even distribution in the b_2 molecular orbital of the resulting two equivalent copper atoms. Analogous reasoning leads to the assignment of the Jahn-Teller distorted trigonal bipyramidal skeleton to the Ag_5 cluster. Unpaired isotropic spin populations here amounted to 62% for two equatorial ^{107}Ag nuclei (30%, Cu); 1.7% for the unique equatorial ^{107}Ag nuclei (1.4%, Cu); and 0.8% for the remaining two axial ^{107}Ag (0.7%, Cu). For both clusters, the remaining *ca.* 40% spin density in the singly occupied MO must comprise p- and/or d-orbital contributions.

Van Zee *et al.* [2] inferred a structure of pentagonal planar geometry for Mn_5. The molecule is apparently high spin, having a total electronic spin S = 25/2, corresponding to five unpaired spins on each Mn atom. This research group has also been successful in detecting the ESR of matrix-isolated bimetallic CrCu, CrAg, and CrAu [53]. Isotopically enriched ^{53}Cr (I = 3/2) aided in the assignment of a $^4\Sigma$ ground state to CrCu. The heavier molecules have the expected $^6\Sigma$ ground states. Pre-dating Van Zee's studies of the mixed metal dimers were experiments by Kasai and McCleod [54], who examined AgM (M = Mg, Ca, Sr, Be, Zn, Cd, or Hg). For all of these dimers the hyperfine coupling to the Ag nucleus was found to be isotropic. Howard and co-workers [49] successfully isolated the CuAgCu, $^2B_2$ obtuse angled trimer by codepositing $^{63}Cu/^{107}Ag/C_6D_6$ at 77 K. For comparison with the Ag_3 hyperfine parameters, we note the large hyperfine splitting from two equivalent ^{63}Cu nuclei [I = 3/2, a(2) = 880.5 G] and a smaller splitting from the one ^{107}Ag center [I = 1/2, a(1) = 35.5 G], with g = 1.962.

Concluding the examples of this section is an outstanding observation of Knight and coworkers [4]. This concerns Sc atom neon matrix codepositions which yield an ESR spectrum centered around the free-electron value (g = 1.9914) showing more than 60 observable hyperfine lines with a constant splitting of 1.68 ± 0.02 G (Fig. 6). The line shape has been analyzed in terms of a homogeneously broadened signal of a well isolated scandium cluster containing at least nine equivalent ^{45}Sc (I = 7/2) atoms containing a single unpaired electron. Interestingly, the g value for bulk scandium metal via conduction ESR (see next section) has been calculated at around 1.95. On the basis of possible structures for transition metal clusters and their relationships to bulk metals, Knight *et al.* [4] considered the ESR spectrum and unusual structural stability of their Sc_x

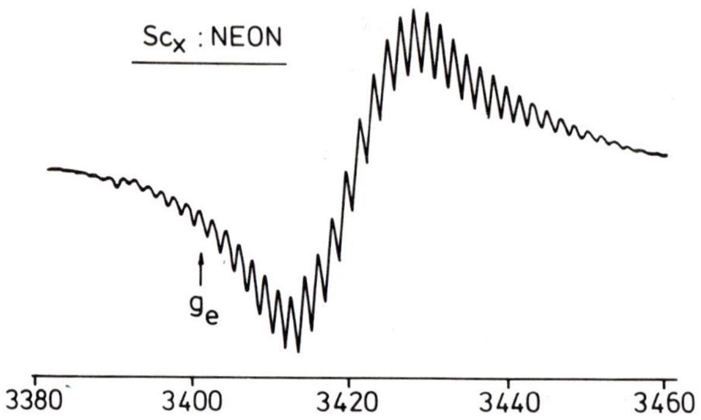

Fig. 6A. ESR spectrum of the Sc_x molecule isolated in solid neon at 2 K. The magnetic field is given in G for spectrometer frequency = 9.532 GHz.

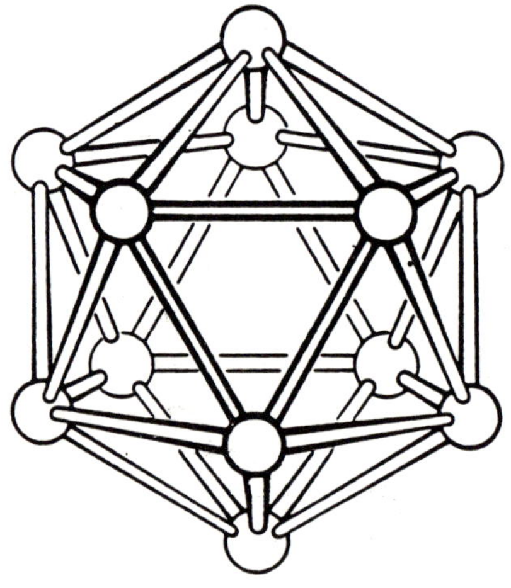

Fig. 6B. Proposed icosahedral structure for Sc_x, where x = 13. The central Sc atom is not shown [4].

cluster in solid Ne in terms of a little piece of hcp Sc metal, i.e., a 3-7-3 ABA truncated hexagonal bipyramidal array of 13 atoms having D_{3h} symmetry. Close in energy are the 3-7-3 ABC cubooctahedral and icosahedral arrays of 13 atoms having O_h and I_h symmetries, respectively. Each of these models contains 12 equivalent surface atoms, consistent with the rigid-structure ESR spectrum observed for Sc_x at 2-4 K, with the requirement that the odd electron be in an orbital spread over at least nine equivalent Sc nuclei. As one model, Knight and coworkers considered Sc_{13} as a 39-electron icosahedron (Fig. 6B), by analogy with the pervasiveness of such electron-deficient icosahedral fragments in borides and boron compounds. With $A_{\parallel}$ = 1.681 G and $A_{\perp}$ = 0.0 G, a satisfactory fit to all of the observed lines in Fig. 6A was obtained. Thus 12 A_{iso} is small relative to A_{iso} = 1007 G for the Sc atom and A_{dip} = 12 x 1.6 = 20 G, which compares favorably with $A_{dip} \cong 25$ G of the free atom. These data indicate that the odd electron has little s character and probably resides in a bonding or nonbonding sd hybrid orbital. A qualitative analysis of the lowest σ-hybrid orbitals for icosahederal Sc_{13} having their largest amplitudes pointing inwards towards the central Sc atom (with reasoning from a crystal field model), yields the electronic configuration

$$a_g{}^2\ t_{1u}{}^6\ h_g{}^{10}\ t_{2u}{}^6\ t_{1u}{}^6\ g_u{}^8\ a_g{}^1$$

with a ^2A_g ground state which would be in accord with the observed hfs for Sc_x described above.

In view of the extensive information that has accrued from matrix ESR spectroscopic studies of ligand-free unimetallic metal clusters over the past few years, a summary of their relevant magnetic parameters, electronic ground states, and geometries—with comments and references—has been collected in Table 2. The growing ESR data bank of corresponding information for matrix-entrapped bimetallic clusters has not been included in this tabulation because of space limitations; the reader is referred to the original literature for this information [53-55].

7.2.2.b Electron spin resonance spectroscopy of high-nuclearity metal clusters. Experimental studies of metal clusters in rare gas matrices are subject to an inherent difficulty; generally it is not possible to produce a species of specific cluster nuclearity to the exclusion of all others. This situation is exemplified by the coisolated Na_3 and Na species in the ESR

TABLE 2. Matrix isolation ESR data for ligand-free metal clusters.

Molecule	Electronic G.S.	Matrix Material	g tensor	A_{iso} (MHz or G)	A_{dip} (MHz or G)	Spin(ρ_s) density	Comments	Ref.
$^{6}Li_3$	$^{2}E'$	Ar	2.00282(1)	12.21(1)(G)		0.225	pseudorotating	[3g]
$^{7}Li_3$	$^{2}E'$	Ar	2.00262(1)	32.19(8)(G)		0.225	at 4.2 K	
$^{23}Na_3$	$^{2}B_2$ or $^{2}\Sigma_u^+$	Ar	2.0012	-60(25)(G)		-0.07	three sites, rigid	[3a, b, f]
				424(40)(G)		0.47	at 4.2 K;	
	$^{2}E'$			261.0(6)(G)			pseudorotating	
$^{39}K_3$	$^{2}B_2$ or $^{2}\Sigma_u^+$	Ar	1.9985(5)	-13.1(6)(G)		-0.06	rigid	[3a, b, f]
				109.1(9)(G)		0.47		
	$^{2}E'$	Ar	1.9990(5)	68.5(20)(G)			pseudorotating	
$^{23}Na_7$	$^{2}A_2''$	Ar	1.9994(1)	6.66(12)(G)		0.021	D_{5h} pentagonal	[3d]
				118.16(4)(G)		0.374	bipyramid - 3 sites	
$^{7}Li_7$	$^{2}A_2''$	Ar	2.00045	2.19(2)(G)		0.015	D_{5h} pentagonal	[3d]
				36.39(5)(G)		0.254	bipyramid	
$^{39}K_7$	$^{2}A_2''$	Ar	2.0017(1)	1.79(2)(G)		0.022	D_{5h} pentagonal	[3h]
				30.55(G)		0.371	bipyramid	
$^{45}Sc_2$	$^{5}\Sigma_u^-$	Ne	2.002	252(11)	20(11)	0.23	Hyperfine parameters	[56]
			2.000	(MHz)	(MHz)		favor $s\sigma^1\ d\sigma^1\ d\pi^2$	
							electronic configuration	
$^{45}Sc_3$	$^{2}A_1'$	Ar	1.993(1)	24(4)(MHz)	24(2)(MHz)	0.022(4)	D_{3h} static equilateral	[4]
			2.993(4)				triangle at 2 K	
$^{89}Y_3$	$^{2}B_2$	Ar	1.9603(17)	18.9(3)(MHz)	4.4(2)(MHz)	0.0174(3)	C_{2v} bent	[4]
			1.9578(17)	22.4(3)(MHz)	2.7(2)(MHz)	0.260(3)		
$^{45}Sc_{13}$	$^{2}A_g$	Ne	1.9914(4)	1.6±2.0(MHz)	1.6±1.0(MHz)	0.0014(18)	I_h icosahedron	[4]

TABLE 2 (Continued)

Molecule	Electronic G.S.	Matrix Material	g tensor	A_{iso} (MHz or G)	A_{dip} (MHz or G)	Spin(ρ_s) density	Comments	Ref.
$^{55}Mn_2$	$^1\Sigma_g^+$	Ar(S=1) Ar(S=2) Ar(S=3)	2.0 1.9986 1.9983	14(G) 14(G) 14(G)			Antiferromagnetic exchange coupled dimer; S = 1, 2, 3 Landé states accessible from 4-70 K; J = 9 ± 3 cm^{-1}; $\lvert D_1\rvert$ = 0.49; $\lvert D_2\rvert$ = 0.128; $\lvert D_3\rvert$ = 0.064 cm^{-1}	[57]
$^{55}Mn_5$		Kr	1.997(1) 1.979(2)				S = 25/2; D = 0.013(1) cm^{-1}; D_{5h} planar pentagon	[2]
$^{63}Cu_3$	$^2B_2$	Adamantane	1.9925	-55.6(G) 625.5(G)		-0.026 0.29	Deposited at 77 K; bent C_{2v} G.S.	[49]
$^{107}Ag_3$	$^2B_2$	C_6H_6	1.960 1.962 1.966	-105.7(9)(G) 810.1(9)(G)		-0.06 0.44	Deposited at 77 K; bent C_{2v} G.S.	[47]
$^{197}Au_3$	$^2B_2$	C_6D_6	1.855	-58(G) 382.9(G)		-0.06 0.42	Deposited at 77 K; bent C_{2v} G.S.	[58]
$^{63}Cu_5$	$^2B_2$	C_6D_{12}	2.055	15(G) 30(G) 608.6(G)		0.01 0.3	Distorted trigonal bipyramid at 77 K; obtuse C_{2v}; static Jahn Teller	[49]
$^{107}Ag_5$	$^2B_2$	Adamantane	2.081	5.5(G) 11(G) 204.6(G)		0.01 0.3	Distorted trigonal bipyramid at 77 K; obtuse C_{2v}, static Jahn Teller	[49]

spectrum of Lindsay's sodium heptamer. With the possible exception of the recently reported Sc_{13} molecule [4], the synthesis and identification of aggregates larger than 10 atoms by matrix isolation methods has not been realized. This lack is likely due to the low probability associated with bringing together larger and larger numbers of atoms to form precisely the desired cluster during the cocondensation step whereby most of these $n > 3$ clusters are grown. Part of the difficulty may also reside in the necessity of having to resort to high metal atom:gas ratios in order to favor formation of the larger clusters [3d, 59, 60]. Lindsay has noted that at Na:Ar ratios larger than the 1:100 ratio used to generate Na_7, a broad resonance centered near the free-electron g value was observed. Howard and coworkers [48, 49, 58] and Ozin [61] have also observed similar effects with Cu and Ag.

This feature may be attributed to a conduction electron spin resonance (CESR). A CESR is a resonant transition between Zeeman levels of the conduction electrons at frequency

$$\omega_z = \frac{g\mu_B B}{K} \tag{1}$$

in the applied field B, where μ_B is the Bohr magneton. The observed g factor is shifted from the free electron value by some amount Δg_{so}, approximately related to the strength of spin orbit coupling. CESR is a much-studied phenomenon, receiving attention because of the relevance of CESR as a gauge of the onset of geometric and quantum size effects [18a, 62]. Hermerschmidt and Haul [63] observed a CESR from silver, aggregated after reduction of Ag_6^{n+} clusters within the cubooctahedral sodalite cages of AgA and AgX zeolites. Ozin [61] observed similar effects for Ag atoms aggregated in rare gas solids. The thermally induced growth of the CESR (~3320 G) from silver atoms isolated in squalene [64] [a triterpene, small-molecule analogue of poly(isoprene)] is depicted in Fig. 7. This system is interesting, as the substantially higher melting points of many hydrocarbons relative to the melting points of the noble gases has suggested to some researchers [25d] the possibility of using hydrocarbon matrices as variable-temperature reaction media for observing incipient metal atom aggregation processes over a much wider temperature range than is possible with the rare gas solids. Use of the hydrocarbon supports has met with some limited success. It has been possible to establish the atomic spectrum of Ag atoms in, for example, the n-$C_{22}H_{46}$ wax [25d]. Whereas the onset of aggregation of silver atoms in methane and short-chain hydrocarbons appeared to follow the Tamann predictive diffusion temperature of $0.3T_{mp}$, diffusion in the paraffin wax is extremely facile at temperatures as low as 30 K. This observation is explained in terms of diffusion in voids in the very disordered and open structures characteristic of vapor-condensed wax films. These hydrocarbons are therefore not very useful materials for maintaining atomic dispersions at higher temperatures; however, observations of weak silver atom-olefin interactions [65] and a trend towards

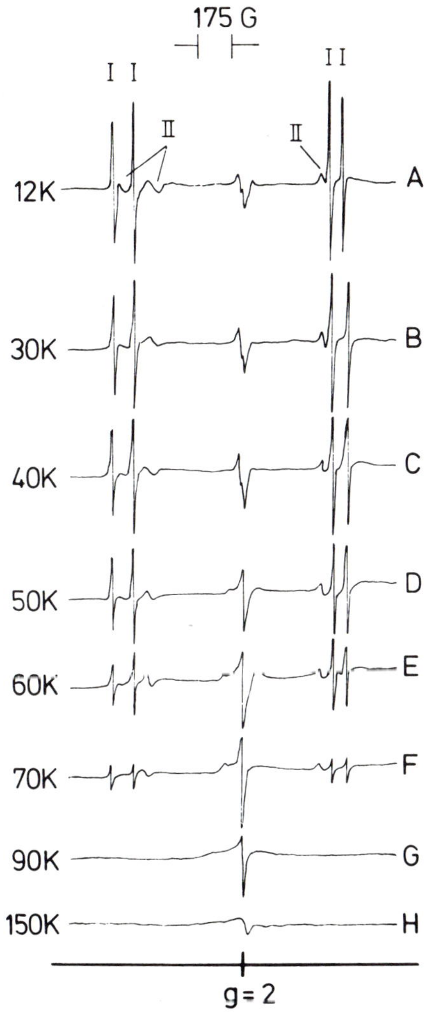

Fig. 7. Silver atoms cocondensed with squalene at 12 K and annealed at 30, 40, 50, 60, 70, 90 and 150 K showing the agglomeration of silver atoms in sites I and II (see text) to Ag_n^0 clusters capable of supporting a CESR spectrum (central feature around g = 2) [64].

increasing thermal stability in the series [25d]: Ag(but-1-ene)$_n$ > Ag(propylene)$_n$ > Ag(ethylene)$_n$ suggests the idea of using squalene as a "hybrid matrix support" with alkene traps in an alkane matrix. The signals in Fig. 7 confirm that silver atoms can be isolated in squalene. The observed splitting pattern of Ag for the $m_s = \pm 1/2$ transitions is readily accounted for by the hyperfine interaction with the nuclei ^{107}Ag (natural abundance = 51%, I = 1/2 and $\mu_B = -0.1130\beta_N$) and ^{109}Ag (natural abundance = 49%, I = 1/2 and $\mu_B = -0.1299\beta_N$) nuclei. The respective hyperfine splittings of 628 and 723 G and g values (2.0018, 2.0021) are in line with those for $^{107,109}Ag$ in frozen adamantane, cyclohexane [47], and methane [66]. The weak features (indicated as site II in Fig. 7), which persist to 70 K while the atom lines decrease, are thought to be due to Ag-olefin 1:1 pseudocomplexes. The g value (1.9998) and hyperfine splitting [a(^{107}Ag) = 542.5] are very similar to those values derived from analysis of the spectrum of the weak Ag.....C_2D_4 or Ag.....C_3D_6 complexes diluted in rare gas matrices [65]. If the site II ESR transitions in squalene do represent interactions between Ag and the backbone carbon double bonds of squalene, then it follows that silver atoms partition themselves in both saturated and unsaturated regions of the squalene matrix, and the function of a hybrid matrix support would seem to be defined. Thus in a sense the spectra A-F of Fig. 7, evidencing coexisting silver atoms and aggregates, represent a linking of the circles of SCHEME 1 through the central chain of particles varying from atoms to low- and high-nuclearity clusters.

Under conditions for which the mean free path of a conduction electron exceeds the particle diameter, a simple geometric effect on the linewidth is predicted, proportional to d^{-1}, where d is the particle diameter. In Fig. 8 the linewidth of the CESR at ~3320 G is decreasing with increasing annealing temperature. This is the expected behavior, given that matrix annealing should lead to growth in numbers and sizes of particles. Also depicted is the loss of atomic silver (triangles) through decay of the silver $m_s = \pm 1/2$ transitions. The initial increase of the CESR signal intensity (at constant linewidth) with increasing temperature is probably due to the growing number of similarly sized particles having the required size to show the effect. At temperatures greater than 70 K silver aggregate growth likely dominates (silver atoms have vanished); consequently the portion of sufficiently small particles decreases until the signal finally disappears. The decrease in linewidth at temperatures greater than 30 K is also caused by large particle growth.

Conduction electrons may modulate a number of surface-sensitive phenomena of relevance to catalysis. It is thought, for example, that the

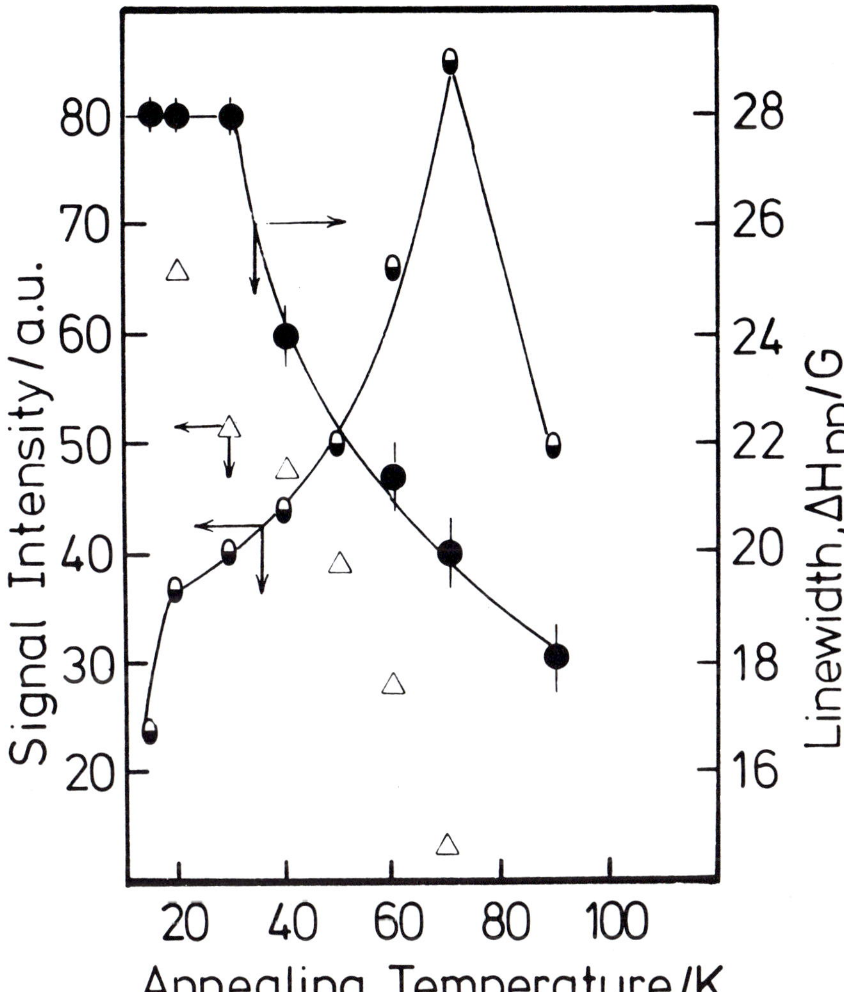

Fig. 8. Behavior of the intensity (◒), and line width ΔH_{pp} (peak to peak, ●) of the Ag_n^o CESR signal and the intensity of the Ag^o ESR signal (Δ) on annealing Ag^o/squalene matrices at 12-90 K [64].

broadening of the infrared absorption of CO chemisorbed at low coverages on a metal may be enhanced through the intermediacy of the conduction electrons. Such non-local effects must be removed in the limit of particles sufficiently small to disrupt the band structure of the solid.

7.2.2.c **Mössbauer spectroscopy of metal clusters.** Mössbauer spectroscopy shows potential for probing subtle changes in electronic and bonding interactions among metal atoms. The technique is sensitive to changes in s-electron density at the metal nucleus, changes which can be influenced by polarization by d-electrons. Among the most important results to be gleaned from such experiments are (a) *absorption peak areas* giving effective Debye temperatures (phonon cut-off frequencies), a measure of the strength of binding of a Mössbauer atom to a solid matrix, the probability of occurrence of the Mössbauer effect and relative abundances of individual species (concentrations); (b) *isomer shifts* (resulting from differences in electron density at the source and absorber Mössbauer atom) reflecting the electronic configuration of the Mössbauer atom ground state, which relates to oxidation state, covalency, coordination number, and ligand type; (c) *quadrupole splittings* reflecting the electric field gradient and hence the symmetry of the charge distribution around the Mössbauer atom; and (d) *magnetic dipole splittings*, the temperature and magnetic field dependence of which provide detailed information concerning the nature of magnetic ordering and particle size effects in ferromagnetic, antiferromagnetic, and superparamagnetic materials.

An experimental demonstration of the marked deviation of the properties of metal clusters from the bulk phase is represented in the study by Dyson and Montano [67]. They have observed the evolution of iron clusters from atom to bulk using Mössbauer matrix spectroscopy (Fig. 9). In the size range Fe_n (n = 1-8) the isomer shift undergoes a large change on proceeding from the atom to the dimer, with the maximum value of the isomer shift occurring at the dimer and trimer level, gradually diminishing to the value of α-iron for a cluster of eight atoms. The magnitude of the quadrupole splitting is also reduced with cluster size.

Much attention has been paid to the study of local electronic structure in metals and alloys to better understand their physical and catalytic properties. The alloys of course enjoy a special place in catalysis [68]. FeCo has a $^6\Sigma$ ground state according to analysis of matrix isolation Mössbauer spectra [69]. Other Fe_xCo_y molecules, including Fe_2Co and $FeCo_2$, have been identified and their Mössbauer parameters determined, although geometries were not assigned. It might be expected that in metal vapor matrix isolation cocondensations, the product distributions themselves might be amenable to some sort of statistical treatment. Monte Carlo methods have been used to estimate cluster size and composition. In this way the variation of the Fe isomer shift with the number of Co nearest neighbors in the Fe_xCo_y clusters has been examined [69]. The isomer shift approaches the value for iron in a cobalt matrix for about 12 Co nearest neighbors. For one Co nearest neighbor, a maximum in the isomer shift of

Fe indicated a change in the configuration of the iron atom and a decrease in the s-electron density at the iron nucleus. Increasing the number of surrounding Co atoms then decreased the isomer shift, implying an increase in the s-electron density at the nucleus as a result of 3d-electron

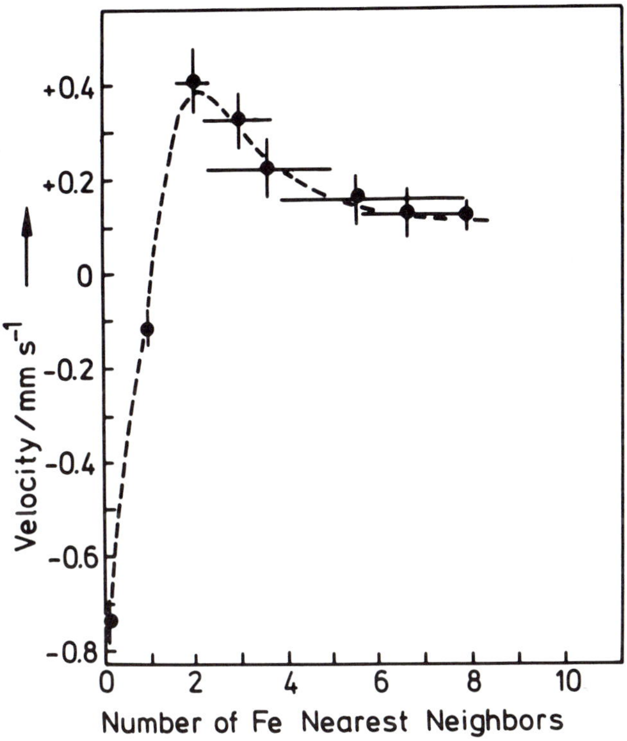

Fig. 9. Variation of IS (at 4.2 K) with the number of nearest neighbor iron atoms for argon-entrapped Fe^{o}_{n} [67].

polarization effects. Monte Carlo analysis of these 5 atom % (Fe:Co = 1:3) matrices predicted the formation of only four- and five-atom clusters. At these concentrations the Mössbauer spectra already display the six lines characteristic of magnetic hyperfine splitting. At 25 atom % the observed magnetic hyperfine field is the same as that observed for disordered iron-cobalt alloys of the same composition.

Separate Mössbauer studies have been made of matrix-isolated Fe_xSn_y molecules [70, 71]. Dimers, heterodimers, and heteromultimers of 3-6 atoms were detected. In bulk metal FeSn alloys the isomer shifts are linearly related to the number of Fe-Sn bonds through the reduced electron density at both nuclei due to bond formation. Bonding is essentially covalent and unpolarized. These are characteristics also of the Fe_xSn_y matrix-isolated molecules. The increased isomer shift of FeSn over Fe_2 reflects the reduced electron density at the Fe nucleus [70a], while the decrease observed in the 5s atomic orbital population at the Sn site compared with that of Sn_2 is consistent with the decrease observed in the tin IS in the heterobimetallic [70a].

7.2.2.d **Vibrational spectroscopy of metal clusters.** With the recent successes in Raman and resonance Raman observations of matrix-entrapped metal diatomics, the vibrational frequencies and anharmonicities of the first transition series are now complete [50]. A systematic assessment of their observed and calculated properties, such as M-M bond stretching force constants, bond dissociation energies, and bond lengths is now possible (Table 3). In particular, comparison of the properties of M_2 and bulk M as one moves across the periodic table yields an interesting insight into the similarities and differences between the nature of the bonding interactions responsible for the cohesive forces in these seemingly different states of matter.

Diatomic metal-bulk metal correlations of this kind have been of considerable interest for sometime [70a]. Thus, for example, it has been noted previously that the dissociation energies of the transition metal diatomics vary across a period in an irregular manner, somewhat akin to the variation of certain solid-state properties of the bulk metals [70a and references cited therein]. This has led to the suggestion that the valence bond model used for bulk metallic systems might be applicable to the gaseous diatomics as well as various clusters with nuclearities intermediate between those of the bulk and diatomic molecules [70a]. By considering various degrees of $3d^n4s^2 \longrightarrow 3d^{n+1}4s^1$ and $3d^n4s^2 \longrightarrow 3d^n4s^14p^1$ atomic promotion to optimize the bonding between unpaired valence electrons, in conjunction with basic physical concepts, it is found that quite a few solid-state properties (e.g., crystal structures, cohesive energies, conductivity, superconductivity, phase stability, and alloying) can be examined by reference to the properties of the respective diatomic molecule [70a].

A most striking feature of the trend of both the bond stretching force constants and dissociation energies of the first-row transition metal diatomics as a function of the number of valence electrons in the way they shadow the corresponding binding energies of the respective transition elements (Fig. 10). Trends in the depth (D_0^0) and narrowness at the minimum (k_e) of the diatomic metal ground state potential energy curves appear to follow one another, as well as the cohesive energies (E_c) of the bulk metal. In all cases, these properties at first increase, reach a maximum at vanadium, then fall off at manganese, rise again up to nickel, and steadily drop off to zinc. This behavior for the bulk metals may be rationalized on the basis of band theory, which at a zeroth-order level of approximation shows resemblences to the bonding in the respective homonuclear diatomics, in the spirit that one is monotonically filling a set of bonding and antibonding orbitals with a maximum in binding energy occurring near the center of the series.

TABLE 3. Selected Observed and Calculated Properties of Metal Diatomics and Bulk Metals.[a]

Metal	k_e, mdyn/nm[a]	ΔE_c, eV[b]	D_0^0, eV[c]	r_e, calc[d], nm	r_e, reported, nm
K	1.0	0.941	0.56	0.355	0.392
Ca	0.74	1.825	0.12±0.01	0.365	
Sc	7.6	3.93	1.3	0.245	
Ti	23.5	4.855	1.31±0.18	0.177	
V	43.4	5.30	2.47±0.22	0.147	
Cr	28.0	4.18	1.56±0.22	0.169	0.1685*
Mn	02.5	2.98	0.44±0.3	0.313	0.340
Fe	14.8	4.29	1.30±0.22	0.204	0.203*
Co	14.6	4.387	1.73±0.22	0.204	
Ni	24.8	4.435	2.38±0.22	0.175	
Cu	13.0	3.50	1.97±0.06	0.211	0.22197*
Zn	1.2	1.35	0.19±0.07	0.348	

[a] M. Moskovits, D. P. DiLella and W. Limm, J. Chem. Phys., 80 (1984) 626.

[b] C. Kittel, Introduction to Solid Physics, 3rd ed. (Wiley, New York, 1967).

[c] The bond dissociation energy of the diatomics were taken from K. A. Gingerich, Faraday Disc. Chem. Soc., 14 (1980) 109, for all the species with the exception of that of Sc_2, for which the value was taken from G. Verhaegen, S. Smoes and J. Drowart, J. Chem. Phys., 40 (1964) 239.

[d] Calculated with the formula $k_e = -0.22 \pm 14.35/r_e^3$, which was obtained by fitting the three measured r_e values marked with asterisks in the last column to the observed k_e values.

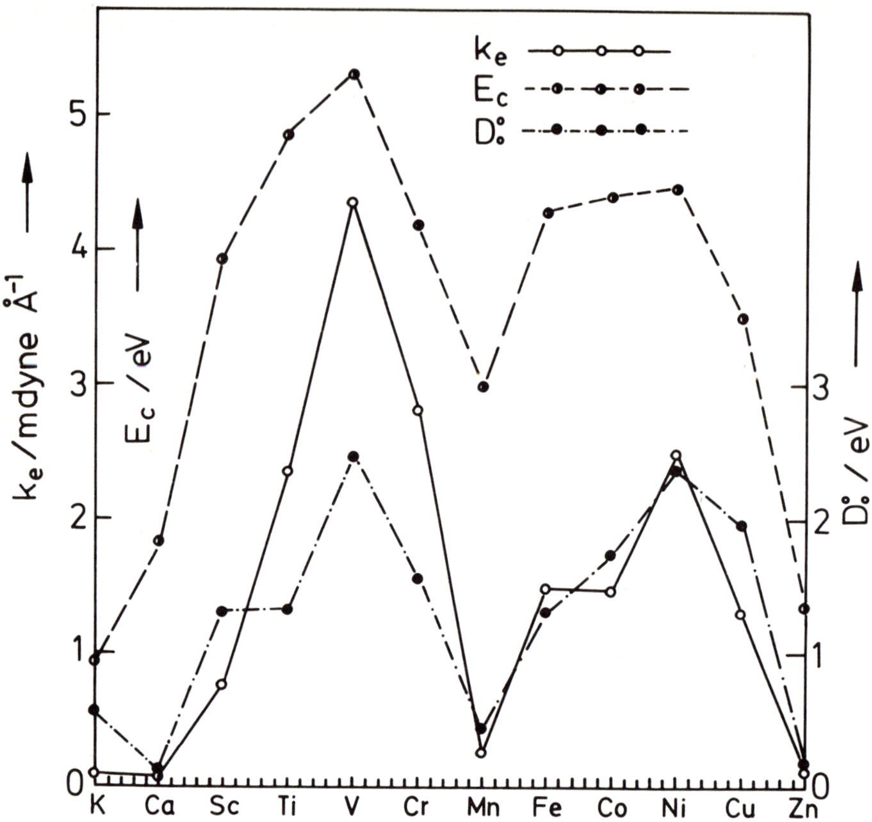

Fig. 10. Bond stretching force constants k_e and bond dissociation energies D_0^0 for M_2 and cohesive energy E_c for bulk M on passing from K through to Zn [50].

Clearly, this is an oversimplification of the picture that exists in practice for dimers and bulk metals. One may consider the d-band of the metal and its s-type conduction band (as well as the mixing between these two bands) as having counterparts in the diatomics. In the latter, the energy difference between the bonding and antibonding counterparts of the σ, π, and δ orbitals deriving from the d-electrons relates to the d-band width, while the σ_g and σ_u orbitals and their energy separation originating from the s-electrons relate to the conduction band width. Besides these contributions to the binding energies of M_2 and bulk M, one also needs to consider the atomic promotional energy term that describes the energy required to take M from its ground electronic $3d^n4s^2$ state to the $3d^{n+1}4s^1$ and $3d^n4s^14p^1$ states appropriate for optimum binding to neighboring metal atoms in either diatomic or extended metal arrays [70a]. As one traverses left to right across the first transition series, the d orbitals stabilize in energy and contract, and the d-band width of the bulk metal (as well as the separation of antibonding and bonding counterparts of the d-orbitals of M_2) are expected to decrease. Simultaneously, the d orbital electron population increases.

These two effects alone should lead to a monotonic increase in binding energy from Sc_2 to Cr_2 (where at Cr one has completely filled bonding orbitals), subsequently dropping off monotonically on the way to Zn. That this is not exactly the case in practice implies the existence of other terms and effects (including spd hybridization, electron correlation, and spin-orbit coupling effects) not well accounted for on the basis of an s, d-bonding model alone. Mn_2 is clearly a case in point, where the $^1\Sigma_g^+$ ground state turns out to be very weakly bound, with five unpaired electrons localized on each Mn atom, exchange coupled in an antiferromagnetic fashion [50, 57]. The more strongly bonded nature of bulk Mn, on the other hand, underscores the existence of interactions in the extended lattice which defy explanation in terms of two-atom forces alone.

V_2 and Cr_2 nicely exemplify the delicate balance that exists between orbital overlap and metal-metal bond strength [71a]. Here, d-orbital contraction effects are already showing up because, although the bond length of Cr_2 (0.168 nm) is considerably shorter than that of V_2 (0.177 nm), the bond-stretching force constant of the latter is much higher. In Cr_2 one has the situation where the atoms must approach more closely to maximize d orbital overlap (Cr_2 is more contracted than V_2), but this contraction creates additional repulsive interactions of the filled core orbitals as well as the nuclear Coulomb repulsions, which lead to a shallower minimum in Cr_2 than in V_2. This results in a lower vibrational frequency of Cr_2 (470 cm^{-1}) than of V_2 (529.5 cm^{-1}). The disproportionate contraction of 3d orbitals relative to the 4s orbital thus forces Cr_2 to have a shorter bond length than V_2, but at the expense of a lower dissociation energy. The 3d contraction therefore is anticipated to play a major role in controlling bond lengths and electronic properties of the first transition metal dimers, and most likely their higher-nuclearity clusters as well [71a].

Raman and resonance Raman data have been published for tri- and multi-atomic metal clusters, exemplified by Sc_3, Cr_3, Ni_3, Cu_3, Ag_3, Pb_3,

Sb_3, Sb_4, Bi_4, and Pb_4 [50, 70, 72-76]. Much structural information has been gleaned from the spectra. We briefly mention only Cu_3 because of its cross reference with the ESR data. The resonance Raman spectrum of Cu_3 in solid Ar (excitation center *ca.* 16810 cm^{-1}) displays an irregularly spaced progression of lines (major series 355, 710, 1063, and 1430 cm^{-1}; minor series 404, 760, and 1115 cm^{-1}) with vibrational components possessing an unexpected form of isotopic structure thought to be diagnostic of a pseudorotating Cu_3 molecule (Fig. 11) [74].

The assignment of fluxionality is subtle. Vibrational features are observed to arise only from $^{63}Cu_3$ and $^{65}Cu_3$, but not from the isoptomers, $^{63}Cu^{65}Cu_2$ or $^{63}Cu_2{}^{65}Cu$. The explanation invokes lifting of the degeneracy of the vibrational ground (and possibly excited) state of $D_{3h}(^2E')^{63}Cu_3$ in the $C_{2v}(^2B_2 + {}^2A_1)^{63}Cu_2{}^{65}Cu$ isoptomer. Such splitting

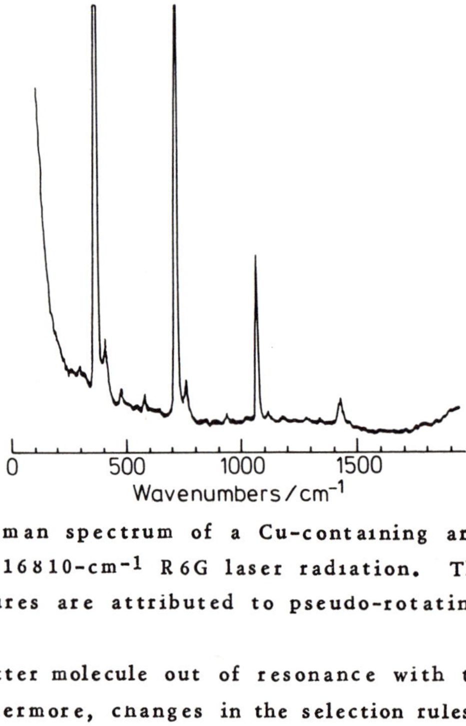

Fig. 11. Resonance Raman spectrum of a Cu-containing argon matrix, excited with 16810-cm^{-1} R6G laser radiation. The observed spectral features are attributed to pseudo-rotating Cu_3 [74].

could possibly put the latter molecule out of resonance with the exciting laser frequency. Furthermore, changes in the selection rules could make the dipole-allowed electronic excitation in $^{63}Cu_3$ forbidden in $^{63}Cu_3{}^{65}Cu$. Either or both of these effects could result in the absence of $^{63}Cu_2{}^{65}Cu/{}^{63}Cu^{65}Cu_2$ vibronic components in the resonance Raman spectrum of the Cu_3 molecule undergoing dynamic Jahn-Teller distortions.

Raman and resonance Raman spectra have also been obtained for Cr_3 in argon matrices, and a bent structure with an estimated apex angle of *ca.*

60° was proposed [72]. Complementary information was provided by Fourier transform far infrared spectroscopy by Ozin *et al.* [77]. Trichromium, immobilized in argon and xenon matrices, displayed molecular geometries ranging from that of an equilateral triangle to that of an almost linear triatomic molecule (cf. Fig. 6 and Table 1 in reference 77). From these data it was concluded that the trimer must have a rather shallow potential for the framework deformation [77]. These results reinforce a more general observation that even a "weak" interaction between the solute and the support can markedly influence the electronic, vibrational, and geometric properties of the cluster. Some additional insight into such matrix effects is furnished, in selective cases, by magnetic circular dichroism spectroscopy.

7.2.2.e Magnetic circular dichroism (MCD) of metal clusters. This technique utilizes measurements of the differential absorption of left- and right-circularly polarized light propagating along the direction of an externally applied magnetic field [78]. Information is obtained concerning the degeneracy of the electronic states involved in optical transitions. Because the MCD spectra of certain paramagnetic molecules show a temperature dependence that is diagnostic of the presence of unpaired spins, Grinter *et al.* [79] were able to confirm the assignment of the absorption band at 246 nm attributed previously [80] to Ag_3 in an argon matrix. The MCD results here pointed to an orbital degeneracy in the excited electronic state, consistent with a $^2\Pi \longleftarrow {}^2\Sigma$ assignment for the electronic transition, i.e., a linear Ag_3 molecule. This geometry agrees with an interpretation [81] of the Raman spectrum of the trimer (in solid krypton), but is at variance with the bent structure favored by Howard and coworkers [48] in their ESR study of this molecule in solid benzene at 77 K (probably a matrix effect). Other dimers and clusters that have been examined by MCD include $Cu_{2,3}$, Ag_2, $Mg_{2,3,4}$, Ca_2, Sr_2, Mn_2, and Ni_n [79, 82, 83]. Clusters of uncertain nuclearity have also been studied by photoemission spectroscopy, [84], to which we now turn.

7.2.2.f Photoemission spectroscopy of metal clusters. Photoemission spectroscopy can be used to measure the electronic structure of metal cluster particles. The technique is capable of mapping cluster density of states, ionization threshold, core-level positions, and adsorbate structure. Studies here have also emphasized details of the convergence of electronic properties with cluster size from atom to bulk metal. Much success has been achieved in experiments with metal clusters grown at high and low coverage on supports under ultra high vacuum conditions [41]. The consensus here is that even clusters in the size range 100-150 atoms do not always reproduce bulk spectral cluster features.

Jacobi *et al.* [84] have successfully used UPS to study Ni_n and Cu_n clusters isolated in rare gas matrices. A broad emission at *ca.* 4eV for Cu/Kr $\cong 1/10^2$ was assigned to entrained Cu_n (Fig. 12). The position of this emission is quite distinct from that of bulk copper (*ca.* 2.5 eV). Discrepancies between the UPS results and the predictions of theory seem to indicate a need to reassess those theoretical calculations [31] which indicate that copper clusters with as few as 6 to 13 atoms display d-band energies similar to those of bulk Cu.

As a logical extension of this work, the UPS studies were applied to Ni_n and Cu_n clusters in reactive O_2 and CO matrices [84]. Comparisons were made between adsorption of O_2 and CO on Cu and Ni films with that on the small Cu_n and Ni_n clusters. Thus, for Cu_n in CO a new emission was observed at 1.5 eV which was not present in the UPS of Cu_n isolated in rare gas matrices. The new signal was assigned to a photoemission from the $2\pi^*$ orbital of CO, which is unoccupied in the free molecule but partly occupied by charge transfer from Cu_n. A similar peak is observed for CO adsorbed on Cu films and surfaces. This peak has been interpreted as diagnostic of a $Cu \longrightarrow CO(2\pi^*)$ charge transfer, an interaction which plays a central role in Cu-CO bond formation.

7.2.2.g Ultraviolet-visible absorption and emission spectroscopy of metal clusters. A rather large amount of data has been collected for the ligand-free metal clusters, providing additional spectroscopic bases for examining cluster-bulk transformations, matrix effects, theoretical calculations, etc. Matrix-isolated metal cluster species characterized by their UV/vis absorption spectra include $Na_{3,4}$ [58, 85, 86], $Mg_{3,4}$ [82], Co_3 [87], Ni_3 [88], $Cu_{3,4}$ [89], Cr_3, Mo_3, Cr_nMo_{n-3} [90a], Ag_{3-6} [60, 91] and Cr_nAg_m [90b]. Silver clusters rank highly among those favored for study in absorption and emission. Of interest at the low-nuclearity cluster scale is the lowest-energy HOMO $\longrightarrow$ LUMO transition, which decreases monotonically in value with increasing cluster nuclearity [25d, 41, 94]. This trend was predicted theoretically [41], the theory showing good agreement with the observed transitions. For copper clusters of increasing size, it seems that the HOMO-LUMO transition should recede into the infrared, broadening, until in the bulk it is replaced by the spectrum due to free electrons [89].

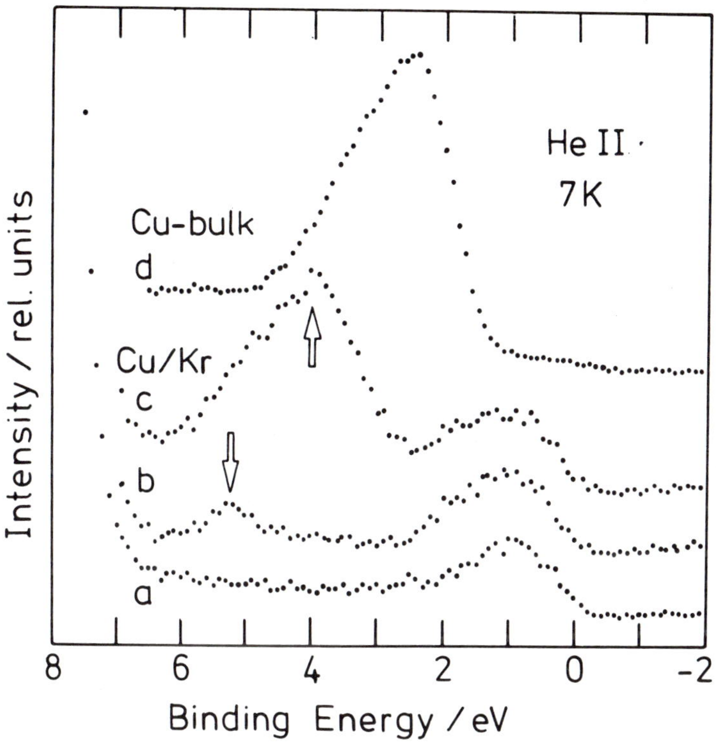

Fig. 12. UPS of (a) a pure Kr matrix, (b) Cu atoms isolated in Kr, and (c) Cu clusters isolated in Kr. For comparison, the spectrum of bulk Cu is also shown in (d). Excitation: He II; T = 7K [84].

Synthesis of matrix-isolated metal clusters can be photo-assisted. The study of Ag_n (n = 2-6) clusters by Ozin and Huber [91] introduced a technique for photoaggregating metal atoms. Since then a number of experiments, notably the ESR cluster studies referred to earlier, have utilized light to specify cluster growth. Matrix-isolated Na, Ni, Cu, Cr, Mo, and Rh atoms also show the effect [25a, c, d, 85, 87, 92]. The energetics and dynamics of this process in rare gas solids have been reported [32], analysis of the absorption and emission spectra indicating strong guest-host interactions in the low-lying excited states of the Ag, Ag_2, and Ag_3 species. SCHEMES 3-5 summarize the photoprocesses thought to be responsible for their photoactivated growth and decay. The kinetics of the photo-induced silver atom aggregation is apparently adequately

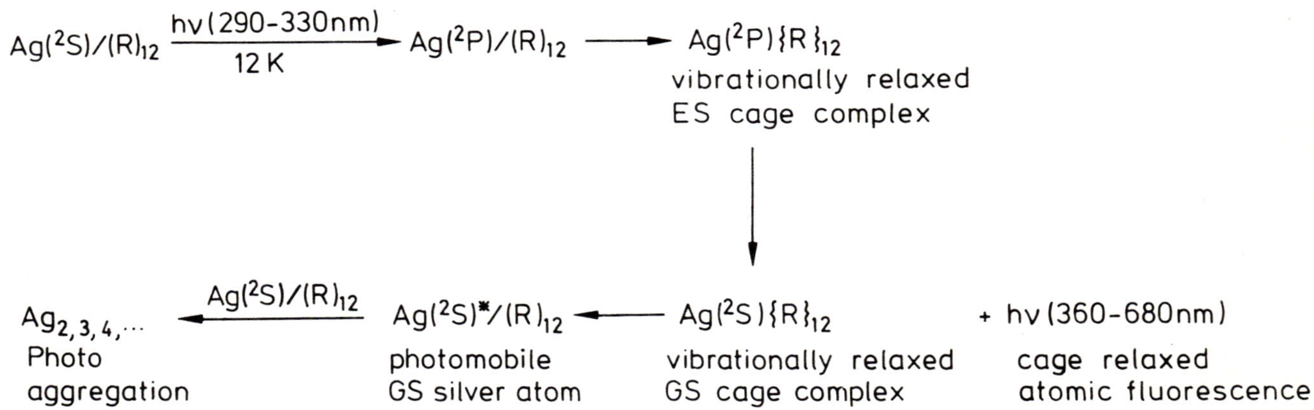
Ag(²S)/(R)₁₂
hν(290-330nm)
12 K
Ag(²P)/(R)₁₂
Ag(²P){R}₁₂
vibrationally relaxed
ES cage complex
Ag(²S){R}₁₂
vibrationally relaxed
GS cage complex
+ hν(360-680nm)
cage relaxed
atomic fluorescence
Ag(²S)*/(R)₁₂
photomobile
GS silver atom
Ag(²S)/(R)₁₂
Ag₂,₃,₄,...
Photo
aggregation

SCHEME 3

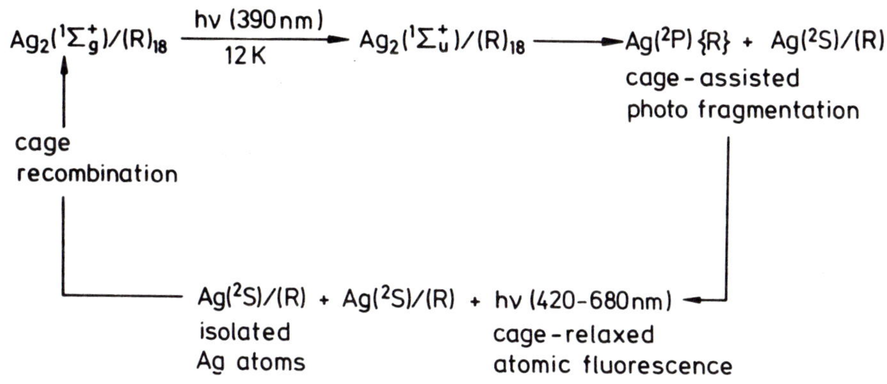

SCHEME 4

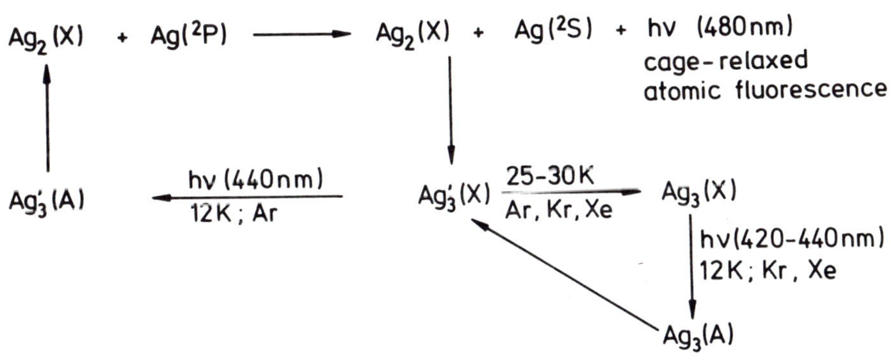

SCHEME 5

described by a simple statistical cluster growth model [93]. At the high-nuclearity end of the scale in SCHEME 1, the surface plasmon absorption of small silver particles has been intensively studied, both theoretically and experimentally, because of the sensitivity of this collective phenomenon to particle size and shape [18, 94].

We observe that an underlying theme of some UV/vis matrix-isolation studies remains the details of the transition from single electronic transitions to collective oscillations of conduction-band electrons in cluster systems of increasing size [18d, 59, 89, 91, 94]. Hoffman *et al.* [86] introduced a procedure for directly preparing and isolating Na_n molecules: the clusters are preformed in a molecular beam from a supersonic nozzle and deposited into Kr matrices at 4.2 K. By this means, the relative concentrations of atoms and molecules could be simultaneously mass analyzed. The experiments demonstrated that preformed Na_n clusters could be matrix isolated as such. Optical absorption assignments of the products could therefore be made directly, and the results confirm other [25d, 59] assignments. They found the Na_2 molecule to be rather stable to matrix aggregation in annealing experiments. They also observed that molecular aggregates having $n > 2$ evolved from very small molecules (Na_{2-4}). Sodium atoms in sites causing the "red" absorption were rapidly mobilized and combined to give stable clusters having extended overlapping absorption bands. There appeared to be some selectivity towards the growth of a species of a given nuclearity, depending on the size of the aggregates already present in the matrix, with the largest clusters developing from the larger aggregates predisposed in the matrix.

As Thompson *et al.* [3d] had found in the ESR of Na_n in Ar, alkali metal-rare gas ratios in excess of 1/100 gave rise to high-nuclearity clusters which appeared to show incipient metallic character through a conduction electron spin resonance. In optical absorption, the nearly structureless band recorded by Hoffmann *et al.* [86] was argued to result from a number of overlapping absorption bands due to the Na_n clusters. In some respects this band resembles that produced by optical transitions between surface states of small metal particles (surface plasmon absorption)—care must be taken to distinguish the origins of the observed processes in the rare gas solid. For high metal concentrations (in this case exceeding 10 mole%), the sharp molecular lines abruptly disappeared and a single broad plasmon absorption, characteristic of microcrystals, appeared. The authors explained the shape of the absorption band using a continuum theory of dielectric response, taking into account the microcrystal size, shape, and mutual interaction.

It is worth pointing out the similarity between metal atoms and clusters dispersed in a rare gas matrix and expanded states of simple metals. In one sense the metal/rare gas matrix isolation experiments can be considered as models of a metal of extremely low density at low temeratures. A sharp metal-insulator transition has been observed in sodium-argon mixtures at 15% Na [95]. Analysis of these results gave a density of sodium at the transition of approximately 0.14 g/cm^3, close to the density at the liquid-vapor critical point of pure sodium. Similarly, Cu/Ar compositions [96] underwent dramatic changes in the electrical properties at a critical concentration corresponding to a density of about 2.3-2.7 g/cm^3. Berggren [38] argued on the basis of Herzfeld's [97] criterion of metallization that the metal-nonmetal transition in the Na/Ar mixtures is due to changes in the dielectric screening properties of the system, rather than phase separation or percolation to give contacts between individual atoms or clusters.

A word of caution is in order, however. Experience gained from matrix isolation studies of the growth of the Na and Cu clusters from the atoms in rare gas solids [86, 89] suggests that these "expanded" metal/rare gas systems are actually rather aggregated. In calculating the critical density and interatomic spacing at which the metal-insulator transition occurs, one cannot reliably assume that (all) the metal atoms enter the rare gas host sites substitutionally.

7.2.3 Reactions of ligand-free metal clusters

Having considered some features of ligand-free metal clusters in matrices, one can proceed around the hermeneutic circles of SCHEME 1 and begin to examine what happens when organic molecules and fragments thereof react with low-nuclearity clusters in rare gas solids. By synthesizing and characterizing such species, researchers in the field of matrix isolation have sought to clarify aspects of the chemisorption bond. These studies [25a] have dealt, for example, with the following molecules:

$M_n(CO)_x$ (where n = 1, 2, 3, 4; x = 1, 2, ...)

$M_n(O_2)_x$ (where n = 1, 2, 3: x = 1, 2, ...)

$M_n(N_2)_x$ (where n = 1, 2; x = 1, 2, ...)

$M_n(C_2H_x)_x$ (where n = 1, 2, 3, 4; x = 1, 2, ...)

From this array of products, we restrict attention primarily to $M_n(CO)_x$ low-nuclearity clusters, as these overlap at one end with the Chini-type cluster carbonyls (Chapter 2), and at the other end with some high-nuclearity clusters $[M_n(CO)_x]$ to be presented below.

Thus in SCHEME 1, ML_x (x = 1) represents the metal atom synthesis of the series of triatomic monocarbonyls, MCO (M = V, Cr, Mn, Fe, Co, Ni, or Cu) [98]. Eischens, as a proponent of the localized chemisorptive bond, has observed both two-center and multicenter bonding of CO to surface metal atoms. From the matrix isolation perspective [98], the argument proceeds from Blyholder's proposal [99], fortified by extended Hückel calculations, that the degree of π-overlap between metal d orbitals and the π^* orbitals of terminal CO on a (Ni_9) cluster "mini-surface" increases as one proceeds from central to corner sites, implying a concomitant shift of the CO stretching frequency to lower values. This is reasonable because corner atoms have few nearest neighbors with which to share d electrons and therefore would be expected to have a greater residue of electron density to donate to the $2\pi^*$ orbital, weakening the CO π bond. A monocarbonyl metal species should therefore absorb at a lower frequency than the lowest observed for terminally bonded CO in the bulk metal. A superposition of MCO spectra (Fig. 1 in Ref. 25c) and spectra reported by Blyholder [99] and coworkers for CO chemisorbed on metal particles reveals an envelope of bands *below* that of MCO, implying that these bands are most likely bridging groups, probably involving two- and multi-center bonds.

By using high Ni/rare gas matrix ratios, "low coverage" CO chemisorption model compounds were synthesized [26b]. These clusters included monocarbonyls having two-center and three-center bridge-bonded CO, as well as linear CO. The structures shown below were assigned on the basis of metal-concentration studies using kinetics analyses, mixed $^{12}CO/^{13}CO$ isotope experiments, and normal coordinate analyses of the vibrational data.

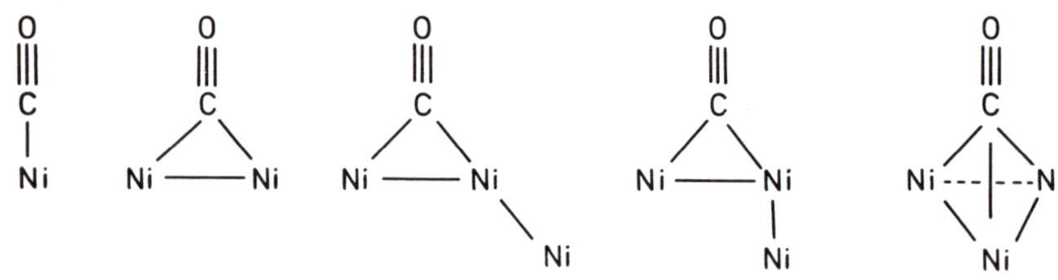

I II III IV V

Although MO calculations [100] predict that the stable form of Ni_3 is linear, recent Raman studies of Ni_3 in Ar [72] have been interpreted in terms of a bent C_{2v} form for the cluster, probably trapped in different angular forms. Therefore, the alternate trinuclear geometries shown above may be products determined by kinetics rather than thermodynamics. A plot of ν_{CO} versus 1/n for the three-center bonded CO molecule extrapolates to 1950 cm^{-1} for $n \longrightarrow \infty$ (the chemisorbed limit) and corresponds to the center of the spectrum of CO on Ni. A different extrapolation, from NiCO to Ni_3CO to Ni_mCO yields a value of 1905 cm^{-1}, close to that observed for chemisorbed CO on Ni at low coverage. Such convergence behavior may mean that CO, interacting with bulk nickel, bonds in a manner similar to that found in stable metal carbonyls; the results may also be considered to support both terminal and multi-center-bonded CO on nickel surfaces, although such bonding may not be easily distinguished [101].

We have already referred to the UPS studies of Jacobi *et al.* [84] of matrix-isolated CO/Ni_n and CO/Cu_n. Extending matrix isolation cluster chemistry to $Cu_n(CO)_x$ molecules, Moskovits and Hulse [26b, d] were able to assign frequencies at 2128, 2117, 2103, and 2011 cm^{-1} to complexes of stoichiometry Cu_2CO, Cu_3CO, Cu_4CO, and CuCO. Matrix annealing experiments resulted in CuCO aggregates of indeterminate stoichiometry having ν_{CO} = 2090 cm^{-1}, and an infrared spectrum closely resembling that of CO chemisorbed on bulk copper [102]. Extrapolations like those given above for $Ni_n(CO)_x$ provided a basis for inferring localized Cu-CO bonding. In fact Cu_4CO was offered as a good model for CO chemisorption on bulk copper.

There are also some other factors to consider. Baerends and Post [103] published theoretical calculations which sought to probe chemisorption of CO on (100) and (111) Cu surfaces by means of small model clusters. They found that properties such as M-CO distance, M-CO vibration, and C-O distance can be obtained from small-cluster calculations, as these are determined by "local" bonding aspects. Small clusters are, however, unable to mimic the polarization of conduction electron density, the latter process expressing the capability of an extended metal to act as an electron source and sink (nonlocal effects). These authors argued that the adsorption energy is a delocalized property which cannot be adequately handled within the metal/small-cluster analogy. Careful appraisal of the metal cluster-surface analogy [1, 26d, 27, 104] gives evidence that as a didactic instrument the analogy can provide insight into chemisorption presumed by the various surface-adsorbate interaction theories. Still, much remains to be done before comparisons between the chemistry occurring on surfaces and on clusters can be put to predictive use. (For further development of these ideas, see Chapter 11.)

Infrared studies of NO and CO adsorbed on Ar matrix-isolated "massive" silver clusters (mean diameter 30-10 nm) by Frank *et al.* [105] are paradigmatic for high-nuclearity M_bL_x clusters in SCHEME 1. Frequencies of vibration of CO on Ag particles prepared by a gas aggregation technique showed a coverage dependence similar to that observed for CO on silver films. Reducing the CO coverage on the Ag colloid (by reducing the amount of CO dopant in the argon) resulted in a CO frequency shift from 2110 to 2140 cm^{-1}. Since CO binds very weakly to silver surfaces, the authors suggested that the upward frequency shift results primarily from dipole-dipole coupling [99a]. Alternatively, for an essentially σ-bonded CO (as expected on Ag; cf. matrix-isolated $Ag(CO)_{1,2,3}$ and $Ag_2(CO)_6$ [106]), an increase in σ charge transfer per CO ligand with decreasing CO coverage could also cause the same blue shift in ν_{CO}.

Moving to still larger particles, we consider a fascinating paper by Sheets and Blyholder [78] describing a technique for investigating competitive chemisorption and interaction between the coadsorbed species on metal colloids prepared by evaporation of metal atoms under 0.5 Torr of helium into a paraffin oil matrix in a spectroscopic cell. Interactions between the hydrocarbon and the metal particles are sufficiently strong to inhibit sintering. The interactions are also sufficiently weak that in most cases there is only slight competition between the oil and the adsorbate gas for sites on the metal surface. Numerous studies showed that spectra of various gases chemisorbed on such particles did not differ substantially from spectra of gases adsorbed on supported metals or vacuum evaporated films. The hydrocarbon-colloid interaction can, however, be sufficiently strong that neither tetrafluoroethylene/Ni nor CO/Cu interactions (known to be weak) could be observed in this medium. The technique has been used to establish those electronic factors (simple molecular orbital ideas of electron donation and withdrawal) responsible, for example, for shifts in the infrared spectra of ND_3 and NH_3 coadsorbed with PF_3 on Fe, Ni, and Cu, and for alkylamines with CO on Fe and Ni.

These experiments by Blyholder lead us to consider a range of experimental techniques for producing low- and high-nuclearity metal clusters (as defined in SCHEME 1) in matrices. The review by Perenboom *et al.* [18a] contains a succinct summary of the imaginative procedures used by various researchers to produce small metal particles suitable for study of such size-dependent phenomena as Knight shifts in NMR, conduction electron spin resonance, magnetic susceptibility, superconductivity, and infrared and UV/visible extinction, to mention some of the specific issues. The articles by Papavassiliou [18d] and Hughes and Jain [107], and a monograph by Bohren and Huffman [94] give details of colloidal and subcolloidal metal particles and molecules, accenting their interaction with light. Most of these studies, implicitly or explicitly, address the question of density of states, localized and delocalized electronic/magnetic phenomena, and the metal-adsorbate interface, which in turn have a bearing on the broad area of catalysis. In the last section of this chapter, we consider low- and high-nuclearity clusters in matrices and on supports, with emphasis on some of the less traditional methods of preparation.

7.3 METAL CLUSTERS IN POLYMERIC AND OLIGOMERIC SUPPORTS

Literally hundreds of discrete metal-metal bonded cluster compounds have been synthesized (Chapter 1). Their structural characterization (Chapter 2), reactivity (Chapter 4), and opportunities offered in catalysis are detailed elsewhere in this book. In the context of SCHEME 1, the cycle of interconversions second from the top represents the low-nuclearity cluster regime. In the counterclockwise direction, $M_nL_x \longrightarrow M_nL_{x-y}S$ represents in the first step, the process of depositing a coordinatively saturated or unsaturated organometallic cluster onto a support S and implies some sort of interaction (chemical bonding) with the support. In the second step, the reaction $M_nL_{x-y} \rightleftarrows M_nS$ represents reversible (it need not be) ligand stripping on the support. In keeping with the metal vapor/cluster spirit of this chapter, we shall be concerned with procedures for placing metal particles in and on supports (as symbolized in this and the subsequent reaction cycles) that lead progressively to larger and larger metal particles.

Polymers have received a great deal of attention regarding their use as supports for anchoring catalysts. The term "heterogenized homogeneous catalysts" emphasizes the relationship between a macroscopically insoluble crosslinked polymer and its locally "soluble" intra-crosslink chain segments (resulting in a swellable macromolecule) giving microenvironments within which catalysis comparable to solution-phase (homogeneous) catalysis can take place. Polymer systems are interesting from this viewpoint because of the possibility for controlling the local molecular environment in ways that can result in tuned structure-property relationships leading, for example, to enhanced catalytic activity through cooperative, local reactive functional group mobility linked to main and side chain dynamics [108]. Approaches to low- and high-nuclearity metal clusters through the polymer routes are indicated in SCHEME 6.

Section 7.3 is predominantly concerned with metal atoms in polymers, outlining specific features of their interactions which can be viewed as significant in promoting the growth and stabilization of metal clusters prepared directly from the atomic vapor. Some of the conclusions reached may serve as a prescription for successful preparation of metal clusters in other reactive polymers. Before proceeding to these studies, it will serve our purpose to indicate different approaches to combining polymers with metal atoms obtained from the bulk. Shuttleworth [109] preparated metal-polymer dispersions by a radiofrequency (rf) plasma process. In his experiments, the monomer is plasma polymerized and codeposited with either a rf-sputtered metal or a plasma-decomposed organometallic onto a quartz surface. The result is a rather complex metal-polymer composition. Boonthanum and White [110] evaporated solid polymer pellets at 650 K (with substantial thermolysis) and codeposited the volatile products (mainly radicals) with metal atoms to give, for example, copper islands *ca.* 10 nm in diameter in a "polyethylene" matrix. In essence, the polymer prevented coalescence of the copper islands during deposition. Films prepared in this manner showed sheet resistivity changes from 10^{13} to 10^{7} Ω/□ over a temperature range of 200 K. Lunina and Novozhilov [111] prepared colloidal dispersions of metal in organic media using a device which generated a pulsed high-frequency AC spark discharge across electrodes fabricated from the metal to be dispersed.

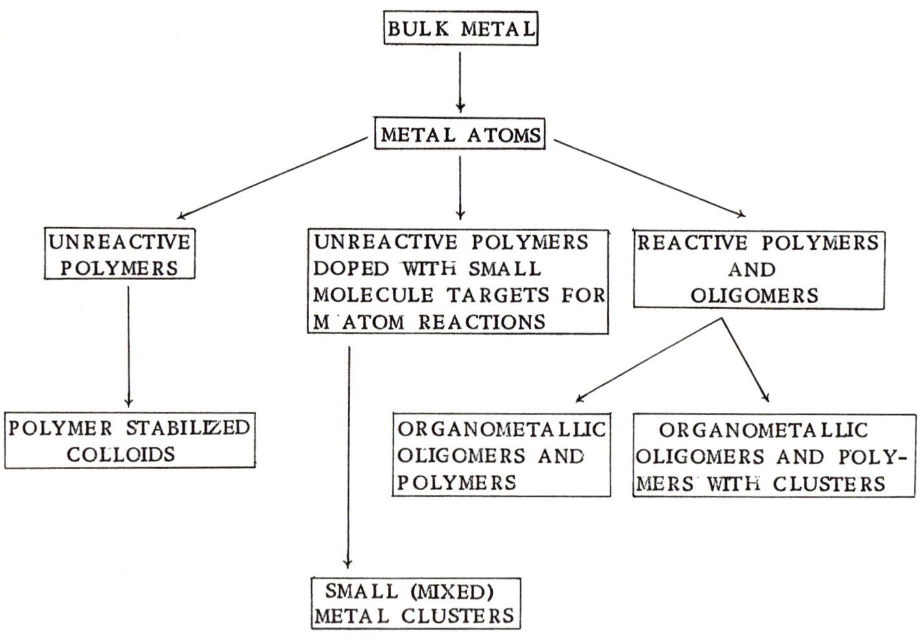

SCHEME 6

7.3.1 Metal clusters in reactive polymers and oligomers

The aforementioned metal atom/polymer experiments produce rather large metal aggregates and metal/support (matrix) interactions of considerable chemical complexity. It is axiomatic that transition metal atoms can be isolated from each other only if the interaction of the atom with the support exceeds that of the atom in the aggregate. If the atom-support interaction is primarily van der Waals in nature, then a kinetic barrier to aggregation must be imposed. This principle extends also to low-nuclearity clusters.

To those interested in stabilizing atoms and clusters in environments equilibrated near or at room temperature, a judicious choice of materials and preparation techniques can produce the desired results. In a thematic exposition, Ozin *et al.* [112, 113] has emphasized some of the conditions required to undertake metal vapor chemistry to generate and stabilize small metal clusters at or near room temperature. Rigid macromolecules are appealing matrices for use in segregating reactive species in room-temperature environments. Unfortunately, it has not yet been possible to introduce metal atoms directly into rigid polymers without first dissolving them in organic solvents. Here, too, the ease with which metal vapor reactions with polymers can be conducted is often hindered by the tendency for chain molecules to phase separate and precipitate from solution when polymer/solvent systems are cooled to provide the tolerable background vapor pressures required for these experiments. Since 1978, our research group has been gathering information concerned with the effect of different matrix supports and reaction media for controlling the growth and stabilization of metal atomic, cluster, and colloidal dispersions. The utility of such reaction media for metal vapor synthesis can be considered to be determined by the following constraints:

(1) the vapor pressure/viscosity/solubility characteristics of the medium over the working liquid range, as they will define the temperature regime useful for synthetic preparations;

(2) the reactivity of the solvent, ligand and/or support toward the deposited metal atoms and/or cluster and/or colloid, as well as during any subsequent thermal or photolysis treatment;

(3) the nature of the metal atom and/or cluster, ligand, solvent and/or support interactions, as these will influence the extent to which metal atom aggregation processes occur and the degree to which the spectroscopic and chemical properties of the atoms and clusters are perturbed by the medium; and

(4) the thermal stability of metal atomic/cluster/colloidal dispersion,

indicated by the temperature above which desired or undesired diffusion/agglomeration processes become noticeable.

The Dow Corning polymer, DC510, a copolymer of dimethyl- and methylphenylsiloxanes, fits many of the above criteria, being a low-vapor-pressure liquid with sufficient fluidity at 250 K to make the material an appealing target for direct metal vapor reactions. Francis and Timms [114] first reported that *bis*(arene)M transition metal complexes could be formed from reaction of M = Ti, V, Cr, Mo, or W with DC510 at 273 K using a rotary metal vapor reactor similar to that shown in Fig. 13. (The equipment is described in the caption.) These reactions produced organometallic polymers:

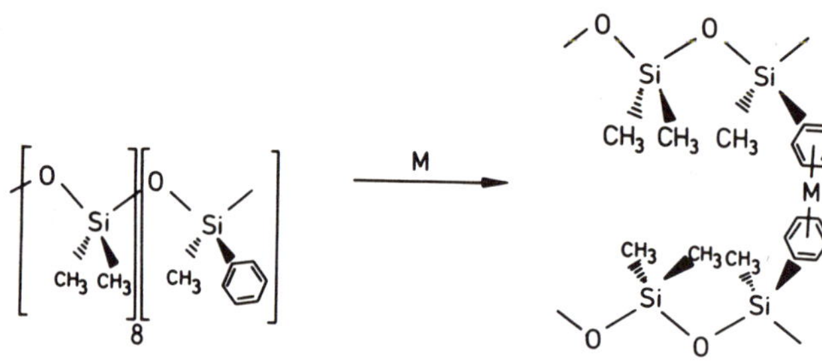

M = Ti, V, Cr, Mo, W

Following these studies, techniques were developed [115] (see next section) which allowed one to examine this reaction in more detail under a variety of experimental conditions.

7.3.2 Metal clusters in DC510; microscale, quiescent liquid film metal vapor chemistry

Adapting the techniques of nanogram-microgram matrix isolation synthesis and spectroscopy, a microscale metal vapor technique was introduced [115] for assessing the outcome of controlled depositions of one to three different metal atoms into thin polymer or oligomer films or organic solvent/solute systems. The idea of these experiments is to provide an informed basis from which to design gram-scale metal vapor reactions in polymers. The apparatus is shown in three different functional orientations in Fig. 14, and its operation is described in the caption. Using this apparatus, the principle of the Francis-Timms reaction was extended to quantitative depositions (effectively, titrations of reactive phenyl groups) of transition metal atoms into a thin DC510 film in a series of experiments that essentially established an intermediate link between matrix isolation and macroscale rotary reactor synthesis.

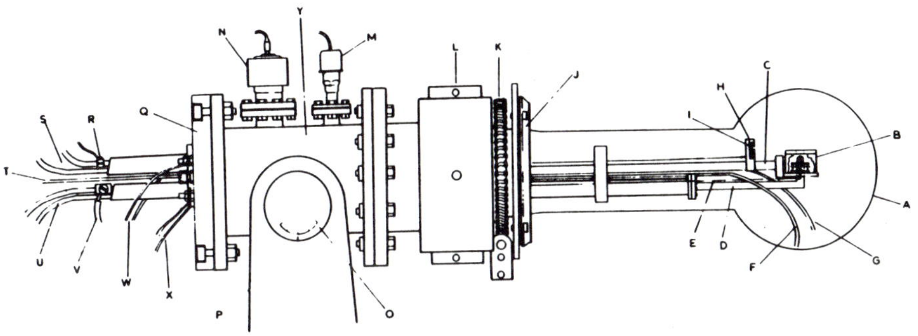

Schematic of reverse-polarity electron gun-metal vapor rotary reactor equipment (Torrovap) used for preparing liquid poly(dimethyl-co-methylphenylsiloxane)-supported divanadium: (A) 2-5-L glass reaction vessel; (B) electron gun; (C) liquid-N_2-cooled electron-gun support; (D) water-cooled hearth (anode); (E) thoriated tungsten filament power leads; (F) rotatable liquid-product outlet tube; (G) rotatable liquid-reagent inlet tube; (H) quartz crystal support; (I) 5 MHz AT cut shear-mode quartz crystal; (J) rotary vacuum seal-reaction vessel mating flange; (K) motor-driven worm and gear assembly; (L) rotary vacuum seal; (M) thermocouple vacuum gauge; (N) cold cathode vacuum gauge; (O) pivot point for setting angle of inclination of reaction vessel; (P) reactor support; (Q) plexiglass back flange; (R) external dc power supply for *in situ* electron beam focusing; (S) liquid nitrogen feed through; (T) liquid-reagent inlet tube; (U) electron-gun water supply; (V) high-voltage connection to electron-gun hearth; (W) quartz crystal microbalance oscillator-power supply; (X) filament power supply; (Y) main vacuum manifold connected to angle-adjustable vacuum seal and diffusion pump-rotary backing pump assembly.

Fig. 13. Rotary metal vapor reactor (M. P. Andrews, C. G. Francis and G. A. Ozin, Inorg. Synthesis, 22 (1983) 116).

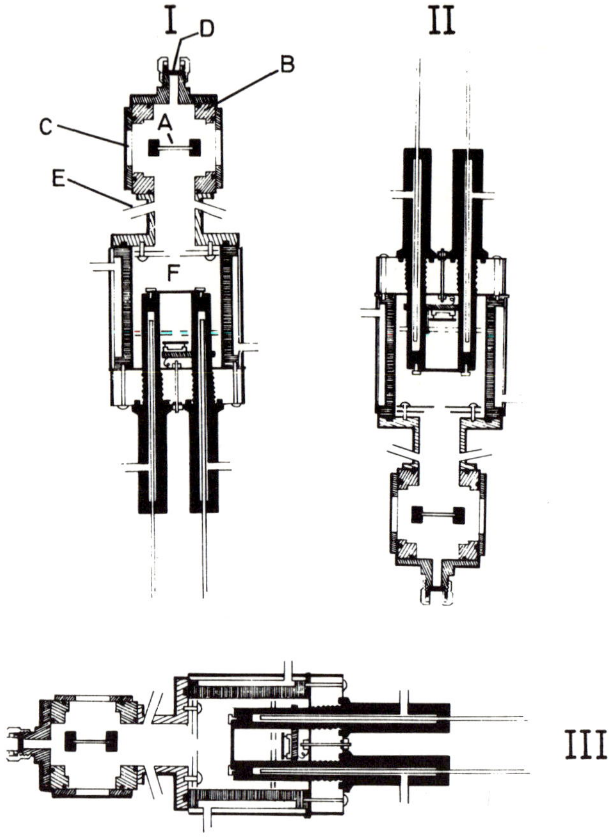

Fig. 14. Schematic representation of vacuum furnace-closed cycle helium refrigeration system used for metal vapor microliquid or micro-solution optical spectroscopy as well as conventional metal vapor-matrix experiments: (A) NaCl or Suprasil optical window, horizontal configuration; (B) stainless-steel vacuum-shroud; (C) NaCl or Suprasil optical viewing ports; (D) Cajon-rubber septum, liquid or solution injection port; (E) gas deposition ports; (F) vacuum furnace and quartz crystal microbalance assembly. With the optical window in a fixed horizontal configuration, liquid or solution sample injection onto the window at any desired temperature in the range 12-300 K is performed in position I, metal deposition is conducted in position II, and optical spectra are recorded in position III [112].

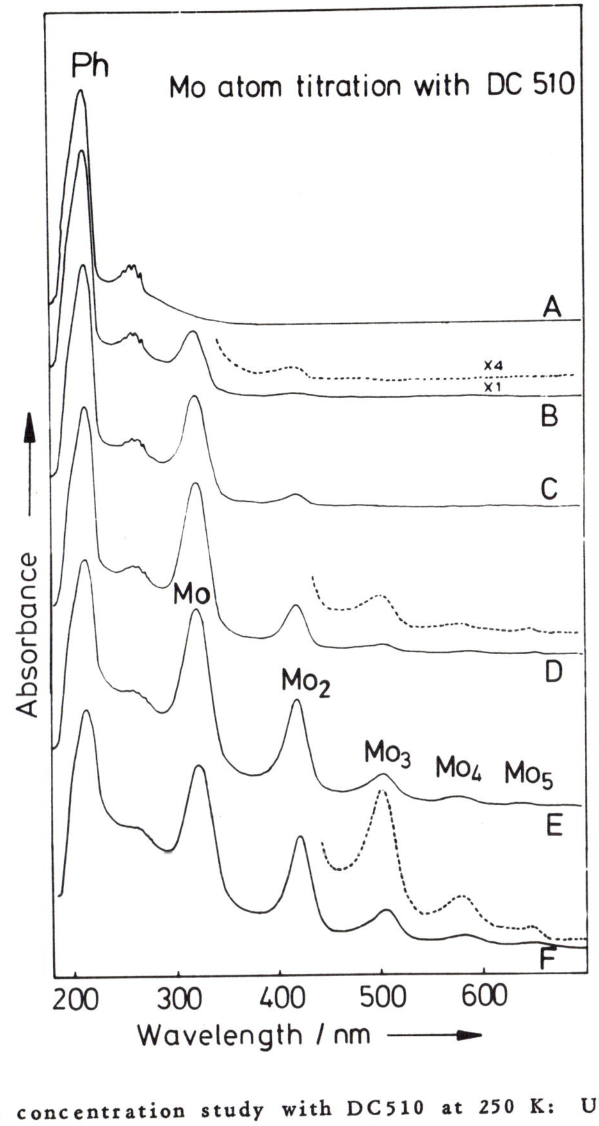

Fig. 15. Mo atom concentration study with DC510 at 250 K: UV-visible study. (A), neat liquid polymer film; (B)-(F), increasing concentration of Mo atoms [116].

In Fig. 15 are shown the results [116] of depositing increments of Mo atoms into 10-100-μm-thick 250 K films of DC510. The initial Mo deposition represented in trace B produced the polymer-supported *bis*(arene)Mo, as identified by its intense metal-ligand charge transfer (MLCT) band at 318 nm. Increasing the metal loading (at approximately

2×10^{-2} mmol h^{-1} deposition rate) to roughly 10^{-8} mol resulted in the growth of a new absorption at 418 nm (Fig. 15C). As the amount of Mo was gradually increased, more transitions could be observed at 502, 578, and 640 nm (Fig. 15D-F), which have been assigned to Mo_n (n = 2-5) clusters [116].

At this point it is useful to distinguish some differences expected in the outcome of metal vaporizations into quiescent or turbulently mixed DC510 films. In the macroscale metal vapor rotary reactor synthesis of poly-(siloxane)-anchored *bis*(arene)chromium the maximum concentration of *bis*(arene)chromium that can be produced in the polymer corresponds to 45% conversion of the appended phenyl ligands to the complex [114], whereas in the metal vapor microscale fluid matrix technique, the maximum complexation corresponds to 25-30% of the phenyl groups. The reduction in the maximum complexation in the latter instance is most likely a reflection of simple mixing dynamics. In the former case, the polymer is continuously mixed in the spinning glass reactor, a fresh surface of polymer being exposed to the flux of metal vapor. Maximum coordination is known to be associated with a dramatic increase in the measured macroscopic viscosity of the polymer [114]. Presumably at this maximum, intra- and intermolecular crosslinking severely reduces the frequency of reactive encounters among phenyl substituents and metal atoms, since the diffusion of any two arene ligands together is strongly correlated to the segmental motions of the polymer, and these will be dependent on the density and distribution of crosslinks. In the early stages of metal deposition, the effect of turbulent mixing is seen as exposing unreacted and partially reacted polymer to the incident metal atoms. In other words, mixing suffices to maintain a sufficiently high local concentration of unreacted ligand in the vicinity of a diffusing metal atom and short circuits the competitive route to metal atom polymerization. On the other hand, metal atoms deposited into a quiescent liquid polymer film are more likely to quickly exhaust the surface and near subsurface supply of arene. The path length for diffusion of a single metal atom to a site favorable for association of two arenes and a metal atom must rapidly increase during the deposition period. Surface renewal with unreacted polymer-bound ligand is not likely to be an efficient process, since this requires the collective motion of large groups of atoms (translation of polymer segments containing pendant arenes), a process which is likely to be disfavored both by very low temperature (220-250 K) and the presence of the ramifying, crosslinked polymer network through which the supply of ligand and metal atoms must pass. Then either colloid formation becomes the dominant or competing reactive fate of a diffusing metal atom, or the newly formed *bis*(arene)transition metal complexes themselves become targets for reaction. In this view, more arene will remain unreacted in the static film than would remain if the polymer were continuously agitated.

The problem of diffusion with chemical reaction is an interesting one in relation to metal vapor chemistry. Consideration of the problem within a simplified quantitative theoretical model can give some qualitative insights into the chemical and physical processes accompanying a metal vapor/thin film microsolution experiment. This problem can be treated within the framework of diffusion accompanied by a pseudo-first order reaction for the case of a constant flux boundary condition at the surface x = 0 of a semi-infinite medium, in which the concentration of atom trapping sites is random and fixed [64]. Two quantities of interest emerge. These are the bivariate concentration profiles of the unbound and bound metal atoms, given, respectively, in Equations (2) and (3):

Unbound metal atom concentration profile:

$$C(x,t) = \frac{R}{2\sqrt{Dk}}\left[\exp\left(-x\sqrt{\frac{k}{D}}\right)\operatorname{erfc}\left(\frac{x}{2\sqrt{Dt}}-\sqrt{kt}\right)-\exp\left(x\sqrt{\frac{k}{D}}\right)\operatorname{erfc}\left(\frac{x}{2\sqrt{Dt}}+\sqrt{kt}\right)\right] \quad (2)$$

Bound metal atom concentration profile:

$$\begin{aligned} S(x,t) = \frac{R}{2}\sqrt{\frac{k}{D}}\Bigg[&\left(t-\frac{x\sqrt{\frac{k}{D}}-1}{2k}\right)\exp\left(-x\sqrt{\frac{k}{D}}\right)\operatorname{erfc}\left(\frac{x}{2\sqrt{Dt}}-\sqrt{kt}\right) \\ &-\left(t+\frac{x\sqrt{\frac{k}{D}}-1}{2k}\right)\exp\left(x\sqrt{\frac{k}{D}}\right)\operatorname{erfc}\left(\frac{x}{2\sqrt{Dt}}+\sqrt{kt}\right) \\ &+2\sqrt{\frac{t}{\pi k}}\exp\left(-kt-\frac{x^2}{4Dt}\right)\Bigg] \end{aligned} \quad (3)$$

These expressions have been plotted in Fig. 16 for a one-hour deposition period, with the parameters $k = 1000\ s^{-1}$ and various values of the diffusion coefficient D. The concentrations of the unbound and bound atoms are normalized to their respective surface values, S^o and C^o. Also plotted is the diffusion profile for a freely diffusing atom in the absence of chemical reaction. Some important features of these curves are the following:

(1) For all combinations of k and D, the concentration of free and bound metal atoms is restricted to regions very near the surface of the film, there being little penetration beyond 10^{-6} to 10^{-2} cm, where the concentration of species declines rapidly to zero.

(2) The larger the value of D, the greater the penetration depth for a given value of k.

(3) In the absence of chemical reaction, the diffusant penetrates the film substantially, compared with the actual film thickness. Clearly, the effect of chemical reaction on diffusion is to restrict the concentrations of free and bound species to a near-subsurface region of the film in a metal vapor/thin, static polymer film experiment.

On the basis of the foregoing analysis, some remarks can be made regarding the fact that the pendant phenyl groups in the liquid DC510 copolymer cannot be quantitatively converted to *bis*(arene) metal complexes in a static, thin-film metal vapor experiment. According to the computed

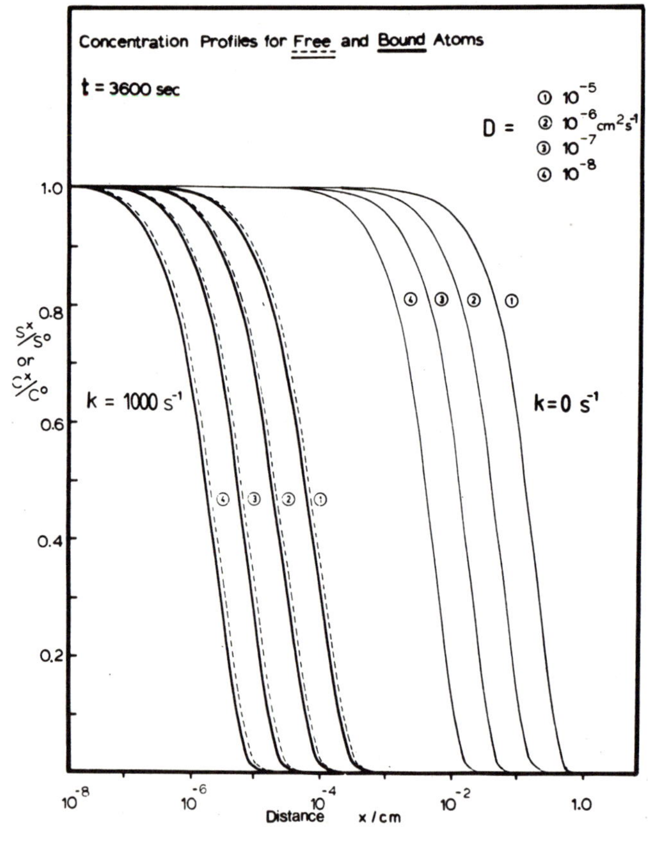

Fig. 16. Calculated concentration profiles for unbound and bound metal atoms formed when metal atoms are deposited into a nonreactive and reactive semi-infinite liquid medium for various values of the diffusion coefficient D and pseudo-first order rate constant k (s^{-1}) as a function of the penetration depth x into the medium. See text and Ref. 64 for details of the method.

concentration profiles, the metal atoms simply do not penetrate the film sufficiently to consume the available ligand. In spite of the simplicity of the model, the restrictions that chemical reaction are expected to impose on the extent of penetration of a diffusant as reactive as a metal atom makes sense. Evidently, if high conversion of uncomplexed arene to mononuclear *bis*(arene)M complexes is desirable, this would best be accomplished by means of turbulent mixing. Formation of metal sites containing two or more metal atoms is therefore expected to be favored in quiescent films where conversion of *bis*(arene) mononuclear transition metal complexes will occur competitively with their production. Reasonably, it would seem that the static thin film experiment is biased towards the generation of higher-nuclearity metal species (including colloids), since free atoms are not converted efficiently into mononuclear complexes by mixing, but must be converted solely on the basis of diffusion; consequently additionial reaction pathways (metal cluster formation) can become competitive with *bis*(arene) complex formation. Qualitatively, this can account for the observation that it is much more difficult in a macroscale experiment to prepare DC510-supported metal multimers—i.e., considerably higher quantities of deposited metal are required to achieve significant metal cluster growth in a rotary reactor experiment.

We return to the microscale reaction between Mo atoms and DC510; the growth characteristics of the Mo_n species from Fig. 15 are given in Fig. 17. The order of stability of the species giving rise to the bands red-shifted from the 318-nm MLCT of $(arene)_2Mo$ is the same as the order of production. Except for the MLCT absorption, all the absorptions gradually disappear on warming the polymer medium to 300 K. In part, this thermal instability also accounts for why these species went undetected in the macroscale M/DC510 syntheses reported by Francis and Timms. With an awareness of this thermal lability, it has been possible to prepare macroscale quantities of the Mo and V dimers in DC510 when the samples, carried to high enough metal loadings, are worked up at low temperatures.

Growth curves similar to those of Fig. 17 have been obtained for Ti, V, and Cr in DC510. In addition, mixed metal depositions on the microscale have led not only to DC510-supported *bis*(arene) mixed metal organometallics, but also to what can be described as alloy dimers $VCr(arene)_n$, where n is probably 2 (see below).

7.3.3 Assignment of metal cluster nuclearities

A number of heuristic arguments have been advanced [117] to rationalize the absorption spectra such as those of Mo/DC510 at high metal loadings

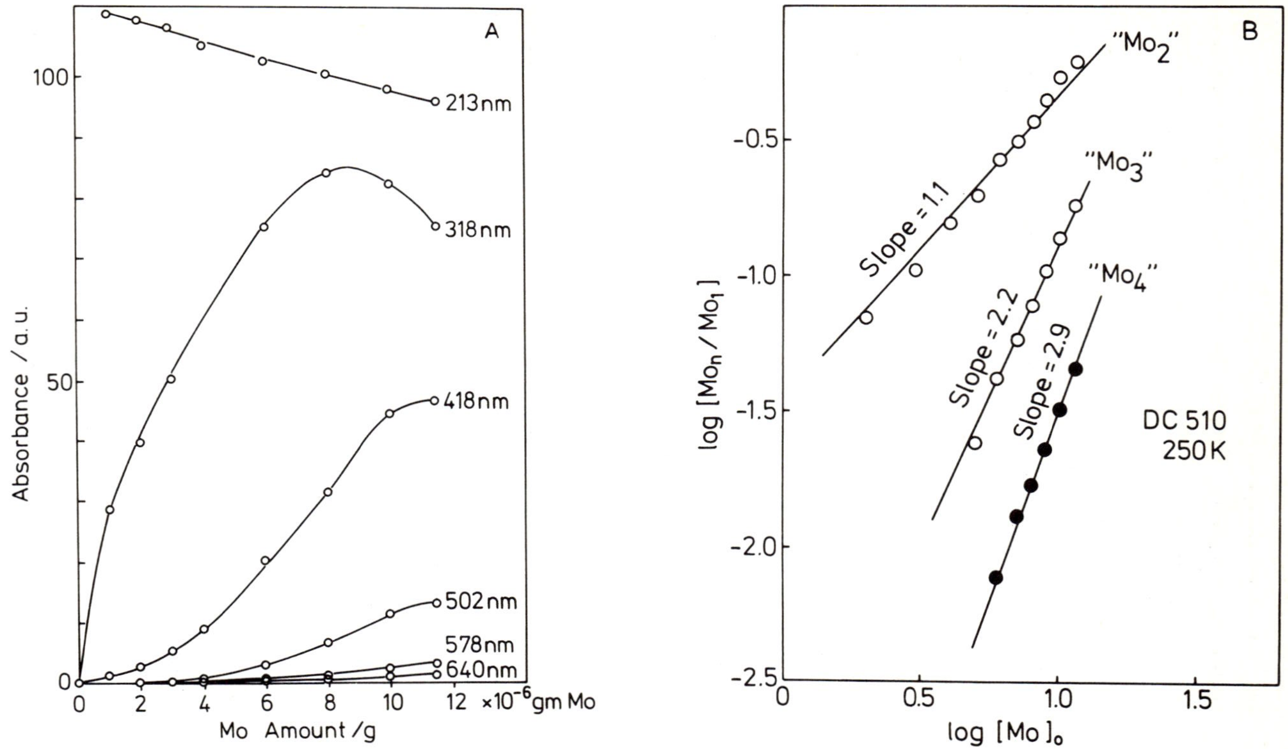

Fig. 17. (A) Growth curves for the production of polymer-supported Mo_n cluster complexes resulting from depositing Mo atoms into DC510 at 250 K. (B) Linear kinetics plots used to determine cluster nuclearity referenced to the mononuclear Ar_2Mo center [121].

(Fig. 15E). Additional support for the assignments derives from kinetics and chemical studies. A brief return to the matrix isolation experiments is helpful in establishing the identity and nuclearity of the cluster compounds and in elucidating those structural, chemical, and electronic properties of metal atoms and multimers which can be viewed as coextensive with properties of species isolated at ambient and near-ambient temperatures in the rather different gross polymeric environments. In this manner, it can be seen that the microsynthetic methods of matrix isolation chemistry express themselves continuously through to the macrosynthetic methods of gram-scale metal vapor chemistry by way of the preparation of species common to both synthetic scales.

Fig. 18 (curve A) shows a portion of the UV-visible absorption spectrum of V atoms and dimers co-isolated in argon at 12 K. The multiplets of absorptions at 478, 578, and 740-840 nm belong to V_2 ($^3\Sigma_g^-$ ground state), as established by experiment and theory [118]. The formally pentuply bonded dimer has a short 0.177-nm bond length. The remaining multitude of lines in this spectrum arises from the many spin-allowed transitions from the vanadium atom $^4F_{3/2}$ Russel-Saunders ground state. This spectrum serves as a background against which to calibrate the outcome of reactions of V atoms and dimers in benzene-doped argon matrices.

To understand cluster growth in the oligomer and polymer matrices, it is important to be able to follow step by step the growth sequence itself. Using combination of matrix isolation and microsolution methods, it was possible to stablish the following sequence of events:

$$V + C_6H_6 \longrightarrow V(\eta^6\text{-}C_6H_6) \qquad \text{(i)}$$

$$V(C_6H_6) + C_6H_6 \longrightarrow V(\eta^6\text{-}C_6H_6)_2 \qquad \text{(ii)}$$

$$V + V(\eta^6\text{-}C_6H_6)_2 \longrightarrow V_2(C_6H_6)_2 \qquad \text{(iii)}$$

Fig. 18 (curve E) shows the absorption of the half-sandwich complex at 550 nm. The existence of this species is confirmed in Fig. 19, where an ESR spectrum for this molecule (Fig. 19, curve A) is clearly distinguishable from that of the *bis*(arene)V complex (Fig. 19, curve C), which is formed when the mixed system containing unreacted C_6H_6 is carefully annealed [119]. Here the hyperfine splitting is governed by the eight-line vanadium coupling (vanadium has a nuclear spin of $I = 7/2$). The resolved seven- and 13-line superhyperfine inserts [(a) and (b)] are diagnostic of a single unpaired electron interacting with six and 12 equivalent protons, respectively. Just as it is possible to infer the existence of V_2 in 12-K argon matrices on the basis of quantitative metal atom studies,

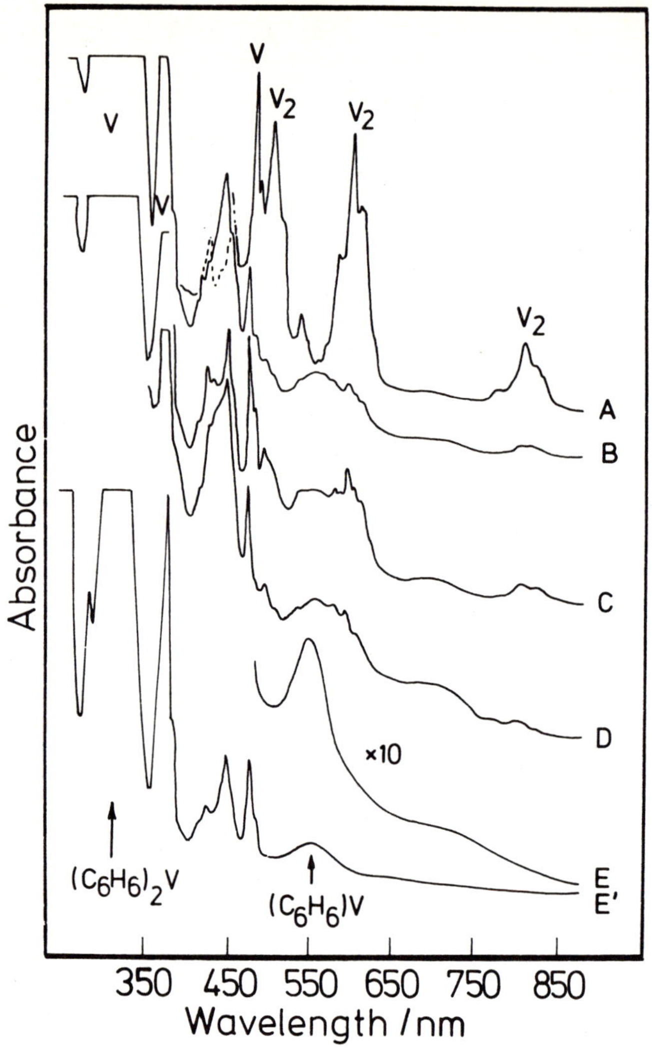

Fig. 18. Results of cocondensing vanadium atoms under binuclear (curves A-D) and mononuclear (curves E, E') conditions, with Ar or C_6H_6/Ar. Curve A: V/Ar at 12 K (V, 1.5 x 10^{-3} mmol h^{-1}; Ar, 0.5 mmol h^{-1}). Curve B: V + C_6H_6/Ar (1/10) (V, 7 x 10^{-4} mmol h^{-1}; C_6H_6/Ar, 0.8 mmol h^{-1}). Curve C: V + C_6H_6/Ar (1/25) (V, 9.5 x 10^{-4} mmol h^{-1}; C_6H_6/Ar, 0.8 mmol h^{-1}). Curve D: as in C, but annealed 5 min. at 28 K. Recorded at 12 K. Curve E': V + C_6H_6/Ar (1/25) (V, < 10^{-4} mmol h^{-1}; C_6H_6/Ar, 0.8 mmol h^{-1}). Curve E: 10 x expansion of E' [64].

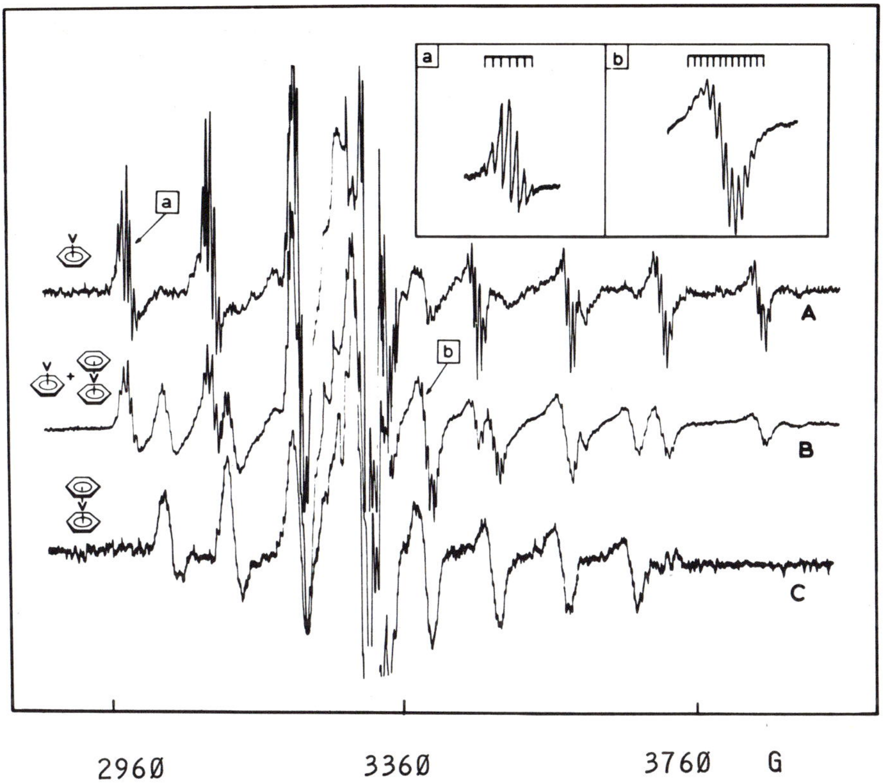

Fig. 19. Electron paramagnetic resonance spectra for the products of the cocondensation of V atoms with C_6H_6/Ar matrices at 12 K. (A) $C_6H_6/Ar \cong 1/250$; (B) $C_6H_6/Ar \cong 1/50$; and (C) the same as (B) but after annealing to 40 K. Inserts (a) and (b) show superhyperfine splitting on selected vanadium hyperfine components of $(\eta^6\text{-}C_6H_6)V$ in (a) and $(\eta^6\text{-}C_6H_6)_2V$ in (b) [119].

complementary information obtained from spectroscopic studies, and associated kinetics analysis of the growth of the dimer [118], so it is also possible to infer the existence of V_2 interacting with benzene rings in 12 K V/C_6H_6 deposits. The basis of the kinetics experiments is the analysis of metal cluster formation in reactive and unreactive matrices as proposed by Moskovits and Hulse [120]. This method for assigning cluster nuclearities proceeds from the network of reactions shown in SCHEME 2. A quenched reaction model assumes that the nucleation sequence of reactions shown in the SCHEME 2 evolves to some time τ_q, then abruptly ceases. Solution of the set of simultaneous differential equations based on SCHEME 2 established that the concentration ratio was roughly proportional to the metal concentration raised to the power (n - 1). Therefore, the slope of a log ($[M_nL_m]/[ML_p]$) vs. log $[M_o]$ plot provides a direct analytical route to metal cluster size.

Now, suppose that ML_p is M(arene)$_2$; then, under appropriate conditions of ligand and metal atom deposition rate, it is possible to assign the nuclearity of the M_nL_m complexes (also presumed to exist in DC510). For example, when vanadium or molybdenum atoms are cocondensed with benzene or benzene/argon, metal atom concentration-dependent absorptions arise at 455 nm ("V_2") or 418 nm ("Mo_2") to the red of the respective MLCT absorptions. Plots of these kinetics data (Fig. 20) point to the dimer absorption in each case. For molybdenum there is evidence of trimer formation in neat benzene matrices at the higher Mo concentrations (absorption at 502 nm). Significantly, these dimer and trimer bands fall at precisely the same positions as those observed for the DC510 polymer.

Leaving the matrix isolation experiments and shifting to somewhat higher temperatures, we learn that dimers can be grown in 145-K methylcyclohexane solutions of toluene [Fig. 21, curve (a)] or in 250-K *bis*(arene)M-doped DC200 fluid. The latter fluid is a poly(dimethyl-siloxane), containing no phenyl groups [121]. Apart from the different atomic number of the metal atoms, there is a distinction to be made between the MCH/toluene and DC200 microsolution experiments. In the former study, the (vanadium) dimer can be prepared by depositing the atoms into the arene-MCH solution, the results showing that the dimers synthesized by both the cocondensation and microsolution routes are the same. The DC200 experiment involves taking *bis*(toluene)Mo prepared in a separate reaction and subsequently dissolved in the polymer and adding Mo atoms to the complex, essentially titrating the starting organometallic. In this manner, the spectrum of Fig. 21, curve (f), is obtained, shown here in the late stages of the titration. This experiment exemplifies the case (SCHEME 6) in which an unreactive polymer hosts a small molecule dopant which is a target for metal atom reactions leading to metal clusters.

When these results are combined with results of a similar study involving the addition of M = V or Cr atoms to (arene)$_2$M (SCHEME 7) in MCH (microsolution), it appears that the dimer molecule is formed by direct addition of a metal atom to the metal center of (arene)$_2$M. It was noted [122] that when (1) or (2) in SCHEME 7 reacts with Cr or V atoms, respectively, the heterobimetallic species 4 is produced, which absorbs at 420 nm. This wavelength is the same as that attributed to CrV produced by cocondensation of atomic Cr and V into DC510 at 250K [123]. Since no evidence was found for free arene or metal displacement reactions for the bimetallic species (3-5), a one-step addition reaction has been proposed to account for the building up of bimetallics and higher-nuclearity clusters in

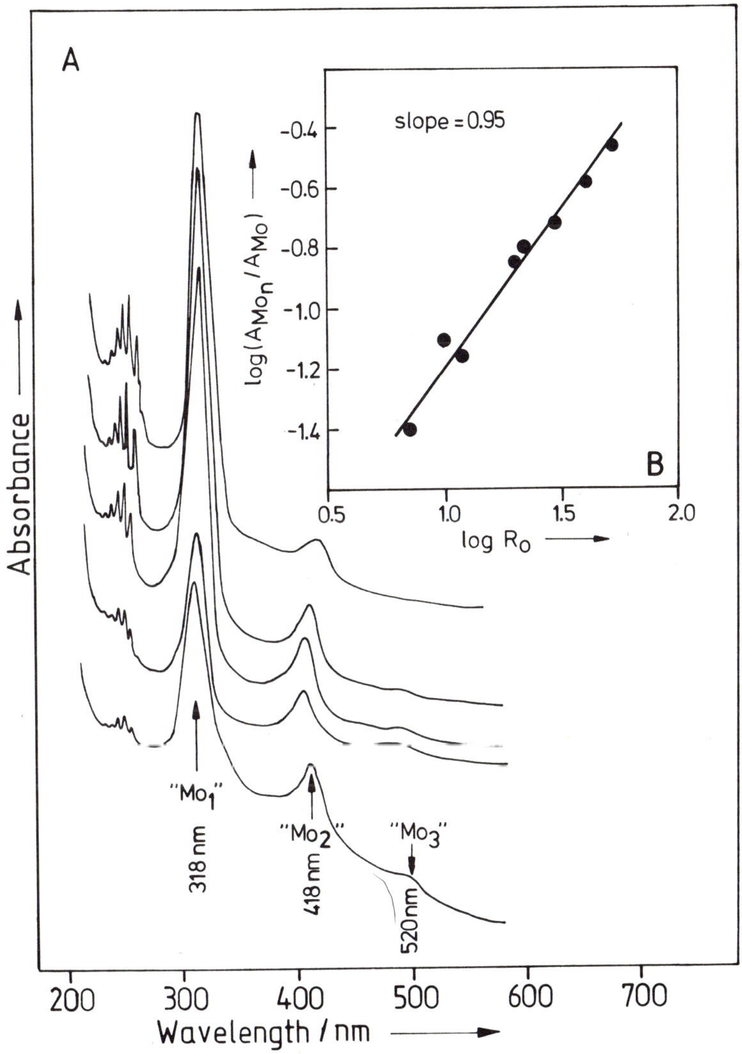

Fig. 20. (A) Results of UV-visible spectroscopic Mo atom concentration study with neat benzene matrices at 12 K, with increasing metal concentration from top to bottom trace. (B) plot of the data in (A) based on the cluster agglomeration kinetics analysis of Ref. 120, where A_n = $(C_6H_6)_mMo_n$; A_1 = $(C_6H_6)_2Mo$; R_o = rate of deposition of Mo atoms [64].

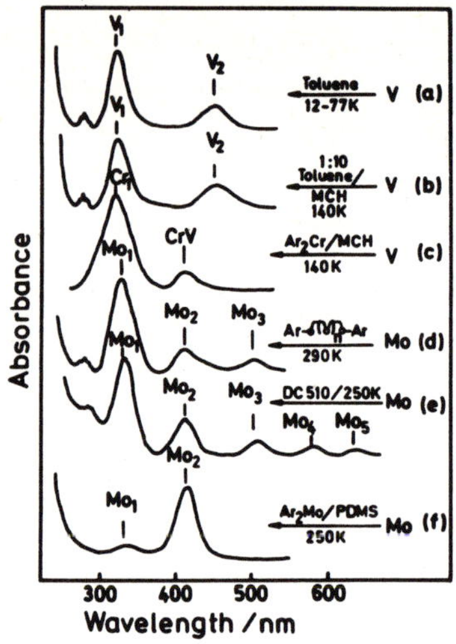

Fig. 21. Optical absorption spectra of low-nuclearity metal clusters formed in the presence of arene functionalities (monomers, oligomers, or polymers) in a series of metal vapor experiments. In each case the high-energy absorption (310-330 nm) is characteristic of the MLCT band of the *bis*(arene)metal center [121].

the DC510 polymers shown in SCHEME 8. Mass spectrometry [117], which has established the existence of $[(C_6H_6)_2Cr_2]^+$, and the gas phase direct addition of a transition metal monocation to the 18-electron $Cr(CO)_6$ complex [124] can be considered to support the dimer formulation and initial

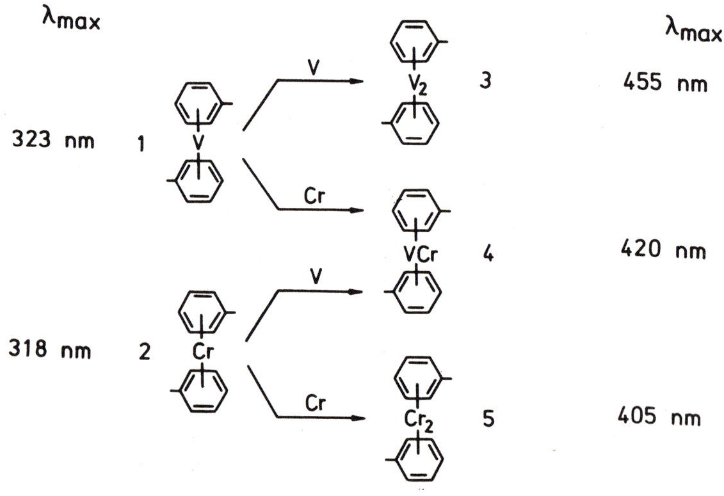

SCHEME 7

cluster growth process in the M/DC510 reactions.

The mechanism of addition of the diffusing atom to the *bis*(arene)M complex is not known, but it could involve $\eta^6 \longrightarrow \eta^{n<6}$ haptotropic ring excursions [125] or simply ring tilting as possible deformations which expose the central metal atom. The sequence in SCHEME 8 is amenable to treatment by a set of kinetics expressions describing series-parallel competitive addition reactions. The problem in the early stages is formally analogous to a batch chemical reaction in which a small quantity of reagent is added a little at a time to a reactor containing initially a single

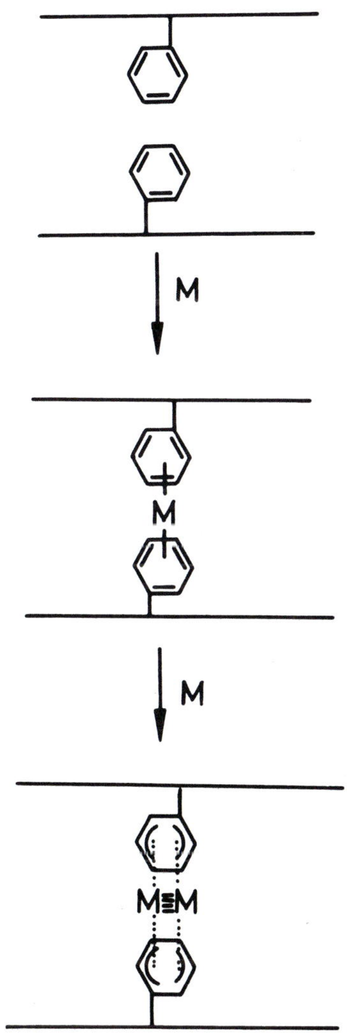

SCHEME 8

reactive component. Ratios of rate equations describing the sequence of irreversible, bimolecular reactions,

$$M + L \xrightarrow{k_1} ML$$

$$ML + L \xrightarrow{k_2} ML_2$$

$$ML_2 + M \xrightarrow{k_3} M_2L_2 \tag{4}$$

$$\begin{matrix} \cdot & & \cdot \\ \cdot & & \cdot \\ \cdot & & \cdot \\ \cdot & & \cdot \\ \cdot & & \cdot \end{matrix}$$

lead to a series of linear differential equations of the form

$$\frac{d[ML_2]}{d[L]} = \frac{k_3}{2k_1}\frac{[ML_2]}{[L]}$$

$$\frac{d[M_2L_2]}{d[L]} = -\frac{k_3}{2k_1}\frac{[ML_2]}{[L]}$$

$$\frac{d[ML_2]}{d[M_2L_2]} = \frac{k_1}{k_3}\frac{[L]}{[ML_2]} - 1 \tag{5}$$

when the steady-state approximation is applied to the half-sandwich complex, ML, and the absence of a competitive channel to metal colloid is assumed.

According to the analysis, direct differentiation of the growth curves such as those in Fig. 15 should lead to an arrangement of data points verifying the linearity predicted by these equations. These predictions have been confirmed in kinetics studies involving M/DC510 (M = Cr, V, Mo) [64]. However, the kinetics does not strictly obey the predictions for series-parallel competitive addition reactions in the late stages of reaction where the absorbance of all species [not shown in Fig. 15; but see Fig. 1 of reference [117] becomes independent of the amount of deposited metal atoms, a large fraction of the deposited atoms being directed into colloid.

A quenched reaction model for the kinetics of the growth of DC510 supported Mo_n in Fig. 15 also appears to confirm the stepwise aggregation model (SCHEME 8) and is useful for predicting cluster nuclearities. The clustering process can be represented by the following set of differential equations:

$$\frac{d[M]}{dt} = R - \sum_{j=0}^{p} k_m[M][M_jL_2] \tag{6}$$

$$\frac{d[M_nL_2]}{dt} = k_m[M]\{[M_{n-1}L_2] - [M_nL_2]\} \tag{7}$$

In solving the rate equations, cluster development is envisaged to evolve continuously throughout the metal atom deposition time τ at rate R in the film of polymer. Interruption of the deposition quenches the evolution of clusters and allows one to examine the product distribution spectroscopically.

The general solution for the concentration of a cluster species $[M_nL_2]$ referenced to the mononuclear species $[ML_2]$ on the polymer is as follows:

$$\frac{[M_nL_2]}{[ML_2]} = [M_o]^{n-1} \frac{2^{n-1}}{n![L_o]^{n-1}} = K[M_o]^{n-1} \tag{8}$$

where $[M_o]$ is the total mass of metal atoms deposited into the polymer film in time τ. The slopes of log ($[M_nL_2]/[ML_2]$) *vs.* log $[M_o]$ plots should yield values of n-1 and hence a spectroscopic probe into cluster nuclearity n. The series of experiments performed on the V_n/DC510, Cr_n/DC510, and Mo_n/DC510 systems appeared to conform reasonably well to this simple model (see, for example, Fig. 17B for Mo, n = 2, 3, 4).

The above experiments involving reactions of metal atoms with DC510 are thought to involve the consecutive growth of successively higher-nuclearity metal clusters in a process whereby the transition metal site is itself the target for cluster expansion. In Smith and Wychik's terminology [126], the DC510 macromolecules provide functional loci (pendant arenes) for the controlled growth of metal clusters. The function of the macromolecular domain is to restrict metal atom clustering to regions where the nucleated particles will be unable to accrete and flocculate.

More information regarding the ability of the DC510 polymer to stabilize low-nuclearity clusters can be gained by examining in some detail the effect of crosslinking on the thermally enhanced diffusion-controlled loss of the clusters. For a second-order decay process, bimolecular encounters leading to loss of cluster complexes should be increasingly inhibited as the molecular weight decreases between crosslinks derived from the *bis*(arene)M complexes. This idea is examined below.

7.3.4 Thermal agglomeration kinetics for Mo_n/DC510

Viscometry and thermal agglomeration studies pertaining to the Mo/DC510 system elucidate, among other factors, the role played by inter- and intramolecular crosslinking via *bis*(arene)molybdenum complex formation in stabilizing higher-nuclearity metal clusters in the order n = 2 > 3 > 4 > 5 [121]. Loss of the binuclear (arene)$_2$Mo$_2$ species supported on DC510 has been found to follow second-order kinetics (Fig. 22A). The rate of the diffusion-controlled aggregation and the activation energy for the process (Fig. 22B) depend on the metal loading through the number of crosslinks.

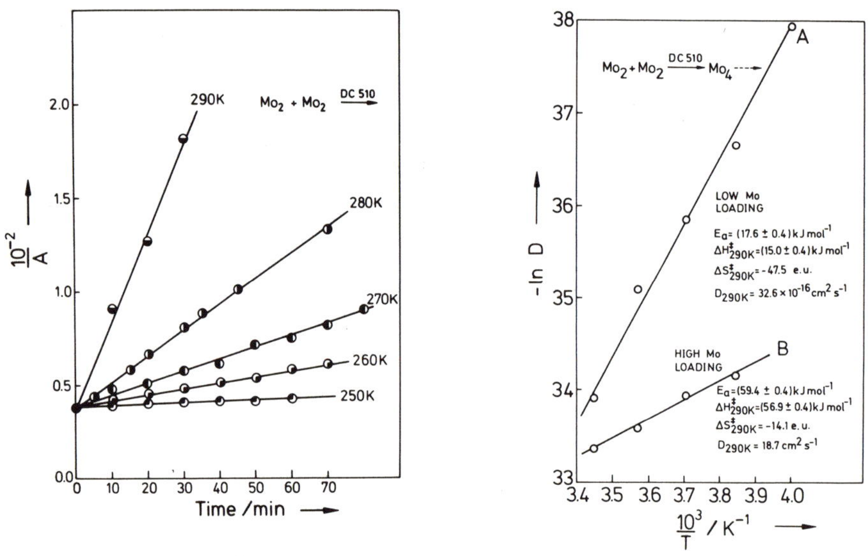

A B

Fig. 22. A. Second-order kinetics plot for the aggregation reaction (arene)$_2$Mo$_2$ + (arene)$_2$Mo$_2$ supported in DC510 at various temperatures. B. Arrhenius dependence of the diffusion coefficients extracted from kinetics data for a high metal loading limit, A, and a low metal loading limit, B [121].

Thus the activation energy for the loss of $(arene)_2Mo_2$ in the limit of low metal loading [note that the $(arene)_2Mo$ is unreactive under these conditions] is 17.6 kJ mol^{-1}. The respective diffusion coefficients at 290 K of 32.6×10^{-16} and 18.7×10^{-16} cm^2/s reflect the microscopic mobility of the polymer constrained by crosslinking to mediate aggregation. These diffusion coefficients are extremely low, of the same order of magnitude as those extracted for the dimerization of various transition metal atoms in weakly interacting solid matrices in the cryogenic range 10-40 K [127].

It should be noted that diffusional encounters of the "Mo_2" species in DC510 require approach of the actual crosslink sites. These movements are therefore very different from those of a small molecule diffusing in a crosslinked matrix. The mobility of crosslinks will be strongly correlated with the motions of the free and crosslinked portions of other chain segments. The fact that the rate of agglomeration at a given temperature was found to decrease with increasing metal loading supports the idea that an increase in the crosslink density impairs local macromolecular mobility in these metal-polymer systems. Information gained by incorporating ESR-active vanadium atoms as *bis*(arene)V complexes in DC510 confirms that motion in the vicinity of the crosslink is quite restricted. For comparison, not until a $V(C_6H_6)_2$/pentane solution has reached 138 K does its ESR spectrum resemble that of V/DC510 at 300 K [64, 112].

7.3.5 **Mode of metal cluster formation, bonding, and stabilization in poly-(dimethyl-co-methylphenylsiloxane)**

Mo atoms concatenate liquid α,ω-diphenyloligoethylene oxides at 275-290 K [121]. Like DC510, these ligands are able to harbor mono-, bi-, and trinuclear Mo species, showing growth characteristics and optical absorption spectra similar to those of DC510. However, no clusters of nuclearity greater than three can be prepared from any transition metal vapor under given conditions in these oligomers. Notably, "tadpole" molecules such as $C_6H_6(X(CH_2)_2)_nXCH_3$ (where X = CH_2 or O), would support only *bis*(arene)M complexes and $(arene)_2M_2$ species. Similarly, only $(arene)_2Mo_2$ was observed when $(arene)_2Mo$ dissolved in DC200 was titrated with Mo atoms, showing that mere existence of a liquid macromolecular matrix is insufficient to guarantee stabilization of higher-nuclearity clusters. In fact, low-nuclearity clusters stable to agglomeration cannot be grown in the absence of arene functionalities in liquid media.

How the metal atom multimers are bonded within the DC510 matrix is not known. Some plausible structures can be proposed for the dimers based on known multiply bonded binuclears such as $(\eta^6\text{-}C_6H_6)_2M_2(CO)_x$ (M = V, x = 3; M = Cr, x = 4) [128, 129]. In these molecules, the dimer lies perpendicular to the planes of the capping rings. Structures having the dimers parallel to the ring planes are also reasonable [130, 131]. Xα scattered wave molecular orbital calculations of simplified half-sandwich C_{6v} and C_{2v} mono(benzene)divanadium model compounds suggest that the optical transitions, red-shifted from the main MLCT band in M/DC510, are essentially metal localized [64].

An idealized physical chemical picture (Fig. 23) emerges from these studies. Here the *bis*(arene)M complexes are viewed as crosslinking and/or chain propagating agents, as well as metal nucleation sites for the production of clusters through an array of parallel, competitive metal atom addition reactions. The stable aggregates in the case of V, Cr, and Mo appear to require two arene rings and are unsolvated for n = 2 (Fig. 23 III) and perhaps solvated for nuclearities n = 3, 4, or 5 (Fig. 23 I) by oxygens of the oligomer or polymer backbone. The evidence suggests that this interpretation is more likely than an alternate one involving solely crown ether or cryptand-like interactions (Fig. 23 II). It is possible that the order of development of the species carrying the absorption bands to the red of the MLCT band in, say, Mo/DC510 may reflect not only the series-parallel growth kinetics (higher-nuclearity species ought to form

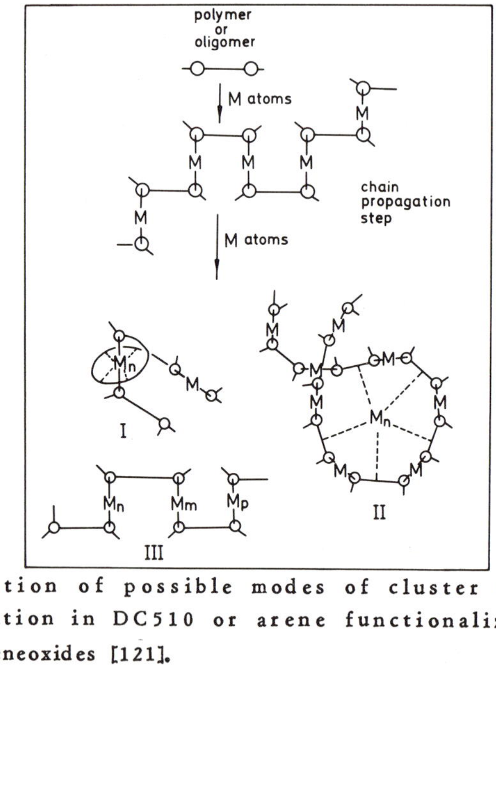

Fig. 23. Idealization of possible modes of cluster formation and stabilization in DC510 or arene functionalized oligo- or polyethyleneoxides [121].

later), but also the changing physical impediments to aggregation of the clusters by an increasingly crosslinked and therefore less locally mobile polymer network. It is conceivable that higher-nuclearity Mo clusters (with nuclearity 3, 4, 5,...) will be formed only if the polymer microenvironment is sufficiently rigid to prevent desolvation of the cluster. Thus the most highly crosslinked or networked systems should present the greatest kinetic barriers to cluster aggregation. This may account for the fact that no new absorptions are observed in the Mo_2L_2 dimerization kinetics experiments, where the Mo loading was intentionally kept low to favor the production of MoL_2 and Mo_2L_2 alone. Nor are new absorptions detected when DC510 systems with high Mo loading are warmed. In this case it is possible that the cluster species are ejected from the stable sites into more open regions of the polymer matrix, where rapid aggregation takes place.

One can suppose that the cluster growth and agglomeration characteristics in the polymer are determined by a number of factors, including the following:

(1) The presence or absence of covalently bound arene groups;

(2) changes in the microenvironment in the vicinity of the metal aggretate (cluster size and stability may be viewed as being determined in part by the dimensions and arene density of stabilizing cavities);

(3) steric interaction between the metal aggregate (immobilized), the polymer support, and the diffusant (atom, covalently bound arene, nucleation site) (diffusional encounters are controlled by polymer micro-dynamics, which are dependent on both temperature and crosslink density);

(4) diffusional limitations related to points (2) and (3); such limitations will affect the rate at which metal atoms penetrate into the polymer to contribute to cluster growth or be removed by a colloid sink; cluster aggregation is diffusion controlled;

(5) structural changes or constraints imposed on metal aggregates or the polymer as a result of binding or stabilization of macromolecules by crosslinking—such changes will control both the number and size distribution of clusters, determining a maximum metal loading limit beyond which colloid formation becomes the favored process for metal atom consumption.

7.3.6 Metal atom reactions with oligosilastyrenes and polydimethylcyclopentadienesiloxanes

Since there is an immense variety of functional groups that can be incorporated into oligomers and polymers, the range of new materials that can now be accessed using metal vapor synthesis methods is quite enormous.

One simply has to extend the data bank of known MV vapor reactions involving "single" ligands to the more "complex" polymer-bound ligand. For example, in the case of arene functionalities, the metal vapor synthesis method was extended to oligo- and polysilastyrenes in the liquid and solution phase [64, 113], as illustrated in SCHEME 9 for M = Ti, V, Cr, and Mo. Interesting differences in reactivity resulted when iron atoms were brought in contact with trimethylsilylcyclopentadiene and polysiloxane-bound cyclopentadiene (SCHEMES 10 and 11). The 16-electron complex of the free ligand was green and stable up to 243 K. At higher temperatures it transformed (in the presence or absence of O_2) into the corresponding 18-electron orange ferrocene derivative. The polymer-bound ligand reacted with iron atoms to produce green polymers the color of which was not discharged at room temperature. Fe Mössbauer spectra verified the presence of a 16-electron *bis*(cyclopentadiene)-iron(0) complex appended to the polymer backbone [112, 132]. However, the properties of this iron(0) site on the polymer differed from those of the site on the monomer. For example, reaction with O_2 yielded colloidal iron oxide in the polymer matrix rather than a polymer-anchored ferrocene. These results again point to the active role that a polymeric medium can play in determining chemical reactivity.

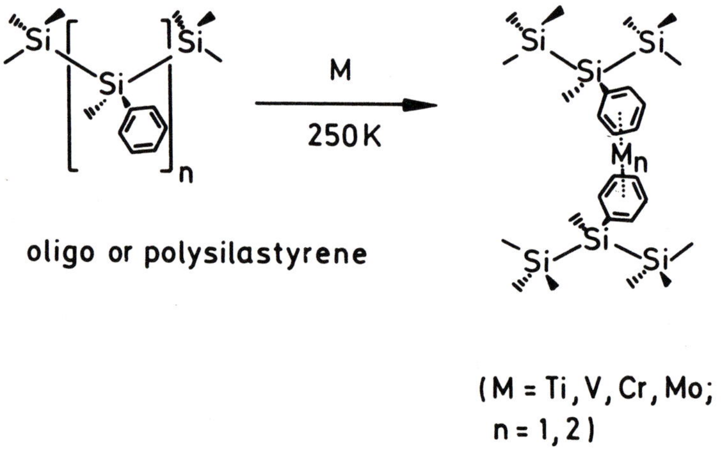

SCHEME 9

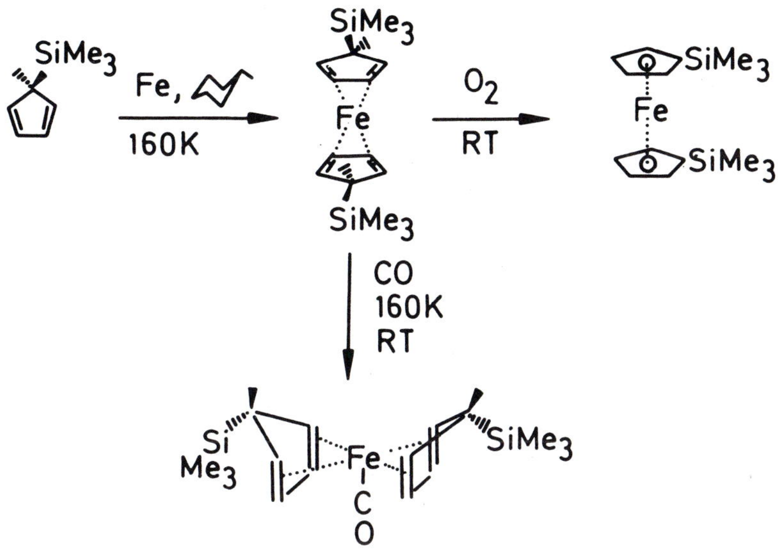

SCHEME 10

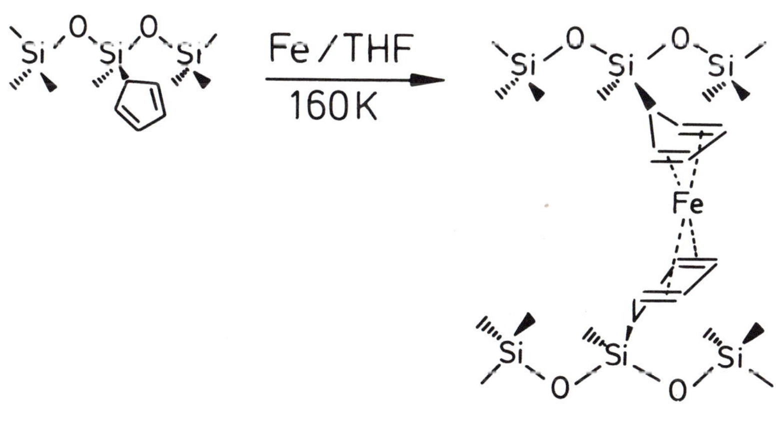

SCHEME 11

7.3.7 Metal atoms in unreactive polymers and oligomers

Francis and Timms [114, 133] evaporated Cr, Fe, Co, and Ni into a poly-(dimethylsiloxane) (DC200), the phenyl-free analog of DC510. These evaporations produced polymer-dispersed metal colloids and no metal-polymer insertion products. In contrast to Klabunde's and Tanaka's catalytically active SMAD slurries [134], the metal-DC200 slurries were found to be inactive. It does appear that some iron atoms can be solvated by the siloxane polymer backbone ether oxygen atoms when the reaction is carried out with DC200 (diluted with methylcyclohexane) at 150 K. Depositions of iron atoms into DC200 at 273 K produced only colloidal iron.

Metal vapors can be used to prepare colloidal silver directly in weakly interacting liquid supports such as polyolefins, oligo-olefins, vinyl and phenylsiloxane polymers, and crown ether and low-molecular-weight ether solvents [64, 112, 135]. What is interesting about this technique is that silver particles (1.0 nm < R < 20 nm) develop with a size distribution that remains independent of silver loading at a fixed temperature during the deposition period. The effect manifests itself optically in the linear growth of the surface plasmon absorption with increasing concentration of deposited Ag and in the invariance of both λ_{max} and the half bandwidth. An example is given in Fig. 24 for the system silver/polybutadiene. Particles generated in oligomeric and polymeric olefinic supports showed greater stability towards thermally induced agglomeration than did the easily desolvated species produced in low-molecular-weight liquid solvent media.

The growth behavior of silver particles in these liquid media can be exploited to produce metal-carbon composites suitable for fuel cell electrodes [113, 135]. The technique combines Ag, Pd, or Pt atoms, for example, with weakly coordinating ether and aromatic solvents, liquid oligo- and poly-ethers, and olefins, simultaneously or subsequently contacted with various forms of pretreated carbon powders. These experiments are informed by independent microsolution studies of metal atom reactions with thin liquid films of the above type. For reactions in SCHEME 1 proceeding to M_mL_x and M_bL_x, the growth and stabilization of monodispersions of metal clusters up to a critical saturation loading of the liquid with metal atoms could be demonstrated [64, 112, 135]. Consistent with this information, ligand stripping reactions, $M_mL_x \longrightarrow M_mL_{x-y}S \longrightarrow M_mS$ (for example, SCHEME 12) deposited sub-5-nm metal atom aggregates on the surface and within the pores of the carbon. Control of both particle size and size distribution depends sensitively on the preparative conditions, as outlined in section 7.3.1. This approach is likely to generate interest in any application requiring ultrafinely divided supported metal particles. For example, Ag_n^o/C, Pd_n^o/C, and Pt_n^oC compositions produced in this way were fabricated into porous oxygen gas electrodes (backed onto porous nickel) for use in an alkaline H_2/O_2 fuel cell [64, 113, 135].

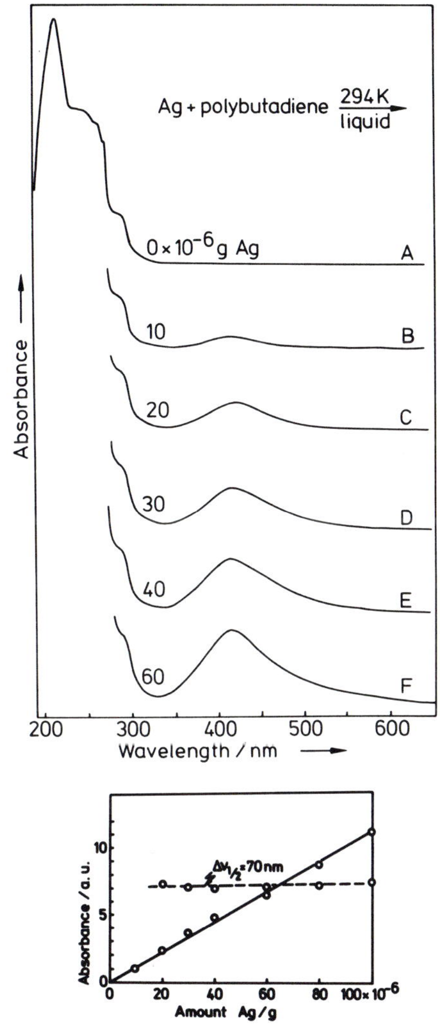

Fig. 24. Silver atom deposition into liquid polybutadiene monitored by *in situ* optical transmission spectroscopy performed and recorded at 294 K. The insert shows a graphical representation of the silver atom concentration dependence of the plasma resonance absorption and respective band width. Note the unchanging λ_{max} and $\Delta\nu_{1/2}$ (bwhh) of the plasma resonance absorption with increasing silver atom concentration [64, 113, 135].

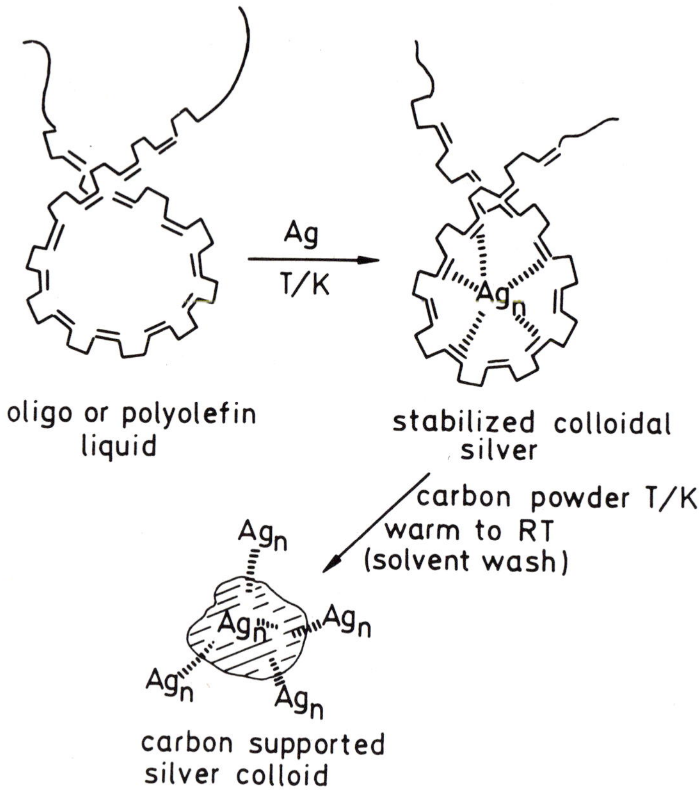

SCHEME 12

7.4 OXIDE- AND ZEOLITE-SUPPORTED METAL CLUSTERS

7.4.1 Solvated metal atom dispersion methods

In many of the aforementioned studies of the metal/polymer and metal/oligomer systems, control over diffusion and aggregation of the clusters is achieved by "solvating" or weakly binding the cluster species in a chemically interacting matrix. This idea is not new. Wilke wrote of "naked nickel" in the exceptionally reactive (1,5-cycloctadiene)$_2$Ni complex [136]. The idea is reinforced by Stone's reference [137] to the olefin complexes of Pd(0) and Pt(0) as sources of "solvated" metal atoms, and by Klabunde and Tanaka's approach [134] to metal particles through the use of "weakly stabilized" or "solvated metal atom solutions". Klabunde and Tanaka emphasized the building up of a catalytic metal particle one atom at a time, comparable to the clustering of metal atoms in DC510. Here we are reminded of the approaches adopted by Ichikawa [138] and Smith *et al.* [139] to lay down on a catalyst support clusters that carry the nuclearity of the parent compound, e.g., Ni_3 from $(C_5H_5)_3Ni_3(CO)_2$ or Rh_6 from $Rh_6(CO)_{16}$. Perhaps somewhat less control is achieved by thermal decomposition of organometallic clusters chemically bound to or adsorbed onto a support surface. Ligand stripping here leads to cluster-derived, dispersed metal and alloy particles [140]. Klabunde and Tanaka varied this cluster-derived dispersion technique by using weakly stabilized or solvated metal atom solutions [134]. The solutions are brought in contact with catalyst supports, and upon warming of the mixture to room temperature, the solvates decompose, depositing metal atoms and clusters (nucleation of metal atoms). The technique goes by the acronym SMAD (solvated metal atom dispersion), illustrated in SCHEME 13. The dispersing solvent affects the catalyst activity and selectivity, while the support affects metal particle nucleation and deposition differently for metals like Co and Ni. Furthermore, the dispersing reagent profoundly affects the shape, size, and magnetic properties of the resulting metal particles [134]. Carbonaceous fragments, which are often incorporated by the metal particles produced during the solvent meltdown, may be important in attaching the particles to support surfaces. Production of carbonaceous fragments is minimized by use of a solution phase metal atom technique (see below) for metal deposition rather than a cocondensation meltdown, method. This method apparently leads to more highly dispersed, more active catalysts than the SMAD procedure. In Klabunde and Tanaka's work also, arenes feature prominently in the solvated metal atom approach. An illustration of the competing processes in cluster nucleation and reaction is given in SCHEME 14.

Klabunde and Imizu [141] have extended the SMAD method to the preparation of unsupported and supported bimetallic clusters, for example, Fe/Mn, Fe/Co, Ni/Mn, Co/Mn, and other combinations. Significantly, the catalytic properties of these bimetallic materials are quite distinct from those of their respective monometallic counterparts. In particular, small dopings of Mn were found to modify dramatically the catalytic selectivity and activity patterns of Fe-, Co-, and Ni-supported clusters for alkene and diene hydrogenation and isomerization. In fact, when Mn was incorporated into Co particles on silica, the increase in activity was sufficient to create a catalyst active enough to hydrogenate but-1-ene at 213 K in a diffusion-controlled fashion (Fig. 25). Noteworthy also was the effect of

Mn to decrease the tendency of these metals to incorporate carbon from the carrier solvent (C-H and C-C activation). These are pivotal observations for the SMAD catalyst designer.

The discovery that Mn can activate SMAD Co/SiO_2 catalysts to such a large extent is timely, as there is a great deal of interest currently in both homogeneous cluster chemistry and heterogeneous catalysis in making "early-late" transition metal clusters and catalysts. This impetus is mainly based on the supposition that the combination of two such metals may lead to unusual CO activation mechanisms and hence affect selectivity in Fischer-Tropsch reduction products.

The fact that *bis*(toluene)iron or *bis*(toluene)cobalt synthesized by metal atom reactions in solution liberates a metal atom and two toluene molecules at 243 and 203 K for the iron and cobalt complexes, respectively, has motivated experiments which have resulted in the production of $\leq$ 1.2-nm metal particles on the surface and in the cavities of dehydrated Na-Y zeolite [23]. These metal particle-zeolite compositions are Fischer-Tropsch catalysts, showing selectivity (cluster size dependence) for low carbon-number olefins. Two approaches have been taken to prepare the zeolite. The first is essentially that of SCHEME 14, pathway II, in which iron or cobalt metal is evaporated into a slurry of dehydrated sodium zeolite Y and toluene at low temperatures in a metal vapor rotary

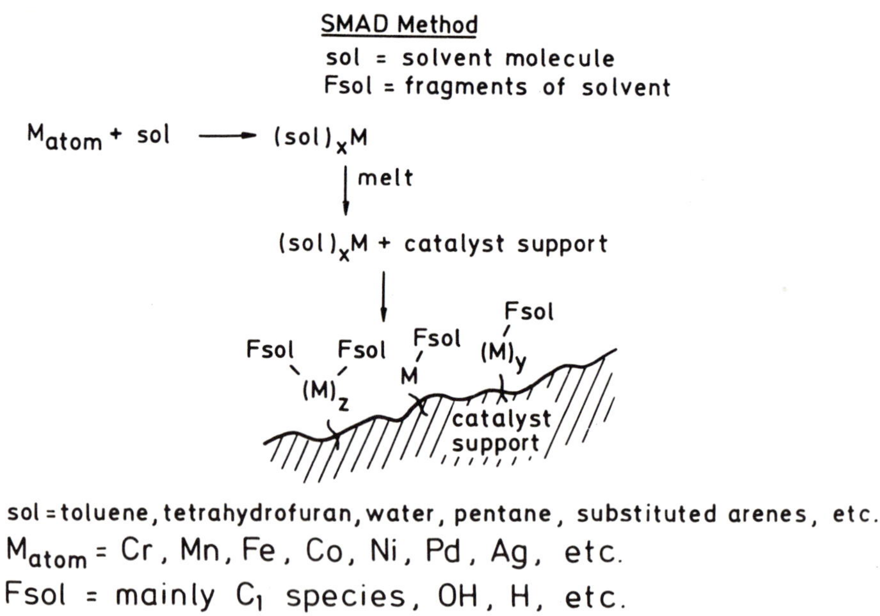

SCHEME 13

(no support) (support) (support) (support)

(I) Cluster growth competes with reaction of the forming clusters with the host solvent (toluene). Low-temperature aging aids the solvent-cluster reaction channel.

(II) *In-situ* Solution Phase Metal Atom Technique

Using the solution technique the support immediately and continuously is in contact with the metal solvate. Nucleation commences immediately, leading to very small metal particles with little carbonaceous buildup (solvent-cluster reaction channel is disfavored).

(III) Cocondensation Metal Atom Technique

Using the codeposition technique, the M-solvate is given a chance both to begin cluster growth and to react. Upon contact with the support later, the solution contains larger clusters with more carbonaceous material buildup (solvent-cluster reaction channel is allowed). The cluster's attachment to the support is aided by these carbonaceous fragments, yielding a more stable cluster deposit.

(IV) *In-situ* Solution Phase Metal Atom Technique

Using a rotary reactor, preformed solutions of solvated metal atoms are transferred below their decomposition temperature to a slurry of the support in the same solvent. Following impregnation and cold filtration (and cold washing for porous supports) the metal-support combination is warmed to room temperature under dynamic vacuum, causing desolvation, and metal atom aggregation on the surface or in the pores of the support.

SCHEME 14

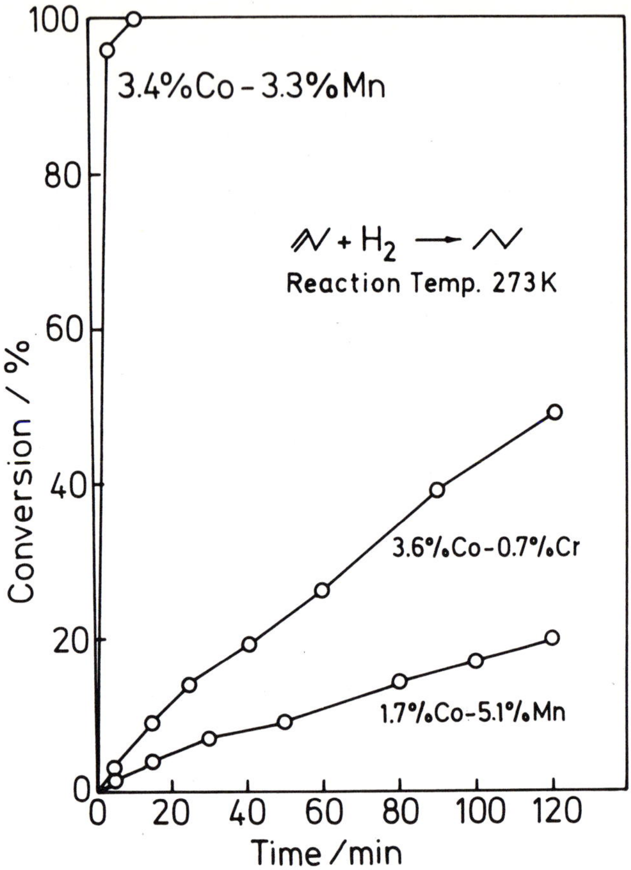

Fig. 25. Effect of Mn addition on hydrogenation activity for a Co/SiO_2 SMAD catalyst [141].

reactor. The second approach, pathway IV, takes the previously prepared *bis*(arene) complex and subsequently brings it in contact with the zeolite. In both cases the slurry is carefully warmed to a temperature below the decomposition temperature of the complex to enhance diffusion of the monatomic metal atom carrier into the zeolite α-cages. The slurry is then transferred at the same low temperature directly into a precooled filtration apparatus. At this stage, the solid zeolite is washed with cooled toluene (to facilitate removal of molecules adsorbed on the outer surface) and then allowed to warm slowly to room temperature under dynamic vacuum. The zeolite-entrapped *bis*(toluene)M complex then gently "desolvates" well below room temperature, thereby releasing metal atoms which can diffuse and agglomerate to give small metal aggregates mainly within the zeolite α-cages.

(The subject of metals in zeolite cages is discussed in detail in Chapter 8 of this book.)

Thus, the idea behind the method is to provide an exceedingly "mild" preparative route for impregnating the supercages of zeolites with small, formally zerovalent metal clusters. By selecting a "metal atom solvate" of the molecular dimensions which are such that it can access the channel structure of the zeolite Y ($\lesssim$ 0.8 nm) and enter the α-cage (~1.25 nm) under conditions that also maintain the integrity of the solvate, one has the opportunity to inquire into the fate of a "single" zerovalent metal atom when placed into the supercage of zeolite Y under extremely gentle conditions (contrast the rather severe methods traditionally used for preparing metal aggregates in zeolites, as discussed in Chapter 8).

A study of this type has been reported for Fe and Co utilizing Mossbauer, magnetization, and ferromagnetic resonance methods for metal particle size characterization and CO hydrogenation as a structure-sensitive catalytic probe reaction [23, 142]. In brief, Mössbauer spectra demonstrated for $^{57}Fe_n$/Y that the iron clusters were superparamagnetic, which for zerovalent iron translates into a particle size less than 3 nm. A low- and high-field Langevin analysis of the magnetic field (0-52 KOe) dependence of the magnetization curves at 4.2 K and 300 K for Fe_n/ZY (Fig. 26) confirmed that at low Fe loading, samples could be prepared containing between 80 and 95% superparamagnetic iron clusters with

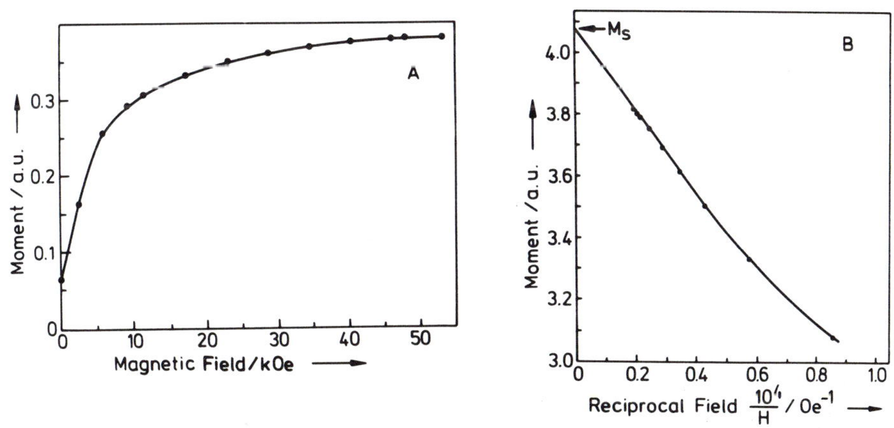

Fig. 26. (A) Magnetization of solvated metal-atom-generated Fe_n/NaY zeolite at 4.2 K (magnetic moment in arbitrary units) as a function of applied field; (B) saturation magnetization of Fe_n/NaY zeolite at 4.2 K [23, 142].

diameters (assuming spherical symmetry) in the range 0.5-1.2 nm. It was significant that for all Fe_n/Y samples generated from toluene-solvated iron atoms, the major portion (> 75%) of the iron clusters generated did not exceed the diameter of the α-cage (1.25 nm). This observation provides support for the idea that the majority of superparamagnetic iron clusters reside within the pore structure of the zeolite Y rather than on the external surface. These results imply that when a single, neutral iron atom finds itself in the α-cage of zeolite Y around room temperature, it readily diffuses and agglomerates with other iron atoms in adjoining α-cages, growing up to a maximum cluster dimension controlled by the initial iron atom loading, as well as the size of the α-cage and its 12-ring entrance channels.

Further evidence for the existence of very small metal clusters entrapped in the α-cages of zeolite Y for the Fe_n/Y and Co_n/Y compositions comes from the hydrocarbon product distribution formed when these materials are brought in contact with CO/H_2 mixtures in a catalytic flow reactor at atmospheric pressure [23, 142]. For both metals, the pattern of olefinic and paraffinic products in the C_1 to C_6 range studied was not typically Schulz-Flory. In particular, the more stable Co_n/zeolite Y samples (compared to Fe_n/Y) at 520 K displayed a definite propensity towards the formation of C_4 hydrocarbons (around 70 mole % but-l-ene) (Fig. 27). Although the selectivity for C_4 products is pronounced and the product distribution unchanged after 60 h on stream, the conversion was less than 1%. The metal loading of these samples was only 1-2 wt%, and higher loadings prepared with multiple impregnations might give higher activities. These same Co_n/Y samples, which are selective for C_4 products at atmospheric pressure, give C_5 selectivity when the CO/H_2 reaction is conducted at 15 atm. A judicious choice of metal, solvent, zeolite type, metal loading, and impregnation conditions can lead to metal cluster-zeolite compositions of novel types. These catalysts are considered further in Chapter 8.

The solvated metal atom approach has been exploited by Berry [143] and Scott and coworkers [144] to prepare metal particles according to various routes outlined in SCHEME 1. Ozin, Power, and coworkers [29, 33] had identified nickel-olefin cluster complexes of the type $Ni_n(olefin)_x$ (n = 1,2,3; x = 1,2,....) in matrix isolation experiments; Berry [143] was stimulated to synthesize new labile organometallics capable of depositing metallic films. $(PF_3)_3Ni(olefin)$ compounds so prepared did decompose between 257 and 276 K to give $Ni(PF_3)_4$, olefin, and Ni metal. The structural, morphological, and magnetic properties of small

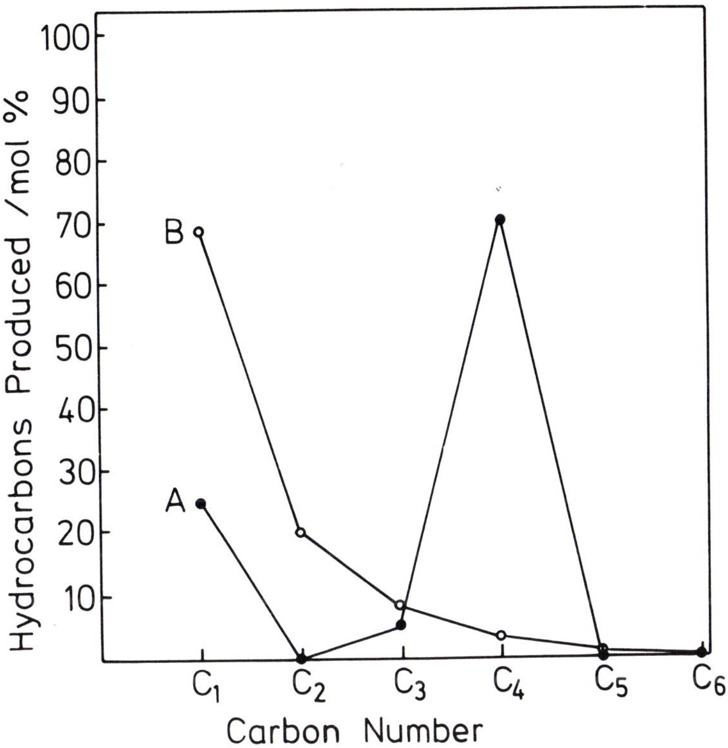

Fig. 27. Hydrocarbon product distribution for solvated-metal-atom-generated Co_n/NaY catalysts (mole %). (A) 520 K, CO/H_2 = 1/2, 0.02% conversion; 1-C_4H_8, 33%; iso-C_4H_8, 3%; 2t-C_4H_8, 18.5%; 2c-C_4H_8, 15.5% (B) 563 K, CO/H_2 = 1/2, 0.04% conversion [23, 142].

nickel particles prepared by the metal atom technique in toluene, SF_6, and xenon matrices have been reported by Scott *et al.* [144]. Careful consideration was given to the role of the solvent matrix in controlling the growth and nucleation of metal microparticles.

7.4.2 Modification of well-defined hydroxylated oxide surfaces with solvated metal atoms

7.4.2.a Iron on silica. A systematic study of the genesis of iron clusters supported on silica with different concentrations of surface hydroxyl groups has been carried out with solution-phase solvated iron atom deposition techniques [145]. This can be considered to be a precursor study into the effect of the physicochemical properties of oxide support materials on the crucial stage of metal cluster formation using solvated metal atom reagents.

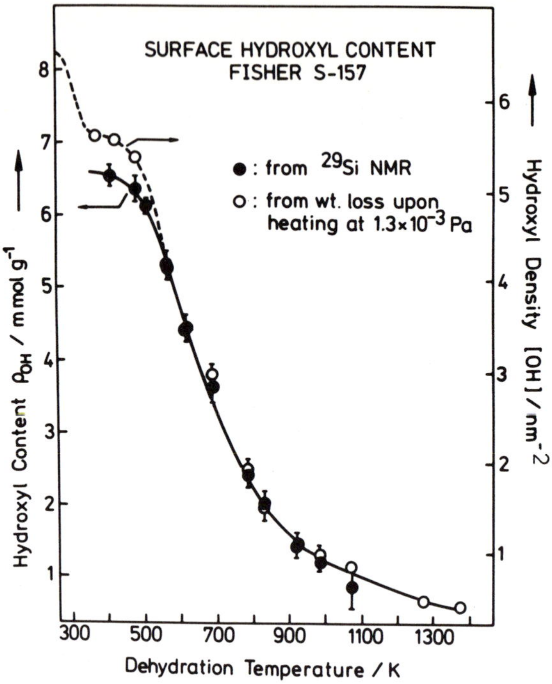

Fig. 28. Bulk hydroxyl density determined by the NMR approach (ρ_{OH}^{NMR}) and by the weight-change approach (ρ_{OH}^{d}) as a function of dehydration temperature [146].

The surface concentration of hydroxyl groups on silica varies according to the dehydroxylation temperature, as depicted in Fig. 28 [146]. Thus at 473 and 673 K, the surface of silica has approximately 5 and 2 hydroxyl groups per nm^2, respectively. Using *in-situ* microsolution UV-visible spectroscopy, *bis*(toluene)iron (formed by depositing Fe atoms into toluene held at 150 K on a thin self-supporting wafer of silica dehydroxylated at 473 K) was found from careful warm-up experiments to decompose at about 180 K. The fact that this temperature is roughly 60 K less than the temperature of solution-phase decomposition of *bis*(toluene)iron in the absence of the silica support indicates that the decomposition process is facilitated by the surface hydroxyl groups of the silica. In line with this proposal is the observation that the rate of decomposition of *bis*(toluene)iron(0) is much faster on silica treated at 473 K than on silica treated at 673 K.

The average pore radius $\bar{r}$ of the silica used in this study was calculated from the standard equation

$$\bar{r} = 2\ v_g/S_g \tag{9}$$

where v_g is the void volume per gram and S_g is the surface area per gram. This radius was found to be 1.5 nm. Thus most of the *bis*(toluene)iron(0) can diffuse into pores quite easily and interact with most of the surface hydroxyl groups on the silica.

The color of silica (pretreated at 473 K) impregnated with *bis*(toluene)iron(0) at 195 K was pale brown, which changed to reddish brown upon slow warming of the sample to room temperature under vacuum. Infrared studies of this process showed that most of the surface hydroxyl groups were consumed in the reaction with *bis*(toluene)iron(0); bands appearing at 972 and 672 cm^{-1} are characteristic of toluene coordinated to group VIII metal atoms (Fig. 29A). Further evacuation of the sample I at 10^{-5}-10^{-6} Torr for an extended period changed the color to a still redder brown, indicating the presence of yet another surface compound. This species was investigated by Mössbauer spectroscopy using ^{57}Fe enriched *bis*(toluene)$^{57}Fe(0)$. A room-temperature Mössbauer spectrum of a 0.044 wt%-loaded (10% ^{57}Fe-enriched) sample is shown in Fig. 30A, displaying a well-resolved doublet. As a first approximation the curve was treated as an asymmetric quadrupole doublet with isomer shift δ = 0.347 mm/s, quadrupole splitting Δ = 0.99 mm/s with area ratio A_2/A_1 = 0.681 and respective bwhh ratio of 1.12. This confirms that the iron is no longer present in the form of *bis*(toluene)iron(0), which has Mössbauer parameters of δ = 0.39 mm/s and Δ = 0.5 mm/s [147].

The isomer shift is, however, in good agreement with those expected for Fe(III). When *bis*(toluene)iron was brought in contact with silica pretreated at 673 or 873 K, the observed color changes were different from those observed with the silica pretreated at 473 K. Thus the *bis*(toluene)iron(0) filtrate from silica pretreated at 673 and 873 K remained green-brown (rather than colorless for silica pretreated at 473 K), indicating that the decomposition of *bis*(toluene)iron at 195 K is not complete due to the lower concentration of hydroxyl groups on the silica samples pretreated at 673 and 873 K. Furthermore, after prolonged evacuation at room temperature, the color of the samples pretreated at 673 and 873 K changed to dark gray, suggesting the presence of colloidal iron on the surface of the silica. However, the color of the sample changed to grayish brown on storage under Ar for one month. A representative Mössbauer spectrum of the sample prepared from silica pretreated at 873 K (0.1 wt% loading, 10% ^{57}Fe enriched) is shown in Fig. 30B. It displays a similar asymmetric quadrupole doublet (δ= 0.33 mm/s, Δ = 1.00 mm/s, A_2/A_1 = 0.934, and bwhh ratio =

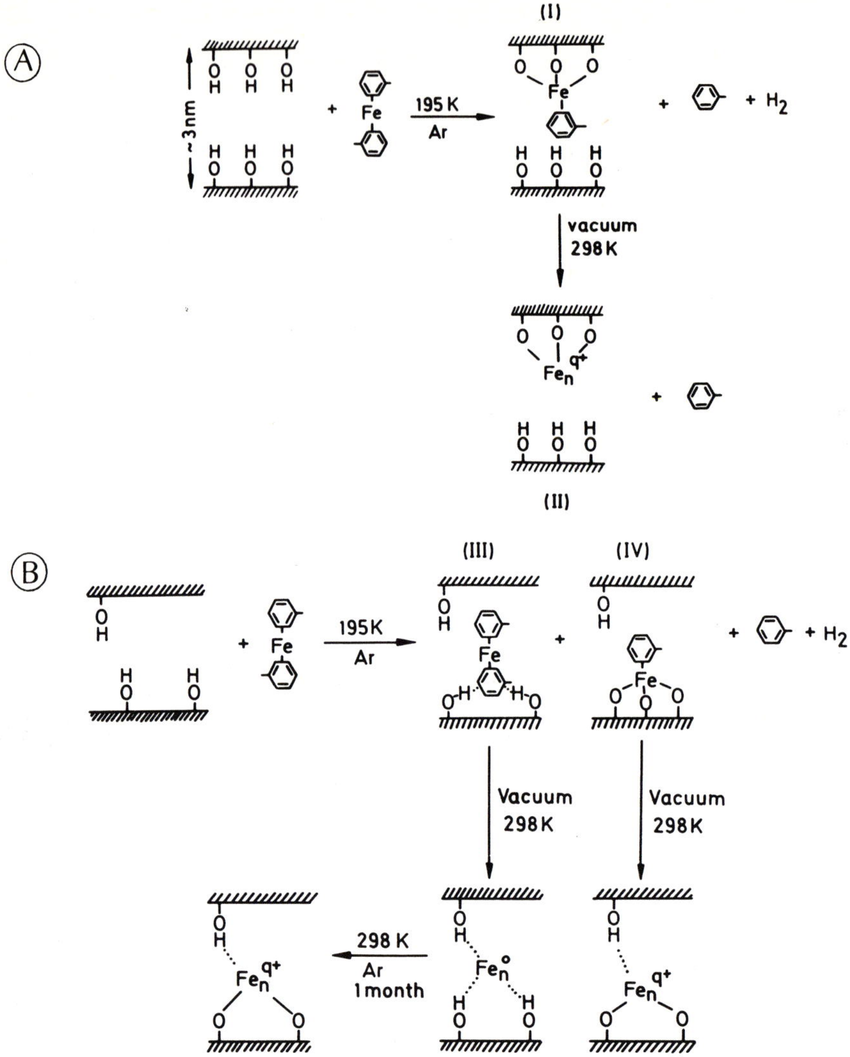

Fig. 29. A. Proposed steps for the formation of iron clusters on the surface of partially dehydroxylated silica gel at 473 K; B. proposed steps for the formation of iron clusters on the surface of silica gel dehydroxylated at 673 K.

1.09), also indicative of Fe(III) surface species. For the silica pretreated at 873 K, the number of Fe atoms corresponding to 0.1 wt% Fe loading is 1.07×10^{19} atoms per gram of silica. This is still less than the number of available surface hydroxyl groups, 5.5×10^{20} per gram of silica. It is

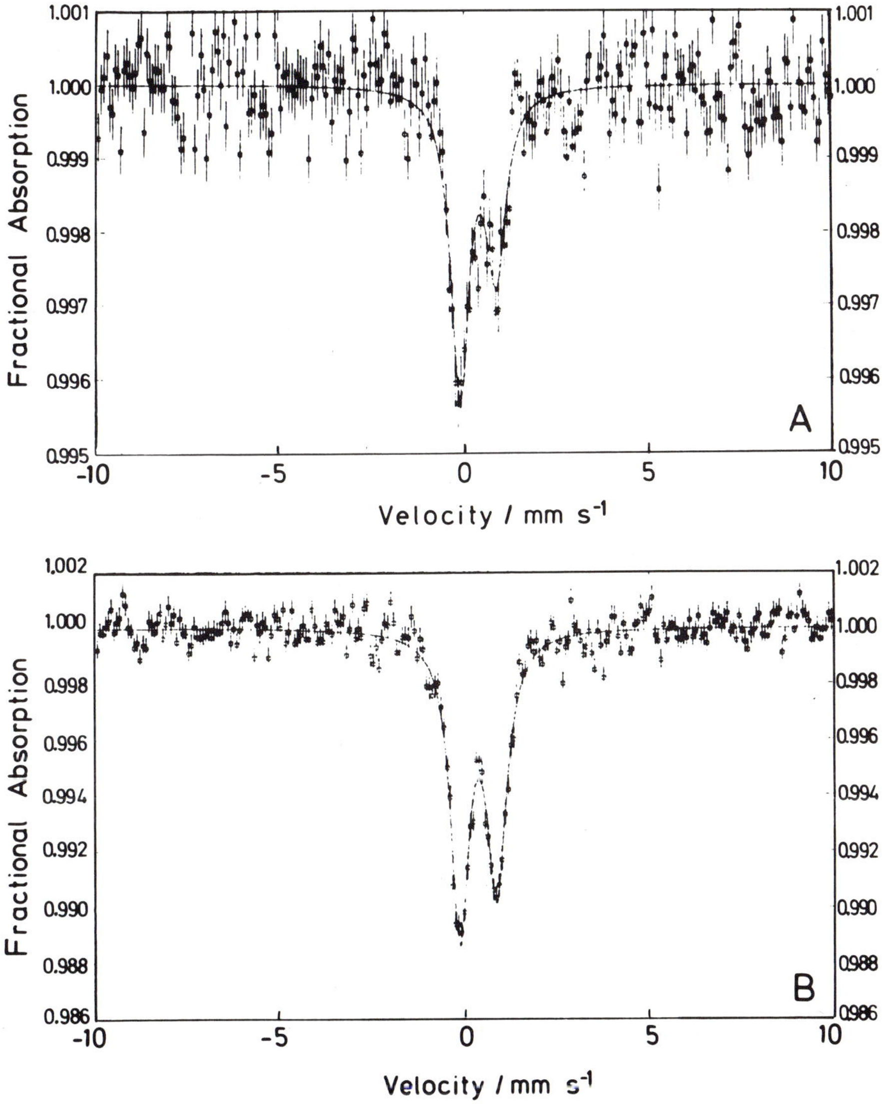

Fig. 30. Room-temperature Mössbauer spectra of (A) 0.044 wt% iron (10% ^{57}Fe enriched) on silica pretreated at 473 K and (B) 0.1 wt% iron (10% ^{57}Fe enriched) on silica pretreated at 873 K generated by solution phase solvated iron atom deposition techniques [145].

clear, therefore, that all of the impregnated *bis*(toluene)iron(0) can be oxidized by surface hydroxyl groups, as indicated by the color of the samples pretreated at 473, 673, and 873 K as well as the ^{57}Fe Mössbauer spectra of 0.1 wt% (or less) iron-loaded samples.

The asymmetric doublets of Fig. 30A and B show unequal line intensities and widths. As discussed by Garten [148], this can arise from one of the following effects (or a combination): a Gol'Danskii-Karyagin effect, relaxation phenomena, and a multiplicity of iron environments. Transmission electron micrographs of *bis*(toluene)iron(0) decomposed rapidly and slowly on silica pretreated at 473 and 873 K show that a larger population of smaller iron crystallites between 1.0 and 2.0 nm in diameter is immobilized on both silicas for the latter mode of impregnation, whereas the former leads to iron crystallites with diameters between 5 and 10 nm. Therefore, the rapid desolvation of *bis*(toluene)iron(0) appears to favor the formation of large iron clusters in the solution phase, which can be trapped on the exterior surface of silica after removal of the toluene under vacuum. In contrast, slow desolvation of *bis*(toluene)iron(0) on the surface of silica enhances the formation of smaller iron clusters. The microscopy data suggest that the corresponding Mössbauer asymmetric quadrupole doublet probably arises from either Fe(III) occupying a number of sites with different symmetries and coordination numbers or Fe(III)-containing particles with a distribution of sizes. This would give rise to a distribution of isomer shifts and quadrupole splittings. Since the iron clusters on silica pretreated at 473 and 673 K are not monodisperse in size and since the line intensities and widths of the Mössbauer quadrupole doublets are different, fitting the spectrum with a range of isomer shifts and quadrupole splittings would better approximate the physical situation (recall that the isomer shift has been shown to be a strongly varying function of particle size for very small iron aggregates [67], and this probably holds true for iron oxide aggregates) than relaxation and Gol'Danskii-Karyagin effects, although these cannot be discounted as contributing to line asymmetry.

With this information it is possible to propose steps for the formation of iron clusters on the surface of silica dehydroxylated at 473 and 673 K, as illustrated in Fig. 29. Thus the red-brown product formed from sample I on silica dehydrated at 473 K is best described as either a surface-anchored Fe^{3+} ion or superparamagnetic Fe_2O_3. The results of transmission electron microscopy support the latter assignment. Mössbauer measurements at liquid helium temperatures would help clarify the matter; we anticipate a characteristic six-line magnetic hyperfine splitting pattern for the latter but not the former. In the case of the silica dehydroxylated at 673 K, the final state of the surface iron species is similar to that of sample II on the silica dehydroxylated at 473 K, as indicated in Fig. 29.

7.4.2.b **Iron on titania.** The solution-phase SMAD approach has been used to produce heterogeneous photocatalysts [149]. Materials of this type, particularly titanium dioxide powders, predominantly in the anatase crystal form, have been reported to photocatalyze a variety of reactions and are attracting increasing attention for solar energy splitting of water and selective organic photooxidations [150]. It is known that doping of TiO_2 with small metal crystallites enhances its photoactivity [151].

Although many routes have been devised for the deposition of Pt metal onto these powders, few general techniques are available for incorporating other metals into photosemiconducting oxides. An interesting metal vapor synthesis application in this context stems from the work of Radford and Francis [149], who dispersed toluene-solvated iron atoms onto finely divided TiO_2 powders and used the resulting material for the aqueous photoreduction of N_2:

$$N_2 + 3H_2O \xrightarrow{Fe/TiO_2} 2NH_3 + 3/2\ O_2 \qquad (10)$$

$$\Delta G^o = 678\ kJ\ mol^{-1}$$

This is a reaction that has been shown to proceed on Fe_2O_3/TiO_2 particles [151].

Results for iron-doped anatase and rutile illuminated under a N_2 atmosphere are shown in Fig. 31. For comparison, undoped samples demonstrated minimal or no activity when irradiated under either N_2 or Ar. In addition, the dark reactions yielded no observable activity for doped or undoped samples of anatase or rutile. From Fig. 31 one learns that Fe-doped anatase is much more active than doped rutile, consistent with earlier investigations of water photo-splitting reactions on TiO_2. Also, reducing the Fe/anatase concentration by a factor of about 10 caused only halving of the initial rate. This result was rationalized in terms of inefficient use of active sites owing to poor penetration of photons in the more concentrated sample. The quantum efficiency for the formation of NH_3 from Fig. 31b was estimated to be 1%. Interestingly, lower Fe loadings led to higher initial activities but to loss of stability with increasing irradiation time, which implies that samples with higher iron dispersions, although exposing a larger number of active sites, are more susceptible to deactivation, possibly via poisoning by evolved O_2 or by

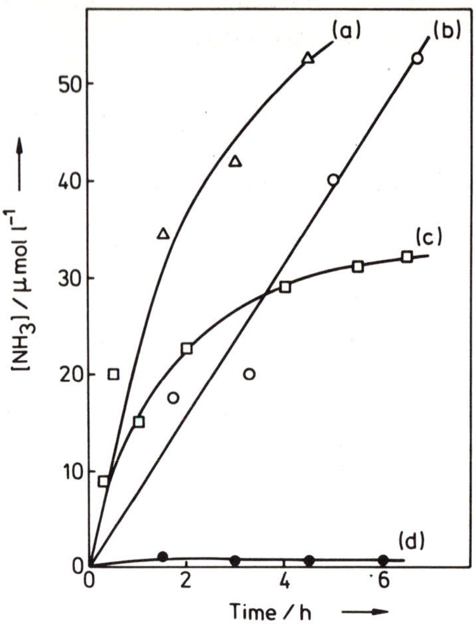

Fig. 31. Evolution of NH_3 from illuminated aqueous suspensions of Fe/TiO_2 under a nitrogen atmosphere: (a) 26 mg of anatase (2.3% Fe) per ml of water; (b) 1.6 mg of anatase (2.3% Fe) per ml of water; (c) 1.6 mg of anatase (0.9% Fe) per ml of water; (d) 8.1 mg of rutile (1.9% Fe) per ml of water [149].

reaction with surface hydroxyl groups. According to this interpretation, the behavior illustrated in Fig. 31 [curves (a) and (c)] may be diagnostic of a photostoichiometric reaction approaching completion.

The metal vapor synthesis route for doping powders to produce photoactive materials can be extended to other metals and supports, provided that a solvated metal atom precursor can be prepared that is relatively unstable to desolvation. Materials of this type generated by metal vapor synthesis methods might be expected to find applications as microphotoelectrochemical devices and photocatalysts for H_2 production from water, selective C-H bond activation, nitrogen fixation, and a wide range of other light-driven processes.

These few examples should serve to illustrate the potential of the metal vapor synthesis technique in new materials science. Much is yet to be learned about the mechanisms by which metal atoms are liberated from ligand carrier molecules and how the atoms subsequently aggregate in various environments. Matrix isolation and microsolution metal vapor techniques will continue to contribute to this understanding as chemical and physical techniques for manipulating matter at the atomic level are progressively refined.

ACKNOWLEDGMENTS

We acknowledge the invaluable assistance of the graduate students, postdoctoral fellows, and colleagues whose names appear in this chapter and who have made enormous contributions to the development of various aspects of metal atom chemistry, spectroscopy, and catalysis over the past decade. We also express our special indebtedness to Ted Huber, Karl Molnar, and Wolfgang Panning for their invaluable technical assistance with the instrumental aspects of the field. Helen Sarkissian's expert typing of the preliminary manuscript is also appreciated. The generous financial assistance of the Natural Sciences and Engineering Research Council of Canada's Operating and Strategic Grants Programmes and the Connaught Foundation of the University of Toronto are all gratefully acknowledged. The award of a 3M Corporation Research Fellowship to GAO for the period 1982-85 is also gratefully acknowledged.

REFERENCES

1. (a) P. Chini, Gazz. Chim. Ital., 109 (1979) 225; (b) P. Chini, J. Organomet. Chem., 200 (1980) 37.
2. R. J. Van Zee, C. A. Baumann, S. V. Bhat and W. Weltner, Jr., J. Chem. Phys., 76 (1982) 5636.
3. (a) G. A. Thompson and D. M. Lindsay, J. Chem. Phys., 74 (1981) 959; (b) D. M. Lindsay, D. R. Herschbach and A. L. Kwiram, Mol. Phys., 32 (1976) 1199; (c) G. A. Thompson, F. Tischler, D. Garland and D. M. Lindsay, Surf. Sci., 106 (1981) 408; (d) G. A. Thompson, F. Tischler and D. M. Lindsay, J. Chem. Phys., 78 (1983) 5946; (e) D. A. Garland and D. M. Lindsay, to be published; (f) D. M. Lindsay and G. A. Thompson, J. Chem. Phys., 77 (1981) 1114; (g) D. A. Garland and D. M. Lindsay, J. Chem. Phys., 78 (1983) 2813; (h) D. A. Garland and D. M. Lindsay, J. Chem. Phys., 80 (1984) 4761.
4. L. B. Knight, Jr., R. W. Woodwood, R. J. Van Zee and W. Weltner, Jr., J. Chem. Phys., 79 (1983) 5820.
5. C. E. Briant, B. R. C. Theobald, J. W. White, L. K. Bell, D. M. P. Mingos and A. J. Welch, J. Chem. Soc. Chem. Commun. (1981) 201.
6. D. M. Washecheck, E. J. Wucherer, L. F. Dahl, A. Ceriotti, G. Longoni, M. Manassero, M. Sansoni and P. Chini, J. Am. Chem. Soc., 101 (1979) 6110.
7. B. K. Teo and K. Keating, J. Am. Chem. Soc. 106 (1984) 2224.
8. P. Fantucci, J. Koutecky and G. Pacchioni, J. Chem. Phys., 80 (1984) 325. 9. J. Muhlbach, K. Sattler, P. Pfau and E. Rechnagel, Phys. Lett. 87A (1982) 415.
10. G. Hortig and M. Müller, Z. Phys., 221 (1969) 119.
11. M. R. Hoare, P. Pal and P. P. Wegener, J. Colloid Interface Sci., 75 (1980) 126; A. L. Mackay, Acta Crystallogr., 15 (1962) 916.

12. S. Ino, J. Phys. Soc. Japan, 21 (1966) 346; K. Kimoto and I. Nishiyida, J. Phys. Soc. Japan, 22 (1967) 940; J. G. Allpress and T. V. Saunders, Surf. Sci., 7 (1967) 1; J. Farges, B. Raoult and G. Torchet, J. Chem. Phys., 59 (1973) 3454; J. Farges, M. F. de Feraudy, B. Raoult and G. Torchet, J. Chem. Phys., 78 (1983) 5067.
13. S. Martinego, G. Ciani and A. Sironi, J. Am. Chem. Soc., 102 (1980) 7564.
14. J. L. Vidal, R. C. Schoening and J. M. Troup, Inorg. Chem., 20 (1981) 227.
15. R. W. Broach, L. F. Dahl, G. Longoni, P. Chini, A. J. Schulz and J. M. Williams, Advan. Chem. Ser., 167 (1978) 93.
16. P. F. Jackson, B. F. G. Johnson, J. Lewis, W. J. H. Nelson and M. McPartlin, J. Chem. Soc. Dalton Trans. (1982) 2099.
17. C. T. Hayward, J. R. Shapley, M. R. Churchill, C. Bueno and A. L. Rheingold, J. Am. Chem. Soc., 104 (1982) 7347.
18. (a) J. A. A. J. Perenboom, P. Wyder and F. Meier, Phys. Reports, 78, (1981) 173; (b) International Meeting on Small Particles and Inorganic Clusters, J. Phys. (Paris), 38, Colloque C-2 (1977); (c) Second International Meeting on Small Particles and Inorganic Clusters, Surf. Sci., 106 (1981); (d) G. C. Papavassiliou, Progr. Solid State Chem., 12 (1979) 185; (e) Conference on Metal Clusters, Ber. Bunsenges. Phys. Chem., 88 (1984) 188.
19. Y. Takasu, T. Akimaru, K. Kasahara, Y. Matsuda, H. Miura and I. Toyoshima, J. Am. Chem. Soc., 104 (1982) 5249.
20. J. M. Dartigues, A. Chambellan and F. G. Gault, J. Am. Chem. Soc., 98 (1976) 856.
21. (a) H. E. Rhodes, P. K. Wang, H. T. Stokes, C. P. Slichter and J. H. Sinfelt, Phys. Rev., B26 (1982) 3559; (b) H. E. Rhodes, P. K. Wang, C. D. Makowka, S. L. Rudaz, H. T. Stokes, C. P. Slichter and J. H. Sinfelt, ibid., 3569; H. T. Stokes, H. E. Rhodes, P. K. Wang, C. P. Slichter and J. H. Sinfelt, ibid., 3575.
22. D. L. Doering, H. Popper and J. T. Dickinson, J. Vac. Sci. Technol., 18 (1981) 460; J. Catal., 73 (1982) 91, 104.
23. L. F. Nazar, G. A. Ozin, F. Hugues, J. Godber and D. Rancourt, J. Mol. Catal., 21 (1983) 313.
24. P. A. Jacobs and H. H. Nijs, J. Catal., 65 (1980) 328.
25. (a) G. A. Ozin, Catal. Rev.—Sci. Eng., 16 (1977) 191; (b) Coord. Chem. Rev., 28 (1979) 117; (c) G. A. Ozin, Acc. Chem. Res., 10 (1977) 21; (d) G. A. Ozin, Faraday Disc. Chem. Soc., 14 (1980) 1.
26. (a) M. Moskovits and G. A. Ozin, "Cryochemistry," Wiley-Interscience, New York, 1976; (b) M. Moskovits and J. Hulse, Surf. Sci., 57 (1976) 125; (c) M. Moskovits and J. Hulse, Surf. Sci., 61 (1976) 302; (d) M. Moskovits, Acc. Chem. Res., 12 (1979) 229.
27. (a) R. Ugo, Catal. Rev. 11 (1975) 225; (b) E. L. Muetterties, T. N. Rhodin, E. Band, C. F. Brucker and W. R. Pretzer, Chem. Rev., 79 (1979) 91; (d) J. W. Lauher, J. Am. Chem. Soc., 101 (1979) 2604; (e) J. C. Green, D. M. P. Mingos, and E. Seddon, Inorg. Chem., 20 (1981) 2595.
28. R. P. Eischens and W. A. Pliskin, Advan. Catal., 10 (1958) 1.
29. G. A. Ozin, W. J. Power, T. Upton and W. A. Goddard, III, J. Am. Chem. Soc., 100 (1980) 4750.
30. R. P. Messmer, "Cluster Model Theory" in T. N. Rhodin and G. Ertl (eds.), "The Nature of the Surface Chemical Bond," North-Holland, Amsterdam, 1979.
31. N. Rösch, D. Menzel, Chem. Phys., 13 (1976) 243; R. P. Messmer, S. K. Knudsen, K. H. Johnson, J. B. Diamond and C. Y. Yang, Phys. Rev., B13 (1976) 1396.

32. (a)G. A. Ozin and S. A. Mitchell, Angew. Chem. Int. Ed. Engl., 22 (1983) 674; (b) G. A. Ozin and S. A. Mitchell in M. H. Chisholm, Ed. "Inorganic Chemistry Towards the 21st Century," Amer. Chem. Soc. Symp. Ser. 211 (1983) 303.
33. W. J. Power and G. A. Ozin, Advan. Inorg. Chem. Radiochem., 23 (1980) 79.
34. J. R. Blackborow and D. Young, "Metal Vapor Synthesis in Organometallic Chemistry," Springer, New York (1979).
35. V. E. Bondybey and L. E. Brus, Advan. Chem. Phys., 41 (1980) 269.
36. (a)A. Herrmann, E. Schumacher and L. Wöste, J. Chem. Phys., 68 (1978) 2327; (b) A. Herrmann, M. Hoffmann, S. Leutwyler, E. Schumacher and L. Wöste, Chem. Phys. Lett., 62 (1979) 216; (c) T. G. Dietz, M. A. Duncan, D. E. Powers and R. E. Smalley, J. Chem. Phys., 74 (1981) 6511; (d) D. L. Michalopoulos, M.E. Geusic, S. G. Hansen, D. E. Powers and R. E. Smalley, J. Phys. Chem., 86 (1982) 3915.
37. P. P. Edwards and M. J. Sienko, Int. Rev. Phys. Chem., 3 (1983) 83.
38. K. F. Berggren, J. Chem. Phys., 60 (1974) 3399.
39. P. P. Edwards and M. J. Sienko, Phys. Rev., B17 (1978) 2575.
40. P. P. Edwards amd M. J. Sienko, J. Chem. Educ., 60 (1983) 691.
41. R. C. Baetzold and J. F. Hamilton, Progr. Solid State Chem., 15 (1983) 1.
42. P. P. Edwards and M. J. Sienko (ref. 37) pointed out that Silsbee first made this point in 1968.
43. (a)G. Ertl, Catal. Rev.–Sci. Eng., 21 (1980) 201; (b) W. Y. Wen, Catal. Rev.–Sci. Eng., 22 (1980) 1; (c) W. D. Ross, Catal. Rev.–Sci. Eng., 25 (1983) 591.
44. See, for example, G. A. Ozin and H. Huber, Inorg. Chem., 18 (1979) 1402.
45. W. H. Gerber and E. Schumacher, J. Chem. Phys., 69 (1978) 1692; (b) W. H. Gerber, E. Schumacher, H. P. Harri, M. Hoffmann and E. Scholl, Amer. Chem. Soc. Symp. Ser., 179 (1982) 83.
46. P. Fantucci, J. Koutecky and F. Pacchioni, J. Chem. Phys., 80 (1984) 325.
47. A. J. Buck, B. Mile and J. A. Howard, J. Am. Chem. Soc., 105 (1983) 3381.
48. J. A. Howard, K. F. Preston and B. Mile, J. Am. Chem. Soc., 103 (1981) 6226.
49. (a) J. A. Howard, K. F. Preston, R. Sutcliffe, B. Mile and J. S. Tse, Chem. Phys. Lett., 94 (1983) 561; (b) J. A. Howard, R. Sutcliffe and B. Mile, J. Phys. Chem., 87 (1983) 2268.
50. M. Moskovits, D. F. DiLella and W. Limm, J. Chem. Phys., 80 (1984) 626.
51. See for example, W. Weltner, Jr., "Magnetic Atoms and Molecules," Van Nostrand Reinhold, New York, 1983.
52. (a) C. Bachman, J. Demuynck and A. Veillard, Faraday Symp. Chem. Soc., 14 (1980) 170; (b) H. Tatewaki, E. Miyoshi and T. Nakamura, J. Chem. Phys., 76 (1982) 5073.
53. C. A. Baumann, R. J. Van Zee and W. Weltner, Jr., J. Chem. Phys., 79 (1983) 5272.
54. P. H. Kasai and D. McCleod, Jr., J. Phys. Chem. 82 (1978) 1554.
55. J. A. Howard, R. Sutcliffe and B. Mile, J. Am. Chem. Soc., 105 (1983) 1394.
56. L. B. Knight, R. J. Van Zee and W. Weltner, Jr., Chem. Phys. Lett., 94 (1983) 296.
57. R. K. Nesbet, Phys. Rev., A 135 (1964) 460; R. J. Van Zee, C. A. Baumann, S. V. Bhat and W. Weltner, Jr., J. Chem. Phys., 78 (1983) 190.
58. J. A. Howard, R. Sutcliffe and B. Mile, J. Chem. Soc. Chem. Commun. (1983) 1449.

59. T. Welker and T. P. Martin, J. Chem. Phys., 70 (1979) 5683.
60. W. Schulze, H. W. Becker and H. Abe, Chem. Phys., 35 (1978) 177.
61. G. A. Ozin, J. Am. Chem. Soc., 102 (1980) 3301; Faraday Soc. Symp. Chem. Soc., 14 (1980) 7.
62. R. F. Marzke, Catal. Rev.–Sci. Eng., 19 (1979) 43.
63. D. Hermerschmidt and R. Haul, Ber. Bunsenges, Phys. Chem., 84 (1980) 902.
64. G. A. Ozin and M. P. Andrews, J. Phys. Chem., in press.
65. (a) P. H. Kasai, J. Am. Chem. Soc., 106 (1984) 3069 and references cited therein; (b) P. Kasai, J. Phys. Chem., 86 (1982) 3684; (c) P. H. Kasai and D. McCleod, Jr., J. Am. Chem. Soc., 97 (1975) 6602.
66. G. A. Ozin, S. A. Mitchell and J. G. Prieto, unpublished results.
67. W. Dyson and P. Montano, Solid State Commun., 33 (1980) 191.
68. L. Guczi, Catal. Rev.–Sci. Eng., 23 (1981) 329.
69. W. Dyson and P. Montano, Phys. Rev., B20 (1979) 3619.
70. (a) S. Shamai, M. Pasternak and H. Micklitz, Phys. Rev., B26 (1981) 3031; (b) L. M. Pecora and P. Ficalora, J. Solid State Chem., 27 (1979) 239; L. Brewer and J. S. Winn, Faraday Symp. Chem. Soc., 14 (1980) 126.
71. H. M. Nagarathna and P. Montano, J. Chem. Soc. Faraday Trans. II, 78 (1982) 923; (a) M. D. Morse and R. D. Smalley, Ber. Bunsenges. Phys. Chem., 88 (1984) 228.
72. M. Moskovits and D. DiLella, J. Chem. Phys., 72 (1980) 2267; D. P. DiLella, W. Limm, R. H. Lipson, M. Moskovits and K. V. Taylor, J. Chem. Phys., 77 (1982) 5263.
73. D. D. Stanz and R. K. Khanna, J. Chem. Phys., 74 (1981) 2116.
74. D. P. DiLella, K. V. Taylor and M. Moskovits, J. Phys. Chem., 87 (1983) 524.
75. V. E. Bondybey and J. H. English, J. Chem. Phys., 73 (1980) 42.
76. H. Sontag ant R. Weber, Chem. Phys., 70 (1980) 23.
77. G. A. Ozin, M. D. Baker, S. A. Mitchell and D. F. McIntosh, Angew. Chem. Int. Ed. Engl., 22 (1983) 166.
78. S. B. Piepho and P. N. Schatz, Group Theory in Spectroscopy with Applications to Magnetic Circular Dichroism, Wiley, New York, 1983.
79. R. Grinter, S. Armstrong, U. A. Jayasooriya, J. McCombie, D. Norris and J. P. Springall, Faraday Symp. Chem. Soc., 14 (1980) 94.
80. G. A. Ozin, H. Huber and S. A. Mitchell, Inorg. Chem., 18 (1979) 2932.
81. W. Schulze, H. U. Becker, R. Minkwitz and K. Manzel, Chem. Phys. Lett., 55 (1978) 59.
82. J. C. Miller, R. L. Mowery, E. R. Krausz, S. M. Jacobs, H. W. Kim, P. N. Schatz and L. Andrews, J. Chem. Phys., 74 (1981) 6349.
83. J. C. Rivoal, J. S. Emampour, K. J. Zeringue and M. Vala, Chem. Phys. Lett., 92 (1982) 313.
84. K. Jacobi, D. Schmeisser and D. M. Kolb, Chem. Phys. Lett., 69 (1980) 113; J. Chem. Phys., 75 (1981) 5300.
85. G. A. Ozin and H. Huber, Inorg. Chem., 18 (1979) 1402.
86. M. Hoffmann, S. Leutwyler and W. Schulze, Chem. Phys. Lett., 40 (1979) 145.
87. G. A. Ozin and A. J. L. Hanlan, Inorg. Chem., 18 (1979) 1781.
88. M. Moskovits and J. E. Hulse, J. Chem. Phys., 67 (1977) 4271.
89. M. Moskovits and J. E. Hulse, J. Chem. Phys., 66 (1977) 3988.
90. (a) W. E. Klotzbucher and G. A. Ozin, J. Am. Chem. Soc., 100 (1978) 2262; (b) Inorg. Chem., 18 (1979) 2101.
91. G. A. Ozin and H. Huber, Inorg. Chem., 17 (1978) 155.
92. G. A. Ozin, S. A. Mitchell and J. Garcia-Prieto, J. Phys. Chem., 86 (1982) 473.
93. S. A. Mitchell and G. A. Ozin, J. Phys. Chem., 88 (1984) 1425.

94. C. F. Bohren and D. R. Huffman, "Absorption and Scattering of Light by Small Particles," Wiley, New York, 1983.
95. R. C. Cate, J. G. Wright and N. E. Cusack, Phys. Lett., 32 (1970) 467.
96. N. Endo, A. I. Eatah, J. C. Wright and N. E. Cusack, J. Phys. Soc. Japan, 34 (1973) 666.
97. K. F. Herzfeld, Phys. Rev., 29 (1927) 701.
98. M. Moskovits and G. A. Ozin in "Vibrational Spectra and Structure," J. Durig, Ed., Elsevier, Amsterdam (1976).
99. (a) G. Blyholder, J. Phys. Chem., 68 (1964) 2772; (b) G. Blyholder and M. C. Allen, J. Am. Chem. Soc., 91 (1969) 3158.
100. A. B. Anderson, J. Chem. Phys., 66 (1977) 5108.
101. D. W. Bullett and E. P. O'Reilly, Surf. Sci., 89 (1979) 274.
102. J. Pritchard and M. L. Sims, Trans. Faraday Soc., 70 (1969) 427.
103. D. Post and E. J. Baerends, J. Chem. Phys., 78 (1983) 5663.
104. N. D. S. Canning and R. J. Madix, J. Phys. Chem., 88 (1984) 2437.
105. F. Frank, W. Schulze, B. Tesche and F. W. Froben, Chem. Phys. Lett., 103 (1984) 336.
106. D. McIntosh and G. A. Ozin, J. Am. Chem. Soc., 98 (1976) 3167.
107. A. E. Hughes and S. C. Jain, Advan. Phys., 28 (1979) 717.
108. G. Challa, J. Mol. Catal; 21 (1983) 1; C. U. Pittman, "Comprehensive Organometallic Chemistry," Vol. 8, p. 553, Pergamon, Oxford (1982).
109. D. Shuttleworth, J. Phys. Chem., 84 (1980) 1629.
110. S. Boonthanum and M. White, Thin Solid Films, 24 (1974) 295.
111. M. A. Lunina and Yu A. Novozhilov, Colloid J., USSR, (1969) 370.
112. G. A. Ozin, C. G. Francis, H. X. Huber, M. Andrews and L. F. Nazar, Coord. Chem. Rev., 48 (1983) 203.
113. G. A. Ozin, CHEMTECH, in press, 1985; G. A. Ozin and M. P. Andrews, J. Phys. Chem., in press.
114. C. G. Francis and P. L. Timms, J. Chem. Soc., Dalton (1980) 1401.
115. G. A. Ozin, C. G. Francis, H. X. Huber, M. Andrews and L. F. Nazar, J. Am. Chem. Soc., 103 (1981) 2453.
116. C. G. Francis, H. Huber and G. A. Ozin, Inorg. Chem., 19 (1980) 219.
117. C. G. Ozin and C. G. Francis, J. Mol. Struct., 59 (1980) 55.
118. T. A. Ford, H. Huber, W. Klotzbucher, E. P. Kundig, M. Moskovits and G. A. Ozin, J. Chem. Phys., 66 (1977) 524; C. Cosse, M. Fouassier, T. Mejean, M. Tranquille, D. P. DiLella and M. Moskovits, J. Chem. Phys., 73 (1980) 6076; M. P. Andrews and G. A. Ozin, J. Phys. Chem., in press.
119. M. Andrews, G. A. Ozin and S. Mattar, J. Am. chem. Soc., 105 (1983) 6170.
120. M. Moskovits and J. E. Hulse, J. Chem. Soc. Faraday II, 73 (1977) 471.
121. G. A. Ozin and M. P. Andrews, Angew. Chem. Suppl., (1982) 1255.
122. G. A. Ozin and M. P. Andrews, Angew. Chem., Suppl. (1982) 381.
123. G. A. Ozin and C. G. Francis, J. Macromol. Sci., A 16 (1981) 167.
124. M. M. Kappes and R. H. Staley, J. Phys. Chem., 86 (1982) 1332.
125. E. L. Muetterties, J. R. Bleeke, E. J. Wucherer and T. A. Albright, Chem. Rev., 82 (1982) 499.
126. T. W. Smith and D. Wychik, J. Phys. Chem., 84 (1980) 1621.
127. Ref. 26a, p. 335.
128. J. D. Atwood, T. S. Janick, J. L. Atwood and R. D. Rogers, Synth. React. Inorg. Met-Org. Chem., 10 (1980) 397.
129. L. Knoll, K. Reiss, J. Schoefer and P. Klupfers, J. Organomet. Chem., 193 (1980) C40.
130. D. J. Brower and C. Kruger, Inorg. Chem., 15 (1976) 2511; J. Muller, W. Holtzinger and F. Kohler, Chem. Ber., 109 (1976) 1222.

131. A. W. Duff, K. Jonas, R. Goddard, H. J. Kraus and C. Kruger, J. Am. Chem. Soc., 105 (1983) 5479; K. Jonas, G. Kope, L. Schieferstein, R. Mynott, C. Kruger and Y. H. Tsay, Angew. Chem. Suppl. (1983) 920.
132. L. F. Nazar, Ph.D. Thesis, University of Toronto, 1985.
133. C. F. Francis, Ph.D. Thesis, University of Bristol, 1978.
134. K. Klabunde and Y. Tanakata, J. Mol. Catal., 21 (1983) 57.
135. G. A. Ozin and M. P. Andrews, U. S. patent application, filed Dec., 1982.
136. G. Wilke, Angew. Chem., Int. Ed. Engl., 2 (1963) 105; B. Bogdanovic, M. Kroner and G. Wilke, Justus Liebigs Ann. Chem., 1 (1966) 699.
137. M. Green, J. A. K. Howard, J. L. Spencer and F. G. A. Stone, J. Chem. Soc., Dalton Trans. (1977) 364; F. G. A. Stone, Inorg. Chim. Acta, 50 (1981) 33; Acc. Chem. Res., 14 (1981) 318.
138. M. Ichikawa, J. Chem. Soc. Chem. Commun. (1976) 11.
139. A. K. Smith, F. Hughes, A. Theolier, J. M. Basset, R. Ugo, G. M. Zanderighi, J. L. Bilhou, V. Bilhou-Bougnol and W. F. Graydon, Inorg. Chem., 18 (1979) 3104.
140. J. M. Basset and A. Choplin, J. Mol. Catal., 21 (1983) 95.
141. K. J. Klabunde and Y. Imizu, J. Am. Chem. Soc., 106 (1984) 2721.
142. G. A. Ozin, F. Hugues, and L. F. Nazar, U. S. patent application, filed in Dec., 1982.
143. A. D. Berry, Organometallics, 2 (1983) 895.
144. B. A. Scott, R. M. Plecenik, G. S. Cargill, J. R. McGuire and S. R. Herd, Inorg. Chem., 19 (1980) 1252.
145. G. A. Ozin and S. Woo, J. Catal., submitted.
146. D. W. Sindorf and G. A. Maciel, J. Phys. Chem., 87 (1983) 5516.
147. G. A. Ozin, L. F. Nazar and P. Montano, unpublished observations.
148. R. L. Garten, J. Catal., 43 (1976) 18.
149. P. R. Radford and C. G. Francis, J. Chem. Soc. Chem. Commun., 6 (1983) 1520.
150. S. N. Frank and A. J. Bard, J. Phys. Chem., 81 (1977) 1484; J. Am. Chem. Soc., 99 (1977) 303; B. Aurian-Blajeni, M. Hallmann and J. Manassen, Solar Energy, 25 (1980) 165.
151. G. N. Schrauzer and T. D. Guth, J. Am. Chem. Soc., 99 (1977) 7189.

Chapter 8

METAL CLUSTERS AND ZEOLITES

P. A. JACOBS

8.1 INTRODUCTION

This chapter is concerned with metal clusters and related structures in the molecular-scale pores of zeolites. The preparation, modification, physical, physico-chemical, chemical, and catalytic characterization of metals associated with zeolite phases have been thoroughly investigated and reviewed regularly:

(1) An overview of the catalytic properties of metal-containing zeolites by Minachev and Isakov [1] includes older work with emphasis on Russian literature. The same authors have also reviewed recent developments in the field [2].

(2) Reviews of the preparation, modification, and catalytic properties of metal-containing zeolites up to 1977 were reported by Jacobs [3] and Uytterhoeven [4].

(3) The formation and potential uses of nickel clusters in zeolites have received a great deal of attention, as summarized by Delafosse in 1980 [5].

(4) In 1982 a symposium was devoted to a review of the formation, properties, and applications of metal microstructures in zeolites [6].

(5) The physical characterization of Pt- and Pd-faujasites is in large measure the result of fundamental work by Gallezot *et al.*; this aspect of metals in zeolites has also been reviewed [7].

(6) The characterization of iron and nickel zeolites by physical methods has been reviewed by Schmidt *et al.* [8].

(7) The potential uses of metal-containing zeolites in synthesis gas conversion and in carbonylatien reactions have been examined by Jacobs [10] and Ben Taarit *et al.* [11, 12], and a summary of the U. S. patents dealing with this subject appears in the book by Scott [13].

(8) All aspects of nonacid catalysis with zeolites, including metal catalysis, have been reviewed by Maxwell [14].

This short enumeration of reviews treating various aspects of metals in zeolites shows that the matter is complex and the approach is always interdisciplinary in nature. It also follows that it will be difficult for a newcomer to enter the field and almost impossible for him to judge the relevance of a particular piece of research in the area. Therefore, it is the aim of this chapter to approach the subject in a basic and general way. Consequently, the literature survey is not exhaustive, but intended to highlight key references.

8.2 IMPORTANCE OF METAL-CONTAINING ZEOLITES IN INDUSTRIAL CATALYSIS

Metal-containing catalysts are used in large-volume petroleum refining as well as in the production of petrochemicals. Typical catalysts are listed in Table 1. From this table the following conclusions can be drawn:

(1) In industrial practice, alumina is the preferred support for metal catalysts.

(2) Metal-containing zeolites are used exclusively in refinery applications and still find no use in petrochemical applications.

Table 1 also shows that industrial uses of metal-containing zeolites are confined to acidic zeolites loaded with Pt or Pd. The industrial applications of bifunctional catalysis with zeolites has been reviewed by

TABLE 1. Catalysts for Petroleum Refining and Petrochemical Processes

Process	Typical catalyst
Hydrodesulfurization	Cobalt-molybdenum or nickel molybdenum on alumina
Naphtha hydrogenation	Group VIII metals on supports
Reforming	Platinum/rhenium on alumina
Hydrocracking	Platinum on acid wide pore zeolites (faujasite-type)
Hydrodewaxing	Platinum on acidic medium-pore zeolites (ZSM-5 type)
Octane upgrading	Group VIII metals on faujasite zeolites
Ammonia synthesis	Iron
HCN synthesis	Platinum on corundum
Alkene oligomerization	Nickel on alumina
Methanation	Nickel
Fischer-Tropsch synthesis	Cobalt, iron, ruthenium
Acetone from isopropanol	Copper
Cyclohexane	Nickel on alumina
Linear alkylbenzene synthesis	Platinum on alumina

Bolton [15] and Jacobs [3]. The only function of the metal phase in this type of catalyst is to establish a fast but unselective hydrogenation/dehydrogenation equilibrium. In the case of hydrocarbon feedstocks, alkenes are readily generated this way, and the feedstock becomes susceptible to proton attack. Simple metal catalysis with zeolites as supports is not yet industrial practice.

Mechanistically, metal catalysis in industrial applications involves the following rearrangements:

(1) hydrogenation of organic functions
(2) dehydrogenation of paraffinic C-C bonds
(3) double bond or skeletal isomerization in hydrocarbons
(4) hydrogenolysis or carbon-carbon bond scission in hydrocarbons.

Each of these catalytic acts can be performed on metals associated with any support. In the case of structure-insensitive reactions, a gain in catalytic activity is to be expected when zeolites are used as supports, provided that the support is able to generate higher and more stable metal dispersions. For structure-sensitive reactions, not only a better activity but also changes in selectivity may be expected when zeolites are used as supports.

8.3 ASSOCIATION OF METAL PRECURSORS WITH A ZEOLITE MATRIX

8.3.1 Generation of transition-metal ion zeolites by stoichiometric ion-exchange processes

Zeolites are three-dimensional inorganic cation exchangers. It is therefore a straightforward matter to prepare metal containing zeolites by exchange with reducible transition-metal ions. Depending on the zeolite structure, the cation exchange capacity (CEC) and the number and nature of the cation sites in the structure are to be considered as variable parameters. All available structure types are represented in the atlas of zeolite structure types shown by stereo drawings [16]. All the possible cationic extra-framework sites have been compiled by Mortier [17].

In order to be able to fix a predetermined amount of transition metal ion in a zeolite structure, it is very useful to know the equilibrium ion-exchange isotherms. Typical isotherms are shown in Fig. 1.

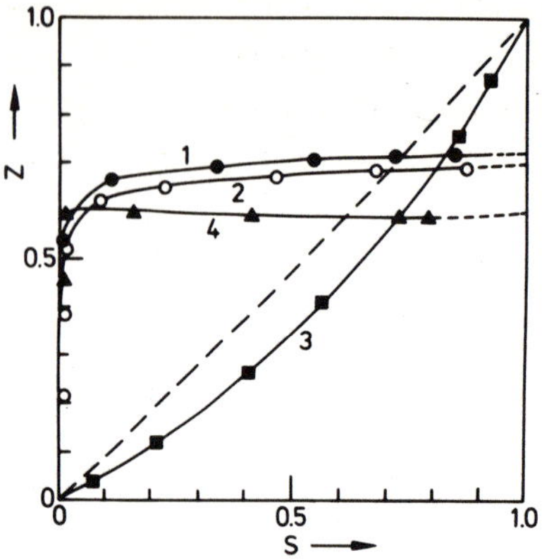

Fig. 1. Ion-exchange isotherms of transition-metal ions and their complexes in zeolites at 0.1 total molarity and a solid:liquid ratio of 1 gdm^{-3}: (1) Ni^{2+} in NaY; (2) $Pt(NH_3)_4^{2+}$ in NaY; (3) Cu^{2+} in Na-ZSM-5; and (4) $Ru(NH_3)_6^{3+}$ in NaY.

Exchange isotherms with other ions and zeolites are available [18-25]. This information can be summarized as follows:

(1) In all the zeolites [e.g., A, X, Y, mordenite (MOR), and chabasite (CHA)] except one (ZSM-5), transition-metal ions show a high affinity for the zeolite; for degrees of ion exchange not exceeding 50% of the CEC, all ions are fixed in the matrix, as can be observed from the rectangular form of the isotherm. The same is true with the ammine complexes of these ions.

(2) Polyvalent transition-metal ions show lower selectivity for ZSM-5 than the Na ions originally present [25]. This observation has been attributed to the high hydration energy of polyvalent ions and to the high degree of dilution of the exchange sites in the zeolite matrix.

(3) The exchange isotherms in Y, X, and MOR zeolites are incomplete, which is related to the existence of sites which are hidden from direct exchange and which place a limit on the total loading of metal which can be introduced in this way.

(4) The isotherms indicate a reversible process and show that under these conditions a stoichiometric cation exchange has taken place. For the preparation of metal-zeolite catalysts, the knowledge of these limiting values is of primary importance. As shown in Table 2, the maximum metal loading which can be achieved via stoichiometric ion exchange amounts to

about 10% for metals such as Ni, Co, and Cu. For the Group VIII metals, the maximum loading even on ZSM-5 is sufficient for most catalytic applications, provided that the metal ions are completely reducible. It is stressed that this technique allows preparation of metal-containing zeolites in a reproducible way, irrespective of the amount of zeolite used. This is much less the case for the other methods, which are discussed below.

TABLE 2. Characteristics of Transition Metal-containing Zeolites prepared by Cation Exchange at Room Temperature with dilute Aqueous Solutions.[a]

Transition metal ion or complex	Zeolite	Maximum exchange,%CEC	Maximum metal loading[b], wt%	Ref.
Co^{2+}	NaX	65	12.0	[19]
$Co(NH_3)_6^{3+}$	NaMOR	46	8.5	[22]
Ni^{2+}	NaY	76	9.4	[21]
	ZSM-5[c]	100	0.9	[25]
Cu^{2+}	NaY	74	9.9	[21]
	ZSM-5[c]	100	1.1	[25]
$[Cu(NH_3)_4(H_2O)_2]^{2+}$	NaMOR	48	3.7	[22]
Ag^+	NaY	100	45.4	[21]
	ZSM-5[c]	100	3.6	[26]
$Ag(NH_3)_2^+$	NaMOR	75	34.0	[22,23]
$Pt(NH_3)_4^{2+}$	NaY	70	28.7	[24]
	NaMOR	62	14.5	[25]
$Pd(NH_3)_4^{2+}$	NaY	60	13.4	[24]
	NaX	70	23.7	[24]
	NaMOR	60	7.6	[24]
		60	8.6	—

[a]0.05<u>M</u> for the ions and 0.1 equiv. dm^{-3} for the complexes.
[b]On a water-free basis.
[c]With SiO_2/Al_2O_3 = 400.

8.3.2 Preparation of transition metal containing zeolites by nonstoichiometric ion exchange

Attempts to increase the amount of a transition metal in a zeolite by successive exchanges, by increasing the exchange temperature, or by working with more concentrated solutions or higher solid-to-liquid ratios often result in at least partial hydrolysis of the transition metal ions in the zeolite. This hydrolysis is reflected by the irreversible nature of the ion-exchange process. The hydrolyzed metal species have an increased reducibility compared with that of metal ions located in lattice sites, and their existence gives rise to sample properties which are not easily reproduced [26].

The influence of preparation conditions on the excess of Ni and Cu retained in zeolites X and Y has been studied systematically [26]. The most significant effects are summarized in Table 3. Thus, it seems that stoichiometric ion exchange is possible only under well-defined conditions; however, some authors [27] even claim that small amounts of hydrolyzed transition metal ions are always formed on the outer surfaces of the zeolite crystals.

TABLE 3. Influence of Ion-exchange Conditions on the Amount of Nonstoichiometrically held Transition metal Ions (adapted from ref. [26]).

Sample	Concentration of exchange solution, mol dm^{-3}	Zeolite/liquid ratio, g dm^{-3}	Exchange pH	Temperature, K	$(TI)^{2+}$ + Na^{+}, % of CEC
NiY	0.1	6	5.0	293	100.7
NiY	0.1	6	6.0	293	101.4
NiY	0.1	6	7.0	293	104.9
	0.1	1	5.0	293	96.7
	0.1	1	6.0	293	100.0
	0.1	1	7.0	293	109.2
	0.1	6	5.0	363	125.9
	0.5	12	5.0	363	112.5
NiX	0.1	6	6.0	293	109.2
	0.1	6	5.0	363	102.9
CuY	0.01	1	5.2	293	103.3
	0.1	20	4.6	293	137.9
	0.1	6	5.0	reflux	155.5
	1.0	5	2.9	293	244.5

8.3.3 Competitive ion exchange

Even with stoichiometrically exchanged transition-metal-containing zeolites, after drying and reduction with molecular hydrogen, an inhomogeneous distribution of the metal phase may be formed [31, 32]. This nonuniformity was demonstrated for bifunctional zeolite catalysts containing relatively small amounts of Pd and/or Pt. Indeed, during exchange of $Pt(NH_3)_4^{2+}$ ions in NH_4Y or NH_4-MOR carried out under conditions such that the number of available ion-exchange sites is in excess, the rate of cation exchange is much higher than the rate of diffusion of the complex through the zeolite pores [31, 32]. After termination of the exchange, the complex will therefore be concentrated near the outer rims of the zeolite crystals, and, after reduction, the Pt will be concentrated in the same locations.

It has been shown that this is no longer the case when an excess of a second cation competes with the $Pt(NH_3)_4^{2+}$ for exchange sites. Indeed, when either NH_4^+ [31] or Ag^+ [32] ions are added in excess during the ion-exchange process, a homogeneous distribution of the Pt metal throughout the zeolite crystals is obtained. A degree of competition (x/α) has been defined as [31, 32]

$$\frac{N_o + N_z}{mM_o{}^{n+}}$$

in which N_o is the total concentration of the competing ions, N_z, the concentration of the competing ions initially present on the solid, and M_o^{n+} the total number of metal ion equivalents present in the system. The distribution of reduced Pt in the zeolite crystals is shown in Fig. 2 for different values of x/α. Optimum values of this parameter are between 200 and 300 and provide a nearly homogeneous distribution of metal in the zeolite.

8.3.4 Loading of zeolites with metal carbonyls

Metal carbonyls with sufficient vapor pressure can be adsorbed at room temperature on a previously degassed zeolite. This technique has been used several times for the preparation of Fe-Y and Ni-containing X and Y

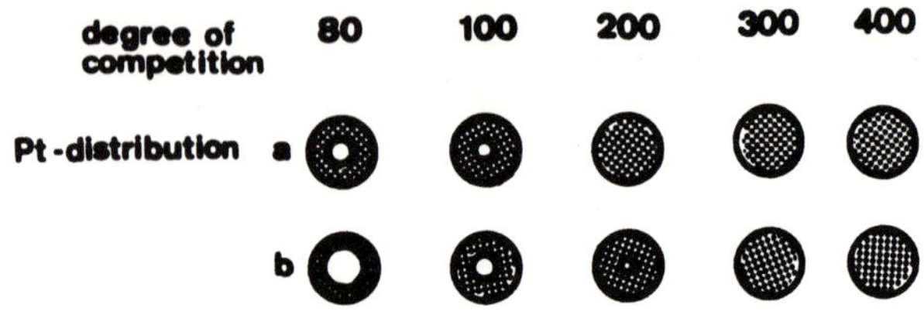

Fig. 2. Distribution of reduced Pt in zeolite crystals for different degrees of competition between $Pt(NH_3)_4^{2+}$ and NH_4^+ in zeolite Y; a represents a static ion-exchange process and b an ion-exchange process with continuous stirring.

zeolites [33-36], although only recently have adsorption isotherms become available [35,36]. From Fig. 3 it is clear that these isotherms are of a Langmuir type and that the adsorption is not completely reversible after a room-temperature degassing.

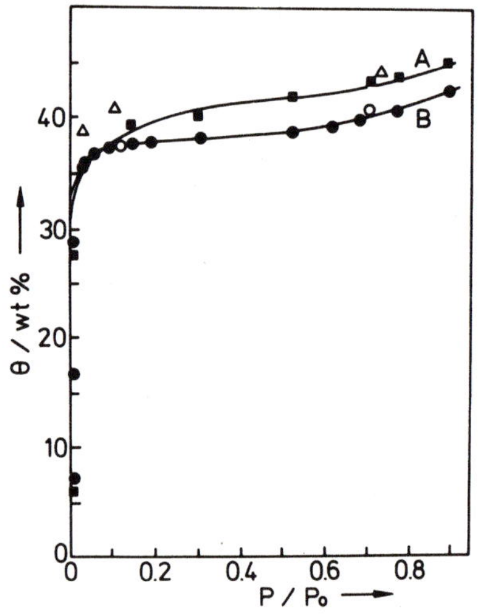

Fig. 3. Adsorption isotherms of $Fe(CO)_5$ on HY (A) and NaY (B) zeolites at 293 K. The full points were measured during adsorption and the open points during desorption [35].

TABLE 4. Adsorption of $Fe(CO)_5$ on Zeolites at Saturation Loading and Room Temperature (from ref. [36]).

Zeolite	Amount sorbed in zeolite		pore/cage dimensions[a], nm
	wt %	% of pores filled	
NaY	38.3	0.48	1.3
HY	38.9	0.46	1.3
KL	11.7	0.89	0.71 x 0.75
NaΩ	9.4	0.72	0.74 x 0.76
NaMOR	3.4	0.16	0.67 x 0.70
silicalite	0.4	0.02	0.54 x 0.56

[a] from ref. 16.

The saturation loadings of several zeolite structures which $Fe(CO)_5$ are shown in Table 4, together with the pertinent channel or pore dimensions of the zeolites. In the faujasite structure (X or Y zeolite), the large cages are filled with 3 molecules, which is consistent with the known geometries. All other cages are inaccessible to the $Fe(CO)_5$ because the molecule is too large to enter.

In zeolites L and Ω, the one-dimensional channels are filled completely, which is no longer the case for MOR. In this zeolite, the amount adsorbed remains too high for coverage of the external surface of the crystals, so that pore-mouth blocking was invoked to explain the amounts sorbed. Monolayer coverage of the external surface explains the silicalite behavior.

As an industrial preparation technique, this has serious disadvantages; $Ni(CO)_4$ is extremely toxic, and $Fe(CO)_5$ is highly sensitive to photochemical decomposition. On the other hand, heavy metal carbonyls such as $Mo(CO)_6$, $Re_2(CO)_{10}$, $Ru_3(CO)_{12}$, $Fe_2(CO)_4$, and $Fe_3(CO)_{12}$ can be easily loaded into the wide-pore zeolites via sublimation [34, 37, 38].

8.3.5 Sorption of labile organic complexes

Attempts to support neutral metal complexes on zeolites and to keep the metal encapsulated in the pores or cages have also been reported. Reaction of triallylrhodium with the lattice hydroxyl groups of the zeolite has been asserted to lead to the formation of zeolite-supported rhodium hydrides, as shown in SCHEME 1 [39, 40]. Here, ZOH stands for a protonated zeolite matrix.

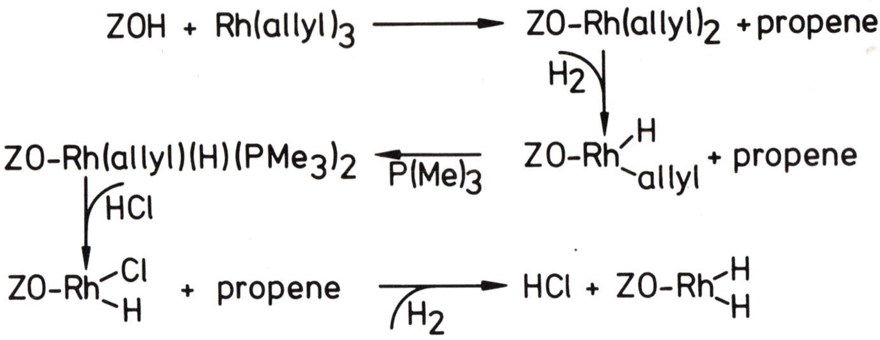

SCHEME 1

When Ni(II) dithiophosphates are refluxed in toluene or carbon tetrachloride with NH_4Y, they are irreversibly retained by the zeolite [41]:

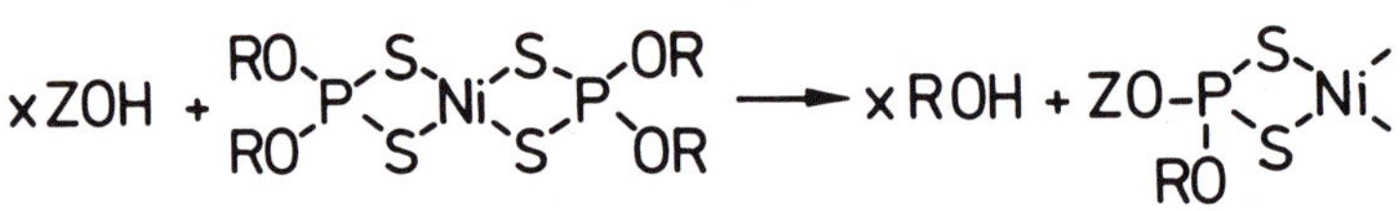

(1)

Labile organometallics can be used to transport metal atoms into zeolites [42, 43]. Bis-(toluene) metal(0) complexes were formed at temperatures <223 K and then adsorbed on a zeolite. The temperature was then gradually raised above the decomposition point of the complex, with continuous removal of the volatile ligands. The method has been demonstrated with cobalt and iron; in principle, all metals can be used, provided that the complex is small enough to enter the zeolite pores and the complex is thermally unstable. A more detailed account of this subject is provided in Chapter 7.

8.3.6 Formation of polycyano inclusion compounds

Scherrer *et al.* [44, 45] showed recently that a transition metal exchanged zeolite reacts with anionic and water-soluble cyanide complexes to form insoluble compounds distributed throughout the zeolite crystals. The chemistry is as follows [44]:

$$Fe - ZO + (NH_4)_3Fe(CN)_6 \longrightarrow (NH_4,Fe)ZO + (Fe,NH_4)Fe(CN)_6 \qquad (2)$$

A whole variety of inclusion compounds can be prepared using this method, including $Co^{II}(NH_4)_2Fe^{II}(CN)_6$ [44] and $KFe^{III}Co^{II}(CN)_6$ [46].

8.3.7 Association of large amounts of metal precursors with ZSM-5 zeolites

The deposition of significant amounts of metal precursors on ZSM-5 is not possible by ion exchange (Table 2), and therefore metal incorporation has been carried out by impregnation of metal salts by the incipient wetness technique [47] or with metal carbonyls of low volatility using a Soxhlet extraction technique [48]. For metal loadings between 5 and 10 wt%, the major amount of the metal precursor does not enter the pores of the ZSM-5, but remains external to the zeolite crystals [49]. When a physical mixture of an oxide phase (e.g., α-Fe_2O_3 [47]) was used with this zeolite and subsequently reduced again, most of the metal remained external to the zeolite, but the catalyic properties of the resulting zeolite were distinctly different from those of the same metal on a classical support [47].

8.3.8 Preparation of precursors of shape-selective metal catalysts

Small amounts of metal precursors can be brought into shape-selective zeolites by simple ion exchange: up to 3% of Pt can be brought into ZSM-5 by ion exchange with $Pt(NH_3)_4^{2+}$ [50, 51]. A more general method involves the addition of a transition metal ion or complex during the synthesis of the zeolite. In all the reported examples, the major amount of the transition metal added to the synthesis gel was occluded by the zeolite after crystallization. The following specific examples illustrate this method of preparation:

(1) Pt in CaA, 0.6 wt% [52].

(2) Pt in zeolite alpha (a high-silica analog of A), 0.35 wt% [53].

(3) Pt in silicalite, 1 wt% [54].

The crystallization-nuclei method [55] is a variation of the preceding method: a metal on alumina (or on another support) is added to provide nucleation centers for the synthesis of a zeolite. For example, Ru or Rh on alumina is added to a gel from which zeolite ZSM-34 ultimately crystallizes. Although the method looks promising, it is at this time difficult to decide whether the product is just an intimate physical mixture of the metal on the support with the zeolite or whether part of the metal becomes occluded in the zeolite.

8.3.9 Comparison of methods and their simultaneous utilization

The crystallization-nuclei method, after appropriate activation and reduction of the samples, leads to catalysts with lifetimes superior to those prepared by more conventional methods, such as physical blending, impregnation, or ion exchange [55]. Also for Ru on NaY, it was reported that catalyst behavior was strongly dependent on the preparation method of the metal precursor [56]: impregnation of aqueous $RuCl_3$ via incipient wetness, ion exchange with Ru (III)-hexammine, and vapor impregnation with $Ru_3(CO)_{12}$ produced three completely different catalysts.

Some of the methods mentioned can be applied successively or simultaneously for the preparation of multimetallic zeolite precursors. Precursors for Pt-Mo bimetallic Y zeolites are easily made by application of ion exchange with $Pt(NH_3)_4^{2+}$ followed by sublimation of $Mo(CO)_6$ into the precalined zeolite [57]. Alternatively, mixed-metal zeolite Y precursors have been made by ion exchange with mixed-metal complexes of the following type [58]:

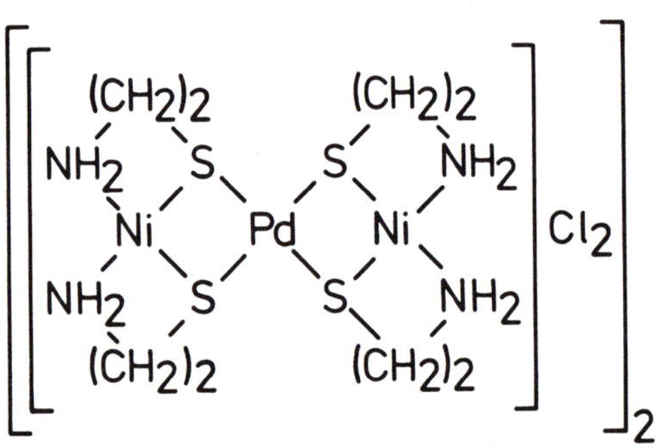

Other mixed-metal compositions which can be prepared using the same complexes are Pd-Ni-Pd, Ni-Pt-Ni, and Pt-Ni-Pt.

8.3.10 Ball-milling of elemental tellurium with zeolite X

This technique allows loading of zeolite X with up to 10% of metal [59] and produces active dehydroyclization catalysts. The retainment of Te metal is increased on a Cs-exchanged [59] NaX zeolite or on a NaX zeolite in which NaCl has been preadsorbed [60]. Although the system has some catalytic potential, the solid state chemical behavior is far from being understood.

8.4 ACTIVATION OF METAL-ZEOLITE PRECURSORS

8.4.1 Dehydration of cation-exchanged zeolites

A detailed overview of transition metal ion locations in hydrated zeolites was presented by Mortier [17]. To summarize, the hydrated metal ions are always located in the accessible sites. In order to avoid hydrolysis of waters of hydration during degassing, this treatment should consist of a slow heating so that the bare metal ions distribute themselves over the different sites, the site populations being determined by the following:

(1) Energetic and coordinative differences: In the hexagonal prism of a faujausite, octahedral coordination to lattice oxygen atoms is possible, whereas in a six-membered ring, only a one-sided trigonal coordination may exist. The stability at this site is improved when extra-lattice species formed from water hydrolysis remain in the zeolite, thereby changing the site symmetry from trigonal to tetrahedral.

(2) Competition for a site with other cations: Since most ion exchanges are not complete, at least one other cation is competing with the transition metal ion. A typical example has been reported for Ca^{2+}/Ni^{2+} exchange in NaX [61]: Consecutive or competitive ion exchange of the two ions ultimately gives catalysts with distinctly different properties. The same is also true for La^{3+}/Ni^{2+} and Ce^{3+}/Ni^{2+} combinations [5, 64, 65]. Since even for a given structure, the energy differences among the different sites are dependent upon the chemical composition [62], this whole matter will be extremely difficult to rationalize.

The degree of hydrolysis of hydrated transition metal ions in zeolites will depend not only upon the rate of temperature increase; for very low rates of temperature increase it may be determined by the geometry of the bed of zeolite. Hydrolysis of metal ions will affect the nature and location of the species which will be reduced afterwards. Under some conditions, even the zeolite matrix may become partially hydrolyzed, which ultimately may cause collapse of the structure.

A nonstoichiometrically exchanged zeolite will, upon dehydration, produce occluded oxide clusters. This may be a desired objective and can be achieved by treatment of degassed stoichiometrically exchanged zeolites with boiling aqueous Na_2CO_3-K_2CO_3 solutions [63] followed by dehydration.

8.4.2 Autoreduction of transition metal ions in zeolites

In some cases the zeolite framework itself can actively participate in solid-state redox reactions. Upon slow heating of stoichiometrically exchanged Ag^+ [66-69] and Cu^{2+} zeolites [70, 71], oxygen is desorbed from the framework, the metal ion is partially reduced, and Lewis acid sites (I) are formed. In the case of CuY zeolite, this process may be visualized as follows [71]:

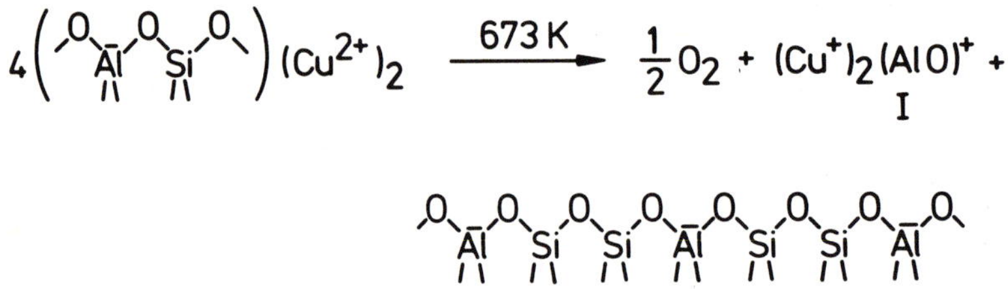

(3)

In Ag^+-zeolites of the A, X, Y, CHA, or MOR type, a whole range of reduced species is formed during autoreduction, as is evidenced by the drastic color changes of the samples [66-69]. The formation of colored centers in AgA has been known since 1962 [72], but it has only recently been associated with the existence of an autoreduction process [66]:

$$2(Ag^+ \; ZO^-) \longrightarrow \frac{1}{2}\, O_2 + 2\; Ag^{6+} + ZO^- + Z^+ \qquad (4)$$

Here, ZO^- stands for a zeolite lattice and Z^+ for a Lewis acid site. With X-ray diffraction of such samples, it was possible to locate an excess of Ag ions in sites I' [68], which was interpreted first in terms of the existence of a linear Ag_3 cluster and later a Ag_3^{2+} cluster [80]:

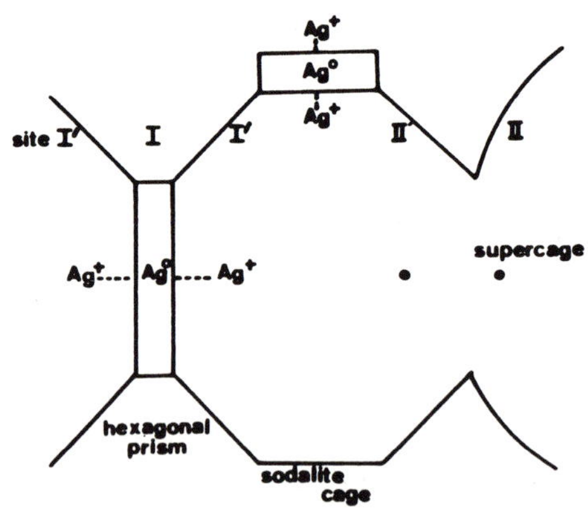

This cluster is responsible for the yellow sample color. The concentration increases with the cation concentration initially in the sample, since in AgX and Ag zeolites, a larger number of clusters is present than in AgY. Eventually or red sample color is developed, which indicates that interaction between the clusters occurs.

8.4.3 Autoreduction of transition metal ions in zeolites exchanges with ammine complexes

Noble metal ions such as Ru^{3+}, Pd^{2+}, Pt^{2+}, Rh^{2+} and Ir^{2+} are highly susceptible to hydrolysis under ion-exchange conditions. Fortunately, their ammine complexes are stable under these conditions and can therefore be used for ion-exchange purposes. For Pt- and Pd-ammine complexes in Y zeolites, it is an established procedure to perform an air oxidation prior to reduction [73,74], since this treatment produces samples with increased metal dispersions. With Ru-ammine, such an oxidizing treatment produces bulk RuO_2 located external to the zeolite [75]. Even slow degassing to 673 K of a $Ru(NH_3)_6{}^{3+}Y$ zeolite results in a degree of reduction of 0.80, caused by dissociated ammonia ligands [76]. With Pt- and Pd-ammine complexes in zeolites, the same autoreduction reaction occurs and can be quantified as follows:

$$Pt(NH_3)_6Y \longrightarrow Pt^o + 4\ HY + 2/3\ N_2 + 4\tfrac{2}{3}\ NH_3 \quad [77] \qquad (5)$$

or as follows:

$$Pt(NH_3)_4Y \xrightarrow{>473\ K} (4 - x)\ NH_3 + Pt(NH_3)_xY$$

$$Pt(NH_3)_xY \xrightarrow{473\ K} Pt^o\text{-}NH + 2\ HY + (x - 1)\ NH_3$$

$$2Pt\text{-}NH \longrightarrow 2\ Pt^o + N_2 + H_2 \quad [78] \qquad (6)$$

With Pt and Pd, degassing in oxygen seems to delay such autoreductions [77]. There is even an optimum calcination temperature which is believed to be the minimum temperature at which, in a reasonable calcination time, the ammine complex is completely decomposed [77]. Since there is evidence that intermediates such as $Pt(NH_3)_2H_2$ are responsible for sintering of Pt clusters [74], the effect of oxygen is to suppress their concentration. It follows that the optimum pretreatment conditions will also depend on sample amount and bed geometry and that reproducibility in attainment of high dispersions will be difficult. Indeed, in contrast to the above results of

Reagen *et al.* [77], Gallezot *et al.* [81] reported the complete oxidation of $Pt(NH_3)_4^{2+}$ to Pt^{2+} ions.

8.4.4 Thermal decomposition of sorbed metal carbonyls

The infrared spectra in the CO stretching region of metal carbonyls sorbed in zeolites, although quite complex [34-38], are closely similar to those of the metal carbonyls in solution and therefore allow the conclusion that most metal carbonyls sorbed at room temperature [34, 35] (or even at elevated temperature in the case of $Re(CO)_{10}$ [37]) retain their molecular structures.

When the zeolite contains residual acidity, however, a metal reoxidation reaction occurs when the metal carbonyl/zeolite adduct is heated [34, 35]:

$$[Fe(CO)_5 \cdot 2\ ZOH] \longrightarrow H_2 + Fe(II) + 5\ CO \qquad (7)$$

This chemistry is illustrated by the data of Fig. 4.

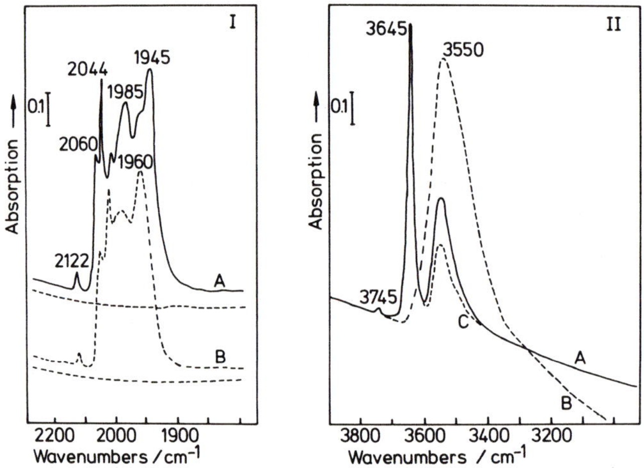

Fig. 4 I. Infrared spectra of $Fe(CO)_5$/zeolite adducts at 293 K for A, NaY and B, HY zeolites.
II. Hydroxyl spectra of the $Fe(CO)_5$/HY adducts: A, OH groups on the zeolite; B, saturated at 293 K with carbonyl; C, after heating of the adduct to 423 K [35].

The carbonyl forms weak hydrogen bonds with the 3650-cm^{-1} supercage -OH groups, shifting them to 3550 cm^{-1}. This interaction is therefore the cause of the restricted mobility found for these species [33]. At the same time, the half-band width of the adduct in the carbonyl stretching region is smaller, and the ν_1 vibration (>2100 cm^{-1}) becomes infrared active [36].

After complete decarbonylation of the sample at 423 K, the original intensity of the supercage hydroxyls is only incompletely restored [Fig. 4B], and it is estimated that up to 25% of the available iron may become oxidized [35]. The phenomenon is inferred to be quite general in nature, since it has been observed for other carbonyls as well [34, 37].

In the presence of zeolite water of hydration, a chemistry closely related to that in aqueous solution occurs involving metal carbonyls and nonacidic zeolites:

$$Fe_3(CO)_{12} \xrightarrow[NaY \cdot H_2O]{297\ K} [HFe_3(CO)_{11}] \qquad [35] \qquad (8)$$

$$[HFe(CO)_{11}]^- \xrightarrow[NaY \cdot H_2O]{333\ K} [Fe_3(CO)_{11}]^{2-} \qquad [35] \qquad (9)$$

$$3\ Fe_2(CO)_9 \xrightarrow[NaY \cdot H_2O]{333\ K\ 2} [HFe_3(CO)_{11}]^- + 3\ CO + 2CO_2 \qquad [38] \qquad (10)$$

During these reactions, the zeolite water of hydration should exhibit basic properties [35]:

$$2\ H_2O + Na^+ZO^- \longrightarrow OH^- + Na^+ + H_3O^+ + ZO^- \qquad (11)$$

Upon heating of metal carbonyl/zeolite adducts, decarbonylation occurs in discrete steps, as shown by volumetric and thermoanalytic results [34, 35, 37]. In the acidic HY zeolite, multinuclear intermediates are formed starting from mononuclear carbonyls, as is shown by the appearance of bridged CO absorbing at $<1900\ cm^{-1}$ in the infrared spectrum [34, 35]:

$$Fe(CO)_5 \xrightarrow{slow} Fe_3(CO)_{12} \xrightarrow{fast} Fe(CO) \xrightarrow{fast} Fe \qquad (12)$$

For $NaY/Fe(CO)_5$ adducts, however, only stepwise decarbonylation is observed [35]:

$$Fe(CO)_5 \xrightarrow{slow} Fe(Co)_2 \xrightarrow{fast} Fe(CO)_{0.25} \xrightarrow{slow} Fe \qquad (13)$$

Decomposition of $(CO)_9Co_3CCH_3/NaY$ adducts in inert or oxidizing atmospheres seems to occur along the same lines [79]:

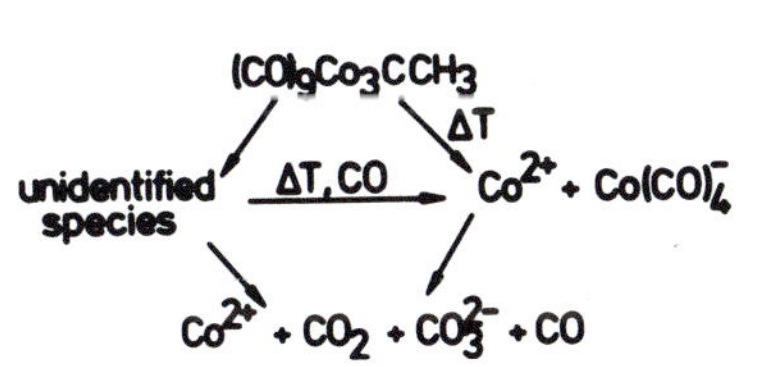

(14)

Here, successive decarbonylations are accompanied by autoxidation. When samples are prepared from adducts of $Mo(CO)_6$ [82] or $W(CO)_6$ [83] with NaY zeolite in the presence of water of hydration, the intermediate formation of the following multinuclear species is proposed:

$$[W_2(CO)_x]^+,\ [W_2(CO)_{10}]^{2-},\ [Mo_2(CO)_{10}]^{2-}.$$

In HY zeolites [82, 83], similar subcarbonyl species are also formed; in these zeolites, the metal is directly bonded to the zeolite framework, but at the same time another part of the carbonyl is completely oxidized with irreversible consumption of lattice hydroxyl groups.

8.4.5 Thermal activation of metal-ZSM-5 zeolite precursors

Activation of metal precursors associated in large quantities with the zeolite ZSM-5 gives rise to profound rearrangements in the nature and number of acid sites present on the acidic zeolite. This is easily quantified with the infrared spectra of adsorbed pyridine [47, 49]. Data characterizing this behavior are collected in Table 5. After calcination of

TABLE 5. Effect of Preparation and Activation of Metal-Zeolite ZSM-5 Precursors on the Nature of the Acid Sites in H-ZSM-5.

Catalyst	Preparation	Pretreatment	Ratio: Brønsted sites/ Lewis sites	Ref.
ZSM-5	-	Calcination	2.10	[47]
2.8 Co/ZSM-5[a]	Impregnation (Co nitrate)	Calcination	0.24	[49]
5.1 Co/ZSM-5	"	"	0.27	[49]
3.7 Co/ZSM-5	Sorption $[Co_2(CO)_8]$	"	0.38	[47]
3.8 Fe/ZSM-5	Sorption $[Fe_3(CO)_{12}]$	Calcination + Reduction	2.07	[47]
7.0 Fe/ZSM-5	"	Calcination	3.30	[47]
7.0 Fe/ZSM-5	"	Reduction	0.28	[47]

[a] wt% metal associated with the zeolite.

salt-impregnated H-ZSM-5, the surface hydroxyl groups (i.e., the Brønsted acid sites) disappear and are replaced by transition metal ions. The results suggest that an ion-exchange reaction of the following nature has occurred [47]:

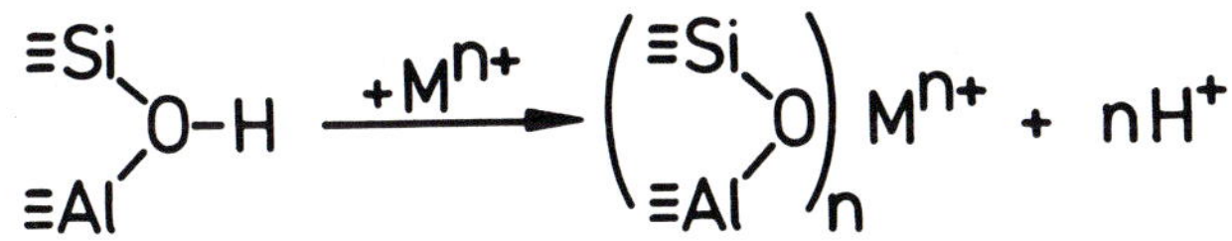 (15)

The disappearance of lattice hydroxyl groups during decarbonylation suggests that part of the Co metal is oxidized and becomes trapped in ion-exchange positions of the zeolite, which is confirmed by XPS data [47, 49]. It is presently not clear why sorbed $Fe_3(CO)_{12}$ behaves in the same way only after reduction. During calcination of such samples, the transition metal ions become fixed in cation positions. They seem to be irreducible during a hydrogen treatment [49].

8.4.6 Thermal activation of polycyano inclusion compounds

During calcination of polycyano inclusion compounds, amorphous "patches" are formed, as shown by electron microscopy [84], suggesting that an agglomeration of these compounds occurs at the external surface of the zeolite. During reduction in hydrogen, ferromagnetic metal is formed outside the zeolite crystals, as shown by Mössbauer spectroscopy, but some 30% of the total iron is superparamagnetic in nature and is either present as small metal particles in the zeolite pores or as Fe(II) ions in ion-exchange positions.

Associated with a ferro-silicate of ZSM-5 structure, these inclusion compounds do not give rise to X-ray diffraction peaks, which suggests that they are highly dispersed and therefore present at pore entrances or in grain boundaries in the crystals [46].

8.5 REDUCTION OF ACTIVATED METAL-ZEOLITE PRECURSORS

8.5.1 Reduction with molecular hydrogen

8.5.1. a Overall stoichiometry. The overall reduction stiochiometry of transition metal ions located at cation positions in zeolites is well documented and understood on a quantitative basis, at least for the most important zeolite structures such as X, Y, and MOR [3-5]. On the basis of the chemical analysis of the transition metal-containing zeolites and the amount of hydrogen consumed and water formed during reduction, the process can be pictured quantitatively. The pioneering work in this field was reported by Riekert [85]. Selected data in Table 6 illustrate the behavior.

TABLE 6. **Illustration of Hydrogen Consumption by Transition metal-containing Zeolites.**

Sample	Reduction temperature,	Degree of reduction	H_2O formed, mmol g^{-1}	Ref.
AgNaY-71[a]	623	1.00	0.00	[112]
AgNaMOR-75	643	0.93	0.00	[113]
AgNaCHA-87	343	0.55	0.00	[114]
CuNaY-10	673	0.52	0.09	[115]
CuNaY-68	673	0.65	0.32	[115]
NiNaY-68	723	0.33	--	[116]

[a]Degree ot cation exchange.

It is seen that hydrogen is consumed and only trace amounts of water are formed, except at the high reduction temperatures. Infrared spectra show the development of lattice hydroxyl groups [112-116]. These observations are of a general nature; irrespective of the nature and oxidation state of the transition metal ion (Me^{n+}) and the nature and chemical composition of the zeolite [Table 6], they imply the following overall reduction stoichiometry:

$$Me^{n+}(ZO^-)_n + \frac{n}{2}H_2 \longrightarrow Me^o + n\ ZOH \tag{16}$$

The amount of water desorbed is then a measure of the degree of zeolite dehydroxylation, which will increase with the reduction temperature. In aqueous solutions a metal chloride is reduced according to a similar reduction stoichiometry [86]:

$$MeCl_{x(s)} + \frac{n}{2}\ H_{2(g)} \longrightarrow Me_{(s)} + x\ HCl_{(g)} \tag{17}$$

Here, the subscripts s and g represent the solution and gas phase, respectively.

For metal oxides this chemistry is distinctly different:

$$MeO_x + x\ H_{2(g)} \longrightarrow Me + x\ H_2O \tag{18}$$

It follows that part of the water formed during hydrogen reduction [$(H_2O)_r$] of nonstoichiometrically exchanged transition metal-containing zeolites may be indicative of the presence of such species, so that the following relation holds:

$$(H_2O)_r = \frac{1}{x}MeO_x + L \tag{19}$$

The amount of MeO_x species can then be determined, since L (the concentration of true Lewis acid sites) can be determined from the infrared spectra of the zeolites saturated with pyridine [47, 48].

8.5.1.b Differences in reducibility of transition metal-containing zeolites

Supported oxide species in zeolites are easier to reduce than the corresponding transition metal ions in cation-exchange positions. The following experimental evidence supports this assertion:

(1) Co^{2+} and Fe^{2+} ions in zeolites in activated metal-oxide/ZSM-5 associations are irreducible, but the oxide phase is entirely reduced under the same conditions to give a metal phase external to the zeolite crystals [47-49].

(2) Fe^{2+} ions in zeolite Y are irreducible, but after a high-temperature oxidation, converting them into an oxide, this reduction occurs readily [87].

(3) Ni^{2+} ions in Y zeolites can only be 50% reduced to Ni^{0} at 673 K, whereas NiO dispersed in X or Y can be reduced completely under the same conditions [88].

This difference in behavior can be explained by thermodynamic considerations. Favorable values for the standard free energy differences in the two reduction processes are determined mainly by the ratios of the partial pressures, P_{H_2O}/P_{H_2} and P_{HCl}/P_{H_2} [86, 87]. These ratios are always lowest in the case of P_{H_2O}/P_{H_2} [86-88].

8.5.1.c Correlations in reducibility of different transition metal ions in zeolites

Differences in reducibility between transition metal ions in a given zeolite under a fixed set of conditions can also be explained by free energy considerations [87]. Uytterhoeven [4] reported the analogy between transition metal ions in zeolites and in aqueous solutions with respect to their reducibility. A qualitative relation seems to exist between the electrochemical potential of the metal ion and its reducibility in zeolite Y [4]. Below a potential of around - 0.4 V, metal ion reduction is not possible with hydrogen under easily accessible conditions of temperature and hydrogen partial pressure. Klier *et al.* [83] explained differences in reducibility on a more quantitative basis. The potential for reduction to neutral atoms [ΔE] is determined as follows [89]:

$$\Delta E = I_1 + I_2 + E_c[Me^{2+}(D_{3h})] \qquad (20)$$

I_n represents the n-th ionization potential of the metal under consideration, and E_c is the stabilization energy of the divalent metal ion Me^{2+} in a site with D_{3h} symmetry, i.e., coordinated to the three nearest oxygen atoms of a six-membered ring in faujasite zeolites. Similar data are available for bulk transition metal oxides [90], and a comparison of the data confirms the easier reducibility of the oxides.

A comparison of the electrochemical approach and the one using the reduction potential is given in Fig. 5. Except for the case of Zn^{2+}, the two approaches lead to consistent predictions. The experimental results characterizing the reducibility of Zn^{2+} in zeolites are conflicting: some authors report the reduction of Zn^{2+} ions [97] and others mention their irreducibility [47].

The reduction of Fe^{2+} ions is a limiting case. Indeed, external factors determine the reducibility of Fe^{2+}: for faujasite, the Al content and therefore the charge density in the zeolite are critical. In zeolite X (with about 85 Al atoms per unit cell), Fe^{2+} is reducible to Fe^{o} [92], whereas in zeolite Y (with only 54-56 Al atoms per unit cell) this reduction no longer occurs [87, 93].

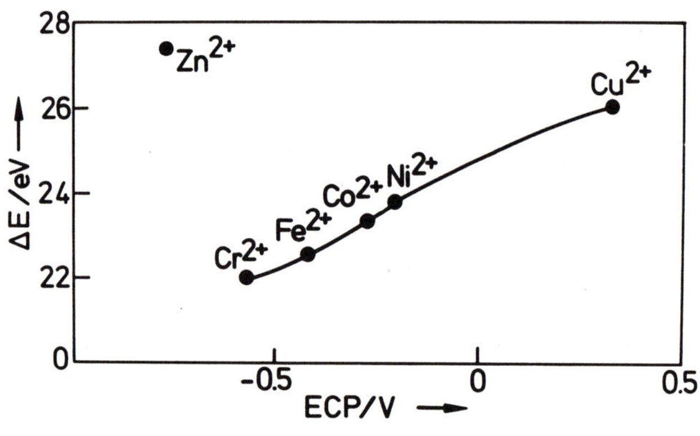

Fig. 5. Comparison of the reducibilities of transition metal ions in zeolites as measured by their electrochemical potentials, ECP [4] and their reduction potentials ΔE [89].

8.5.2 Reduction with atomic hydrogen

Provided that the reduction is not kinetically controlled, the reducibilities of all transition metal ions in a zeolite will be enhanced for thermodynamic reasons if atomic rather than molecular hydrogen is used as the reducing agent. Indeed, this will make the free energy change more negative by 360-440 kJ mol^{-1} of hydrogen involved when the reduction temperature is in the range 300-700 K [87].

In situ generation of H atoms is possible as follows:

(1) By use of zeolite-occluded Pt or Pd aggregates on which molecular hydrogen is activated, then spilling over to the metal ion to be reduced. In this way complete reduction of Ni^{2+} in faujasites is possible at lower temperatures than with molecular hydrogen, so that less sintering occurs and a better dispersion of Ni is obtained [5, 64, 94, 95].

(2) With a microwave discharge in hydrogen, H atoms can be generated *ex-situ* [96, 97]. Their lifetimes seem to be long enough to enhance reduction of the metal ions (Table 7).

TABLE 7. Reducibility of NiX Zeolite with Molecular and Atomic Hydrogen [88].

Reducing agent	Reduction temperature, K	Degree of reduction	Particle size, nm			Ref.
H_2	573	0.19	1.5	6.0	10.0	[5, 64]
H	273	0.60	1.0	-	-	[97]

8.5.3 Reduction with carbon monoxide

Naccache and Ben Taarit [98] were the first to show that Cu^{2+} ions in Y zeolite can be reduced with CO. Adding to this work the results of later studies [99, 100], the following chemistry can be written as in SCHEME 2, where the structure in square brackets is considered to be metastable.

CO treatment is also a gentle method for preparation of metal carbonyl species occluded in the zeolite starting from a transition metal-complex-exchanged material. Treatment of $Rh(NH_3)_6^{3+}$ ions exchanged into zeolite NaY with a CO/H_2 mixture at 300 K and 80 bar generates occluded rhodium carbonyl [28]. This reaction has been well characterized by infrared spectroscopy of the CO stretching region, by XP spectroscopy of the transition metal, by volumetric measurements, and by UV-visible spectroscopy [12, 101-105].

$Ru(NH_3)_6^{3+}$ exchanged into zeolite NaY is transformed in the presence of CO into the following species:

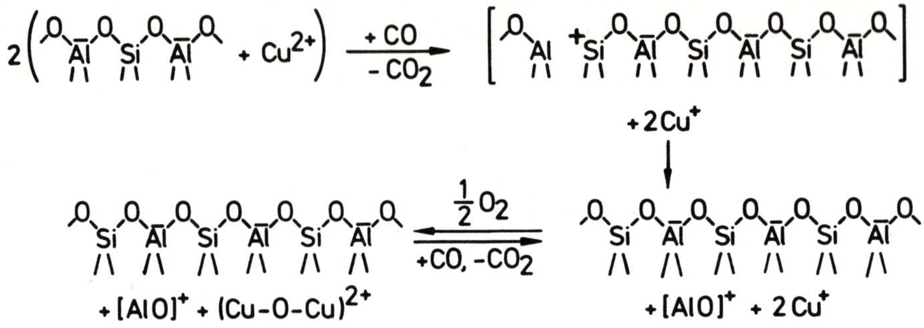

SCHEME 2

(1) Ru^{II} pentammine carbonyl formed at temperatures <373 K with

ν_{CO} at 1950 cm^{-1}

$^1A_{1g} \longrightarrow {}^1T_{1g}$ in Ru^{II} at 360 nm

$d\pi(Ru) \longrightarrow \pi^*(CO)$ and $^1A_{1g} \longrightarrow {}^1T_{1g}$ at 265 nm

with uptake of one molecule of CO and formation of one molecule of CO_2 per Ru atom.

This change occurs as follows:

$$2[Ru^{III}(NH_3)_6]^{3+} + 3CO + H_2O \xrightarrow{T \leq 373\ K} 2[Ru^{II}(NH_3)_5CO]^{2+} + CO_2 + 2NH_4^+ \qquad (21)$$

(2) Ru^I triscarbonyl coordinated to the zeolite lattice formed at temperatures between 423 and 523 K with

ν_{CO} at 2055, 2005, and 1960 cm^{-1}

UV-visible absorption at 405 nm

with uptake of 2.8 CO molecules and formation of 0.4 CO_2 molecules per atom of Ru.

Rh^{III} ions in zeolites in the presence of CO undergo the same transformations as in aqueous solution [101, 103]. An oxidized $[Rh(NH_3)_5Cl]^{2+}$-exchanged Y zeolite undergoes reduction according to the following reactions:

$$Rh^{3+} + 3CO + H_2O \longrightarrow Rh^+(CO)_2 + CO_2 + 2H^+ \quad [101] \qquad (22)$$

$$Rh_2O_3 + 7CO \longrightarrow 2Rh^+(CO)_2 + 3CO_2 \quad [103] \qquad (23)$$

with ν_{CO} at 2101 and 2022 cm^{-1}.

When the sample is treated in the presence of water at temperatures between 400 and 443 K, there is spectroscopic evidence for the following structure [103]:

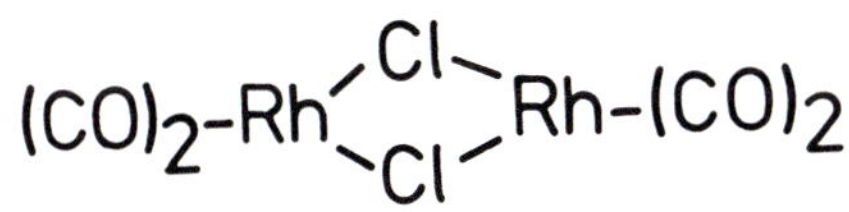

The chemistry of $[Rh(NH)_3)_5Cl]^{2+}$-exchanged faujasites is different in some details [102], and $Rh^{I}(CO)_3$ species are formed. In the presence of H_2O/CO mixtures, these monovalent metal carbonyls are further reduced, as shown by the appearance of cluster carbonyls. This cluster formation is found for Ir as well as Rh [12]:

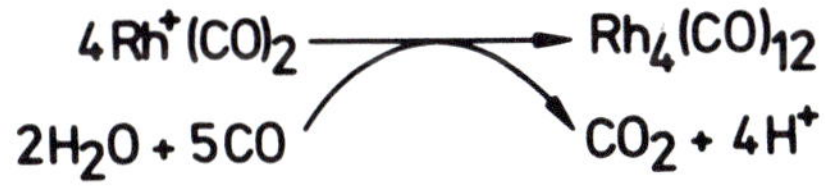

(24)

The nature of these polynuclear carbonyls is well established, since:

(1) CO_2 appearance was observed;

(2) the metal carbonyl spectra agree with those of the molecular cluster in a Nujol mull; and

(3) the ^{13}C MAS NMR shifts are identical for the cluster in the zeolite and in Nujol.

It is stressed that polynuclear carbonyls are formed under milder conditions in zeolites than in solution [12], which illustrates the excellent solvating abilities of the faujasite structure.

At temperatures >343 K, further reduction was observed for the zeolite-encaged tetrarhodium carbonyl [12]:

$$3\ Rh_4(CO)_{12} \longrightarrow 2\ Rh_6(CO)_{16} + 4\ CO \qquad (25)$$

the occurrence of this reaction is definitely not a chemical effect; it seems rather that the Rh carbonyl fits perfectly in the zeolite supercage, which is indicative of a cage effect.

8.5.4 Reduction with other reactants such as ammonia and metal vapors

Reduction of NiY zeolites with ammonia rather than with molecular hydrogen gives an increased degree of reduction but also a higher degree of metal sintering [88, 106]. A possible explanation of the difference is as follows:

(1) Ammonia removes cations from hidden sites, and zeolite impurities generate activated hydrogen through ammonia decomposition.

(2) Sintering of aminated intermediates which are also formed is fast.

(3) Reduction of Ni and Co ions in zeolites Y or A with metal vapors has also been reported [107-111]. Under these conditions, sintering of metal also occurs [108], although this may be decreased by doing the reduction in liquid ammonia [109]. In this way, metal clusters of the reducing metal are formed, but contamination of the metal phase in the presence of residual zeolite water of hydration probably results from formation of alkali metal hydroxides. A typical reaction occurs as follows [107]:

$$Ni^{2+}(ZO^-)_2 + 4\ Na \longrightarrow Ni^o + Na_4^{2+}(ZO^-)_2 \qquad (26)$$

8.5.5 Factors determining the reducibility of transition metal ions in zeolites

There is a firm body of experimental evidence demonstrating that the reducibility of transition metal ions in zeolites is determined by the following:

(1) the structure and chemical composition of the zeolite matrix,
(2) the nature and amount of co-cation,
(3) the site locations in the structure,
(4) the presence of oxidizing sites such as surface hydroxyl groups, and
(5) the presence of residual water of hydration.

However, the effects are not all understood in detail, and the available evidence is sometimes conflicting. One general warning: it is important to recognize that the metals in the zeolite may be in a dynamic equilibrium; as metal ions are reduced at certain sites, there will be a continuous rearrangement of all the residual cations together with newly formed protons over the different exchange sites, thereby keeping the system in its thermodynamically most favored condition. Since there are several parameters determining the reducibility of the metal ions and since the reduction reaction itself is exothermic (possibly causing overheating), it will not always be easy for one to reproduce the results of other research groups or to scale up the reduction process.

8.5.5. a **Effects of residual lattice hydroxyl groups.**

The reaction representing the overall reduction stoichiometry when molecular hydrogen is used as the reducing agent has been written above as an irreversible reaction. However, it seems now that the presence of surface hydroxyl groups suppresses this degree of reduction [117]. Pertinent data illustrating this behavior for HNi-MOR are shown in Fig. 6.

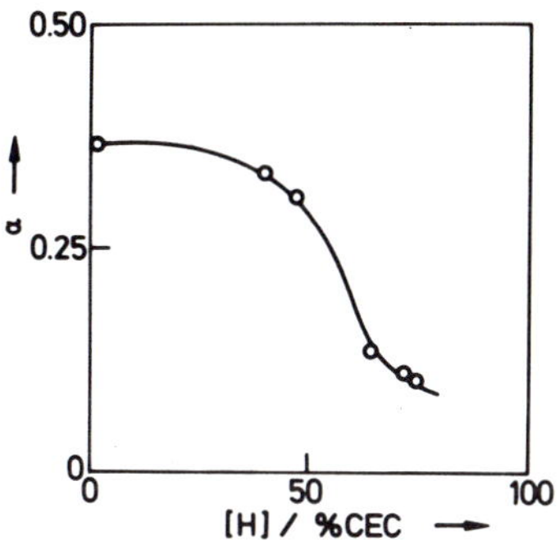

Fig. 6. Influence of the H^+ content of a HNi(23-30)-MOR sample on the degree of reduction (α) of Ni^{2+} with molecular hydrogen [117].

For this sample there seems to be a critical hydroxyl group concentration in the lattice (40 to 50% of the CEC), above which the degree of Ni^{2+} reduction is markedly decreased. This is not an effect of location of Ni^{2+} at sites which are energetically or sterically different. Indeed, Ni^{2+} ions cannot be exchanged into hidden sites of the mordenite structure, as shown by results of ion-exchange experiments [118] or CO adsorption measurements [117]. When a Ni(27)H(71)-MOR sample is gradually heated, the hydroxyl group concentration decreases steadily, but the degree of reduction of Ni^{2+} as well as the initial reduction rate remain unaltered up to a given temperature and then suddenly increase [117]. This result is illustrated in Fig. 7. The conclusion provided by these data is that a critical -OH group concentration exists, above which the Ni^{2+} ions become almost irreducible.

The oxidation of metal clusters in zeolites by hydroxyl groups has already been reported for zeolite-loaded metal carbonyls. It is clear that this phenomenon may be at the origin of the observation that zeolites exchanged to a high degree with transition metal ions do not easily undergo complete reduction, and the reduction should be considered as an equilibrium reaction [120]:

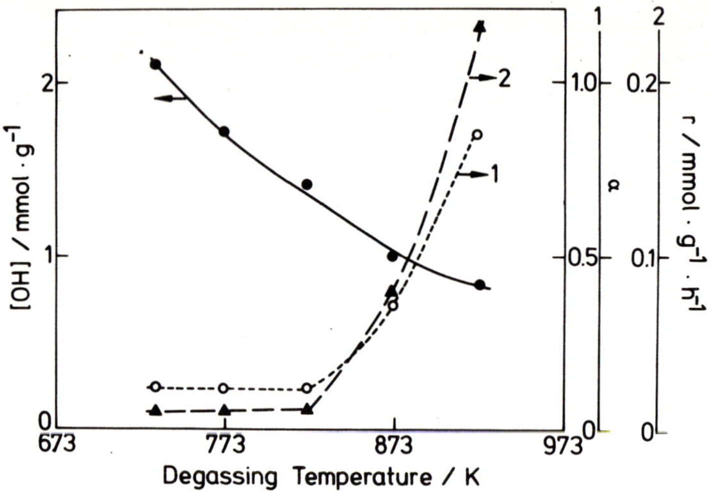

Fig. 7. Change of -OH group concentration, reducibility (r), and degree of Ni^{2+} ion reduction (α) in Ni(27)H(71)-MOR samples [117].

$$Me^{x+}(ZO^-)_x + H_2 \rightleftharpoons \frac{x}{2} ZOH + Me^0 \tag{27}$$

The back reaction had already been taken into account in one of the early kinetics studies of this subject [119].

8.5.5.b Influence of cation location. Using ferromagnetic resonance (FMR) techniques to characterize the Ni^0 species, UV-visible spectroscopy to characterize Ni^+, and infrared spectroscopy of CO sorbed on different cationic species, Lausch *et al.* [21] characterized the distribution of the various species over the cation sites in zeolite Y. However, it is emphasized that the reduction was started with an incompletely dehydrated Ni^{2+} Y zeolite, so that the results cannot be generalized.

Fig. 8 shows the changes in concentration when the reduction temperature was further increased. The results in this figure clearly indicate that during reduction there is a dynamic equilibrium of the cationic species over the different cation sites. Indeed, the Ni^{2+} which is initially abundantly present in the supercages of the zeolites either moves as such to the sodalite cages and hexagonal prisms, is hydrolyzed in the supercages, or is reduced to Ni^+ at the same location. The curves further show that the supercage species are reduced first, and this reduction is followed by reduction of Ni^{2+} in the hidden sites (i.e., in sodalite units or hexagonal prisms). The data, however, do not allow a distinction between the reducibilities of metal ions at these hidden sites.

Conflicting ideas have recently been published as to the ease of reduction of Ni^{2+} ions in hexagonal prisms of faujasites, in comparison with the other sites. It is stressed that these studies all start from completely dehydrated samples and that the irreducible co-cation was Na^+, which has a lower selectivity for sites I than Ni^{2+} [122]. For dehydrated $Ni^{2+}Y$ zeolites with variable total contents (N_T is the total number of Ni^{2+} ions per unit cell), the number of Ni^{2+} ions in the hexagonal prisms, N, was found to be related to the parameter a_0 of the cubic unit cell by the following equation [122]:

$$N = 1368 - 54.945\ a_0 - 0.808\ N_T \tag{28}$$

with limiting values of N_T between 7 and 20, values of N between 3 and 12 are obtained. It follows that on a relative basis, the sites in the hexagonal prisms are always preferentially occupied in comparison with the sodalite and supercage sites.

Using XRD techniques, Delafosse *et al.* [123] showed that these Ni^{2+} ions in the hexagonal prisms were hardly reduced in comparison with those in the other sites. However, it is expected that this selectivity will be less pronounced for faujasites with higher Al contents (zeolite X), since it is known [62] that the energy differences of the different sites gradually disappear with increasing Al content of the faujasite matrix. This expectation has been experimentally confirmed [123] (Table 8).

For Ag^+Y zeolites, upon partial reduction, X-ray diffraction shows that the hexagonal prisms are emptied selectivity (Table 8). However, the changes are the result of the formation of charged clusters which preferentially fill the sites I' in the sodalite cages [69]. Egerton *et al.* [124], using magnetic susceptibility measurements, could not confirm the preferential location of Ni^{2+} ions in the hexagonal prisms of NaY. Tetrahedral coordination of Ni^{2+} in the sodalite cages to three oxygen atoms of a six-membered ring and one oxygen atom from residual water or hydrolyzed water seems preferable to the octahedral coordination in the hexagonal prisms. These tetrahedrally coordinated cations are then less susceptible to reduction, since they are able to bind to the anions more closely than in the hexagonal prisms. These results are therefore not in contradiction to those of Gallezot and Imelik [122] and Briend-Fauré *et al.* [123], since the presence of residual water in the samples investigated by Egerton [124] is the perturbing factor.

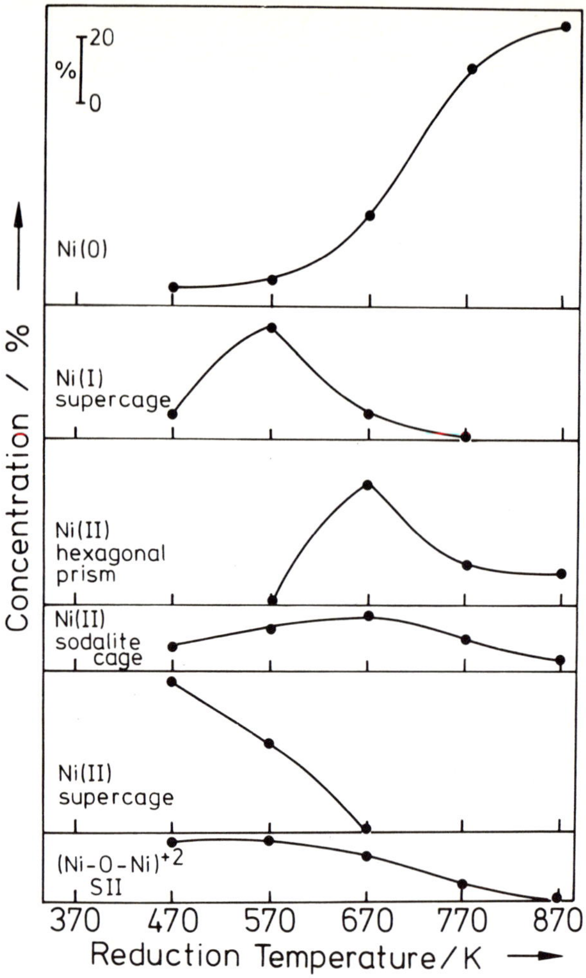

Fig. 8. Changes in concentration and location of Ni species during reduction of partially hydrated Ni^{2+} Y zeolite at increasing temperatures [121].

Also, Mintchev *et al.* [106], in a very indirect way, claimed to have demonstrated that in NiY, the ions in the hexagonal prism are easiest to reduce. Residual water may have caused this easy reduction, although the methods used to come to this conclusion may be critized. Indeed, these authors claimed that ammonia sorbed at 613 K on partially reduced faujasites is able to titrate selectively Ni^{2+} ions in supercage positions; they did not take into consideration an interaction with the lattice hydroxyl groups formed during reduction. They also presumed that

TABLE 8. Cation locations in Y zeolites before and after (partial) reduction of the transition metal ions.

Samle	Treatment temperature, K	Cation locations[a] Hexagonal prism (SI)	Sodalite SI' SII'	Supercage SIII'	Unlocalized cations	Ref.
Ag(55.5)Y	ox[b] 873	13.1	13.2 4.8	19.7	4.7	[69]
	red[c] 348	5.4	25.7 5.8	18.0	1.0	[69]
Ni(13)Na(27)Y	vac[d] 773	10.9(Ni)	1.7(Ni)	25.0(Na)	2.0(Na)	[123]
	red 733	9.0(Ni)		22.0(Na)	5.0(Na) 4.0(Ni)	[123]
Ni(31)Na(24)X	vac 773	11.7(Ni)	9.6(Ni)	24.0(Na)[e]	9.7(Ni)	[123]
	red 713	9.6(Ni)	7.6(Ni)	25.0(Na)	13.8(Ni)	[123]

[a]Cations per unit cell.
[b]Oxidation.
[c]Reduction.
[d]Under vacuum.
[e]In the original article, it was stated that 34 Ni ions are located here; this is clearly a printing error.

ammonia is unable to induce cation migration from hidden sites to supercage sites, although there is experimental evidence of this [125].

Thus it seems as if the dependency of the reducibility of a transition metal ion in a faujasite structure on its location may be obscured by factors such as thermal history of the sample before reduction and the presence of residual water before or during the reduction.

8.5.5.c **Influence of irreducible polyvalent cations.** This influence may be rationalized in terms of competition for the same sites between transition metal ions and polyvalent ions which are irreducible, or in terms of formation of mixed-oxide species having lowered mobility (which may be hidden in, e.g., sodalite cages) or having decreased reducibility. This influence has been investigated mostly with nickel-exchanged faujasite-type zeolites, and the co-cations considered are Ca^{2+}, La^{3+}, and Ce^{3+} [5, 61, 64, 65, 123, 126-129].

In mixed Ni^{2+}-La^{3+} cationic forms of NaX, the Ni^{2+} ions under completely anhydrous conditions preferentially fill the hexagonal prisms [65, 128]. During reduction, the hexagonal prisms are progressively emptied and La^{3+} ions then occupy the SI' sites in the sodalite cages, since for electrostatic reasons both types of sites cannot be fully occupied at the same time. Comparison of these cation location data with kinetics data teaches that the reduction of Ni^{2+} in the hexagonal prisms occurs with an apparent activation energy (E_a) of 117 kJ mol^{-1}, whereas the reduction in the supercages requires an energy of 200 kJ mol^{-1} [65]. In NaNiX, all ions are reduced with E_a = 117 kJ mol^{-1}. It is assumed that the formation of mixed cation forms of the type

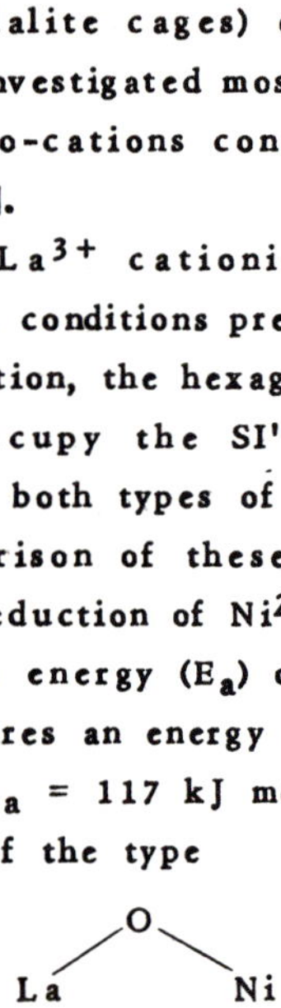

during the sample dehydration is responsible for the decreased reducibility.

On the other hand, in mixed Ni^{2+}-Ce^{3+} cationic forms of the same zeolite, such species do not seem to be formed [127]; upon dehydration, the Ce^{3+} ions are preferentially fixed in the hexagonal prisms, thus forcing the Ni^{2+} ions to be located to a greater extent in the supercages and increasing the reducibility of the latter ions. First, it is strange that with Ce^{3+} a distinctly different chemistry occurs in the same zeolite. Second, the conclusions derived for NiLaNaX are true only if the E_a values reflect differences in the kinetics and not in diffusional behavior, as will be discussed later. Third, it is strange that such pronounced differences in cation location are observed, since the same authors [122] as well as others [62] have shown that in X zeolites all sites are energetically equivalent, so that no such site preferences would be expected. Furthermore, still other authors [130] have shown that in NaY zeolites, where competition of the ions for the different sites is to be expected, the presence ot Ce^{3+} ions has exactly the opposite effect on the reducibility of Ni^{2+} ions. It follows that the effect of Ce^{3+} observed in NiNaX zeolites cannot be a mere location effect, although it is not clear by which other parameter the reduction might have been dominated.

For CaNiNaX [129] as well as Y zeolites [130], the reducibility of Ni^{2+} is enhanced in the presence of Ca^{2+} ions. In the absence of any species of the form Ni-OH in tetrahedral coordination [124], the authors' [129] explanation that Ni^{2+} in the hexagonal prisms of X zeolites is easier to reduce is again difficult to rationalize. The presented relation between the occupation of site I and the relative activity of the reduced sample in the CO/H_2 reaction provides us only with indirect arguments. Indeed, it requires that there be no particle size effect on the activity and that the specific metal surface be linearly related to the Ni content in SI before reduction. The second relation has not yet been demonstrated, and in CO hydrogenation reactions the reaction rate generally increases with increasing metal particle size [10].

It seems, therefore, more attractive at this stage to ascribe preferential reduction in the hexagonal prisms to the formation of mixed cation oxide species in the sodalite cages and supercages. The extent to which this occurs should be intimately related to the amount of water present at high temperatures, either during activation or reduction.

8.5.6 Attempt to rationalize reducibilities of transition metal ions in zeolites

It emerges from the above consideration that reduction of transition metal ions in zeolites is a result of the simultaneous action of several parameters, each of which is not always easily controllable. In an attempt to rationalize this reduction behavior, reducibility is quantified in two different ways: (1) as the initial rate of reduction, r_i, at a given temperature, i.e., the slope of the tangent line through the origin of the degree ot reduction vs. time curve, and (2) as the degree of reduction, α, at a given temperature and after a given reduction time.

The two parameters do not necessarily vary in a parallel way. In order to take into account the differences in cation composition, the average electronegativity of the zeolite according to Sanderson (S) has been used

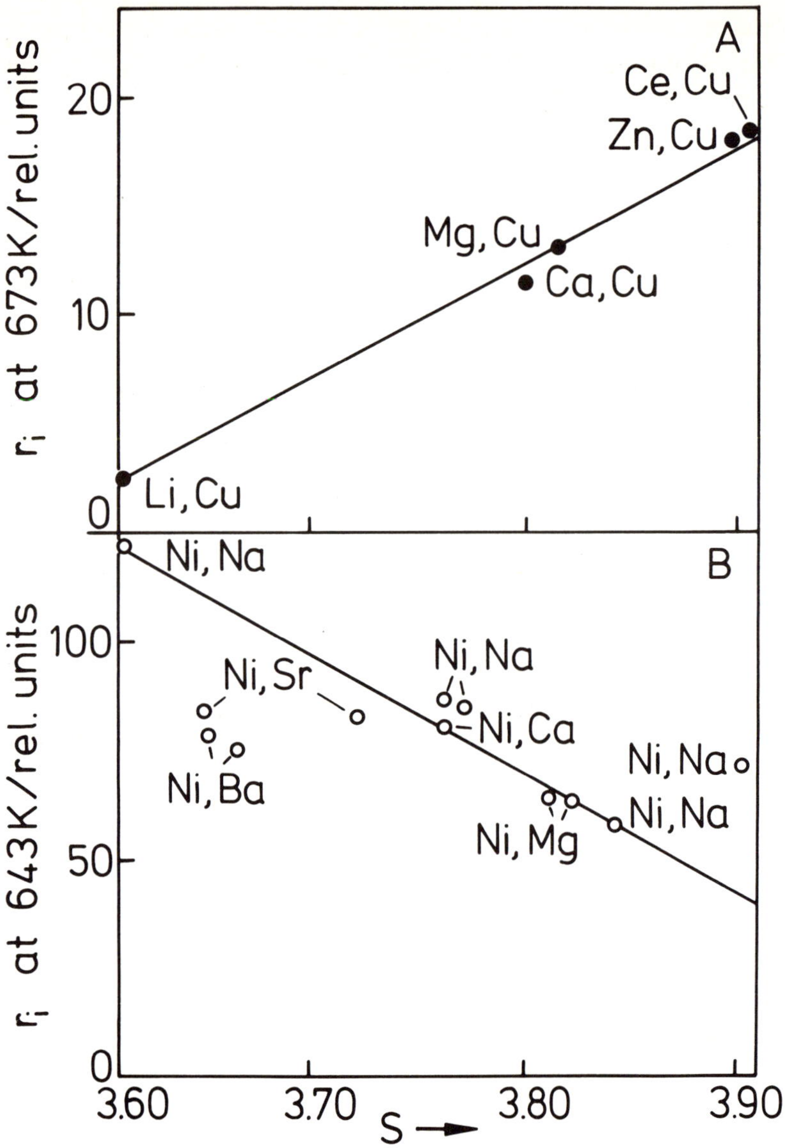

Fig. 9. Dependence of the initial rate of reduction r_i on the Sanderson electronegativity S of multi-cationic forms of zeolite Y: A, the reduction of Cu^{2+} to Cu^+ [10]; B, the reduction of Ni^{2+} to Ni^o; values are from ref. [131].

[132-134]. The interdependence of r_i and S can be understood as follows: an isolated transition-metal ion is assumed to be present in a large zeolite cage, while its reducibility is mainly determined by the chemical composition of the surrounding zeolite matrix. This is shown in Fig. 9 for the reduction of Cu^{2+} to Cu^{+} and of Ni^{2+} to Ni^{0}.

When the end product of the reduction is cationic, it is stabilized by increasing electrostatic interaction with the lattice, and the increased initial reducibility of Cu^{2+} to Cu^{+} with higher overall electronegativity of the matrix (Fig. 9A) may be understood. However, when no such electrostatic interaction will determine the stability of the end product, the reduction will be increasingly difficult for higher values of S. This reasoning explains the decrease in reducibility of Ni^{2+} to Ni^{0} with increasing electrostatic interaction of Ni^{2+} with the zeolite matrix (Fig. 9B).

The overall reducibility as expressed by the degree of reduction, α, varies in a complex manner with S (Fig. 10). Only beyond a certain value of S is the expected decrease of α with S observed. It is striking that two different sets of values from independent laboratories confirm this relation. However, for low values of S, the compositional parameter is of much less

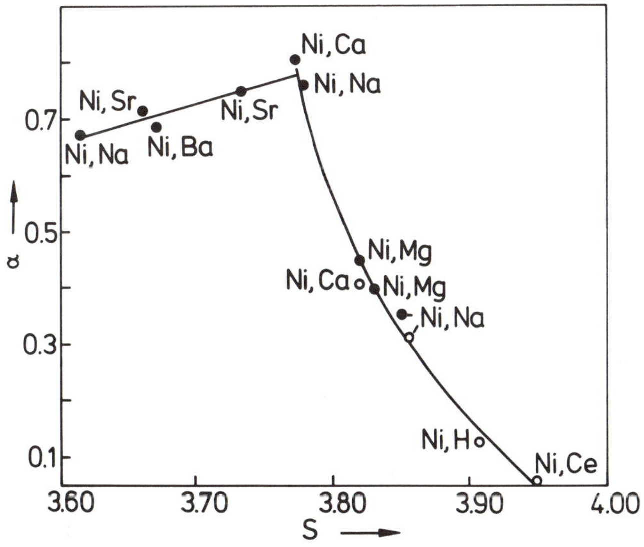

Fig. 10. Variation of the degree of reduction, α, of Ni in Y zeolite with the number and nature of the irreducible cations as quantified by the Sanderson electronegativity, S; the α values are from Ref. [131] (full points) and from Ref. [130] (open points).

influence. It is logical that for these samples with lower contents of Ni^{2+} and the second polyvalent cation, other parameters such as site location can be of greater influence on the degree of reduction, although this is not apparent when only the initial rates of reduction (Fig. 9B) are considered.

The decreased reducibility of transition metal ions in zeolites in general, irrespective of the structure [3, 4, 130, 136], namely,

$$A > X > Y \text{ zeolite}$$

can also be rationalized, at least qualitatively and in the absence of accurate data, in terms of the same electronegativity concept.

8.5.7 Kinetics of reduction of transition metal ions in zeolites with molecular hydrogen

The curves indicating hydrogen uptake by zeolites containing transition metal ions, irrespective of the nature of the cation or the zeolite structure [66, 99, 114, 119, 123, 137-139], show a fast hydrogen uptake followed by a much slower reaction. Even when pre-reduced Pt aggregates are present, the final uptake remains slow, and only initially is the uptake much faster. The curves in all cases are different from those found for corresponding bulk oxides, for which nucleation and growth processes occur in sequence. Typical curves are shown in Fig. 11 for NiY and bulk NiO with and without Pt. Experimentally, caution must be exercised to remove rapidly the heat generated by these exothermic reduction processes [139, 140]. For isothermal reactions, special furnaces have been used, and with temperature-programmed reduction, the use of small quantities of solid gives improved resolution.

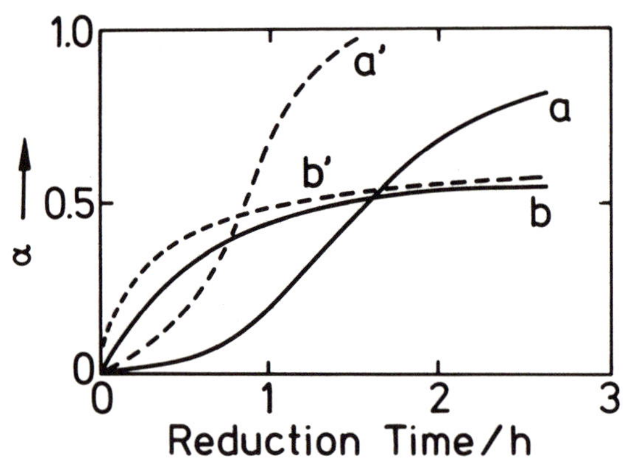

Fig. 11. Hydrogen uptake by NiY at 773 K (b) and NiO at 437 K (a). The dashed lines represent data for samples doped with 0.5 wt% Pt [10].

The rate equations used to fit the hydrogen uptake curves are representative either of simple chemical kinetics [66, 114, 99, 138] or of a mass-transport-influenced process with the diffusion of reactants and products through and between phases being rate limiting [123, 137, 138]. It is striking that chemically controlled reductions are observed only for low-temperature reductions such as $Ag^+ \longrightarrow Ag^0$ and $Cu^{2+} \longrightarrow Cu^+$. In NiX and NiY zeolites and for the reduction $Cu^+ \longrightarrow Cu^0$ in zeolite Y, the equations always indicate diffusion-controlled reduction. Two typical equations are the following:

$$3(1-\alpha)^{-1/3} + \log(1-\alpha) - 3 = kt \qquad [99, 123, 137] \tag{29}$$

$$0.5 - \alpha/3 - \frac{(1-\alpha)^{2/3}}{2} = kt \qquad [99, 138] \tag{30}$$

The first equation describes a diffusion mechanism with steady formation of a reaction zone in which the diffusion coefficient (D) shows a linear dependence on the degree of reduction (α):

$$D = D_o (1 - \alpha) \tag{31}$$

The decrease of D with increasing α can be explained by assuming that the reaction is retarded by products. The second equation is also based on the assumption that the reaction is retarded by a product zone.

In view of the preceding discussion, it is inferred that this retarding species is probably water or zeolite hydroxyl groups. It should be noted that these processes are characterized by increasing values of the apparent activation energies. In view of this, the original interpretation that the two processes in the Cu^{2+} to Cu^+ reduction in zeolite X and Y correspond to reduction in supercage and sodalite positions [138-140] has to be reconsidered. In any case of diffusion-controlled reduction the $\alpha = f(t)$ curves are superimposible after a $[\alpha(t) \longrightarrow k\alpha(t)]$ transformation, in which k is constant [137].

In the case of a chemical control of the reduction, this transformation is no longer applicable, and a stepwise reduction is observed [66, 114], irrespective of the zeolite structure. This behavior is shown in Fig. 12 for Ag^+ ions in chabasite.

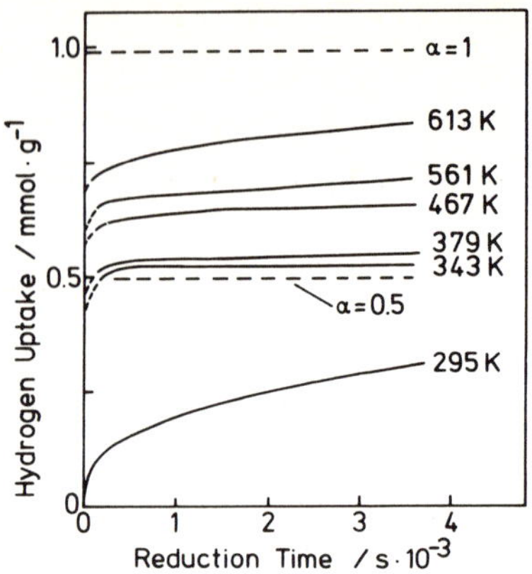

Fig. 12. Stepwise uptake of hydrogen by Ag-chabasite at different temperatures [114].

The low-temperature reduction curve can be interpreted mechanistically only when charged silver clusters are formed: the hydrogen activation is either heterolytic (on iron impurities in zeolites Y and MOR [66, 113]) or homolytic (as in chabasite [114]). In either case, the rate-determining event is the collision of neutral atoms with cations leading to formation of a charged cluster.

From the kinetics results, the following average cluster compositions are inferred:

Y:	Ag_5^{2+}	[114]
MOR:	Ag_5^{+}	[113]
CHA:	Ag_3^{+}	[114]

The further reduction of these clusters occurs at higher temperatures via a distinct mechanism. This interpretation of the kinetics data no longer requires the assumption that the Ag^+ ions in the hexagonal prisms are difficult to reduce [66], which is in agreement with the X-ray diffraction data characterizing these systems [69]. Since the reduction of other transition-metal ions with H_2 requires higher temperatures, these charged clusters, if they are formed, cannot be more than metastable intermediates. In principle, a careful kinetics study of the reduction of Ni^{2+} in X zeolites with molecular hydrogen should be able to show these charged clusters, if they are indeed formed.

8.6 METAL SINTERING IN ZEOLITES AND STABILIZATION OF THE SYSTEM

Metals in zeolites can in principle be located in the different structural positions, or external to the zeolite crystals. The details of location and migration during reductive or sintering treatments will be easiest to unravel for transition metal ions that can be reduced under rather mild conditions, such as Ag^+ and Pt^{2+}. To characterize the metallic phase associated with the zeolite, almost every chemical, physical, and physico-chemical technique available has been used.

8.6.1 Silver zeolites

As discussed in earlier sections, and in Chapter 7, upon treatment of samples with molecular hydrogen, charged clusters are fixed in the small cages of zeolites A, X, Y, and chabasite [66-69, 71, 80, 141]. Chemical evidence as well as evidence from optical spectroscopy and X-ray methods supports this conclusion. It is also established that the charged entities ultimately agglomerate in the large zeolite cages to form neutral metal aggregates (or particles) [124]. When the metals are first reduced to neutral particles, reoxidation under much milder conditions is possible [113, 114].

The neutral aggregates of Ag located initially in the zeolite crystals migrate gradually to the external surfaces of the crystals, where they agglomerate into large crystals (up to 40 nm); these are easily detectable by X-ray line broadening techniques, allowing determination of the average particles size [66, 113]. The latter particles (Ag^o_o), are more inert towards oxidation than the neutral internal particles (Ag^i_o), and consequently temperature-programmed oxidation techniques allow quantification of the amount of Ag^o located as aggregates located inside the zeolite (Ag^o_i) as here or particles located outside the zeolite (Ag^o_o) [113].

Since the Ag_2O phase is thermally unstable [66], the silver can be redispersed completely in cationic positions as Ag^+ [66, 67, 113, 114]. As a result, the reduction reaction is entirely reversible:

$$Ag^+ZO^- \underset{\frac{1}{4}O_2,\ -\frac{1}{4}H_2O}{\overset{+\frac{1}{2}H_2}{\rightleftarrows}} Ag^o + ZOH \tag{32}$$

The fate of a silver ion located in the hexagonal prism of a faujasite crystal during reduction and subsequent reoxidation is schematically represented in Fig. 13.

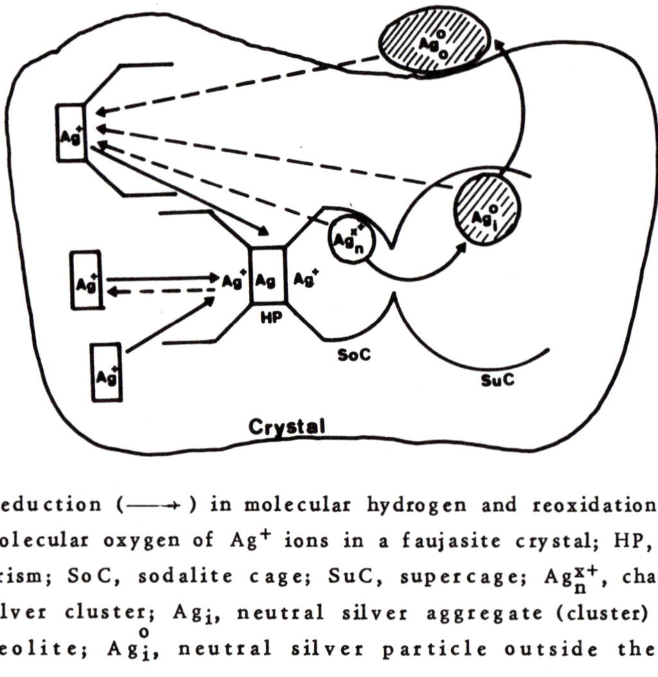

Fig. 13. Reduction (——→) in molecular hydrogen and reoxidation (←– – –) in molecular oxygen of Ag^+ ions in a faujasite crystal; HP, hexagonal prism; SoC, sodalite cage; SuC, supercage; Ag_n^{x+}, charged small silver cluster; Ag_i, neutral silver aggregate (cluster) inside the zeolite; Ag_i^o, neutral silver particle outside the zeolite.

8.6.2 Pt and Pd zeolites

Platinum aggregates or particles associated with X or Y zeolites have been extensively studied, mainly by Gallezot and coworkers, probably because of the industrial importance of the system and of the ease of application of physical techniques to characterize the heavy Pt particles.

An overview of physical methods used for the characterization of Pt zeolites (almost exclusively 10 wt% Pt in Y zeolite) is given in Table 9, together with the typical information obtained by a particular technique. As a result of this work, the location, particle size distribution, and structure of the particles are reasonably well understood.

The size and location of Pt particles in zeolite Y have been determined by crystal structure analysis [81]. Pertinent results are shown in Table 10.

TABLE 9. Methods used for physical characterization of Pt particles in zeolites.

Technique	Information	Ref.
Crystal structure analysis (of powders)	Location of cations, isolated atoms or pairs of atoms	[81]
X-ray line broadening[a]	Aggregates or large particles (>2 nm)	[81, 143]
SAXS[b]	Particle size distribution down to atomic size	[81, 144]
Radial electron distribution (RED)	Structures of aggregates or clusters	[195, 148]
Transmission electron microscopy (TEM)	Particle sizes >0.5 nm	[146, 147, 151]
EXAFS[c]	structures of aggregates or clusters	[149, 150]
XPS[d]	Metal location inside or outside zeolite	[152, 153]
NMR chemical shift of adsorbed xenon	Average number of atoms per aggregate or particle	[154, 155]

[a]From the Scherrer equation.

[b]Small angle X-ray scattering.

[c]Extended X-ray absorption fine structure spectroscopy.

[d]X-ray photoelectron spectroscopy.

TABLE 10. Size, location, and dispersion of Pt aggregates in Y zeolite after various treatments [81].

Treatment/ temperature, K	Aggregate size, nm	Aggregate location	Average dispersion, %
Oxygen/573	Pt^{+} ions	supercage	-
Hydrogen/573	0.6-1.3	supercage	100
Vacuum/1073	0.6-1.3	supercage	100
Oxygen/773	Pt^{+} ions	sodalite cage	-
Hydrogen/573	Pt^{o} atoms;	sodalite cage	25
	0.6-20	?	
Vacuum/1073	1.5-2.0	in bulk of crystal	65

It follows that (1) the location and size of the Pt particles is determined by the temperature of oxidation before reduction is carried out and (2) that Pt agglomeration in the zeolite crystals can take place to give aggregates with sizes exceeding that of the supercages. It was pointed out earlier that during this oxidizing period kinetics parameters intervene and the proposed behavior is dependent on external factors and is therefore not a general occurrence [77]. The preferential agglomeration of Pt in cracks or holes of the crystals of faujasite zeolites has been confirmed directly by transmission electron microscopy [196, 156] and by XPS [151, 153]. These 2.0-nm particles are homogeneously distributed throughout the zeolite crystals as a monodispersed phase. Possibly these particles have a structure resembling a cluster of grapes. If neighboring supercages are filled with metal, the electron microscope might view these as single particles. Nevertheless, since electron micrographs and oxygen chemisorption experiments agree rather well when used to characterize similar samples, the microscopy was thought to be negative evidence as far as the existence of these grape-like structures is concerned [156].

The structure of 1.0-nm aggregates of Pt in Y zeolite has been determined by the RED-method from X-ray diffraction data [145, 148]. The interatomic distances are contracted with respect to the f.c.c. (face-centered-cubic) structure of bulk Pt, although upon admission of hydrogen, relaxation towards the normal F.C.C. structure occurs. Adsorption of CO, ethylene, or butene also induces a displacement disorder of the Pt atoms, which can be removed upon hydrogen sorption. From an EXAFS analysis of he first-neighbor distances [149], it was concluded that for the 0.7-0.8-nm aggregates, a mixture of icosahedra and cuboctahedra with contracted lattice parameters existed, whereas for the 2.0-nm particles in the bulk, agreement with a cubic model is better.

The electron-deficient nature of Pt aggregates in zeolites has also become a firmly established property [150, 157-161]. The early evidence for this behavior stems from infrared frequency measurements of adsorbed probe molecules such as CO [153, 160] and NO [160]: ν_{CO} shifts to higher frequencies when fewer electrons are available for back-donation of filled d-metal orbitals into empty π*-antibonding orbitals of CO, while ν_{NO} increases with decreasing particle size rather than with the degree of dehydroxylation of the zeolite matrix.

These observations are consistent with an electronic effect rather than with a support effect [7]. XPS as well as EXAFS gives more direct evidence for this difference from bulk-like metal behavior. With a PtNaY zeolite, it was found that electron deficiency of the Pt was higher for the 1.0- than for the 3.0-nm particles, which points to an intrinsic size effect [162]. However, EXAFS at the same time shows that the number of unoccupied states in 1.0-nm aggregates increases when multivalent ions or residual acidity are present [162].

It can be concluded that the aggregates of Pt in supercages of faujasites are electron deficient, but it is not clear to what extent this is the result of intrinsic size effects or electron transfer from metal to support.

When Mo atoms from decomposed $Mo(CO)_6$ are added to such a Pt zeolite, the electron-deficient Pt particles gradually lose their special electronic properties and finally even lose their chemisorptive properties, which suggest that the Mo atoms cover the Pt aggregates, shielding them from an interaction with the support [163].

Pd zeolites behave similarly in all these respects, showing the same electron-deficient behavior [9] and also forming particles in the bulk outside the supercages of the zeolite crystals [164]; structures have not been studied in great detail, but the redispersion of the Pd has been studied.

Atomically dispersed Pd^o is easily reoxidized at 500 - 780 K to give Pd^{2+} ions, but the 2.0-3.5-nm particles in the bulk of the zeolite crystals are then transformed into palladium oxide [164]. These particles can be dissolved at room temperature in nitric oxide according to the following overall reaction stoichiometry [165]:

$$Pd^o + 2NO + 2\ ZOH \longrightarrow N_2O + H_2O + (ZO^-)_2\ Pd^{2+} \qquad (33)$$

8.6.3 Ni zeolites

Since the reducibility of Ni^{2+} ions in zeolites by molecular hydrogen requires high temperatures, metal aggregates, if they are initially formed, will sinter very rapidly and at least partially agglomerate at the external surfaces of the crystals or in holes or defects in the bulk of the zeolite crystals. All treatments or modifications which enhance the Ni reducibility will therefore also decrease the degree of metal sintering and provide superior dispersions. Such possibilities are the following (also see section 8.5):

(1) replacement of molecular hydrogen by atomic hydrogen;

(2) preparation of hydrolyzed Ni species;

(3) incorporation of co-cations such as Ca^{2+}, Ce^{3+}, Mg^{2+}, Ba^{2+} and Sr^{2+};

(4) incorporation of Pt or Pd aggregates.

In most cases a bi- or multinodal distribution of metal particles is formed in this way. The distinction between supercage Ni aggregates (Ni_i) in faujasite and external Ni particles (Ni_o) can easily be made by a modified temperature-programmed reduction [166, 167]. This reduction procedure takes advantage of the fact that NiO phases are easier to reduce than Ni^{2+} ions in cation positions. Also, for Ni in general, the following holds true [166]:

$$Ni_i + 2\ ZOH + 1/2\ O_2 \longrightarrow 2\ ZO^- + Ni^{2+} + H_2O \qquad (34)$$

$$Ni_o^o + 1/2\ O_2 \longrightarrow (NiO)_o, \qquad (35)$$

where the subscripts i and o refer to particles located inside and outside the zeolite, respectively.

The bidisperse character of Ni in such samples is confirmed by ferromagnetic resonance [167], as illustrated in Table 11. The latter technique clearly shows that the aggregates in the supercages are either charged or interact strongly with the support.

The sintering mechanism of Ni in zeolites has been characterized recently with magnetic measurements [168]. Since the amount and volume of the particles exceeding the cage dimensions increases in the sequence

$$MOR < Y < X < A,$$

it was concluded that the growth of the particles is mainly controlled by the concentration of lattice defects in the matrix. Fusion of the particles is considered to be rate determining.

TABLE 11. Characterization of Ni^0 in Y Zeolite by Temperature-programmed Reduction (TPR) and Ferromagnetic Resonance (FMR) Methods [166, 167].

	Sample reduction temperature, K	
	723	823
Degree of reduction	44	56
% Ni^0 by TPR[a]	30	57
% Ni^0 by FMR	38	53
% Ni_i by TPR	37.8	45.3
H/Ni^0 by TPR	0.60	0.17

[a]From total amount of Ni.

8.6.4 Fe and Ru zeolites

The specific characteristics of Pt and Pd in zeolites have also been observed for ruthenium. It is possible to form a whole range of metal particles in faujasite zeolites, including atomically dispersed Ru^0 in zeolite cages [169], 1.0-nm aggregates in supercages with particles outside the zeolite [170], and 2.4 - 4-nm particles in holes or cracks of the zeolite crystals; the formation of the latter is increased when water is present

TABLE 12. Catalytic Activities of Supported Pt with Well-characterized Particle Sizes or Metal Dispersions.

Catalytic reaction	Temperature, K	Catalyst	Pt content wt%	10^3 x Rate: Turnover number, s^{-1}	10^3 x Rate: mol h^{-1} (m^2 of Pt)$^{-1}$ x 10^{-4}	Average particle size, nm	H/Pt	Ref.
Ethylene hydrogenation	189	Pt/NaY	0.54	5.3	-	-	1.4	[74]
		Pt/CaY	0.59	25.0	-	-	1.3	[74]
		Pt/SiO_2	0.53	6.3	-	-	0.6	[74]
Cyclopropane hydrogenation	295	Pt/NaY	3.8	944	-	1.0	-	[147]
		Pt/SiO_2	2.6	25.0	1.5	1.5	-	[147]
		Pt/HY	3.5	1170	-	1.0	-	[147]
Neopentane hydrogenolysis	545	Pt/CaY	-	12.0	-	-	1.0	[74]
		Pt/Al_2O_3	-	0.3	-	-	0.6	[147]
Ethane hydrogenolysis	663	Pt/NaY	3.8	-	50.0	1.0	-	[147]
		Pt/NaY	3.8	-	13.0	1.5	-	[147]
		Pt/SiO_2	2.6	-	14.0	1.5	-	[147]
		Pt/HY	10.8	-	60.0	1.0	-	[147]

during reduction [170].

Under certain reduction conditions (reduction of a previously degassed sample), however, mononodal dispersions of metal that are very stable in hydrogen can be achieved [170-172]. A metal-support interaction exists which may give rise to electron-deficient metal aggregates in the supercages [173].

The behavior of the Ru in oxygen-containing atmospheres is entirely exceptional, because of the high volatility of ruthenium oxides as well as their high stability. Upon oxygen treatment, all Ru^0 is oxidized to RuO_2, which easily agglomerates outside the zeolite as a bulk oxide phase [170-171].

Iron zeolites also typically have multinodal metal distributions [174, 175], unless particular treatment and preparation procedures are followed [168]. Besides the methods already mentioned, Mössbauer spectroscopy is a valuable tool in the characterization of these zeolites [176].

8.6.5 Summary

The results summarized in sections 8.3-8.6 lead to a few general conclusions that facilitate the detailed discussion of the unusual catalytic properties of metal-containing zeolites:

(1) The characterization of a metal-containing zeolite after a particular treatment is a complex matter, but the structure can be determined in detail when a battery of chemical, physical, and physico-chemical techniques is used in concert.

(2) Although general lines are available along which a particular metal-zeolite association can be designed, it remains unclear to what extent scale-up of the procedures is possible. Therefore, correlations with catalytic results or properties will always be subject to uncertainties, unless characterization of the material taken from the catalytic reactor is done.

8.7 CATALYTIC PROPERTIES OF METAL-CONTAINING ZEOLITES

The catalytic properties of metal-zeolite combinations have been reviewed on several occasions [1-7, 11-15]. Most reactions known to be catalyzed by a metal can also be catalyzed by a metal-zeolite combination. The activity and selectivity of such a combination depend on the application. The specific catalytic properties of metals located in zeolite pores and/or cages are often obscured by the presence of a metal phase external to the zeolite. It is clear that unusual reaction selectivities to be attributed to the metals in the zeolite may be at least partially obscured when the metal is also present external to the zeolite. Conclusions regarding unusual catalytic properties of metals in zeolites can therefore be made only for well-characterized materials. Some of these properties are now fairly well established.

8.7.1 Particle size and support effects

Typical literature data characterizing these topics are collected in Table 12. The activity of zeolite-supported Pt aggregates is considerably greater than that of Pt aggregates on classical supports such as silica and alumina. This behavior is true for reactions considered to be structure-insensitive (hydrogenation of alkene) as well as structure-sensitive (hydrogenolysis of hydrocarbons). This enhanced activity per unit of available metal surface has been ascribed to particle-size and to electronic effects. Electron deficiency of the metal resulting from the interaction with the support causes Pt aggregates to resemble bulk Ir in their catalytic nature [7, 74, 174], because Ir has a higher percentage of d-band character and therefore a higher activity. However, the statement of electron deficiency as presented here is rather primitive and does not provide a basis for understanding the enhanced catalytic properties in detail [7].

The observed increase in turnover number with increased metal surface area for a structure-insensitive reaction has also been observed for benzene

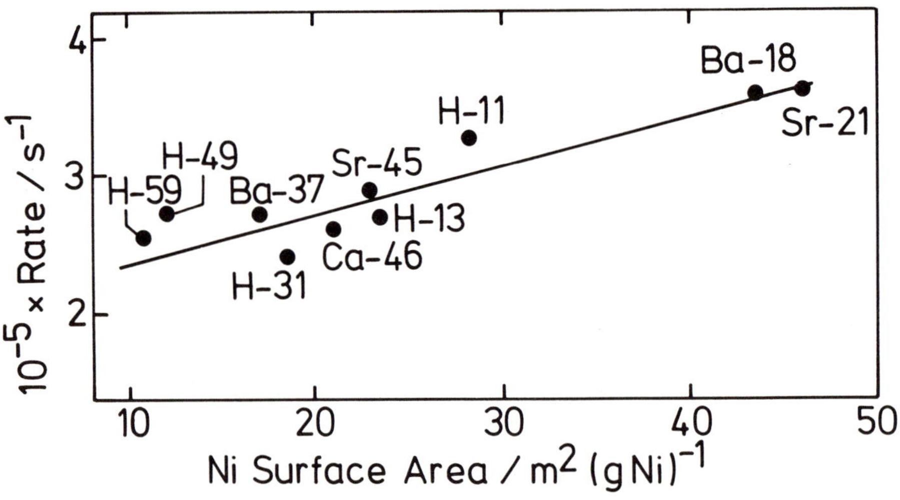

Fig. 14. Rates of benzene hydrogenation catalyzed by NiMeNaY zeolites with various nickel surface areas measured by hydrogen chemisorption [177]. The figures represent the degree (percentage) of cation exchange of the irreducible second cation (Me).

hydrogenation catalyzed by NiMeY zeolites [177]. Data illustrating this behavior are shown in Fig. 14.

Also for this reaction, the turnover frequency N does not remain constant with changing degree of Ni dispersion. Since N increases linearly with the dispersion, irrespective of the nature of the second cation and the degree of cation exchange, the present data tend to indicate that the particle size effect dominates over the effect exerted by the electron deficiency of the metal aggregates. However, the data of Table 12 for Pt indicate that this conclusion is not general.

Effects of particle size on catalyst selectivity have also been established for hydrodechlorination and oligomerization of carbon tetrachloride catalyzed by NiY zeolites [178]. It seems that the hydrodechlorination reaction is favored at Ni particles located outside the zeolite crystals on which an excess of activated hydrogen is present. On Ni aggregates inside the supercages, much less activated hydrogen is apparently present, resulting in a preferential oligomerization of CCl_4. Therefore, the catalyst can be tailored by changing the distribution between internal and external Ni particles.

Support effects are in general more important when the acidity of the zeolite has increased and consequently when the electron-deficient character of the metal is more important. In catalysis this results in a decreased Fischer-Tropsch or methanation activity [179], a decreased alkene yield in the same reaction [180], or a decreased activity for dehydrocyclization of *n*-hexane [181]. Consequently, it is not abnormal to find that KL zeolite, which contains K^+ ions in the small cages and therefore exhibits a strongly basic behavior, has superior dehydrocyclization properties [181, 182]. Similarly, the presence of residual acidity switches the selectivity of supported Pd from methanol formation to methane formation [183]. In all cases, however, the explanations advanced are always in terms of the general and ill-defined concept of electron deficiency of zeolite-supported metal aggregates.

In other reactions, however, the properties of the zeolite-supported metal aggregates are in no way different from those of metal aggregates on other supports. The Kölbel-Engelhardt reaction proceeds via a sequence of water gas shift and methanation reactions [9]:

$$3\,[\,CO + H_2O \rightleftarrows CO_2 + H_2\,] \quad (36)$$

$$3\,H_2 + CO \rightleftarrows CH_4 + H_2O \quad (37)$$

$$4\,CO + 2\,H_2O \rightleftarrows CH_4 + 3\,CO_2 \quad (38)$$

For Ru on Y zeolite, alumina, or silica, the support is observed not to play any role in this reaction [184]. Further, the properties of metal aggregates as water gas shift catalysts were shown not to be influenced by the zeolite but rather by the electronegativity or redox potential of the element itself [185].

In the oxidation of ethylene, all selectivity for ethylene oxide is absent, and only CO_2 is formed when residual acidity is present in addition to silver particles [186]. When the silver-zeolite Y combination is prepared in such a way that residual acidity is absent, epoxide selectivities equal to those of other unpromoted silver catalysts are observed [186].

Taking into account all the available knowledge, it is impossible to rationalize this behavior. It seems, therefore, that the link between general mechanistic information for a given reaction and the specific physical properties of metal aggregates in cages of zeolites is totally missing.

8.7.2 Sensitivity to poisoning of metal aggregates in zeolites

For electron-deficient metal aggregates, it may be anticipated that the bonding energy of electronegative atoms with the electron-deficient surface atoms will be decreased, and thus it is expected that electron-deficient metal aggregates in zeolites will show enhanced resistance to sulfur poisoning. The experimental results clearly show that this is the case:

(1) Dealuminated PdHNaY zeolite shows enhanced resistance to sulfur poisoning as well as further exchange of H^+ for Na^+ [187].

(2) PtNaY is more sulfur-resistant than PtKL [188].

(3) Addition of Ce^{3+} to PtHNaY leads to enhanced sulfur resistance [189].

In every case, enhanced sulfur resistance occurs when the acidity of the support increases and thus when the electron-deficient character of the metal aggregates becomes more pronounced.

It is evident tnat the greater the electron-deficient character of metal particles, the more pronounced will be their poisoning with electron-donor molecules. This was indeed shown to be the case for poisoning of Pt aggregates in Y zeolite with ammonia [7, 160].

8.7.3 Shape-selective metal catalysis

The shape-selective hydrogenation and oxidation properties of Pt aggregates in zeolite A were discovered by Weisz *et al.* in the early 1960's [190, 191]. Such a catalyst is able to discriminate between normal and methyl-branched alkanes for hydrogenation. With the same catalyst, normal aggregates in zeolite A were discovered by Weisz *et al.* in the early 1960's [190, 191]. Such a catalyst is able to discriminate between normal and methyl-branched alkanes for hydrogenation. With the same catalyst, normal alkanes can be completely oxidized in the presence of methyl-branched alkanes [192]. It is of course critical that metal external to the zeolite should be absent or poisoned by treatment with large molecules unable to reach the intrazeolitic metal.

Newer and thermally more stable zeolites such as alpha [193] and ZSM-5 [50, 51, 194, 195] have been proposed as supports for metals in shape-selective hydrogenation and oxidation reactions. Pertinent data are shown in Table 13. The results shown in this table are just what one would expect. In some cases, however, unexpectedly high selectivities have been obtained. Indeed, from acetone, high yields of methylisobutyl ketone can be obtained with a bifunctional Pd/H-ZSM-5 catalyst [194]:

$$2(CH_3)_2C{=}O + H_2 \longrightarrow H_2O + (CH_3)_2CH{-}CH_2{-}C({=}O){-}CH_3 \qquad (39)$$

On the acid function, acetone undergoes successive aldol condensations. On the metal, some intermediates are then hydrogenated and can thereby be desorbed. The selectivity for this product is determined by the geometry of the zeolite cages; in the more open Y-type zeolite, the selectivity amounts only to one third of the value observed with ZSM-5.

Metals encaged in zeolites also impose characteristic product selectivities in the Fischer-Tropsch reaction. Indeed, instead of products with carbon numbers conforming to a polymerization distribution law (Schulz-Flory distribution), the carbon chain length beyond a certain carbon number is drastically decreased. This distribution has been attributed to a particle size effect [42, 43, 197] or to diffusion effects [34, 110, 196, 198].

8.8 CONCLUSIONS

A wide variety of techniques can be used for the preparation of metal-zeolite combinations, but preparation of a mononodal distribution of metal particle sizes for a given metal and a specific zeolite structure requires that highly specific procedures be followed. In most cases, metal aggregates encaged in the zeolite can indeed be prepared.

The factors determining metal reducibility are fairly well understood and consequently can be used to tailor dispersion and particle size. The same

TABLE 13. Selective catalysis by metal-containing zeolites.

Reaction type	Reactants	Zeolite catalyst	Reaction temperature, K	Conversion, %	Refs.[b]
Hydrogenation	but-1-ene	Pt/alpha[a]	511	74.4	[193, ex. 6]
	isobutene			0.0	
Hydrogenation	hex-1-ene	Pt/CoZSM-5	578	25.6	[50, ex. 14]
	6-methyl-hep-1-ene			2.0	
Hydrogenation	styrene	Pt/CsZSM-5	698	50	[50, ex. 19]
	2-methylstyrene			1.8	
Oxidation	o-xylene	Cu/ZSM-5	683	10.0	[51]
	p-xylene			79.0	
Hydrogenation	*p*-methylstyrene	Pt/ZSM-5	673	69.1	[195, ex. 1]
	m-methylstyrene			42.2	

[a]External Pt on this zeolite was poisoned with triphenylphosphine; *b*, number in the reference list followed by the example number in case of patent literature.

factors help to determine the specific characteristics of metals in zeolites, such as structure and surface properties. However, to characterize a metal-zeolite combination in full detail, the application of several physical and physico-chemical techniques is a prerequisite.

Catalytically, the specific properties of these materials can be rationalized in terms of support or particle-size effects, although the concept of electron-deficient particles and their influence on the intimate chemistry is not fully understood and requires further study.

A promising area for application seems to be the shape-selective metal-catalyzed conversions with metal aggregates encaged in zeolites.

ACKNOWLEDGMENTS

The author acknowledges the National Fund of Scientific Research for a research position as Senior Research Associate and for financial support of his research. A research grant from the Belgian Government in the frame of a concerted action on catalysis is gratefully acknowledged. The author is also grateful to Prof. Uytterhoeven for his constant interest in and stimulation of this work.

REFERENCES

1. Wm. M. Minachev and Ya. I. Isakov, in Zeolite Chemistry and Catalysis, J. A. Rabo, ed., ACS Monograph, 171 (1976) 552.
2. Ya. I. Isakov, and Wm. M. Minachev, Russ. Chem. Rev., 51 (1982) 1188.
3. P. A. Jacobs, Carboniogenic Activity of Zeolites, Elsevier, 1977.
4. J. B. Uytherhowen, Acta. Phys. Chem., 24 (1978) 53.
5. D. Delafosse, in Catalysis by Zeolites, Stud. Surf. Sci. and Catal., 5 (1980) 235, Elsevier.
6. Stud. Surf. Sci. and Catal., 12 (1982), Elsevier.
7. P. Gallezot, Catal. Rev. Sci. Eng., 20 (1979) 121.
8. P. Gallezot and G. Berghet, ref. 6, p. 167.
9. F. Schmidt, ref. 6, p. 172.
10. P. A. Jacobs, ref. 5, p. 293.
11. Y. Ben Taarit and M. Che, ref. 5, p. 167.
12. P. Gelin, F. Lejebvre, B. Elleuch, C. Naccache, and Y. Ben Taarit, Intrazeolite Chemistry, G. D. Stucky and F. G. Dwyer, eds., ACS Symp. Ser. 218 (1983) 469.
13. J. Scott, Zeolite Technology and Application, Noyes Data Corp., 1980.
14. I. E. Maxwell, Advan. Catal., 31 (1982) 2.
15. A. P. Bolton, ref. 1, p. 714.
16. W. M. Meier, and D. H. Olson, Atlas of Zeolite Structure Types, Polycrystal Books, Pittsburgh, 1978.
17. W. J. Mortier, Compilation of Extra Framework Sites in Zeolites, Butterworth, 1982.
18. A. Cremers, ACS Symp. Ser., 40 (1977) 179.
19. A. Maes and Cremers, J. Chem. Soc. Faraday I, 71 (1975) 265.
20. B. H. Wiers, R. J. Grosse and W. A. Cilley, Environ. Sci. Technol., 16 (1982) 617.
21. A. Maes and A. Cremers, Advan. Chem. Ser., 121 (1973) 231.

22. P. Fletcher and R. P. Townsend, J. Chem. Soc. Faraday I, 77 (1981) 497.
23. R. M. Barrer and R. P. Townsend, JCS Faraday I, 72 (1976) 2650.
24. P. Fletcher and R. P. Townsend, Zeolites, 3 (1983) 129.
25. P. Chu and F. G. Dwyer, ACS Symp. Ser., 218 (1983) 59.
26. R. A. Schoonheydt, L. Y. Vandamme, P. A. Jacobs, and J. B. Uytterhoeven, J. Catal., 43 (1976) 292.
27. K. G. Ione, V. N. Romannikov, A. A. Davydov, and L. B. Orlova, J. Catal., 57 (1979) 126.
28. E. Mantovani, N. Palladino and A. Zanobi, U.S. Patent 4,334,101, (1982) assigned to Snamprogetti S.P.A.
29. P. Gelin, G. Couduier, Y. Ben Taarit and C. Naccache, J. Catal., 70 (1981) 32.
30. C. A. Clausen and M. L. Good, Inorg. Chem., 16 (1977) 816.
31. F. Ribeiro and C. Marcilly, Rev. Inst. Frans. Petrole, 3 (1979) 40.
32. M. El Malki, J. P. Franck, C. Marcilly, and R. Montarnal, C. R. Acad. Sci. Paris, 288 (1979) 173.
33. J. B. Nagy, M. van Enoo, and E. G. Derouanc, J. Catal., 58 (1979) 230.
34. D. Ballivet-Tkatchenko and G. Coudurier, Inorg. Chem., 18 (1979) 558.
35. T. Bein, P. A. Jacobs and F. Schmidt, Stud. Surf. Sci. Catal., 12 (1982) 111, Elsevier.
36. T. Bein and P. A. Jacobs, J. Chem. Soc. Faraday I, 79 (1983) 1819.
37. P. Gallezot, G. Coudurier, M. Primet, and B. Imelik, ACS Symp. Ser., 40 (1977) 144.
38. M. Iwamoto, H. Kusano, and S. Kagawa, Chem. Lett., (1983) 1483.
39. M. D. Ward, and J. Schwartz, Organometallics, 1 (1982) 1030.
40. T. N. Huang and J. Schwartz, J. Am. Chem. Soc., 104 (1982) 5244.
41. G. M. Woltermann, and V. A. Durante, Inorg. Chem., 22 (1983) 1954.
42. L. F. Nazar, G. A. Ozin, C. G. Francis, M. P. Andrews and H. X. Huber, J. Am. Chem. Soc., 103 (1981) 2453.
43. L. F. Nazar, G. A. Ozin, F. Hugues, J. Godber and D. Rancourt, Angew. Chem., 95 (1983) 645.
44. J. Scherzer and D. Fort, J. Catal., 71 (1981) 111.
45. J. Scherzer, U.S. Patent 4,397,164 (1982) assigned to Filtrol Corp.
46. L. E. Iton, R. B. Beal, and D. T. Hodul, J. Mol. Catal., 21 (1983) 151.
47. K. H. Rhee, F. R. Brown, D. H. Finseth and J. M. Stencel, Zeolites, 3 (1983) 394.
48. K. H. Rhee, V. U. S. Rao, J. M. Stencel, G. A. Melson, and J. E. Crawford, Zeolites, 3 (1983) 337.
49. J. M. Stencel, V. U. S. Rao, J. R. Diehl, K. H. Rhee, A. G. Dhere, and R. J. De Angelis, J. Catal., 84 (1983) 109.
50. R. M. Dessau, U.S. Patent 4,377,503 (1983), assigned to Mobil Oil Corp.
51. R. M. Dessau, J. Catal., 77 (1982) 304.
52. P. B. Weisz, Erdöl Kohle, 18 (1965) 525.
53. G. H. Kuehl, U.S. Patent 4,191,663 (1980) assigned to Mobil Oil Corp.
54. P. A. Jacobs, M. Tielen, and J. Martens, J. Mol. Catal., 27 (1984) 11.
55. T. Iniu, G. Takeuchi, and Y. Takegami, Appl. Catal., 4 (1982) 21.
56. Y. W. Chen, H. T. Wang and J. J. Goodwin, J. Catal., 83 (1983) 415.
57. T. M. Tri, J. P. Candy, P. Gallezot, J. Massardier, M. Primet, J. C. Vedrine and B. Imelik, J. Catal., 79 (1983) 396.
58. B. Engler, C. Vruger, W. Keim, and J. C. Sekutowski, Erdöl Kohle Erdgas Petrochem. Brennst. Chem., 3 (1978) 87.
59. For a literature survey of this system see G. L. Price and C. Egedy, J. Catal., 84 (1983) 461.
60. B. E. Langner and J. H. Kagon, Stud. Surf. Sci. Catal., 16 (1983) 619.

61. N. Jaeger, U. Melville, R. Nowak, H. Schrubbers and G. Schulz-Ekloff, Stud. Surf. Sci. Catal., 5 (1980) 335.
62. W. Mortier, J. Phys. Chem., 79 (1975) 1497.
63. P. A. Jacobs, Stud. Surf. Sci. Catal., 12 (1982) 71.
64. M. Briend Faure, M. F. Guilleux, J. Jeanjean, D. Delafosse, G. Djega Mariadassou, and M. Bureau-Tardy, Acta. Phys. Chem., 24 (1978) 19.
65. B. Briend Faure, J. Jeanjean, D. Delafosse and P. Gallezot, J. Phys. Chem., 84 (1980) 875.
66. P. A. Jacobs, J. B. Uytterhoeven and H. K. Beyer, J. Chem. Soc. Faraday I, 75 (1979) 56.
67. Y. Kim and K. Seff, J. Phys. Chem., 82 (1978) 921.
68. L. R. Gellens, W. Y. Mortier, and J. B. Uytterhoeven, Zeolites, 1 (1981) 11.
69. L. R. Gellens and R. A. Schoonheydt, Stud. Surf. Sci. Catal., 2 (1982) 87.
70. P. A. Jacobs, W. De Wilde, R. A. Schoonheydt, J. B. Uytterhoeven, and H. J. Beyer, J. Chem. Faraday I, 72 (1976) 1221.
71. P. A. Jacobs and H. K. Beyer, J. Phys. Chem., 83 (1979) 1124.
72. M. Ralek, P. Jiru, O. Grubner and H. Beyer, Coll. Czech. Chem. Commun., 27 (1962) 142.
73. T. Kubo, H. Arai, H. Tominaga and T. Kunuzi, Bull. Chem. Soc. Japan, 45 (1972) 607, 613.
74. R. A. Dalla Betta and M. Boudart, Proc. 5th Int. Congr. Catal., 2 (1973) 1329.
75. L. A. Pedersen and J. H. Lunsford, J. Catal., 61 (1980) 39.
76. J. J. Verdonck, P. A. Jacobs, M. Genet and G. Poncelet, J. Chem. Soc. Faraday I, 76 (1980) 403.
77. W. J. Reagan, A. W. Chester and G. T. Kerr, J. Catal., 69 (1981) 89.
78. D. Exner, N. Jaeger, K. Möller, and G. Schulz-Ekloff, J. Chem. Soc. Faraday I, 78 (1982) 3537.
79. R. L. Schneider, R. F. Howe and K. L. Watters, J. Catal., 79 (1983) 298.
80. L. R. Gellens, W. J. Mortier, R. A. Schoonheydt and J. B. Uytterhoeven, J. Phys. Chem., 85 (1981) 2783.
81. P. Gallezot, A. Alcaron-Diaz, J. A. Dalmon, A. J. Renouprez, and B. Imelik, J. Catal., 39 (1975) 334.
82. S. Abdo and R. F. Howe, J. Phys. Chem., 87 (1983) 1713.
83. J. Gosbee, S. Abdo, and R. F. Howe, J. Chem. Phys. Chem. Biol., 78 (1981) 885.
84. J. Scherzer, J. Catal., 80 (1983) 465.
85. L. Riekert, Ber. Bunsenges, Phys. Chem., 73 (1969) 331.
86. J. R. Anderson, "Structure of Metallic Catalysts", Academic Press, 1975, pp. 166-167.
87. Y. Y. Huang and J. R. Anderson, J. Catal., 40 (1975) 143.
88. P. A. Jacobs, Stud. Surf. Sci. Catal., 12 (1982) 71.
89. K. Klier, P. J. Hutta and R. Kellerman, ACS Symp. Ser., 40 (1977) 108.
90. K. Klier, J. Catal., 8 (1967) 14.
91. K. M. Minachev, G. V. Antoshin, E. S. Shapiro and Y. A. Yusifov, Proc. 6th Int. Congr. Catal., 2 (1976) 621.
92. L. A. Morice and L. V. C. Rees, Trans. Faraday Soc., 64 (1968) 1388.
93. R. L. Garten, W. N. Delgass and M. Boudart, J. Catal., 18 (1970) 20.
94. M. F. Guilleux, M. Kermarec and D. Delafosse, J. Chem. Soc. Chem. Commun. (1977) 102.
95. J. Jeanjean, D. Delafosse and P. Gallezot, J. Phys. Chem., 83 (1979) 2761.
96. M. Che, M. Richard and D. Oliver, J. Chem. Soc. Faraday I, 76 (1980) 1526.

97. D. Oliver, M. Richard, L. Bonnerot and M. Che, "Growth and Properties of Metal Clusters", J. Bourdon, ed., Elsevier (1980) p. 165.
98. C. M. Naccache and Y. Ben Taarit, J. Catal., *22* (1971) 171.
99. F. C. Bravo, J. Dwyer and D. Zambouli, Proc. 5th Int. Conf. Zeolites, L. V. C. Rees, ed., Heyden (1980) p. 749.
100. P. A. Jacobs and H. K. Beyer, J. Phys. Chem., 83 (1979) 1174.
101. M. Primet, D J. C. Vedrine and C. Naccache, J. Mol. Catal., 4 (1978) 411.
102. P. Gelin, G. Coudurier, Y. Ben Taarit and C. Naccache, J. Catal., 10 (1981) 39.
103. H. van Brabant, R. A. Schoonheydt and J. Pelgrims, Stud. Surf. Sci. Catal., 12 (1981) 61.
104. J. Verdonck, D. R. A. Schoonheydt and P. A. Jacobs, J. Phys. Chem., 87 (1983) 683.
105. P. A. Jacobs, R. Chantillon, P. De Laet, J. Verdonck and M. Tielen, ACS Symp. Ser., 218 (1983) 440.
106. C. Mintchev, V. Kanazirev, L. Kosova, V. Penchev, W. Gunsser, and F. Schmidt, Proc. 5th Int. Conf. Zeolites, L. V. C. Rees, ed., Heyden (1980) 335.
107. J. A. Rabo, C. L. Angell, P. H. Kasai and V. Schomaker, Disc. Faraday Soc. (1966) 329.
108. F. Schmidt, D W. Gunsser and J. Adolph, ACS Symp. Ser., 40 (1977) 291.
109. F. Steinbach and H. Minchev, Z. Phys. Chem. NF, 99 (1976) 223.
110. D. Fraenkel and B. C. Gates, J. Am. Chem. Soc., 102 (1980) 2480.
111. D. J. Yates, J. Phys. Chem., 69 (1965) 1676.
112. P. A. Jacobs, J. B. Uytterhoeven and H. K. Beyer, J. Chem. Soc. Faraday I, 75 (1979) 56.
113. H. K. Beyer and P. A. Jacobs, ACS Symp. Ser., 40 (1977) 493.
114. H. K. Beyer and P. A. Jacobs, Stud. Surf. Sci. Catal., 12 (1982) 95.
115. R. G. Herman, J. H. Lunsford, H. K. Beyer, P. A. Jacobs, and J. B. Uytterhoeven, J. Phys. Chem., 79 (1975) 2388.
116. P. A. Jacobs, H. Nijs and J. Verdonck, J. Chem. Soc. Faraday I, 75 (1979) 1196.
117. M. Suzuki, U. Tsutsumi and H. Takahashi, Zeolites, 2 (1982) 87.
118. R. M. Barrer and R. P. Townsend, J. Chem. Soc. Faraday I, 72 (1976) 661.
119. A. C. Herd and C. G. Pope, J. Chem. Soc. Faraday I, 69 (1973) 833.
120. C. Mirodatos, J. A. Dalmon, E. D. Garbowski, and D. Barthomeuf, Zeolites, 2 (1982) 125.
121. H. Lausch, W. Morke, F. Vogt and H. Bremer, Z. Anorg. Allg. Chem. (Leipzig), 499 (1983) 213.
122. P. Gallezot and B. Imelik, J. Phys. Chem., 77 (1983) 625.
123. M. Briend-Faure, J. Jeanjean, M. Kermarec, and D. Delafosse, J. Chem. Soc. Faraday I, 74 (1978) 1538.
124. T. A. Egerton and J. C. Vickerman, J. Chem. Soc. Faraday I, 69 (1973) 39.
125. P. Gallezot, Y. Ben Taarit and B. Imelik, J. Phys. Chem., 77 (1973) 2556.
126. M. F. Guilleux, D. Delafosse, G. A. Martin, and J. A. Dalmon, J. Chem. Soc. Faraday I, 75 (1979) 165.
127. J. Jeanjean, S. Djemel, M. F. Guilleux and D. Delafosse, J. Phys. Chem., 85 (1981) 4145.
128. M. Briend-Faure, J. Jeanjean, G. Spector, D. Delafosse, and F. Bozon-Verduraz, J. Chem. Phys., 79 (1982) 489.
129. H. Schrubbers, G. Schulz-Ekloff, and G. Wildeboer, Stud. Surf. Sci. Catal., 12 (1982) 261.
130. K. H. Bager, F. Vogt, and H. Bremer, ACS Symp. Ser., 40 (1977) 528.
131. M. Suzuki, K. Tsutsumi, and H. Takahashi, Zeolites, 2 (1982) 51.

132. W. J. Mortier, J. Catal., 55 (1978) 138.
133. P. A. Jacobs, W. J. Mortier and J. B. Uytterhoeven, J. Inorg. Nucl. Chem., 40 (1978) 1919.
134. P. A. Jacobs, Catal. Rev.-Sci. Eng., 24 (1982) 415.
135. W. Romanovski, Rocz. Chem., 45 (1971) 427.
136. A. Tungler, J. Petro, T. Mathe, G. Besenyei and Z. Csuros, Acta Chem. Acad. Sci. Hung., 82 (1974) 183.
137. M. Kermarec, M. Briend-Faure and D. Delafosse, J. Chem. Spc. Chem. Commun. (1975) 272.
138. P. A. Jacobs, M. Tielen, J. P. Linart, J. B. Uytterhoeven and H. Beyer, J. Chem Soc. Faraday I, 72 (1976) 2793.
139. S. J. Gentry, N. W. Hurst, and A. Jones, J. Chem. Soc. Faraday I, 75 (1979) 1688.
140. N. W. Hurst, S. J. Gentry, A. Jones and B. D. McNicol, Catal. Rev.-Sci. Eng., 24 (1982) 233.
141. G. A. Ozin and F. Hugues, J. Phys. Chem., 87 (1983) 94.
142. D. Hermerschmidt and R. Haul, Ber. Bunsenges, Phys. Chem., 84 (1980) 902.
143. P. H. Lewis, J. Catal., 11 (1968) 162.
144. A. Renouprez, C. Hoang Van and P. Compagnon, J. Catal., 34 (1974) 411.
145. P. Gallezot, A. Bienenstock, and M. Boudart, Nouv. J. Chim., 2 (1978) 263.
146. P. Gallezot, I. Mutin, G. Dalmai-Imelik, and B. Imelik, J. Microsc. Spectrosc. Electron., 1 (1976) 1.
147. C. Naccache, N. Kaufherr, M. Dufaux, J. Bandiera, and B. Imelik, ACS Symp. Ser., 40 (1977) 538.
148. P. Gallezot, Zeolites, 2 (1982) 103.
149. B. Moraweck and A. J. Renouprez, Surf. Sci., 106 (1981) 35.
150. R. S. Weber, M. Boudart, and P. Gallezot, "Growth and Properties of Metal Clusters", J. Bourdon, ed., Elsevier, 1980, p. 415.
151. D. Exner, N. Jaeger and G. Schulz-Ekloff, Chem. Ing. Tech., 52 (1980) 734.
152. J. H. Lunsford and D. S. Treybig, J. Catal., 68 (1981) 192.
153. G. Schulz-Ekloff, D. Wright and M. Grunze, Zeolites, 2 (1982) 70.
154. L. C. de Menorval, J. P. Fraissard and T. Ito, J. Chem. Soc. Faraday I, 78 (1982) 403.
155. J. Fraissard, T. Ito, L. C. de Menorval, and M. S. Springuel-Huet, Stud. Surf. Sci. Catal., 12 (1982) 179.
156. S. Briese-Gulban, H. Kompa, H. Schrubbers and G. Schulz-Ekloff, React. Kinet. Catal. Lett., 20 (1982) 7.
157. M. Primet, P. Fouilloux and B. Imelik, J. Catal., 61 (1980) 553.
158. R. A. Dalla Betta, M. Boudart, P. Gallezot, and R. S. Weber, J. Catal., 69 (1981) 514.
159. J. C. Vedrine, M. Dufaux, C. Naccache and B. Imelik, J. Chem. Soc. Faraday I, 74 (1978) 440.
160. P. Gallezot, J. Datka, J. Massardier, M. Primet and B. Imelik, Proc. 6th Int. Congr. Catal., 2 (1976) 696.
161. F. Figueras, R. Gomez, and R. Primet, Advan. Chem. Ser., 121 (1973) 480.
162. P. Gallezot, R. Weber, R. A. Dalla Betta and M. Boudart, Z. Naturforsch., 34A (1979) 40.
163. T. M. Tri, J. P. Candy, P. Gallezot, J. Massardier, M. Primet, J. C. Vedrine and B. Imelik, J. Catal., 79 (1983) 396.
164. G. Bergeret, P. Gallezot and B. Imelik, J. Phys. Chem., 85 (1981) 411.
165. M. Che, J. F. Dutel, P. Gallezot, and M. Primet, J. Phys. Chem., 80 (1976) 2367.
166. P. A. Jacobs, J. P. Linart, H. Nijs and J. B. Uytterhoeven, J. Chem. Soc. Faraday I, 73 (1979) 1745.

167. P. A. Jacobs, H. Nijs, J. Verdonck, E. G. Derouane, J. P. Gilson, and A. J. Simoens, J. Chem. Soc. Faraday I, 75 (1979) 1196.
168. F. Schmidt, T. Bein, V. Ohlerich and P. A. Jacobs, Proc. 6th Int. Zeolite Conf., paper B-14, Butterworths, 1984.
169. J. R. Pearce, W. J. Mortier, and J. B. Uytterhoeven, J. Chem. Soc. Faraday I, 75 (1979) 1395.
170. J. J. Verdonck, P. A. Jacobs, M. Genet and G. Poncelet, J. Chem. Soc. Faraday I, 76 (1980) 403.
171. L. A. Pederson and J. H. Lunsford, J. Catal., 61 (1980) 39.
172. C. H. Yang and J. G. Goodwin, J. Catal., 78 (1982) 182.
173. D. G. Blackmond and J. G. Goodwin, J. Chem. Soc. Chem. Commun. (1981) 125.
174. F. Schmidt, W. Gunsser, and J. Adolph, ACS Symp. Ser., 90 (1977) 291.
175. W. Gunsser, J. Adolph, and F. Schmidt, J. Magn. Magnet. Meter., 15-18 (1980) 1115.
176. S. L. Suib, K. C. McMahon and D. Psaras, ACS Symp. Ser., 218 (1982) 301.
177. M. Suzuki, K. Tsutsumi and H. Takahashi, Zeolites, 2 (1982) 185.
178. A. H. Weiss, S. Valinski and G. V. Antoshin, J. Catal., 74 (1982) 136.
179. D. G. Blackmond and J. G. Goodwin, J. Chem. Soc. Chem. Commun. (1981) 125.
180. I. R. Leith, J. Chem. Soc. Chem. Commun. (1983) 93.
181. J. R. Bernard, Proc. 5th Int. Conf. Zeolites, L.V.C. Rees, ed., Heyden, 980, p. 686.
182. C. Besoukhanova, M. Breysse, J. R. Bernard, and D. Barthomeuf, Proc. 7th Int. Cong. Catal. (1980) 1410.
183. F. Fajula, R. G. Anthony, and J. H. Lunsford, J. Catal., 73 (1982) 237.
184. B. L. Gustafson and J. H. Lunsford, J. Catal., 74 (1982) 393.
185. M. Iwamoto, T. Hasuwa, H. Furukawa and S. Kagawa, J. Catal., 79 (1983) 291.
186. N. Giordano, J. C. J. Bart, and R. Maggiore, Z. Phys. Chem. NF, 127 (1981) 109.
187. G. D. Chukin, M. V. Landau, V. Kruglikov, D. A. Agievskii, B. V. Smirnov, A. L. Belozerov, V. D. Asrieva, N. V. Goncharova, E. D. Radchenko, O. D. Konovalcherov, and A. V. Agafonov, Proc. 6th Int. Congr. Catal., 1 (1977) 668.
188. C. Besoukhanova, M. Breysse, J. R. Bernard and D. Barthomeuf, in "Catalyst Deactivation", B. Delmon and G. F. Froment, eds., Elsevier, 1980, p. 201.
189. Tran Manh Tri, J. Massardier, P. Gallezot and B. Imelik, Stud. Surf. Sci. Catal., 5 (1980).
190. P. B. Weisz and V. J. Frilette, J. Phys. Chem., 64 (1960) 382.
191. P. B. Weisz, V. J. Frilette, R. W. Maatman and E. B. Mower, J. Catal., 1 (1962) 307.
192. See ref. 52.
193. G. H. Kuehl, U.S. Patent 4,191,663 (1980) assigned to Mobil Oil Corp.
194. T. J. Huang and W. O. Haag, U. S. Patent 4,339,606 (1982) assigned to Mobil Oil Corp.
195. J. M. Klosek and M. M. Wu, U. S. Patent 4,379,027 (1983) assigned to Mobil Oil Corp.
196. D. Ballivet-Tkatchenko and I. Tkatchenko, J. Mol. Catal., 13 (1981) 1.
197. P. A. Jacobs and D. van Wouwe, J. Mol. Catal., 17 (1982) 145.
198. D. Vanhove, Z. Zhuyong, L. Makambo and M. Blanchard, Appl. Catal., 9 (1984) 327.

Chapter 9

SUPPORTED METAL CATALYSTS PREPARED FROM MOLECULAR METAL CLUSTERS: ORGANOMETALLIC SURFACE CHEMISTRY

Section 9.1

METAL CLUSTERS ON FUNCTIONALIZED SUPPORTS

B. C. GATES

9.1.1 INTRODUCTION

Metal clusters anchored to solid supports have been prepared by a variety of routes, some inferred from organometallic solution chemistry combined with the functional group chemistry of the support and some determined empirically. Most attempts at synthesis have led to structurally complex materials; in this section, we consider only those materials with demonstrated catalytic activity which appear to be structurally unique, i.e., with the supported metal species being analogues of molecular clusters. This section is concerned specifically with clusters on supports functionalized with ligands common in molecular cluster chemistry (e.g., phosphines). The following section (9.2) is concerned with analogues of molecular clusters and their reactions on unfunctionalized metal-oxide supports.

Part of the motivation for investigation of supported metal clusters derives from the opportunities they offer for fundamental understanding of catalysis. A compelling goal is the preparation of supported metals having simple, uniform structures with neighboring metal centers; variation of cluster structures and compositions may allow determination of relations between structure and catalytic properties and, ultimately, rational catalyst design. With conventional supported metals, in contrast, there is no unique metal structure, and researchers are limited to determining relations between catalytic properties and some average structural properties.

A second, more practical goal motivates research with metal clusters: the discovery of catalysts with new structures and new activities and, especially, selectivities. Metal clusters offer many opportunities for new combinations of metal centers and surrounding ligands, and some of the common ligands (e.g., H, CO, small hydrocarbons) are important building blocks in the chemical industry. The initial high hopes for application of metal clusters as catalysts have not been realized, however; the primary difficulty is the lack of stability of the clusters.

The potential advantages of anchoring metal clusters to supports are the same as for acids, bases, and simpler metal complexes; when the catalytic species are isolated in a solid phase, separation from products is facilitated and corrosion is minimized. Whether anchoring can lead to increased stability of metal clusters is an open question; supported clusters may sometimes be even less stable than their analogues in solution because the support may react with the cluster.

9.1.2 SYNTHESIS

9.1.2.1 Clusters on polymers

An early attempt to prepare supported metal clusters with a functionalized support was reported by Collman *et al.* in 1972 [1]; these authors brought solutions containing $Rh_6(CO)_{16}$ and $Rh_4(CO)_{12}$ in contact with phosphine-functionalized poly(styrene-divinylbenzene), forming polymer-supported arene hydrogenation catalysts that are still not well characterized structurally [2]. Supported metal clusters with relatively well defined structures were described in 1978 [3]; these are tetrairidium carbonyls supported on phosphine-functionalized poly(styrene-divinylbenzene). A large fraction of the known supported metal clusters with well-defined structures have been prepared with functionalized poly(styrene-divinylbenzene). These are listed in Table 1.

Hydrocarbon polymers have been used frequently as catalyst supports for many acids, bases, and organometallic groups [4-6]. The polymers offer the advantages of easy functionalization, e.g., with phosphine, amine, hydroxyl, and thiol groups, and virtual inertness of the remainder of the support [21]. Consequently, a number of straightforward syntheses of supported metal clusters have been developed by simple analogy to solution organometallic chemistry (Chapters 1 and 4). The synthetic routes are summarized in the following sections.

It is emphasized that a major disadvantage of the polymer supports is their lack of stability at high temperatures (the exact limits being poorly defined) and their lack of mechanical stability--they tend to be pulverized in a stirred reactor.

9.1.2.1.a **Ligand association.** Cluster attachment has been carried out by ligand association involving a coordinatively unsaturated metal cluster and an electron-donor ligand of the support, as follows [11, 12]:

$$H_2Os_3(CO)_{10} + Ph_2P\text{-}\textcircled{P} \longrightarrow H_2Os_3(CO)_{10}(Ph_2P\text{-}\textcircled{P}) \qquad (1)$$

This method is limited by the small number of stable coordinatively unsaturated clusters (Chapter 4).

TABLE 1. Metal Clusters with Well-defined Structures Supported on Functionalized Poly(styrene-divinylbenzene).

Metal Framework Composition	Structures	Synthesis Method	Principal Characterization Method	Refs.
Ir_4	$Ir_4(CO)_{12-y}(Ph_2P{-}Ⓟ)_y$ y = 1 or 2	*In situ* synthesis from $Ir(CO)_2Cl(p\text{-toluidene})$	IR	[3, 7]
Ru_4	$H_4Ru_4(CO)_{12-x}(Ph_2P{-}Ⓟ)_x$ x = 1, 3, or 4	Ligand exchange	IR	[8]
Fe_2Pt	$Fe_2Pt(CO)_8(Ph_2P{-}Ⓟ)_2$	Ligand exchange	IR	[9]
$RuPt_2$	$RuPt_2(CO)_5(Ph_2P{-}Ⓟ)_3$	Ligand exchange	IR	[9]
$AuOs_3$	$HAuOs_3(CO)_{10}(Ph_2P{-}Ⓟ)$	Ligand exchange	IR	[10]
$AuOs_3$	$ClAuOs_3(CO)_{10}(Ph_2P{-}Ⓟ)$	Ligand exchange	IR	[10]
$PtOs_3$	$H_2PtO_3(CO)_{10}(Ph_2P{-}Ⓟ)_2$	Ligand exchange	IR	[10]
Os_3	$H_2Os_3(CO)_{10}Ph_2P{-}Ⓟ$	Ligand association	IR	[11, 12]
Os_3	$H_2Os_3(CO)_9Ph_2P{-}Ⓟ$	Ligand association and decarbonylation	IR	[11, 12]
Os_3	$Os_3(CO)_8(\mu\text{-}Cl)_2(Ph_2P{-}Ⓟ)_2$	Ligand exchange	IR	[13]
Os_3	$[HOs_3(CO)_{11}]^-Et_3\overset{+}{N}\text{-}CH_2{-}Ⓟ$	Ion exchange	IR	[14]
Os_3	$HOs_3(CO)_{10}O(CH_2)_2{-}Ⓟ$	Oxidative addition	IR	[15]
$RhOs_3$	$H_2RhOs_3(CO)_{10}(acac)Ph_2P{-}Ⓟ$	Ligand exchange	IR	[16]
Fe_3	$[HFe_3(CO)_{11}]^-Et_3\overset{+}{N}\text{-}CH_2{-}Ⓟ$	Ion exchange	IR	[17]
Fe_4S_4	$[Fe_4S_4(St\text{-}Bu)_4]^-Et_3\overset{+}{N}\text{-}CH_2{-}Ⓟ$	Ion exchange	far IR	[18]
Co_3C	$Co_3C(CO)_9\text{-}CH_2{-}Ⓟ$	?	(far) IR	[19]

9.1.2.1.b **Ligand exchange.** Most of the known polymer-supported clusters have been prepared by ligand exchange, e.g., as follows [8, 10]:

$$H_4Ru_4(CO)_{12} + Ph_2P\text{-}ⓟ \longrightarrow CO + H_4Ru_4(CO)_{11}(Ph_2P\text{-}ⓟ) \xrightarrow{Ph_2P\text{-}ⓟ} CO + H_4Ru_4(CO)_{10}(Ph_2P\text{-}ⓟ)_2, \text{ etc.} \quad (2)$$

$$ClAuOs_3(CO)_{10}(PPh_3) + Ph_2P\text{-}ⓟ \longrightarrow ClAuOs_3(CO)_{10}(Ph_2P\text{-}ⓟ) + PPh_3 \quad (3)$$

This method is widely applicable, failing, however, when side reactions interfere, as for example with $Ir_4(CO)_{12}$ and phosphine-functionalized polymer [3, 7].

9.1.2.1.c **Oxidative addition.** In a synthesis designed to mimic that occurring on oxide surfaces (discussed in Section 9.2), polymers functionalized with -OH groups have undergone oxidative addition with $Os_3(CO)_{12}$ to give $HOs_3(CO)_{10}O(CH_2)_2\text{-}ⓟ$ [15]. The analogous structure on silica, alumina, and other oxides is discussed in Section 9.2.

9.1.2.1.d **Ion exchange.** Cluster anions have been anchored to anion-exchange resins by simple ion exchange, e.g., as follows [17]:

$$[HFe_3(CO)_{11}]^-[HN(Et)_3]^+ + ⓟ\text{-}CH_2\text{-}\overset{+}{N}(Et)_3Cl^- \xrightarrow{CH_2Cl_2 \text{ solvent}} [HN(Et)_3]^+Cl^- + ⓟ\text{-}CH_2\text{-}\overset{+}{N}(Et)_3[HFe_3(CO)_{11}]^- \quad (4)$$

9.1.2.1.e **Synthesis variables.** Attempted syntheses often give mixtures of supported clusters, but there are a number of variables available for maximizing yields of the desired structures. Both block and random copolymers of phosphine-functionalized poly(styrene-divinylbenzene) have been prepared; these offer contrasting reactivities in the synthesis of supported metal clusters [7, 8]. Random copolymers with low concentrations of functional groups are best for synthesis of monosubstituted clusters [e.g., $Ir_4(CO)_{11}PPh_2\text{-}ⓟ$], whereas block copolymers (having regions with high concentrations of functional groups) are best for multiple substitution of the clusters. The flexible polymer networks of lightly crosslinked polymers have the ability to chelate a metal cluster; as many as four phosphine groups can be bonded to each tetraruthenium cluster in the polymer [8]. Polymers with high degrees of crosslinking [e.g., most macroporous poly(styrene-divinylbenzenes)] are rigid and favor monosubstitution. The degree of substitution can also be regulated with synthesis conditions such as time, temperature, and solvent; for example, manipulation of these variables allowed preparation of high yields of tetrairidium clusters having one or two phosphine ligands [7] and tetraruthenium clusters having one, three, or four phosphine ligands [8] (Table 1).

To sum up, the syntheses of polymer-supported clusters are often direct extensions of organometallic chemistry in solution, and manipulation of the synthesis conditions and physical properties of the support allows systematic variations in the ligand surroundings of the cluster. Extension of this class of supported metal clusters is expected to be straightforward. The straightforwardness of this chemistry stands in contrast to much of the chemistry of metal clusters on unfunctionalized oxides, discussed in Section 9.2.

9.1.2.2 Clusters on oxides

Functional groups such as those mentioned above can be tethered to oxides such as silica, e.g., by the condensation reaction of surface Si-OH groups with $PPh_2(CH_2CH_2)Si(OEt)_3$, a commercially available reagent [22]. Pendant diphosphine ligands have also been prepared [23], as well as mono- and diamine ligands [24], thiols [25], and others [26].

The oxide supports offer several advantages over the polymers: being the common industrial catalyst supports, they are inexpensive and available with wide ranges of physical properties such as surface area and pore size distribution; they have the stability required for use at high temperatures. It must be borne in mind, however, that the supported organometallics themselves are likely to be unstable at high temperatures.

Synthetic routes for attachment of metal clusters through these pendant ligands are essentially the same as those mentioned above for polymer supports. Castrillo *et al.* [27] developed alternative routes for anchoring of tetrairidium clusters by reaction of $Ir_4(CO)_{12}$ with pendant phosphine groups. The syntheses involved a base-catalyzed substitution and a trimethylamine-N-oxide-promoted ligand exchange. Mixtures of molecular clusters were formed on the support, however, and these are not the preferred synthetic routes.

Several authors have also succeeded in attaching metal clusters with tether ligands already bonded to the clusters (Table 2). For example, supported triosmium clusters were prepared by bringing $HOs_3(CO)_{10}S(CH_2)_2Si(CH_3)_2(OCH_3)$ in contact with silica; the resulting condensation reaction proceeded as with simple trimethoxysilanes [25].

Complications sometimes arise in the syntheses with functionalized oxide supports, since the surfaces of the oxides themselves (including -OH groups

TABLE 2. Supported Metal Clusters on Functionalized Oxides.

Metal Framework Composition	Structures	Synthesis Methods	Principal Characterization Methods	Refs.
Os_3	$H_2Os_3(CO)_9(Ph_2P\text{-SIL})^a$	Ligand association and decarbonylation	IR	[9, 11, 29-31]
	$H_2Os_3(CO)_{10}(Ph_2P\text{-SIL})$, among others	Ligand association		
Os_3	$HOs_3(CO)_9(\mu\text{-Cl})_2PPh_2(CH_2)_2\text{-SIL}$	Ligand exchange	IR, UV-visible	[13]
Os_3	$HOs_3(CO)_{10}S(CH_3)_2\text{-SIL}$	Condensation at $HOs_3(CO)_{10}\text{-}S(CH_2)_3(OCH_3)$ with passivated SiO_2 and reaction of thiol-functionalized SiO_2 with $Os_3(CO)_{10}(C_6H_8)$	IR	[25]
Ru_3	$HRu_3(CO)_{10}S(CH_3)_2\text{-SIL}$	Ligand exchange	IR	[25]
Ru_3	$H_2Ru_3(CO)_9\{\mu_3\text{-P}(CH_2)_2\text{-SIL}\}^b$	Condensation with $H_2Ru_3(CO)_9\{\mu_3\text{-P}(CH_2)_2Si(OEt)_3\}$	IR, UV-visible	[32]
Ru_3	$Ru_3(CO)_9[Ph_2P(CH_2)_2\text{-SIL}]_3$	Condensation with $Ru_3(CO)_9[PPh_2(CH_2)_2Si(OEt)_3]_3$	IR	[33, 34]
Ir_4	$Ir_4(CO)_9(PPh_3)_2[Ph_2P(CH_2)_3\text{-SIL}]$ $Ir_4(CO)_{10}(PPh_3)[Ph_2P(CH_2)_3\text{-SIL}]$, $Ir_4(CO)_{11}[Ph_2P(CH_2)_3\text{-SIL}]$	*In situ* synthesis from $Ir(CO)_2Cl(p\text{-toluidene})$	IR	[27, 35]
$AuOs_3$	$HAuOs_3(CO)_{10}(Ph_2P\text{-SIL})$	Ligand exchange	IR	[28]
Ru_4	$H_4Ru_4(CO)_{12-x}(Ph_2P\text{-SIL})_x$ x = 1, 2, or 3	Ligand exchange	IR	[9]

(continued on following page)

TABLE 2. (Continued)

Metal Framework Composition	Structures	Synthesis Methods	Principal Characterization Methods	Refs.
Ru_4	$H_4Ru_4(CO)_{11}[Ph_2P(CH_2)_2\text{-SIL}]$	Ligand exchange	IR, UV-visible	[36, 37]
Ru_4	$H_4Ru_4(CO)_{11-x}[(Ph_3P)_xPh_2P(CH_2)_2\text{-SIL}]$ x = 1 or 2	Ligand exchange	IR, UV-visible	[36, 37]
Ru_4	$H_4Ru_4(CO)_8[Ph_2P(CH_2)_2\text{-SIL}]_4$	Condensation of $H_4Ru_4(CO)_8[Ph_2P(CH_2)_2Si(OEt)_3]_4$ with SiO_2	IR, UV-visible	[34, 38]
$ReCo_3$	$[FeCo_3(CO)_{12}]^-[\overset{+}{H}_3N(CH_2)_3\text{-SIL}]$	Ion exchange	IR, UV-visible	[39]

[a]SIL is silica.
[b]Also prepared on Al_2O_3, TiO_2, MgO, and ZnO.

with acidic and/or basic character and Lewis acid sites such as coordinatively unsaturated Al^{3+} ions) often react with the cluster in competition with the tethered functional groups (Section 9.2). To minimize these complications, some workers have passivated a fraction of the silica surface (by reaction with methanol, giving pendant methoxyl groups) prior to incorporation of the tethered ligand or clusters [25]. This procedure is at best partially successful [28]; the more nearly inert hydrocarbon polymers often appear to be the preferred supports if the goal is to prepare analogues of molecular clusters.

A summary of the metal clusters anchored to functionalized metal-oxide supports is given in Table 2. Many of these are entirely analogous to the polymer-supported clusters listed in Table 1.

9.1.3 STRUCTURAL CHARACTERIZATION

The principal characterization technique for each of the supported clusters listed in Tables 1 and 2 is infrared spectroscopy. All the listed clusters have been identified by comparison of their carbonyl stretching bands with those of molecular analogues, as illustrated in Section 6.2. Some of these identifications are quite firm, especially when a family of variously substituted clusters has been prepared (e.g., phosphine-substituted tetrairidium and tetraruthenium). The UV-visible spectra are of only modest value; the far-infrared spectra are quite useful, but this technique is limited, as discussed in Section 6.2. A much broader range of techniques has been applied to characterize metal clusters on unfunctionalized oxides, as discussed in Sections 6.2 and 6.3.

9.1.4 REACTIVITY

The reactivities of the supported metal clusters listed in Tables 1 and 2 are for the most part poorly defined. It has been difficult to observe changes in the ligand environments of supported clusters without significant accompanying changes in the cluster frameworks; the most commonly observed reactions involve drastic restructuring of the clusters, typically with the formation of metal particles and sometimes mononuclear metal complexes.

The most thorough set of reactivity data has been determined by infrared spectroscopy with triosmium clusters anchored to poly(styrene-divinylbenzene) and to silica through pendant phosphine ligands [11, 12, 29, 31]. The supported cluster $H_2Os_3(CO)_{10}Ph_2P$–Ⓟ (or its silica-supported analogue), upon refluxing in hexane, loses a CO ligand to give the coordinatively unsaturated cluster $H_2Os_3(CO)_9Ph_2P$–Ⓟ [11, 12].

The unsaturated clusters react in the same way as their soluble analogue, e.g., undergoing ligand association with alkynes and alkenes to give stable adducts [11]. These triosmium clusters are the only supported clusters known to give stable coordinatively unsaturated species by simple ligand dissociation. The analogous silica-supported clusters $Os_3(CO)_9(\mu\text{-}Cl)_2PPh_2CH_2CH_2$-[SIL] also have labile CO ligands, and one CO ligand appears to be easily and reversibly exchanged with but-1-ene [13].

All the supported metal clusters are subject to degradation processes, especially at elevated temperatures. The most typical products are aggregates (or crystallites) of metal; mononuclear metal complexes may also be formed. These processes are complex and for the most part poorly characterized. They have been investigated more fully with unfunctionalized oxide supports (Section 9.2).

A systematic investigation of metal aggregation was reported with rhodium cluster precursors in phosphine-functionalized poly(styrene-divinylbenzene) [40]. The reduction and aggregation took place in the presence of hydrogen and were manifested by changes in the infrared spectrum of adsorbed CO and in the kinetics of catalytic alkene and arene hydrogenation reactions. The aggregation was apparently inhibited by polymer-bound phosphine groups chelating the rhodium clusters and aggregates. Similarly, samples of rhodium clusters on phosphine- or amine-functionalized silica were found to give small aggregates of metal in the presence of H_2 or H_2 + ethylene [41]; the rhodium aggregates were characterized by transmission electron microscopy and found to have regularities of structure, suggesting that they consisted of aggregates of a few hexarhodium units [42].

It is important to be aware of the complications that these cluster decomposition reactions can cause. They may occur even during synthesis of the supported clusters, perhaps only to a small degree; if the new structures go undetected, then incorrect conclusions about cluster reactivity and catalytic activity can easily be inferred.

In summary, there have been only a few systematic investigations of reactivities of metal clusters on functionalized supports. Often the reactions are complex; even if a structurally well-defined cluster can be prepared on a support, it will often react to give complex mixtures. Understanding of surface reactivity is expected to develop most rapidly for clusters with stable metal frameworks (such as tetrairidium and triosmium); characterization by the concerted application of a variety of spectroscopic techniques (plus determination of stoichiometries of reactions involving gain or loss of ligands) will lead to the most rapid progress.

Catalytic activites of some of the supported clusters mentioned here are discussed in Section 9.3.

REFERENCES

1. J. P. Collman, L. S. Hegedus, M. P. Cooke, J. R. Norton, G. Dolcetti and D. N. Marquardt, J. Am. Chem. Soc., 94 (1972) 1789.
2. M. S. Jarrell and B. C. Gates, J. Catal., 54 (1978) 81.
3. J. J. Rafalko, J. Lieto, B. C. Gates and G. L. Schrader, Jr., J. Chem. Soc. Chem. Commun. (1978) 1063.
4. P. Hodge and D. C. Sherrington (Eds.), Polymer-supported Reactions in Organic Synthesis, Wiley, New York, 1977.
5. C. U. Pittman, in G. Wilkinson, F. G. A. Stone and E. W. Abel (Eds.), Comprehensive Organometallic Chemistry, Vol. 8, Pergamon, Oxford, 1982, p. 553.
6. B. C. Gates and J. Lieto in Encyclopedia of Polymer Science and Engineering, Vol. 2, Wiley, New York, 1985, p. 708.
7. J. Lieto, J. J. Rafalko and B. C. Gates, J. Catal., 62 (1980) 149.
8. Z. Otero-Schipper, J. Lieto and B. C. Gates, J. Catal., 63 (1980) 175.
9. R. Pierantozzi, K. J. McQuade, B. C. Gates, M. Wolf, H. Knözinger and W. Ruhmann, J. Am. Chem. Soc., 101 (1979) 5436.
10. R. Pierantozzi, K. J. McQuade, and B. C. Gates, in T. Seiyama and K. Tanabe (Eds.), New Horizons in Catalysis, Part B, Kodansha, Tokyo, Elsevier, New York, 1981, p. 941.
11. M. B. Freeman, M. A. Patrick and B. C. Gates, J. Catal., 73 (1982) 82.
12. J.-B. N'Guini Effa, J. Lieto and J.-P. Aune, J. Mol. Catal., 15 (1982) 367.
13. M. Wolf, J. Lieto, B. A. Matrana, D. B. Arnold, B. C. Gates and H. Knözinger, J. Catal., 89 (1984) 100.
14. H. Marrakchi, J.-B. N'Guini Effa, J. Lieto and J.-P. Aune, to be published.
15. J. Lieto, M. Prochaska, D. B. Arnold, and B. C. Gates, J. Mol. Catal., 31 (1985) 89.
16. J. Lieto, M. Wolf, B. A. Matrana, M. Prochazka, B. Tesche, B. C. Gates and H. Knözinger, J. Phys. Chem., 89 (1985) 991.
17. J.-B. N'Guini Effa, J. Lieto and J.-P. Aune, Inorg. Chim. Acta, 65 (1982) L105.
18. M. D. Monteil, J.-B. N'Guini Effa, J. Lieto, P. Verlaque and D. Benlian, Inorg. Chim. Acta, 76 (1983) L309.
19. H. Atif, J.-B. N'Guini Effa, J. Lieto and J.-P. Aune, to be published.
20. H. Menzel, W. P. Fehlhammer and W. Beck, Z. Naturforsch., 37B (1982) 201.
21. J. Lieto, D. Milstein, R. L. Albright, J. V. Minkiewicz and B. C. Gates, CHEMTECH, 13 (1983) 46.
22. K. G. Allum, R. D. Hancock, S. McKenzie and R. C. Pitkethly, Proc. 5th Int. Congr. Catal., North-Holland, Amsterdam, 1973, p. 477.
23. L. L. Murrell, in J. J. Burton and R. L. Garten (Eds.), Advanced Materials in Catalysis, Academic Press, New York, 1977.
24. H. Knözinger and E. Rumpf, Inorg. Chim. Acta., 30 (1978) 51.
25. T. Castrillo, K. Knözinger, J. Lieto and M. Wolf, Inorg. Chim. Acta, 44 (1980) L239.
26. H.-P. Boehm and H. Knözinger, in J. R. Anderson and M. Boudart, (Eds.), Catalysis - Science and Technology, Vol. 4, Springer, Berlin, 1983.

27. T. Castrillo, H. Knözinger, M. Wolf and B. Tesche, J. Mol. Catal., 11 (1981) 151.
28. M. Wolf, H. Knözinger and B. Tesche, J. Mol. Catal., submitted for publication.
29. S. C. Brown and J. Evans, J. Chem. Soc. Chem. Commun., (1978) 1063.
30. J. Evans and B. P. Gracey, J. Chem. Soc. Chem. Commun., (1980) 852.
31. S. C. Brown and J. Evans, J. Mol. Catal., 11 (1981) 143.
32. S. L. Cook and J. Evans, J. Chem. Soc. Chem. Commun., (1983) 713.
33. D. K. Liu and M. S. Wrighton, J. Am. Chem. Soc., 104 (1982) 898.
34. Y. Doi, K. Yano, A. Yokota, H. Miyake and K. Soga, Proc. 8th Int. Congr. Catal., DECHEMA, Frankfurt, 1984, Vol. 4, p. 689.
35. T. Castrillo, H. Knözinger and M. Wolf, Inorg. Chim. Acta, 45 (1980) L235.
36. T. Castrillo de Castro, Dissertation, University of Munich, 1982.
37. T. Castrillo and H. Knözinger, to be published.
38. Y. Doi, K. Soga, K. Yano and Y. Ono, J. Mol. Catal., 19 (1983) 359.
39. R. Hemmerich, W. Keim and M. Röper, J. Chem. Soc. Chem. Commun. (1983) 428.
40. M. S. Jarrell, B. C. Gates and E. D. Nicholson, J. Am. Chem. Soc., 100 (1978) 5727.
41. E. W. Thornton, H. Knözinger, B. Tesche, J. J. Rafalko and B. C. Gates, J. Catal., 62 (1980) 117.
42. B. Tesche, H. Knözinger and B. C. Gates, J. Catal., 64 (1980) 232.

Section 9.2

METAL CLUSTERS ON UNFUNCTIONALIZED INORGANIC OXIDES

R. PSARO and R. UGO

9.2.1 INTRODUCTION

Molecular metal clusters of increasing nuclearity constitute a new generation of organometallic complexes which almost fill the gap between the molecular state (simple metal complexes) and the metallic state (small metallic particles) [1-12].

Today, the sizes of very large transition metal clusters approach those of very small metal particles, so that it is possible to propose a progressive link between molecular clusters and metal particles. In the one case we are dealing with a kind of very small arrangement of metal atoms stabilized by appropriate ligands, whereas in the other, the very small crystallites are stabilized by interaction with the surface of an oxide or are trapped in the frame of a zeolite or in an inert matrix.

Several kinds of relationships between metal clusters and metallic arrangements (e.g., comparison of the electronic characteristics and of the reactivity of ligands and chemisorbed molecules, etc.) have been discussed by a number of authors [13-28].

It was pointed out early that a real comparison between very small particles (VSP) and molecular clusters (MC) would require reversible poisoning of the surface of a VSP by a kind of ligand L (such as CO) that easily forms a low-valent transition metal cluster frame. In this respect there is a great interest in the investigation of thermal treatments which could regenerate the VSP according to the following equilibrium:

$$xL + VSP \rightleftharpoons MC \quad (1)$$

Such a decomposition was observed originally in some mass spectra of metal clusters [29].

Therefore, a large amount of work (based on these few elegant speculations) has been devoted in the last few years to the investigation of the thermal and chemical behavior of metal clusters of various nuclearities supported on various metal oxides or trapped in the frameworks of zeolites [30-35]. The main purpose of this approach was threefold:

(1) To prevent an easy aggregation into large particles of the original metal framework of the cluster to obtain a reversible ligand dissociation from the metallic arrangement [36].

(2) To investigate and reach direct evidence of the existence of chemical

interactions between the original metal clusters and the so-called inert supports [37].

(3) To prepare, by thermal treatment of supported clusters, very highly dispersed and tailor-made metallic particles [31].

In the course of the various investigations it has been shown that by interaction of metal clusters with surface groups, surface organometallic-like compounds are easily produced as intermediates; the solid surface therefore behaves as a real ligand [38] toward the cluster species, which may maintain the original nuclearity, or aggregate to give higher nuclearities, or often break up to give monomeric metallic species [26, 30, 31, 39]. This surface reactivity is the origin of a kind of surface organometallic chemistry and of a new class of hybrid catalysts which quite often exhibit unusual activities and selectivities in comparison with conventional homogeneous and supported catalysts [26, 30, 31, 33, 40-50].

The purpose of this chapter is to describe the organometallic chemistry associated with transition metal clusters supported on inorganic oxides.

Provided that substantial aggregation can be prevented, this application of metal cluster compounds can add a further dimension and an alternative method to the preparation of highly dispersed supported metal catalysts, approaching the situation where many atoms are surface atoms [25, 31, 51, 52]. Moreover, clusters can offer the following differences (advantages) in comparison with conventionally prepared [52-55] supported metal catalysts:

(1) Nonaqueous methods of catalyst preparation can be employed since metal cluster compounds are usually soluble (although often sparingly) in organic solvents.

(2) Clusters do not usually contain halogen atoms so that the resulting catalyst can be halide-free.

(3) The high-temperature reduction by hydrogen, used in conventional catalyst preparation, should be unnecessary, since in many cases metal clusters already have the metal formally in the zerovalent state. Cluster-derived catalysts may therefore by activated under mild conditions [33, 45, 56-59].

(4) The discovery of an ever-increasing number of mixed-metal or heteronuclear cluster compounds [9] offers the possibility of preparing

bimetallics or alloys of precisely known stoichiometry (composition) in a controlled manner [25, 60-62].

9.2.2 METHODS FOR PREPARATION AND CHARACTERIZATION

Before going into a detailed description of the organometallic chemistry occurring on the surfaces of different supports, we deal briefly with the methodologies of preparation and characterization of supported metal clusters. It is important to point out that the surface organometallic chemistry of a cluster deposited on a support is affected by the following: (1) the techniques of preparation of the original samples containing the supported cluster; (2) the preliminary chemical and thermal treatments of the surface of the support; and (3) the conditions of the surface transformations (e.g., resulting from heating in vacuum or in an atmosphere of an inert gas, the presence of water in the gas phase or on the support, etc.). Finally, the degree of characterization and detailed evidence of the surface chemistry from the initial physisorption stage up to the formation of metallic particles is dependent upon the sophistication of the methodologies used in the characterization of the surface species (see Chapter 6). Therefore, we devote only a few words to these subjects.

9.2.2.1 Techniques for preparation of supported or entrapped metal clusters

Clusters are usually supported by conventional techniques such as impregnation from solutions (wet or dry) [63-65] or dry grinding [66]. These processes may be carried out in air, but sometimes an inert atmosphere is necessary [64, 65]. One interesting technique, which provides clusters trapped in the cavities of the structure of a zeolite, requires ion-exchange methods, followed by controlled reduction [67] (see Chapter 8). Another technique, which has not been used often but which has substantial potential, is the sublimation of the cluster directly onto the support [68, 69]. The potential of this method follows from the fact that many neutral transition metal clusters are sufficiently volatile to be sublimed in high vacuum without decomposition; moreover, with this technique the presence of solvents and of other impurities such as water can be completely avoided. In any case, the surface organometallic chemistry of a cluster, when deposited on a well-defined support, can be quite dependent on the methodology of impregnation [70].

In addition, the treatment of the support surface, such as the control of the degree of dehydration or dehydroxylation, the decarbonation or deoxygenation (accomplished by thermal or vacuum treatments), is extremely important [45]. Also, the surface area and the intrinsic nature of the basic or acidic sites (Lewis, Brønsted, etc.) [38, 72-74] are important in defining the surface organometallic chemistry. Moreover, the surface of a support can be chemically modified by functionalization with, for example, phosphine or amine groups. In this latter case, chemical bond formation between the polynuclear organometallic species and the functional groups of the surface produces an "anchored" cluster in the manner followed extensively in recent years for immobilizing soluble catalysts [30, 71]. This approach is discussed in Section 9.1.

9.2.2.2 **Methodologies of characterization of surface species.** The utilization of appropriate methods for the characterization of surface species (see Chapter 6) poses severe problems and underlines the shortcomings of physical techniques [75-77]. In fact, in this area of surface organometallic chemistry the required information often falls outside the range of sensitivity of traditional surface science techniques and beyond that obtainable by molecular structure methods such as X-ray crystallography, vibrational and NMR spectroscopy, etc. No one technique is satisfactory in direct isolation and characterization, and, for example, until ultra-high resolution electron microscopy becomes available on a nearly routine basis, reliance on a combination of techniques and/or indirect evidence is necessary.

One interesting indirect technique is the preparation of so-called molecular models of surface species, which are complexes with special model ligands (like silanols) [59, 78-81] having structures that can be determined by classical direct methodologies. The comparison of some properties (e.g., vibrational spectra) of well-known models and surface species provides indirect support for the definition of possible structures of the surface species [92]. The methodologies most frequently used for the preparation and characterization of surface organometallic species are referred to in the following pages.

Metal particles, which may finally be produced either thermally or chemically, are characterized by those methodologies normally used in this area of surface science [83].

9.2.3 SURFACE ORGANOMETALLIC CHEMISTRY OF FIRST-ROW TRANSITION METAL CARBONYL CLUSTERS

We describe only the organometallic chemistry which involves molecular carbonyl clusters, although often a related surface chemistry is obtained starting from simple mononuclear metal carbonyls [30, 31, 66, 84-88]. We first discuss the behavior of clusters of the first-row transition metals because these typically provide higher reactivity in comparison with those of the second- and third-row transition metal clusters [89-91].

9.2.3.1 Iron clusters

A large series of investigations concerned the interaction of $Fe_3(CO)_{12}$ with inorganic oxides such as silica [92, 93], alumina [92, 93, 94], magnesia [92, 95], and zinc oxide [92]. The physisorption of $Fe_3(CO)_{12}$ occurs when it is sublimed onto a silica surface which is thermally treated under vacuum at temperatures between 298 and 773 K. Some $Fe(CO)_5$ which is formed on the surface (probably during the sublimation of $Fe_3(CO)_{12}$) can be removed by evacuation at 209 K (SCHEME 1):

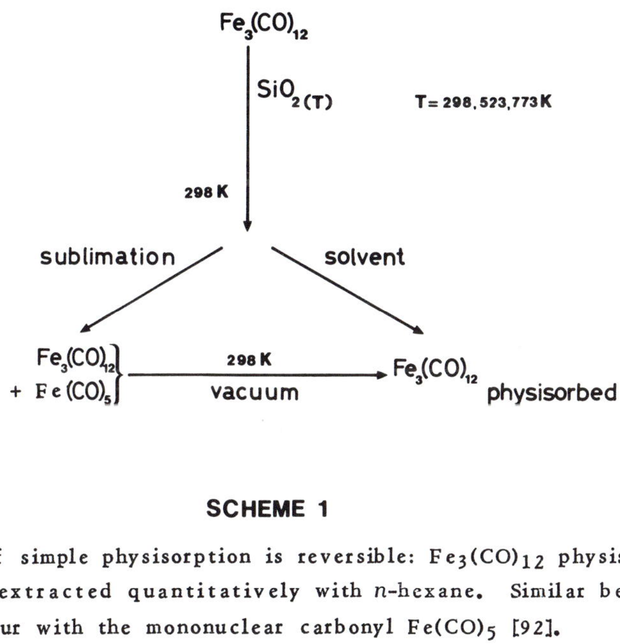

SCHEME 1

The process of simple physisorption is reversible: $Fe_3(CO)_{12}$ physisorbed on silica can be extracted quantitatively with *n*-hexane. Similar behavior is reported to occur with the mononuclear carbonyl $Fe(CO)_5$ [92].

Thermal treatment under vacuum, at temperatures between 323 and 373 K, of physisorbed $Fe_3(CO)_{12}$ produces a complete decomposition of the original cluster. Partial oxidation of the iron atoms with parallel hydrogen evolution occurs, together with the formation of metallic particles. In summary, the interaction of the silanol groups of the silica surface with both $Fe_3(CO)_{12}$ and $Fe(CO)_5$ is rather weak; however, increasing temperature leads to partial oxidation of Fe(0) with hydrogen evolution by electrophilic attack of silanol groups.

The behavior of these organometallics on alumina is much more complex.

When a hexane solution of $Fe_3(CO)_{12}$ is stirred with alumina (η or γ) pretreated under vacuum at 298 K, some interaction of the cluster occurs as demonstrated by the formation of a pink coloration of the alumina. A similar process occurs when $Fe_3(CO)_{12}$ is supported by sublimation. The analysis of the gas phase during this process indicates no significant evolution of CO or CO_2. This process corresponds to a real chemisorption because the surface species cannot be extracted with *n*-hexane. When the surface metal loading is higher than 2%, a simple physisorption of $Fe_3(CO)_{12}$ adds to the initial chemisorption process. In fact some unaltered $Fe_3(CO)_{12}$ can be extracted with *n*-hexane.

The chemisorbed iron carbonyl species present on the surface of the pink-colored sample was characterized, in particular by infrared and UV spectroscopies. The spectra are in good agreement with the presence on the surface of the anion $[HFe_3(CO)_{11}]^-$, although some differences are observed in the infrared spectrum with respect to that of the corresponding molecular species $[NR_4]$ $[HFe_3(CO)_{11}]$ [96]. In fact, the pattern of the carbonyl bands is quite in agreement but the stretching frequencies of the linear carbonyl ligands are shifted by about 10 cm^{-1} to higher frequencies and those of the bridging carbonyl ligand by about 100 cm^{-1} to lower frequencies. This kind of shift can be explained as an effect produced by the interaction of the bridging carbonyl of the anionic cluster with a surface aluminum cation acting as Lewis acid [72, 97-99], by analogy with the known interaction of some molecular metal carbonyl clusters with a series of Lewis acids [100, 101]:

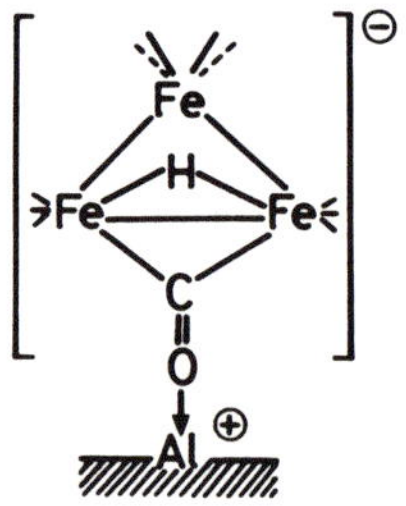

The anionic nature of the surface species (which cannot be extracted with simple solvents) is confirmed by its extraction with a hexane solution of $[NEt_4]Cl$. The chemistry occurring during the chemisorption process and the further transformations of the species $[HFe_3(CO)_{11}]^-$ bound to the surface are summarized in SCHEME 2.

The formation of the anionic hydrido cluster implies the nucleophilic attack of the original $Fe_3(CO)_{12}$ by surface -OH groups with simultaneous elimination of CO_2 (which remains strongly adsorbed on the alumina surface [102]). This kind of surface organometallic reaction is quite similar to the reaction which occurs between $Fe_3(CO)_{12}$ or $Fe(CO)_5$ and H_2O in alkaline solutions [103, 104]. On the alumina surface, the formation of $[HFe_3(CO)_{11}]^-$ is observed also in the absence of physisorbed molecular water under more drastic conditions. After pretreatment of γ-alumina either at 573 or at 773 K in vacuum, physisorption of $Fe_3(CO)_{12}$ first occurs; after evacuation at 298 K, heating at 333 K induces the formation of $[HFe_3(CO)_{11}]^-$ on the surface (SCHEME 2).

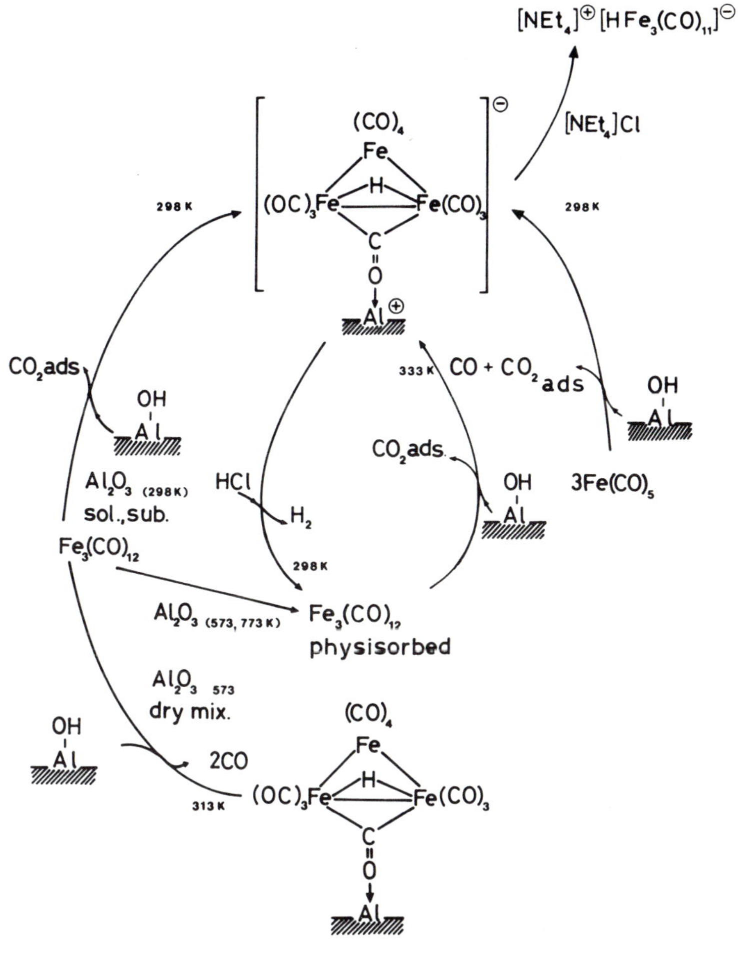

SCHEME 2

This latter observation and the fact that the anionic surface species can be extracted only by ion-exchange techniques would suggest that the counter-cation is an aluminum acidic site (and not a surface H_3O^+ species), in agreement with the hypothesis that the strongly basic character of some surface -OH groups of alumina [105, 106] is responsible of the original nucleophilic attack.

Recently, Iwasawa *et al.* [93] reported a different reactivity of $Fe_3(CO)_{12}$ with the surface of γ-Al_3O_3. On the basis of EXAFS, infrared, and UV spectra (together with analysis of the gas evolved during the chemisorption process), they proposed an oxidative addition of a surface -OH group to a Fe-Fe bond of $Fe_3(CO)_{12}$, with the formation of the following surface cluster:

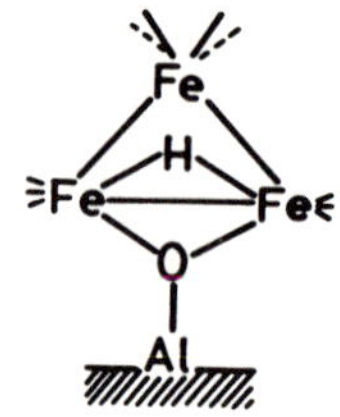

The chemisorption of $Fe_3(CO)_{12}$ on γ-Al_2O_3 (thermally treated at 573 K for only 1.5 h) was performed by dry mixing, followed by treatment under vacuum at 313 K for 2 h.

Two groups have investigated in detail the thermal decomposition of physisorbed and chemisorbed species originating from the interaction of $Fe_3(CO)_{12}$ with the alumina surface [51, 92, 94, 107, 108]. The decomposition pattern depends strongly on the pretreatment of the support. Decarbonylation is complete at about 383 K (independent of the pretreatment of the support). This process is irreversible. If the surface of the alumina is hydrated, CO is evolved together with H_2 and minor amounts of CO_2. The amount of hydrogen evolved (about 2.5 moles for each mole of cluster) cannot be fully accounted for by the surface reaction

$$[HFe_3(CO)_{11}]^- + H^+ \rightarrow H_2 + [Fe_3(CO)_{11}] \qquad (2)$$

An electrophilic attack of surface protons on the zerovalent unsaturated species $[Fe_3(CO)_{11}]$, with some oxidation of Fe(0) to Fe(III), must also occur.

On strongly dehydrated alumina (pretreated at 773 K), only CO evolution is observed, whereas H_2 evolution is a minor process. In this case oxidation of Fe(0) can be excluded, so that much of the iron is agglomerated to form metallic particles covered with CO. The evidence for formation of carbon in this latter thermal decomposition indicates the presence of iron(0) particles, which are known to favor CO dissociation to carbon [109, 110].

Much less work has been done on the interaction between $Fe_3(CO)_{12}$ and zinc oxide, because ZnO is less easily accessible to infrared spectroscopy. It appears that hydrated zinc oxide behaves like alumina, in that it instantaneously produces the surface species $[HFe_3(CO)_{11}]^-$ [92]. On dehydrated zinc oxide (pretreatment at 573 K under vacuum), simple physisorption of $Fe_3(CO)_{12}$ occurs, but after removal of the solvent at 298 K by vacuum treatment, the green color of the surface changes to yellow-brown. Unfortunately, the corresponding surface species could not be characterized due to the difficulty of getting good and reproducible infrared spectra.

It appears that ZnO behaves quite similarly to alumina. However, the interaction of $Fe(CO)_5$ with hydrated zinc oxide produces a simple physisorption, whereas on hydrated alumina the instantaneous condensation to form $[HFe_3(CO)_{11}]^-$ occurs (SCHEME 2), in agreement with a higher basicity of the alumina surface and a high mobility of the adsorbed organometallic species. This kind of condensation is observed in alkaline solutions of $Fe(CO)_5$ [111].

A support of high basicity such as magnesia [112] behaves in the same way. Both $Fe_3(CO)_{12}$ and $Fe(CO)_5$ react instantaneously with the surface of fully hydroxylated magnesia to form the surface anion $[HFe_3(CO)_{11}]^-$ (SCHEME 3) [92, 95].

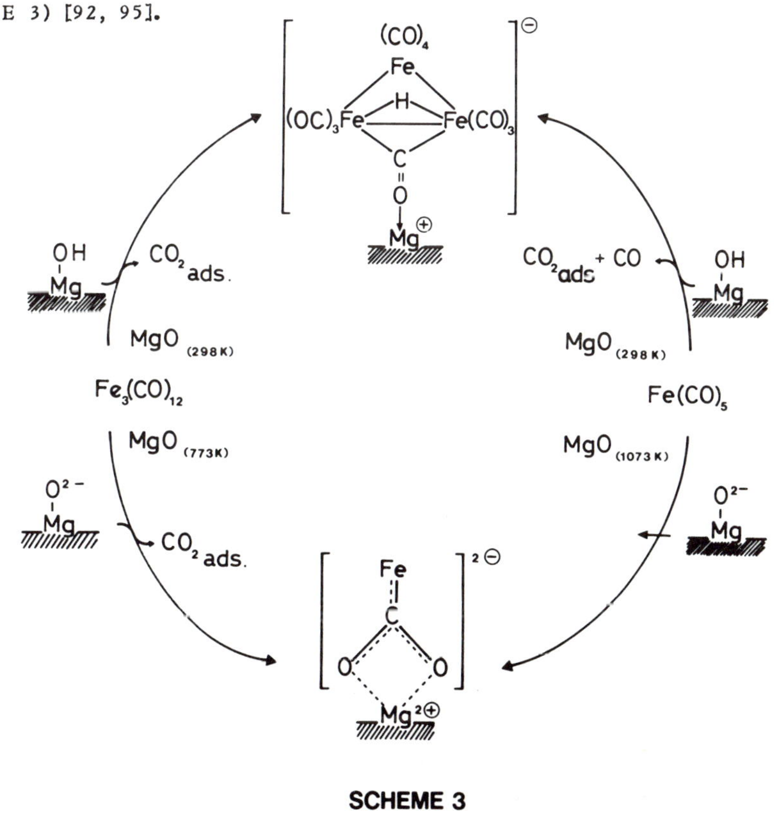

SCHEME 3

On dehydrated magnesia (pretreated at 673 K under vacuum), the sublimation of $Fe_3(CO)_{12}$ or its adsorption from *n*-hexane solution leads to a complex mixture of surface organometallic species which has not been completely characterized [92, 95]. According to the authors, many species are present, namely physisorbed $Fe_3(CO)_{12}$, chemisorbed $[HFe_3(CO)_{11}]^-$ (interacting with a magnesium acid center as counter-cation), and other mononuclear carbonyl species bound to the magnesia surface by complex ionic interactions.

Theolier and Basset [113] proposed that on highly dehydroxylated magnesia, $Fe_3(CO)_{12}$ is chemisorbed as $Fe(CO)_4(COOMg)$, in agreement with the behavior of $Fe(CO)_5$ on the same support [88]. On this support, the strong basic sites [114] attack coordinated CO by a nucleophilic mechanism with parallel breakup of the metallic cluster frame to form monomeric species (SCHEME 3).

The thermal decomposition of such complex surface mixtures has been studied by means of infrared and Mössbauer spectroscopies, ferromagnetic resonance (FMR), magnetic measurements, and analysis of the gas evolved [95]. Complete decarbonylation occurs at 383 K on hydrated magnesia, whereas on dehydrated magnesia the complete decarbonylation is achieved at about 403 K.

On hydrated magnesia, some hydrogen evolution is observed, as on hydrated alumina, with formation of monomeric Fe(II) ions, which probably occupy interstitial positions in the magnesia lattice. These surface-oxidized entities are responsible for a water gas shift reaction which occurs between the surface water and the free carbon monoxide evolved during the thermal treatment. As a consequence, only a fraction of the original iron atoms aggregate to form highly dispersed metal particles, which unfortunately cannot be fully characterized due to the simultaneous presence of Fe(II) [95].

On dehydrated magnesia, the irreversible decarbonylation is not accompanied by corresponding amounts of hydrogen, so that small particles of metallic iron, which exhibit superparamagnetic properties, are produced. Obviously, the original cluster nuclearity is not preserved; however, the resulting particles have a rather narrow size distribution between 0.6 and 1.7 nm, corresponding to aggregates of about 100-150 atoms. It must be pointed out that such aggregation on the support surface is more pronounced than that achieved during the simple pyrolysis of low-nuclearity clusters (3 or 4 atoms) to generate clusters of higher nuclearity (usually up to 6-12 atoms) [115-119]. Magnetic measurements indicate that only 50% of the original iron is transformed into metal particles, while the other 50% is present as Fe(II). Since the corresponding amount of hydrogen is not evolved, the oxidation of the iron(0) atoms must be ascribed to a process such as CO dissociation occurring on the metal particles [108, 109].

An interesting, clear example of organometallic interactions is given by the tetrameric cluster $[(\eta^5\text{-}C_5H_5)Fe(CO)]_4$, characterized by a high thermal stability [120, 121] and by carbonyl ligands bridging the Fe_3 face of a tetrahedron [122]. This kind of bridging ligand is strongly basic towards molecular Lewis acids [123, 124]; in addition, the carbon atom should not be susceptible to attack by nucleophiles on both electronic and steric grounds [125]. In agreement with such statements, the interaction between a highly dehydroxylated surface of alumina and a dichloromethane solutions of $[(\eta^5\text{-}C_5H_5)Fe(CO)]_4$ produces a simple molecular adsorption [126]. The infrared spectrum of the final sample (see Section 6.2) shows two carbonyl bands (a fairly sharp band at higher frequency and a broad band at lower frequency), compared with the single band of the original cluster, supporting a donor-acceptor interaction between only one carbonyl ligand and a cationic aluminum surface site (as shown below); similar spectral changes are observed when this cluster interacts with molecular Lewis acids in solution [123, 124].

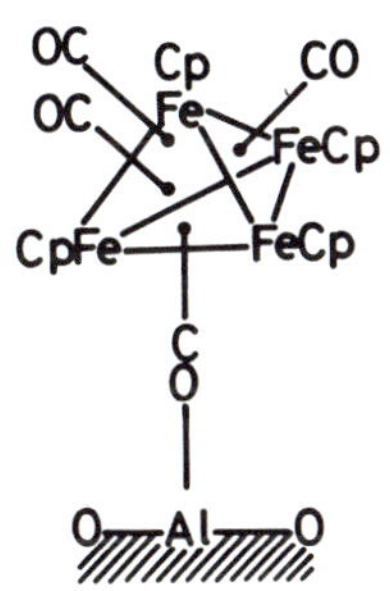

The adsorption is quite strong; desorption occurs only by addition of an excess of a strong base like pyridine [126]. The infrared frequency shifts observed when $[\eta^5\text{-}C_5H_5Fe(CO)]_4$ interacts with a series of Lewis acids or with the surface of dehydroxylated alumina have been compared. The magnitude of the shifts for the alumina is intermediate, indicating that the relative electron pair acceptor strength follows the sequence:

$$AlBr_3 > \text{dehydroxylated } Al_2O_3 > AlR_3$$

9.2.3.2 Cobalt clusters

The interaction of a series of cobalt carbonyl clusters with alumina, silica, and zinc oxide surfaces was investigated by temperature-programmed decomposition (TPDE) to produce metallic catalysts [48, 63, 108, 123, 128]. However, the knowledge of the intermediate surface organometallic chemistry is poor.

Nakamura *et al.* [129] studied the behavior of $Co_2(CO)_8$ supported on γ-Al_2O_3 and SiO_2 by adsorption experiments, analysis of gas evolution, infrared spectroscopy, and the TPDE technique. When a pentane solution of $Co_2(CO)_8$ is adsorbed on silica previously treated at 773 K, 2 moles of CO per mole of $Co_2(CO)_8$ are evolved at 298 K. In agreement with earlier infrared data [130], this observation indicates the formation of $Co_4(CO)_{12}$ on the silica surface. The usual synthesis of $Co_4(CO)_{12}$ from $Co_2(CO)_8$ is achieved by refluxing of a petroleum ether solution [131]; but the reaction is easily reversed at atmospheric pressure [132], showing that the thermodynamically more stable product is $Co_2(CO)_8$. It is quite relevant that the same organometallic reaction can be performed on the silica surface even under mild conditions. The nature of the interaction of $Co_2(CO)_8$ with an alumina surface depends on the state of dehydration of the support and on the metal loading. When a pentane solution of $Co_2(CO)_8$ interacts at 298 K with alumina surfaces having different hydroxyl contents, the solution becomes colorless and some evolution of CO occurs, suggesting a chemisorption process. Quantitative experiments revealed that the surface saturation of partially and fully dehydroxylated aluminas corresponds to metal loadings of 1.5 and 1.3%, respectively. During the impregnation followed by a treatment in flowing helium at 298 K, only two moles of CO per mole of $Co_2(CO)_8$ are evolved. These data suggest a behavior corresponding to that observed on silica surfaces.

The authors [129] assume, only on the basis of TPDE spectra (the infrared data are very poor), that the formation of physisorbed $Co_4(CO)_{12}$ is less easy than the formation of oxidized surface species, such as $[Co_2(CO)_6(OAl{\equiv})_2]$. Obviously this system requires further systematic investigation.

The surface chemistry of the cluster $[Co_3(CO)_9CCH_3]$, supported on silica and alumina, was investigated in some detail recently [133]. $[Co_3(CO)_9CCH_3]$, in contrast to other cobalt carbonyl clusters, is thermally robust [134, 135] (it can be easily sublimed) and resistant to air oxidation [136]; the cluster has therefore been brought in contact with the surface by sublimation. Independent of the thermal pretreatments of the support surface, simple physisorption occurs in a first step, giving rise to a surface species which is quite stable in vacuo. Only upon addition of oxygen does a gradual disappearance of all carbonyl stretching bands take place, probably due to the destruction of the original cluster. This latter observation is quite interesting, because it is reported that $[Co_3(CO)_9CCH_3]$ is insensitive to oxygen, either as a solid or in solution [136].

The surface reactivity of this cluster supported on alumina is strongly dependent on the preliminary thermal treatment of the surface. Whereas only physisorption occurs on a partially dehydroxylated alumina surface (thermal treatment at 525 K), a surface reaction takes place in the presence of surface water (thermal treatment at 295 K), which leads to a mixture of uncharacterized carbonyl species with carbonyl stretching frequencies between 2020 and 1850 cm^{-1}. Treatment under vacuum for several hours results in a new infrared spectrum which retains some features of the original carbonyl stretching frequencies, but which shows a new strong peak at about 2020 cm^{-1}. Admission of carbon monoxide increases the infrared carbonyl bands related to those of the original cluster, while the bands in the region 1970-1850 cm^{-1} remain unaffected. The authors interpreted these infrared data as an indication of the formation on the surface of the anion $[Co(CO)_4O]^-$ [133]. It is known that $[Co_3(CO)_9CCH_3]$ yields the anion $[Co(CO)_4]^-$ in alkaline solutions [137]; consequently, its formation on the surface can be related to the intrinsic basicity of a hydrated alumina surface [138] (it does not occur on silica). The parallel formation of other uncharacterized polynuclear carbonyl species cannot be excluded, however.

Admission of oxygen causes the rapid disappearance of all the carbonyl bands, while the parallel formation of surface carbonates or bicarbonates [102] is observed. These latter species can be removed by evacuation at room temperature. Obviously the formation of surface carbonates must be ascribed to a facile oxidation process leading to surface decarbonylated species [probably Co(II) ions incorporated in the alumina surface] together with oxidation of CO to CO_2. Again, such an easy oxidation process is noteworthy, the more so as the original molecular cluster is highly stable towards oxidation.

9.2.3.3 Nickel clusters

The interaction of the cluster $[(\eta^5-C_5H_5)_3Ni_3(CO)_2]$ with silica and alumina surfaces is the only example in this category reported in the literature so far [126, 139]. The infrared spectra observed after its adsorption on silica suggest a simple physisorption. During heating, the physisorbed species are inferred to react with the silanol groups of the surface, since stoichiometric amounts of both CO and cyclopentadiene were detected in the gas phase. By carbonylation, the original cluster is reformed. Studies by infrared spectroscopy of the reversible carbonylation process together with some ESR studies of the recarbonylated samples suggest that the surface species produced by thermal decomposition of the original cluster retains the trinuclear arrangement similar to that of the parent molecular compound.

Comparable investigations have been carried out with a highly dehydroxylated alumina by means of infrared spectroscopy, analysis of gas evolution, and other chemical techniques. Simple impregnation from a dichloromethane solution produces a sample in which the original cluster is probably strongly adsorbed by an adduct formation between a bridging carbonyl ligand and an acidic Lewis center of the surface:

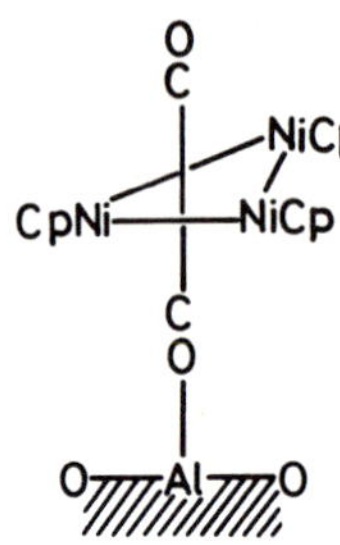

Due to the intrinsic lower stability of $[(\eta^5\text{-}C_5H_5)_3Ni_3(CO)_2]$ in comparison with that of $[(\eta^5\text{-}C_5H_5)Fe(CO)]_4$, this particular surface species is less stable than the corresponding iron surface species. In fact, some decomposition occurs after one hour at room temperature, and a broad carbonyl stretching band, typical of terminal carbonyls, appears in the infrared spectrum. The new band could be attributed to $Ni(CO)_4$ adsorbed on the alumina surface [140]; interaction in solution of $[(\eta^5\text{-}C_5H_5)_3Ni_3(CO)_2]$ with $AlBr_3$ leads to the expected adduct $[(\eta^5\text{-}C_5H_5)_3Ni_3(CO)(COAlBr_3)]$, together with some $Ni(CO)_4$ and $[Ni(\eta^5\text{-}C_5H_5)_2]$.

9.2.4 SURFACE ORGANOMETALLIC CHEMISTRY OF SECOND- AND THIRD-ROW TRANSITION METAL CARBONYL CLUSTERS

This group of clusters has been investigated preferentially because they provide relatively low reactivity. It has therefore been possible to define a series of intermediate steps during the interactions of these clusters with surfaces, and the corresponding organometallic surface species have been characterized by a variety of techniques. In particular, a complex surface organometallic chemistry has been reported to occur in the case of Rh, Os, and, to some degree, Ru clusters adsorbed on many oxides.

9.2.4.1 Rhodium clusters

The particular interest in the surface organometallic chemistry of rhodium clusters originated from the interesting and often unusual catalytic activity of such systems for the selective reduction of carbon monoxide

[141-143]. The major work was carried with $Rh_6(CO)_{16}$ [32, 45, 64, 128, 144-149] and $Rh_4(CO)_{12}$ [32, 128, 150, 151].

9.2.4.1.a **$Rh_6(CO)_{16}$ and $Rh_4(CO)_{12}$ on silica.** The behavior of both $Rh_6(CO)_{16}$ [146] and $Rh_4(CO)_{12}$ [150] has been extensively investigated, mainly by infrared studies. Although this technique is usually not sufficient to provide a clear picture of the surface chemistry, the relative simplicity of the infrared spectra of the surface species formed on this support permitted interesting hypotheses and conclusions.

Impregnation (in the complete absence of air or moisture [65]) of $Rh_6(CO)_{16}$ on silica (treated under vacuum at 573 K for 16 h) produces a surface species which retains the original Rh_6 cluster framework; it has been proposed that hydrido carbonyl species are formed according to the reaction:

$$Rh_6(CO)_{16} + nH_2O \rightarrow Rh_6(CO)_{16-n}H_{2n} + nCO_2 \qquad (3)$$

The probable formation of these hydrido species is supported by the decrease in the intensity of the infrared bands indicative of terminal carbonyls, together with a small frequency shift. Interestingly, such spectral trends are observed in the couples $Ir_4(CO)_{12}/Ir_4(CO)_{11}H_2$ [152, 153] and $Os_6(CO)_{18}/Os_6(CO)_{17}H_2$ [115, 154]. Moreover, the spectrum of $Rh_6(CO)_{16}$ can be regenerated simply by treatment with CO at 473 K for a few hours. The $Rh_6(CO)_{16}$ cluster thus formed on the silica surface is stable indefinitely under argon or carbon monoxide, but it decomposes slowly under vacuum even at room temperature; at higher temperatures, the decomposition process is easier and produces metallic particles.

When the surface-supported $Rh_6(CO)_{16}$ is treated with oxygen at 373 K, a neat reaction occurs to produce a new carbonyl species (SCHEME 4) characterized by two sharp carbonyl bands of equal intensity at 2093 and 2038 cm^{-1}. The same surface species can be obtained from the metallic particles produced by decomposition under vacuum, after simple treatment with oxygen at room temperature followed by reaction with carbon monoxide (SCHEME 4). This new carbonyl species was described as a $Rh^I(CO)_2$ surface complex by analogy of its infrared spectrum with those of molecular $Rh^I(CO)_2$ species such as $[Rh(CO)_2Cl]_2$ [155].

A weak band at 2108 cm^{-1} is always observed simultaneously with these two strong bands; this band was also observed in the very similar infrared spectrum of the dinuclear complex $Rh_2(CO)_4(OSiPh_2)_2O$ [81], which can be considered a good molecular model of these carbonyl surface species.

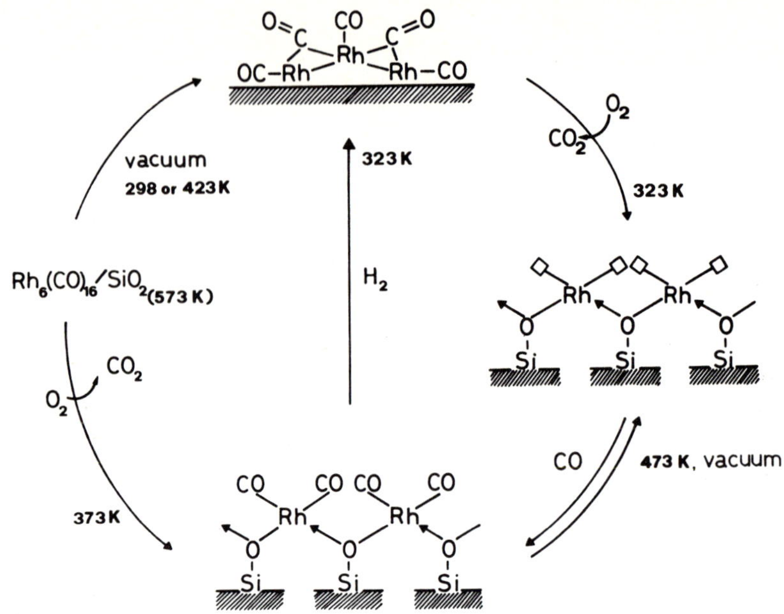

SCHEME 4

Recently, Evans *et al.* [156] made a careful analysis of the infrared spectra and arrived at the conclusion that the band at 2108 cm^{-1} must be associated with another band obscured by one of the two strong absorptions at lower frequencies. At the moment the real nature of this surface species is not fully established, although monomeric or dimeric $Rh^{I}(CO)_2$ species have been proposed.

These oxidized rhodium surface species show an interesting organometallic chemistry (SCHEME 4); a decarbonylation process occurs in vacuo at 473 K in about half an hour, but the simple exposure to CO at 100 kPa at temperatures between 298 and 353 K regenerates the orginial species in a few hours; these species can be reduced with H_2 at 323 K to give metallic rhodium particles. Interestingly, by treatment of the dicarbonyl surface species with CO at pressures <26 kPa in the presence of water (P_{H_2O} = 2.6 kPa), first $Rh_4(CO)_{12}$ and then $Rh_6(CO)_{16}$ is formed; such a surface synthesis of rhodium clusters does not occur if the partial pressure of water is too low (SCHEME 5). It is quite interesting to note that $Rh_4(CO)_{12}$ is an intermediate in the regeneration of the original $Rh_6(CO)_{16}$; when $Rh_4(CO)_{12}$ is supported (in the absence of air or moisture)

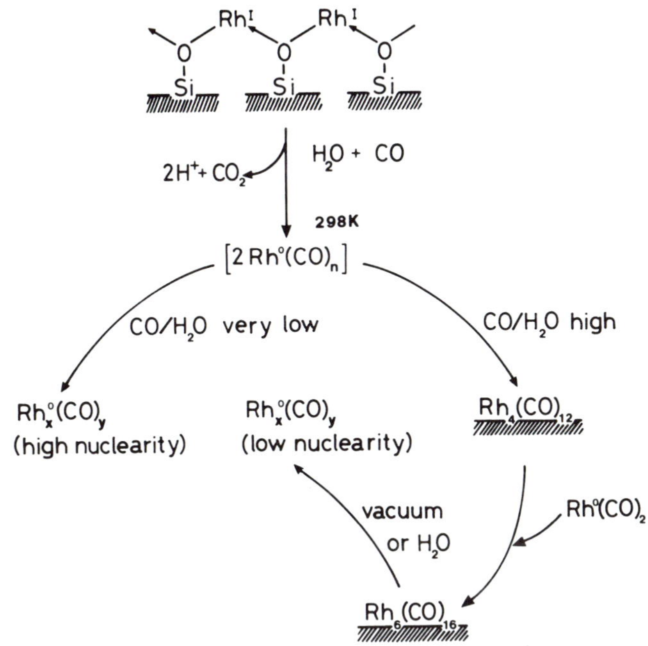

SCHEME 5

on silica (pretreated at 773 K under vacuum for few hours), $Rh_6(CO)_{16}$ is suddenly formed; the same transformation occurs on hydrated silica. $Rh_4(CO)_{12}$ is adsorbed as such only when a CO partial pressure is introduced just after adsorption of this cluster.

The reported easy rearrangements of the nuclearity of rhodium clusters adsorbed on surfaces are superb examples of the surface mobility of rhodium carbonyl species, corresponding to a solutionlike behavior [157]. Of course the same kind of transformations reported for $Rh_6(CO)_{16}$, when adsorbed on silica, occurs with $Rh_4(CO)_{12}$ under similar conditions (SCHEME 5).

9.2.4.1.b **$Rh_6(CO)_{16}$ and $Rh_4(CO)_{12}$ on alumina.** In contrast to the behavior of $Rh_6(CO)_{16}$ on silica, the behavior of $Rh_6(CO)_{16}$ on alumina is strongly dependent upon the mode of impregnation and the water content of the surface [45, 147]. When the impregnation is carried out in the presence of air, as reported by Watters *et al.* [64], a fast, complete decarbonylation occurs. Admission of a CO partial pressure produces surface species with an infrared spectrum characterized by two strong carbonyl bands, similar to those reported for Rh^I dicarbonyl species on silica. The behavior is completely different when the impregnation is carried out in the absence of air [65]. In this case, the surface chemistry is dependent on the water or hydroxyl content of the surface (SCHEME 6).

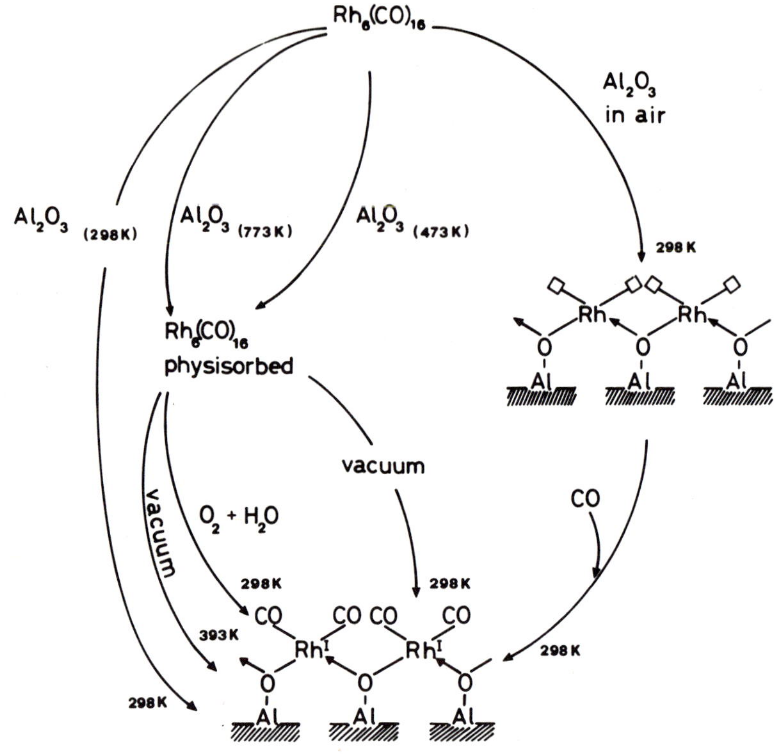

SCHEME 6

With highly dehydroxylated η-Al_2O_3 (treated at 773 K under vacuum overnight), the original $Rh_6(CO)_{16}$ is only physisorbed; it can be easily extracted from the surface by simple treatment with a solvent. The physisorbed cluster is stable at room temperature under vacuum or even in the presence of oxygen. However, if traces of water are added together with oxygen, instantaneous decarbonylation and oxidation of the original cluster occur; surface species with two strong carbonyl bands at 2105 and 2040 cm^{-1} are formed. The same result can be obtained in the absence of water, but after heating in vacuum at 393 K for 50 h.

If the alumina has been pretreated at only 473 K (giving a higher density of surface hydroxyl groups), the stability of $Rh_6(CO)_{16}$ adsorbed on the surface is lower, and the cluster decomposes even as a result of simple treatment under vacuum at 298 K for 4 h (SCHEME 6).

If the alumina is pretreated under vacuum at room temperature (giving a high surface water content), decomposition occurs within a few minutes. The following surface reaction leading to $Rh^{I}(CO)_2$ surface species was proposed:

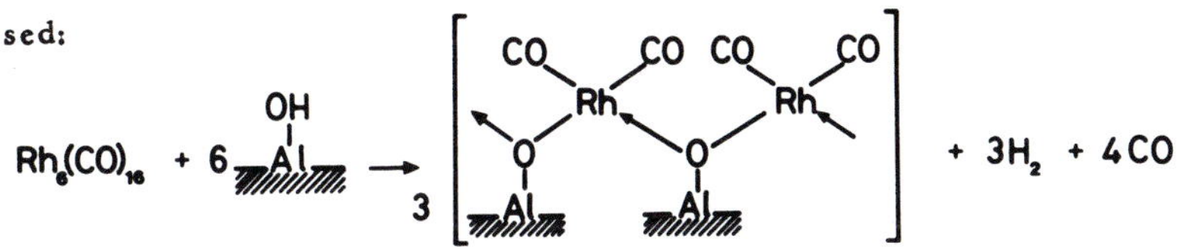

(4)

However, since hydrogen evolution was not observed, and more than the stoichiometric amount of carbon monoxide was found in the gas phase, there are various opinions reported in the literature as regards the nature of the rhodium surface species characterized by the two strong carbonyl absorption bands [158-161]. Suggestions such as formation of a surface $Rh_6(CO)_{12}O_{12}$ entity or of metallic particles with CO adsorbed on particular sites have been made [148, 149]. However, recently Basset *et al.* [162] confirmed unambiguously that the decomposition of $Rh_6(CO)_{16}$ on a highly hydroxylated alumina surface is a real oxidation of the cluster framework. Hydrogen could not be detected in the gas phase because the oxidation, corresponding to an electrophilic attack by surface protons, produces both $Rh^{I}(CO)_2(OAl\leftleftarrows)$ and $Rh^{III}(H)(H)(OAl\leftleftarrows)$ species (SCHEME 7).

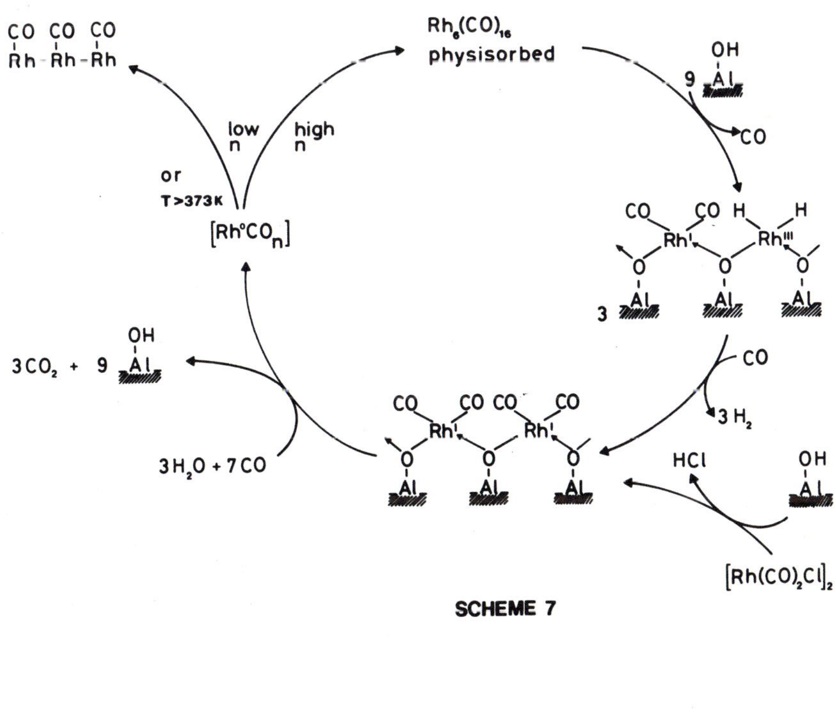

SCHEME 7

While the formation of Rh^{I} carbonyl surface species could be proved by infrared spectroscopy, this was not yet possible for the *bis* hydrido surface species $Rh^{III}(H)(H)(OAl\leqslant)$.

The presence of such a species (which can be also obtained by interaction of the *tris* η^3-allyl rhodium species chemisorbed on silica (SCHEME 8) with hydrogen [163-165] was provided by an elegant catalytic

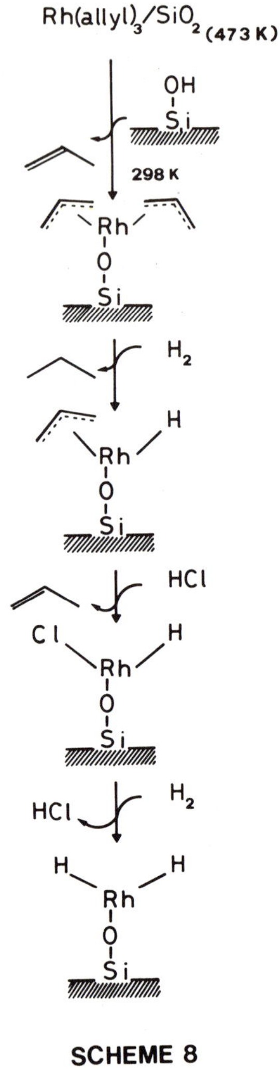

SCHEME 8

experiment. The mixture of oxidized rhodium species (formed on a deuterated alumina surface by decomposition of $Rh_6(CO)_{16}$) was reacted with ethylene at 353 K, and the stoichiometric amount of $C_2H_4D_2$ could be detected in the gas phase according to the following surface reaction [ref. 162]:

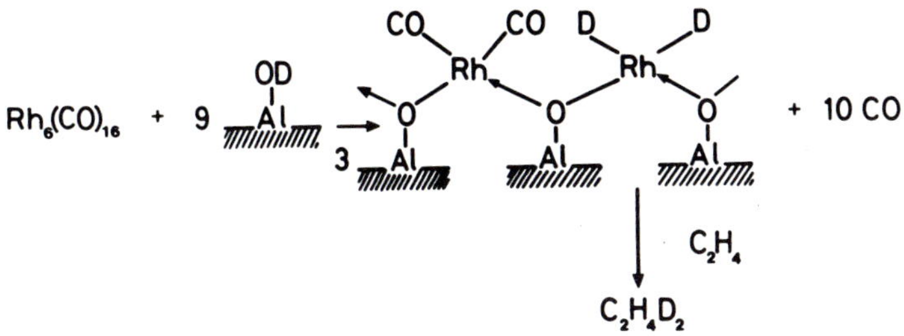

(5)

This interpretation of the surface species formed by decomposition of $Rh_6(CO)_{16}$ also explains why the intensity of the two strong carbonyl bands increases upon admission of carbon monoxide. In this case, all the rhodium atoms are transformed into $Rh^I(CO)_2$ surface species, with parallel evolution of molecular hydrogen from the *bis* hydrido surface species (SCHEME 7).

The dicarbonyl surface compound can be decarbonylated thermally to produce coordinatively unsaturated species which then react with either H_2 or CO according to the following surface organometallic chemistry:

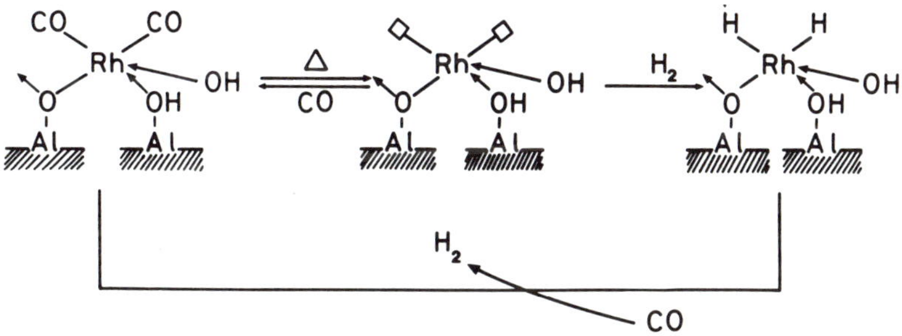

(6)

In addition, these $Rh^I(CO)_2$ surface species show an interesting organometallic chemistry (SCHEME 7): treatment with water at temperatures >373 K produces metallic particles as a result of reduction of Rh^I or Rh^{III} by coordinated CO. The particles were characterized by electron microscopy and their sizes were found to be dependent on the water partial pressure [45]. However, if the reduction process is carried out in the presence of an excess of CO and a low amount of water, the quantitative regeneration of $Rh_6(CO)_{16}$ occurs on the surface (SCHEME 7); under these conditions, metallic particles are formed only at temperatures >373 K.

It is noteworthy that the surface synthesis of $Rh_6(CO)_{16}$ from surface $Rh^I(CO)_2$ species

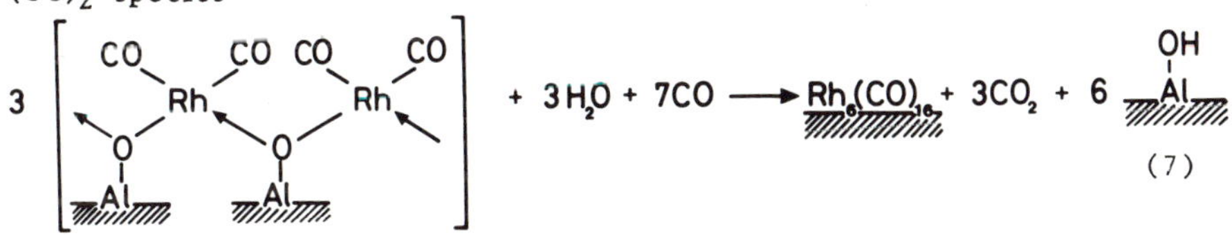

(7)

corresponds to the typical synthesis of $Rh_6(CO)_{16}$ in homogeneous solution starting from $[Rh(CO)_2Cl]_2$ [157, 166] under a CO atmosphere and in the presence of traces of water:

$$3[Rh(CO)_2Cl]_2 + 3H_2O + 7CO \longrightarrow Rh_6(CO)_{16} + 3CO_2 + 6HCl \quad (8)$$

In summary, the surface behavior of Rh^I and Rh^{III} units (SCHEME 7) is a clear example of what we call surface organometallic chemistry. However, there is still a controversy about the real nature of the above-mentioned oxidized rhodium surface species. Even XPS [149] and EPR [148] studies did not elucidate this point, although their direct formation by simple chemisorption of $[Rh^I(CO)_2Cl]_2$ (SCHEME 7) or of $(\eta^3\text{-}C_5H_5)_3Rh$ (SCHEME 8) is in support of their characterization as Rh^I and Rh^{III} surface species.

9.2.4.1.c **$Rh_6(CO)_{16}$ on other oxides.** Very scattered data are available characterizing the interaction of $Rh_6(CO)_{16}$ with the surfaces of other oxides such as magnesia, titania, zinc oxide, and silica-alumina [147].

On magnesia pretreated at 773 K under vacuum, both the parallel physisorption of $Rh_6(CO)_{16}$ and its decomposition to give oxidized rhodium species are reported to occur. Addition of water (2.6 kPa) leads to the complete disappearance of $Rh_6(CO)_{16}$, but the carbonyl bands due to the oxidized $Rh^I(CO)_2$ surface species shift to lower frequencies, and some metallic particles are formed. Only after three days under these conditions (P_{H_2O} = 2.6 kPa) does the complete transformation into metallic particles occur. With magnesia treated under vacuum at only 298 K, the instantaneous formation of oxidized surface species is observed upon adsorption of $Rh_6(CO)_{16}$.

In summary, on magnesia it is very difficult to maintain molecular $Rh_6(CO)_{16}$, in agreement with the important role played by the basicity of the surface in the process of decomposition and oxidation of the parent cluster [157, 167]. Further work is necessary to elucidate the effect of water, because it is known that the basic properties of the surface of hydrated magnesia are diminished [106].

On the acidic surface of silica-alumina, pretreated at 773 K under vacuum, simple physisorption of $Rh_6(CO)_{16}$ occurs (only a small shift of the carbonyl bands to higher frequencies is reported); the physisorbed cluster is stable for several days at room temperature under vacuum [147].

A few studies have been carried out with $Rh_4(CO)_{12}$ supported on titania and zirconia-silica [143]. The original Rh_4 cluster is first physisorbed on these supports, but after a few minutes in vacuo, transformation to $Rh_6(CO)_{16}$ takes place; moreover, with oxygen an easy oxidation occurs to form $Rh^I(CO)_2$ surface species. Reduction of these latter entities by CO (26 kPa) in the presence of some water (1.3-2.6 kPa) again reforms Rh_4 and Rh_6 clusters, as occurs on silica and alumina surfaces.

9.2.4.1.d **Rhodium carbonyl anions on various oxides.** $[Rh_6(CO)_{15}]^{2-}$ and $[Rh_6(CO)_{14}]^{4-}$ are strongly adsorbed on TiO_2 or ZrO_2-SiO_2 [143], producing surface species with infrared spectra similar to those of the corresponding hydrides such as $[Rh_6(CO)_{15}H]^-$ [168]. In contrast, $[Rh_7(CO)_{16}]^{3-}$ and $[Rh_{13}(CO)_{23}H_{5-n}]^{n-}$ (n = 2 or 3), when adsorbed on TiO_2-SiO_2 [143], produce samples with infrared carbonyl bands similar to those of the original cluster anions but shifted by about 20-30 cm^{-1} to higher energy, thus suggesting an interaction with Lewis acid centers of the support [169].

The original surface cluster carbonyl species are almost completely decomposed and oxidized by treatment with oxygen to form the well-known $Rh^I(CO)_2$ species. These latter surface species can be reduced by CO (26 kPa) and H_2O (1.3-2.6 kPa) to form, first, $Rh_4(CO)_{12}$ and then $Rh_6(CO)_{16}$, adsorbed on the surface. However, it has been reported that when the $Rh^I(CO)_2$ species are formed from $[Rh_7(CO)_{16}]^{3-}$ or $[Rh_{13}(CO)_{23}H_{5-n}]^{n-}$ (n = 2, 3), the reported reduction process produces surface species different from $Rh_6(CO)_{16}$. The final infrared spectra suggest the reformation of the original Rh_7 or Rh_{13} clusters. Although these observations need to be confirmed, it is possible that the $Rh^I(CO)_2$ species bind to the surface in such a way as to be related to the original cluster arrangement (a kind of surface "island"). In this respect, it is noteworthy that the reduction of $Rh^I(CO)_2$ surface species to form $Rh_6(CO)_{16}$ would have to be quite fast if the oxidized species were formed from the adsorbed $Rh_6(CO)_{16}$ cluster, whereas the same process would have to be very slow if the same species were produced by chemisorption of $[Rh^I(CO)_2Cl]_2$ [147].

9.2.4.2 **Iridium clusters**

A large amount of work has been devoted to the preparation of supported metallic iridium particles by decomposition of $Ir_4(CO)_{12}$ adsorbed on silica and alumina [63, 170-172]. However, although the final metallic catalysts so produced are well characterized, minor attention has been paid to the investigation of the intermediate organometallic chemistry [63, 172-174].

It appears that $Ir_4(CO)_{12}$ is not physisorbed on hydrated silica; evaporation of a slurry of such hydrated silica in a *n*-hexane solution of $Ir_4(CO)_{12}$ produces only crystallites of the cluster dispersed throughout the support. In fact, the infrared spectrum of the final sample shows an intense negative absorption at about 2130 cm^{-1}, which can be associated with a strong Christiansen effect of the broad band at 2066 cm^{-1} of $Ir_4(CO)_{12}$ [175]. This effect is clearly due to the crystalline cluster. Evacuation of these samples at higher temperatures causes the progressive reduction of the intensity of the carbonyl bands, in parallel with a decarbonylation process, which is almost complete at 623 K, as confirmed by temperature-programmed desorption measurements [174].

An accurate infrared investigation has shown that two related bands at 2086 and 2008 cm^{-1} are formed in the intermediate steps; they disappear at higher temperatures more easily than single broad band at 2054 cm^{-1}. This latter band shifts to lower frequencies when the temperature is increased.

This body of information was interpreted as indicating two parallel modes of irreversible decomposition [175]. The first, which is by far the more important, is associated with the strong and broad band located at 2054 cm^{-1}. It corresponds to the formation of metallic particles covered by CO [176-182], which increase through aggregation of the initial Ir_4 units. The second, which is associated with the two weak bands at 2086 and 2008 cm^{-1}, corresponds to a fragmentation and oxidation of the original cluster into $Ir^I(CO)_2$ units, probably incorporated in the structure of the silica surface [183].

The behavior of $Ir_4(CO)_{12}$ on fully hydroxylated γ-Al_2O_3 is completely different. $Ir_4(CO)_{12}$ is easily adsorbed on this support, as indicated by the fading of the yellow color of the initial solution after addition of γ-Al_2O_3 to form a slurry. The infrared spectrum in air of a freshly prepared sample is quite similar to that of the parent cluster, suggesting that $Ir_4(CO)_{12}$ is simply physisorbed on the surface. After a treatment at 373 K under vacuum for one hour, the final infrared spectrum shows three bands in the carbonyl stretching region, at 2069, 2039, and 1997 cm^{-1}. The bands at 2069 and 1997 cm^{-1} are clearly associated, and they behave as the corresponding pair of bands on silica. However, these bands, which in the case of alumina are dominant, are completely removed only by heating at 623 K under vacuum for one hour. In contrast, the broad band at 2039 cm^{-1} is removed even by thermal treatment at 423 K for one hour.

This is a behavior completely opposite to that observed on silica. Both decomposition processes are irreversible, as inferred from the observation that addition of CO after the thermal treatments did not restore the original spectra. As in the case of silica, two parallel processes of decomposition of the original $Ir_4(CO)_{12}$ were proposed, but on alumina the fragmentation into $Ir^I(CO)_2$ units is the predominant pathway of decomposition. By analogy with the corresponding $Rh(CO)_2$ surface species, it has been proposed that the $Ir(CO)_2$ units contain the metal in the oxidation state Ir(I). Evidence for partial oxidation during the thermal decomposition is given by TPDE results which show some hydrogen evolution [49, 174]. In addition, the frequencies of the two carbonyl bands of the surface species are very close to those of some Ir(I) dicarbonyl complexes such as $[Ir(CO)_2Cl_2]^-$ [184, 185].

The further loss of CO from these $Ir(CO)_2$ units probably produces individual iridium ions which form surface iridium aluminates of very low reactivity [68]. In fact, metallic iridium particles with dispersion comparable to that obtained on silica by simple thermal treatment are formed only by hydrogen reduction at 623 K.

9.2.4.3 Platinum clusters

There are only two reports describing the use of polynuclear platinum complexes such as $(NEt_4)[\{Pt_3(CO)_6\}_n]$ (n = 1-5) as precursors of highly dispersed metallic platinum on γ-Al_2O_3 or on silica gel [172, 186]. The dispersed cluster anions, deposited on the support from tetrahydrofuran solutions, were pyrolyzed in vacuo (or under He or H_2) to produce the stoichiometric amounts of CO and NEt_3. The platinum aggregates formed in this way consist of crystallites with diameters of about 1 nm [186]. The intermediate formation of surface organometallic species and the pathway of the thermal decomposition were not investigated.

9.2.4.4 Osmium clusters

This class of clusters has been thoroughly investigated by several groups; the high kinetic stability of osmium carbonyl clusters implies that surface organometallic intermediates are relatively stable, and a complex surface organometallic chemistry has been established with the help of various chemical and physical techniques. The major effort has been devoted to investigation of the behavior of $Os_3(CO)_{12}$ on inorganic supports [32, 39, 45, 59, 78, 82, 187-197].

9.2.4.4.a **$Os_3(CO)_{12}$, $H_2Os_3(CO)_{10}$, and $Os_6(CO)_{18}$ on silica.** With these clusters, the type and the technique of impregnation (in air or in the absence of air) is not so important as in the case of rhodium, ruthenium, or other carbonyl clusters. Whatever the nature of the silica surface, simple physisorption of $Os_3(CO)_{12}$ occurs [59]. Only by thermal treatment is argon or under vacuum at about 373-423 K, does a reaction take place between $Os_3(CO)_{12}$ and the silanol groups of the surface, corresponding to an oxidative addition of a surface Si-OH group to an Os-Os bond as in the following surface reaction:

$$Os_3(CO)_{12} + HOSi\equiv \rightarrow HOs_3(CO)_{10}(OSi\equiv) + 2CO \tag{9}$$

The stoichiometry of the reaction has been confirmed by gas evolution measurements and by the formation of the same surface species by simple adsorption at 298 K of $Os_3(CO)_{10}(CH_3CN)_2$ (SCHEME 9). This latter complex is known to react easily with HX species to form $HOs_3(CO)_{10}(X)$ compounds with release of CH_3CN [198].

The triangular metal framework of the new surface cluster, which is covalently bonded to the surface, has been confirmed by direct techniques such as EXAFS [82] (see, however, Section 6.7) or by indirect techniques such as infrared [59] and Raman spectroscopies. In particular, the complete agreement of the infrared spectrum of the surface species with that of the model compound $HOs_3(CO)_{10}(OSiPh_3)$ [59, 80, 199, 200] is in support of the proposed structure (SCHEME 9).

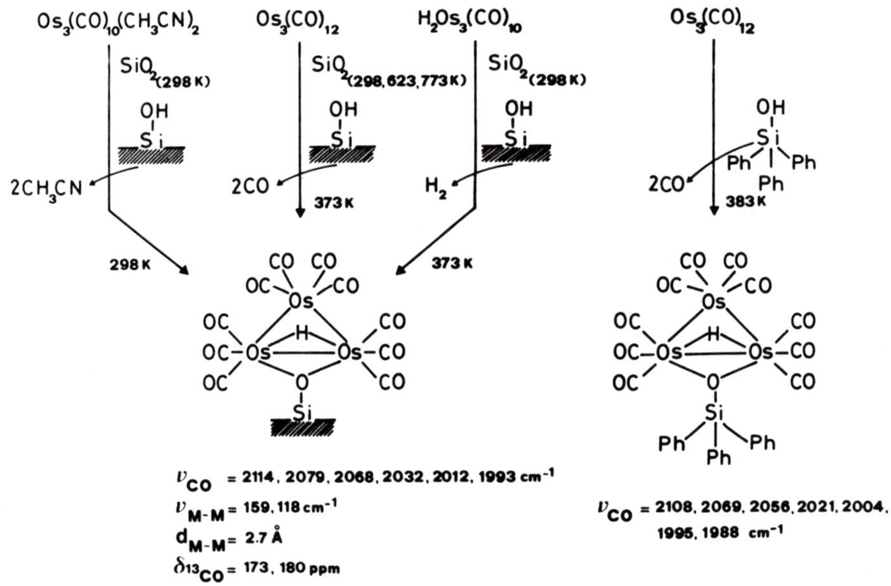

SCHEME 9

As for rhodium carbonyl clusters, the presence of water adsorbed on the surface plays a significant (although not so important) role. In fact, the presence on the surface of water in equilibrium with the gas phase favors the selective formation of the above-mentioned surface species.

The metal loading of the surface is also an important factor in controlling the surface reactions; for instance, with high loadings (amount of cluster $\geqslant 2$ wt %) three parallel transformations take place during the thermal treatment at 423 K: surface reaction of chemisorbed $Os_3(CO)_{12}$, sublimation of $Os_3(CO)_{12}$ (if the treatment is carried out under vacuum), and simple pyrolysis of $Os_3(CO)_{12}$ (which is physisorbed due to the high loading of the surface) to thermally generate osmium clusters of higher nuclearity [115].

When adsorption is carried out under proper conditions (complete absence of air), the cluster $H_2Os_3(CO)_{10}$ is physisorbed on silica [59]. Upon introduction of a partial pressure of CO, a complete transformation into physisorbed $H_2Os_3(CO)_{11}$ occurs even at room temperature; upon evacuation, the original cluster $H_2Os_3(CO)_{10}$ is reformed according to the surface equilibrium:

$$H_2Os_3(CO)_{10} + CO \rightleftharpoons H_2Os_3(CO)_{11} \quad (10)$$

This behavior is equivalent to that reported to occur in solution [201]. By heating under argon at about 373-423 K, the known surface species $HOs_3(CO)_{10}(OSi\equiv)$, covalently grafted to the surface, is formed with H_2 evolution (SCHEME 9).

All the surface cluster species, upon heating to temperatures >423 K, lead to rather similar surface complexes characterized by well-defined infrared spectra (SCHEME 10) and by the white color of the final sample. The appearance of these rather similar species starting from either Os_3 or Os_6 clusters suggests a complete rupture of the original metal cluster framework and the formation of surface species that are basically mononuclear. In fact, the absence of strong direct osmium-osmium bonds was confirmed by EXAFS [82]. Infrared studies at different temperatures (under vacuum or in the presence of an atmosphere of CO) suggest the existence of a reversible carbonylation process.

The experimental evidence is mentioned above; the similarities with the infrared spectra of known complexes of Os^{II}, such as $[Os(CO)_3X_2]_2$ (X = Cl, Br, I) and $[Os(CO)_2I_2]_n$ [202]; and the observation that an oxygen atmosphere favors this kind of surface transformation (which under these conditions occurs even at 373 K) all suggest the following surface reaction (starting from $HOs_3(CO)_{10}(OSi\equiv)$:

$$\begin{array}{l} HOs_3(CO)_{10}(OSi\equiv) + 5HO\text{-}Si\equiv \longrightarrow 3Os(CO)_3(OSi\equiv)_2 + CO + 3H_2 \\ \qquad\qquad\qquad\qquad\qquad\qquad\qquad \Delta \uparrow\downarrow CO \\ \qquad\qquad\qquad\qquad\qquad\qquad\qquad 3Os(CO)_2(OSi\equiv)_2 + 3CO \end{array} \quad (11)$$

The oxidation of the surface clusters to Os^{II} surface species is indirectly supported also by the fact that comparable infrared spectra and a similar reversible carbonylation behavior are obtained by adsorption of preformed Os^{II} species (such as $[Os(CO)_3Cl_2]_2$) on the silica surface (see SCHEME 10).

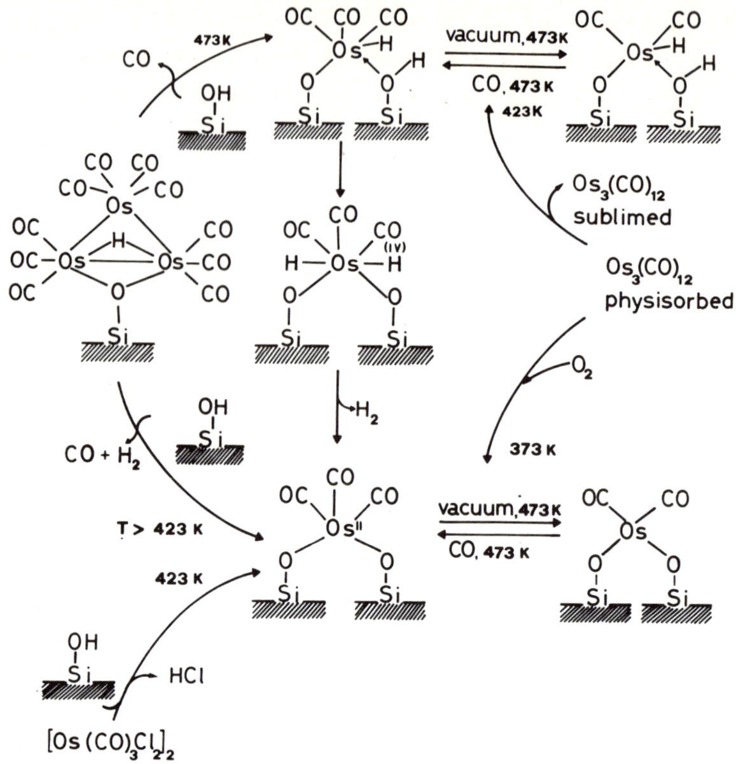

SCHEME 10

However, a careful investigation of the mass balance could not support the above stoichiometry of the reaction because the amount of the evolved hydrogen was less than 3 moles per mole of $Os_3(CO)_{12}$ in all cases.

Basset *et al.* [45] recently suggested that the oxidation process occurs with complete breaking of the cluster framework but with formation of hydrido carbonyl osmium species incorporated into the silica surface by covalent bonds (SCHEME 10). In this case the stoichiometry of the surface reaction could be the following (starting from $HOs_3(CO)_{10}(OSi\equiv)$:

$$HOs_3(CO)_{10}(OSi\equiv) + (n-1)HOSi\equiv \longrightarrow 3HOs(CO)_3(OSi\equiv)_n + CO + \frac{n-3}{2}H_2 \quad \Delta\uparrow\downarrow CO \quad 3HOs(CO)_2(OSi\equiv)_n + 3CO \qquad (12)$$

The oxidation state of the final osmium species is unknown. The previously proposed species $Os^{II}(CO)_3(OSi\leq)_2$ are formed only to a very small extent with evolution of H_2 (SCHEME 10). They are the predominant species formed in the presence of oxygen. Obviously, these different surface species must have rather similar infrared spectra in the carbonyl region (carbonyl bands are in fact rather broad, indicating the presence of mixtures). The nature of these surface species is still a matter of controversy and requires further investigation; in particular, the thermal behavior of these species is not well established.

Basset *et al.* [45] reported the formation of metallic particles only by treatment of the sample under vacuum at temperatures >523 K. However, other authors [56] described a higher thermal stability; in particular, upon increasing the temperature to 673 K under vacuum, a complete decarbonylation was observed, without formation of metallic particles. In fact, subsequent treatment with a partial pressure of CO (at 373 K) produces the reversible formation of the original osmium carbonyl species. Formation of metallic particles occurs only at temperatures >673 K or in the presence of H_2 at lower temperatures. These discrepancies can be ascribed to factors such as different surface conditions, for instance the concentration of active surface ⋛Si-OH groups, or different loadings of osmium on the surface. In addition, a matrix effect can be suggested, particularly when nonporous silica is used; this effect can increase the temperature of decomposition of carbonyl species, thus favoring the formation of oxidized species over metallic particles. Such an effect was observed during the thermal decomposition of many neutral carbonyl clusters in nonporous systems [37].

Two groups [45, 58] reported that the original grafted surface cluster $HOs_3(CO)_{10}(OSi⋚)$ can be reconstituted from the carbonyl mononuclear osmium fragments by controlled recarbonylation. However, their evidence was based only on infrared investigations.

Rather different experimental results and interpretations of the surface transformations were reported by Collier *et al.* [196]. These authors described a thermal treatment up to 523 K in a flow of helium with parallel analysis of the gas evolution; the samples were obtained by impregnation of $Os_3(CO)_{12}$ or $Os_6(CO)_{18}$ from a solution in an organic solvent. Their infrared spectra are in agreement with those reported and interpreted by different authors as being due to mononuclear carbonyl oxidized osmium species. In contrast, Collier *et al.* interpreted these spectra as being indicative of a process of cluster growth which occurs upon heating and produces species with nuclearity between 12 and 18 metal atoms. The retention of osmium-osmium bonds was demonstrated only by UV/visible reflectance spectroscopy, and the upper limit of the nuclearity of the clusters was suggested by electron microscopy. These authors also proposed an alternative interpretation for the decarbonylation-carbonylation process, which they claimed was not reversible. In summary, they described the final infrared spectra as the result of the spectra of mixtures of species having the osmium atoms in zerovalent and partially positive or negative oxidation states [176, 203-205]. Although the major infrared bands reported by all the different authors are located at very similar wavelengths, it is possible, due to their broadness, that they correspond to different mixtures.

The formation of high-nuclearity species claimed by Collier *et al.* may be ascribed to their high surface loading of $Os_3(CO)_{12}$ in comparison with those used by other authors. It is possible that at high loadings a large amount of $Os_3(CO)_{12}$ is not interacting chemically with the active surface groups, which may already be saturated. This, in fact, seems to be the case, as indicated by the infrared spectra reported by the authors for the freshly prepared materials. These spectra are typical solid state spectra and indicate the presence of crystal-like $Os_3(CO)_{12}$ (also see Chapter 6.2). This unreacted part is usually sublimed when heating in vacuo under controlled conditions. However, the thermal treatment reported by Collier *et al.* (steady rate of 10 K min^{-1} in the helium flow) is so fast that complete sublimation cannot easily occur. Therefore, a thermal pyrolysis of $Os_3(CO)_{12}$ probably occurs in parallel to thermally form high-nuclearity clusters; this latter is a well known process [115, 117]. In addition, we can suggest also that such a fast thermal treatment of physisorbed $Os_3(CO)_{12}$ would favor in any case a simple pyrolysis, because the reactions with the surface reactive groups are reported to occur rather slowly (requiring at least one or more hours).

In conclusion, we infer that great care must be taken in the comparison of results obtained under rather different conditions. In any case this area, although investigated in detail by many authors, still deserves more experimentation together with a well-controlled experimental approach in order to provide more precise conclusions regarding the chemistry that takes place between the original clusters and the silica surface.

9.2.4.4.b **$Os_3(CO)_{12}$ and $Os_6(CO)_{18}$ on alumina.** The surface organometallic chemistry of osmium carbonyl clusters deposited on alumina [45, 59] is slightly different from that just described in the case of silica. As on the silica surface, $Os_3(CO)_{12}$ is simply physisorbed at room temperature by any type of alumina pretreated in different ways. However, just a mild thermal treatment (about 3 h at 373 K under vacuum) induces a chemisorption process corresponding to the oxidative addition of a surface $\rightarrow$AlOH group to an osmium-osmium bond of the cluster frame of $Os_3(CO)_{12}$; in this way, the cluster $HOs_3(CO)_{10}(OAl\lessdot)$ covalently anchored to the alumina surface is generated (SCHEME 11).

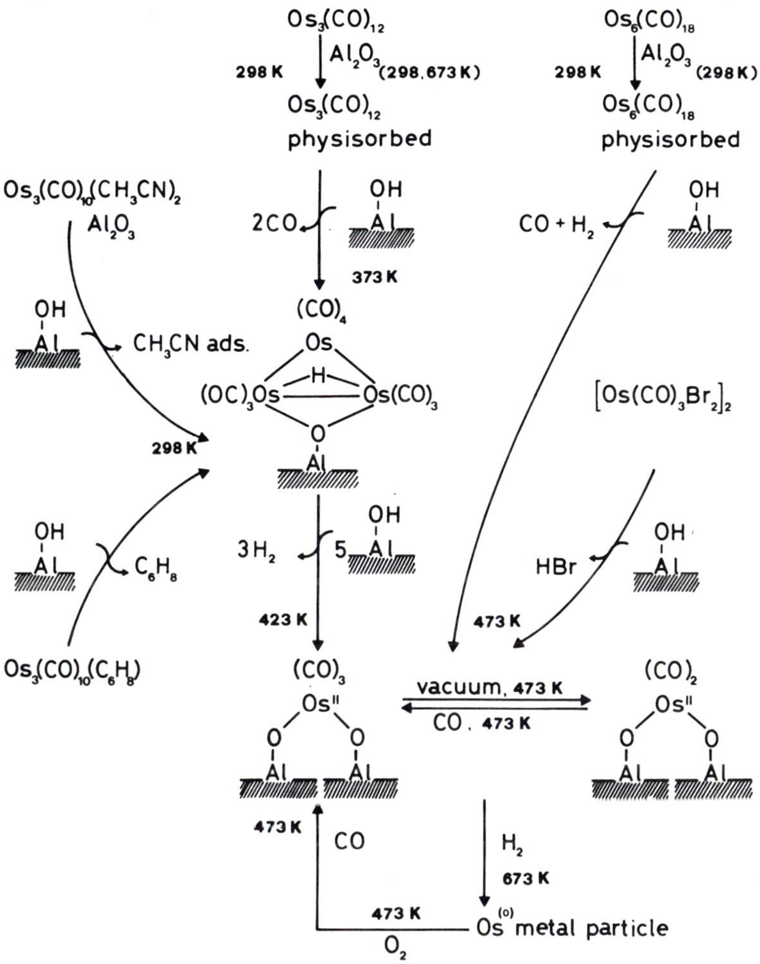

SCHEME 11

The definition of the nature of this new species, which is similar to that previously described in the case of silica, is based on a series of different experimental results. Chemical evidence indicates the formation of this species by adsorption under mild conditions of reactive "$Os_3(CO)_{10}$" fragments such as are formed from $Os_3(CO)_{10}(C_6H_8)$ [187] or $Os_3(CO)_{10}(CH_3CN)_2$ [59]; the mass balance is in agreement with the surface reaction

$$Os_3(CO)_{12} + HOAl\lessdot \rightarrow HOs_3(CO)_{10}(OAl\lessdot) + 2CO \quad (13)$$

Physical evidence is based on infrared and Raman [59, 188, 189, 191, 192,], UV/visible [189], XPS [192], and EXAFS [82, 193, 206] spectroscopic data (see Chapter 6).

This anchored cluster is thermally less stable than the analogous species on silica. A prolonged treatment under vacuum at temperatures between 373 and 423 K induces a complete decomposition to form new surface species, characterized by infrared spectra rather similar to those obtained on silica by a more vigorous thermal treatment. The same species are obtained by a similar thermal treatment of physisorbed $Os_6(CO)_{18}$. This result suggests that also on alumina a complete breakup of the original cluster cage takes place to form mononuclear osmium species, incorporated by chemical bonds into the alumina surface. The formation of similar surface species by chemisorption of oxidized Os(II) complexes, such as $[Os(CO)_3Br_2]_2$ (SCHEME 11) is in support of this picture. Evidence for reversible carbonylation/decarbonylation suggest the following surface reactions [starting from $HOs_3(CO)_{10}(OAl\lessdot)$]:

$$\begin{array}{l} HOs_3(CO)_{10}(OAl\lessdot) + 5HOAl\lessdot \longrightarrow 3Os(CO)_3(OAl\lessdot)_2 + CO + 3H_2 \\ \qquad\qquad\qquad\qquad\qquad\qquad\qquad \Delta \Updownarrow CO \\ \qquad\qquad\qquad\qquad\qquad\qquad\qquad 3Os(CO)_2(OAl\lessdot)_2 + 3CO \end{array} \qquad (14)$$

The mass balance is completely consistent with the above surface reactions because by controlled thermal treatment of preformed $HOs_3(CO)_{10}(OAl\lessdot)$ from 373 K up to 523 K, about 1.6 moles of CO per mole of cluster and 3 moles of H_2 per mole of cluster are evolved [59]. It is interesting that on alumina there is no evidence for the formation of hydrido carbonyl osmium mononuclear species as reported for silica. A detailed infrared analysis of the structure of these surface carbonyl species and of the reversible carbonylation/decarbonylation process was carried out by Knözinger and Zhao [191]. They arrived at the conclusion that the Os-Os distance is 0.59 nm, a value much higher than the distance of 0.2877 nm reported for $Os_3(CO)_{12}$ [207]. This conclusion was confirmed by EXAFS studies [206]. However, both EXAFS and transmission electron microscopy [192, 194] led to the suggestion that discrete ensembles of osmium atoms were formed reproducing in a certain way the original cluster arrangement, on the surface, but without direct Os-Os bonds. These "islands" are different in size when starting from $Os_3(CO)_{12}$ or $Os_6(CO)_{18}$. They are thought to be formed by Os^{II} atoms (XPS evidence [192]) and are highly resistant to sintering and in general to either disintegration or agglomeration [59, 192].

The behavior at high temperatures was not fully investigated; even under rather drastic conditions (heating under vacuum at 673 K) evidence was not reached for the formation of metallic particles, which are formed only by prolonged treatment with H_2 at 673 K (SCHEME 11). Interestingly, these metallic particles are easily oxidized by O_2 at 473 K to form again the same mononuclear Os^{II} surface species which can then be carbonylated to $Os^{II}(CO)_3$ surface species. If this treatment is followed by exposure to a CO atmosphere at increasing temperatures, the original spectrum is easily restored [192]. These experiments could be interpreted either in terms of the formation of unstable hydrido carbonyl osmium species (similar to those observed on silica) or in terms of a reduction process (the existence of Os^{I} species in reduced samples was recently supported by ESR spectra [195], although these low-valent species comprised a small fraction of the total osmium).

9.2.4.4.c $Os_3(CO)_{12}$ or $H_2Os_3(CO)_{10}$ on titania and zinc oxide. Preliminary results have been reported characterizing the behavior of $Os_3(CO)_{12}$ adsorbed on TiO_2 and ZnO [188]. The surface reactivity appears to be rather similar to that reported for the surface of alumina, because the first step is the oxidative addition of OH surface groups to the cluster to form the covalently anchored osmium cluster $HOs_3(CO)_{10}(O\text{-}M)$ (where M = Ti≡ or Zn-). Thermal decomposition under vacuum produces the complete destruction of the cluster frame with formation of Os^{II} surface mononuclear carbonyl species. In contrast to the behavior observed with silica, the adsorption of $H_2Os_3(CO)_{10}$ on TiO_2 [188] (but also on $\gamma\text{-}Al_2O_3$) even at 298 K produces the destruction of the Os_3 framework with parallel formation of the oxidized osmium carbonyl species.

9.2.4.4.d $Os_3(CO)_{12}$ on magnesia. This system has been investigated in some depth [190, 197]. When magnesia was not specially treated, the behavior of the surface toward $Os_3(CO)_{12}$ is similar to that observed on silica or on alumina. In fact, first simple physisorption occurs followed by the formation of the anchored species $HOs_3(CO)_{10}(OMg)$ by treatment under argon at 334 K (see SCHEME 12). Deeba *et al.* [190] suggested that this latter species could correspond to the anchored structure $Os_3(CO)_{10}(OMg)_2$. It is quite difficult to support one or the other formulation only on the basis of very small differences of the stretching frequencies of carbonyls as in the very similar spectra of the $HOs_3(CO)_{10}Y$ and $Os_3(CO)_{10}Y_2$ molecular species (Y = OR, OPh) [199, 200].

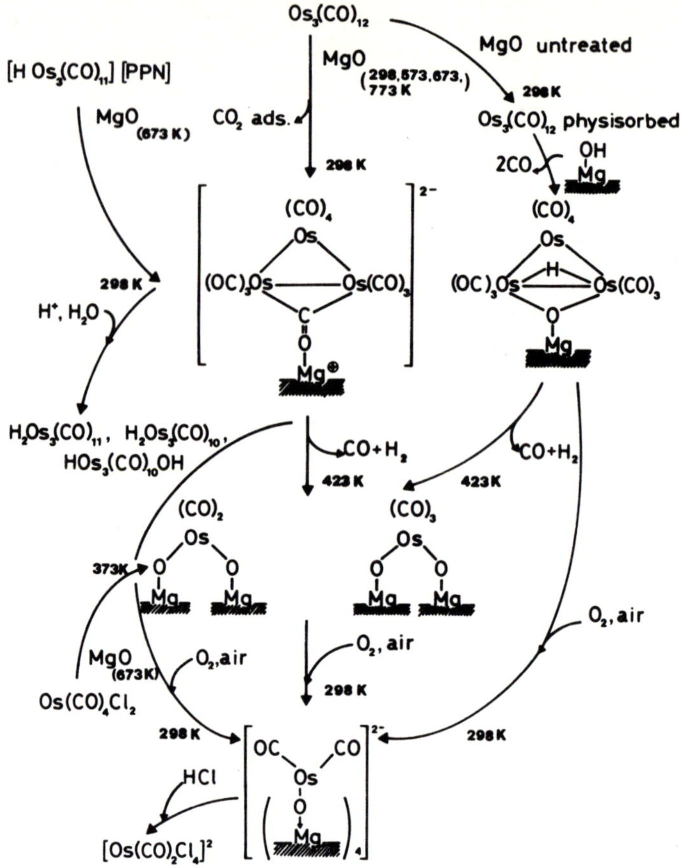

SCHEME 12

The anchored species is quite stable under a CO atmosphere, but by simple exposure to air or oxygen, a new surface species, characterized by an infrared spectrum in the carbonyl region very similar to that of anionic molecular complexes of Os^{II} such as $[Os(CO)_2X_4]^{2-}$ (X = Cl, Br, I) [207], is quickly formed. The following structure for this new surface organometallic species was suggested [190]:

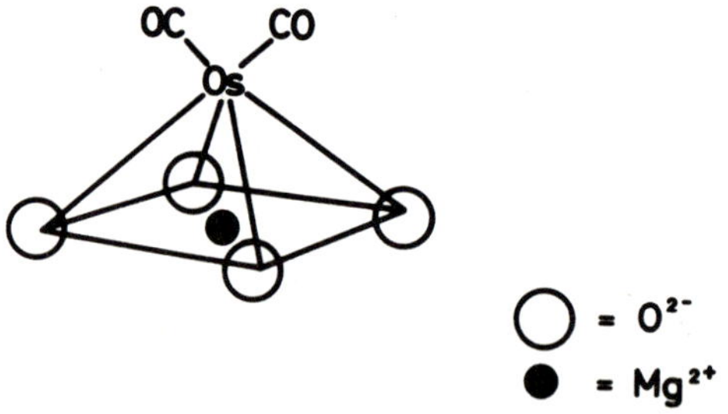

The anionic nature of such a surface species was confirmed by other authors [198] since anionic species such as $[Os(CO)_2Cl_4]^-$ and some $[Os(CO)Cl_5]^-$ could be isolated by simple treatment with HCl. Interestingly, this unusual anionic surface species is particularly stable up to 673 K even in the presence of oxygen. It was claimed that by treatment with CO at 473-573 K, an agglomeration occurs to again form osmium carbonyl clusters. However, this observation, which is based only on infrared evidence, requires additional investigation (see Section 9.5).

It was shown later [198] that the nature of the interaction between $Os_3(CO)_{12}$ and the surface of magnesia depends strongly upon the pretreatment of the magnesia itself. If the magnesia is treated in such a way as to avoid surface carbonates and to diminish the content of surface water and hydroxyl groups, a quick reaction occurs when $Os_3(CO)_{12}$ is deposited on the surface under well controlled conditions. The anion $[Os_3(CO)_{11}]^{2-}$ is easily and quantitatively formed (see SCHEME 12). This anion is strongly bonded, suggesting a complex ionic-covalent interaction with the surface. The shift of the infrared bands is quite in agreement with a tight ion-pair complex, where a bridging carbonyl ligand interacts with a magnesium ion via an acid-base bond [100, 101]:

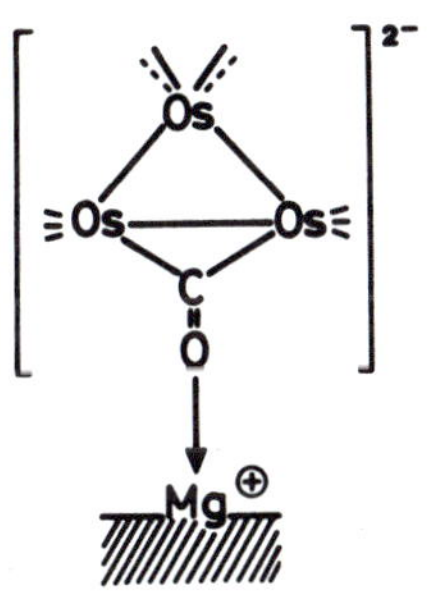

The presence of the $[Os_3(CO)_{11}]^{2-}$ anion on the surface is supported not only by infrared spectroscopy but also by chemical evidence. Controlled reaction of the sample with H_3PO_4 or H_2SO_4 produces $H_2Os_3(CO)_{10}$ and $H(OH)Os_3(CO)_{10}$, in agreement with the known reactivity of Os_3 carbonyl anions [199, 208]. In addition, the same surface species is quickly formed by simple adsorption of the preformed anion $[HOs_3(CO)_{11}]^-$ (SCHEME 12). In homogeneous solution, the anion $[Os_3(CO)_{11}]^{2-}$ can be obtained only by treatment of $Os_3(CO)_{12}$ with a strong reducing base such as potassium benzophenone [199, 209, 210].

As in the case of $Fe_3(CO)_{12}$ supported on alumina or magnesia, the nature of the adsorption is related to the surface loading. At high loadings (osmium content > 1.5%), a simple physisorption of $Os_3(CO)_{12}$

occurs in addition to the formation of $[Os_3(CO)_{11}]^{2-}$. Obviously the organometallic reactivity of the surface is controlled by a few special sites, which can be saturated. The thermal behavior of physisorbed or chemisorbed osmium clusters still deserves a deeper investigation. When physisorbed $Os_3(CO)_{12}$ or surface clusters such as $HOs_3(CO)_{11}(OMg)$ or $[Os_3(CO)_{10}(COMg)]^-$ are heated under argon or in vacuo at 373-423 K, a facile transformation occurs on pretreated magnesia, while on untreated magnesia the same transformation requires more time. The final sample shows an infrared spectrum rather similar to that described for Os^{II} surface carbonyl species formed, for instance, on alumina. Consequently, it has been suggested that a mixture of $Os(CO)_3(OMg)_2$ and $Os(CO)_2(OMg)_2$ species is formed. In agreement with this suggestion, rather similar spectra are shown by species obtained when preformed Os^{II} carbonyl complexes such as $[Os(CO)_3Cl_2]_2$ or $Os(CO)_4Cl_2$ are chemisorbed on magnesia [197]; however, their reactivity (for instance, decarbonylation/carbonylation) is much less than that of the related surface species formed from $Os_3(CO)_{12}$.

All these species react with oxygen to form the very stable surface complex $[Os(CO)_2(OMg)_4]^{2-}$, discussed above (SCHEME 12). The high stability of this unusual anionic surface species, particularly toward thermal decarbonylation or oxidative decarbonylation is relevant; probably the strong basicity and nucleophilicity of a magnesia surface [112, 114] produce anionic sites which kinetically stabilize the bonding with carbonyl ligands by increasing the electron density on the metal, thus making difficult any carbonyl dissociation. In any case, so far this appears to be a characteristic unique to the magnesia surface.

Evidence for the formation of metallic osmium by thermal treatment up to 573 K has not been reported.

In summary, a clean magnesia surface produces a surface organometallic chemistry that is unique, being markedly different from that described for the surfaces of other oxides.

9.2.4.5 Ruthenium clusters

Unlike osmium clusters, although studied by many authors [39, 45, 68, 69, 192, 211-214], supported ruthenium clusters have never been investigated in sufficient depth. $Ru_3(CO)_{12}$ is usually much more reactive than $Os_3(CO)_{12}$, so that the characterization of surface intermediates and consequently the definition of the corresponding surface organometallic chemistry are difficult tasks. Much attention has been devoted to the catalytic aspects, because ruthenium catalysts are very active in various catalytic reactions, such as olefin isomerization or hydrogenation [215, 216] and CO reduction [32, 39, 45, 192, 217-224].

9.2.4.5.a $Ru_3(CO)_{12}$ on silica. In contrast to the behavior of $Os_3(CO)_{12}$, that of $Ru_3(CO)_{12}$, when adsorbed on silica, is strongly dependent upon the amount of water present on the surface and upon the presence of dioxygen. When adsorption is carried out under well controlled conditions (under vacuum or under argon), simple physisorption takes place, irrespective of the content of water of the surface [213]. By mild heating at 343-353 K under vacuum or in an argon atmosphere, a chemisorption process occurs corresponding to the anchoring to the surface of the triangular clusters via a covalent bond. The infrared and Raman spectra [213] of the new surface species are very similar to those reported for the related anchored surface cluster $HOs_3(CO)_{10}(OSi\equiv)$. Moreover, these infrared spectra are typical of molecular clusters such as $HRu_3(CO)_{10}X$ (X = SC_2H_5, etc.) [225] of similar structure. The stoichiometry of the surface reaction:

$$Ru_3(CO)_{12} + HOSi\equiv \longrightarrow HRu_3(CO)_{10}(OSi\equiv) + 2CO \qquad (15)$$

is supported also by the mass balance (1.7 ± 0.2 moles of CO per mole of cluster are evolved between 343 and 393 K, without a significant production of H_2 or CO_2).

The thermal and chemical stability of this grafted ruthenium cluster is much lower than that of the corresponding osmium species. Mild thermal treatment at 393 K under vacuum leads to a complete transformation, giving a mixture of metallic particles and of surface-oxidized mononuclear carbonyl species [$Ru^{II}(CO)_2$ and $Ru^{II}(CO)_3$ fragments incorporated into the silica surface] (SCHEME 13). The partial oxidation process is confirmed by the evolution of a certain amount of H_2, together with the evolution of 10 moles of CO per mole of cluster when the temperature is progressively raised to 523 K.

The relative amounts of metallic particles and oxidized species are dependent upon the loading of the surface. At low loadings (0.2%), oxidized Ru(II) species are predominantly formed, and at high loadings (5-12%) mainly metallic particles are formed [68, 226]. These metallic particles, characterized by electron microscopy, exhibit a very narrow size distribution around 1.4 nm [213, 227]. This is a clear example of how the loading of the surface can influence the total reactivity toward the added cluster. It seems also from many other examples [$Os_3(CO)_{12}$ on magnesia, $Fe_3(CO)_{12}$ on alumina] that a low surface loading is a necessary condition in order to obtain only a pure organometallic interaction with the reactive groups of the surface. The formal oxidation state of the ruthenium in these new surface carbonyl species is suggestive of a variety of observations. The surface species can be easily and quantitatively obtained by treatment with oxygen even at room temperature, with the starting

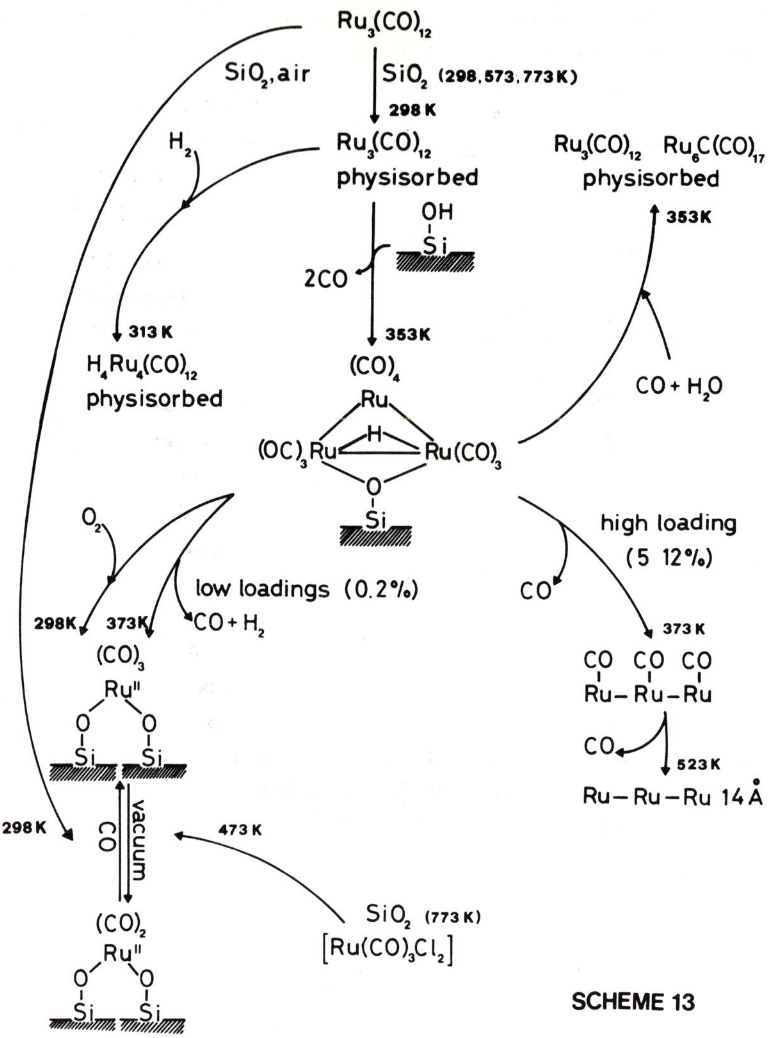

SCHEME 13

material being any ruthenium cluster adsorbed on the surface. Surface species characterized by similar infrared spectra are also obtained by direct chemisorption of molecular Ru^{II} species such as $[Ru(CO)_3Cl_2]_2$ under mild conditions and on various types of surfaces (treated in vacuum from 298 up to 773 K) (SCHEME 13). Further, the infrared spectra are similar to those of the corresponding well-characterized Os^{II} surface species and also to those of molecular Ru^{II} carbonyl complexes such as $[Ru(CO)_3Cl_2]_2$ [228] or $[Ru(CO)_2I_2]_n$ [229].

In summary, all this evidence suggests monomeric or dimeric species incorporated in the structure of the silica surface.

Recently, Evans and McNulty [230] proposed a different interpretation for the nature of these species, claiming that they were related to mononuclear carbonyl species with the ruthenium atom in oxidation states higher than II. Interestingly, it was reported that the typical carbonyl bands are present even in samples containing 6% Ru metal supported on silica; obviously their presence must be dependent upon the presence of oxygen [231]. We have confirmed that supported metallic particles of ruthenium can easily be oxidized and carbonylated to generate the oxidized surface carbonyl ruthenium species just discussed.

The chemical behavior of the grafted cluster $HRu_3(CO)_{10}(OSi\equiv)$ has been investigated in detail [226]. By reaction with CO at room temperature, the original $Ru_3(CO)_{12}$ (physisorbed on the surface) is restored. It seems that the reversible process is favored by water; this same reversible process (but occurring much more slowly) was also reported to occur with the corresponding osmium surface species [59]. When $Ru_3(CO)_{12}$, so reformed, is extracted with n-pentane, the presence of small amounts of $Ru_6C(CO)_{17}$ is evidenced by a series of different techniques [226]. The formation of this Ru_6 cluster from $Ru_3(CO)_{12}$ under very mild conditions is worthy of attention, because it is known that in the solid state the same transformation can be obtained only by prolonged pyrolysis of $Ru_3(CO)_{12}$ at 423 K [115].

The grafted cluster $HRu_3(CO)_{10}(OSi\equiv)$ is rather stable in the presence of water. However, water favors its decomposition by oxygen; in fact, when the impregnation is carried out in the presence of air, immediate decomposition of $Ru_3(CO)_{12}$ occurs with formation of the oxidized surface ruthenium species. This high surface reactivity stands in sharp contrast to the relative stability of $Os_3(CO)_{12}$, which under similar conditions always produces the air-stable grafted cluster $HOs_3(CO)_{10}(OSi\equiv)$.

By reaction with H_2 under mild conditions (20 h at room temperature), $Ru_3(CO)_{12}$, physisorbed on dehydrated silica (pretreated at 773 K overnight under vacuum), transforms cleanly into $H_4Ru_4(CO)_{12}$ [226] (SCHEME 13). In solution, the same transformation requires more drastic conditions (363-393 K) [232]. This is another clear example of an easy change of cluster nuclearity by a mild surface reaction, suggesting a high surface mobility of organometallic fragments [233]. It is noteworthy how organometallic reactions can take place on the surface under conditions which are often different (usually milder) than those reported for the corresponding reactions in solution.

9.2.4.5.b **$Ru_3(CO)_{12}$ and other ruthenium clusters on alumina.** Several authors [68, 69, 211, 212], working under rather different conditions, have shown that $Ru_3(CO)_{12}$ interacts with alumina surfaces, keeping a cluster arrangement. Kuznetsov *et al.* [68] reported that the interaction of $Ru_3(CO)_{12}$ with the surface of γ-Al_2O_3 is dependent upon the pretreatment of the support. With highly dehydroxylated γ-Al_2O_3, simple physisorption occurs, whereas with a hydroxylated alumina, a surface reaction occurs, as indicated by a shifting to lower frequencies of the carbonyl infrared bands. These authors suggested that the new species are anionic ruthenium clusters generated by a reductive nucleophilic attack of surface water (or -OH surface groups) facilitated by the intrinsic basic character of some sites of the alumina surface [97-99]. A completely similar kind of reaction occurs in solution when $Ru_3(CO)_{12}$ is treated with alkali [234, 235].

On alumina free of water and highly dehydroxylated (treatment under vacuum in the range of temperatures between 573 and 923 K), the physisorbed $Ru_3(CO)_{12}$ is quite stable, independent of the manner of adsorption (impregnation from a solution in an organic solvent such as *n*-pentane or sublimation) [68]. It was shown that the physisorption process probably occurs via an acid-base interaction as it occurs between $Ru_3(CO)_{12}$ and $AlBr_3$ [124]; a Lewis acid center of the alumina surface [98] may interact with a carbonyl ligand of $Ru_3(CO)_{12}$.

All the above surface species [physisorbed $Ru_3(CO)_{12}$ or chemisorbed carbonyl ruthenium anions] decompose as a result of thermal treatment at 353-363 K under vacuum, producing surface species which were claimed to be oxidized, such as $[Ru(CO)_n(OAl\langle)_2]_y$ (n = 2, 3, 4) (SCHEME 14). Their nature is still a matter of debate; Knözinger *et al.* [192] discussed the infrared spectra, suggesting the existence of $Ru(CO)_3$ and two different types of $Ru(CO)_2$ surface species, probably corresponding to different oxidation states or to different coordination geometries of the ruthenium atoms.

These oxidized species (they are easily obtained when impregnation is carried out in the presence of air or when physisorbed $Ru_3(CO)_{12}$ is treated with oxygen) show interesting chemical properties, which have not been fully investigated. They can be decarbonylated by thermal treatment under vacuum to form ruthenium ions incorporated into the lattice of the alumina surface (ruthenium surface aluminates) [236-238]. Simple treatment with CO restores the original surface carbonyl complexes. It was also reported that these aluminates show a high thermal resistance; ruthenium metal particles cannot be obtained by prolonged heating under vacuum at

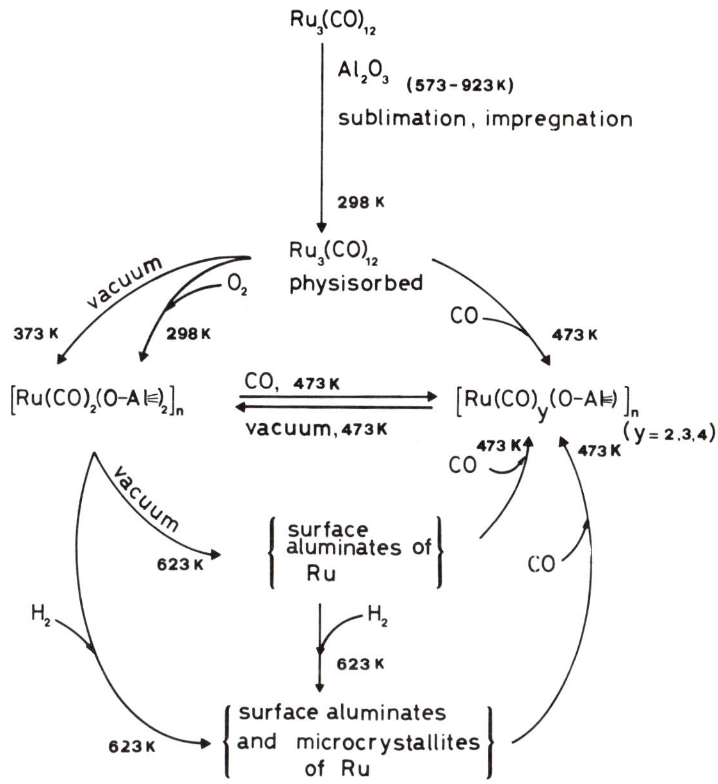

SCHEME 14

relatively high temperatures. Microcrystallites of metallic ruthenium are obtained only by reduction with H_2 at temperatures > 673 K [68, 239]. These metallic particles can be easily reoxidized by oxygen to give the original surface ions [68]. Zecchina *et al.* [69] again pointed out the importance of the surface loading. At low loading (0.33% of Ru metal), a variety of completely decarbonylated ruthenium species are incorporated in the surface structure, mainly as oxidized species, after thermal treatment under vacuum at 623 K. At higher loading (1.5% of Ru metal), ruthenium surface complexes formed by thermal decomposition have been investigated by ^{13}CO-^{12}CO isotopic substitution; oxidation states II and III have been suggested, together with the presence of some metallic ruthenium [211].

In conclusion, despite the extensive work, additional studies of the surface organometallic chemistry of $Ru_3(CO)_{12}$ on Al_2O_3 are required. The behavior of $H_4Ru_4(CO)_{12}$ [68, 214] adsorbed on highly dehydroxylated alumina is interesting; the infrared evidence and extraction with a solution of the cation PPN$^+$ or $(PPh_3)_2N^+$ suggest an easy formation of the corresponding anion $[H_3Ru_4(CO)_{12}]^-$ [240], chemisorbed by ionic bonds to the surface. The surface reaction could be the following:

$$H_4Ru_4(CO)_{12} + HOAl\leqslant \longrightarrow [H_3Ru_4(CO)_{12}]^{\ominus}[Al\leqslant]^{\oplus} + H_2O \quad (16)$$

This anionic surface species is thermally unstable; decomposition takes place in the same way as previously described (SCHEME 15) for $Ru_3(CO)_{12}$.

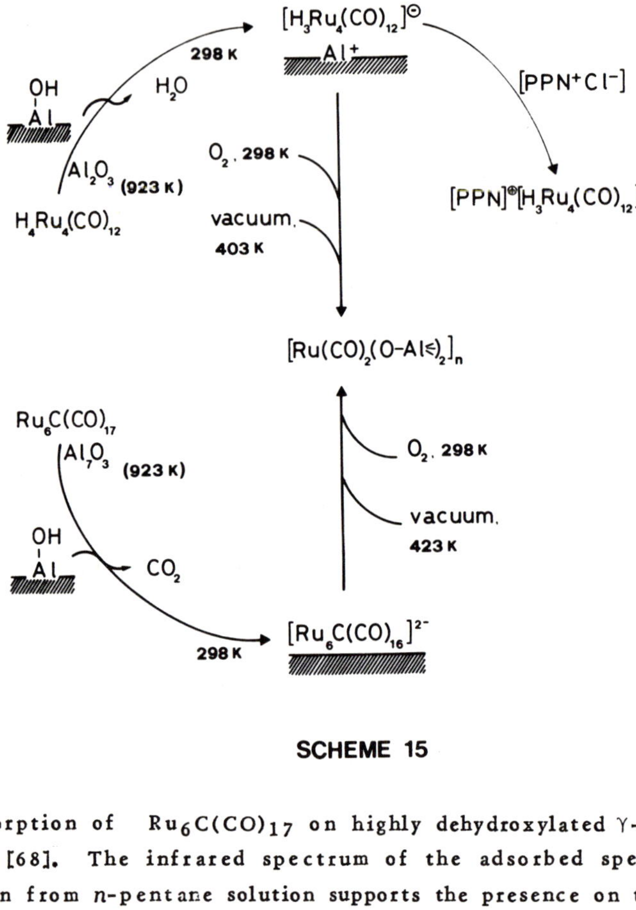

SCHEME 15

The adsorption of $Ru_6C(CO)_{17}$ on highly dehydroxylated γ-Al_2O_3 was also studied [68]. The infrared spectrum of the adsorbed species after impregnation from *n*-pentane solution supports the presence on the surface of the anion $[Ru_6C(CO)_{16}]^{2-}$ [241]. This is an unexpected result, because anionic species are generally obtained by electron transfer with surface water or -OH groups mediated by basic centers of the surface; but on the surface of highly dehydroxylated alumina, the acidic centers are predominant [98]. The thermal decomposition of such an anion (under vacuum at 423 K) produces results similar to those obtained by thermal treatment of the surface species generated by adsorption of $Ru_3(CO)_{12}$ and $H_4Ru_4(CO)_{12}$ (SCHEME 15) on a similar alumina.

9.2.4.5.c $Ru_3(CO)_{12}$ and $H_4Ru_4(CO)_{12}$ on magnesia. As observed with other clusters [particularly $Os_3(CO)_{12}$], the reactivity of the magnesia surface is dependent upon its thermal pretreatment [112, 114]. On magnesia pretreated at 473 K under vacuum, the adsorption of $Ru_3(CO)_{12}$ corresponds to the formation of the anion $[HRu_3(CO)_{11}]^-$, a behavior similar to that shown by $Fe_3(CO)_{12}$; but $Os_3(CO)_{12}$ behaves differently [242]. The surface anion was not only characterized by infrared spectroscopy, it was also extracted from the surface by anion exchange with a solution of the salt $[N(PPh_3)_2]^+Cl^-$. As in the case of the anion $[HFe_3(CO)_{11}]^-$, an acid-base interaction probably occurs between the bridging carbonyl of the anion and a cationic Mg^+ surface center. In fact, the infrared band due to bridging carbonyls shifts from 1702 to 1608 cm^{-1}.

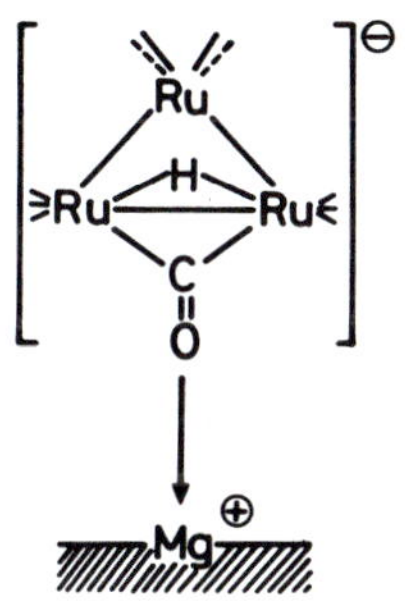

Recently, Pierantozzi *et al.* [214] (working under rather different conditions) reported that other anions are also formed when $Ru_3(CO)_{12}$ is adsorbed on magnesia pretreated at 473 K under vacuum. These authors were able to extract with a CH_2Cl_2 solution of $[N(PPh_3)_2]Cl$ a mixture of $[N(PPh_3)_2]_2[Ru_6(CO)_{18}]$ (major product) [243] and $[N(PPh_3)][HRu_3(CO)_{11}]$ [235]. It is known from solution studies that $Ru_3(CO)_{12}$ reacts with bases to form, first, Ru_3 anions, and then the reaction continues to form Ru_6 anions [234, 235, 243, 244]. The results just reported would suggest that the same series of reactions occurs on a magnesia surface.

$H_4Ru_4(CO)_{12}$ interacts with the magnesia surface in the same way as on alumina; the anion $[H_3Ru_4(CO)_{12}]^-$ is quickly formed [214, 240]. The thermal and chemical reactivity of these anionic species have not been investigated in any detail. However, Pierantozzi *et al.* [214] reported on the chemical transformations which occur under catalytic conditions for the $CO + H_2$ reaction. The surface species catalyze the formation of oxygenated products (particularly methanol); after the catalytic reaction the only species identified on the surface is the anion $[Ru_6C(CO)_{16}]^{2-}$. The authors suggested that intact metal clusters might be associated with the high catalytic selectivity to oxygenated products. These results are discussed further in Section 9.5.

9.2.5 SUPPORTED BIMETALLIC CLUSTERS

Bimetallic transition metal clusters hold promise as highly selective catalysts [60, 61]; in addition, supported bimetallic clusters may be considered to be models of and precursors of bimetallic (alloy) crystallites deposited on supports, which are used industrially in reforming of petroleum distillates [245, 246] (see Chapter 10). In recent years many authors [62, 63, 128, 246-248] reported on the catalytic properties of supported bimetallic carbonyl clusters, but the understanding of the related surface organometallic chemistry is still quite limited.

9.2.5.1 Bimetallic Rh-Co carbonyl clusters on metal oxides

Work has been carried out with $Co_2Rh_2(CO)_{12}$ and $Co_3Rh(CO)_{12}$ adsorbed on γ-alumina and silica (Aerosil) [62, 63]. Infrared evidence indicates that $Co_2Rh_2(CO)_{12}$ is chemisorbed on γ-alumina with total loss of bridging carbonyls; the remaining carbonyl ligands disappear only by treatment at temperatures higher than 300 K. The decarbonylation process is not reversible, as indicated by the observation that admission of CO (about 200 Pa) at room temperature on the decarbonylated sample does not induce the reversible formation of the original cluster. The irreversible nature of this process suggests a strong chemisorption process.

Unfortunately, the available experimental results are not sufficient to clarify the organometallic pathways leading to the final surface species. A more accurate study was carried out in the characterization of the final bimetallic catalysts, which generally show higher dispersions than those prepared by classical techniques [249]. Of particular interest are the unusual properties of hydroformylation catalysts prepared from mixed rhodium-cobalt carbonyl clusters supported on ZnO (pretreated by evacuation at 433 K) [128]. The surface species behave as highly dispersed alloys; the same results could not be achieved by impregnation with a mixture of $Rh_4(CO)_{12}$ and $Co_4(CO)_{12}$.

9.2.5.2 $HFeCo_3(CO)_{12}$ and $H_2RuOs_3(CO)_{13}$ on alumina

A preliminary note appeared concerning the surface organometallic chemistry of $HFeCo_3(CO)_{12}$ on alumina [250]. Infrared evidence suggests that immediately after the impregnation, $HFeCo_3(CO)_{12}$ leads to an anion according to the surface reaction

$$HFeCo_3(CO)_{12} + HOAl\equiv \longrightarrow [FeCo_3(CO)_{12}]^{\ominus}[Al\leqslant]^{\oplus} + H_2O \quad (17)$$

In this case the surface behaves as a Brønsted base which deprotonates the hydride, a reaction well known in solution [251].

A rather similar process occurs when $H_2RuOs_3(CO)_{13}$ [252] is adsorbed on γ-Al_2O_3, previously heated at 475 K under vacuum. Infrared evidence and the absence of gas evolution suggest the following surface reaction:

$$H_2RuOs_3(CO)_{13} + HOAl\leqslant \longrightarrow [HRuOs_3(CO)_{13}]^{\ominus}[Al\leqslant]^{\oplus} + H_2O \quad (18)$$

Treatment of this surface species at 373-473 K under equimolar H_2 + CO leads to the formation of $[H_3RuOs_3(CO)_{12}]^-$. This anion can be extracted from the surface by ion exchange with $[AsPh_4]^+Cl^-$ and fully characterized by classical solution techniques. This is a relevant transformation, because a supported cluster maintains the same bimetallic frame after a rather drastic treatment (H_2 + CO at 473 K). However, the surface anionic bimetallic species are stable only up to 373 K when heated under vacuum; at higher temperatures extensive decomposition occurs, probably giving mononuclear oxidized carbonyl species of osmium and ruthenium.

9.2.5.3 $H_2FeOs_3(CO)_{13}$ on silica

By impregnation on silica, pretreated under vacuum at 623 K, simple physisorption of $H_2FeOs_3(CO)_{13}$ takes place. This inference is based on infrared and Raman spectra and on the absence of any relevant gas evolution [253]. By thermal treatment at 403 K under argon, the Fe-Os bonds are cleaved to produce the known grafted Os_3 cluster $HOs_3(CO)_{10}(Si\leqslant)$. The infrared and Raman spectra and the mass balance are in support of the surface organometallic reactions reported in SCHEME 16. At higher temperatures the behavior is that already reported for such surface-anchored clusters of osmium. In summary, it appears that the surface organometallic chemistry of the $FeOs_3$ metallic frame cannot be too different from that of an Os_3 frame.

Interestingly, metallic particles of osmium (about 1.6 nm) are formed by heating the mixed Fe-Os cluster under argon at temperatures higher than 523 K. The properties (for the CO + H_2 reaction) of the catalyst obtained by thermal decomposition of this mixed cluster are between those of the two monometallic catalysts, as expected on the basis of the surface organometallic chemistry just discussed (complete rupture of Fe-Os bonds).

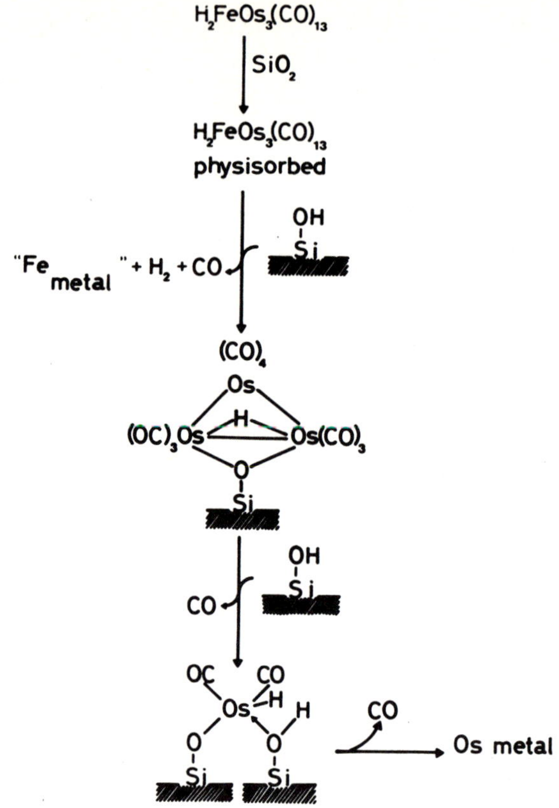

SCHEME 16

9.2.5.4 Other mixed metal clusters on γ-alumina

Clusters such as $(\eta^5\text{-}C_5H_5)WOs_3(CO)_{12}H$ and $(\eta^5\text{-}C_5H_5)MoOs_3(CO)_{12}$ have been adsorbed on γ-alumina; the selectivity of these systems in the reduction of CO has been compared with that of the system prepared by $Os_3(CO)_{12}$ adsorbed on alumina [254].

Also $(\eta^5\text{-}C_5H_5)WIr_3(CO)_{11}$ and $(\eta^5\text{-}C_5H_5)W_2Ir_2(CO)_{10}$ have been adsorbed on γ-alumina [255]; these systems were compared with those obtained from $Ir_4(CO)_{12}$ and $(\eta^5\text{-}C_5H_5)_2W_2(CO)_6$ adsorbed on γ-alumina. Their thermal decomposition was studied by programmed heating (TPDE) in a flow of hydrogen. Accurate analysis of the gas evolved (CO and CH_4) shows differences for all these systems. This result indicates that different sites or a different distribution of sites (which catalyze CO reduction to CH_4) are formed during the decomposition of the two bimetallic clusters.

The intermediate organometallic chemistry which probably leads to the breakup of the bimetallic framework was not investigated, but the dispersion of the final supported metal is very high, with particle sizes probably less than 1 nm [256].

9.2.6 SUMMARY AND CONCLUSIONS

We have reported evidence for surface organometallic species prepared from a wide range of samples, summarized in Table 1. These materials define a new area of molecular chemistry resulting from the overlap between organometallic (and/or coordination) chemistry and surface science. This field deals today with the interaction of organometallic complexes and/or coordination compounds (in particular, carbonyls and cluster carbonyls) with specific surfaces of inorganic oxides.

It has been possible to understand at a molecular level the reactions occurring between an organometallic species and the reactive chemical groups located on the surface of an oxide or a zeolite, etc. The new surface complexes resulting from such surface reactions can be characterized in few cases; the transition metal is linked to the support by "rigid" ligands which belong to the support framework, but conventional ligands, such as carbon monoxide, are also present in the coordination spheres of these unusual molecular coordination compounds. In our opinion there is still much work to do, not only on the systems described in this review (namely, those starting from carbonyl clusters), but also on other systems starting from other organometallic entities.

The main problem is the development of well-defined chemical and physical techniques for characterization of the surface molecular species. Knowledge of the ligand environment of the transition metal and its stoichiometric reactivity will probably illuminate the potential catalytic properties of such surface species. As a result, there is hope, via this approach, of more effectively bridging the concepts of coordination chemistry and homogenous catalysis to surface science and heterogeneous catalysis [15].

9.2.7 ADDENDUM

During the preparation of this manuscript a number of relevant new results came to our attention. These are summmarized below to indicate recent development and current research directions. This addendum, with a few exceptions, is limited to June, 1984.

The behavior of $Fe_3(CO)_{12}$ on Cab-O-Sil was investigated by infrared and Mössbauer spectroscopies and by accurate mass spectrometric

TABLE 1. **Preparation and Characterization of Surface Organometallic Species**

Metal Carbonyl Cluster	Support and Surface Area, $m^2\ g^{-1}$	Thermal Pretreatment of Support, T/K	Metal Loading, wt%	Techniques of Impregnation	Methodologies of Characterization of Surface Species	Refs.
<u>IRON</u>						
$Fe_3(CO)_{12}$	SiO_2 (200)	298,523,773	1.0-2.0	sublimation, hexane	IR	[92]
	SiO_2 (500)	473	---	dry mixing	IR, DRS, EXAFS, TPDE	[93]
	Cab-O-Sil (580)	570	1.0	hexane	IR, TPDE Mössbauer	[257]
	η-Al_2O_3 (315)	298	2.0	sublimation, hexane	IR, DRS, ^1H NMR	[92]
	γ-Al_2O_3 (100)	298	2.5	sublimation, hexane	IR, DRS, ^1H NMR	[92]
	γ-Al_2O_3 (150)	573	---	dry mixing	IR, DRS, EXAFS, TPDE	[93]
	γ-Al_2O_3 (250)	773	1.0	dry mixing, pentane	TPDE	[51]

	MgO (96)	298, 673	2.7	sublimation, hexane	IR, DRS, Mössbauer, Magnetism, FMR	[92,95]
	ZnO (75)	298	1.0	hexane	IR, DRS	[92]
$[(\eta^5\text{-}C_5H_5)Fe(CO)]_4$	γ-Al_2O_3 (160)	1223	---	dichloromethane	IR	[126]
<u>RUTHENIUM</u>						
$Ru_3(CO)_{12}$	SiO_2	373-1073	---	sublimation, pentane	IR	[68]
	SiO_2	723	1.5	sublimation	IR	[212]
	SiO_2 (200)	773	1.0	pentane	IR, Raman	[213]
	γ-Al_2O_3 (100)	373-1073	0.3-1.2	pentane, sublimation	IR	[68]
	Al_2O_3 (100)	298	0.33, 1.57	*n*-heptane	IR, DRS, MS	[69]
	γ-Al_2O_3 (100)	673	---	*n*-octane	IR	[192]
	γ-Al_2O_3	723	1.5	sublimation	IR	[212]
	MgO	473	---	pentane	IR	[242]

	MgO	(29)	473	0.42	hexane	IR	[214]
$H_4Ru_4(CO)_{12}$	γ-Al_2O_3		923	1.1-1.2	pentane	IR	[68]
	γ-Al_2O_3	(257)	473	0.9	hexane/toluene	IR	[214]
	γ-Al_2O_3	(274)	673	0.31	pentane	XPS	[271]
	MgO	(29)	473	0.43	hexane/toluene	IR	[214]
$Ru_6C(CO)_{17}$	γ-Al_2O_3	(274)	923	1.1-1.3	pentane	IR	[68]
<u>OSMIUM</u>							
$Os_3(CO)_{12}$	SiO_2	(200)	298, 773	1.0	dichloromethane	IR, EXAFS	[59,82]
	SiO_2	(382)	573	---	refluxing *n*-octane	IR	[188]
	Cab-O-Sil	(110)	773	2.0	dichloromethane	IR, DRS, TPD, EM	[196]
	Cab-O-Sil		673	0.83	refluxing *n*-octane	IR, DRS	[56]
	η-Al_2O_3	(300)	298, 773	1.0	dichloromethane	IR, EXAFS	[59,82]
	γ-Al_2O_3	(185)	573	1.0	retluxing	IR, TEM	[188,195]

	γ-Al_2O_3 (415)	673	---	refluxing *n*-octane	IR, Raman	[189]
	Al_2O_3 (100)	670	0.78	"	IR, TPD, XPS, EPR	[191,192,195]
	Al_2O_3 (97)	773	2.0	dichloromethane	IR, DRS, TPD, EM	[196]
	TiO_2 (120)	573	---	refluxing *n*-octane	IR	[188]
	TiO_2 (43)	773	2.0	dichloromethane	IR, DRS, TPD, EM	[196]
	ZnO (20)	573	---	refluxing *n*-octane	IR	[188]
	MgO (30)	473	1.0	cyclohexane	IR	[190]
	MgO (200)	298, 573, 673, 773	1.0-1.5	dichloromethane	IR	[197]
$H_2Os_3(CO)_{10}$	SiO_2 (200)	298	---	dichloromethane	IR	[59]

	γ-Al_2O_3 (185)	573, 673	---	cyclohexane	IR	[188]
	TiO_2 (120)	573, 673	---	"	IR	[188]
$Os_6(CO)_{18}$	SiO_2 (200)	298, 773	---	heptane	IR, EXAFS	[59,82]
	Cab-O-Sil (100)	773	2.0	1,2-dichloro-ethane	IR, DRS, TPD, EM	[196]
	η-Al_2O_3 (300)	298, 773	---	heptane	IR, EXAFS	[59,82]
	γ-Al_2O_3 (100)	673	---	refluxing *n*-octane	IR, TEM	[192]
	Al_2O_3 (97)	773	2.0	1,2-dichloro-ethane	IR, DRS, TPD, EM	[196]
	TiO_2 (43)	773	2.0	1,2-dichloro-ethane	IR, DRS, TPD, EM	[196]
$H_4Os_4(CO)_{12}$	SiO_2 (200)	298, 773	---	dichloromethane	IR	[59]
	Cab-O-Sil (110)	773	2.0	cyclohexane	IR, DRS, EM	[268]
	Al_2O_3 (97)	773	2.0	"	IR, DRS, EM	[268]

	γ-Al_2O_3	(250)	673	---	*n*-octane	IR	[192]
	TiO_2	(43)	773	2.0	cyclohexane	IR, DRS, EM	[268]
<u>COBALT</u>							
$Co_2(CO)_8$	SiO_2	(190)	773	0.1-4.5	pentane	IR, TPDE	[129,261]
	SiO_2	(500)	473	---	dry mixing	IR, Raman, EXAFS, TPDE	[259]
	SiO_2	(600)	400	0.5-2.0	sublimation	IR, MS	[262]
	Cab-O-Sil	(200)	400	0.5-2.0	sublimation	IR, MS	[262]
	γ-Al_2O_3	(160)	295, 475, 525, 775	0.5-2.0	sublimation	IR, MS	[262]
	γ-Al_2O_3	(175)	773, 1223	1.5-3.0	pentane	IR, TPDE	[129,260,261]
	γ-Al_2O_3	(150)	573	---	dry mixing	IR, Raman, DRS, EXAFS, TPDE	[259]
	MgO	(105)	773	0.5	pentane	IR, TPDE	[261]
	MgO	(28)	673	0.15	sublimation	IR, XPS	[263]

$[Co_3(CO)_9CCH_3]$	SiO_2 (600)	475	---	sublimation	IR, MS	[133]
	Cab-O-Sil (200)	400	---	sublimation	IR, MS	[133]
	γ-Al_2O_3 (160)	295,525,700	---	sublimation	IR, MS	[133]
$Co_4(CO)_{12}$	SiO_2 (600)	400	0.5-2.0	sublimation	IR, MS	[262]
	Cab-O-Sil (200)	400	0.5-2.0	sublimation	IR, MS	[262]
	γ-Al_2O_3 (160)	295, 475, 525, 775	0.5-2.0	sublimation	IR, MS	[262]
	MgO (28)	673	0.15	sublimation	IR, XPS	[263]
RHODIUM						
$Rh_4(CO)_{12}$	SiO_2 (200)	298, 773	---	hexane	IR	[150]
	SiO_2 (200)	433	---	hexane	IR, MS	[156]
	η-Al_2O_3 (315)	73	---	hexane	IR	[150]
	Al_2O_3	433	---	hexane	IR, MS	[156]
	TiO_2	433	---	hexane	IR, MS	[156]

	TiO_2 (40)	433	0.1-0.5	hexane or tetrahydrofuran	IR, TPD	[143]
	ZnO (26.7)	433	0.1-0.5,1.5	hexane or tetrahydorturan	IR, XPS, TPD	[128,151]
	ZrO_2-SiO_2 (300)	433	0.1-0.5	hexane or tetrahydroturan	IR, XPS TPD	[143]
$Rh_6(CO)_{16}$	SiO_2 (200)	298	---	chloroform	IR, UPS	[145]
	SiO_2 (200)	433	---	hexane	IR, MS	[156]
	SiO_2 (200)	573	5.0	chlorotorm	IR	[146]
	Al_2O_3	433	---	hexane	IR, MS	[156]
	η-Al_2O_3 (315)	298, 473, 773	0.4-2.0	chlorotorm or tetrahydrofuran	IR	[147]
	Al_2O_3 (200)	298	0.67.0	chlorotorm	IR, TEM, EPR, XPS	[148,149]
	SiO_2-Al_2O_3 (500)	773	0.4-2.0	chlorotorm	IR	[147]
	TiO_2	433	---	hexane	IR, MS	[156]

$Rh_6(CO)_{16}$	MgO (245)	773	0.4-2.0	chloroform	IR	[147]
	ZnO (26)	433	0.1-0.5	hexane or tetrahydrofuran	IR, TPD	[128]
	γ-Al_2O_3	650-773				[172]
	MgO (43)	650-773				[172]
	TiO_2 (42)	650-773	0.1-0.5	hexane or tetrahydrofuran	IR, TPD	[172]
	La_2O_3 (23)	650-773	0.1-0.5	hexane or tetrahydrofuran	IR, TPD	[172]
	ZrO_2 (35)	650-773	0.1-0.5	"	IR, TPD	[172]
$[Rh_6(CO)_{15}]^{2-}$	TiO_2 (42)	673	---	tetrahydrofuran	IR	[143]
	ZrO_2-SiO_2 (300)	673	---	tetrahydrofuran	IR	[143]
$[Rh_6(CO)_{14}]^{-4}$	TiO_2 (42)	673	---	"	IR	[143]
	ZrO-SiO_2 (300)	673	---	"	IR	[143]
$[Rh_7(CO)_{16}]^{3-}$	TiO_2-SiO_2 (300)	673	---	"	IR	[143]

$[Rh_{13}(CO)_{23}H_{5-n}]^{n}$ (n = 2 or 3)	TiO_2-SiO_2 (300)	673	---	tetrahydrofuran	IR	[143]
IRIDIUM						
$Ir_4(CO)_{12}$	SiO_2 (340)	298	0.18-1.05	refluxing cyclohexane	IR, TPD, TEM	[63,174]
	γ-Al_2O_3 (200)	298	0.23-1.34	refluxing cyclohexane	IR, TPD TEM	[63,174]
NICKEL						
$[(\eta^5$-$C_5H_5)_3Ni_3(CO)_2]$	SiO_2 (200-380)	298	0.1-0.5	tetrahydrofuran	IR, EPR	[139]
	γ-Al_2O_3 (160)	1223	---	dichloromethane	IR	[126]
PLATINUM						
$[NET_4)_2][\{Pt_3(CO)_6\}_n]$ (n = 1-5)	SiO_2 (250), γ-Al_2O_3 (215)	650-773	0.05-1	tetrahydrofuran	IR, TPD	[172,186]
BIMETALLIC CLUSTERS						
$Co_2Rh_2(CO)_{12}$	SiO_2 (200)	400-630	1	hexane	IR, EM	[62,63]
$Co_3Rh(CO)_{12}$	γ-Al_2O_3 (200)	400-630	1	hexane	IR, EM	[62,63]

	ZnO	(26)	433	0.25	hexane	IR	[128]
$Fe_2Ru(CO)_{12}$ / $H_2FeRu_3(CO)_{13}$	Cab-O-Sil		570	1	hexane	IR, MS, TPR, Mössbauer	[269,270]
	γ-Al_2O_3		570	1	hexane	IR, MS, TPR, Mössbauer	[269,270]
$H_2RuOs_3(CO)_{13}$	γ-Al_2O_3	(100)	473	0.89	hexane	IR	[252]
	γ-Al_2O_3	(274)	673	2.0	hexane	XPS, IR, EM	[271]
$H_2FeOs_3(CO)_{13}$	SiO_2	(200)	623	---	dichloromethane	IR, Raman, TPD, TEM	[253]
$(\eta^5$-$C_5H_5)_2WIr_3(CO)_{11}$ / $(\eta^5$-$C_5H_5)_2W_2Ir_2(CO)_{10}$	γ-Al_2O_3		773	---	cyclohexane	IR, TPD	[255]

determinations [257]. The authors suggested, in contrast to the current literature data [92], that a strong interaction occurs during the physisorption process. This kind of interaction probably involves hydrogen bonding between the -OH surface groups and the bridging carbonyl ligands of the cluster. By thermal decomposition at 370 K, the original metal framework must be retained, because the original carbonyl cluster $Fe_3(CO)_{12}$ is regenerated by treatment with a CO atmosphere. At temperatures exceeding 420 K, the thermal decomposition corresponds to a total oxidation of the iron atoms.

Guglielminotti and Zecchina [258] investigated the interaction of $Fe(CO)_5$ with a surface of completely dehydroxylated magnesia. The formation of the mononuclear anion $[(CO)_4Fe(CO)_2]^{2-}$, bound ionically to cationic magnesium centers of the surface, has been suggested. Such a species is formed by direct interaction of $Fe(CO)_5$ with an active $Mg^{2+}O^{2-}$ pair of the surface. It is worth mentioning that other authors [113] proposed that on highly dehydroxylated magnesia, $Fe_3(CO)_{12}$ is chemisorbed on the support as $[(CO)_4Fe(COOMg)]$.

Iwasawa *et al.* [259] reported extensively on the interaction of $Co_2(CO)_8$ with silica or γ-Al_2O_3 by a dry mixing method. They confirmed by different techniques the easy rearrangement of $Co_2(CO)_8$ to give $Co_4(CO)_{12}$. This latter tetranuclear cluster decomposed upon exposure to oxygen at 293 K; subsequent thermal treatment at 473 K under vacuum induced a complete oxidation of cobalt atoms to Co^{II} species incorporated in the silica surface. Thermal treatment at 773 K in the presence of oxygen led to the formation of Co_3O_4 on the surface. Similar results were obtained with γ-Al_2O_3; only the intermediate species formed by decomposition with oxygen at 293 K could be characterized by EXAFS as a $[CoO]_n$ species (probably Co_4O_4).

Nakamura *et al.* [260, 261] reported on the interaction of $Co_2(CO)_8$ with the surface of a highly dehydroxylated alumina upon impregnation from pentane solution. They suggested the easy formation of $Co_6(CO)_{16}$ adsorbed on the surface.

Schneider *et al.* [262] described the transformations occurring when $Co_2(CO)_8$ or $Co_4(CO)_{12}$ is sublimed onto silica, γ-Al_2O_3, or silica-alumina. On silica, as reported by other authors, $Co_2(CO)_8$ is converted under vacuum and at room temperature into $Co_4(CO)_{12}$. This rearrangement is completely inhibited by a low pressure of CO. On γ-Al_2O_3, such a transformation occurs quite rapidly; the behavior is independent of the original carbonyl compound [$Co_2(CO)_8$ or $Co_4(CO)_{12}$]. On a surface treated under vacuum at 285 K, a fast reaction takes place, and the anion

$[Co(CO)_4]^-$ was identified among the many surface species formed.

On silica-alumina thermal treatment at temperature > 475 K or simple exposure to oxygen resulted in an irreversible decarbonylation process with parallel oxidation of the cobalt atoms.

The behavior of $Co_2(CO)_8$ and $Co_4(CO)_{12}$ on magnesia was also investigated [263]. An easy decarbonylation and oxidation of the cobalt atoms occurred either by reaction with oxygen or by thermal treatment up to 725 K. On addition of CO, reduction occurred to form again the original carbonyl compounds. Such a reversible behavior is worthy of attention because it does not occur with other inorganic oxides [264, 265].

The cobalt carbonyls bound to the surface exchange very easily with ^{13}CO. The final products can be extracted in some cases from the support, providing an excellent method for the preparation of fully exchanged cobalt carbonyls [266].

The state of rhodium in supported metallic catalysts prepared by adsorption of $Rh_4(CO)_{12}$ on silica, γ-Al_2O_3, or La_2O_3 was investigated by XPS, EXAFS, and other techniques [267]. On silica, rhodium is in the metallic state, but on γ-Al_2O_3, a mixture of $Rh^I(CO)_2$ species, metallic rhodium, and probably Rh_2O_3 was found.

Some authors studied the impregnation of $H_4Os_4(CO)_{12}$ on alumina, silica, and titania [268a]. A cluster structure is maintained even during a thermal treatment in a stream of pure helium up to 523 K at a rate of 10 K min^{-1}. Other authors [268d] found that the cluster was deprotonated on the alumina, giving $H_3Os_4(CO)_{12}^-$; the chemistry parallels that observed with $H_4Ru_4(CO)_{12}$ [214] and $H_4RuOs_3(CO)_{12}$ [252].

Schay *et al.* [269] studied the interaction of $Ru_3(CO)_{12}$ with Cab-O-Sil. The simple impregnation is always accompanied by some oxidation to give $Ru^{II}(CO)_n$ (n = 2, 3, 4) surface species. The thermal decomposition of the supported ruthenium clusters is faster in hydrogen than in helium or in vacuum. Surprisingly, this decomposition was reported to be reversible at temperatures up to 400 K. The authors demonstrated the formation of highly dispersed surface carbon by CO dissociation.

An investigation of the infrared spectra of $Fe_2Ru(CO)_{12}$ and $H_2FeRu_3(CO)_{13}$ physisorbed on Cab-O-Sil [270] suggested the presence of mild interactions. On alumina, extensive decomposition occurs even with treatment under vacuum at room temperature. Mössbauer spectra suggest the formation of Fe^{II} and Fe^{III} ions incorporated in the alumina surface. In any case, $H_2FeRu_3(CO)_{12}$ seems to be more stable than $Fe_2Ru(CO)_{12}$, when supported.

Metallic catalysts on alumina have been prepared by decomposition of supported $H_4Ru_4(CO)_{12}$ or $H_2RuOs_3(CO)_{12}$ [271]. The catalysts, which have been applied for catalytic reduction of carbon monoxide, were characterized by XPS, infrared spectroscopy, and electron microscopy. Evidence for the presence of oxidized ruthenium species was presented. It was claimed that these species are related to the production of oxygenated products (particularly methanol), as occurs homogeneously [272-275]. Driessen *et al.* [277] provided evidence that the methanol synthesis catalyzed by palladium supported on silica is associated with the presence of palladium ions on the surface.

Alumina-supported catalysts were also prepared from $H_2RhOs_3(CO)_{12}(acac)$ and $Rh_4(CO)_{12}$ [277]. Evidence provided by infrared spectroscopy, XPS, and electron microscopy indicated that the bimetallic clusters broke apart, initially giving the surface-bound triosmium clusters $[HOs_3(CO)_{10}OAl\leqslant]$ and mononuclear rhodium species, and at higher temperatures giving mononuclear osmium complexes and crystallites of rhodium metal. The catalytic activity for hydrogenation of CO to give hydrocarbons was attributed to the rhodium metal; a slight activity for methanol synthesis was suggested to be associated with ionic rhodium complexes.

A few results were also reported for $H_2FeOs_3(CO)_{13}$ on γ-Al_2O_3 [277]. Recent data indicating the formation of high-nuclearity osmium clusters from $HOs(CO)_4^-$ on magnesia are discussed in Section 9.5.

REFERENCES

1. B. R. Penfold, Perspect. Struct. Chem., 7 (1968) 71.
2. M. C. Baird, Progr. Inorg. Chem. 9 (1968) 1.
3. B. P. Biryukov and Y. T. Struchov, Russ. Chem. Rev., 39 (1970) 789.
4. R. D. Johnston, Advan. Inorg. Chem., 14 (1971) 471.
5. R. B. King, Progr. Inorg. Chem., 15 (1972) 288.
6. (a) P. Chini, Inorg. Chim. Acta Rev., 2 (1968) 31; (b) P. Chini, Pure Appl. Chem., 23 (1970) 489; (c) P. Chini, G. Longoni and V. G. Albano, Advan. Organometal. Chem., 14 (1976) 285; (d) P. Chini and B. Heaton, Topics Curr. Chem., 71 (1977) 1; (e) P. Chini, Gazz. Chim. Ital., 109 (1979) 225; (f) P. Chini, J. Organometal. Chem., 200 (1980) 37.
7. (a) H. D. Kaesz, Chem. Brit., 9 (1973) 334; (b) H. D. Kaesz, in Y. Ishii and M. Tsutsui (Eds.), Organotransition Metal Chemistry, Plenum, New York, 1975, p. 291; (c) A. P. Humphries and H. D. Kaesz, Progr. Inorg. Chem., 26 (1978) 146.
8. P. R. Raithby, in B. F. G. Johnson (Ed.), Transition Metal Clusters, Wiley, New York, 1980, p. 5.
9. (a) W. L. Gladfelter and G. L. Geoffroy, Advan. Organometal. Chem., 18 (1980) 207; (b) G. L. Geoffroy, J. R. Fox, E. Burkhardt, H. C. Foley, A. D. Harley and R. Rosen, in J. P. Fackler, Jr. (Ed.), Inorganic Syntheses, Vol. XXI, Wiley, New York, 1982, pp. 57-65.
10. J. S. Bradely, Advan. Organometal. Chem., 22 (1983) 1.
11. M. Tachikawa and E. L. Muetterties, Progr. Inorg. Chem., 28 (1982) 203.

12. E. L. Muetterties, J. Organometal. Chem., 200 (1980) 177.
13. L. Vaska and J. W. Luzio, J. Am. Chem. Soc., 83 (1961) 1262; and 84 (1962) 679.
14. G. C. Bond, Faraday Disc. Chem. Soc., 41 (1966) 200.
15. (a) R. Ugo, Proc. 5th Intern. Congr. Catal., Palm Beach 1972, vol. 1 (1973) 19; (b) R. Ugo, Catal. Rev., 11 (1975) 225.
16. J. Lewis and B. F. G. Johnson, Pure Appl. Chem., 44 (1975) 43.
17. (a) E. L. Muetterties, Bull. Soc. Chim. Belg., 84 (1975) 959; (b) *ibid.*, 85 (1976) 451; (c) *ibid.*, Science, 196 (1977) 839; (d) *ibid.*, Angew. Chem. Int. Ed. Engl., 17 (1978) 545.
18. T. N. Rhodin and G. Ertl (Eds.), The Nature of the Surface Chemical Bond, North Holland, Amsterdam, 1979.
19. J. M. Basset and R. Ugo, in R. Ugo (Ed.), Aspects of Homogeneous Catalysis, Vol. III, Reidel, Dordrecht, 1977, p. 137.
20. E. L. Muetterties, J. C. Hemminger and G. A. Somorjai, Inorg. Chem., 16 (1977) 3385.
21. E. L. Muetterties, T. N. Rhodin, E. Band, C. F. Brucker and W. R. Pretzer, Chem. Rev., 79 (1979) 91.
22. E. Band and E. L. Muetterties, Chem. Rev., 78 (1978) 639.
23. E. L. Muetterties, Chem. Eng. News. Aug. 30, 1982, 28.
24. J. R. Anderson, The Structure of Metallic Catalysts, Academic Press, London and New York, 1975, p. 244.
25. J. H. Sinfelt, Rev. Mod. Phys., 51 (1979) 569.
26. J. Evans, Chem. Soc. Rev., 10 (1981) 159.
27. K. Wade, in B. F. G. Johnson (Ed.), Transition Metal Clusters, Wiley, New York, 1980, p. 193.
28. J. A. Connor, in B. F. G. Johnson (Ed.), Transition Metal Clusters, Wiley, New York, 1980, p. 345.
29. C. R. Eady, B. F. G. Johnson, J. Lewis and T. Mathenson, J. Organometal. Chem., 57 (1973) C 82.
30. D. C. Bailey and S. H. Langer, Chem. Rev., 81 (1981) 109 and references cited therein.
31. J. Phillips and J. A. Dumesic, Appl. Catal., 9 (1984) and references cited therein.
32. R. Ugo, R. Psaro, G. M. Zanderighi, J. M. Basset, A. K. Theolier, and A. K. Smith, in M. Tsutsui (Ed.), Fundamental Research in Homogeneous Catalysis, Vol. 3, Plenum, New York and London, 1979, p. 579.
33. B. C. Gates and J. Lieto, CHEMTECH (1980) 195 and 248.
34. D. Ballivet, I. Tkatchenko, G. Coudurier and H. Mazzanega, in M. Tsutsui (Ed.), Fundamental Research in Homogeneous Catalysis, Vol. 3, Plenum, New York and London, 1979, p. 257.
35. T. Bein and P. A. Jacobs, J. Chem. Soc. Faraday Trans. I, 79 (1983) 1819.
36. B. F. G. Johnson and R. E. Benfield, in B. F. G. Johnson (Ed.), Transition Metal Clusters, Wiley, New York, 1980, p. 471.
37. R. Psaro, A. Fusi, R. Ugo, J. M. Basset, A. K. Smith and F. Hugues, J. Mol. Catal., 7 (1980) 511.
38. H. P. Boehm and H. Knözinger, in J. R. Anderson and M. Boudart (Eds.), Catalysis--Science and Technology, Vol. 4, Springer, Berlin and New York, 1983, p. 39.
39. J. M. Basset and A. Choplin, J. Mol. Catal., 21 (1983) 95, and references cited therein.
40. (a) J. M. Basset and A. K. Smith, in M. Tsutsui and R. Ugo (Eds.), Fundamental Reserch in Homogeneous Catalysis, Vol. 1, Plenum, New York and London, 1977, p. 69, and references cited therein; (b) A. K. Smith and J. M. Basset, J. Mol. Catal., 2 (1977) 229.
41. E. L. Muetterties and M. J. Krause, Angew. Chem. Int. Ed. Engl., 22 (1983) 135.
42. E. L. Muetterties, Chem. Soc. Rev., 11 (1982) 83.

43. A. F. Simpson and R. Whyman, J. Organometal. Chem., 213 (1981) 157.
44. R. Whyman, in B. F. G. Johnson (Ed.), Transition Metal Clusters, Wiley, New York and London, 1980, p. 547, and references cited therein.
45. J. M. Basset, B. Besson, A. Choplin, F. Hugues, M. Leconte, D. Rojas, A. K. Smith, A. Theolier, Y. Chauvin, D. Commereuc, R. Psaro, R. Ugo and G. M. Zanderighi, in M. Giongo and M. Graziani (Eds.), Fundamental Research in Homogeneous Catalysis, Vol. 4, Plenum, New York and London, 1984, p. 19, and references cited therein.
46. J. F. Hamilton and R. C. Baetzold, Science, 205 (1979) 1213, and references cited therein.
47. J. Haggin, Chem. Eng. News, Feb. 8, 1982, 13; *ibid.*, June 28, 1982, 31.
48. A. Brenner and D. A. Hucul, J. Am. Chem. Soc., 102 (1980) 2484.
49. D. A. Hucul and A. Brenner, J. Am. Chem. Soc., 103 (1981) 217.
50. D. A. Hucul and A. Brenner, J. Phys. Chem., 85 (1981) 496.
51. A. Brenner and D. A. Hucul, Inorg. Chem., 18 (1979) 2836.
52. J. R. Anderson, Structure of Metallic Catalysts, Academic Press, London and New York, 1975, p. 163.
53. J. H. Sinfelt, Ann. Rev. Mater. Sci., 2 (1972) 641.
54. J. R. Anderson, Structure of Metallic Catalysts, Academic Press, London and New York, 1975, p. 451.
55. R. L. Moss, in R. B. Anderson and P. T. Dawson (Eds.), Experimental Methods in Catalytic Research, Vol. 2, Academic Press, New York and London, 1976, p. 43.
56. R. Barth, B. C. Gates, Y. Zhao, H. Knözinger and J. Hulse, J. Catal., 82 (1983) 147.
57. X.-J. Li and B. C. Gates, J. Catal., 82 (1984) 55.
58. B. Besson, A. Choplin, L. D'Ornelas and J. M. Basset, J. Chem. Soc. Chem. Commun., (1982) 843.
59. R. Psaro, R. Ugo, G. M. Zanderighi, B. Besson, A. K. Smith and J. M. Basset, J. Organometal. Chem., 213 (1981) 215.
60. J. H. Sinfelt, Acc. Chem. Res., 10 (1977) 15, and references cited therein.
61. J. H. Sinfelt and G. H. Via, J. Catal., 56 (1979) 1, and references cited therein.
62. J. R. Anderson and D. E. Mainwaring, J. Catal., 35 (1974) 162.
63. J. R. Anderson, P. S. Elmes, R. F. Howe and D. E. Mainwaring, J. Catal., 50 (1977) 508.
64. G. C. Smith, T. P. Chojnacki, S. R. Dasgupta, K. Iwatate and K. L. Watters, Inorg. Chem., 14 (1975) 1419.
65. M. Primet, J. M. Basset, M. V. Matthieu and M. Prettre, J. Catal., 28 (1973) 368.
66. R. F. Howe, D. E. Davidson and D. A. Whan, J. Chem. Soc. Faraday Trans. I, 68 (1972) 2266.
67. P. Gelin, F. Lefebvre, B. Ellench, C. Naccache and Y. Ben Taarit, in G. D. Stucky and F. G. Dwyer (Eds.), Intrazeolite Chemistry, ACS Symposium Series 218, Am. Chem. Soc., Washington D.C., 1983, p. 455.
68. V. L. Kuznetsov, A. T. Bell and Yu. I. Yermakov, J. Catal., 65 (1980) 374.
69. A. Zecchina, E. Guglielminotti, A. Bossi and M. Camia, J. Catal., 74 (1982) 225.
70. R. Psaro, C. Dossi and R. Ugo, in preparation.
71. Yu. I. Yermakov, B. N. Kuznetsov and V. A. Zakharov, Studies in Surface Science and Catalysis, Vol. 8, Catalysis by Supported Complexes, Elsevier, Amsterdam, 1981, Chapter 1 and references cited therein.

72. K. Tanabe, in J. R. Anderson and M. Boudart (Eds.), Catalysis-Science and Technology, Vol. 2, Springer, Berlin and New York, 1981, p. 231.
73. B. C. Lippens and J. J. Steggerda, in B. G. Linsen (Ed.), Physical and Chemical Aspects of Adsorbents and Catalysts, Academic Press, London and New York, 1970, p. 172.
74. T. L. Brown, J. Mol. Catal., 12 (1981) 41.
75. G. A. Somorjai and F. Zaera, J. Phys. Chem., 86 (1982) 3070.
76. H. W. K. Tom, C. M. Mate, X. D. Zhu, J. E. Crowell, T. F. Heinz, Y. R. Shen and G. A. Somorjai, Phys. Rev. Lett., in press (1984).
77. S. P. Wolsky and A. W. Czanderna (Eds.), Methods and Phenomena: Their Application in Science and Technology, Vol. 1, Methods of Surface Analysis, Elsevier, Amsterdam, 1975.
78. A. Theolier, A. K. Smith, M. Leconte, J. M. Basset, G. M. Zanderighi, R. Psaro and R. Ugo, J. Organometal. Chem., 191 (1980) 415.
79. A. Vizi-Orosz and L. Marko, Transition Met. Chem., 7 (1982) 216.
80. J. Evans and B. P. Gracey, J. Chem. Soc. Dalton Trans. (1982) 1123.
81. R. Psaro and N. Ravasio, 3rd Int. Symp. Homog. Catal., Milan, 1982, Fr. 19.
82. B. Besson, B. Moraweck, A. K. Smith, J. M. Basset, R. Psaro, A. Fusi and R. Ugo, J. Chem. Soc. Chem. Commun. (1980) 569.
83. R. Whyman, in B. F. G. Johnson (Ed.), Transition Metal Clusters, Wiley, New York, 1980, pp. 577-585.
84. A. Brenner and R. L. Burwell, Jr., J. Am. Chem. Soc., 97 (1975) 2565.
85. E. G. Derouane, J. B. Nagy and J. C. Vedrine, J. Catal., 46 (1977) 434.
86. R. F. Howe, Inorg. Chem., 15 (1976) 486.
87. E. Guglielminotti and A. Zecchina, J. Chim. Phys., 78 (1981) 891.
88. E. Guglielminotti and A. Zecchina, F. Boccuzzi and E. Borello, in J. Bourdon (Ed.), Growth and Properties of Metal Clusters, Elsevier, Amsterdam, 1980, p. 165.
89. F. Calderazzo, R. Ercoli and G. Natta, in I. Wender and P. Pino (Eds.), Organic Syntheses via Metal Carbonyls, Wiley, New York and London, 1968, pp. 45-63.
90. J. A. Connor, Topics Current Chem., 71 (1977) 71.
91. J. A. Connor, in B. F. G. Johnson, Transition Metal Clusters, Wiley, New York and London, 1980, pp. 347-359.
92. F. Hugues, J. M. Basset, Y. Ben Taarit, A. Choplin, M. Primet, D. Rojas and A. K. Smith, J. Am. Chem. Soc., 104 (1982) 7020.
93. Y. Iwasawa, M. Yamada, S. Ogasawara, Y. Sato and H. Huroda, Chem. Lett., (1983) 621.
94. F. Hugues, A. K. Smith, Y. Ben Taarit, J. M. Basset, D. Commereuc and Y. Chauvin, J. Chem. Soc. Chem. Commun. (1980) 68.
95. F. Hugues, J. A. Dalmon, P. Bussiere, A. K. Smith, J. M. Basset and D. Olivier, J. Phys. Chem., 86 (1982) 5136.
96. J. R. Wilkinson and L. J. Todd, J. Organometal. Chem., 118 (1979) 199.
97. J. B. Peri, J. Phys. Chem., 69 (1965) 211; 220; 231.
98. J. M. Parera, Ind. Eng. Chem. Prod. Res. Develop., 15 (1976) 234.
99. H. Knözinger and P. Ratnasamy, Catal. Rev.--Sci. Eng., 17 (1978) 31.
100. D. F. Shriver, J. Organometal. Chem., 94 (1975) 259.
101. F. Correa, R. Nakamura, R. E. Stimson, R. L. Burwell and D. F. Shriver, J. Am. Chem. Soc., 102 (1980) 5113.
102. (a) N. D. Parkyns, J. Chem. Soc. A. (1969) 102; (b) N. D. Parkyns, J. Phys. Chem., 75 (1971) 526; (c) M. P. Rosynek, J. Phys. Chem., 79 (1975) 1280.
103. W. McFarlane and G. Wilkinson, in H. F. Holtzclaw, Jr., (Ed.), Inorganic Syntheses, Vol. VIII, McGraw-Hill, New York and London, 1966, p. 181.

104. T. Kruck, M. Hofler and M. Noack, Chem. Ber., 99 (1966) 1153.
105. H. Knözinger and B. Stübner, J. Phys. Chem., 82 (1978) 1526.
106. M. Che, C. Naccache and B. Imelik, J. Catal., 24 (1972) 328.
107. D. Commereuc, Y. Chauvin, F. Hugues, J. M. Basset and D. Olivier, J. Chem. Soc. Chem. Commun. (1980) 154.
108. A. Brenner, J. Chem. Soc. Chem. Commun. (1979) 251.
109. H. P. Bonzel and H. J. Krebs, Surface Sci., 91 (1980) 499.
110. B. S. Brøden, T. N. Rhodin, C. Brucker, R. Benbow and Z. Hurych, Surface Sci., 59 (1976) 593.
111. K. Farmery, M. Kilner, R. Greatey and N. W. Greenwood, J. Chem. Soc. A (1969) 2339.
112. H. Knözinger, Advan. Catal., 25 (1976) 184.
113. A. Theolier and J. M. Basset, to be published.
114. O. V. Krylov, Kinet. Katal., 6 (1965) 128.
115. C. R. Eady, B. F. G. Johnson and J. Lewis, J. Chem. Soc. Dalton Trans. (1975) 2606.
116. J. M. Fernandez, B. F. G. Johnson, J. Lewis and P. R. Raithby, J. Chem. Soc. Dalton Trans. (1981) 2250.
117. P. F. Jackson, B. F. G. Johnson, J. Lewis, M. McPartlin and W. J. H. Nelson, J. Chem. Soc. Commun. (1980) 224.
118. M. J. Mays and P. D. Gavens, J. Organometal. Chem., 177 (1979) 443.
119. M. I. Bruce, J. G. Matisons, J. R. Rodgerd and R. C. Wallis, J. Chem. Soc. Chem. Commun. (1981) 1070.
120. R. Psaro, A. Fusi and M. Sabbatini, J. Mol. Catal., 16 (1982) 117.
121. J. A. Ferguson and T. J. Meyer, J. Am. Chem. Soc., 94 (1972) 3409.
122. M. A. Neuman, L. F. Dahl and R. B. King, quoted in R. Doedens and L. F. Dahl, J. Am. Chem. Soc., 88 (1966) 4847.
123. A. Alich, N. J. Nelson, D. Strope and D. F. Shriver, Inorg. Chem., 11 (1972) 2976.
124. J. S. Kristoff and D. F. Shriver, Inorg. Chem., 13 (1974) 499.
125. M. Y. Darensbourg, H. L. Cereder, D. J. Darensbourg and C. Hasday, J. Am. Chem. Soc., 95 (1973) 5919.
126. C. Tessier-Youngs, F. Correra, D. Pioch, R. L. Burwell, Jr. and D. F. Shriver, Organometallics, 2 (1983) 898.
127. D. Vanhove, P. Makambo and M. Blanchard, J. Chem. Soc. Chem. Commun. (1980) 719.
128. M. Ichikawa, J. Catal., 56 (1979) 127; 59 (1979) 67.
129. R. Nakamura, A. Oomura, N. Okada and E. Echigoya, Chem. Lett. (1982) 1463.
130. A. S. Lisitsyn, V. L. Kuznetsov and Yu. I. Yermakov, React. Kinet. Catal. Lett., 14 (1980) 445.
131. G. Bor, G. Sbrignadello and F. Mercati, J. Organometal. Chem., 45 (1972) 357.
132. P. Chini, A. Ceriotti and S. Martinengo, unpublished observations.
133. R. L. Schneider, R. F. Howe and K. L. Watters, J. Catal., 79 (1983) 298.
134. G. Palyi, F. Piacenti and L. Markó, Inorg. Chim. Acta Rev., 4 (1970) 109.
135. R. C. Ryan and C. U. Pittman, Jr., J. Am. Chem. Soc., 99 (1977) 1986.
136. B. R. Penfold and B. H. Robinson, Acc. Chem. Res., 6 (1973) 73.
137. R. Markby, I. Wender, R. A. Friedel, F. A. Cotton and H. W. Steinberg, J. Am. Chem. Soc., 80 (1958) 6529.
138. K. Tanabe, in J. R. Anderson and M. Boudart (Eds.), Catalysis-Science and Technology, Vol. 2, Springer, Berlin and New York, 1981, p. 250.
139. M. Ichikawa, J. Chem. Soc. Chem. Commun. (1976) 11.
140. N. D. Parkyns, Proc. 3rd Int. Congr. Catal., Amsterdam, Part II, 1965, p. 914.

141. M. Ichikawa, J. Chem. Soc. Chem. Commun. (1978) 566.
142. M. Ichikawa, Bull. Chem. Soc. Japan, 51 (1978) 2268; 2273.
143. M. Ichikawa, K. Sekizawa, K. Shikakura and M. Kawai, J. Mol. Catal., 11 (1981) 167.
144. P. Gelin, Y. Ben Taarit and C. Naccache, J. Catal., 59 (1979) 357.
145. H. Conrad, G. Ertl, H. Knözinger, J. Küppers and E. E. Latta, Chem. Phys. Lett., 42 (1976) 115.
146. J. L. Bilhou, V. Bilhou-Bougnol, W. F. Graydon, J. M. Basset, A. K. Smith, G. M. Zanderighi and R. Ugo, J. Organometal. Chem., 153 (1978) 73.
147. A. K. Smith, F. Hugues, A. Theolier, J. M. Basset, R. Ugo, G. M. Zanderighi, J. L. Bilhou, V. Bilhou-Bougnol and W. Graydon, Inorg. Chem., 18 (1979) 3104.
148. K. L. Watters, R. F. Howe, T. P. Chojnacki, C. Fu, R. L. Schneider and N. Wong, J. Catal., 66 (1980) 424.
149. S. L. T. Anderson, K. L. Watters and R. F. Howe, J. Catal., 69 (1981) 212.
150. A. Theolier, A. K. Smith, M. Leconte, J. M. Basset, G. M. Zanderighi, R. Psaro and R. Ugo, J. Organometal. Chem., 191 (1980) 415.
151. P. S. Wehner, P. N. Mercer and G. Apai, J. Catal., 84 (1983) 244.
152. L. Malatesta, G. Caglio and M. Angoletta, in F. A. Cotton (Ed.), Inorganic Synthesis, Vol. XIII, McGraw-Hill, New York, 1972, p. 95.
153. M. Angoletta, L. Malatesta and G. Caglio, J. Organometal. Chem., 94 (1975) 99.
154. C. R. Eady, B. F. G. Johnson and J. Lewis, J. Chem. Soc. Dalton Trans. (1976) 838.
155. D. N. Lawson and G. Wilkinson, J. Chem. Soc. (1965) 1900.
156. J. Evans and G. S. McNulty, J. Chem. Soc. Dalton Trans. (1984) 587.
157. (a) P. Chini and S. Martinengo, Inorg. Chim. Acta, 3 (1969) 21; (b) *ibid.*, 3 (1969) 315.
158. J. T. Yates, Jr., T. M. Duncan and R. W. Vaughan, J. Chem. Phys., 71 (1979) 3908.
159. D. J. C. Yates, L. L. Murrell and E. B. Prestridge, J. Catal., 57 (1979) 41.
160. (a) M. Primet, J. Chem. Soc. Faraday Trans. I, 74 (1978) 2570; (b) M. Primet, J. C. Vedrine and C. Naccache, J. Mol. Catal., 4 (1978) 411.
161. F. Solymosi, A. Erdohelyi and M. Kocsis, J. Catal., 65 (1980) 428.
162. (a) J. M. Basset, B. Besson, A. Choplin and A. Theolier, Phil. Trans. R. Soc. London A, 308 (1982) 125; (b) J. M. Basset and A. Theolier, to be published.
163. M. D. Ward and J. Schwartz, J. Mol. Catal., 11 (1981) 397.
164. S. J. DeCanio, H. C. Foley, C. Dybowski and B. C. Gates, J. Chem. Soc. Chem. Commun. (1982) 1372.
165. H. C. Foley, S. J. DeCanio, K. D. Tau, K. J. Chao, J. H. Onuferko, C. Dybowski and B. C. Gates, J. Am. Chem. Soc., 105 (1983) 3074.
166. B. R. James, G. L. Rempel and W. K. Teo, in F. Basolo (Ed.), Inorganic Synthesis, Vol. XVI, McGraw-Hill, New York, 1976, p. 49.
167. S. Martinengo, P. Chini, G. Giordano, A. Ceriotti, V. G. Albano and G. Ciani, J. Organometal. Chem., 88 (1975) 375.
168. S. Martinengo and P. Chini, Gazz. Chim. Ital., 102 (1972) 344.
169. K. Tanabe, in J. R. Anderson and M. Boudart (Eds.), Catalysis-Science and Technology, Vol. 2, Springer, Berlin and New York, 1981, pp. 255-260.
170. J. R. Anderson and R. F. Howe, Nature, 28 (1977) 129.
171. K. Foger and J. R. Anderson, J. Catal., 59 (1979) 325.

172. M. Ichikawa and K. Shikakura, in T. Seiyama and K. Tanabe (Eds.), Studies in Surface Science and Catalysis, Vol. 7, New Horizons in Catalysis, Elsevier, Amsterdam, 1981, Part. B, p. 925.
173. R. F. Howe, J. Catal., 50 (1977) 196.
174. K. Tanaka, K. L. Watters and R. F. Howe, J. Catal., 75 (1982) 23.
175. W. C. Price and K. S. Tetlow, J. Chem. Phys., 16 (1948) 1157.
176. C. R. Guerra and J. H. Schulman, Surface Sci., 7 (1967) 229.
177. R. P. Eischens and W. A. Pliskin, Advan. Catal. 10 (1958) 1.
178. R. A. Shigeishi and D. A. King, Surface Sci., 58 (1976) 379.
179. A. M. Bradshaw and F. M. Hoffman, Surface Sci., 72 (1978) 513.
180. D. Reinalda and V. Ponec, Surface Sci., 91 (1980) 113.
181. J. S. Wang, Proc. R. Soc. London Ser. A, 161 (1937) 127.
182. D. A. King, CRC Crit. Rev. Solid State Mater. Sci., 7 (1978) 167.
183. P. Gelin, Y. Ben Taarit and C. Naccache, in T. Seiyama and K. Tanabe (Eds.), Studies in Surface Science and Catalysis, Vol. 7, New Horizons in Catalysis, Elsevier, Amsterdam, 1981, Part. B, p. 898.
184. R. Whyman, J. Chem. Soc. Dalton Trans. (1972) 2294.
185. J. E. Crawford, G. A. Melson, L. E. Makovsky and F. R. Brown, J. Catal., 83 (1983) 454.
186. M. Ichikawa, J. Chem. Soc. Chem. Commun. (1976) 11.
187. P. L. Watson and G. L. Schrader, J. Mol. Catal., 9 (1980) 129.
188. M. Deeba and B. C. Gates, J. Catal., 67 (1981) 303.
189. M. Deeba, B. J. Streusand, G. L. Schrader and B. C. Gates, J. Catal., 69 (1981) 218.
190. M. Deeba, J. P. Scott, R. Barth and B. C. Gates, J. Catal., 71 (1981) 373.
191. H. Knözinger and Y. Zhao, J. Catal., 71 (1981) 337.
192. H. Knözinger, Y. Zhao, B. Tesche, R. Barth, R. Epstein, B. C. Gates and J. P. Scott, Faraday Disc. Chem. Soc., 72 (1981) 53.
193. S. L. Cook, J. Evans and G. N. Greaves, J. Chem. Soc. Chem. Commun. (1983) 1287.
194. J. Schwank, L. F. Allard, M. Deeba and B. C. Gates, J. Catal., 84 (1983) 27.
195. V. A. Shvets, A. L. Tarasov, V. B. Kazansky and H. Knözinger, J. Catal., 86 (1984) 2.
196. G. Collier, D. J. Hunt, S. D. Jackson, R. B. Moyes, I. A. Pickering, P. B. Wells, A. F. Simpson and R. Whyman, J. Catal., 80 (1983) 154.
197. R. Psaro, C. Dossi and R. Ugo, J. Mol. Catal., 21 (1983) 331.
198. M. Tachikawa and J. R. Shapley, J. Organometal. Chem., 124 (1977) C 19.
199. B. F. G. Johnson, J. Lewis and P. A. Kilty, J. Chem. Soc. A (1968) 2859.
200. K. A. Azam, A. J., Deeming, R. E. Kimber and P. R. Shukla, J. Chem. Soc. Dalton (1976) 1853, and references cited therein.
201. A. J. Deeming and S. Hasso, J. Organometal. Chem., 114 (1976) 313.
202. (a) L. A. W. Hales and R. J. Irving, J. Chem. Soc. A (1967) 1932; (b) M. I. Bruce, H. Cooke, M. Green and D. J. Westlake, J. Chem. Soc. A (1969) 987.
203. P. A. Breeze, J. K. Burdett and J. J. Turner, Inorg. Chem., 20 (1981) 3369.
204. R. Mason, K. M. Thomas and D. M. L. Mingos, J. Am. Chem. Soc., 95 (1973) 3082.
205. (a) P. S. Braterman, Metal Carbonyl Spectra, Academic Press, London and New York, 1975; (b) L. A. W. Hales and R. J. Irving, Spectrochim. Acta A, 23 (1967) 2981.
206. F. B. M. Duivenvoorden, D. C. Koningsberger, Y. S. Uh and B. C. Gates, J. Am. Chem. Soc., in press.

207. F. H. Johansen, W. Pretzer and A. Scheffler, J. Organometal. Chem., 102 (1975) 527.
208. C. R. Eady, B. F. G. Johnson, J. Lewis and M. C. Malatesta, J. Chem. Soc. Dalton Trans. (1979) 1358.
209. C. R. Eady, P. F. Jackson, B. F. G. Johnson, J. Lewis, M. C. Malatesta, M. McPartlin and W. J. H. Nelson, J. Chem. Soc. Dalton Trans. (1975) 383.
210. C. C. Nagel, J. C. Bricher, D. G. Alway and S. Shore, J. Organometal. Chem., 219 (1981) C 9.
211. (a) E. Guglielminotti, A. Zecchina, A. Bossi and M. Camia, J. Catal., 74 (1982) 240; (b) *ibid.*, 74 (1982) 252.
212. (a) J. G. Goodwin, Jr., and C. Naccache, Appl. Catal., 4 (1982) 145; (b) *ibid.*, J. Mol. Catal., 14 (1982) 259.
213. A. Theolier, A. Choplin, L. D'Ornelas, J. M. Basset, C. Sourisseau, G. M. Zanderighi, R. Ugo and R. Psaro, Polyhedron, 2 (1983) 119.
214. R. Pierantozzi, E. G. Valagene, A. F. Nordquist and P. N. Dyer, J. Mol. Catal., 21 (1983) 189.
215. J. Robertson and G. Webb, Proc. R. Soc. A, 341 (1974) 383.
216. R. A. Sanchez-Delgado, I. Duran, J. Montfort and E. Rodriguez, J. Mol. Catal., 11 (1981) 193.
217. A. K. Smith, A. Theolier, J. M. Basset, R. Ugo, D. Commereuc and Y. Chauvin, J. Am. Chem. Soc., 100 (1978) 2550.
218. G. B. McVicker and M. A. Vannice, J. Catal., 63 (1980) 25.
219. L. Guczi, Z. Schay, K. Lázár, A. Vizi-Orosz and L. Markó, Surface Sci., 106 (1981) 516.
220. C. S. Kellner and A. T. Bell, J. Catal., 70 (1981) 418.
221. D. Ballivet-Tkatchenko and I. Tkatchenko, J. Mol. Catal., 13 (1981) 1.
222. T. Okuhara, K. Kobayashi, T. Kimura, M. Misono and Y. Yoneda, J. Chem. Soc. Chem. Commun. (1981) 1114.
223. H. E. Ferkul, D. J. Stanton, J. D. McCowan and M. C. Baird, J. Chem. Soc. Chem. Commun. (1982) 955.
224. D. J. Hunt, S. D. Jackson, R. B. Moyes, P. B. Wells and R. Whyman, J. Chem. Soc. Chem. Commun. (1982) 82.
225. G. R. Crooks, B. F. G. Johnson, J. Lewis and I. G. Williams, J. Chem. Soc. A (1969) 97.
226. G. M. Zanderighi, R. Ugo, *et al.*, in preparation.
227. A. F. Simpson and R. Whyman, J. Organometal. Chem., 213 (1981) 157.
228. E. Bendetti, G. Braca, G. Sbrana, F. Salvetti and B. Grassi, J. Organometal. Chem., 153 (1978) 73.
229. M. J. Cleare and W. P. Griffith, J. Chem. Soc. A (1969) 372.
230. J. Evans and G. McNulty, J. Chem. Soc. Dalton Trans. (1984) 1123.
231. M. F. Brown and R. D. Gonzales, J. Phys. Chem., 80 (1976) 1731.
232. S. A. R. Knox, J. W. Koepke, M. A. Andrews and H. D. Kaesz, J. Am. Chem. Soc., 97 (1978) 3942.
233. Ref. 21, pp. 124-133, and references cited therein.
234. B. F. G. Johnson, R. D. Johnston, J. Lewis and B. H. Robinson, J. Chem. Soc. A (1968) 2856.
235. B. F. G. Johnson, J. Lewis, P. R. Raithby and G. Suss, J. Chem. Soc. Dalton Trans. (1979) 1356.
236. H. C. Yao, Y. F. Yu Yao and K. Otto, J. Catal., 56 (1979) 21.
237. H. C. Yao, S. Japar and M. Shelef, J. Catal., 50 (1977) 407.
238. J. T. Yates, Jr., T. M. Duncan, S. D. Worley and R. W. Vaughan, J. Chem. Phys., 70 (1979) 1219.
239. R. A. Dalla Betta, J. Phys. Chem., 79 (1975) 2519.
240. (a) J. W. Koepke, J. R. Johnson, S. A. R. Knox and H. D. Kaesz, J. Am. Chem. Soc., 97 (1975) 3947; (b) S. A. R. Knox, J. W. Koepke, M. A. Andrews and H. D. Kaesz, J. Am. Chem. Soc., 97 (1975) 3942.
241. J. S. Bradley, G. B. Ansell and E. W. Hill, J. Organometal. Chem., 184 (1980) C 33.

242. Reference 39, p. 97.
243. C. R. Eady, B. F. G. Johnson, J. Lewis, M. C. Malatesta, P. Machin and M. McPartlin, J. Chem. Soc. Chem. Commun. (1976) 945.
244. A. A. Bhattacharyya, C. C. Nagel and S. Shore, Organometallics, 2 (1983) 1187.
245. J. H. Sinfelt and J. Cusumano, in J. J. Burton and R. L. Garten (Eds.), Advanced Materials in Catalysis, Academic Press, New York 1977, pp. 1-31.
246. J.H. Sinfelt, Advan. Catal., 23 (1973) 91.
247. H. Matsumoto, Y. Saito and Y. Yoneda, J. Catal., 22 (1971) 182.
248. J. R. Anderson and D. E. Mainwaring, Ind. Eng. Chem. Prod. Res. Devel., 17 (1978) 202.
249. J. H. Sinfelt, J. Catal., 29 (1973) 308.
250. A. S. Lysitsyn, V. L. Kuznetsov, E. K. Mamaeva and Yu. I. Yermakov, to be published.
251. P. Chini, L. Colli and M. Peraldo, Gazz. Chim. Ital., 90 (1960) 1005.
252. J. R. Budge, J. P. Scott and B. C. Gates, J. Chem. Soc. Chem. Commun. (1983) 342; J. P. Scott, J. R. Budge, A. L. Rheingold and B. C. Gates, to be published.
253. A. Choplin, M. Leconte, J. M. Basset, S. Shore and W. L. Hsu, J. Mol. Catal., 21 (1983) 389.
254. J. R. Shapley, S. J. Hardwick, D. S. Foose and G. D. Stucky, Prepr. Div. Petrol. Chem. Am. Chem. Soc., 25 (1980) 780.
255. J. R. Shapley, S. J. Hardwick, D. S. Foose, G. D. Stucky, M. R., Churchill, C. Bueno and J. P. Hutchinson, J. Am. Chem. Soc., 103 (1981) 7383.
256. G. B. McVicker, R. T. K. Baker, R. L. Garten and E. L. Kugler, J. Catal., 65 (1980) 207.
257. K. Lázár, K. Matusek, J. Mink, S. Dobos, L. Guczi, A. Vizi-Orosz, L. Markó and W. M. Reiff, J. Catal., 87 (1984) 163.
258. E. Guglielminotti and A. Zecchina, J. Mol. Catal., 24 (1984) 331.
259. Y. Iwasawa, M. Yamada, Y. Sato and H. Kuroda, J. Mol. Catal., 23 (1984) 95.
260. R. Nakamura, N. Okada, A. Aomura and E. Echigoya, Chem. Lett., (1984) 119.
261. R. Nakamura, A. Aomura and E. Echigoya, Proc. 8th Int. Congr. Catal., Berlin, 1984, Vol. V, p. 51.
262. R. L. Schneider, R. F. Howe and K. L. Watters, Inorg. Chem., in press.
263. P. G. Gopal and K. L. Watters, Proc. 8th Int. Congr. Catal., Berlin, 1984, Vol. V, p. 75.
264. M. J. Heal, E. C. Leisegang and R. C. Torrington, J. Catal., 51 (1978) 314.
265. R. A. Gardner and R. H. Petrucci, J. Phys. Chem., 67 (1963) 1376.
266. (a) K. L. Watters, Surf. Interface Anal. 3 (1981) 55; (b) K. L Watters, R. Schneider and R. F. Howe, Prepr. Div. Petrol. Chem. Am. Chem. Soc., 25, (4) 1980.
267. V. L. Kusnetzov, A. V. Romanenko, I. L. Mudrakovskii, V. M. Mastikhin, V. A. Shmachkov and Yu. I. Yermakov, Proc. 8th Int. Congr. Catal., Berlin, 1984, Vol. V, p. 3.
268. (a) D. J. Hunt, R. B. Moyes, P. B. Wells, S. D. Jackson and R. Whyman, Proc. 8th Inter. Congr. Catal., Berlin, 1984, Vol. V, p. 27; (b) T. R. Krause, M. E. Davies, J. Lieto and B. C. Gates, J. Catal., 94 (1985) 195.
269. Z. Schay, K. Lázár, J. Mink and L. Guczi, J. Catal., 87 (1984) 179.
270. I. Böszormenyi, S. Dobos, L. Guczi, L. Markó, K. Lázár, V. M. Reiff, Z. Schay, L. Takáas and A. Vizi-Orosz, Proc. 8th Intern. Congr. Catal., Berlin (West), 1984, Vol. V, p. 183.
271. J. R. Budge, B. F. Lücke, J. P. Scott and B. C. Gates, Proc. 8th Int. Congr. Catal., Berlin, 1984, Vol. V, p. 89.

272. J. S. Bradley, J. Am Chem. Soc., 101 (1979) 7419.
273. B. D. Dombek, J. Am. Chem. Soc., 103 (1981) 6508.
274. J. F. Knifton, J. Am. Chem. Soc., 103 (1981) 3959.
275. B. K. Warren and B. D. Dombek, J. Catal., 79 (1983) 334.
276. J. M. Driessen, E. K. Poels, J. P. Hindermann and V. Ponec, J. Catal., 82 (1983) 26.
277. J. R. Budge, B. F. Lücke, B. C. Gates and J. Toran, J. Catal., 91 (1985) 272.

Section 9.3

SUPPORTED METAL CLUSTER CATALYSTS FOR ALKENE CONVERSION

B. C. GATES

9.3.1 INTRODUCTION

A large fraction of the reactions known to be catalyzed by metal clusters in solution are alkene conversions (Chapter 5), and, not surprisingly, most of the catalytic reactions that occur in the presence of the supported analogues also involve alkenes, the reactions including isomerization, hydrogenation, and hydroformylation.

The synthesis, characterization, and reactivity of supported metal clusters have been considered in Sections 9.1 and 9.2; here the discussion is extended to the catalytic performance in alkene conversion. For the most part, we consider only well-defined catalysts incorporating anchored "molecular" species; in the following sections some discussion of cluster-derived supported metals not having these simple structures is also included; the pertinent catalytic reactions include hydrocarbon hydrogenolysis and skeletal isomerization (Section 9.4) and CO hydrogenation (Section 9.5).

9.3.2 CATALYTIC ACTIVITY

Metal complexes, to be catalytically active, require sites of coordinative unsaturation for bonding of reactants. The few relatively well-documented examples of catalysis by supported metal clusters involve clusters which are easily stabilized in a state of coordinative unsaturation, e.g., triosmium clusters bonded to phosphine-functionalized supports (Section 9.1). The catalytic reactions involve alkenes and are analogous to those occurring in solution in the presence of the coordinatively unsaturated cluster $[H_2Os_3(CO)_{10}]$. Evidently, coordination of an alkene to $H_2Os_3(CO)_9Ph_2P$-Ⓟ or $H_2Os_3(CO)_9Ph_2P(CH_2)_2$-SIL initiates a cycle similar to that shown in SCHEME 10 of Chapter 5. The rate equation describing the reaction of pent-1-ene to give pent-2-ene at 348 K catalyzed by the supported cluster $H_2Os_3(CO)_9Ph_2P(CH_2)_2$-SIL in a toluene slurry has been found to be

$$r = kC_{\text{pent-1-ene}} \tag{1}$$

where k is 0.08 ± 0.02 L/(mol of Os_3·s) and r has dimensions of molecules/(Os_3 cluster·s) [1].

For comparison, rates of the same reaction at the same temperature in the presence of $[H_2Os_3(CO)_{10}]$ in toluene at 348 K were measured, the results determining the rate equation

$$\frac{r}{C_{Os_3\ cluster}} = k\ C_{pent\text{-}1\text{-}ene} \tag{2}$$

where k is 0.026 L/(mol•s) [1]. The approximate agreement of the kinetics supports the conclusion of a close similarity between the soluble cluster $[H_2Os_3(CO)_{10}]$ and its supported analogue [1].

The supported triosmium clusters, like the soluble cluster $[H_2Os_3(CO)_{10}]$, are also active for alkene hydrogenation, although higher pressures are required, and again there are close similarities between the soluble and supported clusters [1, 2]. The supported clusters, like their soluble analogue, are unstable during the alkene conversions, ultimately forming saturated (and catalytically inactive) triosmium species that presumably incorporate bridging vinyl ligands [1, 2].

Other coordinatively unsaturated clusters on supports have been prepared, including the butterfly cluster $[ClAuOs_3(CO)_{10}(Ph_2P\text{-}ⓟ)]$, which is an active and stable catalyst for ethylene hydrogenation at 346-365 K. The rate equation is the following:

$$r = \frac{k\ P_{H_2} P_{C_2H_4}}{1 + K\ P_{C_2H_4}} \tag{3}$$

Rates were found to be in the range 10^{-3}-10^{-2} molecules/cluster•s at 346 to 365 K [3]. Infrared spectra of the functioning catalyst indicated the above-mentioned bimetallic cluster as the only detectable organometallic species. The catalyst was stable during steady-state operation in a flow reactor, and its synthesis was reproducible, suggesting that the cluster itself provided the catalytic sites. Support for this suggestion is provided by the contrast with the similar--but coordinatively saturated--supported metal cluster $HAuOs_3(CO)_{10}(Ph_2P\text{-}ⓟ)$, which has the same metals in a closed tetrahedral framework structure and an immeasurably low catalytic activity under the same conditions [3]. It would be inappropriate, however, to exclude fully the possibility that the alkene hydrogenation in the presence of the butterfly cluster was catalyzed by small amounts of metal formed from the clusters; undetectably small amounts of metal could be sufficiently active to account for the observed catalysis. However, with metal-catalyzed ethylene hydrogenation, a reaction order in ethylene is usually near zero, in contrast to the observations.

A number of coordinatively saturated supported metal clusters appear to be precursors of catalysts for alkene hydrogenation. These include the polymer-supported tetrairidium and tetraruthenium clusters listed in Table 1 of Section 9.1. The nature of the catalytic species remains obscure, and at higher temperatures than those applied ($\lesssim$ 370 K), metal aggregates formed in all the samples; it is possible that in some instances small metal aggregates were the true catalytic species. This possibility seems to be highly likely for catalysts containing tetrairidium clusters on phosphine-functionalized SiO_2 [4], and it is also considered likely for catalysts containing tetraruthenium clusters on polymers [5], which were apparently not as stable as the polymer-supported catalysts containing tetrairdium clusters.

The likelihood of decomposition of a cluster in the hydrocarbon matrix of the nearly inert polymer supports is clearly less than that of the analogous cluster on the oxide surfaces. The stability of the polymer-supported tetrairidium clusters suggests that the clusters themselves provided the catalytic sites. This suggestion is supported by the reproducibility of the syntheses; the indications from infrared spectroscopy that the clusters were the only detectable metal species under catalytic reaction conditions; the form of the kinetics (different from what would be expected for metal catalysis); and the influence of changes in the ligand environment on the catalytic activity (Fig. 1) [6].

If the clusters themselves were catalytically active, then some undetermined mechanism, possibly metal-metal, metal-phosphine, or metal-CO bond dissociation, provided the sites of coordinative unsaturation in a small fraction of the clusters [6]. It has been speculated that the mechanism involves metal-metal bond dissociation, for example that converting a closed tetrahedral cluster framework into an open butterfly framework [3]:

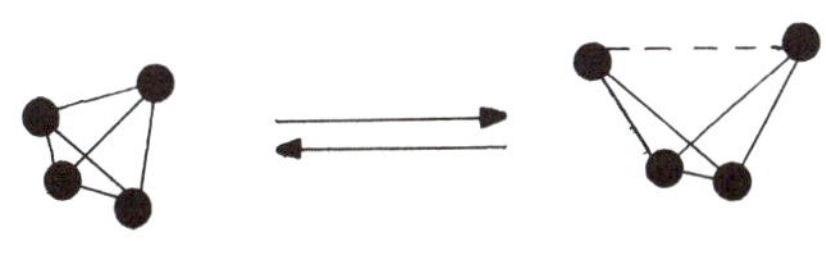

There are literature reports [7] attesting to the importance of metal-metal bond cleavage in the activation of organic reactants by metal clusters.

Triosmium clusters anchored directly to SiO_2 [8], γ-Al_2O_3 [9], and poly(styrene-divinylbenzene) functionalized with -OH groups [10] have been shown to be catalytically active for alkene isomerization (Fig. 2, Table 1). Infrared spectra of each sample were measured during catalysis in a flow

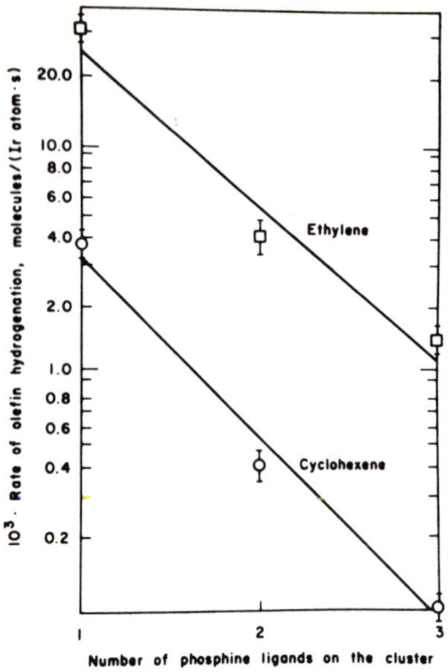

Fig. 1. Catalysis by polymer-supported tetrairidium clusters: effect of ligand substitution on the rate of hydrogenation of ethylene (□) (P_{H_2} = 0.8 bar, $P_{C_2H_2}$ = 0.2 bar, T = 323 K) and cyclohexene (O) (P_{H_2} = 0.72 bar, $P_{C_6H_{10}}$ = 0.15 bar, T = 353 K) [6].

reactor/infrared cell. The spectra of the SiO_2-supported catalyst indicated that the predominant form was the cluster $HOs_3(CO)_{10}$-O-Si≡ (mentioned in Section 9.2), which is coordinatively saturated. As flow of but-1-ene was initiated over the sample at temperatures in the range 323-363 K, the catalytic activity slowly increased to a maximum and then remained constant; simultaneously, there were small changes in the infrared spectrum in the carbonyl region: two new bands indicated the formation of small amounts of a cluster of different symmetry, inferred to have been formed by coordination of reactants (but-1-ene and/or hydrogen) [8]. When CO was introduced into the feed stream, the changes in the spectrum were reversed and the catalytic activity ceased. The results point to catalysis by a small fraction of the surface-bound clusters--intact--at these low temperatures [8].

Similar results were observed with the γ-Al_2O_3-supported cluster $HOs_3(CO)_{10}$-O-Al≤ [9], the polymer-supported analogue [10], and the silica-supported cluster $Os_3(CO)_9(\mu$-$Cl)_2PPh_2(CH_2)_2$-Si≤ [11]. Catalytic reaction rate data for isomerization of hex-1-ene in the presence of H_2 and the former sample are shown in Fig. 2. The reaction was zero-order in alkene and in H_2, and the rate constants are shown in the plot. Again, all

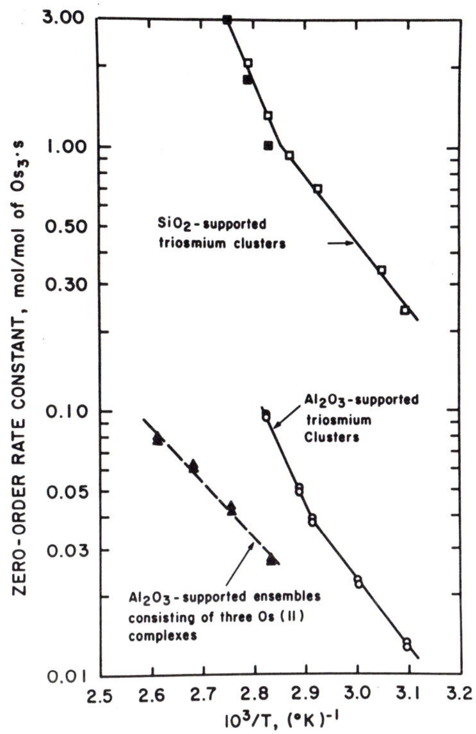

Fig. 2. Alkene isomerization activities of catalysts prepared from triosmium clusters on oxide supports. The upper curve [8] is for isomerization of but-1-ene and the lower curves [9] are for isomerization of hex-1-ene.

the data point to catalysis by a small fraction of the clusters themselves at these low temperatures. When the temperature was raised to 393 K, however, the catalytic activity declined, reaching a new steady state in about two hours; during this period, the cluster broke up into the mononuclear complexes $Os^{II}(CO)_x\{OAl\}_3$ (x = 2 or 3) (described in Section 9.2) as shown by the infrared spectrum [9, 13]. Catalytic activity data for the broken-up clusters (i.e., the mononuclear complexes) are also shown in Fig. 2 (again, the isomerization reaction was zero-order in the reactant and in H_2). The complexes are less active than the clusters; this comparison is important in confirming that the clusters themselves were catalytically active--we can minimize the liklihood of the possibility that small amounts of cluster degradation products were actually responsible for the activity attributed to the clusters, since the degradation products are significantly less active than the clusters themselves.

TABLE 1. Comparison of Activities of Supported Osmium Cluster Catalysts for Alkene Isomerization at 363 K and atmospheric Pressure.

Predominant form of catalyst	Reactant	Rate of isomerization (molecules $(cluster)^{-1}$ s^{-1})	Ref.
$HOs_3(CO)_{10}$-O-Si	but-1-ene	0.028[a]	[8]
$HOs_3(CO)_{10}$-O-$(CH_2)_2$-(P)	but-1-ene	0.00015[a]	[10]
$HOs_3(CO)_{10}$-O-Al	hex-1-ene	0.25[a,b]	[9]
$H_3Os_4(CO)_{12}^{-}$ $^{+}$Al	but-1-ene	0.52[a]	[13]
$H_3RuOs_3(CO)_{12}^{-}$ $^{+}$Al	but-1-ene	0.053[a]	[14]

[a] P_{H_2} = 0.5 bar, P_{alkene} = 0.4 bar.
[b] Extrapolated value.

Supported tetraosmium clusters were prepared by the reaction of $[H_4Os_4(CO)_{12}]$ with the surface of γ-Al_2O_3 in refluxing *n*-octane [13]. The surface-bound organometallic species was identified as $[H_3Os_4(CO)_{12}]^-$ by its infrared spectrum; this anion could be extracted from the surface as the tetraphenylarsonium salt. This solid was also catalytically active for isomerization of but-1-ene at 363 K and atmospheric pressure (Table 1). The only organometallic species on the catalyst surface observed by infrared spectroscopy was $[H_3Os_4(Os)_{12}]^-$, which was suggested to be the catalyst precursor. At higher temperatures, the clusters broke up and were oxidized by -OH groups on the surface to give $Os(II)(CO)_x\{OAl\}_3$ (x = 2 or 3); these surface-bound complexes had a lower catalytic activity than the clusters. Quite similar results have been obtained with $[H_3RuOs_3(CO)_{12}]^-$ clusters supported on γ-Al_2O_3 [14] (Table 1).

It is not possible to determine exactly the activity of any of these clusters involved in the catalytic cycles; evidently the fraction of clusters engaged in the catalysis is small in each sample. The data indicate that the silica- and alumina-bound clusters are more active for alkene isomerization than the above-mentioned clusters anchored through phosphine ligands, but it is not known whether the differences are to be attributed to different fractions of the clusters being active, or to different intrinsic activities of the clusters on the different supports, or to both.

The silica-supported triosmium clusters have also been reported to be active catalysts for ethylene hydrogenation at 343-373 K, the reaction orders in ethylene and hydrogen being 0 and 1, respectively [15]. Infrared spectra indicated that ethylene and hydrogen coordinated to the cluster, and a somewhat speculative catalytic cycle has been suggested, involving reversible breaking of Os-O bonds (Fig. 3) [15]. The reaction orders are the values expected for metal catalysis, but the observed activation energy (32 kJ mol^{-1}) is different from that expected with Os metal [15]. The possibility that metal particles were responsible for the catalysis should not be completely ruled out.

Osmium cluster anions $[HOs_3(CO)_{11}]^-$ have been bonded to crosslinked polystyrene functionalized with $\overset{+}{N}R_3$ groups and tested as catalysts for the hydroformylation of hex-1-ene at 373 K and atmospheric pressure [16]. Some of the cluster remained intact on the polymer during catalysis over a period <48 h, as indicated by infrared spectroscopy. The selectivity was different from that of the cluster analogue in solution, and the stability was greater. It was speculated that the cluster itself was involved in the catalysis.

In summary, there are still only few data documenting catalysis of alkene conversions by supported metal clusters--all for low-temperature reactions. The lack of stability of the supported clusters is a major limitation to their application as catalysts. The available data suggest several tentative working hypotheses for maximizing stability of cluster catalysts on supports:

1. Use of stable metal frameworks, e.g., triosmium and tetraosmium.
2. Use of optimum support/metal combinations, e.g., MgO, to stabilize cluster anions.
3. Use of high CO partial pressures to stabilize the clusters.

These ideas are considered in Section 9.5, which is concerned with CO hydrogenation catalysis.

9.3.3 SUPPORTED METALS WITH SIMPLE STRUCTURES DERIVED FROM SUPPORTED METAL CLUSTERS

Reactions leading to changes in the metal framework structure of a supported cluster usually give complex mixtures of surface structures. There is an intriguing exception, however, involving triosmium clusters on oxides. When $HOs_3(CO)_{10}$-O-Al$\leqslant$ is heated to roughly 370-470 K, cluster breakup into mononuclear $Os^{II}(CO)_2$ and/or $Os^{II}(CO)_3$ complexes takes place,

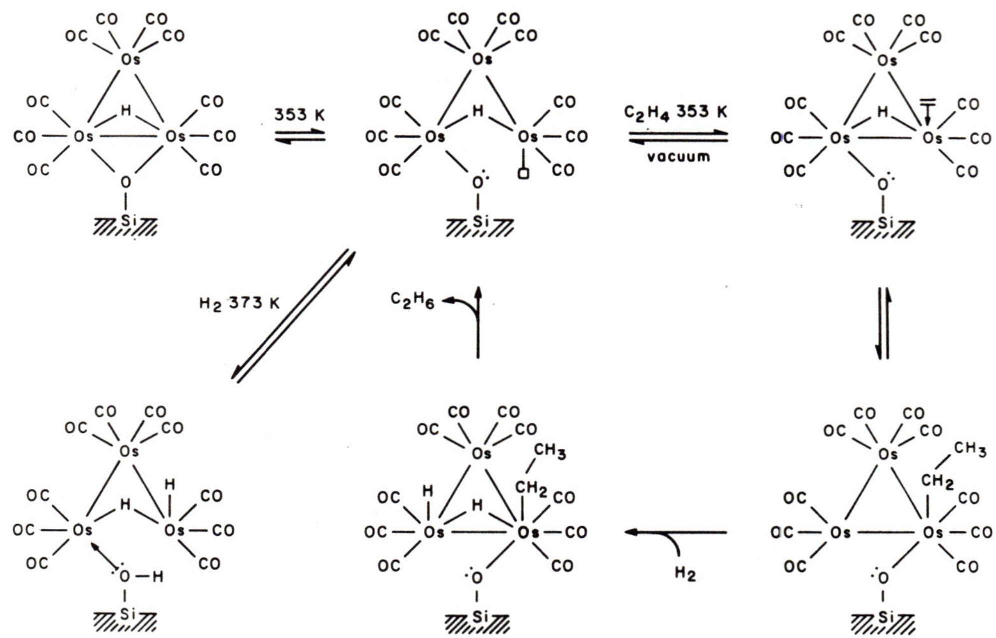

Fig. 3. Suggested mechanism for the hydrogenation of ethylene in the presence of the silica-supported cluster $HOs_3(CO)_{10}$-OSi≡ [15].

as mentioned in Section 9.2. Knözinger *et al.* [17] concluded that the mononuclear complexes were present in the form of ensembles, each consisting of three Os ions. These ensembles have been observed by electron microscopy [17, 18]--the micrographs exhibit uniform scattering centers, all the same size (about 0.7 nm), on the Al_2O_3 surface. Evidence for the formation of the ensembles on MgO was inferred by Deeba *et al.* [19], who observed that cluster reconstruction from the complexes occurred in the presence of CO; a similar observation was made for the SiO_2-supported species [8], among others (Section 9.2).

The ensembles appear to be unique among supported metals with respect to their uniformity of structure; the ensembles on γ-Al_2O_3 have been observed to be stable in the high-energy electron beam of the electron microscope [18], consistent with the strong ionic bonding of the osmium to the oxide surface. When the sample was treated in hydrogen at 673 K, reduced osmium ensembles of the same size were observed--these constitute perhaps the first example of a zero-valent metal supported on an amorphous oxide and having a single size of metal aggregate. These aggregates, in contrast to the ions, were not stable in the electron microscope, consistent with the weaker metal-support interaction.

Triosmium clusters have been anchored to a wide range of oxide supports (SiO_2, γ-Al_2O_3, MgO, TiO_2, ZrO_2, Cr_2O_3, and ThO_2) and broken up into complexes, which likely took the form of three-ion ensembles [20]. The catalysts supported on MgO, Al_2O_3, and SiO_2 had activities related to the basicity of the surface oxygens. The activities of the osmium supported on the other transition metal oxides were less (by an order of magnitude or more) than the activities of the osmium supported on the nontransition metal oxides. Infrared spectra indicated that coordinatively unsaturated mononuclear osmium complexes were the catalytically active species. The activity is apparently related to the ease of electron transfer between the osmium and the support.

To summarize, metal clusters offer new opportunities for preparation of supported metals with unique, well-defined structures—not just clusters themselves, but also ensembles formed by cluster breakup. These materials offer opportunities for determination of the effect of various metal-support interactions, ensemble size, and ensemble composition on reactivity and catalytic activity. The opportunities parallel those provided by intact metal clusters on supports--but the complications associated with lack of stability of the samples are expected to be less with some of the cluster-derived materials than with the supported clusters themselves.

9.3.4 SUMMARY AND EVALUATION

Supported metal clusters are a new and little explored class of materials, most of them lacking stability; consequently, there are still only a few examples of their involvement in catalysis. Nonetheless, the available results portend rapid advances in the understanding of relations between structure and catalytic activity of surface-bound organometallics. The key is the simplicity of structures of the materials--metal clusters and species derived from them. The structural simplicity allows characterization with a powerful and expanding arsenal of techniques available to the surface scientist (Chapter 6). The most rapid advances in our understanding of "molecular" catalysis on surfaces are expected to result when these techniques are used in concert, especially with catalysts in the functioning state.

There is a need for discovery of more stable surface organometallic structures, not only to allow characterization of more catalysts over wider ranges of conditions, but to provide candidate catalysts with new activities and selectivities. The prospects of wholly new catalytic properties associated with structurally unique metal clusters are still unrealized; the prospects are valid, but much work remains to develop stable species which are structurally and catalytically novel.

REFERENCES

1. M. B. Freeman, M. A. Patrick and B. C. Gates, J. Catal., 73 (1982) 82.
2. S. C. Brown and J. Evans, J. Mol. Catal., 11 (1981) 143.
3. R. Pierantozzi, K. J. McQuade and B. C. Gates, in Proceedings of the Seventh International Congress on Catalysis, Kodansha, Tokyo, Elsevier, New York, 1981, p. 941.
4. T. Castrillo, H. Knözinger, M. Wolf and B. Tesche, J. Mol. Catal., 11 (1981) 151.
5. Z. Otero-Schipper, J. Lieto and B. C. Gates, J. Catal., 63 (1980) 175.
6. J. Lieto, J. J. Rafalko and B. C. Gates, J. Catal., 62 (1980) 149.
7. G. Huttner, J. Schneider, H. D. Müller, G. Mohr, J. von Seyrl and L. Wohlfahrt, Angew. Chem. Int. Ed. Engl., 18 (1979) 76.
8. R. Barth, B. C. Gates, Y. Zhao, H. Knözinger and J. Hulse, J. Catal., 82 (1983) 147.
9. X.-J. Li, J. H. Onuferko and B. C. Gates, J. Catal., 85 (1984) 176.
10. J. Lieto, M. Prochazka, D. B. Arnold and B. C. Gates, J. Mol. Catal., 31 (1985) 89.
11. M. Wolf, J. Lieto, B. A. Matrana, D. B. Arnold, B. C. Gates and H. Knözinger, J. Catal., 89 (1984) 100.
12. X.-J. Li and B. C. Gates, J. Catal., 84 (1983) 55.
13. T. R. Krause, M. E. Davies, J. Lieto and B. C. Gates, J. Catal., 94 (1985) 195.
14. J. P. Scott, J. R. Budge, A. L. Rheingold and B. C. Gates, to be published.
15. J. M. Basset, B. Besson, A. Choplin and A. Theolier, Phil. Trans. R. Soc. London A, 308 (1982) 115; B. Besson, Thesis, University Claude Bernard, Lyon, 1984.

16. H. Marrakchi, J.-B. Nguini Effa, M. Haimeur, J. Lieto and J.-P. Aune, J. Mol. Catal., 30 (1985) 101.
17. H. Knözinger, Y. Zhao, B. Tesche, R. Barth, R. Epstein, B. C. Gates and J. P. Scott, Faraday Disc. Chem. Soc., 72 (1981) 54.
18. J. Schwank, L. F. Allard, M. Deeba and B. C. Gates, J. Catal., 84 (1983) 27.
19. M. Deeba, J. P. Scott, R. Barth and B. C. Gates, J. Catal., 71 (1981) 373.
20. X.-J. Li, B. C. Gates, H. Knözinger and E. Alizo Delgado, J. Catal., 88 (1984) 355.

Section 9.4

SUPPORTED METAL CATALYSTS PREPARED FROM MOLECULAR METAL CLUSTERS: ACTIVITIES FOR HYDROGENOLYSIS AND SKELETAL ISOMERIZATION OF HYDROCARBONS

G. MAIRE

9.4.1 INTRODUCTION

The purpose of this section is an assessment of mechanisms of skeletal rearrangements and hydrogenolysis of hydrocarbons on metals, related to the active surface species and the catalytic centers; molecular metal clusters are playing an increasing role in the preparation of supported catalysts for these reactions. The reactions (isomerization, dehydrocyclization, aromatization, and hydrogenolysis) belong to an important class of hydrocarbon conversions having few counterparts in homogeneous catalysis. Nevertheless, as recently demonstrated by Gault [1], metallacarbene and metallacarbyne intermediates play an important role in these reactions, which implies some relation between the surface-catalyzed reactions and organometallic chemistry.

These intermediate species require one or several coordinatively unsaturated surface metal atoms, such as those located on the edges or corners of crystallites. The involvement of metal atoms with several free valencies in skeletal rearrangements has been demonstrated by catalytic studies of single crystals exposing high-Miller-index faces [2-4] and by investigation of particle size effects in supported catalysts [5, 6]. However, the peculiar properties of highly dispersed Pt/Al_2O_3 catalysts have not been simulated by platinum single crystals. They may be explained by the presence of pseudocrystals with irregular symmetry or by electronic properties of very small metal aggregates that differ from those of infinite crystals. Other hypotheses have also been advanced, including strong metal-support interactions or promoter effects due to the surroundings of the crystallites [4]. The small metal particles may have high activities and stabilities, depending on the method of preparation.

New approaches to the preparation of supported metal catalysts by the decomposition of surface organometallics (including bimetallics [7-9]--see Chapter 10) have made possible the preparation of highly dispersed particles of transition metals; these have been used to study the dependence of catalytic properties on the size of the metal particles. The review by Yermakov *et al.* [7] of catalysis by supported complexes exemplifies the interest in the investigation of skeletal rearrangement reactions of hydrocarbons on supported metal catalysts prepared from molecular metal clusters.

9.4.2 APPROACHES TO MECHANISMS OF SKELETAL REARRANGEMENTS OF HYDROCARBONS VIA METALLACARBENE AND METALLACARBYNE CHEMISTRY

There is traditionally a gap between the appraoches to homogeneous (molecular) catalysis and heterogeneous (surface) catalysis. There are opportunities to narrow the gap and to realize improved activities and selectivities in heterogeneous catalysis as a result of research with catalysts intermediate between large metal crystals and molecular metal complexes. At this stage, an attempt will be made to explain skeletal rearrangements on metals (isomerization, hydrogenolysis, dehydrocyclization, and aromatization) with a limited number of species (carbenes, carbynes, and metallacyclobutane) and mechanisms (dicarbene recombination, metallacyclobutane dismutation, and carbene-olefin addition) which all have their equivalents in coordination chemistry. This subject has been reviewed [1, 10, 11]; the emphasis here is placed on the relations to homogeneous catalysis and the structures and reactivities of organometallic compounds and on comparisons between supported metal catalysts prepared by the decomposition of molecular metal clusters and those prepared conventionally.

9.4.2.a Skeletal Isomerization

The ^{13}C-tracer technique allows one to distinguish between two main classes of reactions: (1) a bond-shift that corresponds to the simple displacement of a carbon-carbon bond and (2) cyclic-type isomerization which involves consecutive 1-5 ring closure and ring opening in the adsorbed phase [1, 4]. The initial product distributions obtained from a number of ^{13}C-labeled hexanes and heptanes may all be accounted for by a combination of these two basic processes.

Bond-shift. The reactions of *n*-pentane-2-^{13}C and 2-methylbutane-2-^{13}C on a platinum catalyst of low dispersion may be classified into four groups, according to their apparent activation energies [4]. The reaction with the highest activation energy is isomerization of *n*-pentane proceeding by a cyclic mechanism (E_a = 294 kJ/mol). A second group includes interconversions between isopentane and *n*-pentane (E_a = 231 kJ/mol). The third group includes the methyl shift, isomerization of isopentane to neopentane, and hydrogenolysis of internal C-C bonds (E_a = 189 kJ/mol). In a fourth group demethylations proceed with the lowest activation energy (E_a = 147 kJ/mol). This classification shows that two different bond-shift mechanisms and two different types of hydrogenolysis should be considered. Metallacyclobutane was suggested as an intermediate in all the reactions of the third group, bond-shift and internal fission, especially on platinum (SCHEME 1):

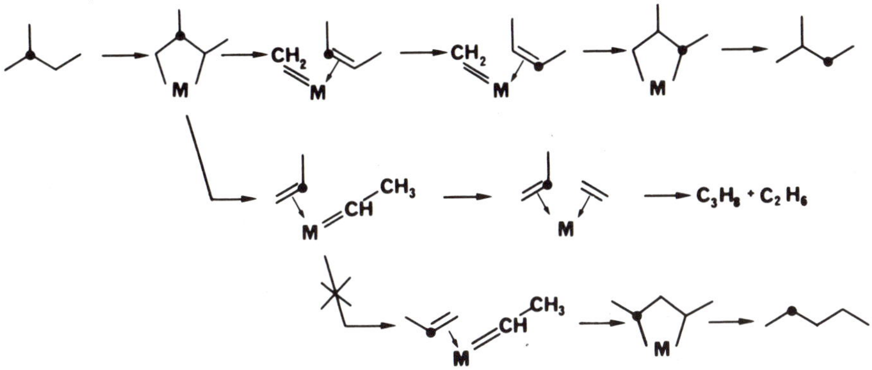

SCHEME 1

Dismutation of this species into metallacarbene and π-adsorbed olefin, followed by isomerization or formation again of a metallacyclobutane, would account for hydrogenolysis and isomerization, respectively [1]. Thus it is explained why, in isopentane, hydrogenolysis of the internal C-C bond has the same activation energy as the methyl shift and why chain lengthening does not take place by this mechanism but requires a mechanism with a higher activation energy.

Since the metallacyclobutane mechanism requires rotation of the π-adsorbed olefin, replacement of hydrogen and methyl groups in this species by more bulky substituents would increase the barrier to rotation and decrease significantly the reaction rate. One can thereby explain the decreasing contribution of the bond-shift mechanism in isomerization when passing from pentanes to hexanes and to heptanes [1]. A further demonstration of the validity of this mechaism is the occurrence of homologation on platinum catalysts at very low hydrogen partial pressures. This reaction, consisting of the addition of one carbon unit at a time, takes place to a very small extent because the formation of metallacyclobutane again is much faster than the required methylene migration [13].

On platinum catalysts of very high dispersion, the methyl shift in isopentane has the same activation energy as the *n*-pentane ⟶ isopentane interconversion; isomerization to neopentane becomes negligible; hydrogenolysis of the internal C-C bonds has the same kinetics parameters as demethylation. Clearly, all the reactions of the third group occurring on catalysts of low dispersion become negligible, and the only bond-shift mechanism left is characterized by the higher activation energy [4]. Since this mechanism does not allow rearrangement at a quaternary carbon atom, it has been suggested that it results from the attack on an allylic species by hydrogen to form a metallacarbene and π-adsorbed olefin. This mechanism, which again involves a metallacyclobutane transition state, had been proposed previously to explain isomerization of isobutane on palladium catalysts [14]:

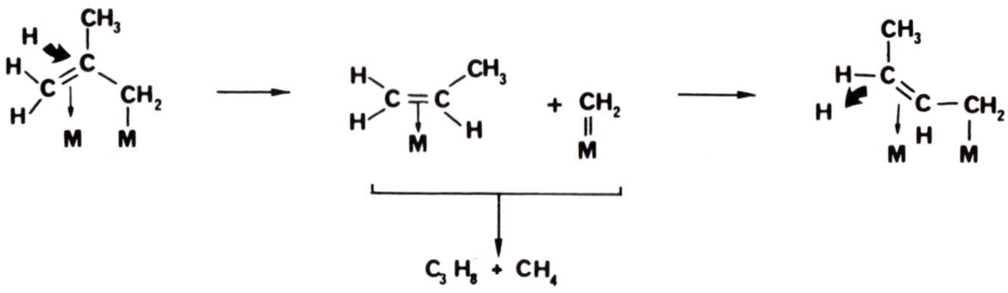
(1)

Cyclic mechanisms. Two cyclic mechanisms may be distinguished for hexane isomerization. According to the first "nonselective" cyclic mechanism, all the cyclic carbon-carbon bonds of the methylcyclopentane intermediate are ruptured with equal probability. The nonselective hydrogenolysis of cyclopentanes was observed with a number of cyclopentanes on highly dispersed Pt/Al_2O_3 catalysts [1, 4]. Under the same conditions, however, quaternary-secondary C-C bonds are not ruptured [15], and 2-methylhexane is not isomerized to give 3,3-dimethylpentane [1]. The reaction mechanism for 1-5 ring closure, therefore, involves only primary and secondary carbon atoms, and that for ring opening involves only carbon atoms bearing at least one hydrogen atom. In this case, 1-5 dehydrocyclization is best described as a dicarbene recombination to form a π-adsorbed olefin [1]:

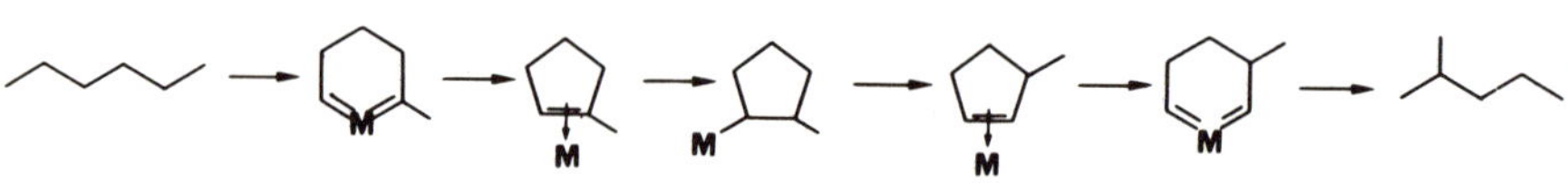
(2)

Another mechanism, taking place on iridium [1], does not allow the isomerization of *n*-hexane, but accounts for rapid interconversion of 2-methylpentane and 3-methylpentane via a methylcyclopentane intermediate. Conversely, on iridium only the CH_2-CH_2 bonds are ruptured during the hydrogenolysis of substituted cyclopentanes. This selective hydrogenolysis had already been observed on platinum catalysts of very low dispersion at low temperature [4]. Since two methyl groups in the 1 and 5 positions are required for dehydrocyclization, and two neighboring methylene groups are required for cyclopentane hydrogenolysis, it was suggested that this "selective" cyclic mechanism consists of a dicarbyne recombination to form a dicarbene species [1]. All the intermediate species are necessarily attached to two metal atoms, and it is possible that the reaction involves a first-step addition of two hydrogen atoms to form a cyclic dicarbene:

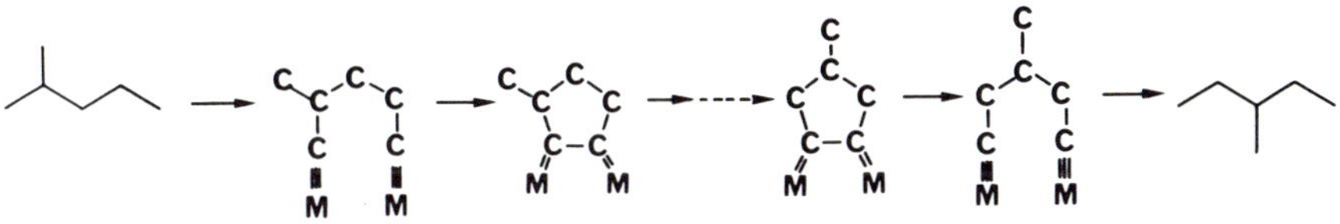
(3)

Analysis of the product distributions in methylcyclopentane hydrogenolysis on platinum catalysts showed that there was at least a third mechanism for ring rupture [1], and therefore a third mechanism for cyclic type isomerization. This, it was believed, is the same as the metallacyclobutane mechanism involved in the bond shift. It would explain the rupture on platinum, to a very small extent, of quaternary-secondary C-C bonds in gem-dimethylcyclopentane [16], the cleavage under severe conditions of the cyclohexane ring, and also some abnormal product distributions in the hydrogenolysis of 1,2-dimethylcyclopentane [1]. Whereas in the hydrogenolysis of most cyclopentanes the product distributions are accounted for by a combination of selective and nonselective mechanisms only, on a Pt/Al_2O_3 catalyst of low dispersion the rupture of the tertiary-tertiary C-C bond in 1,2-dimethylcyclopentane is enhanced relative to the cleavage of the tertiary-secondary C-C bonds (SCHEME 2):

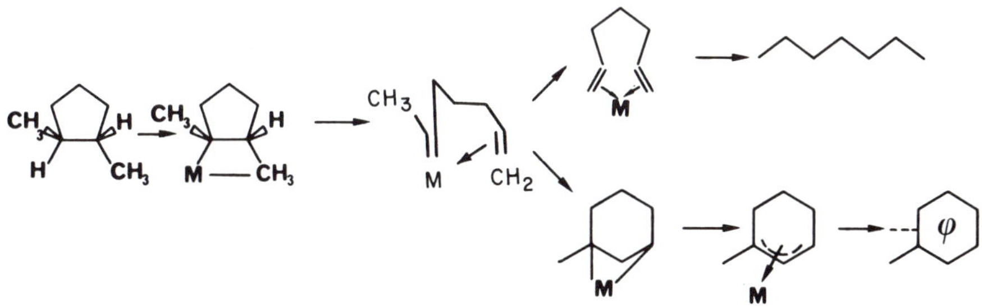

SCHEME 2

This could be accounted for by an exocyclic metallacyclobutane which, after dismutation, would yield either *n*-heptane or methylcyclohexane. Indeed, toluene and *n*-heptane are both obtained as initial products, and the amounts of each respond in the same way to changes in the hydrogen partial pressure.

9.4.2.b Aromatization

According to the preceding scheme, 1-6 ring closure (the major process in aromatization of alkanes) would consist of a carbene-olefin addition:

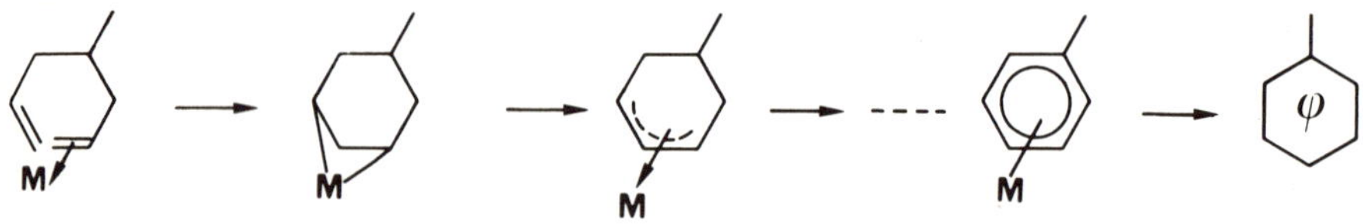 (4)

Similarly, dehydroaromatization of substituted cyclopentanes such as 1,2-dimethyl and 1,1,3-trimethylcyclopentanes [17] would involve ring cleavage followed by 1-6 ring closure. An alternative route is then offered for the aromatization of alkanes on platinum, consisting of the occurrence of three consecutive steps in the adsorbed phase: 1-5 ring closure, ring opening, and 1-6 ring closure. Gault [1] thereby explained the distribution of the aromatics obtained from a number of unlabeled and labeled hydrocarbons by competition between these one-step and three-step mechanisms initiated by 1-5 and 1-6 ring closure, respectively:

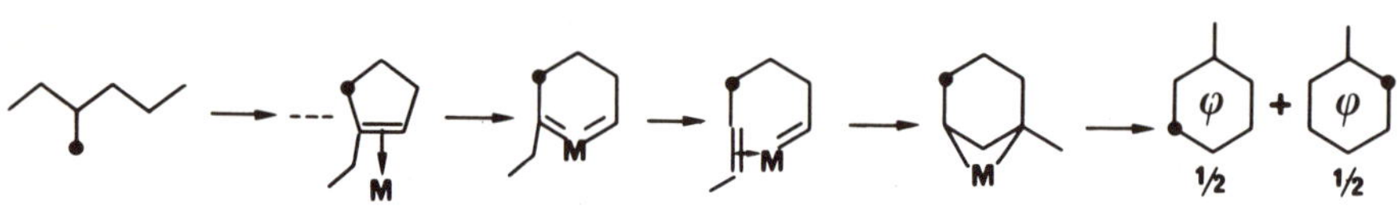 (5)

In a multistep mechanism, ring cleavage may occur in two different ways, either, as in the case of isooctane [17], by a metallacyclobutane mechanism, which provides the carbene-olefin species immediately suitable for 1-6 ring closure, or, as in the case of methylhexanes [1], via a dicarbene species.

9.4.2.c Hydrogenolysis of alkanes

In the case of platinum, several mechanisms compete in hydrogenolysis of alkanes, one of which may be considered as a degenerate bond-shift type isomerization [4]. For a number of branched hydrocarbons, the C-C bonds in the position β to the tertiary carbon atom are preferentially ruptured, which suggests the involvement of an $\alpha\alpha\gamma$ triadsorbed species [18], yielding a metallacarbene and a metallacarbyne:

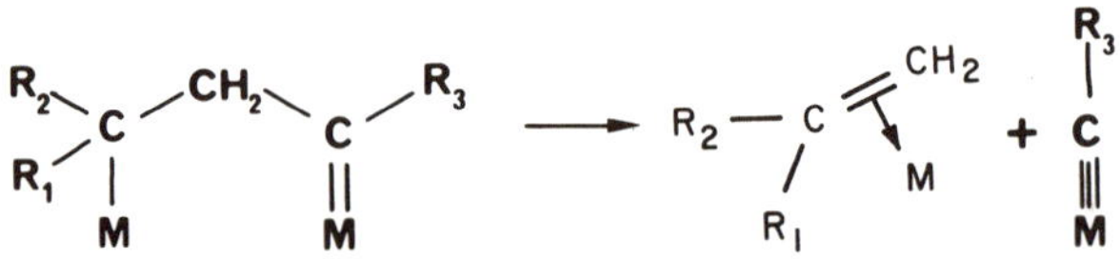 (6)

On iridium, the rupture of bonds between secondary or primary carbon atoms is highly favored, and that is best explained by dismutation of 1,2-dicarbene species [1]:

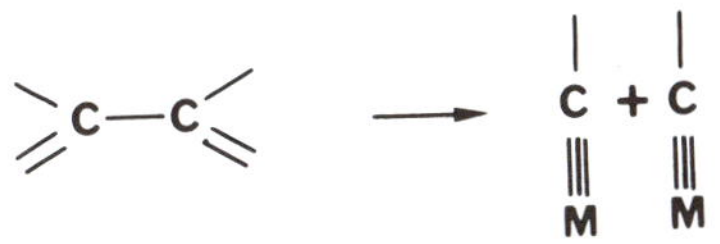

(7)

Therefore, in hydrogenolysis of acyclic hydrocarbons, as in isomerization and aromatization, metallacarbenes and metallacarbynes play an important role. The unambiguous demonstrations which were given for their participation in reactions such as metathesis [19] (olefin-carbene addition) and dimerization (dicarbene recombination) [20-22] now allow a better understanding of the mechanisms of skeletal isomerization on metals. On the other hand, double bond migration, hydrogenation, and exchange on metals also have their counterparts in homogeneous catalysis, so that the involvement of σ-alkyl, π-olefinic, and π-allylic species in these reactions could be demonstrated [1, 10, 11].

9.4.3 SUPPORTED MONOMETALLIC CATALYSTS PREPARED BY DECOMPOSITION OF MOLECULAR METAL CLUSTERS

In this section, a survey is given of the data accumulated in the study of reactions of hydrocarbons on catalysts prepared by decomposition of molecular metal clusters deposited on metal oxides. The methods of preparation of such catalysts are not presented here. It is now well established that many catalysts obtained by decomposition of molecular metal complexes are characterized by "ultradispersed particles" containing atoms of active metal. Here we are concerned mainly with the reactions of hydrocarbons such as hydrogenolysis, isomerization, dehydrocyclization, and aromatization taking place on catalysts derived from molecular monometallic clusters. Introduction of a second metal element into the supported catalyst by use of a bimetallic cluster precursor offers opportunities for modifying the catalysts, as discussed in Chapter 10.

Summarized in Table 1 are the catalytic reactions of hydrocarbons investigated with supported monometallic catalysts prepared by decomposition of molecular metal clusters on metal-oxide supports. The data characterizing the catalytic activities and selectivities are summarized for hydrogenolysis and skeletal rearrangements of hydrocarbons.

TABLE 1. Catalytic Reactions of Hydrocarbons on Supported Monometallic Catalysts prepared by Decomposition of Molecular Clusters on Metal-oxide Supports.

Precursor molecular metal complex	Cluster-derived catalyst	Catalytic Reactions[a]	Refs.
$[Pt_3(CO)_6](Et_4N)_2$	0.5% Pt/SiO_2	Hydrogenolysis, Isomerization of n-C_7	[26]
$[\{Pt_3(CO)_6\}_n](Et_4N)_2$ (n = 1-5)	0.15-0.7% Pt/SiO_2	Dehydrocyclization, Isomerization of n-C_6	[27]
$[\{Pt_3(CO)_6\}](Et_4N)_2$ (n = 2-5)	0.4-5.5% Pt/γ-Al_2O_3	Hydrogenolysis of MCP Isomerization of MP	[4, 6, 28]
$Pt(C_3H_5)_2$	2.3-3.9% Pt/SiO_2	Hydrogenolysis of C_2H_6, Dehydrocyclization of n-C_5, n-C_7	[7, 29]
$Pt_3(CO)_3L_4$	1.6-3% Pt/γ-Al_2O_3	Hydrogenolysis of MCP, poly MCP; cyclohexane	[4, 28]
(L = PMe_3, PEt_3, PPh_3)		Isomerization of MP	
$Pt(PPh_3)_3$	1% Pt/γ-Al_2O_3	"	[30, 31]
$Pd(C_3H_5)_2$	4% Pd/SiO_2	Hydrogenolysis of C_2H_6, Aromatization of n-C_6	[7, 29]
$Pd(C_3H_5)_2\ Cl_2$	3% Pd/γ-Al_2O_3	Hydrogenolysis of MCP Isomerization of MP	[32, 33]
$Pd(PPh_3)_4$	1-3% Pd/γ-Al_2O_3	Isomerization of MP	[32, 33]
$Ir_4(CO)_{12}$	0.25% Ir/γ-Al_2O_3	Hydrogenolysis of MCP Isomerization of MP	[25]
$Ni(C_3H_5)_2$	1% Ni/SiO_2	Hydrogenolysis of C_2H_6	[7, 29]
$Ru_3(CO)_{12}$, $Ru_6C(CO)_{17}$	1.2-2.3% Ru/SiO_3	Hydrogenolysis of n-C_7	[26]
$Ru_3(CO)_{12}$	0.5% Ru/SiO_2	Hydrogenolysis of n-C_4	[34]
$Rh_6(CO)_{16}$, $Rh_4(CO)_{12}$	0.3-2.3% Rh/SiO_2	Hydrogenolysis of n-C_7	[26]
$W(\pi$-$C_4H_7)_4$	8.2% W/SiO_2	Hydrogenolysis	[7, 23]
$Mo(\pi$-$C_3H_5)_4$	5.2% Mo/SiO_2	Isomerization of neo-C_5	

[a]MCP = methylcyclopentane; MP = methylpentanes.

The general trend is that the interaction of organometallic compounds of transition elements with the surfaces of oxide supports leads to highly dispersed catalysts which are more active than conventionally prepared catalysts (which have larger particles), both for hydrogenolysis and isomerization. But the activity of a cluster-derived catalyst is sometimes unstable, depending upon the nature of the initial molecular complex [4, 34].

For example, the hydrogenolysis of methylpentanes, methylcyclopentane, and polymethylcycloalkanes and isomerization of labeled hexanes was studied with Pt catalysts prepared from the clusters $Pt_3(CO)_3L_4$ (L = PMe_3, PEt_3, PPh_3) or from $Pt(PPh_3)_3$ deposited on alumina (Tables 2-5). These catalysts showed high selectivities for demethylation and ring opening, which is quite different from the behavior of classical Pt catalysts; the average particle size (~2.0 nm) was similar to that of classical dispersed catalysts, and the differences in the catalytic properties were associated with the presence of phosphorus (from the phosphine ligand). With such catalysts, the same selectivity for demethylation was observed with C_5 and C_6 rings (Tables 2 and 3). Selective rupture of the secondary-quaternary bond of the C_5 ring occurred preferentially on cluster-derived catalysts in comparison with classical Pt/Al_2O_3 catalysts. The selective demethylation was observed for ethylcyclohexane and ortho- and meta-xylenes.

Palladium catalysts prepared by decomposition of $Pd(C_3H_5)_2$ on silica were more active for ethane hydrogenolysis than conventional catalysts [7], and highly dispersed catalysts prepared from $Ir_4(CO)_{12}$ on alumina are highly active for hydrogenolysis of methylcyclopentane [25]. At 433 K, a 0.25% Ir/Al_2O_3 catalyst derived from $Ir_4(CO)_{12}$ had the same activity and selectivity for ring opening as a classical 10% Ir/Al_2O_3 catalyst [37]. Like the Pt-cluster-derived catalysts, these iridium catalysts exhibited low rates of coke formation and were remarkable for their stability and selectivity. Ruthenium catalysts prepared from $Ru_3(CO)_{12}$ and $Ru_6C(CO)_{17}$ on SiO_2 also displayed enhanced activity for the complete hydrogenolysis of straight-chain paraffins to methane and provided a temperature advantage of 150°C relative to conventionally prepared ruthenium catalysts [26]. The increased activity is correlated with the smaller metal crystallite sizes (1.5-2.0 nm) reproducibly obtained with molecular clusters as catalyst precursors. The high activity for the hydrogenolysis of alkanes has been applied for selective cleavage of the alkyl group in ethylbenzene, giving toluene and methane with $Ru_3(CO)_{12}/SiO_2$-derived catalysts. Such catalysts can provide a means of separating ethylbenzene from mixtures with xylenes [26].

TABLE 2. Reactions of 2-methylpentane at 573 K on Supported Pt Catalysts prepared by Decomposition of Metal Clusters on γ-alumina.

Catalyst precursor	catalyst	$\%\alpha_i$	$\%\alpha_c$	$\%\alpha_T$	%S	n-P/iso-P	$\%C_1$	$\%C_5+C_1$	$\%C_4+C_2$	$\%2C_3$	$\%3C_2$	%3-MP	%n-H	%MCP	S(nm) (TEM)	Ref.
$H_2PtCl_6{\cdot}6H_2O$	2% Pt/Al_2O_3	15.8	15.1	30.9	51	2.11	-	23.3	15.7	9.8	-	21.8	24.8	4.5	1.48	[4, 6]
$[\{Pt_3(CO)_6\}_3(Et_4N)_2]$	0.4% Pt/Al_2O_3	14.7	18.6	33.3	44	2.18	0.97	27.2	15.6	11.2	0.7	18.8	22.6	2.8	1.78	[6]
$Pt_3(CO)_3(PEt_3)_4$	2% Pt/Al_2O_3	0.9	19.3	20.2	4.5	4.16	-	58.8	18.3	18.7	-	1.5	1.9	0.73		[30]
$Pt_3(CO)_3(PEt_3)_4$	1.6% Pt/Al_2O_3	0.18	1.1	1.2	14	4.4	4.6	46	14.5	14.3	6.2	7.5	5.6	1.3	1.5-2.8	[30]
$Pt_3(CO)_3(PEt_3)_4$	1.7% Pt/Al_2O_3	0.45	6.3	6.8	6.7	4.7	-	63.5	15.8	14	-	2.5	2.45	1.8		[30]

Notes:

3-MP = 3-methylpentane; n-H = n-hexane; MCP = methylcyclopentane; n-P/iso-P = n-pentane/isopentane; $\%\alpha_T$ = overall conversion; % S = selectivity = $\sum C_6$ isomers x $100/\alpha_T$; $\%\alpha_i$ = % of C_6 isomers; $\%\alpha_c$ = % of hydrocracking = $\sum$ % CH_4 + % C_2H_6 + % C_3H_8 + % C_4H_{10} + % C_5H_{12}.

TABLE 3. Hydrogenolysis of Methylcycloalkanes on 2% Pt Supported on γ-alumina Catalysts prepared by Decomposition of Metal Clusters [30] and by Classical Impregnation [35].

Catalyst precursor	Reactant hydrocarbon	T/K	$\%\alpha_T$	$\%\alpha_C$	Product distributions in %					
	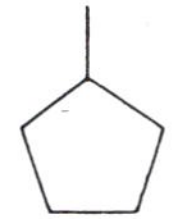				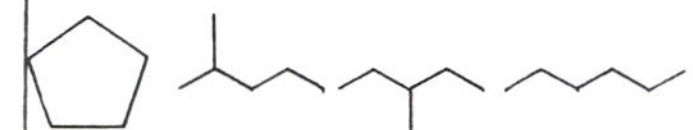					
$Pt_3(CO)_3(PEt_3)_4$		573	4.2	0.4	66.3	14.3	2.6	16.4		
$H_2PtCl_6 \cdot 6H_2O$		493	2	0	0	45.4	21.6	32.4		
	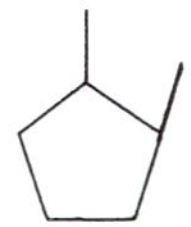				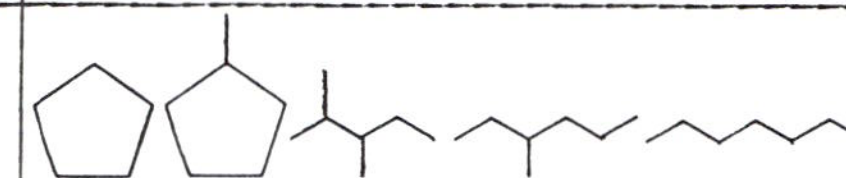					
$Pt_3(CO)_3(PEt_3)_4$		493	4.2	13	1.4	43.8	19	16.7	6	
$H_2PtCl_6 \cdot 6H_2O$		493	14.7	0	0	1.35	37.9	24	36.7	
	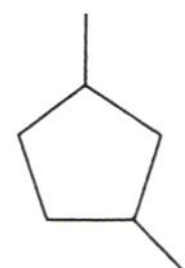				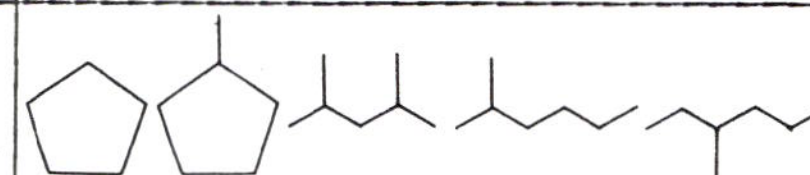					
$Pt_3(CO)_3(PEt_3)O_4$		493	7.2	11.4	4.2	33.1	14	20	17.2	
$H_2PtCl_6 \cdot 6H_2O$		493	3.4	4.5	0	1.50	76.7	10.1	7.1	
					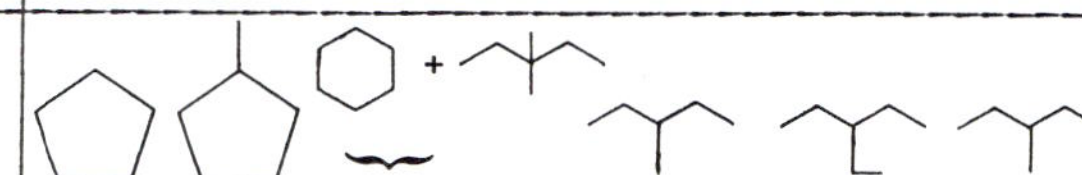					
$Pt_3(CO)_3(PEt_3)_4$		493	14	3.9	3.6	2.2	0	24	11.4	54.9
$H_2PtCl_6 \cdot 6H_2O$		493	12.2	0	0.33	0.4	0.65	41.7	14.3	42.5
					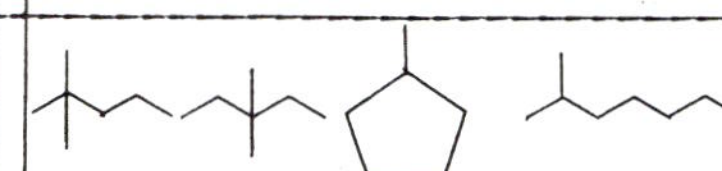					
$Pt_3(CO)_3(PEt_3)_4$		493	25.2	-	45.1	17.1	26.5	11.2		
$H_2PtCl_6 \cdot 6H_2O$		493	8	-	84.2	10.3	-	5.4		

TABLE 3 (Continued)

Catalyst precursor	Reactant hydrocarbon	T/K	%α_T	%α_C	Product distributions in %								
	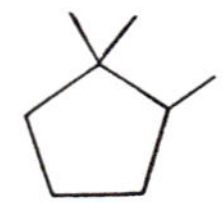				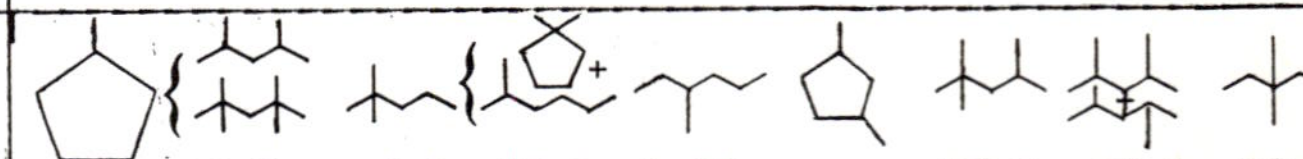								
$Pt_3(CO)_3(PEt_3)_4$		493	37.5		3.8	1.5	0.9	27.8	1.25	-	32.2	22.4	10.2
$H_2PtCl_6{\cdot}6H_2O$		493	13.4		-	0.5	-	2.6	-	0.6	86.6	6.9	2.9
	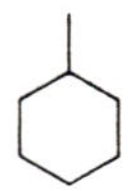				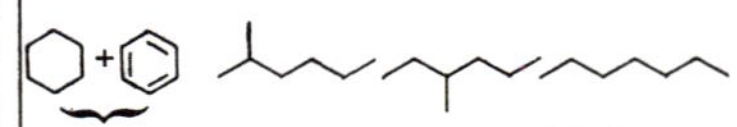								
$Pt_3(CO)_3(PEt_3)_4$		513	8.6	0	72.7	10.6	4	12.5					
$H_2PtCl_6{\cdot}6H_2O$		513	0	-	-	-	-	-					
$Pt_3(CO)_3(PEt_3)_4$	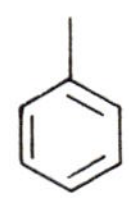	513	11.8	0	81	7.8	3	8.3					
$H_2PtCl_6{\cdot}6H_2O$		513	0	-	-	-	-	-					
					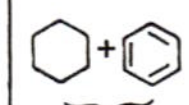	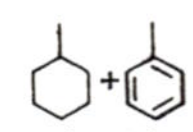	Ring Opening						
$Pt_3(CO)_3(PEt_3)_4$		513	4.7	-	29.9	58.4	11.5						
$H_2PtCl_6{\cdot}6H_2O$	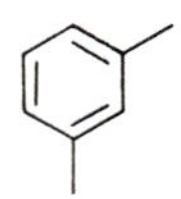	513	0	-	-	-	-						
$Pt_3(CO)_3(PEt_3)_4$		513	7	-	4	95.2	0.7						
$H_2PtCl_6{\cdot}6H_2O$		513	0	-	-	-	-						
$Pt_3(CO)_3(PEt_3)_4$	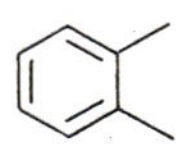	523	0	-	3	85.2	-						
$H_2PtCl_6{\cdot}6H_2O$		523	0	-	-	-							

Notes:

%α_T = overall conversion; %α_C = % of hydrocracking; product distributions in % = demethylation + ring opening.

TABLE 4. Isomerization of ^{13}C-labeled Hexanes and Hydrogenolysis of Methylcyclopentane in the presence of Catalysts derived from $[\{Pt_3(CO)_6\}_n](Et_4N)_2$ (n = 2-5) and of Catalysts prepared by Impregnation from Chloroplatinic Acid.

Catalyst precursor	Catalyst Pt on γ-Al_2O_3	Reaction temperature/ K	a = H/Pt	d_S/nm T.E.M.	Activity $\%\alpha_T$ moles	S	X	Y	R_3	Refs.
	7.1	527	0.35	3.6	-	-	16	90	0.7	
	4.1	527	0.55	1.7	18	56	30	86	0.4	
$H_2PtCl_6 \cdot 6H_2O$	2.5	527	0.70	1.5	15	50	62	97	0.4	[1, 4, 5]
	2.25*	548	0.04	10	-	-	44	92	0.7	
	0.2	527	1	2	-	-	83	100	0.42	
$[\{Pt_3(CO)_6\}_4](Et_4N)_2$	5.5	548	-	2	11	55	44	92	0.57	
$[\{Pt_3(CO)_6\}_2](Et_4N)_2$	1.1	548	-	1.7	3.3	41.5	67	-	0.73	[6]
$[\{Pt_3(CO)_6\}_5](Et_4N)_2$	0.56	548	0.8	2.3	3.3	57	57	90	0.62	
$[\{Pt_3(CO)_6\}_3](Et_4N)_2$	0.4	548	0.54	1.5	7.5	41	80	82	0.7	

Notes:

*Catalyst previously reduced at 473 K sintered at 673 K in H_2.

a = H/Pt : accessibility deduced from chemisorption of H_2; activity $\%\alpha_T$ moles = overall conversion for 100 moles of reactant = % hydrocracking + % ΣC.M. + % ΣB.S. (C.M. = cyclic mechanism; B.S. = bond shift mechanism);

S (selectivity) = $\frac{\Sigma(C_6 \text{ isomers})}{\alpha T}$

X = percentage of cyclic mechanism for isomerization of 2-methylpentane to 3-methylpentane;

Y = percentage of cyclic mechanism for isomerization of 2-methylpentane to *n*-hexane;

R_3 = ratio of 3-methylpentane over *n*-hexane for hydrogenolysis of methylcyclopentane.

TABLE 5. Isomerization of Labeled Hexanes on various Supported Pt Catalysts prepared by Decomposition of Organometallic Complexes on γ-alumina [28, 30, 31].

Catalyst	Support	Reaction temperature/ K	$\bar{d}_H$/nm	$\bar{d}_S$/nm	N	X	Y
Euro-Pt1	SiO_2	527	0.85	1.9	2.2	23	87
0.2% Pt[a]	γ-Al_2O_3	527	0.85	2.0	39.4	83	100
2.5% Pt[b]	γ-Al_2O_3	573	-	2.2	7.6	8	4
2.5% Pt[c]	γ-Al_2O_3	573	-	2.5	-	54	98

Notes:

[a] 0.2% Pt-Al_2O_3 prepared from chloroplatinic acid.

[b] 2.5% Pt-Al_2O_3 prepared from $Pt_3(CO)_3(PPh_3)_4$.

[c] same catalyst treated in air, 16 h, 473 K before the catalytic tests.

$\bar{d}_H$/nm = particle mean size deduced from chemisorption of H_2.

$\bar{d}_S$/nm = particle mean size deduced from T.E.M. measurements.

N = percentage of surface particle sizes in the range 0.5 to 1 nm.

X = percentage of cyclic mechanism for isomerization of 2-methylpentane to 3-methylpentane.

Y = percentage of cyclic mechanism for isomerization of 2-methylpentane to *n*-hexane.

Extensive work has been done to characterize hydrogenolysis catalysts prepared from $Os_3(CO)_{12}$, $H_4Os_4(CO)_{12}$, and $Os_6(CO)_{18}$ supported on silica, alumina, titania, and ceria [40-43]. Decomposition of these clusters on the surfaces under vacuum or in helium leads to the formation of coordinatively unsaturated osmium aggregates of slightly higher nuclearity than the starting clusters. CO and carbon ligands are evidently responsible for preventing sintering of the osmium to give conventional polycrystalline particles.

Ethane hydrogenolysis has been investigated with these catalysts at 395-665 K. Freshly prepared catalysts exhibited an initial period of nonsteady-state activity, during which hydrocarbons were progressively retained, and activity rose and passed through a maximum and declined to a steady value. Most cluster-derived catalysts showed exceptional activity for ethane hydrogenolysis, with some apparent turnover numbers being two orders of magnitude higher than for conventional supported metallic osmium. Owing to the presence of stabilizing CO and C ligands in the supported osmium aggregates, the reactant is inferred to have adsorbed less strongly than on metallic osmium; the weaker reactant adsorption may provide superior catalytic properties. A volcano curve described the dependence of the rate of reaction on the CO stretching frequency of the CO ligands retained by the working catalyst; this may provide a measure of the osmium-support interaction.

Brief mention of molybdenum on alumina is made here since Mo/Al_2O_3 catalysts cannot be prepared from MoO_3/Al_2O_3 by reduction at temperatures $\lesssim$ 900 K [44, 45]. Fully dehydroxylated alumina (pretreated at 1220 K) was brought in contact with $Mo(CO)_6$, yielding Mo^0 after decomposition [46]. Mo/Al_2O_3 prepared in this way from the mononuclear carbonyl complex has been reported to be highly active for propane hydrogenolysis [47, 48] in comparison with catalysts prepared by reduction of MoO_3 [49]. The molybdenum catalyst prepared from $Mo(CO)_6$ ranks among the most active hydrogenolysis catalysts, such as ruthenium.

9.4.3.a **Skeletal rearrangements of hydrocarbons**

Platinum, palladium, rhodium, and iridium catalysts obtained by decomposition of molecular metal clusters on metal-oxide supports have been tested for isomerization, dehydrocyclization, and aromatization of hexanes and heptanes. For example, Ichikawa [27] observed differences in the selectivities for isomerization and dehydrocyclization of *n*-hexane in the presence of supported platinum formed from platinum carbonyl cluster anion salts. He concluded that the small platinum aggregates derived from metal clusters favor the cyclic mechanism for hexanes.

Zahraa *et al.* [6, 28] investigated the skeletal isomerization and dehydrocyclization of 2-methylpentane with a series of catalysts obtained from decomposition of $[Pt_3(CO)_6]_n^{2-}$ complexes. ^{13}C labeling allowed the authors to estimate the relative contributions of the cyclic mechanisms in isomerization. In the isomerization of 2-methylpentane to 3-methylpentane, the relative importance of the selective cyclic mechanism increased with the percentage of metal particles having diameters $\lesssim$1 nm, whatever catalyst was used. In the isomerization of 2-methylpentane to *n*-hexane with catalysts having the same mean size of particles, the cluster-derived catalysts showed differences in selectivity compared with the classical Pt/Al_2O_3 catalysts (Table 4). These differences were attributed to an electronic rather than a geometric effect.

9.4.3.b **Alkene hydrogenation**

Catalysts prepared from $Rh_6(CO)_{16}$ supported on SiO_2 were characterized by Graydon and Langan [50]. The ultradispersed rhodium catalysts were prepared with various metal loadings and particle size distributions, the latter being measured by transmission electron microscopy. The catalytic activity for benzene hydrogenation was inferred to be negligible for rhodium aggregates with diameters less than 1.2 nm, whereas the aggregates with diameters between 1.2 and 2.0 nm were highly active. The relatively inactive rhodium species may have included hexarhodium aggregates.

There are a number of examples of reactions of alkenes apparently catalyzed by intact clusters on oxide supports, as summarized in Section 9.3.

9.4.4 CORRELATIONS BETWEEN THE MECHANISMS OF REACTION OF HYDROCARBONS AND THE STRUCTURE AND COMPOSITION OF THE ACTIVE CENTERS

Decomposition of mono- or multimetallic molecular clusters on oxide supports permits the preparation of highly dispersed supported metal particles (some with high stability) having more uniform distributions of particle sizes <2 nm than conventionally prepared catalysts. Some mechanisms of isomerization, hydrogenolysis, and dehydrocyclization of hydrocarbons are favored by these small particles. The metal aggregates formed by decomposition of molecular clusters may, in some rare instances, retain their initial metal frameworks (see the other sections of Chapter 9); in these cases they represent a unique transition class between mononuclear metal compounds and the classical supported crystallites [31, 51].

As a basis for inferring correlations from the mechanisms of hydrocarbon reactions and the structures and compositions of the catalysts obtained by decomposition of molecular clusters on oxide supports, a summary is presented of the main patterns of activity and selectivity of reactions of skeletal rearrangements of hydrocarbons.

As a general rule, the catalytic activity is enhanced by and directly related to the increase of dispersion of metal on the support.

The selectivity can be greatly modified by varying the chemical composition of the starting molecular cluster and by its interaction with the support. Some ligands (e.g., CO) are not expected to have much influence on the specific catalytic properties of the resulting supported metal, but other ligands are expected to be important. For example, some ligands (e.g., phosphines) might be expected to be catalyst poisons and others might be expected to be promoters. These issues have been discussed by Yermakov *et al.* [7, 23, 52].

We begin with an evaluation of catalysts prepared by decomposition of monometallic molecular clusters. The general trend is that the small aggregates of Pt, Pd, Ir, or Ru formed by decomposition of molecular clusters adsorbed on alumina or silica (when the ligands do not contain phosphine) have roughly the same selectivity for hydrogenolysis, isomerization, or dehydrocyclization as the conventional small aggregates (e.g., catalysts prepared from chloroplatinic acid). On very small particles of platinum obtained from molecular clusters decomposed on alumina ($d <$ 1 nm), the bond shift mechanism is drastically decreased and the nonselective cyclic mechanism becomes predominant, whereas methylcyclopentane hydrogenolysis is nonselective (Table 4). However, it is interesting to note the differences in selectivity for the reaction 2-methylpentane $\longrightarrow$ *n*-hexane for catalysts having the same mean size of aggregates, some prepared conventionally and some from molecular clusters (Table 4). On the other hand, experimental evidence seems to indicate that electronic rather than geometric factors should be considered [1]. Electronic factors, taking on importance mostly when the metal particles are smaller than 1 nm in the case of platinum, regulate the concentrations of the precursor metallacyclobutane and metallacarbene species, and hence the relative contributions of the cyclic and bond shift mechanism.

With catalysts prepared from clusters, the effects of the starting counter ion and of carbon encapsulated in the metal framework [34, 53] may influence the competition between the metallacyclobutane rearrangement (SCHEME 1) and the nonselective mechanism of isomerization described as a dicarbene recombination to form a π-adsorbed olefin [1] [Eq. (2)]. The participation of the metallacyclobutane rearrangement can be correlated with the demethylation occurring at the same time on these catalysts and with the increase in the ratio 3-methylpentane/*n*-hexane for hydrogenolysis of methylcyclopentane via the third mechanism for cyclic type isomerization described in Section 9.4.2.a. On the other hand, the importance of dispersion in hydrogenolysis of ethane, *n*-butane, and *n*-pentane was emphasized by Guczi [54], who interpreted the C-C bond cleavages in terms of the formation of a metallacyclopropane intermediate on Pt catalysts.

Hydrogenolysis and isomerization catalysts derived from $Ir_4(CO)_{12}$ and $Pd(C_2H_5)_2Cl_2$ were found to be similar to classical supported catalysts of similar dispersion; these results are inferred to emphasize the "electronic factor" in view of the drastic changes in mechanisms which arise upon changes in the nature of the metal [1, 4]. It is important that with most of the catalysts prepared from molecular clusters, any influence of chlorine on the reaction mechanisms can be ruled out. It has been shown that traces of chlorine did not influence the properties of small aggregates of platinum for hydrogenolysis of methylcyclopentane and isomerization of hexanes by acidic mechanisms [1, 4].

As a first conclusion, it is remarkable that the model catalysts prepared by the controlled condensation of clusters on a support are directly comparable to industrial catalysts. However, they can also show a specific behavior due to the incorporation of promoters during the preparation (e.g., phosphorus from the phosphine ligands [28, 30, 31]). Selective demethylation and bond-shift mechanisms were evident with catalysts derived from clusters for a series of cyclic hydrocarbons (see Tables 3 and 5) and for isomerization of hexanes.

Mechanisms involving a metallacyclobutane intermediate were proposed to account for all the reactions (demethylation, ring opening, and isomerization), as shown in SCHEME 3. As suggested by Rooney [55], selective demethylation and substantial hydrogenolysis without much cyclization are characteristic of nickel catalysts; it therefore seems that phosphorus in the neighborhood of platinum atoms makes platinum behave like nickel; $\alpha\alpha\beta$-type triadsorbed species could be responsible for demethylation on the P-modified catalysts.

However, toluene led exclusively to benzene and ortho- and para-xylenes to toluene by selective demethylation, which excludes $\alpha\alpha\beta$ species (Table 3). On the other hand, a metallacyclobutane intermediate was suggested by Maire *et al.* [28, 30] as a precursor responsible for the methyl migration and the selective demethylation [28] for hexanes. This hypothesis was reinforced by the results obtained by the authors for isomerization by

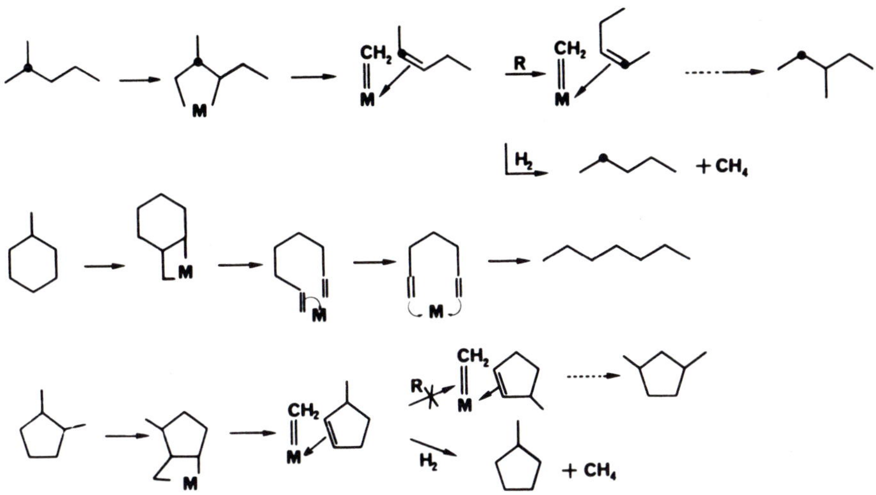

SCHEME 3

hexanes and hydrogenolysis of methylcyclopentane on catalysts prepared by impregnation of alumina by a $Pt(PPh_3)_3$ complex, leading again to the selective demethylation and to enhancement of the bond-shift mechanism [35]. Upon impregnation of alumina with the $Pt(PPh_3)_3$ complex, there was no drastic change in the structure of the complex, as shown by EXAFS measurements [31]. After activation by hydrogen at 573 K, the catalyst consisted of Pt atoms surrounded by two P atoms at a distance of 0.22 nm and two O atoms at a distance of 0.21 nm as first neighbors and four Pt atoms at distances of 0.35 and 0.37 nm as second neighbors. It is appropriate to think of platinacyclobutane rearrangements on such Pt catalysts.

Wilker and Hoffmann [56] recently suggested a new mechanism for metathesis and for understanding how electron count and ligand electronegativity affect the stability of various intermediates in the $L_3Pt(C_3H_6)$ system. On palladium catalysts obtained from $Pd(PPh_3)_4$ complexes adsorbed on alumina, the selective demethylation characteristic of palladium was hindered in favor of the ring opening, but the intrinsic properties of palladium for isomerization of methylpentanes were apparently not affected by the neighboring P atoms.

9.4.5 CONCLUSIONS

Supported metal particles derived from molecular metal clusters have been used in a variety of hydrocarbon conversion reactions. For skeletal rearrangements, such catalysts exhibit various activities and selectivities in isomerization, demethylation, and hydrogenolysis, depending on the sizes of the metal aggregates or particles, which influence both geometric and electronic factors. However, many of the supported catalysts prepared from metal clusters are not stable in their inital forms, and many are scarcely different from supported catalysts prepared in classical ways.

The approach of cluster chemistry also provides a new source of information useful for understanding the mechanisms of reactions on surfaces and a new tool to fill the gap between current ideas in homogeneous and heterogeneous catalysis. These ideas are elaborated in Chapter 11.

ACKNOWLEDGMENTS

The author is grateful to Drs. L. Hilaire and P. Esteban Puges, Universite Louis Pasteur de Strasbourg, for helpful comments and for a critical reading of the manuscript. For the detailed work in the preparation of the data, his sincere thanks go to Mr. S. Wechsler, who drew the schemes and figures, and to Mrs. M. Wechsler, who typed the manuscript.

REFERENCES

1. F. G. Gault, Advan. Catal., 30 (1981) 1.
2. G. A. Somorjai, in D. A. King and D. P. Woodruff (Eds.) The Chemical Physics of Solid Surfaces, vol. 4., Elsevier, New York, 1982, p. 216.
3. F. Garin, S. Aeiyach, P. Legare and G. Maire, J. Catal., 77 (1982) 323.
4. G. Maire and F. Garin, in Catalysis: Science and Technology, vol. 6, Springer Verlag, Eds. M. Boudart and J. R. Anderson (1984).
5. J. M. Dartigues, A. Chambellan and F. G. Gault, J. Am. Chem. Soc., 98 (1976) 856.
6. F. Garin, O. Zahraa, C. Crouzet, J. L. Schmitt and G. Maire, Surf. Sci., 106 (1981) 466.
7. Yu. I. Yermakov, B. N. Kuznetsov and V. A. Zakharov, Catalysis by Supported Complexes, Elsevier, Amsterdam, 1981.
8. D. C. Bailey and S. H. Langer, Chem. Rev., 81 (1981) 110.
9. J. M. Basset, J. Mol. Catal., 21 (1983) 95.
10. J. R. Anderson, Advan. Catal., 23 (1973) 1.
11. J. K. A. Clarke and J. J. Rooney, Advan. Catal., 25 (1976) 125.
12. J. R. Anderson and N. R. Avery, J. Catal., 5 (1966) 446.
13. F. Luck, These d'Etat, Strasbourg (1983).
14. J. M. Muller and F. G. Gault, J. Catal., 24 (1972) 361.
15. B. A. Kazanskii, Ouspiekhii Khimii, 17 (1948) 655.
16. O. V. Bragin, E. G. Khelkoskaya-Sergeeva and A. L. Liberman, Izv. Akad. Naui SSSR Ser. Khim. (1977).
17. a) B. A. Kazanskii, A. L. Liberman, V. T. Aleksanyan and Kh. E. Sterin, Dokl. Akad. Nauk. SSSR, 95 (1954) 281.
    b) B. A. Kazanskii, A. L. Liberman, G. V. Loza and T. V. Vasina, Dokl. Akad. Nauk. SSSR, 128 (1959) 1188.
18. G. Leclercq, L. Leclercq, and R. Maurel, J. Catal., 50 (1977) 87.

19. Y. Chauvin and J. L. Herrisson, Makromol. Chem., 141 (1971) 151.
20. E. O. Fischer and G. G. Kreiter, Angew. Chem. Int. Ed. Engl., 8 (1969) 761.
21. R. L. Anderson and C. P. Casey, J. Chem. Soc. Chem. Commun. (1975) 895.
22. F. Hugues, B. Besson and J. M. Basset, J. Chem. Soc. Chem. Commun. (1980) 719.
23. B. N. Kuznetsov, Y. I. Yermakov, M. Boudart and J. P. Collman, J. Mol. Catal., 4 (1978) 49.
24. Y. Holl, These d'Etat, Strasbourg (1984); Y. Holl, G. Maire, unpublished results.
25. A. Fritsch, P. Legare and G. Maire, unpublished results.
26. A. F. Simpson and R. Whyman, J. Organometal. Chem., 213 (1981) 157.
27. M. Ichikawa, J. Chem. Soc. Chem. Commun. (1976) 11.
28. O. Zahraa, F. Garin and G. Maire, Faraday Disc. Chem. Soc., 72 (1981) 157.
29. N. S. Kozlov, Y. Ivankovitch and V. S. Molodyanova, Izv. Acad. Nauk. B SSSR, Ser. Khim., 3 (1978) 12.
30. G. Maire, L. Hilaire and O. Zahraa, in Metal-support and Metal-additive Effects in Catalysis, B. Imelik *et al.* (Eds.), Elsevier, Amsterdam (1982).
31. G. Maire, L. Hilaire, O. Zahraa and M. F. Ravet, Proc. Inter. Conf. EXAFS and Near Edge Structures, Frascati (Italy), Springer (1982).
32. P. Esteban Puges, These d'Etat, Strasbourg (1984).
33. P. Esteban Puges, P. Girard, M. J. Ledoux and G. Maire, unpublished results.
34. L. Guczi, Z. Schay, K. Matusek, I. Bogyay and G. Steffler, Proc. 7th Int. Congr. Catal., T. Seiyama and K. Tanabe (Eds.), 1 (1981) 211.
35. O. Zahraa, These d'Etat, Strasbourg (1980); O. Zahraa and G. Maire, unpublished results.
36. M. Hajek, G. Maire, A. O'Cinneide, C. Corolleur and F. G. Gault, J. Chim. Phys., 71 (1974) 1329.
37. F. Weisang and F. G. Gault, J. Chem. Soc. Chem. Commun. (1979) 519.
38. L. Guczi, K. Matusek and M. Eszterle, J. Catal., 60 (1979) 121.
39. L. Guczi, M. Matusek, A. Sarkany and P. Tetenyi, Bull. Soc. Chim. Belg., 88 (1979) 497.
40. G. Collier, D. J. Hunt, S. D. Jackson, R. B. Moyes, I. A. Pickering, P. B. Wells, A. F. Simpson and R. Whyman, J. Catal., 80 (1983) 154.
41. D. J. Hunt, S. D. Jackson, R. B. Moyes, P. B. Wells and R. Whyman, J. Catal., 86 (1984) 333.
42. S. D. Jackson, R. B. Moyes, P. B. Wells and R. Whyman, J. Catal., 86 (1984) 342.
43. D. J. Hunt, R. B. Moyes, P. B. Wells, S. D. Jackson and R. Whyman, Proc. 8th Int. Congr. Catal., 5 (1984) 27.
44. P. Gajardo, P. Grange and B. Delmon, J. Chem. Soc. Faraday I, 76 (1980) 929.
45. F. E. Massoth, Advan. Catal., 27 (1978) 265.
46. A. Brenner and R. L. Burwell, Jr., J. Catal., 52 (1978) 353.
47. R. Nakamura, R. G. Bowman and R. L. Burwell, Jr., J. Am. Chem. Soc., 103 (1981) 673.
48. R. L. Burwell, Jr., J. Catal., 86 (1984) 301.
49. J. H. Sinfelt, Advan. Catal., 23 (1973) 91.
50. W. F. Graydon and M. D. Langan, J. Catal., 69 (1981) 180.
51. P. Esteban Puges, F. Garin, P. Girard, P. Bernhardt and G. Maire, Proc. Iberoamerican Congr. Catal., Lisbon, 1984.
52. Y. Yermakov, Proc. 7th In. Congr. Catal., T. Seiyama and K. Tanabe (Eds.), 1 (1981) 57.
53. C. E. Eady, B. F. G. Johnson and J. Lewis, J. Chem. Soc. Dalton Trans. (1975) 2606.
54. L. Guczi, Faraday Disc. Chem. Soc., 72 (1982) 89.

55. J. J. Rooney, Faraday Disc. Chem. Soc., 72 (1982) 87.
56. C. N. Wilker and R. Hoffmann, J. Am. Chem. Soc., 105 (1983) 5285.

Section 9.5

CO HYDROGENATION CATALYSIS AND SUPPORTED METAL CLUSTERS

H. KNÖZINGER AND B. C. GATES

9.5.1 INTRODUCTION

The catalytic hydrogenation of carbon monoxide gives methane and higher hydrocarbons (the Fischer-Tropsch reaction) and/or oxygenated products, including methanol and higher alcohols. Surfaces of many metals, even single crystals, are active for the former reaction. Supported catalysts such as copper on zinc oxide are active for the latter; some authors have implicated ionic species such as Cu^+ [1] and Rh^+ [2], whereas others have implicated Cu(0) [3], and some have implicated both metal particles and cationic sites (promoter sites) adjacent to the metal on the support [4].

The CO hydrogenation catalyst typically performs several functions: if hydrocarbons are produced, CO dissociation occurs, and if higher hydrocarbons are formed, C-C bond formation ensues. The key surface intermediates are evidently CH_x (where x = 2) [4, 5], and a chain-growth process has been postulated involving insertion of these CH_x moieties into surface-bound alkyl groups. As a consequence of this chain-growth process, products typically follow the Schulz-Flory distribution; much effort has been expended in attempts to find catalysts giving different distributions, e.g., higher selectivities to valuable chemical intermediates such as ethylene and propylene. Metal clusters have been suggested as candidate catalysts that might offer different and improved selectivities [6].

Clusters are of interest as candidate catalysts for several reasons: (1) they have neighboring metal centers and in prospect some of the reactivity characteristics of metal surfaces, which might facilitate CO dissociation and hydrocarbon chain growth; (2) many clusters are stabilized by hydride and, especially, carbonyl ligands, which suggests H_2 and CO as reactants; (3) bimetallic clusters might offer neighboring metal centers with complementary functions; (4) clusters with a variety of hydrocarbon ligands are known, which suggests that the required intermediates could form in the coordination sphere of the metal atoms; and (5) a limited number of neighboring metal centers (as in a cluster rather than a metal surface) might limit the chain growth and contribute to a high selectivity for low-molecular-weight products besides methane. The organometallic literature provides ample evidence of elementary reactions (many involving clusters) that provide models for rationalization of CO hydrogenation catalysis on surfaces or clusters [7, 8].

Many attempts have been made to find soluble metal-cluster catalysts for CO hydrogenation (see Chapter 5); rhodium carbonyl clusters have been implicated in the high-pressure synthesis of ethylene glycol, but the nature of the organometallic intermediates remains in doubt. A number of early reports of catalysis of the Fischer-Tropsch reaction by soluble metal clusters remain unconfirmed; often the metal cluster precursors have been unstable, giving colloidal metal which became the actual catalyst. Metal clusters [e.g., $Ir_4(CO)_{12}$] in the presence of molten salts providing Lewis acid groups ($NaCl/AlCl_3$) apparently are catalyst precursors for the Fischer-Tropsch reaction at 433 K and atmospheric pressure, although the activity is extremely low [9, 10].

The motivations for testing metal clusters on supports as CO hydrogenation catalysts have to do with the potential selectivity advantages mentioned above combined with the potential processing advantages of supported catalysts: ease of separation from products and lack of corrosion. Further, some authors have hoped to maintain cluster nuclearity with partial removal of cluster ligands as a consequence of bonding the cluster to the support. The obvious stumbling block is the lack of stability of metal clusters at high temperatures, perhaps especially those on supports. This lack of stability has been documented in the preceding sections of this chapter; at most there are a few well-documented examples of catalysis by supported metal clusters. We might therefore articulate the primary challenge in this field as identification of the optimum combinations of metal, support, and reactive atmosphere to allow stabilization of metal clusters.

9.5.2 MODELING OF THE FISCHER-TROPSCH REACTION BY MOLECULAR METAL CLUSTERS

As mentioned above, CO hydrogenation to form hydrocarbons is considered to involve three fundamental steps, namely, carbon-oxygen bond scission, carbon-hydrogen bond formation, and carbon-carbon bond formation [4, 5]. Numerous examples of organometallic compounds (particularly clusters) that model intermediates in these reaction steps have been synthesized and structurally characterized. This approach towards elucidation of the mechanism of the Fischer-Tropsch reaction has been reviewed by Muetterties and Stein [8] and by Ugo and Psaro [7]; we therefore mention only a few typical examples of related model compounds here.

The clusters [$HFe_4(CO)_{12}(\mu_3\text{-}\eta^2CO)$] [11] and [$(\eta^5\text{-}C_5H_5)_3Nb_3$-$(CO)_6(\mu_3\text{-}\eta^2\text{-}CO)$] [12] each contain a highly activated carbonyl ligand that can be considered as a model for the side-on adsorbed precursor state for C-O bond scission on metal surfaces (see Chapter 11). In the Fe_4 cluster anion, the activated CO ligand bridges four iron atoms and thus resembles sites at the intersection of two close-packed face-centered cubic metal planes, such as (111) and $(\bar{1}11)$. In contrast, the μ_4-η^2-CO ligand in the cluster $[Co_2Mo_2(\mu_2\text{-}CO)_3(\mu_4\text{-}\eta^2\text{-}CO)(\eta\text{-}C_5H_5)_2(\eta\text{-}C_5Me_5)_2]$ [13] resembles CO adsorbed on coordinatively unsaturated sites of stepped metal surfaces, [e.g., at intersections of (111) and (001) planes of face-centered cubic metal]. The structure of this cluster is as follows:

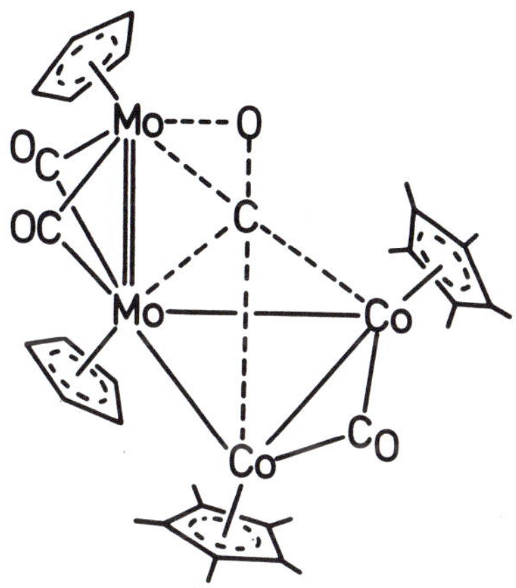

Carbonyl metal clusters can be transformed under various conditions into carbido metal clusters and thus mimic the dissociative chemisorption of CO on metal surfaces; examples are the following [14]:

$$Fe_3(CO)_{12} \longrightarrow Fe_6(CO)_{17}C + Fe_5(CO)_{15}C \qquad (1)$$

$$Ru_3(CO)_{12} \longrightarrow Ru_6(CO)_{17}C + CO_2 \qquad (2)$$

The carbon-hydrogen bond formation by reaction of H_2 with carbido clusters having "exposed" carbon atoms to yield methylidyne clusters has been demonstrated [15, 16]:

$$[Fe_4C(CO)_{13}] \xrightarrow{H_2} [HFe_6(CO)_{12}CH] \qquad (3)$$

The possible interconversion of CH_x species (x = 1, 2, or 3) was shown by Calvert and Shapley [17], who obtained a mixture of the methyl species $H(CH_3)Os_3(CO)_{10}$ and of the bridging methylene species $H_2CH_2Os_3(CO)_{10}$, when they reacted $H_2Os_3(CO)_{10}$ with CH_2N_2. Heating of the methylene and methyl species produced $(H)_3(HC)Os_3(CO)_9$. Steinmetz and Geoffroy [18] have shown the stepwise reduction of a carbonyl ligand in $Os_3(CO)_{12}$ to give CH_4 (SCHEME 1):

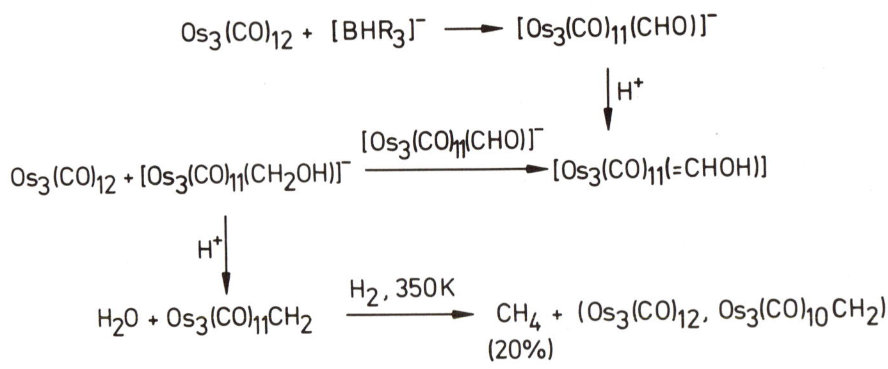

SCHEME 1

Formation of higher hydrocarbon can be envisaged by coupling reactions of CH_x intermediates in which carbenes can be involved, as shown by Brady and Pettit [19]. Coupling of μ-methylene groups in di-(μ-methylene)-dicobalt complexes has been demonstrated by Laws and Puddephat [20] (Scheme 2):

Longoni *et al.* [21] observed the formation of ethane when the cluster anion $[Co_3Ni_9C(CO)_2]^{3-}$ was treated in CO + H_2 at 300 K and 1 bar. They suggested that two carbidic carbon atoms might dimerize inside a metal cage, and the intermediate formation of $[Co_3Ni_7(C\text{-}C)(CO)_5]^{3-}$ was tentatively envisaged.

Alternative chain growth mechanisms could involve CO insertion into metal-alkyl bonds. Numerous examples of this reaction step in the coordination spheres of molecular clusters have been described [7, 8]. A large series of osmium clusters having small oxy ligands and their binding and rearrangements have been discussed by Deeming [22]. Oxygen coordination, which was considered likely to be of importance in catalytic intermediates on metal surfaces, occurs in these compounds.

Formyl complexes have been implicated as key intermediates for the methanol synthesis from CO and H_2. Among a variety of known cluster compounds having formyl ligands, the cluster $[Rh_{17}(CO)_{32-n}(S)_2(CHO)_n]^{(3+n)-}$ is perhaps of special interest as it is derived from $[Rh_{17}(CO)_{32}(S)_2]^{3-}$, which was found to be the most abundant cluster present during Rh-catalyzed homogeneous conversion of carbon monoxide and hydrogen into alcohols [23].

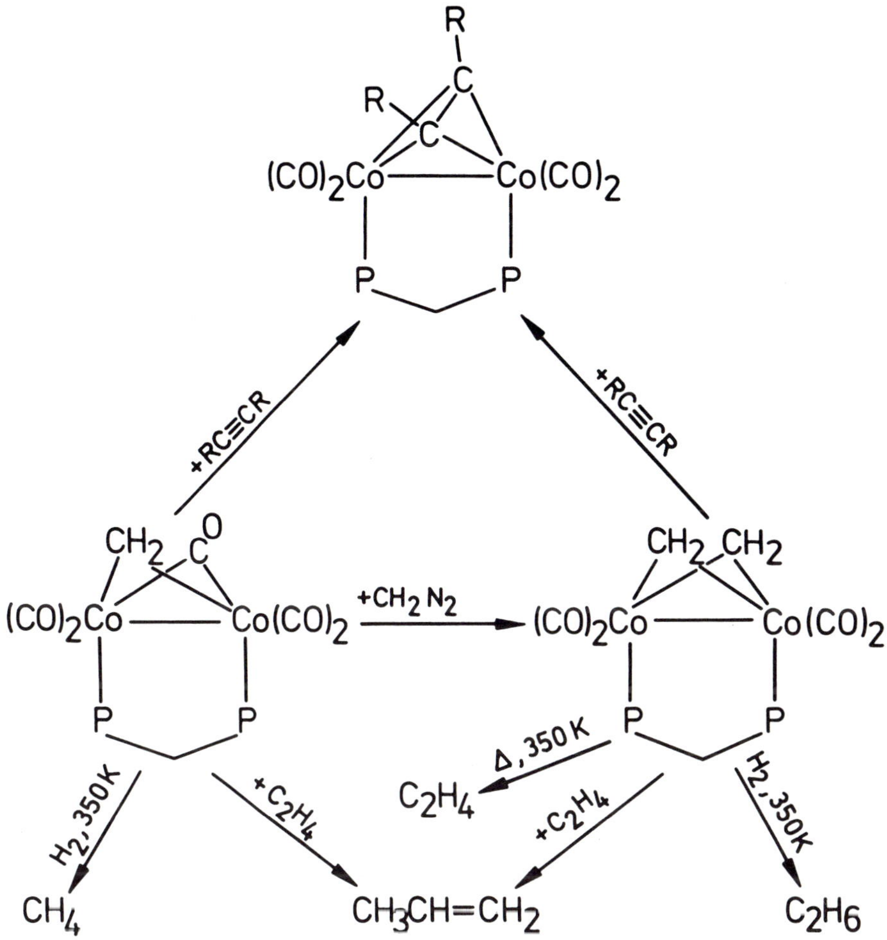

SCHEME 2

In conclusion, there are many examples of well-characterized cluster compounds which can serve as models of intermediates that may occur during catalytic hydrogenation of carbon monoxide.

9.5.3 STOICHIOMETRIC REACTIONS OF METAL CLUSTERS ON OXIDE SURFACES

The reactivities of metal carbonyl clusters with the surfaces of metal oxides are documented in Section 9.2. We mention just a few of the pertinent results that have been used as a basis for the suggestion that these surface reactions are related to the Fischer-Tropsch reaction [10].

The reactions initially lead to destruction of the clusters and formation of methane and higher hydrocarbons (alkenes and alkanes), CO_2, and H_2 [24, 25]. Ethylene is a major product of the initial decomposition of $Rh_6(CO)_{16}$ on alumina, for example, which suggests the possibility that ethylene is a primary product of the catalytic Fischer-Tropsch reaction; a development of these ideas is presented elsewhere [24].

9.5.4 EVIDENCE OF CATALYSIS BY MAGNESIA-SUPPORTED MOLECULAR OSMIUM CLUSTERS

Magnesium oxide is unique as a support for metal clusters (Section 9.2.4.4), and Deeba *et al.* [27] suggested that they had observed molecular catalysis of CO hydrogenation by osmium carbonyl clusters on magnesia. The catalyst was prepared from $Os_3(CO)_{12}$ and magnesia; the infrared spectrum of the initial supported species was suggestive of a triosmium cluster; when this was oxidized, a stable mononuclear complex was formed (Section 9.2.4.4). When either form of the catalyst was introduced into the Fischer-Tropsch reactor, an induction period was observed, followed by a period of nearly steady state activity at 573 K and 32 bar; the activity and product distribution were the same, independent of the initial form of the catalyst. An infrared spectrum of the used catalyst was suggestive of a molecular osmium carbonyl, but a definitive structural inference was not possible. Deeba *et al.* [27] suggested that a molecular osmium cluster was formed on the surface during catalysis and was responsible for the catalytic activity.

A related catalyst was prepared by adsorption of $H_2Os(CO)_4$ on magnesia, the surface species being the anion $HOs(CO)_4^-$, evidently bonded to the surface at a Lewis acid site (Mg^{2+} site) with a carbonyl oxygen being a lone pair electron donor [28a]. The anion could be extracted from the surface as a salt, and its infrared spectrum confirmed its structure. This material was catalytically active for the Fischer-Tropsch reaction at 548 K and 10 bar, giving more-or-less a Schulz-Flory distribution of products at low conversion, and losing activity rapidly for about the first 10 h on stream and slowly thereafter. The initial activity was 1.5×10^{-4} CO molecules/Os atom•s, an order of magnitude less than that of a highly dispersed metal catalyst prepared from $Os(CO)_5$ [29].

The used catalyst derived from $H_2Os(CO)_4$ was red-brown, whereas catalysts having metal particles are typically black; the infrared spectrum was different from that of the initial form. When the used catalyst was brought in contact with a solution of $(CH_3)_4NCl$ in isopropanol, $H_3Os_4(CO)_{12}^-$ was extracted from the surface. The remaining solid was red, and its infrared and UV-visible spectra were nearly identical to those of the cluster $Os_{10}C(CO)_{24}^{2-}$ in solution or on the surface. This large cluster is not appreciably soluble in isopropanol, and it is highly stable, being synthesized from $Os_3(CO)_{12}$ in the presence of sodium metal at 533 K [30].

The authors [28a] inferred that molecular organometallic chemistry occurred on the magnesia surface, paralleling that occurring in solution. The keys to stabilization of the clusters at high temperature were evidently (1) the tendency of osmium to form stable carbonyl clusters, (2) the strongly basic nature of the magnesia, stabilizing anions on the surface, and (3) the presence of CO stabilizing the clusters. The nature of the catalytic species remains to be determined, but the authors speculated that $H_3Os_4(CO)_{12}^-$ might be involved, with the $Os_{10}C(CO)_{24}^{2-}$ being inactive and associated with the catalyst deactivation.

When magnesia-supported osmium catalysts were prepared conventionally (by aqueous impregnation of chloroosmic acid and reduction) and used for the Fischer-Tropsch synthesis, the osmium metal particles initially present on the support were partially converted into $Os_{10}C(CO)_{24}^{2-}$ [28b]. Evidently, this is the thermodynamically favored species; the results suggest that molecular clusters may play a much greater role in supported metal catalysis than has ever been suspected.

The importance of these results is the indication of molecular cluster chemistry on the support at high temperature (548 K) and during catalysis of an industrially important reaction. This is perhaps only the beginning of investigation of stable, supported molecular metal cluster catalysts.

There is some other evidence of simple molecular organometallic chemistry on catalyst surfaces, Pierantozzi *et al.* [31] investigated CO hydrogenation catalysis at 480-600 K and 2 x 10^3 kPa with alumina- and magnesia-supported ruthenium carbonyls [$H_4Ru_4(CO)_{12}$ and $Ru_3(CO)_{12}$] as the precursors. Infrared spectra of used catalysts indicated the presence of the stable cluster $[Ru_6C(CO)_{16}]^{2-}$, in a mixture with other structures, evidently including metal particles. The role of clusters in the catalysis remain to be elucidated.

Meyers and Hall [32] characterized a used supported cobalt catalyst (derived from $Co_4(CO)_{12}$) with laser-induced mass spectrometry (LIMS), finding evidence for cobalt clusters of the type $Co_4(CO)_x$ (x = 2-6) and cluster carbides, $Co_4(CO)_xC$ (x = 0, 1, 2, 4, 11), among other structures. We might suggest that the thermodynamically stable metal carbonyls form on the surfaces of many Fischer-Tropsch catalysts; this suggestion is borne out by the apearance of volatile metal carbonyls in product streams, observed by a number of authors [26].

9.5.5 CONVENTIONAL SUPPORTED METAL CATALYSTS PREPARED FROM METAL CLUSTERS

In contrast to the few examples discussed above, almost all of the supported CO hydrogenation catalysts prepared from metal clusters lose their original metal framework structures and undergo aggregation to give cyrstallites of supported metal; the same generalization pertains to the hydrocarbon conversion catalysts mentioned in the preceding sections of this chapter. We therefore consider this topic only superficially, emphasizing the ties to molecular clusters on oxide supports.

Ichikawa [33] prepared a series of supported rhodium catalysts from metal oxides and clusters including $Rh_4(CO)_{12}$. The catalysts exhibited a wide range of selectivities, some (e.g., Rh/SiO_2) giving Schulz-Flory distributions of hydrocarbons [34], some (e.g., Rh/MgO, Rh/CaO) giving high yields of methanol [34], and some (e.g., Rh/TiO_2, Rh/La_2O_3) giving high yields of ethanol, together with methanol [35]. The metal cluster precursor was not necessarily responsible for the unique catalytic character, since highly dispersed rhodium produced in more conventional ways also gives catalysts with similar selectivities [35]. The support character (e.g., the basicity) seems to be important; some authors [2, 36] have associated activity for methanol formation with the presence of Rh^+ centers. The formation of C_2 and higher oxygenates seems to occur via CO insertion reactions into $M-CH_x$ bonds and thus requires a sufficient supply of CH_x groups, which is typically the case on Rh catalysts but not on Pd catalysts [37]. Catalysts promoted with cations such as Mn^{x+} may facilitate CO dissociation by an interaction of the lone pair on the oxygen with the promoter cation on the support, with the carbon being bonded to a metal particle near its interface with the support [4, 38, 39]:

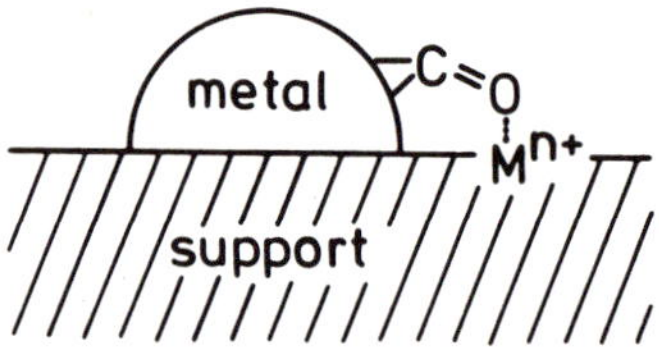

Modification of SiO_2 and ZrO_2 surfaces with alkali metal ions (Li^+, Na^+, K^+) gave selectivities >90% for methanol in the conversion of CO + H_2 on Rh catalysts derived from $Rh_4(CO)_{12}$ [40].

It should be recalled in this context that mononuclear surface Rh^+ carbonyl species can be formed in the presence of gas-phase CO even from small metal aggregates [41] (Section 6.2.3.7).

Catalysts prepared from $Fe_3(CO)_{12}$ on alumina have been inferred to give chemisorbed $HFe_3(CO)_{11}^-$; the infrared spectra suggest that this anionic cluster is bonded to surface Lewis acid centers (Al^{3+} ions) through the lone pair of the bridging carbonyl ligand [42]. The iron clusters are easily decomposed on alumina and magnesia to give very small iron particles; these initially have high selectivities for propylene formation [43], but it is not known whether the catalyst performance can be maintained.

Kuznetsov *et al.* [44] reported increased selectivities towards higher hydrocarbons when iron catalysts (the supports were SiO_2, Al_2O_3, MgO, TiO_2, and ZrO_2) were modified with manganese. Similar effects were observed when the bimetallic cluster $(NEt_4)_2[Fe_2Mn(CO)_{12}]$ was used as a catalyst precursor.

A series of organoosmium complexes and clusters has been used to prepare oxide-supported CO hydrogenation catalysts. The catalysts are typically highly dispersed metals, and, according to recent reports (discussed in Section 9.5.4), they also include molecular clusters.

Alumina-supported catalysts were made by adsorption from paraffinic solvents of $Os(CO)_5$ [29] and $Os_3(CO)_{12}$, $H_4Os_4(CO)_{12}$, and $Os_6(CO)_{18}$ [45, 46]. The catalysts were pretreated in hydrogen prior to the beginning of flow of CO + H_2 to the catalytic reactor; they gave Schulz-Flory distributions of hydrocarbons, being tested only at low conversions. Characterization of the used catalysts by high-resolution electron microscopy typically showed small crystallites of osmium on the support (Table 1) [46]. Os(II) centers (the only ones detected by infrared spectroscopy in used catalysts having low Os loadings) have been suggested to be involved in the catalytic sites; metallic osmium is regarded as the likely catalytic species [24, 46], with Os(II) centers or Al^{3+} sites on the support possibly assisting in the dissociation of CO, as shown schematically above [29, 45]. The larger Os particles were more active than the smaller ones; this result confirms the particle size effects observed for Fe and Ru.

Catalysts prepared from osmium carbonyl clusters on silica also form metal particles, evidently more easily than those prepared with alumina as the support [7, 24].

Effects of the nature of the support and of the pretreatment conditions on the catalytic performance of materials derived from $Os_3(CO)_{12}$ have been observed [26], as shown in Table 2. These catalysts, however, have not yet been sufficiently well characterized to establish firm correlations between support properties and catalytic behavior.

As in the case of rhodium, disintegration of highly dispersed particles in the presence of carbon monoxide has also been observed for alumina-supported osmium catalysts [47]; Os(II) carbonyl complexes were formed.

TABLE 1. γ-Al_2O_3-Supported CO Hydrogenation Catalysts.

Catalyst Precursor	wt % Os	Approx. avg. particle size determined by TEM, nm	$10^4 r_{HC}$[a] mol/(mol of total Os·s)	r_{HC}/r_{ether}[b]	Ref.
$Os(CO)_5$	0.36	≤2.5[c]	10.4	___[d]	[29]
$Os(CO)_5$	0.36	≤1.7[e]	2.5	___[d]	[29]
$Os_3(CO)_{12}$	1.8	1.0-2.0 + large particles	1.9	25	[46]
$H_4Os_4(CO)_{12}$	1.8	1.0	4.1	50	[46]
$Os_6(CO)_{18}$	1.8	1.0-2.0	10	160	[46]
H_2OsCl_6	5	7.0	18	490	[46]

[a]Rate of total hydrocarbon formation at 548 K and 10 bar with a H_2:CO molar ratio of 1.
[b]Rate of total hydrocarbon formation divided by rate of dimethyl ether formation at 548 K and 10 bar.
[c]Appreciable aggregation of Os was reported [29].
[d]Small amounts of ether were observed but not determined quantitatively [29].
[e]Only few aggregates observed [29].

TABLE 2. Product Distributions of $CO + H_2$ Reaction in the Presence of $Os_3(CO)_{12}$-derived Catalysts.[a]

Support	Activated in	Reaction Temp./K	% CO conversion	% HC's[b] formed	% CH_4	% C_2H_4	% C_2H_6
Al_2O_3	He	576	9.3	84.3	75.6	1.1	7.6
	H_2	574	2.5	91.5	60.5	15.5	15.5
SiO_2	He	576	0.6	82.1	82.1	--	--
	H_2	583	1.7	89.9	89.9	--	--
ZrO_2	He	583	1.3	80.9	51.5	29.4	--
	H_2	583	1.5	85.3	37.3	21.3	26.7
TiO_2	He	582	1.7	83.3	27.8	11.1	44.4
	H_2	587	0.4	75.0	25.0	50.0	--
CeO_2	He	580	11.7	38.1	15.9	19.0	3.2
	H_2	580	7.4	27.6	9.2	18.4	--
ThO_2	H_2	589	11.6	10.7			
La_2O_3	He	588	6.2	44.0	32.2	--	11.8
	H_2	581	6.8	46.6	30.1	6.6	9.9
Cr_2O_3	He	588	6.2	44.0	32.2	--	11.8
	H_2	581	6.8	46.6	30.1	6.6	9.9
Nb_2O_5	He	581	1.8	54.3	10.9	43.5	--
	H_2	640	2.8	72.4	72.4	--	--

[a]Catalyst activation temperature-programmed decomposition in He or H_2 up to 770 K; CO hydrogenation at atmospheric pressure with He:H_2:CO = 4:2:1; total flow rate 65 cm^3 NTP/min.

[b]Percentage of CO transformed into hydrocarbons; balance: CO_2 formation.

Bimetallic clusters (also see Chapter 10) have been used to prepare supported CO hydrogenation catalysts: $H_2FeOs_3(CO)_{13}$ on silica [48], $H_2FeOs_3(CO)_{13}$ on γ-alumina [49], $H_2RuOs_3(CO)_{12}$ on γ-alumina [49], and $H_2RhOs_3(CO)_{10}(acac)$ on γ-alumina [50]. Each of these clusters broke apart on the surface prior to or during catalysis; evidently, the stability of the osmium triangle is sufficient to maintain triosmium clusters intact initially after the tetrahedral cluster frames have been destroyed. Each of the used catalysts contained segregated metals, the osmium on alumina typically being detected by infrared and X-ray photoelectron spectroscopy as Os(II) complexes, the rhodium as small metal particles, and the ruthenium as a mixture of cationic complexes and small metal particles. In the latter two catalysts, the hydrocarbon formation activity was attributed predominantly to the metal particles, with the small observed methanol formation activity tentatively associated with the cationic species, which may play the role of promoters suggested above.

In summary, these bimetallic clusters are not sufficiently stable to survive under catalytic conditions, and they appear to offer no advantages in catalyst performance. However, the incorporation of the promoter metal into a cluster with the catalytic metal appears to offer a novel way of producing highly dispersed supported catalysts having the two metals near to each other. McVicker and Vannice [51] prepared alumina- and silica-supported catalysts of Group VIII metals using precursor clusters such as e.g., $K_2Ru_3(CO)_{12}$. They found that several catalysts derived from such salts were more active and exhibited higher selectivities for C_2-C_5 olefins than conventionally prepared potassium-Group VIII metal catalysts (independent two-step impregnation with Group VIII metal and potassium salt) of the same metal stoichiometry. It was suggested that the use of preformed carbonyl complexes or clusters can increase any promotional effect that potassium can exert on supported Group VIII metals.

9.5.6 METAL AGGREGATES IN ZEOLITE CAGES

The idea of limiting chain growth suggests the application of metal clusters or small metal aggregates for CO hydrogenation. The molecular-scale cages of zeolites offer an opportunity for limiting metal aggregate size, and several researchers have prepared such catalysts (containing Ru, Co, and Fe) and found them to have unique selectivities for low-molecular-weight hydrocarbons—at low conversion and low temperature; this work is reviewed by Jacobs (Chapter 8) and by Ozin and Andrews (Chapter 7). The main concern about these catalysts is—again—stability: the metals easily migrate out of the zeolite cages to give metal particles on the surfaces of the zeolite crystals, and these act like conventional (unselective) catalysts.

9.5.7 CONCLUSIONS

Almost all catalysts derived from molecular clusters which have been well characterized after use for CO hydrogenation give evidence for the presence of highly dispersed metal aggregates and/or mononuclear complexes; structurally well-defined clusters (molecular clusters) have been found only rarely. There are a few examples which clearly demonstrate that unique molecular clusters (such as, for example, $[H_3Os_4(CO)_{12}]^-$ and $[Os_{10}C(CO)_{24}]^{2-}$ [28] or $[Ru_6C(CO)_{16}]^{2-}$ [31]) can be formed on suitable support surfaces under synthesis conditions. Nicholls *et al.* [52] have shown that a wide variety of carbonyl, hydrido carbonyl, and carbido clusters of osmium and ruthenium can be formed and interconverted at high CO or H_2 partial pressures as well as in CO + H_2. It has also been shown that small metal aggregates can break up in the presence of CO under rather mild conditions to form mononuclear carbonyl complexes [41, 47] and that molecular clusters can be formed spontaneously on support surfaces [28b]. We infer that under synthesis conditions, delicate equilibria are established between mononuclear complexes, metal clusters, and metal aggregates. The proportions ot individual species are expected to be critically dependent on the nature of the metal and the support, the presence of additives or promoters, the nature of the gas atmosphere, and the pressure and temperature. Catalytic activities and selectivities would then be governed by such equilibria.

REFERENCES

1. K. Klier, in Heterogeneous Catalysis, B. L. Shapiro, Ed., Texas A & M University Press, College Station, Texas, 1984, p. 252.
2. P. R. Watson, and G. A. Somorjai, J. Catal., 66 (1980) 257; 72 (1981) 347; 74 (1982) 282.
3. T. H. Fleisch and R. L. Mieville, J. Catal., 90 (1984) 165.
4. W. M. H. Sachtler, Proc. 8th Int. Congr. Catal. Vol. 1, p. 151, Dechema, Frankfurt, 1984.
5. P. Biloen and W. M. H. Sachtler, Advan. Catal., 30 (1981) 165.
6. E. L. Muetterties, Bull. Soc. Chim. Belg., 84 (1975) 959; 85 (1975) 451.
7. R. Ugo and R. Psaro, J. Mol. Catal., 20 (1983) 53.
8. E. L. Muetterties and J. Stein, Chem. Rev., 79 (1979) 479.
9. G. C. Demitras and E. L. Muetterties, J. Am. Chem. Soc., 99 (1977) 2796.
10. H.-K. Want, H. W. Choi and E. L. Muetterties, Inorg. Chem., 20 (1981) 2661.
11. M. Manassero, M. Sandoni and S. Longoni, J. Chem. Soc. Chem. Commun. (1976) 919.
12. W. A. Herrmann, H. B. Biersack, M. L. Ziegler, K. Weidenhammer, R. Siegel and D. Rehder, J. Am. Chem. Soc., 103 (1981) 1692.

13. P. Bonn, G. M. Dawkins, M. Green, A. D. Miles, A. G. Ozen and F. G. A. Stone, J. Chem. Soc. Chem. Commun. (1982) 926.
14. M. Tachikawa and E. L. Muetterties, Progr. Inorg. Chem., 28 (1981) 203.
15. K. H. Whitemire and D. F. Shriver, J. Am. Chem. Soc., 102 (1980) 1456.
16. E. M. Holt, K. H. Whitemire and D. F. Shriver, J. Organomet. Chem., 213 (1981) 125.
17. R. B. Calvert and J. R. Shapley, J. Am. Chem. Soc., 99 (1977) 5225.
18. (a) G. R. Steinmetz and G. L. Geoffroy, J. Am. Chem. Soc., 103 (1981) 1278; (b) G. R. Steinmetz, E. D. Morrison and G. L. Geoffroy, J. Am. Chem. Soc., 106 (1984) 2559; (c) E. D. Morrison, G. R. Steinmetz, G. L. Geoffroy, W. C. Fultz and A. L. Rheingold, J. Am. Chem. Soc., 106 (1984) 4783.
19. R. C. Brady, III, and R. Pettit, J. Am. Chem. Soc., 102 (1980) 6182; and 103 (1981) 1287.
20. W. J. Laws and R. J. Puddephatt, J. Chem. Soc. Chem. Commun. (1984) 116.
21. G. Longoni, A. Cerotti, R. Della Pegola, M. Manassero, M. Perego, H. Piro and M. Sansoni, Phil. Trans. R. Soc. London (A), 308 (1982) 47.
22. A. J. Deeming, J. Mol. Catal., 21 (1983) 25.
23. R. C. Schoening, J. L. Vidal and R. A. Fiato, J. Mol. Catal., 13 (1981) 83.
24. J. M. Basset, B. Besson, A. Choplin, F. Hugues, M. Leconte, D. Rojas, A. K. Smith, A. Theolier, Y. Chauvin, D. Commereuc, R. Psaro, R. Ugo and G. M. Zanderighi in Fundamental Research in Homogeneous Catalysis, Vol. 4 (M. Graziani and M. Giongo, eds.), Plenum, New York, 1984, p. 19.
25. A. Brenner in Metal Clusters, M. Moskovits, ed., Wiley, New York, 1986.
26. E. Alizo Delgado and H. Knözinger, unpublished results.
27. M. Deeba, J. P. Scott, R. Barth, and B. C. Gates, J. Catal., 71 (1981) 373.
28. (a) H. H. Lamb and B. C. Gates, J. Am. Chem. Soc., 108 (1986) 81; (b) H. H. Lamb, T. R. Krause and B. C. Gates, J. Chem. Soc. Chem. Commun. (1986) in press.
29. E. O. Odebunmi, Y. Zhao, H. Knözinger, B. Tesche, W. H. Manogue, B. C. Gates and J. Hulse, J. Catal., 86 (1984) 95.
30. T. C. Hayward and J. R. Shapley, Inorg. Chem., 21 (1982) 3816.
31. R. Pierantozzi, E. G. Valagene, A. F. Nordquist and P. N. Dyer, J. Mol. Catal., 21 (1983) 189.
32. G. F. Meyers and M. B. Hall, Inorg. Chem., 23 (1984) 124.
33. M. Ichikawa, CHEMTECH, 12 (1982) 674.
34. M. Ichikawa, Bull. Soc. Chem. Jpn., 51 (1978) 2268.
35. M. Ichikawa, Bull. Soc. Chem. Jpn., 51 (1978) 2273.
36. J. R. Katzer, A. W. Sleight, P. Gajardo, J. B. Michel, E. F. Gleason and S. McMillan, Faraday Disc. Chem. Soc., 72 (1981) 121.
37. T. L. F. Favre, G. van der Lee and V. Ponec, J. Chem. Soc. Chem. Commun. (1985) 230.
38. M. Kawai, M. Uda and M. Ichikawa, J. Phys. Chem., 89 (1985) 1654.
39. M. Ichikawa and T. Fukushima, J. Phys. Chem., 89 (1985) 1564.
40. H. A. Dirkse, P. W. Lednor and P. C. Versloot, J. Chem. Soc. Chem. Commun. (1982) 814.
41. H. F. J. van't Blik, J. B. A. D. van Zon, T. Huizinga, G. C. Vis, D. C. Koningsberger and R. Prins, J. Phys. Chem. 87 (1983) 2264.
42. F. Hugues, J. M. Basset, Y. Ben Taarit, A. Choplin, M. Primet, D. Rojas and A. K. Smith, J. Am. Chem. Soc., 104 (1982) 7020.

43. D. Commereuc, Y. Chauvin, F. Hugues, J. M. Basset and D. Olivier, J. Chem. Soc. Chem. Commun. (1980) 154.
44. V. L. Kuznetsov, A. F. Danilyuk, I. E. Kolosova and Yu. I. Yermakov, React. Kinet. Catal. Lett., 21 (1982) 249.
45. H. Knözinger, Y. Zhao, B. Tesche, R. Barth, R. Epstein, B. C. Gates and J. P. Scott, Faraday Disc. Chem. Soc., 72 (1981) 53.
46. E. O. Odenbunmi, B. A. Matrana, A. K. Datye, L. F. Allard, Jr., J. Schwank, W. H. Manogue, A. Hayman, J. H. Onuferko, H. Knözinger and B. C. Gates, J. Catal., 95 (1985) 370.
47. G. Kunzmann, G. Ertl, B. Tesche and H. Knözinger, unpublished results.
48. A. Choplin, M. Leconte, J. M. Basset, S. G. Shore and W.-L. Hsu, J. Mol. Catal., 21 (1983) 389.
49. J. R. Budge, B. F. Lücke, J. P. Scott and B. C. Gates, Proc. 8th Int. Congr. Catal., Vol. 5, p. 89, Dechema, Frankfurt, 1984.
50. J. R. Budge, B. F. Lücke, B. C. Gates and J. Toran, J. Catal., 91 (1985) 272.
51. G. B. McVicker and M. A. Vannice, J. Catal., 63 (1980) 25.
52. J. N. Nicholls, D. F. Farrar, P. F. Jackson, B. F. G. Johnson and J. Lewis, J. Chem. Soc., Dalton Trans. (1982) 1395.

Chapter 10

SUPPORTED BIMETALLIC CATALYSTS DERIVED FROM MOLECULAR METAL CLUSTERS

L. GUCZI

10.1 INTRODUCTION

Bimetallic catalysts have attracted the interest of research workers in heterogeneous catalysis for many years [1, 2]. Effects of alloying on catalytic properties have been shown to be due to geometric (ensemble) and electronic (ligand) effects [3]. Supported bimetallic catalysts are conventionally prepared by impregnation from aqueous salt solutions and H_2-reduction at elevated temperatures. The final dispersion and the composition of bimetallic particles obtained by this approach are already predetermined during the impregnation step, where homogeneous distribution of the two metal precursors can be achieved only if they have similar or identical adsorption properties on the support surface [4, 5]. As this is difficult to a control (cationic and anionic precursors are frequently used in coimpregnation), inhomogeneous compositions of the individual particles can result. Additional complications encountered in the preparation of these catalysts are broad particle size distributions, physical separation of the two metals, surface segregation, incomplete reduction of one or both metal components, etc. Also, as shown by Ollis [6], the thermodynamics, and hence the phase compositions, of small metal particles (< 10 nm) can be significantly different from those of bulk alloys.

These difficulties have prompted research on catalysts prepared from molecular bimetallic clusters. Besides the general advantages of molecular cluster precursors discussed in previous chapters of this book, bimetallic cluster compounds are interesting in prospect since they provide catalyst precursors of well-defined stoichiometry and structure. Even though the cluster framework may break up during thermal treatment and catalysis, there is still the hope that the two metal components remain in intimate contact on the support surface. Moreover, molecular cluster compounds have been synthesized in recent years which combine two metals (e.g., Cu and Ru) [7] that do not form bulk alloys. These materials could offer the opportunity for preparing new catalysts with unique properties.

10.2 PREPARATION OF CLUSTER-DERIVED BIMETALLIC CATALYSTS

10.2.1 Preparation of bimetallic catalysts by deposition of carbonyl clusters on oxides

The organometallic surface chemistry involved in interactions of molecular carbonyl clusters with hydroxylated and dehydroxylated oxide surfaces (interaction with hydroxyl groups, adduct formation, oxidative cluster degradation, etc.) has been discussed in detail in Section 9.2. A few examples have also been reported in which metal particles were first produced by conventional methods and bimetallic catalysts were then obtained by subsequent deposition of carbonyl compounds. Gallezot [15] and Tri *et al.* [16] ion exchanged platinum into synthetic NaY zeolite. Calcination in oxygen and reduction in hydrogen resulted in the formation of Pt particles (1.0 nm average diameter) located in the supercages. Pt-Mo/zeolite catalysts were then prepared by subsequent adsorption and decomposition of $Mo(CO)_6$. Hughes *et al.* [17] deposited platinum pentane-2,4-dionate onto alumina. After reduction in hydrogen, $Fe_3(CO)_{12}$ was adsorbed on this alumina-supported platinum, giving the supported bimetallic catalyst.

10.2.2 Preparation of bimetallic catalysts from organometallic compounds anchored to the support

In preparation of bimetallic catalysts, one of the most interesting methods is to form chemical bonds between surface hydroxyl groups of silica gel or alumina and the reacting metal compounds [18, 19]. There are two fundamental routes to the preparation of bimetallic catalysts:

(1) An organometallic complex containing heteroatomic metal-metal bonds is anchored onto an oxide support as represented by the SCHEME 1 [20].

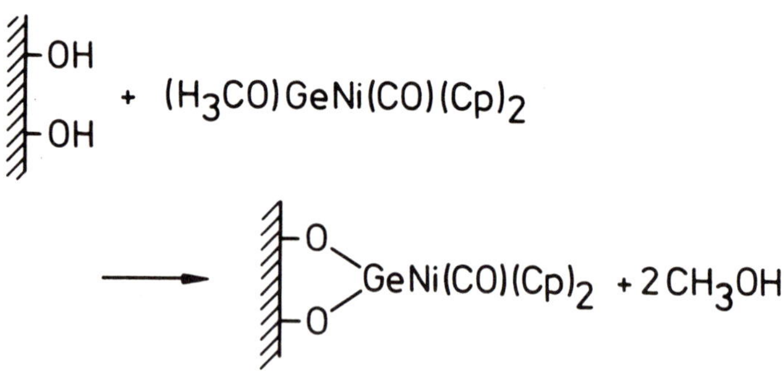

SCHEME 1

The bimetallic surface complex can be decomposed by heating in hydrogen.

(2) Organometallic compounds are reacted with low-valent surface ions (SCHEME 2) [20]:

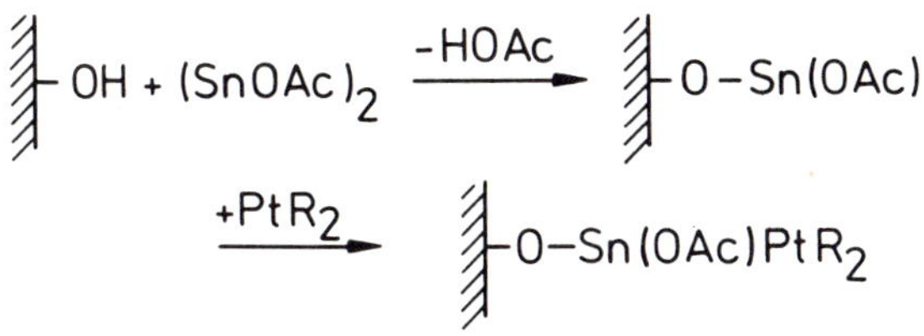

SCHEME 2

Low-valent ions can be prepared either by anchoring a complex in which a metal ion of low oxidation state is initially present or by partial reduction of surface metal ions. Examples follow.

10.2.3 Preparation of bimetallic catalysts on functionalized supports

The synthesis of supported clusters using functionalized oxides or polymers has been described in Section 9.1. The main advantage of this method is that the heteronuclear metal-metal bonds already formed in the molecular cluster may be maintained, and the cluster structure can be stabilized by the ligands in some cases.

10.3 CHARACTERIZATION AND CATALYTIC ACTIVITY OF BIMETALLIC CATALYSTS DERIVED FROM MOLECULAR CLUSTERS

A full characterization of this class of catalysts requires knowledge of the structure of the bimetallic molecular clusters after impregnation, of the processes which control the stability of the supported molecular clusters during mild and high-temperature treatment, and the effect of the support and the ambient atmosphere on their decomposition. The nature of the metal species must be known if correlations between catalytic activity and selectivity and structure and composition of the active species are to be made. Therefore, interactions of molecular clusters with support surfaces after impregnation, mild decomposition, and high-temperature treatment are discussed first, and the catalytic properties follow.

10.3.1 Bimetallic catalysts derived from molecular bimetallic clusters containing iron, ruthenium, or osmium

10.3.1.a Interaction of molecular bimetallic cluster with supports.

Supported clusters have been characterized by several methods such as infrared and Mössbauer spectroscopy and temperature programmed spectroscopies, TPDE, TPR, and TPD (see Sections 6.2, 6.6, and 6.9).

A comparison of the infrared spectra of $Fe_3(CO)_{12}$ supported on Cab-O-Sil and in hexane solution [21] shows a shift by 12 cm^{-1} of the terminal CO stretching frequencies to higher wavenumbers, whereas the frequency of bridging CO is shifted to lower wavenumbers by about 50 cm^{-1}. In the Mossbauer spectra measured at 1.6 K, a superposition of the spectrum for the original iron dodecacarbonyl and that for finely dispersed iron oxide is observed. The oxide is characterized by a hyperfine magnetic field of 454 kOe, which is less than that found for α-Fe_2O_3 (515 kOe) and close to that characteristic of β- or γ-FeOOH (475 and 460 kOe, respectively). On the basis of these results, combined with additional evidence, the following structure has been suggested for the species formed from $Fe_3(CO)_{12}$ on the support surface:

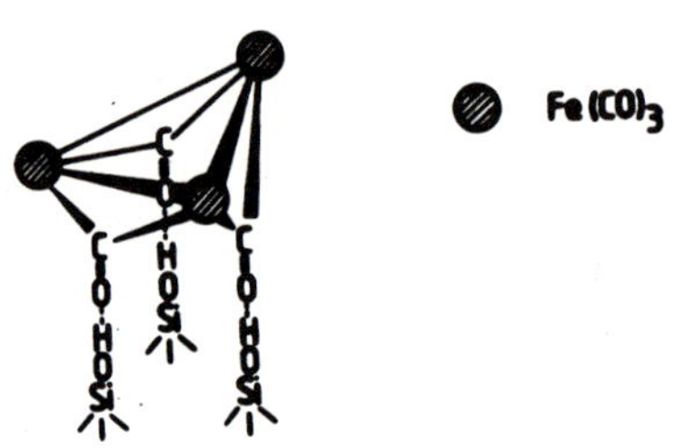

A fraction of the clusters are decomposed, leading to the formation of iron oxide.

With $Ru_3(CO)_{12}$, the interaction is much weaker, as indicated by the small shift of the terminal CO stretching frequencies (about 4 cm^{-1}) to higher wavenumbers. No interaction occurs between $Fe_3(CO)_{12}$ and $Ru_3(CO)_{12}$ when a mixture of these clusters is supported on Cab-O-Sil. The infrared spectrum of this material is simply a superposition of the two individual spectra [22].

When bimetallic molecular clusters such as $Fe_2Ru(CO)_{12}$ and $H_2FeRu_3(CO)_{13}$ were adsorbed on Cab-O-Sil [23], the infrared and Mossbauer spectra showed the molecular structure which is present in solution; hence, it is inferred that no disintegration of the molecular cluster framework occurred upon adsorption. A similar conclusion was reported [24] for $H_2FeOs_3(CO)_{13}$ on silica: the molecular cluster was physisorbed after impregnation as shown by infrared and Raman spectroscopy (see Section 9.2).

Alumina and magnesia supports behave differently. Upon adsorption, cluster anions are often formed, e.g., from $Fe_3(CO)_{12}$ [25], whereas neutral clusters on alumina form from $Ru_3(CO)_{12}$ [26] and $Os_3(CO)_{12}$ [27] (Section 9.2). The bimetallic cluster $H_2RuOs_3(CO)_{13}$ forms an anionic surface species by simple deprotonation under mild conditions, namely $Al^+[HRuOs_3(CO)_{13}]^-$, as shown by infrared spectroscopy [28]. Treatment in an equimolar mixture of H_2 and CO at temperatures of 370-470 K leads to the anion $[H_3RuOs_3(CO)_{12}]^-$ on the surface. This conclusion follows from the fact that this anion was extracted from the surface by ion exchange with $(AsPh_4)^+Cl^-$. Similarly, when $H_2FeRu_3(CO)_{13}$ was brought in contact with MgO, an anionic cluster could be extracted by PPN^+Cl^- (bis-triphenyl phosphine imminium chloride) solution as $[HFeRu_3(CO)_{13}]^-PPN^+$, which was characterized by its infrared and NMR spectra [29]. However, these surface anionic species are stable only up to about 373 K if heated under vacuum; at higher temperatures extensive decomposition occurs (see Section 9.2 and the following paragraphs).

Similarly, the cluster frameworks can be maintained intact when a molecular bimetallic cluster is deposited on phosphine-functionalized polymers, as discussed below and in Section 9.1.

The reactions between molecular bimetallic clusters and hydrated surfaces are fast, occurring in minutes. The standard procedure in preparation of bimetallic catalysts normally includes evacuation to evaporate trace amounts of solvent. During this period, even at room temperature, additional surface reactions can occur which may result in a disintegration of the cluster framework. The transformations of mononmetallic clusters are discussed by Psaro and Ugo (Section 9.2), and some infrared spectra are shown in Section 6.2.

In the following, low-temperature transformations of bimetallic clusters supported on oxide surfaces will be considered. It is interesting to note that regardless of the initial cluster [$Fe_2Ru(CO)_{12}$ or $H_2FeRu_3(CO)_{13}$], the infrared band structures were identical [23, 30], as shown in Fig. 1. When the spectra obtained for alumina are compared with those taken for $Ru_3(CO)_{12}$ on alumina (Section 6.2), it is clear that after evacuation the bimetallic cluster is decomposed in contact with hydrated alumina. Since no infrared bands characteristic of iron carbonyl species could be observed, it was assumed that decomposition of the bimetallic cluster results in oxidation of iron. Mössbauer spectra of the $H_2FeRu_3(CO)_{13}$ deposited on

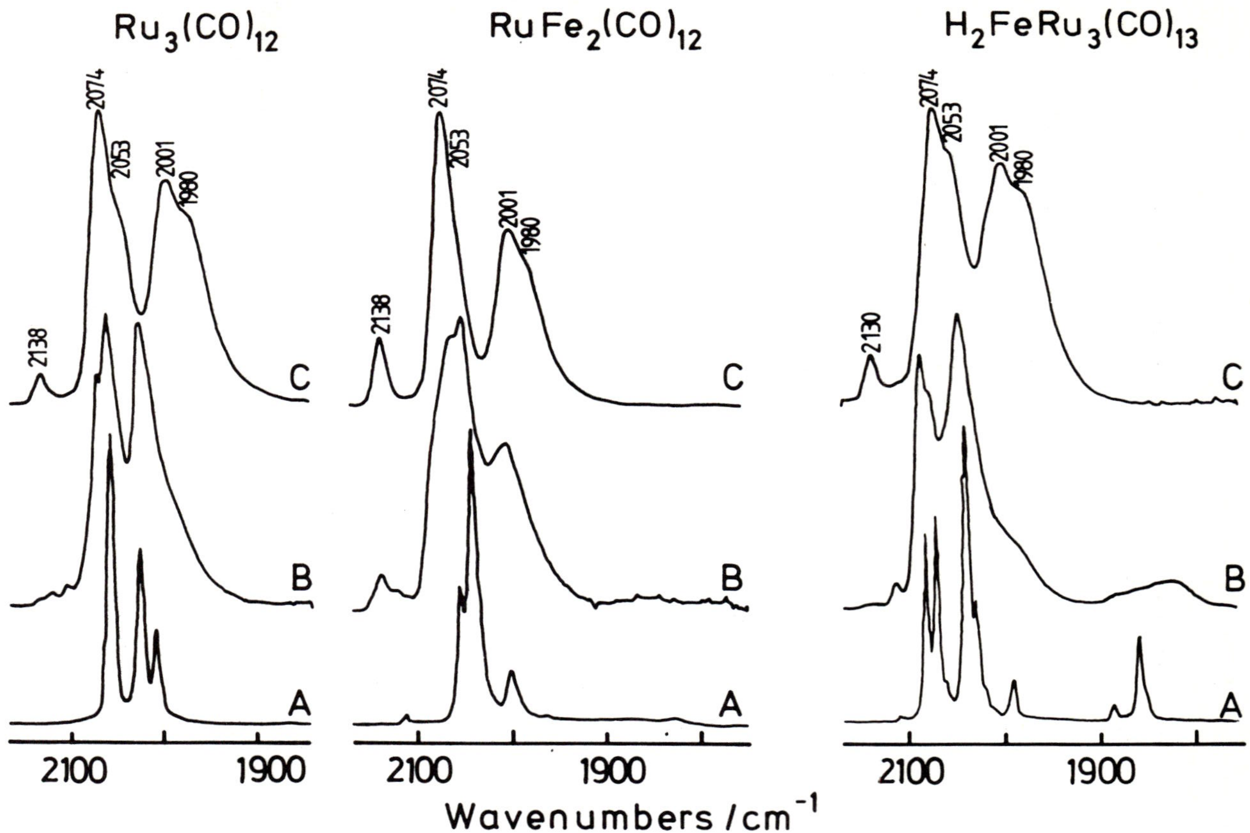

Fig. 1. Infrared spectra of $Ru_3(CO)_{12}$, $RuFe_2(CO)_{12}$, and $H_2FeRu_3(CO)_{13}$ in hexane solution (A), supported on Cab-O-Sil (B), and on Al_2O_3 (C) [23, 30].

alumina show [23] that the main iron species formed are Fe^{3+} and Fe^{2+} in very highly dispersed oxide form, since magnetic hyperfine splitting cannot be observed even at 4.2 K.

On the basis of experimental results showing the CO evolution during cluster adsorption [29] and the exchange between ^{13}CO and ^{12}CO ligands of the supported species [31, 32], it has been suggested that each of these cluster frameworks is more stable on Cab-O-Sil than on Al_2O_3. The stabilities of the clusters on either support were inferred to decrease in the sequence:

$$H_2FeRu_3(CO)_{13} \gg Ru_3(CO)_{12} > Fe_3(CO)_{12} > Fe_2Ru(CO)_{12}$$

Infrared spectra of alumina-supported samples prepared from $Fe_2Ru(CO)_{12}$ [32] indicate the initial formation of $Fe(CO)_5$ characterized by bands at 2022 and 2000 cm^{-1}. These bands disappeared as oxidized iron species formed. The ruthenium spectra observed after long times resemble those of $Ru_3(CO)_{12}$. Thus, the decomposition was suggested to occur as shown in SCHEME 3, where Ly denotes a surface site.

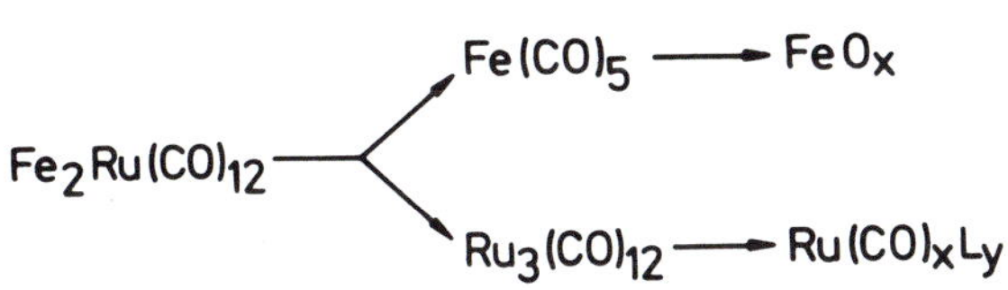

SCHEME 3

This scheme is similar to that described by Fox *et al.* [33] for other bimetallic clusters in solution.

10.3.1.b Decomposition of supported bimetallic clusters with segregation of the metals

$H_2FeRu_3(CO)_{13}$ and $Fe_2Ru(CO)_{13}$ adsorbed on Cab-O-Sil or on alumina break up at temperatures $\leq$400 K; as indicated by temperature programmed decomposition [22], infrared spectroscopy [30], and Mössbauer spectroscopy [23, 34], iron oxide and mononuclear ruthenium complexes are formed.

$H_2FeOs_3(CO)_{13}$ on silica is first physisorbed [24], as indicated by infrared and Raman spectra. However, at about 400 K the iron-osmium bonds are cleaved and Os_3 fragments are formed on the support; at higher temperature mononuclear osmium species are formed (SCHEME 4). Similar decomposition of $(\mu_2\text{-}H)AuOs_3(CO)_{10}PPh_2(CH_2)_2SIL$ and

$(\mu_2\text{-Cl})AuOs_3(CO)_{10}PPh_2(CH_2)_2SIL$ occurred upon prolonged heating at 440 K [35]. At room temperature the former cluster underwent structural changes in the presence of CO, but the osmium framework remained intact.

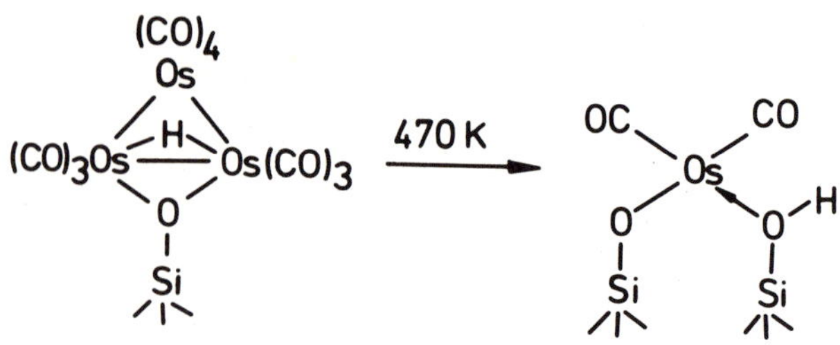

SCHEME 4

At higher temperatures, mononuclear osmium species were formed, as indicated by the infrared bands at 2120, 2040, and 1975 cm^{-1}. The appearance of a band at 2085 cm^{-1} when the sample was treated with CO indicates the formation of gold particles; the size range was 2-4 nm, as determined by TEM. The chloro-osmium cluster is less stable; even at 380 K the same decomposition occurs. Similar behavior has been observed for $(\mu_2\text{-H})_2RhOs_3(acac)(CO)_{10}$ bonded to a polymer matrix [36], whereby Os_3 clusters and Rh particles were formed, and for $H_2RuOs_3(CO)_{13}$ on alumina [37], whereby mononuclear Os complexes Ru particles and mononuclear Ru complexes were formed. One may conclude that numerous supported bimetallic clusters, even if they are stable at moderate temperature, decompose at higher temperatures accompanied by metal-metal bond scission and segregation of the metals.

10.3.1.c Formation of supported bimetallic particles from metal clusters

A mixture of $Fe_3(CO)_{12}$ and $Ru_3(CO)_{12}$ was used to prepare catalysts deposited on Cab-O-Sil with various Ru/Fe ratios. Temperature programmed decomposition in hydrogen and in helium [38] suggests the formation of "alloy" catalysts. The catalytic activity and selectivity in carbon monoxide hydrogenation provide further indication of the interaction of ruthenium and iron [38].

Infrared and Mossbauer spectroscopy [22, 34] give evidence of the interaction between $Fe_3(CO)_{12}$ and $Ru_3(CO)_{12}$. Infrared spectra indicate that the decomposition of $Fe_3(CO)_{12}$ on Cab-O-Sil precedes the decarbonylation of $Ru_3(CO)_{12}$; the decomposition of the $Ru_3(CO)_{12}$ is accelerated by the iron. Preliminary Mossbauer data indicate the effect of ruthenium on the reduction of iron. Whereas after decomposition of

$Ru_3(CO)_{12} + Fe_3(CO)_{12}$ and of $Fe_2Ru(CO)_{12}$ in hydrogen at 420 K the main product was Fe^{2+} (IS = 1.08 mm s^{-1} and QS = 2.06 and 2.4 mm s^{-1}, respectively), no significant reduction was observed for the $Fe_3(CO)_{12}$-derived material.

Although these data indicate interaction between the two metals during cluster decomposition and reduction, additional evidence is required to demonstrate the presence of bimetallic particles. Conventional room-temperature Mössbauer spectroscopy is not adequate for this purpose. After reduction of the supported mixed iron-ruthenium clusters at 700 K, part of the iron existed as small superparamagnetic particles accompanied by the presence of α-iron, as indicated by measurements at 80 K [39]. The superparamagnetic component may be assigned to zerovalent iron associated with ruthenium or to small separate iron particles. Evidence for the formation of iron-ruthenium particles was obtained from low-temperature Mössbauer measurements (carried out at about 1.6 K) in high external magnetic fields [40]. At zero field, the existence of very small Fe_2O_3 particles with a magnetic hyperfine field of 408 kOe was shown, and a single line at zero velocity points to the presence of superparamagnetic iron (Fig. 2). In an external magnetic field of 6T, the single line at zero velocity was split and appeared superimposed in the central doublet. It is possible that the single line corresponds to superparamagnetic iron particles

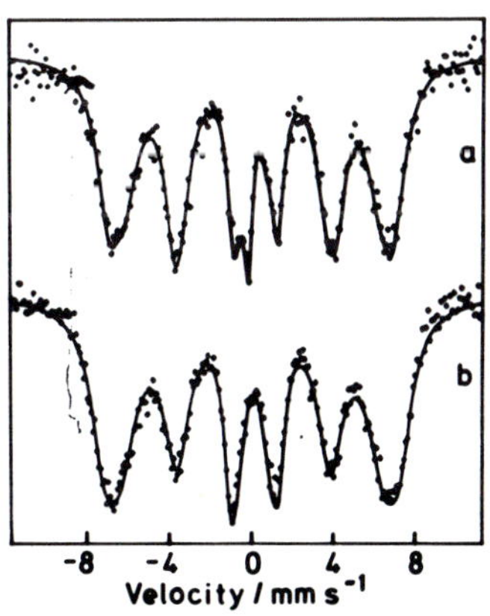

Fig. 2. Low-temperature (1.6 K) Mössbauer spectra of Cab-O-Sil-supported FeRu catalysts prepared from the mixture of $Fe_3(CO)_{12}$ and $Ru_3(CO)_{12}$ and reduced in hydrogen at 770 K: (a) at zero external magnetic field; (b) in 6T external magnetic field [40].

and the splitting occurred even at very low external magnetic field. However, it was found that the hyperfine splitting was proportional to the external magnetic field, which means that the single line can be assigned to iron-ruthenium bimetallic particles which behave as Pauli paramagnets in the concentration range of 12-100 atomic % ruthenium [41]. Mössbauer measurements thus provided evidence for the presence of iron-ruthenium bimetallic particles after high-temperature hydrogen treatment, even though cluster disintegration takes place.

Similar results have been obtained by Kaminsky *et al.* [42], using amorphous carbon black as a support. As no hydroxyl groups exist on the support surface, the metallic state of the bimetallic catalysts is not seriously affected. When the sample was not exposed to air, only a low-temperature pretreatment at 473 K was needed to reduce all the iron to the metallic state. Comparison with catalysts derived from the clusters $Fe_3(CO)_{12}$, $Ru_3(CO)_{12}$, $FeRu_2(CO)_{12}$, $H_2FeRu_3(CO)_{13}$, and $Fe_2Ru(CO)_{12}$ showed that FeRu catalysts made by coimpregnation did not exhibit behavior consistent with that of catalysts derived from the bimetallic clusters, and a lack of homogeneity in these catalysts is indicated. Electron microscopy showed that the metal crystallites were two dimensional in nature and seemed to form rafts within the small pores of the high-surface-area carbon. These experiments demonstrate that the concentration of surface hydroxyl groups on the support plays role in the reducibility of the iron-ruthenium catalysts.

In summary, bimetallic "alloy" catalysts can often be produced by high-temperature treatment in hydrogen, even when a mixture of two monometallic cluster precursors is used.

10.3.1.d **Catalytic activity**

Iron-ruthenium, iron-osmium, and related combinations. The behavior of catalysts derived from Fe-Ru and Fe-Os clusters has been investigated for CO hydrogenation reactions. Activity and selectivity are strongly influenced by the carbon content of the catalyst and the metal-support interaction, which affect the dispersion and oxidation state of the metallic components, the surface hydrogen concentration, and the number of metallic sites. This latter is further influenced by the dispersion: the larger the metal particle, the greater is the deactivation by carbon overlayers. The carbon hinders the migration of metal particles on the surface, which results in highly dispersed catalysts. When carbon is removed, e.g., when cluster decomposition is carried out in hydrogen, larger metal particles are formed.

Activity and selectivity are further controlled by the formation of bimetallic particles and metal-support interactions. It has been established that selectivity to methane in the CO + H_2 reaction on $Ru_3(CO)_{12}$-derived catalysts supported on Cab-O-Sil was nearly 100% [38, 43]. On addition of iron, the selectivity to methane decreased, and the activity was higher than that of a pure iron catalyst [38]. Only a small difference (about twofold) was observed when the catalysts were prepared from a molecular bimetallic clusters as compared with those formed from a mixture of iron and ruthenium clusters [23, 34]. The observed differences could be related to the varying ruthenium contents [14, 47, and 84 atomic % for the mixture, $Fe_2Ru(CO)_{12}$, and $H_2FeRu_3(CO)_{13}$, respectively]. When iron was added to an osmium catalyst, similar changes were observed: catalysts derived from bimetallic clusters had lower selectivities to methane than those prepared from pure osmium dodecacarbonyl [24]. Although there is no reported evidence for the formation of bimetallic particles in these catalysts, the selectivity changes strongly suggest their formation under reaction conditions.

The nature of the support also affects catalytic activity and selectivity. For example, formation of higher hydrocarbons was observed for iron-ruthenium supported on TiO_2 [44] (Table 1); the same holds true for iron-manganese supported on TiO_2 [45].

Ruthenium- and osmium-containing bimetallic catalysts. $H_2RuOs_3(CO)_{13}$ formed the bimetallic cluster anion $H_3RuOs_3(CO)_{12}^-$ on alumina when heated to 370-470 K under a H_2 + CO atmosphere. The material was catalytically active for but-1-ene isomerzation at about 360 K [28]. A more detailed investigation was reported for the surface structure and catalytic properties of Al_2O_3-supported catalysts prepared from $H_4Ru_4(CO)_{12}$, $H_2RuOs_3(CO)_{13}$ and a mixture of $H_4Ru_4(CO)_{12}$ and $H_2Os_3(CO)_{10}$ [37]. In contrast to the low-temperature behavior, the original clusters decomposed under conditions of CO hydrogenation at 543 K. Electron microscopy, XPS, and infrared measurements lend support to the following picture: Metallic ruthenium was present, after reaction, whereas only oxidized osmium species could be identified in the form of Os(II) carbonyl complexes. Oxidized ruthenium was also present, probably as Ru(II) carbonyl complexes. By use of a strongly dehydroxylated alumina support heated to 900 K, the amount of Ru(II) carbonyl complex could be considerably increased, and consequently the catalyst selectivity towards dimethyl ether was significantly enhanced. The authors concluded that metallic and cationic ruthenium are responsible for the formation of hydrocarbons and oxygenated products, respectively.

TABLE 1. Comparison of Iron-Ruthenium Catalysts Supported on Active Carbon, Al_2O_3, Cab-O-Sil, and TiO_2.

Support	Cluster precursor framework	Metal Loading/Wt% Fe	Ru	Activity[a]	C_1[b]	C_{4+}[b]	Refs.
Active carbon	$Fe_3 + Ru_3$	4.2	2.0	1.74	0.5	0.14	[42]
	Fe_2Ru	0.97	0.77	2.52	0.76	0.04	
	$FeRu_3$	0.47	2.6	6.36	0.95	0.01	
Cab-O-Sil	$Fe_3 + Ru_3$	0.5	0.5	0.15	0.70		
	Fe_2Ru	0.52	0.48	0.16	0.73	0.05	[46]
	$FeRu_3$	0.15	0.85	0.63	0.70		
Al_2O_3	$Fe_3 + Ru_3$	0.5	0.5	0.37	0.74		
	Fe_2Ru	0.52	0.48	0.17	0.75	0.1	[23, 46]
	$FeRu_3$	0.15	0.85	1.6	0.72		
TiO_2	$Fe_3 + Ru_3$	0.5	0.5	6.38	0.47		
	Fe_2Ru	0.52	0.48	6.80	0.50	0.3	[44]
	$FeRu_3$	0.15	0.85	1.8	0.54		

[a] Activity is calculated in mol s^{-1} (mol M_T metal)$^{-1}$ x 1000 at 548 K. $H_2/C = 3$, total pressure, 101 kPa; catalyst samples were reduced at 500 and 570 K for carbon and for Cab-O-Sil, Al_2O_3, and TiO_2, respectively.

[b] Selectivites: $C_1 = N_{C_1}/\sum_{i=1}^{7} NC_i$ and $C_4^+ = \sum_{i=4}^{7} NC_i/\sum_{i=1}^{7} NC_i$ with NC_i moles of hydrocarbon molecules having i C atoms.

However, comparison of catalysts prepared from monoruthenium and bimetallic ruthenium-osmium clusters showed the latter to have lower activity. It is believed that an effect is operative here that is similar to what was found in the case of the iron-ruthenium system, i.e., the presence of oxidized osmium decreases the amount of metallic ruthenium.

Additional examples have been reported demonstrating the importance of monometallic particles formed from bimetallic clusters. Budge *et al.* [47] studied the catalytic behavior of $Rh_4(CO)_{12}$, $H_2RhOs_3(CO)_{10}(acac)$, and $H_2FeOs_3(CO)_{13}$ supported on γ-Al_2O_3. They observed by infrared spectroscopy, XPS, and electron microscopy that the $RhOs_3$ cluster decomposed on the surface under CO hydrogenation conditions; segregated Rh metal was formed, which was responsible for the hydrocarbon formation activity. Osmium complexes $Os(CO)_x(O\text{-}Al\lessdot)_3$ were formed on the surface, as indicated by infrared spectra. In agreement with the results mentioned above, the selectivity towards oxygenated products was slightly increased by osmium. The catalyst prepared from $H_2FeOs_3(CO)_{13}$, however, revealed two orders of magnitude lower activity than those formed from Rh_4 and $RhOs_3$ clusters. By physical methods the presence of oxidized iron particles in the size range of 2.0-3.0 nm could be shown along with $Os(CO)_x(O\text{-}Al\lessdot)_3$ species. Due to the absence of metallic particles, it is obvious that not only was the activity lower but the selectivity was also shifted towards oxygen-containing products. The authors compared the activity of this catalyst to that of samples prepared from $Fe_3(CO)_{12}$. The higher activity for this latter catalyst was explained by the enhanced metal particle formation. The authors tentatively attributed the different behavior of the two catalysts to the formation of the relatively stable intermediate $Al^+[H_3FeOs_3(CO)_{12}]^-$. However, Choplin *et al.* [24] have shown that on Cab-O-Sil the breakup of this cluster framework occurs even at 400 K. An alternative explanation may therefore be considered. The $FeOs_3$ framework may break up and small iron particles can be oxidized in the Os(II) environment. Small iron oxide particles may thus be stabilized. The concentration of metallic iron will be low and the catalytic activity consequently small. On the other hand, when osmium is not present, small iron crystallites may be aggregated giving higher activity and hydrocarbon selectivity.

Segregation of two metals was also shown to occur in the tungsten-osmium and molybdenum-osmium system [48]. Alumina-supported catalysts prepared from $(\eta^5\text{-}C_5H_5)WOs_3(CO)_{12}H$ and $(\eta^5\text{-}C_5H_5)MoOs_3(CO)_{12}H$ as well as $Os_3(CO)_{12}$ showed the same activity for CO hydrogenation in the temperature range between 620 and 670 K. The turnover frequencies and the activation energies were near 0.05 s^{-1} and 138 kJ mol^{-1}, respectively. Since separate experiments with catalysts prepared from $(\eta^5\text{-}C_5H_5)_2W_2(CO)_6$ on alumina showed no activity, it was suggested that the bimetallic clusters were decomposed, and the catalytic activity is attributed solely to the surface osmium atoms.

In summary, the ruthenium- and osmium-containing bimetallic clusters discussed here are generally decomposed, with one metal predominantly determining the catalytic performance and the second having a minor influence. In almost all cases osmium is stabilized as a divalent mononuclear carbonyl complex on alumina, and its reduction to zerovalent metal is quite difficult. The only example of bimetallic particles being formed is the iron-ruthenium system after high-temperature reduction.

Catalytic activity of bimetallic clusters with intact frameworks. As mentioned above, the bimetallic cluster framework can be maintained in selected cases if preparation and catalytic reaction are carried out at low temperature. Pierantozzi *et al.* [49] used phosphine-functionalized polymers as supports to prepare $Fe_2Pt(CO)_8(Ph_2P\text{-}Ⓟ)_2$ (Section 9.1). Analogous clusters were obtained with phosphine-functionalized silica supports. The supported bimetallic cluster $RuPt_2(CO)_5(Ph_2P\text{-}Ⓟ)_3$ was also prepared. The supported clusters have been suggested to be catalysts for ethylene hydrogenation (Section 9.3).

A series of polymer-grafted tetranuclear bimetallic clusters has been reported [50-52]. Some of these, namely $ClAuOs_3(CO)_{10}(PhP_2\text{-}Ⓟ)$ and $Co_2Pt_2(CO)_8(PhP\text{-}Ⓟ)_2$, have open butterfly structures, whereas another, namely, $HAuOs_3(CO)_{10}(Ph_2P\text{-}Ⓟ)$ has a closed tetrahedral arrangement. Bimetallic clusters with open butterfly structures were active for ethylene hydrogenation and that having a closed tetrahedral structure had immeasurably low activity. The clusters which provide coordinative unsaturation, provided by formation of the open-butterfly structures, were associated with catalytic activity (Section 9.3).

10.3.2 Bimetallic catalysts derived from molecular cobalt-rhodium clusters

10.3.2.a Interaction with oxide surfaces and cluster decomposition. $Co_2Rh_2(CO)_{12}$ and $Co_3Rh(CO)_{12}$ were adsorbed on γ-alumina and Aerosil [53, 54]. Infrared studies indicated that $Co_2Rh_2(CO)_{12}$ is chemisorbed on γ-alumina. In contrast to Fe-Ru or Fe-Os clusters, for example (Section 10.3.1.a), the Co-Rh clusters lose the bridging carbonyls in this chemisorption step. In addition to adsorption of molecular cluster species with intact metal frameworks, cluster disintegration with Co-Rh metal-metal bond scisson was reported to occur during impregnation [55]. EXAFS data indicated [56] that Co-Rh bonds existed on alumina supports even after treatment in H_2. Rh-Rh and Rh-Co coordination numbers were found to be close to those in the original clusters. Hence highly dispersed bimetallic aggregates were evidently formed on alumina under mild reduction conditions.

10.3.2.b Catalytic activity of CoRh bimetallic catalysts prepared from molecular clusters. Ichikawa reported on the effect of various oxide supports such as ZnO, MgO, and ZrO_2 on the activity and selectivity of cobalt-rhodium bimetallic catalysts in the synthesis of oxygen-containing compounds from synthesis gas [57] and in catalytic hydroformylation of ethylene and propene [58, 59]. Synthesis gas conversion was studied with catalysts prepared from $Rh_4(CO)_{12}$, $Rh_2Co_2(CO)_{12}$, $RhCo_3(CO)_{12}$, and $Co_4(CO)_{12}$ supported on ZrO_2 at low metal loadings (0.2 wt% metal; 2:1 H_2/CO mixture, 470 K). The carbon efficiency {defined by i x C_i/[CO (consumed)], with i = number of C atoms and C_i = concentration of oxygenated products containing i C atoms} in the oxygenated products decreased with increasing CO content, and the proportion of ethanol in the oxygenated products increased (Fig. 3). It was suggested that the $Co_4(CO)_{12}$-derived catalyst preferentially produces methane and other hydrocarbons. Since cobalt catalysts provide sites for dissociative chemisorption of CO, the methylene coverage on the surface increases, which might result in the enhancement of ethanol among the oxygenated products.

The same monometallic and bimetallic rhodium-cobalt molecular clusters were deposited from hexane solution onto ZnO (0.25 wt% metal loading). After decomposition in vacuum at 430 K, a 1:1:1 propene:CO:H_2 mixture was admitted at the same temperature [58, 59]. The specific activity for aldehyde formation decreased in the series $Rh_4 > Rh_2Co_2 > RhCo_3 > Co_4$, but the proportion of linear aldehyde increased as the cobalt content increased. Catalysts prepared from mixtures of the two monometallic molecular clusters were significantly less active than those prepared from bimetallic molecular clusters. This comparison suggests that bimetallic clusters or aggregates were not formed from the monometallic clusters under the conditions applied.

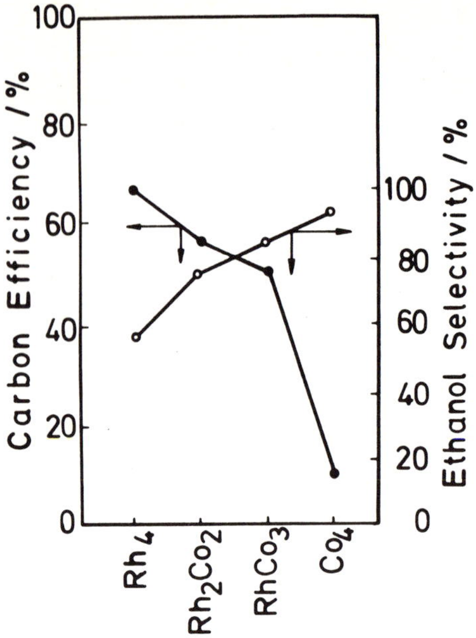

Fig. 3. Carbon efficiencies of the oxygenated products and ethanol selectivities obtained with Rh-Co cluster-derived catalysts supported on ZrO_2 (0.2 wt% metal loading; $CO:H_2$ = 1:2.25; total pressure 8.6 x 10^3Pa; 470 K) (data from Ref. [57]).

10.3.3 Bimetallic catalysts derived from molecular clusters of platinum and palladium

Properties of catalysts derived from platinum and palladium clusters modified by other elements have been reviewed by Yermakov *et al.* [19]. Therefore, only a short summary and some recent data are presented. This topic is less well understood than those stated above, in part because the surface organometallic chemistry of this group of clusters has not been reported.

Yermakov *et al.* [18-20] described catalytic properties of supported bimetallic catalysts prepared by decomposition of surface complexes containing atoms of an active metal M and a promoter element E. For example, Pt-Sn catalysts of various compositions were prepared from $H_4[Pt_3Sn_8Cl_{20}]$ or $(C_8H_{12})_3Pt_3(SnCl_3)_2$ on alumina [19, 60], and Pt-W/SiO_2 and Pt-Mo/SiO_2 catalysts were prepared by adsorption of $Pt(\pi\text{-}C_4H_7)_2$ on silica containing surface Mo^{2+} and W^{2+} ions [19, 61]. In this section, the catalytic properties for hydrogenolysis and skeletal isomerization of hydrocarbons are described for samples prepared by adsorption of Pt-containing bimetallic clusters on supports followed by activation under hydrogen. Table 2 is a summary of some bimetallic catalysts of this type

TABLE 2. Reactions of Hydrocarbons on Supported Bimetallic Catalysts Prepared by Decomposition of Organometallic Complexes on Oxide Support.

Starting organometallic complex	Cluster-derived catalyst	Reactions studied	Refs.
$H_4[Pt_3Sn_8Cl_{20}]$	$Pt\text{-}Sn/Al_2O_3$	Dehydrogenation of Methylcyclopentane	
$[Pt(SnCl_3)]_5[N(C_2H_5)_4]_3$	$Pt\text{-}Sn/Al_2O_3$	Hydrogenolysis and	[19]
$[PtCl_2(SnCl_3)_2][N(C_2H_5)_4]_2$	$Pt\text{-}Sn/Al_2O_3$	dehydrocyclization of n-C_6	
$[PtCl_2(SnCl_3)_2][N(CH_3)_4]_2$	0.5% Pt-0.6% 0n/γ-Al_2O_3	Hydrogenolysis and isomerization of methcyclopentane	[60, 62, 63]
$(C_8H_{12})_3Pt_3(SnCl_3)_2$	8% Pt-3.3% Sn/γ-Al_2O_3	Hydrogenolysis and isomerization of n-hexane	[64, 63, 65]
trans-$Pt[Co(CO)_4]_2[C_6H_{11}NC]_2$	10% Pt-6.6% Co/γ-Al_2O_3	"	
	0.3% Pt-0.2% Co/γ-Al_2O_3	"	
$Pt_2Co_2(CO)_5(\mu_2\text{-}CO)_3[P(C_6H_5)_3]_2$	10% Pt-3% Co/γ-Al_2O_3	"	[63, 64, 66]
$Pt_3Co_2(\mu_2\text{-}CO)_5[P(C_2H_5)_3]_3$	10% Pt-2% Co/γ-Al_2O_3	"	[63, 64, 66]
trans-$Pt[Fe(CO)_3NO]_2[tert\text{-}C_4H_9NC]_2$	10% Pt-5.6% Fe/γ-Al_2O_3	"	[63, 64, 66]
	0.3% Pt-0.17% Fe/γ-Al_2O_3	"	[63, 64, 66]
$Pd_2Cr_2(C_5H_5)_2(CO)_6[P(CH_3)_3]_2$	6.6% Pd-3.3% Cr/ -Al_2O_3	"	[64, 67]
	0.3% Pd-0.15% Cr/ -Al_2O_3	"	
$Pd_2W_2(C_5H_5)_2(CO)_6[P(CH_3)_3]_2$	3.6% Pd-6.4% W/ -Al_2O_3	"	[64, 67]
PtY zeolite + $Mo(CO)_6$	10.6% Pt-5.6% Mo/Y zeolite	"	[16]
$((CH_3)_4N)_2((Pt(SnCl_3)_2Cl_2)$	$Pt\text{-}Sn/Al_2O_3$	"	[68]
$((C_2H_5)_4N)_4(Pt_3Sn_8Cl_{20})$		Hydrogenolysis and	
$((CH_3)_4N)_2((Pt(SnCl_3)_5)$		isomerization of n-heptane	

and the catalytic hydrocarbon reactions reported so far.

The addition of Sn to Pt/Al_2O_3 catalysts leads to increased selectivity and stability in reforming of hydrocarbons. $Pt\text{-}Sn/Al_2O_3$ catalysts exhibit high platinum dispersions and high selectivity towards aromatics [19, 69]. The highest dispersion of metal particles was observed for catalysts prepared by decomposition of platinum-tin clusters adsorbed on alumina [19]. Dehydrogenation is the principal reaction of methylcyclopentane and *n*-hexane on $[Pt(SnCl_3)_2Cl_2]^{2-}/Al_2O_3$-, $[Pt(SnCl_3)_5]^{3-}/Al_2O_3$-, and $[Pt_3Sn_8Cl_{20}]^{4-}/Al_2O_3$-derived catalysts in a hydrogen atmosphere at 670-820 K. These platinum-tin catalysts are characterized by a low rate of hydrogenolysis as compared with aromatization. The total activity decreases on increasing the tin content in the platinum-tin precursor clusters. It was concluded [19] that the C_5 cyclic mechanism (see Section 9.4) for conversion of *n*-hexane was strongly disfavored on these catalysts.

$Pt\text{-}Sn/Al_2O_3$ catalysts prepared from the same bimetallic clusters were also characterized for *n*-heptane conversion [68]. Two types of sites were assumed, M_1, which was active for ethylene hydrogenation, and M_2, which was inactive for this reaction. As the Sn/Pt ratio was increased, the number of M_1 sites decreased and the number of M_2 sites increased; simulataneoulsy, aromatic and hydrogenolysis products decreased whereas isomerization was enhanced. The density of M_2 sites is probably proportional to the number of single Pt atoms in the bimetallic catalyst. The mechanisms were not defined, but the observed effects were explained by the suppression of coke formation.

The bond-shift mechanism for the isomerization of 2-methylpentane $\longrightarrow$ 3-methylpentane and 2-methylpentane $\longrightarrow$ *n*-hexane was observed [69] to be relatively important on these cluster-derived catalysts. In agreement with the conclusions of Yermakov *et al.* [19], the C_5 cyclic mechanism was suppressed.

Supported platinum-cobalt catalysts were obtained by decomposition of linear and nonlinear Pt-Co clusters [64, 65]. The product distributions obtained for the reaction of 2-methylpentane on the cluster-derived catalysts showed (Table 3) a low selectivity for isomerization, while demethylation without extensive degradation ($C_6 \longrightarrow C_5 + C_1$) was favored. On bimetallic Pt-Co catalysts derived from molecular clusters, the bond shift mechanism and the selective demethylation $C_6 \longrightarrow C_5 + C_1$ were enhanced as compared to classical Pt-Co catalysts [70]. The catalytic properties of these cluster-derived materials seem to be dominated by platinum rather than cobalt, which has much more influence in classical Pt-Co catalysts.

TABLE 3. Reaction of 2-Methylpentane on Pt-Co/Al_2O_3 Catalysts.

Catalyst	Precursors	T/K	Isomerization, %	Cracking, %	Refs.
$PtCo_2/Al_2O_3$	*trans*-$Pt[Co(CO_4)]_2[C_6H_{11}NC]_2$	527	14	86	[64, 65]
Pt_2Co_2/Al_1O_3	$Pt_2Co_2(CO)_5(\mu_2\text{-}CO)_3[P(C_6H_5)_3]_2$	573	18	82	[64, 65]
Pt_3Co_2/Al_2O_3	$Pt_3Co_2(CO)_4(\mu_2\text{-}CO)_5[P(C_2H_5)_3]_2$	573	38	72	[64, 65]
10 wt% Pt-6.6 wt% Co/ Al_2O_3[a]	H_2PtCl_6, $Co(NO_3)_3$	573	17	83	[70]
10 wt% Pt-3 wt% Co/ Al_2O_3[b]	H_2PtCl_6, $Co(NO_3)_3$	573	58	42	[70]
10 wt% Pt-2 wt% Co/ Al_2O_3[c]	H_2PtCl_6, $Co(NO_3)_3$	573	59	41	[70]

[a]Atomic ratio Pt:Co $\simeq$ 1:2.
[b]Atomic ratio Pt:Co $\simeq$ 1:1.
[c]Atomic ratio Pt:Co $\simeq$ 3:2.

The synthesis and catalytic activity of bimetallic Pd-Cr catalysts derived from molecular complexes were studied by XPS and EXAFS and by test reactions of ^{13}C-labeled hexanes [71]. The structure of the starting bimetallic cluster $Pd_2Cr_2(C_5H_5)_2(CO)_6(P(CH_3)_3)_2$ determined by EXAFS was in good agreement with the structure determined by X-ray diffraction. The cluster adsorbed on Al_2O_3 maintained its initial structure. The environment of palladium atoms was not modified by adsorption on Al_2O_3. Only after treatment in H_2 at 623 K were drastic changes observed in the radial distribution. The central palladium atom was then surrounded by Pd and Cr atoms and also by Al and O atoms, which suggested a strong interaction of the aggregates with the support. Furthermore, XPS measurements showed a surface enrichment in Cr suggesting that the active centers were blocked by Cr. It was possible to activate these catalysts by treatment in air at 623 K followed by reduction in H_2 at 623 K; the coordination sphere of palladium then consisted mainly of Pd atoms; the Cr atoms were not located in the first shells around Pd.

Esteban Puges *et al.* [71] tested the hydrogenolysis of methylcyclopentane and isomerization of ^{13}C-labeled hexanes on palladium-chromium catalysts derived from $Pd_2Cr_2(\eta^5\text{-}C_5H_5)_2(CO)_6P(CH_3)_3$ clusters on alumina or silica [72]. Prior to any treatment in air, the materials were inactive when the surface complex was decomposed in H_2 at 623 K. It was possible to activate the cluster-derived catalysts, with the results being dependent on the temperature and time of calcination in air. When the catalysts were stabilized, as measured by the formation of methylcyclopentane during isomerization of 2-methylpentane, the total contribution of the cyclic mechanism was the same as for classical well-dispersed palladium catalysts.

Esteban Puges [67] also used palladium-tungsten clusters $Pd_2W_2(C_5H_5)_2(CO)_6(PPh_3)_2$ adsorbed on $\gamma\text{-}Al_2O_3$ to prepare Pd-W/Al_2O_3 catalysts. The reaction of methylcyclopentane on the catalyst formed by decomposition of the surface cluster in H_2 at 623 K led exclusively to the formation of benzene. Isomerization of 2-methylpentane showed a ratio 3-methylpentane/*n*-hexane $\cong$ 30. When the catalyst was treated in air, its properties became similar to those of Pd + WO_3/Al_2O_3 catalysts, where palladium acts only as a hydrogenating agent. These Pd-W catalysts have properties completely different from those of the Pd-Cr cluster-derived catalysts.

With catalysts prepared by decomposition of molecular bimetallic clusters, the general trend for skeletal rearrangements of hydrocarbons is as follows: (1) activity for both hydrogenolysis and isomerization reactions is lower than that on classical dispersed catalysts or on bimetallic catalysts prepared by fixation of organometallic complexes of group VIII metals on low-valent ions, e.g. Mo^{2+} and W^{2+} on SiO_2 [19, 61]; (2) the contributions of the bond-shift mechanism of isomerization and of the selective hydrogenolysis of the C_5 ring are greater than those observed with classical catalysts. When Pt is the base element of the bimetallic catalysts, selective demethylation often occurs due to the increased contributions of the bond shift mechanism.

10.4 COMPARISON OF BIMETALLIC CATALYSTS PREPARED FROM MOLECULAR CLUSTERS AND FROM INORGANIC SALTS

The metal dispersion in supported metal catalysts prepared from molecular clusters is often considerably higher than that observed for catalysts obtained by the incipient wetness method from precursors such as inorganic salts. As a consequence of the high dispersions, support effects may be highly significant in cluster-derived catatalysts.

In this section, differences between cluster-derived bimetallic catalysts and their counterparts prepared conventionally from inorganic salts are summarized briefly.

10.4.1 Iron-ruthenium catalysts

Iron-ruthenium catalysts were first prepared by the incipient wetness method by using $RuCl_3$ and $Fe(NO_3)_3$ on alumina at metal loadings of about 5 wt% [73, 74]. Evidence for the formation of iron-ruthenium bimetallic particles could not easily be found, in contrast to results for iron-platinum [66, 75] and iron-palladium [76] catalysts. The appearance of a singlet and a doublet in the Mossbauer spectrum at room temperature was interpreted as being due to a bimetallic composition in the bulk and on the surface of the particles, respectively [74]. Although a similar assignment had been made for the iron-iridium system [77], the Mössbauer parameters surprisingly resembled those of the high-spin Fe^{3+} ion.

Extensive research with different iron-containing systems [78, 79-82] indicated that at high metal loading iron-ruthenium particles are always present along with some Fe^{3+} oxide. It is established that the bimetallic core is coated by iron oxide, through which hydrogen must diffuse to reach the bimetallic particle.

The structural issue clearly becomes more complicated when the loading is reduced to <1 wt% metal [75, 76, 83, 84]. Under these conditions, nucleation rates may be drastically decreased due to the strong interaction between the metal ions and the support. As a consequence, iron oxide in iron-ruthenium catalysts at low metal loading could not be reduced to metallic particles [85]; only Fe^{2+} ions were formed to small extent. In this case the existence of iron-ruthenium aggregates should be excluded, but the remaining superparamagnetic iron oxide phase exerts a stabilizing effect on the highly dispersed ruthenium particles formed after reduction.

The high dispersion of Ru has been demonstrated not only by X-ray diffraction [75] but also by the catalytic selectivity for *n*-butane hydrogenolysis [85]. The selective rupture of the central C-C bond in *n*-butane gave high yields of ethane, this being a characteristic indication of high dispersion [75]. Since metallic ruthenium particles on which hydrogen dissociation would take place were present [85], there is an opportunity for hydrogen transfer to the iron oxide phase to facilitate the reduction of iron.

If this reduction occurred, formation of bimetallic particles should be expected. However, it was shown [86-88] with Mössbauer and electron spectrosocpy that iron oxide was present and that ruthenium could not be completely reduced either. Consequently, a Fe_2O_3-RuO_x-Ru interphase might exist, possibly providing a link between highly dispersed ruthenium metal particles and iron oxide. The presence of stabilized RuO_x may prevent hydrogen transfer to the iron oxide particles.

In conclusion, formation of the bimetallic iron-ruthenium particles from inorganic salts can be ruled out at low metal loading. The presence of iron oxide results in the stabilization of small monometallic ruthenium particles. As a result of changes in the conditions of catalyst preparation, separate large iron and ruthenium particles can be formed.

In contrast, as shown in Section 10.3.1, bimetallic particles are formed from carbonyl clusters even at very low metal loadings. This statement summarizes the most striking difference between the two systems.

The cluster-derived catalysts also have high dispersions. As shown in Table 4, catalysts prepared from mixtures of $RuCl_3$ and $Fe(NO_3)_3$ have generally lower dispersions at comparable compositions than those prepared from carbonyl clusters. These results are in agreement with data reported by Brenner [89].

10.4.2 Ruthenium- and osmium-containing catalysts prepared from inorganic salts

Ru-Cu and Os-Cu supported on Cab-O-Sil were the first reported examples of bimetallic catalysts containing two metals that do not form binary alloys [90, 91]. As these catalysts contain immiscible metals, the differences in catalytic activity from those of massive bimetallic catalysts

TABLE 4. Dispersions of Fe-Ru Catalysts Supported on Cab-O-Sil (Data from [23, 85]).

Catalyst[a]	Precursor[b]	Treatment	O_2 chemisorbed $mol \cdot g^{-1}$	CO chemisorbed $mol \cdot g^{-1}$	Dispersion[d]
Ru 0 Fe 100	Cluster	H_2, 770 K	---	n.a.[c]	---
		He, 770 K	---	n.a.[c]	---
Ru 14 Fe 86	Cluster	H_2, 770 K	---	17.1	0.3
		He, 770 K	---	38.7	0.6
Ru 28 Fe 72	Cluster	H_2, 770 K	---	25.8	0.44
		He, 770 K	---	60.0	1.0
Ru 100 Fe 0	Cluster	H_2, 770 K	---	10.8	0.75
		He, 770 K	---	36.2	>1.0
Ru 58 Fe 42	Salt	Standard[b]	2.6	---	~0.1
		H_2, 570 K	9.9	---	~0.3
Ru 73 Fe 27	Salt	H_2, 570 K	7.2	---	~0.2
Ru 100 Fe 0	Salt	Standard	Trace	---	---
		H_2, 570 K	8.2	---	~0.2

[a]Total metal loading 1 wt%. Numbers indicate composition in atomic %.

[b]Calcination in O_2 at 570 K for 1 h and reduction in H_2 at 770 K for 2 h.

[c]No adsorption.

[d]Calculated from CO adsorption; a value >1 suggests recarbonylation of the metal.

were attributed to the drastic difference in the particle sizes. The explanation of the catalytic behavior was ascribed to the enrichment of copper on the ruthenium and osmium surfaces and to reduced ensemble sizes, which lead to a decrease in the multiple adsorption considered to be a prerequisite for hydrogenolysis [92, 93]. Electronic interactions between the metals could not be observed by XPS [94]. Electron microscopy revealed structural differences between osmium and ruthenium. Whereas osmium forms small particles with diameters of about 1.0 nm on Cab-O-Sil, ruthenium forms larger particles, some of which have raft-like structures [90].

Although molecular Ru-Cu clusters do exist [7], they have not yet been used for the preparation of supported catalysts. However, Ru-Au catalysts have been prepared from inorganic salts [8, 95-98] and supported Os-Au catalysts from molecular clusters [35, 50]. On a silica support, EDS analysis revealed a bimodal distribution for conventionally prepared catalysts, i.e., large particles (>15 nm) appeared to be pure gold, with smaller particles (>3 nm) being bimetallic and having ruthenium-rich cores and gold-rich surfaces. Because of the gold enrichment on the surfaces of the bimetallic particles, the activity for synthesis gas conversion and ethane hydrogenolysis decreased with increasing gold content over the whole range of compositions.

Infrared spectroscopy indicated the formation of bimetallic particles on magnesia [95]. Recently, a more detailed analysis showed [8] that there is a wide variation in particle size distributions. Materials having large particles contained monometallic ruthenium and gold crystallites, whereas the smaller particles were bimetallic. The gold particles on magnesia were covered by ruthenium. On both supports a large fraction of the gold was separated, being in the form of large monometallic particles. An analogous segregation of the two metals with formation of large gold particles was observed for materials prepared from Os-Au bimetallic clusters on phosphine-modified silica [35].

10.5 CONCLUSIONS

Bimetallic clusters have attracted interest as catalyst precursors primarily because they offer the prospect of preparation of supported metal structures with exactly controlled compositions and high dispersions. Catalysts with these properties have indeed been prepared, and at low temperatures, catalysis by structures similar to those of the molecular precursors even occurs. There are a number of examples illustrating differences in catalytic performance between cluster-derived and classically prepared catalysts; these differences may often be related to the prospective advantages of cluster-derived catalysts stated above, but for the most part the differences remain less than thoroughly explained. Catalysts prepared from supported molecular clusters are almost always unstable at high temperatures and in the presence of hydrogen or other reactive atmospheres; consequently, the high initial dispersion and exact bimetallic compositions of the metal entities are often extremely difficult to maintain under catalytic reaction conditions. For these reasons, and because of their apparent lack of regenerability, catalysts prepared from bimetallic clusters do not offer obvious advantages in application. The principal value of samples made from these clusters has been to provide fundamental understanding of highly dispersed bimetallic catalysts and the influence of one metal on the catalytic behavior of the other.

ACKNOWLEDGMENT

The author thanks G. Maire for his contribution of Section 10.3.3.

REFERENCES

1. J. H. Sinfelt and J. A. Cusumano in J. J. Burton and R. L. Garten, Eds., Advanced Materials in Catalysis, Academic Press, New York, San Francisco, London, 1977, p. 1.
2. J. H. Sinfelt, Bimetallic Catalysts- Discoveries, Concepts, and Applications, Wiley, New York, 1983.
3. W. M. H. Sachtler, Le Vide 164 (1973) 67.
4. H. Knözinger in B. Imelik, C. Naccache, G. Couduries, Y. Ben Taarit and J. C. Vedrine, Eds., Catalysis by Acids and Bases, Elsevier, Amsterdam, Oxford, New York, 1985, p. 111.
5. K. Foger, in J. R. Anderson and M. Boudart, Eds., Catalysis-Science and Technology, Springer, Berlin, 1984, Vol. 6, p. 227.
6. D. F. Ollis, J. Catal., 23 (1971) 131.
7. M. J. Freeman, M. Green, A. Guy Orpen, I. D. Saltes and F. G. A. Stone, J. Chem. Soc. Chem. Commun. (1983) 1332.
8. A. Datye and J. Schwank, Proc. 8th Int. Congr. Catal., Berlin, 1984 Vol. 4., p. 587.
9. J. H. Sinfelt, Rev. Mod. Physics, 51 (1979) 569.
10. F. W. Lytle, G. H. Via and J. H. Sinfelt, J. Phys. Chem., 67 (1977) 3831.
11. G. Meitzner, G. H. Via, F. W. Lytle and J. H. Sinfelt, J. Chem. Phys., 78 (1983) 882.
12. G. Meitzner, G. H. Via, F. W. Lytle and J. H. Sinfelt, J. Chem. Phys., 78 (1983) 2533.
13. C. H. Bartholomew, J. H. Anderson and M. Boudart, J. Chem. Soc. Faraday I, 75 (1979) 257.
14. C. M. Grill, M. L. McLaughlin, J. M. Stevenson and R. D. Gonzales, J. Catal., 69 (1981) 454.
15. P. Gallezot, Surf. Sci., 106 (1981) 459.
16. T. M. Tri, J. P. Candy, P. Gallezot, J. Massardier, M. Primet, J. C. Vedrine and B. Imelik, J. Catal., 79 (1983) 396.
17. I. S. C. Hughes, J. O. H. Newman and G. C. Bond, J. Mol. Catal., 25 (1984) 219.

18. Yu. I. Yermakov, Pure Appl. Chem., 52 (1980) 2075.
19. Yu. I. Yermakov, B. N. Kuznetsov and V. A. Zakharov, Catalysis by Supported Complexes, Elsevier, Amsterdam, 1981.
20. Yu. I. Yermakov and B. N. Kuznetsov, J. Mol. Catal., 9 (1980) 13.
21. K. Lázár, K. Matusek, J. Mink, S. Dobos, L. Guczi, A. Vizi-Orosz, L. Markó and W. M. Reiff, J. Catal., 87 (1984) 163.
22. Z. Schay, K. Lázár, J. Mink and L. Guczi, J. Catal., 87 (1984) 179.
23. I. Böszörmenyi, S. Dobos, L. Guczi, L. Marko, K. Lázár, W. M. Reiff, Z. Schay, L. Tákacs and A. Vizi-Orosz, Proc. 8th Int. Congr. Catal., Berlin, 1984, Vol. 5, p. 183.
24. A. Choplin, M. Leconte, J. M. Basset, S. G. Shore and W.-L. Hsu, J. Mol. Catal., 21, (1983) 389.
25. F. Hugues, J. M. Basset, Y. Ben Taarit, A. Choplin and D. Rojas, J. Am. Chem. Soc., 104, (1982) 7020.
26. A. Theolier, A. Choplin, L. D'Ornelas, J. M. Basset, G. Zanderighi, R. Ugo, R. Psaro and C. Sourisseau, Polyhedron, 2 (1983) 119.
27. R. Psaro, R. Ugo, G. M. Zanderighi, B. Besson, A. K. Smith and J. M. Basset, J. Organomet. Chem., 213 (1981) 215.
28. J. R. Budge, J. P. Scott and B. C. Gates, J. Chem. Soc. Chem. Commun. (1983) 342.
29. A. Choplin, A. Theolier, L. D'Ornelas, P. Defour, L. Huang and J. M. Basset, J. Organomet. Chem., submitted.
30. I. Böszörmenyi, S. Dobos, K. Lázár, Z. Schay and L. Guczi, Surface Sci., 156 (1985) 995.
31. L. Guczi, I. Böszörmenyi, S. Dobos, K. Lázár, L. Tákacs and Z. Schay, Proc. 10th Int. Symp. Reactivity of Solids, Elsevier, Amsterdam, 1985, p. 1087.
32. I. Böszörmenyi, S. Dobos and L. Guczi, to be published.
33. J. R. Fox, W. L. Gladfelter and G. L. Geoffroy, Inorg. Chem., 19 (1980) 2574.
34. L. Guczi, Z. Schay, K. Lázár, A. Vizi and L. Marko, Surface Sci., 106, 516 (1981).
35. H. Knözinger, B. Tesche and M. Wolf, J. Mol. Catal. 25 (1984) 273.
36. J. Lieto, M. Wolf, B. A. Matrana, M. Prochazka, B. Tesche, H. Knözinger and B. C. Gates, J. Phys. Chem., 89 (1985) 991.
37. J. R. Budge, B. F. Lücke, J. P. Scott and B. C. Gates, Proc. 8th Int. Congr. Catal., Berlin, 1984, Vol. 5, p. 89.
38. L. Guczi, Z. Schay, K. Matusek, I. Bogyai and G. Stefler, Proc. 7th Int. Congr. Catal., Tokyo, 1981, Part A, p. 211.
39. L. Guczi, J. Mol. Catal., 25 (1984) 13.
40. K. Lázár, W. M. Reiff and L. Guczi, to be published.
41. H. Fujimori and H. Saito, J. Phys. Soc. Japan, 26 (1969) 1115.
42. M. Kaminsky, K. J. Yoon, G. L. Geoffroy and M. A. Vannice, J. Catal., in press.
43. Z. Schay and L. Guczi, React. Kinet. Catal. Lett., 14 (1980) 207.
44. Z. Schay, L. Bogyay and L. Guczi, Acta Chim. Hung., in press.
45. V. L. Kuznetsov, A. F. Danilyuk, I. B. Kolosova and Y. I. Yermakov, React. Kinet. Catal. Lett., 21 (1982) 249.
46. Z. Schay and L. Guczi, Acta Chim. Hung., 111 (1983) 607.
47. J. R. Budge, B. F. Lücke, B. C. Gates, and J. Toran, J. Catal., 91 (1985) 272.
48. J. R. Shapley, S. J. Hardwick, D. S. Foose and G. D. Stucky, Preprints, ACS Div. Petrol. Chem., 25 (1980) 780.
49. R. Pierantozzi, K. J. McQuade, B. C. Gates, M. Wolf, H. Knözinger and H. Ruhmann, J. Am. Chem. Soc., 101 (1979) 5436.
50. R. Pierantozzi, K. J. McQuade and B. C. Gates, Proc. 7th Int. Congr. Catal., Tokyo, 1980, Part A, p. 941.
51. B. C. Gates and J. Lieto, CHEMTECH 10 (1980) 195.
52. B. C. Gates and J. Lieto, CHEMTECH 10 (1980) 248.
53. J. R. Anderson and D. E. Mainwaring, J. Catal., 35 (1974) 162.

54. J. R. Anderson, P. S. Elmes, R. F. Howe and D. E. Mainwaring, J. Catal. 50 (1977) 508.
55. H. D. Kaesz, R. A. Love, B. A. Matrana, A. P. Humphries and S. Siegel, Preprints, ACS Div. Petrol. Chem. 22 (1977) 1220.
56. T. Yokoyama, K. Yamazaki, N. Kosugi, H. Kuroda, M. Ichikawa and T. Fukushima, J. Chem. Soc. Chem. Commun. (1984) 962.
57. M. Ichikawa, CHEMTECH, 10 (1982) 674.
58. M. Ichikawa, J. Catal., 59 (1979) 127.
59. M. Ichikawa, J. Catal., 59 (1979) 67.
60. O. Zahraa, F. Garin and G. Maire, Faraday Disc. Chem. Soc., 72 (1981) 45.
61. Y. Yermakov, Proc. 7th Int. Congr. Catal., Tokyo, 1984, Vol. 1, p. 57.
62. F. G. Gault, O. Zahraa, J. M. Dartigues, G. Maire, M. Peyrot, E. Weisang and P. A. Engelhard, Proc. 7th Int. Congr. Catal. Tokyo 1981, Part A, p. 199.
63. G. Maire and F. Garin in J. R. Anderson and M. Boudart, Eds., Catalysis - Science and Technology, Springer, Berlin, 1984, Vol. 6, p. 161.
64. O. Zahraa, These d'Etat, Strasbourg, 1980; O. Zahraa and G. Maire, unpublished results.
65. G. Maire, O. Zahraa, F. Garin, C. Crouzet, S. Aeiyach, P. Legare and P. Braunstein, J. Chem. Phys., 78 (1981) 951.
66. M. A. Vannice and R. L. Garten, J. Mol. Catal. 1 (1976) 201.
67. P. Esteban Puges, These d'Etat, Strasbourg, 1984.
68. L. Liwu, Z. Jingling, W. Rongan, W. Chengyu and D. Hongzhang, Proc. 8th Int. Congr. Catal., Berlin, 1984, Vol. 4, p. 565.
69. V. N. Seleznev, Y. V. Fomichev and M. E. Levinter, Neftekhimya, 14 (1974) 205.
70. S. Zyade, These 3° cycle, Strasbourg, 1984; S. Zyade, F. Garin and G. Maire, unpublished results.
71. P. Esteban Puges, F. Garin, P. Girard, P. Bernhardt, and G. Maire, Proc. 9th Iberoamer. Congr. Catal., Lisbon, 1984, in press.
72. R. Bender, P. Braunstein, J. M. Jud and Y. Dusausoy, Inorg. Chem., 22 (1983) 3394.
73. M. A. Vannice, Y. L. Lam and R. L. Garten, Preprints, ACS Div. Petrol. Chem. 23 (1978) 495.
74. J. J. Burton and R. L. Garten in J. J. Burton and R. L. Garten, Eds., Advanced Materials in Catalysis, Academic Press, New York, 1977, p. 33.
75. L. Guczi, K. Matusek and M. Eszterle, J. Catal., 60 (1979) 121.
76. R. L. Garten and D. F. Ollis, J. Catal., 35 (1974) 232.
77. L. L. Murrell and R. L. Garten, paper presented at 8th North American Catal. Soc. Meeting, Philadelphia, May 1-4, 1983.
78. A. T. Bell and P. Winslow, Proc. 8th Int. Congr. Catal., Berlin, 1984, Vol. 3, p. 175.
79. L. Guczi, Catal. Rev.-Sci. Eng. 23 (1981) 329.
80. J. W. Niemantsverdriet, A. M. van der Kraan, J. J. van Loef and W. N. Delgass, J. Phys. Chem., 87 (1983) 1292.
81. H. F. J. van't Blik and J. W. Niemantsverdriet, Appl. Catal., to be published.
82. J. W. Niemantsverdriet, D. P. Aschenbeck, F. A. Fortunato and W. N. Delgass, J. Mol. Catal., in press.
83. L. Guczi, K. Matusek, M. Eszterle and Z. Schay, Proc. Symp. Heterog. Catal., Varna, 1979, p. 117.
84. R. L. Garten, J. Catal., 43 (1976) 18.
85. L. Guczi, K. Matusek, A. Sarkany and P. Tetenyi, Bull. Soc. Chim. Belg., 88 (1979) 497.
86. M. M. Akbarnejad, Ph.D. Thesis, Louisiana State Univ., 1984.
87. J. T. Donner, Ph.D. Thesis, Louisiana State Univ., 1983.

88. M. L. Good, M. Akbarnejad, J. T. Donner, Preprints, ACS Div. Petrol. Chem. 25 (1980) 763.
89. A. Brenner, J. Chem. Soc. Chem. Commun. (1979) 251.
90. E. B. Prestridge, G. H. Via and J. H. Sinfelt, J. Catal., 50 (1977) 115.
91. J. H. Sinfelt, J. Catal. 29 (1973) 308.
92. L. Guczi, A. Frennet and V. Ponec, Acta Chim. Hung., 112 (1983) 127.
93. J. H. Sinfelt, Y. L. Lam, J. A. Cusumano and A. E. Barnett, J. Catal., 42 (1976) 227.
94. D. J. Elliott and J. H. Lunsford, J. Catal., 57 (1979) 11.
95. J. Schwank, G. Parravano, and H. L. Gruber, J. Catal., 61 (1980) 19.
96. S. Galvagno, J. Schwank and G. Parravano, J. Catal., 61 (1980) 223.
97. L. W. Bassi, F. Garbassi, G. Vlaic, A. Marzi, G. R. Tauszik, G. Cocco, S. Galvagno and G. Parravano, J. Catal., 64 (1980) 405.
98. S. Galvagno, J. Schwank, G. Parravano, F. Garbassi, A. Marzi and G. R. Tauszik, J. Catal., 69 (1981) 293.

PART III

RELATIONS BETWEEN MOLECULAR CLUSTERS AND METAL SURFACES

Chapter 11

RELATIONS BETWEEN METAL CLUSTERS AND METAL SURFACES

G. ERTL

11.1 INTRODUCTION

Understanding of bond formation in metal complexes has developed along different lines from that for metal surfaces. A localized picture of interacting orbitals was the natural starting point for the description of systems consisting of ligands and one or only a few metal atoms. On the other hand, the valence electrons in a metal crystal are delocalized, and therefore a series of earlier concepts of the "electron theory of catalysis" was based on itinerant electron models derived from band theory [1].

The first hints for analogies between metal-complex and surface chemistry emerged from infrared spectroscopic studies of adsorbed molecules [2], and as a result researchers started to consider chemical interactions at metal surfaces in terms of localized bonding [3]. Since any electronic disturbance in a metal is completely screened within dimensions of about one to two atomic diameters, it becomes obvious that chemisorption must be considered as a spatially localized effect, even if the metal orbitals involved are not localized. Analogies between the bonding in metal complexes and in chemisorption systems are therefore to be anticipated, and the analogies are expected to be closest when the electronic and structural properties are most similar.

Apart from the infrared spectra, close similarities between carbonyl cluster compounds and CO chemisorption systems with respect to their valence electronic structure as probed by ultraviolet photoelectron spectroscopy (UPS) were recognized only a few years ago [4]. Possible relations between cluster and surface chemistry were probably suggested for the first time by Burwell and Peri [9] and were emphasized in recent years mainly by Muetterties [5]. Recently this idea has been further developed in numerous articles, most written by cluster chemists [6].

On the other hand, "cluster models" are currently en vogue for theoretical treatment of the chemisorption bond [7]: An infinite metal surface is replaced by a small number of metal atoms modelling the adsorption site and probably its immediate vicinity. Despite the apparent successes of this approach, it should be kept in mind that even a system consisting of, say, 50 metal atoms will differ from a bulk crystal with respect to a series of solid-state properties; a similar caution against overstressing the similarity in chemical properties of cluster compounds and chemisorption systems appears to be appropriate here. This caution pertains in particular to chemical reactivity; it does seem, however, as if coupling of a not too strongly bound ligand to a surface is satisfactorily represented by a cluster.

Not only can clusters serve as models of chemisorption systems, but the properties of clusters in turn can in principle be studied by means of metal surfaces. Due to their high thermal stability, clean surfaces (modelling "naked" clusters) can easily by prepared (albeit under ultrahigh vacuum conditions) and the ligand:metal stoichiometry (= coverage) may be continuously varied. The presence of coadsorbates mimics the coupling of different ligands in a cluster, etc.

The applicable experimental tools and therefore the kind of accessible information differ in the two fields: Crystal structure determination by X-ray diffraction can be routinely performed to high precision with clusters, whereas surface crystallography by low energy electron diffraction (LEED) is by no means trivial, and yields accuracies at best of ± 0.05 nm [8], which are too poor for subtle bonding considerations. Metal-ligand bond strengths and their variations can, on the other hand, in general be more accurately determined with metal surfaces. The same holds for the rates of individual reaction steps.

This chapter is devoted to a critical comparison of the fundamental properties of clusters and metal surfaces as they may be relevant to their chemistry and catalytic activity.

11.2 STRUCTURE

11.2.1 General

The structure of the metal skeleton in a cluster compound is frequently very similar to the arrangement of a corresponding ensemble of atoms in a bulk crystal, which is perhaps the main reason for seeking analogies between clusters and surfaces. One should, however, keep in mind that these similarities hold only as long as ligands are attached to the cluster. Although, for example, Ag_6 clusters trapped in the cages of a zeolite exhibit octahedral structure [10], there are also several reports in the literature indicating that the structures of small "naked" crystallites differ from those of the bulk. Gold on mica exhibits fivefold symmetry [11], and calculations by Burton [12] demonstrated that a 13-atom cluster would not exhibit the octahedral geometry characteristic of a fcc crystal, but rather an icosahedral structure with fivefold rotational axes. (Five-fold symmetry may also be present in clusters, e.g. $[(Pt_{19}(CO)_{22}]^{4-}$ [13]). Since the metal-metal bond energy may even be smaller than that of the metal-ligand bond (as will be outlined in more detail later) the removal of ligands in a cluster (which will be necessary for the creation of unsaturated valencies playing the role of catalytically active centers) will most probably also cause a rearrangement of the metal skeleton. This tendency is much less pronounced with metal surfaces, where the bulk acts as a template forcing the surface atoms towards their "normal" lattice positions. But even on surfaces--because of the altered energetics--atomic displacements (relaxation and reconstruction) are frequently observed, and these may be affected by the presence of adsorbates.

Since only little is known about the structure of "naked" clusters, our comparison will necessarily be restricted to stable compounds with all the ligands attached.

11.2.2 Configuration of the Metal Atoms

Fig. 1 shows the configuration of atoms in the three most densely packed surfaces of a fcc crystal, whereby possible reconstruction effects are neglected for the moment. Each surface atom is directly bonded to as many as nine nearest-neighbor metal atoms--a situation encountered only with rather large clusters. Nevertheless, the atomic configuration in smaller clusters often closely resembles a section of the densely packed bulk crystal, and the interatomic distances are often similar. With carbonyl clusters, the following polyhedra are frequently found [14] (also see Chapter 2):

4-atom = tetrahedron (fragment of hcp or fcc crystal)
5-atom = trigonal bipyramid (hcp) or square pyramid (fcc or bcc)
6-atom = octahedron (hcp + fcc)
7-atom = face-capped octahedron (fcc)

a b c

Fig. 1. Top views of the three most denseley packed (unreconstructed) surfaces of a fcc crystal. Full lines: atoms of the topmost layer; dashed lines: atoms of the second layer: a, (111); b, (100); c, (110).

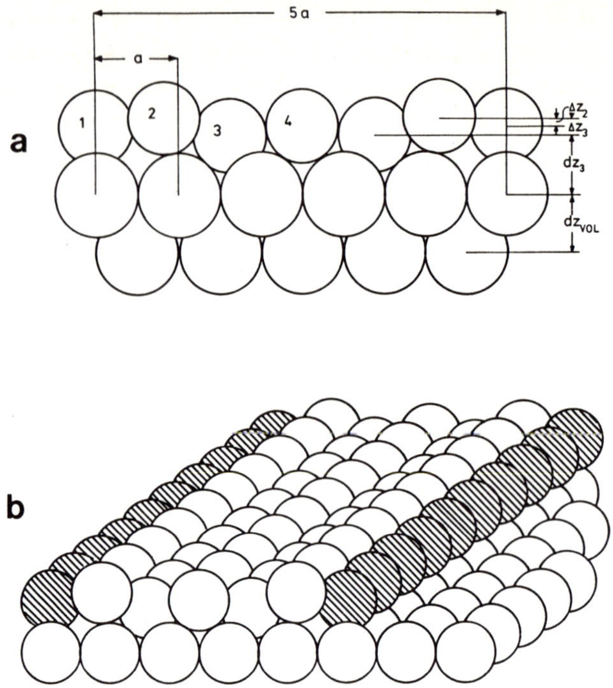

Fig. 2. Model of the reconstructed Ir(100) surface where the topmost atoms form a denseley packed (hexagonal) 1x5 super lattice on the underlying bulk [15]: a, side view; b, top view.

Among the largest known clusters, $[Pt_{38}(CO)_{44}H_x^{2-}]$ exhibits a fcc array of metal atoms. $[Pt_{26}(CO)_{32}]^{2-}$ has a hcp structure; the central plane consists of 12 close-packed Pt atoms, above and below which are planes of seven atoms with hexagonal symmetry [15]. It can be concluded generally that the packing of metal atoms in clusters models that in bulk crystals with only relatively few exceptions.

Clean surfaces of a number of elements exhibit a tendency for reconstruction: The atoms in the outermost layer of Ir(100) or Pt(100) show a hexagonal arrangement, presumably with some buckling [16], as shown in Fig. 2. The (110) surfaces of these elements exhibit a 1x2 superstructure, for which various models have been proposed--so far, however, without conclusive evidence for the actual structure [17].

This reconstruction may be reversed by adsorption. For example, the hexagonal surface of Pt(100) transforms into the "normal" atomic arrangement under the influence of adsorbed CO molecules, which are terminally bonded and form an ordered c2x2 overlayer at a coverage θ = 0.5 [18]. The driving force for this transformation must be sought in the increased adsorption energy of the nonreconstructed surface. This process takes place even far below room temperature, which indicates the existence of a rather low activation barrier-- i.e., displacements of metal atoms may take place quite easily under the influence of ligands.

In turn, adsorption may cause reconstruction of an undistorted surface. If, for example, the coverage of H atoms on a Ni(110) surface exceeds θ = 1.0, a new structure with θ = 1.5 is formed by pairwise displacement of parallel Ni rows, as shown in Fig. 3 [19]. (These structures are formed only at temperatures <220 K. At higher temperatures, activated transformation into a different structure takes place.)

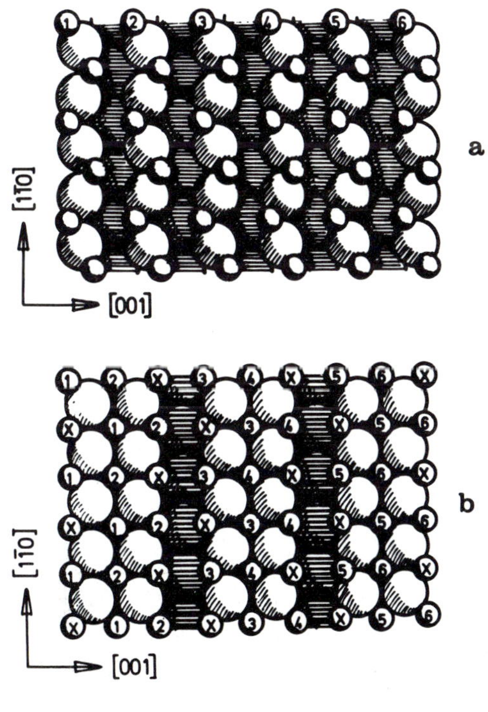

Fig. 3. The Ni(110) surface is unreconstructed up to a coverage θ_H = 1.0 of H atoms (a), but undergoes parallel displacements of the topmost Ni atom rows upon further hydrogen uptake until the structure (b) is reached at θ_H = 1.5 [19].

11.2.3 Distances between Metal Atoms

The spacing between the first and second atomic layers of clean metal surfaces is frequently contracted by up to about 10% with respect to the bulk value, but this relaxation is usually removed in the presence of adsorbates, and then the normal bulk separations are found [8]. A comparison of the metal-metal bond distances in a few selected cluster compounds with the values for the corresponding bulk crystals is given in Table 1; the values are generally found to agree within a few percent.

TABLE 1. Metal-Metal Bond Distances in Bulk Crystals and a few selected Cluster Compounds [61].

Metal or cluster	Metal-metal bond distance, nm
Fe bulk (bcc)	0.249
$Fe_4(CO)_{13}^{2-}$	0.255
Os bulk (hcp)	0.268
$Os_6(CO)_{18}^{2-}$	0.268
Co bulk (hcp)	0.251
$Co_4(CO)_{12}$	0.249
$Co_6(CO)_{15}^{-2}$	0.251
Rh bulk (fcc)	0.269
$Rh_6(CO)_{16}$	0.278
$Rh_4(CO)_{12}$	0.273
Ir bulk (fcc)	0.271
$Ir_4(CO)_{12}$	0.269
$H_2Ir_4(CO)_{10}^{2-}$	0.276
Ni bulk (fcc)	0.249
$(\eta^5\text{-}C_5H_5)_4Ni_4H_3$	0.246
$Ni_4(CO)_6[P(CH_2CH_2CN)_3]_4$	0.251

Therefore, there is a close resemblance between clusters and surfaces not only with respect of the configuration of the metal atoms but also with respect to their distances of separation.

11.2.4 Configuration of Ligands

These structural characteristics will be rationalized in terms of the local geometry of the ligand with respect to the metal atoms to which it is attached (= adsorption site) and to the neighboring ligands. A cluster compound generally exhibits a fixed stoichiometry and thereby also a fixed structure (apart from the occurrence of fluxionality of the ligands). The situation is quite different with metal surfaces, where the stoichiometry (= coverage) may be continuously varied--this range of composition and structure makes surfaces much more suitable as catalysts. Different adsorption sites may be occupied with varying coverage, and the interaction between adsorbed species most frequently gives rise to two-dimensional long-range order, which also changes with coverage. In these respects there are fundamental differences between clusters and surfaces, and our comparison will necessarily be restricted to special situations.

Fig. 4 shows the structure of the skeleton of the $H_3Ni_4(C_5H_5)_4$ cluster, for which the positions of the H atoms were determined by neutron diffraction [20]. The H atoms are located in threefold bridging sites with a Ni-H distance of 0.169 nm. For comparison, the structure of H/Ni(111)

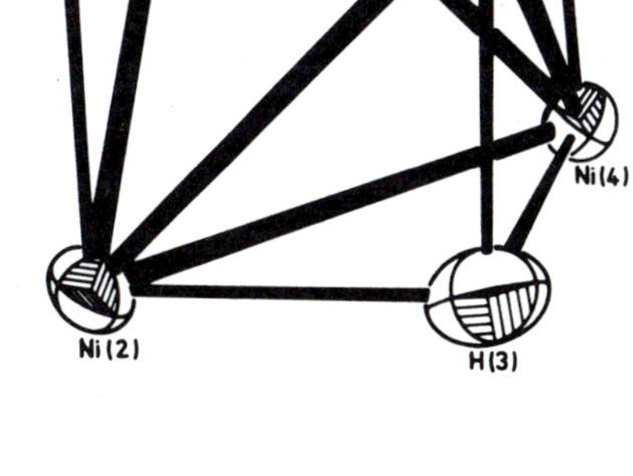

Fig. 4. Structure of the skeleton of the $H_3Ni_4(C_5H_5)_4$ cluster [20].

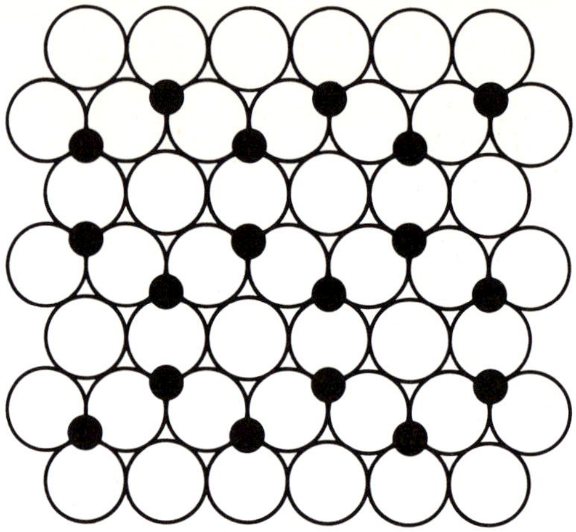

Fig. 5. Structure of the system H/Ni(111) at θ = 0.5 [21].

at θ = 0.5, as determined by LEED, is reproduced in Fig. 5 [21]. Again, the H atoms occupy threefold sites, and the Ni-H distance is 0.184 nm (although known with less accuracy). The close similarity is evident. In general, H atoms have been found to occupy terminal, two-fold and three-fold bridging as well as interstitial sites in cluster compounds [22], and the same presumably holds true for H adsorption on metal surfaces, although only relatively few data are available so far. A counterpart to "interstitial" H-atoms is offered by a subsurface hydrogen species as detected, for example, in the H/Pd(110) system [23].

Atomic adsorbates in general prefer sites with the highest coordination (which can be rationalized theoretically [24]) where the M-A distances are close to the sum of covalent radii [8]; similar conclusions have been reached for the structures of clusters [61].

Among molecular ligands, CO has received by far the most attention. The value of the stretch frequency of adsorbed CO has frequently been used to identify the nature of the adsorption site. For neutral clusters these are in the range of 2150-1950 cm^{-1} for terminal ligands, 1880-1740 cm^{-1} for two-fold (bridge) coordinated ligands, and 1840-1600 cm^{-1} for three-fold coordinated ligands [25] (also see Chapter 6); the same assignments have been applied to adsorbed species. In a few cases, independent structural determinations performed by LEED have confirmed these conclusions: Terminal bonds on Ni(100) and Cu(100) with M-C bond lengths of 0.175 [26] and 0.190 nm [27], respectively, are associated with C-O valence frequencies of 2070 [28] and 2100 cm^{-1} [29].

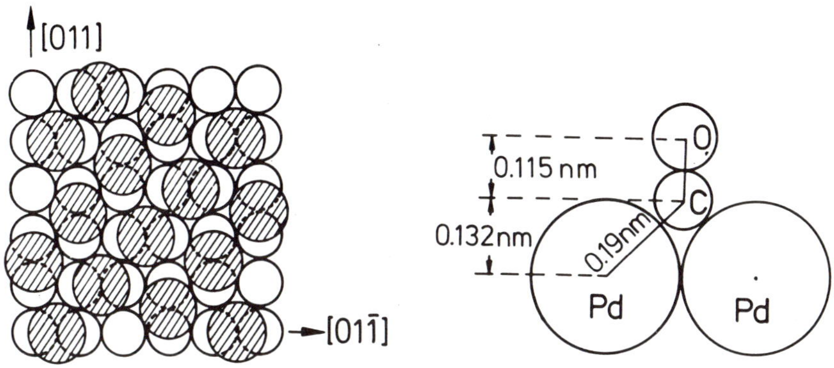

Fig. 6. Structure of CO adsorbed on a Pd(100) surface at θ_{CO} = 0.5 [30].

M-C distances in the range of 0.175-0.189 nm were determined for terminal bonds in Ni carbonyl clusters [6b], and distances of 0.178-0.181 nm were determined for corresponding Cu clusters [36].

The structure of CO adsorbed on Pd(100) at θ = 0.5 is reproduced in Fig. 6 [30]. The bridge-bonded CO molecules exhibit a stretch frequency around 1900 cm^{-1} [30, 31]. The M-C distance of 0.193 nm is slightly smaller (but still equivalent within the experimental accuracy) than the average value of 0.205 nm in the analogous cluster compound $Pd_{10}(CO)_{13}(PBu^n{}_3)_6$ [32].

The stoichiometry of carbonyl clusters is essentially governed by steric constraints. On the basis of a comparison of a large series of polynuclear clusters, Chini [33] concluded that a CO ligand requires a minimum space, as determined by its dimensions depicted in Fig. 7. A similar tendency is observed for CO adsorption on metal surfaces. Saturation of the adlayer occurs at densities of around $1x10^{15}$ CO molecules/cm^2, irrespective of the surface orientation [34], and is characterized by the formation of densely packed layers. As a consequence, at higher coverages equivalent adsorption sites can no longer be occupied, but the frequently observed "compressed" structures might also require the occupation of sites with low symmetry. Fig. 8 shows the configuration adopted by CO adsorbed on a Ni(111) surface at θ = 0.57; CO occupies twofold bridge and terminal sites [35].

In conclusion, there are also close similarities between clusters and surfaces with respect to the configuration of the ligands.

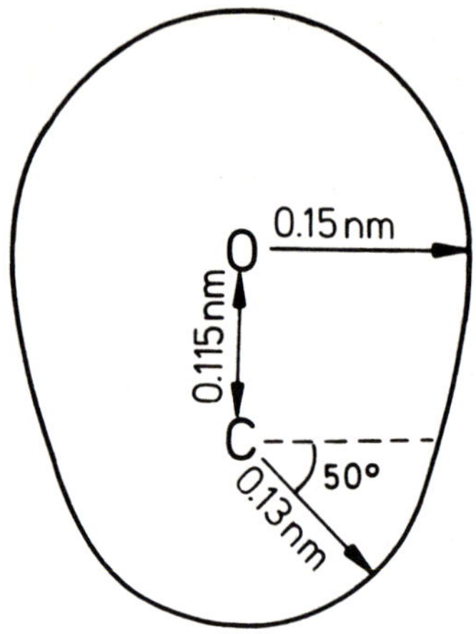

Fig. 7. The effective size of CO in clusters as derived by Chini [33], which holds equally for CO adsorption.

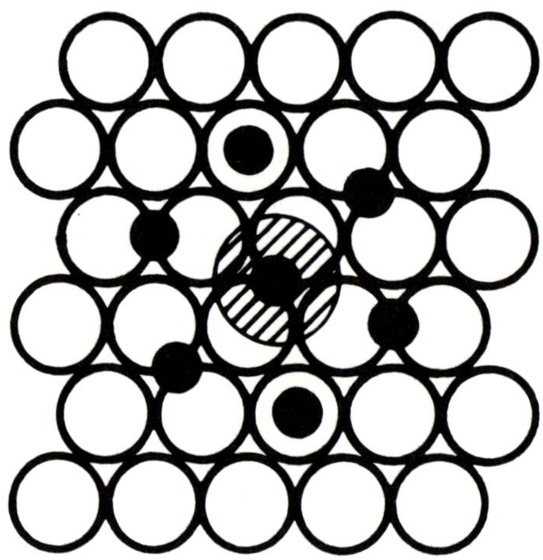

Fig. 8. Structure of CO adsorbed on a Ni(111) surface at saturation coverage [35].

11.3 ENERGETICS

The energy of the bond between two metal atoms is a quantity that is not directly accessible by experiment, either for clusters or for crystals. Approximate values for clusters can be obtained from thermochemical measurements if it is assumed that the average M-M bond energy is constant. This subject is discussed further in Chapter 3. A set of data obtained in this way [37] is listed in Table 2. For bulk metals, so-called "calculated bond energies" (CBE) (which are equal to the heats of atomization per bonding electron per atom) were derived by Brewer [38] and are supposed to reflect the strength of the metal-metal bond. Representative data of this type are also included in Table 2. The comparison shows that these numbers are in some cases very similar for clusters and bulk metals, but the agreement should not be considered to be generally valid quantitatively in view of the uncertainties. What appears to be more important, however, is the relatively small magnitude of the energies of these bonds in comparison with the energies of the metal-ligand bonds (see below). This relative weakness of the metal-metal bonds is considered to be the key to understanding the structural instability of the cluster frameworks when their ligand spheres are perturbed by reactions. With extended surfaces, in contrast, the underlying bulk acts as a template for the surface atoms, which tends to keep them in fixed positions. But even then reconstruction and/or relaxation effects may come into play, as was mentioned above.

TABLE 2. Metal-Metal Bond Energies in Clusters [37] and Bulk Metals [38].

Metal	Metal-metal bond energy, kJ/mol	
	Cluster	Bulk Metal
Mn	67	42
Co	84	142
Ru	117	121
Rh	113	113
Os	130	130
Ir	130	138

With respect to the metal-ligand bond energy, the situation is somewhat different; a limited set of data is available for cluster compounds corresponding to the mean enthalpy for dissociation. In the case of surfaces, much more accurate and extended values have been determined in terms of the adsorption energy. The latter quantity is derived from thermodynamic measurements of the adsorption-desorption equilibrium via the Clausius-Clapeyron equation, or alternatively (and more rapidly but with less accuracy), from the kinetics of desorption by application of thermal desorption spectroscopy (TDS). A compilation of such data can be found in Ref. [39].

Table 3 is a list of data for the heat of CO chemisorption on the most densely packed planes of a few transition metals together with the M-CO dissociation energies of corresponding carbonyl compounds. The values for the compounds appear to be larger by 30-50%, irrespective of whether mono- or multinuclear carbonyls are considered. This result appears to be qualitatively plausible, since a surface atom is surrounded by a larger number of metal atoms and has therefore less "free valency" available. As a consequence, the heat of chemisorption should increase as the surface crystallography is varied from densely packed to more open planes. Table 4 is a list of corresponding data for CO adsorption on different Pd single crystal surfaces [34]; these results confirm the trend for the three low-index planes [(111) < (100) < (110)], but no longer for the more open planes [(210) and (311)].

The variation of CO adsorption energy with surface orientation is well reproduced by a semi-empirical theory [40], the results of which are included in Table 4. As a general result, it should be kept in mind that variation of the surface structure (including defects such as steps, etc.) affects the strength of the metal-ligand bond by about $\pm 10\%$.

Another source of variation of the bond energy on a surface is its stoichiometry. With increasing coverage, interactions between the adsorbed species come into play, and these modify the effective adsorption energy and eventually also cause the occupation of different adsorption sites. For example, Fig 9. shows the variation of the adsorption energy E_{ad} for CO on Pd(111) as a function of coverage. Up to $\theta = 1/3$, E_{ad} remains constant, and the CO molecules occupy equivalent threefold coordinated sites. Beyond that coverage, "compression" of the adlayer starts and additional CO molecules are forced into less favorable sites. Finally, direct van der Waals repulsion causes a rapid decline of $E_{ad}(\theta)$ and saturation of the adlayer. The energetic and structural variations are also nicely reflected

TABLE 3. M-CO Bond Energies.

Metal or metal complex or cluster	M-CO bond energy, kJ/mol
W(110)	113[a]
$W(CO)_6$	180[b]
Ru(0001)	121[c]
$Ru_3(CO)_{12}$	171[d]
Ir(111)	142[e]
$Ir_4(CO)_{12}$	188[d]
Ni(111)	113[f]
$Ni(CO)_4$	146[g]

[a]T. E. Madey and J. T. Yates, J. Chem. Phys., 45 (1966) 1623.

[b]D. S. Barnes, G. Pilcher, D. A. Pettam, H. A. Skinner, and D. Todd, J. Less-Common Met., 38 (1974) 53.

[c]T. E. Madey and D. Menzel, Jap. J. Appl. Phys. Suppl. 2, Pt. (1974) 229.

[d]J. A. Connor, H. A. Skinner, and Y., Virmani, Faraday Symp., 8 (1973) 18.

[e]J. Küppers and A. Plagge, J. Vac. Sci. Techn., 12 (1976) 259.

[f]K. Christmann, G. Ertl, and O. Schober, J. Chem. Phys., 60 (1974) 4719.

[g]A. K. Fischer, F. A. Cotton, and G. Wilkinson, J. Am. Chem. Soc., 79 (1957) 2044.

TABLE 4. Experimental and Theoretical Adsorption Energies for CO at low coverages on various Pd single Crystal Planes.

Pd crystal plane	Adsorption energy for CO adsorption at low coverage, kJ/mol	
	Experimental [34]	Theoretical [40]
(111)	(142)	(142)
(100)	(153)	(150)
(110)	167	167
(210)	146	146
(311)	148	146

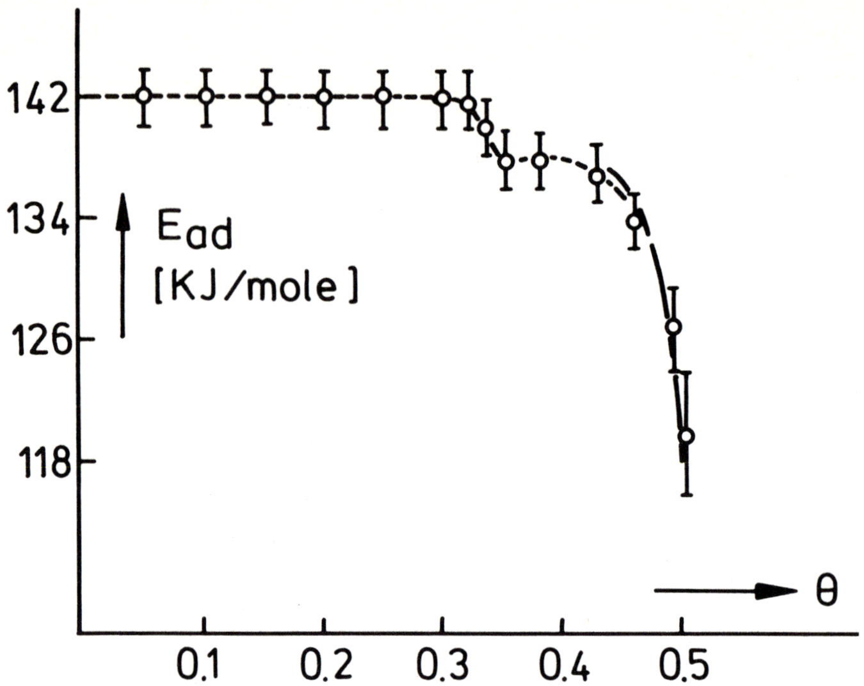

Fig. 9. Variation of the heat of CO adsorption, E_{ad}, with coverage on a Pd(111) surface [41].

by the vibrational spectroscopic data [31] reproduced in Fig. 10. Clearly, similar effects are to be expected for cluster compounds, but these are not accessible to direct experimental observation.

The energetics of dissociative adsorption are illustrated by the Lennard-Jones potential diagram of Fig. 11. The crossing point of the molecular and atomic potential curves determines the effective activation energy E^* for dissociative adsorption. For molecules such as H_2 or O_2, E^* is <0 on typical transition metal surfaces--which constitutes the basis of their high catalytic activities for hydrogenation or oxidation reactions. The M-A bond energy, E_{M-A}, is related to the adsorption energy, E_{ad}, and the dissociation energy of the free molecule, E_{AA}, by the equation $E_{M-A} = 1/2 (E_{ad} + E_{AA})$.

Table 5 is a comparison of a few data determined in this way with dissociation energies of the corresponding diatomic molecules M-H. To our knowledge, no data for cluster compounds are yet available. The close similarity between the two sets of data suggests, however, that similar values should be expected for clusters.

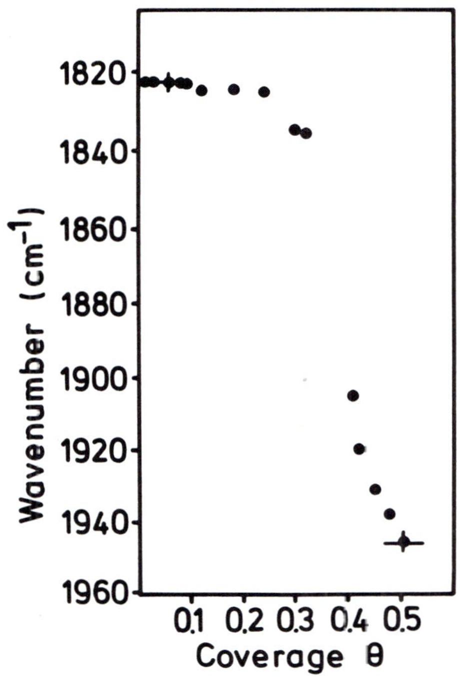

Fig. 10. Variation of the C-O stretch frequency with coverage for CO adsorbed on a Pd(111) surface [31].

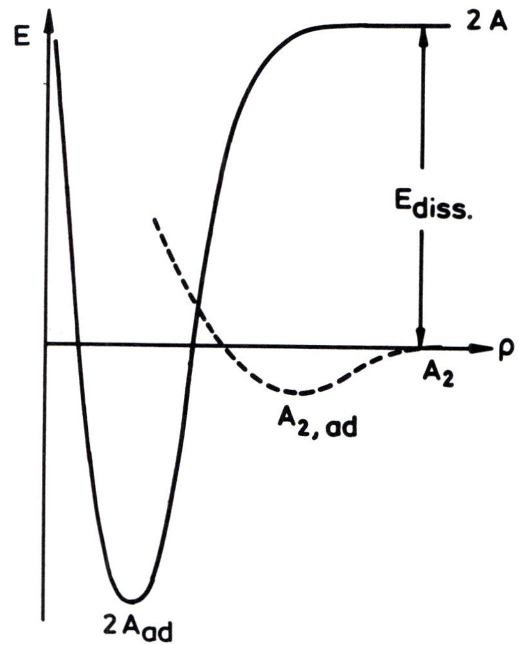

Fig. 11. Potential energy diagram illustrating dissociative chemisorption of a diatomic molecule.

TABLE 5. M-H Bond Energies on Surfaces and in Diatomic Molecules.

	M-H bond energy, kJ/mol	
Metal	Metal surface	MH molecule
Ni	263[a]	251[d]
Cu	234[b]	276[d]
Pt	255[c]	247[d]

[a]K. Christmann, O. Schober, G. Ertl and M. Neumann, J. Chem. Phys., 60 (1974) 4528.

[b]M. Balooch, M. J. Cardillo, D. R. Miller and R. E. Stickney, Surface Sci., 46 (1974) 358.

[c]B. Poelsema, G. Mechtersheimer and G. Comsa, Surface Sci., 111 (1981) 519.

[d]A. G. Goyden, Dissociation Energies and Spectra of Diatomic Molecules, Chapman and Hall, London (1968).

Another aspect of the energetics concerns the variation of the M-A bond energy with respect to the movement of the adsorbed species across the surface, which determines the activation energy for surface diffusion, E^*_{diff}. As a rule of thumb, E^*_{diff} is less than about 20% of the adsorption energy, which is confirmed by theoretical calculations [24, 42]. For example, a value of E^*_{diff} = 29 kJ/mol was determined experimentally for CO/Pt(111) [43], while E_{ad} = 135 kJ/mol. These values imply that an adsorbed species on the average will always undergo many migration steps on the surface (during which it may encounter a reaction partner) before desorbing.

A counterpart to the high mobility of surface species is found in the fluxionality of ligands in cluster compounds as indicated by NMR spectroscopic experiments [44]. These measurements demonstrate facile migration of ligands (e.g., CO) at fairly low temperatures with clusters capable of different types of ligand bonding (e.g., terminal, bridge, and threefold)--i.e., when the energy differences for various geometries are obviously fairly small. Since stable cluster compounds correspond to saturated adsorption layers, however, it should be kept in mind that these phenomena concern the concerted motion of the whole ligand sphere rather than the single species diffusion usually studied on surfaces having low coverages. Probably the best example for comparison with cluster fluxionality is offered by the system CO/Pt(111) as studied by Hayden and Bradshaw [45] by means of infrared spectroscopy. At θ = 0.5, the adsorbed CO molecules were found to be highly mobile between twofold- and threefold-bridge sites even at 100 K, the difference in adsorption energy between these two sites being only about 4 kJ/mol.

In summary, although there exists only a limited set of data for the metal-ligand bond energies in clusters, it appears that these are comparable to the adsorption energies on the more open crystal planes of the corresponding bulk metals. As a general trend, there is a decrease in these energies from left to right in the metals of the periodic table. This trend--at least for H and O--is also reproduced by theory [42]. If bonding of various ligands at the same metal is regarded, then the same trend is observed for both surfaces and clusters:

$$N > O > H > CO \approx NO > NH_3 \approx H_2O > O_2 > N_2 > Xe$$

This sequence follows immediately from general chemical experience. The only remarkable difference is probably offered by H_2. Whereas the adsorption energy of this molecule on metal surfaces has been found to be very small (<20 kJ/mol [46]), a recently discovered metal complex incorporating H_2 as a ligand is apparently so stable that the crystal structure of the complex could be determined [47].

11.4 ELECTRONIC PROPERTIES

A single metal atom exhibits discrete valence electronic levels which split and represent delocalized wave functions when several metal atoms form the skeleton of a cluster. Continuous energy bands of allowed electronic states are formed in macroscopic crystallites [60]. The ionization potentials and electron affinities vary continuously with increasing cluster size and finally converge to the work function, which in turn may vary by up to about 1 eV between different crystal faces of the same bulk metal. Despite these apparent differences between the various systems under discussion, the similarities of the properties outlined in the previous sections suggest that indeed no basic differences exist with respect to the valence electronic properties of the metal-ligand bond.

The latter may be probed most conveniently by means of ultraviolet photoelectron spectroscopy (UPS). It has to be kept in mind, however, that the "binding energies" of valence levels derived in this way represent ionization energies rather than ground-state orbital energies. Although this distinction is often neglected in studies with molecules (= Koopmans' theorem), it must be emphasized that the "extraatomic" relaxation energies for species at metal surfaces are caused mainly by long-range (image charge) effects and are therefore generally quite different (typically by about 2 eV) from the effects observed with molecules. It therefore appears to be more appropriate to consider *relative* binding energies (i.e., differences between various levels) rather than *absolute* binding energies when comparing clusters with adsorbate systems. Close analogies in this respect have been observed first for Ru carbonyls and CO adlayers on Ru [4]. A systematic investigation was later performed with PF_3 adsorbed on various metal surfaces, and the UPS data were compared with those for corresponding mononuclear metal complexes [48].

Bonding of PF_3 to transition metals takes place through a donor-acceptor mechanism similar to that for CO, as illustrated in Fig. 12. Coupling of the highest occupied PF_3 orbital ($8a_1$ = phosphorus lone pair) to the metal is expected to lower its energy and to reduce its energetic separation from the second highest level (6e), which is mainly F lone-pair in character and essentially not involved in bond formation. (Back-donation into the lowest empty level, 7e, is predicted to create another level which

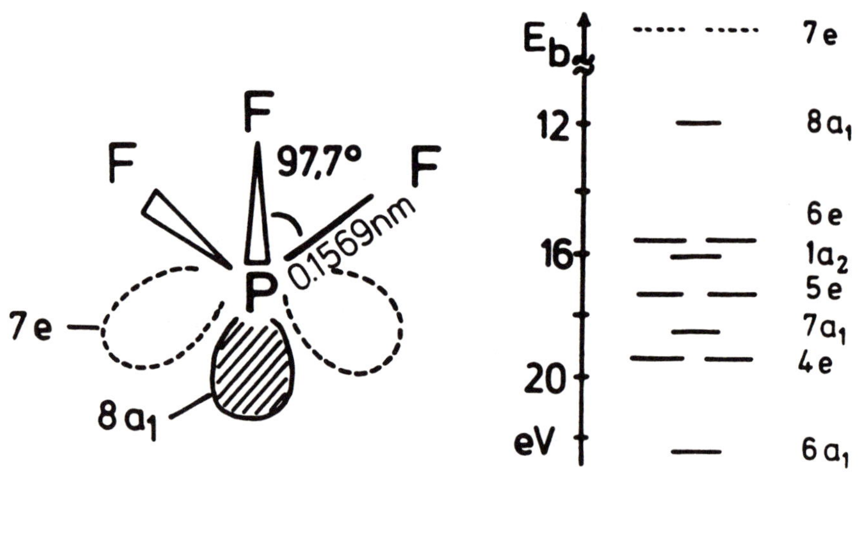

Fig. 12. Orbital scheme for PF_3.

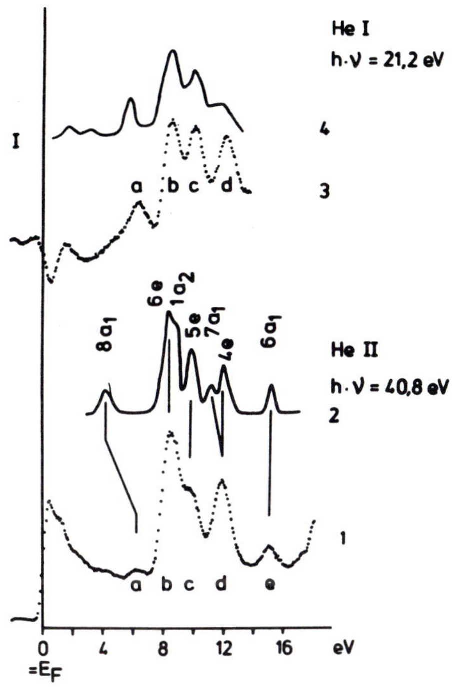

Fig. 13. UPS data of PF_3 systems [48]: 1: He II spectrum of PF_3 adsorbed on a Fe(110) surface; 2: He II spectrum of free PF_3; 3: He I difference spectrum of PF_3/Fe(110); 4: He I spectrum of $Fe(PF_3)_5$. The energy scales for PF_3 and $Fe(PF_3)_5$ were shifted in order to line up peak b with the 6e maximum.

falls into the range of the metallic d-states and will therefore be masked in the UP spectra). The energetic separation $\Delta\varepsilon = b - a$ in the spectra is regarded as reflecting the degree of σ-donor coupling. Fig. 13 is a comparison of spectra for free PF_3 with those for $Fe(PF_3)_5$ and for PF_3 adsorbed on Fe(110), which qualitatively confirms this model. (The absolute energy scales have been adjusted in such a way as to align the peaks arising from levels not involved in the bond for the reasons mentioned above.)

As can be seen from Table 6, the experimental values of $\Delta\varepsilon$ range between 3.85 eV for free PF_3 and 1.50 or 1.35 eV, respectively, for PF_3 adsorbed on Pt(111) or coupled to a single Pt atom. If the data for the

TABLE 6. Difference of Ionization Potentials ($\Delta\varepsilon$) between $8a_1$ and ($6e$ + $1a_2$)-levels of PF_3 either Adsorbed on a Metal Surface or Coupled to a Single Metal Atom [48].

Sample	($\Delta\varepsilon$), ev
Free PF_3	2.85
PF_3/Cr-poly	3.35
$Cr(PF_3)_6$	2.95
PF_3/Fe(110)	2.20
PF_3/Fe(111)	2.75
$Fe(PF_3)_5$	2.75
PF_3/Ni(111)	2.35
$Ni(PF_3)_4$	2.45
PF_3/Ru(0001)	2.35
$Ru(PF_3)_5$	2.50
PF_3/Pd(110)	2.25
$Pd(PF_3)_4$	2.10
PF_3/Ir(100)	1.90
$HIr(PF_3)_4$	1.85
PF_3/Pt(111)	1.50
$Pt(PF_3)_4$	1.25

metal complexes are compared with those for the corresponding adsorbate systems, the agreement is surprisingly good. Note that for Fe the agreement is better for the more open crystal plane! This result is quite remarkable, since the electronic properties of a single metal atom are certainly distinctly different from those of a bulk crystal--nevertheless, the valence electronic structure of a ligand bond appears to be rather similar.

11.5 VIBRATIONAL FREQUENCIES

As a direct consequence of the similar valence electron properties, vibrational frequencies of ligands attached to clusters and to surfaces are also quite analogous. The assignment of the CO adsorption sites on the basis of a comparison with the C-O stretch frequency of corresponding clusters has already been mentioned. This concept is now widely used to identify the chemical nature of surface species. A nice example is the identification of ethylidyne ($\equiv C-CH_3$) formed upon interaction of C_2H_4 with a Pt(111) surface [49], which was based on the comparison of the observed vibrational spectrum with that of the $(CH_3C)Co_3(CO)_9$ cluster, which has a known structure [50]. Table 7 is a list of the vibrational bands and their assignments for the two systems; the agreement is remarkable.

Particularly low C-O stretching frequencies were observed for CO adsorbed either on a stepped Ni surface [51] or on an "open" Fe(111) surface [52]; the surface structures find their analogues in cluster compounds where metal atoms interact directly with the π-electron system of the ligand, as shown in Fig. 14 [53]. Such a geometry is either offered by the step atoms or by adsorption in "deep hollow" sites of the Fe(111) surface.

An extremely low stretch frequency was also found for N_2 adsorbed on Fe(111) (~1500 cm^{-1} versus 2330 cm^{-1} for the free molecule), although the adsorption energy is only about 30 kJ/mol [54]. Similarly low frequencies were also observed with a few dinitrogen cluster compounds which are characterized by interaction of both N atoms with metal atoms of the type M-N=N-M or $M\langle{}^{N}_{\overset{\|}{N}}\rangle M$ [55]. That a similar configuration must indeed also be formed on Fe(111) was confirmed by a detailed analysis of XPS data [54].

11.6 REACTIVITY

The example just described also offers a nice insight into the reactivity of metal surfaces. The low-frequency $N_{2,ad}$ species on Fe(111) undergoes easy dissociation into 2 N_{ad} at temperatures as low as ~150 K [54]. Dissociative adsorption of a diatomic molecule is usually rationalized in terms of a one-dimensional Lennard-Jones potential diagram, as shown in Fig. 11. A more realistic description must be sought in a two-dimensional potential with the $M-A_2$ and the A-A distances as coordinates.

For the N_2/Fe(111) system, such a potential diagram was recently calculated by Tomanek [56] and is reproduced in Fig. 15. This picture shows how the π-bonded N_2 molecule might undergo easy dissociation without an appreciable activation barrier. For this process to occur, it is necessary that neighboring empty adsorption sites for the N atoms be

TABLE 7. Assignment and Frequencies (cm^{-1}) of the Vibrational Bands of Ethylidyne ($\equiv C-CH_3$) Adsorbed on Pt(111) [49] and in the Cluster $(CH_3C)Co_3(CO)_9$ [50].

Assignment	Frequency, cm^{-1}	
	Surface	Cluster
-CCM	310	220
-CM(s)	430	401
-CM(as)	600	555
-CH	980	1004
$-CH_3$(s)	1350	1356
$-CH_3$(AS)	1420	1420
-CC	1130	1163
$-CH_3$(s)	2890	2888
$-CH_3$(as)	2950	2930

Fig. 14. Skeletons of clusters exhibiting very low C-O stretching frequencies: (a) $Nb_3(\eta^5-C_5H_5)_3(CO)_6(\eta^2,\mu_3-CO)$; (b) $Mn_2(CO)_4(\eta^2,\mu_1-CO)(dppm)_2$.

available. Such "ensembles" of free metal atoms providing the possibility for dissociative adsorption will always be available on surfaces as long as the coverage is not too high. With a cluster, on the other hand, such a situation will usually not be encountered. When the ligands are removed, the metal skeleton of the cluster becomes unstable.

Notwithstanding all these similarities in the structural, energetic, and electronic properties, it is believed that this basic difference in reactivity between clusters and surfaces is essentially responsible for the fact that attempts to replace small catalyst particles by clusters (which remain intact during the reaction) have so far not been very successful.

There are numerous hints for the validity of the "ensemble" concept of dissociative adsorption just outlined. The sticking coefficient for

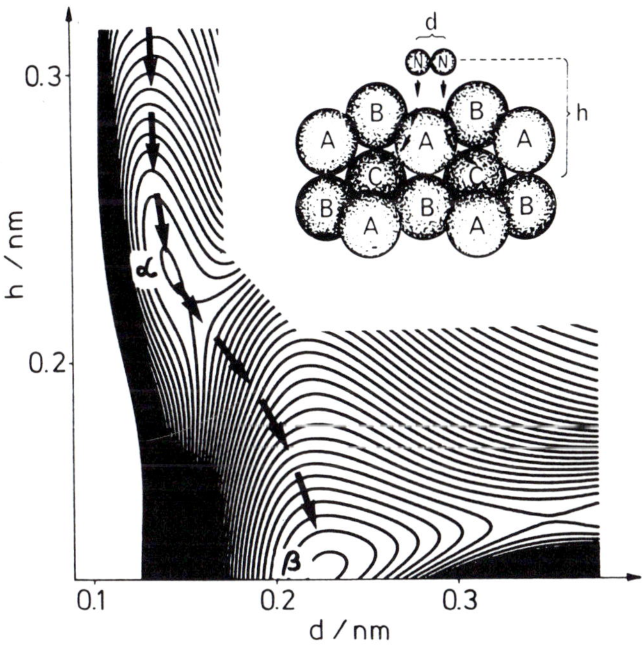

Fig. 15. Theoretical potential diagram showing the energy variation of a N_2 molecule adsorbed on a Fe(111) surface (α-state) and undergoing dissociation (β-state) [56].

dissociative adsorption usually drops rapidly with increasing coverage [57]; the activity and selectivity may be drastically affected by the use of bimetallic surfaces [58] or alloys [59]; and so on.

Consequently, it is concluded that clusters may frequently offer reaction pathways different from those of metal surfaces and/or transform into other species during the reaction. As an example of the latter possibility, a study of the hydrogenation of CO in the presence of $Ir_4(CO)_{12}$ [61] is mentioned, whereby irreversible transformation of the cluster into an unknown species characterized by a C-O stretch frequency of 1630 cm^{-1} was observed. Since this reaction was performed in a NaCl-$AlCl_3$ melt, this band was tentatively attributed to the formation of Ir-C-O-Al groups. On transition metal surfaces, on the other hand, the reaction simply proceeds via dissociation of adsorbed CO [62]. This latter process must involve a transition state in which both atoms of CO interact with the metal. A model for this state can be found in the anionic cluster $[HFe_4(\mu^2\text{-}CO)(CO)_{12}]^-$, where one of the carbonyl ligands is bonded to the Fe framework through both the C and O atoms [65]. Nevertheless, it appears to be highly improbable that this compound can act as a true catalyst.

Another example in which different reaction pathways are most probably followed is offered by the reaction $NO + 5/2H_2 \longrightarrow NH_3 + H_2O$. On noble metal surfaces this reaction proceeds through dissociative adsorption of NO and H_2 which is followed by subsequent hydrogenation of N_{ad} and O_{ad}; in contrast, the reaction appears to be much more complicated with a cluster compound: In a recent study [66] it was found that $[HRu_3(CO)_{10}(\mu_2\text{-}NO)]$ may be reduced by H_2 to give $[HRu_3(CO)_{10}NH_2]$, from which NH_3 is evolved; in this process, however, the starting compound is not restored, but instead $H_4Ru_4(CO)_{12}$ is formed. The formation of the intermediate was postulated to occur through the following steps:

$$HRu_3(CO)_{10}(\mu_2\text{-}NO) \xrightarrow{H_2} H_3Ru_3(CO)_{10}(\mu\text{-}NO) \longrightarrow HRu_3(CO)_{10}(NHOH)$$

$$\xrightarrow{H_2} H_3Ru_3(CO)_{10}(NHOH) \xrightarrow{H_2O} HRu_3(CO)_{10}NH_2 \qquad (1)$$

The fact that another cluster $[H_4Ru_4(CO)_{12}]$ is the end product of this complicated reaction sequence underlines the second conclusion stated above.

A remarkable example of a reaction in which the initial cluster is restored, i.e., acts as a true catalyst, is probably presented by the work of Basset *et al.* [67] on the water-gas shift reaction catalyzed by $Rh_6(CO)_{16}$

adsorbed on alumina. It turns out, however, that in the first step of this reaction the cluster framework is destroyed and Rh^{I} and Rh^{III} ions incorporated into the surface of the support are formed, and these work as the actual catalyst.

11.7 CONCLUSIONS

Structural, energetic, electronic, and vibrational properties of cluster compounds are remarkably similar to those of metal surfaces covered by adsorbed molecules. Nevertheless, one must be warned against considering a small cluster and a metal particle as electronically identical--they are certainly not. The main differences with respect to catalytic activity arise, in my opinion, from the following two points:

(1) Stability: the strength of the metal-metal bond is comparable in magnitude to that of the metal-ligand bond. Clusters therefore undergo structural reorganization quite easily. Even metal surfaces may undergo reconstruction or faceting during a catalytic reaction, but the bulk crystal structure has a strong stabilizing effect.

(2) Reactivity: reactive interaction of a molecule (e.g., dissociative adsorption) frequently requires several neighboring metal atoms which are not blocked by other ligands. In the case of clusters, this will usually be prevented for reasons of stability, whereas a catalytically active metal surface will provide sufficient empty adsorption sites under steady-state reaction conditions--even at high pressures.

Clusters offer excellent models for the surface chemistry on metallic catalysts, primarily because structural information is much more easily accessible for clusters and since possible reaction intermediates may be isolated and spectroscopically characterized.

Principles of cluster coordination chemistry will therefore be invaluable for understanding the mechanisms of metal-catalyzed reactions. On the other hand, there are also aspects which can be more conveniently studied with metal surfaces, such as metal-ligand bond energies, dipole moments, and effects of varying stoichiometry. Future developments will certainly cause further mutually beneficial overlap of these two fields of chemistry.

ACKNOWLEDGMENTS

This chapter is dedicated to the memory of the late E. L. Muetterties, who pioneered this area and stimulated my interest in it during my sabbatical leave in Berkeley in 1981/82. Thanks are also due to B. C. Gates and H. Knözinger, who drew my attention to recent literature on cluster chemistry. The access to work by E. Shustorovitch and D. Tomanek prior to publication is gratefully acknowledged. Financial support was provided by the Deutsche Forschungsgemeinschaft (SFB 128) and by the Fonds der Chemischen Industrie.

REFERENCES

1. a) D. A. Dowden, J. Chem. Soc. (1950) 242.
   b) G. M. Schwab, Angew. Chem., 67 (1955) 433.
2. R. P. Eischens and W. A. Pliskin, Advan. Catal., 10 (1958) 1.
3. See, e.g.,
   a) W. M. H. Sachtler, in E. Drauglis, R. D. Gretz and R. I. Jaffee (Eds.), Molecular Processes on Solid Surfaces, McGraw-Hill, New York, 1969, p. 313.
   b) G. C. Bond, Disc. Faraday Soc., 41 (1966) 200.
4. a) H. Conrad, G. Ertl, H. Knözinger, J. Küppers and E. E. Latta, Chem. Phys. Lett., 42 (1976) 115.
   b) E. W. Plummer, W. R. Salaneck and J. S. Miller, Phys. Rev. B18 (1978) 1673.
5. E. L. Muetterties, Bull. Soc. Chim. Belg., 84 (1975) 959.
6. a) E. L. Muetterties, Angew. Chem., 17 (1978) 545.
   b) E. L. Muetterties, T. N. Rhodin, E. Band, C. Brucker and H. Pretzer, Chem. Rev., 79 (1979) 91.
   c) G. Martino, in J. Bourdon (Ed.), Growth and Properties of Metal Clusters, Elsevier, Amsterdam (1980) p. 399.
   d) E. Borello, Int. J. Quant. Chem., 19 (1981) 1065.
   e) M. Moskovits, Acc. Chem. Res., 12 (1979) 229.
   f) J. M. Basset and A. Choplin, J. Mol. Catal., 21 (1983) 95.
   g) E. L. Muetterties, Pure Appl. Chem., 54 (1982) 83.
   h) J. Evans, Chem. Soc. Rev., 10 (1981) 159.
   i) E. L. Muetterties and R. M. Wexler, Surv. Progr. Chem., 10 (1983) 61.
7. See, e.g.,
   R. P. Messmer, in T. N. Rhodin and G. Ertl (Eds.), The Nature of the Surface Chemical Bond, North Holland, Amsterdam (1979) p. 51.
8. M. A. Van Hove, in T. N. Rhodin and G. Ertl (Eds.), The Nature of the Surface Chemical Bond, North Holland, Amsterdam (1979) p. 275.
9. R. L. Burwell and J. B. Peri, Ann. Rev. Phys. Chem., 15 (1964) 131.
10. Y. Kim and K. Seff, J. Am. Chem. Soc., 99 (1977) 7055.
11. J. G. Allpress and J. V. Sanders, Surface Sci., 7 (1967) 1.
12. J. J. Burton, Catal. Rev., 9 (1974) 209.
13. D. M. Washecheck, E. J. Wucherer, L. F. Dahl, A. Carioti, G. Longoni, M. Manassero, M. Sansoni and P. Chini, J. Am. Chem. Soc., 101 (1979) 6110.
14. P. Chini, J. Organomet. Chem., 200 (1980) 37.
15. L. F. Dahl, personal communication.
16. a) M. A. Van Hove, R. J. Koestner, P. C. Stair, J. P. Biberian, L. L. Kesmodel, I. Bartos and G. A. Somorjai, Surface Sci., 103 (1981) 189.
    b) W. Moritz, personal communication.
17. H. P. Bonzel and S. Ferrer, Surface Sci., 118 (1982) L263.
18. R. J. Behm, P. A. Thiel, P. R. Norton and G. Ertl, J. Chem. Phys., 78 (1983) 7437; 7448.
19. K. Christmann, V. Penka, R. J. Behm, F. Chehab and G. Ertl, Solid State Commun., 51 (1984) 487.
20. T. F. Koetzle, J. Muller, D. L. Tipton, D. W. Hart and R. Bau, J. Am. Chem. Soc., 101 (1979) 5631.

21. K. Christmann, R. J. Behm, G. Ertl, M. A. Van Hove and W. H. Weinberg, J. Chem. Phys., 70 (1979) 4168.
22. R. G. Teller and R. Bau, Struct. Bonding, 44 (1981) 1.
23. R. J. Behm, V. Penka, M. G. Cattania, K. Christmann and G. Ertl, J. Chem. Phys., 78 (1983) 7486.
24. E. Shustorovitch, R. C. Baetzold and E. L. Muetterties, J. Phys. Chem., 87 (1983) 1100.
25. E. L. Muetterties, Isr. J. Chem., 20 (1980) 84.
26. S. Andersson and J. B. Pendry, Surface Sci., 71 (1978) 75.
27. S. Andersson and J. B. Pendry, Phys. Rev. Lett., 43 (1979) 363.
28. S. Andersson, Solid State Commun., 21 (1977) 75.
29. a) J. Pritchard, T. Catterick and R. K. Gupta, Surface Sci., 53 (1975) 1.
b) S. Andersson, Surface Sci., 89 (1979) 477.
c) B. Sexton, Chem. Phys. Lett., 63 (1979) 451.
30. R. J. Behm, K. Christmann, G. Ertl and M. A. Van Hove, J. Chem. Phys., 73 (1980) 2984.
31. A. M. Bradshaw and F. M. Hoffmann, Surface Sci., 72 (1978) 513.
32. E. G. Mednikov, N. K. Eremenko, V. A. Mikhailov, S. P. Gubin, Y. L. Slovokhotov and Y. T. Struchkov, J. Chem. Soc. Chem. Commun., (1981) 989.
33. P. Chini, Gazz. Chim. Ital., 109 (1979) 225.
34. H. Conrad, G. Ertl, J. Koch, J. Küppers and E. E. Latta, Surface Sci., 57 (1976) 475.
35. H. Conrad, G. Ertl, J. Küppers and E. E. Latta, Surface Sci., 57 (1976) 475.
36. M. Pasquali, F. Marshetti and C. Floriani, Inorg. Chem., 17 (1978) 1684.
37. J. A. Connor, Topics Curr. Chem., 71 (1977) 71.
38. L. Brewer, Science, 161 (1968) 115.
39. G. Ertl, in T. N. Rhodin and G. Ertl (Eds.), The Nature of the Surface Chemical Bond, North Holland, Amsterdam, 1979, p. 313.
40. G. Doyen and G. Ertl, Surface Sci. 43 (1974) 197.
41. G. Ertl and J. Koch, Z. Naturforsch., 25a (1970) 1906.
42. J. K. Nørskov, personal communication.
43. B. Poelsema, S. T. de Zwart and G. Comsa, Phys. Rev. Lett., 49 (1982) 578.
44. E. Band and E. L. Muetterties, Chem. Rev., 78 (1978) 639.
45. B. Hayden and A. M. Bradshaw, Surface Sci., 125 (1983) 787.
46. J. P. Cowin, C. F. Yu, S. J. Sibener and J. E. Hurst, J. Chem. Phys., 75 (1981) 1033.
47. G. J. Kubas, R. R. Ryan, B. I. Swanson, P. J. Vergamini and H. J. Wasserman, J. Am. Chem. Soc., 106 (1984) 451.
48. F. Nitschke, G. Ertl and J. Küppers, J. Chem. Phys., 74 (1981) 5911.
49. H. Steininger, H. Ibach and S. Lehwald, Surface Sci., 117 (1982) 685.
50. P. Skinner, M. W. Howard, I. A. Oxton, S. F. A. Kettle, D. B. Powell and N. Sheppard, J. Chem. Soc. Faraday II, 77 (1981) 397.
51. W. Erley, H. Ibach, S. Lehwald and H. Wagner, Surface Sci., 83 (1979) 585.
52. U. Seip, M. C. Tsai, K. Christmann, J. Küppers and G. Ertl, Surface Sci., 139 (1984) 29.
53. a) C. J. Commons and B. F. Hoskins, Australian J. Chem., 28 (1975) 1663.
b) W. A. Herrmann, H. Biersack, M. L. Ziegler, K. Weidenhammer, R. Siegel and D. Rehder, J. Am. Chem. Soc., 103 (1981) 1692.
54. M. Grunze, M. Golze, W. Hirschwald, H. J. Freund, H. Pulm, U. Seip, M. C. Tsai, G. Ertl and J. Küppers, Phys. Rev. Lett., 53 (1984) 850.
55. a) C. Krüger and Y. H. Tsay, Angew. Chem. Int. Ed. Engl., 12 (1973) 998.

b) G. D. Pez, P. Apgar and R. K. Crissey, J. Am. Chem. Soc., 104 (1982) 482.
56. D. Tomanek and K. H. Bennemann, to be published.
57. C. R. Brundle, J. Behm and J. A. Barker, to be published.
58. J. H. Sinfelt, Bimetallic Catalysts, Wiley, New York, 1983.
59. a) W. M. H. Sachtler, LeVide, 164 (1973) 67.
b) R. A. Van Santen, Surface Sci., 53 (1975) 35.
60. For a review of the convergence of cluster calculations to bulk electronic properties see, e.g., J. Demuynck, M. M. Rohmer, A. Strick and A. Veillard, J. Chem. Phys., 75 (1981) 3443.
61. H. K. Wang, H. W. Choi and E. L. Muetterties, Inorg. Chem., 20 (1981) 2661.
62. a) D. W. Goodman, R. D. Kelley, T. E. Madey and J. T. Yates, J. Catal., 63 (1980) 226.
b) A. T. Bell, Catal. Rev.-Sci. Eng., 23 (1981) 203.
c) V. Ponec, Catal. Rev.-Sci. Eng., 18 (1978) 4; 64 (1980) 479.
63. H. Conrad, G. Ertl and E. E. Latta, J. Catal., 35 (1974) 363.
64. T. Engel and G. Ertl, Advan. Catal., 28 (1978) 2.
65. M. Manassero, M. Sansoni and G. Longoni, J. Chem. Soc. Chem. Commun. (1976) 919.
66. B. F. G. Johnson, J. Lewis and J. M. Mace, J. Chem. Soc. Chem. Commun. (1984) 186.
67. J. M. Basset, B. Besson, A. Choplin and A. Theolier, Phil. Trans. Roy. Soc., A308 (1982) 115.

PART IV

SUMMARY AND EVALUATION

Metal clusters constitute an extensive new class of compounds, offering new modes of bonding of ligands to metal and subtle new patterns of reactivity. Most of the literature of metal cluster chemistry still has to do with synthesis and structure. The predictive basis of thermochemistry is still weakly developed, and the understanding of reactivity has only just begun to develop rapidly.

A number of assessments of the state of metal cluster chemistry have emphasized the potential value of the clusters as catalysts. Catalysis has been a major driving force for the advancement of organometallic chemistry in general, and the major technological successes represented by processes such as alkene hydroformylation and methanol carbonylation loom as monuments justifying research in this field.

The forecasts of exciting catalytic routes related to the new bonding and reactivity patterns offered by metal clusters have not been borne out. The clusters are fragile. A rapid and selective cycle of organometallic catalysis requires a complex and delicate dynamic balance; the complication of instability of most cluster frameworks imposes a major limitation on the catalytic opportunities. The typical attempt at catalysis by metal clusters leads to the formation of mononuclear metal complexes and/or colloidal metal. Only a few examples of catalysis can be confidently attributed to metal clusters. One potentially important technological reaction (CO hydrogenation to give ethylene glycol) may involve metal clusters, but the details of the chemistry are lacking.

The prospects for short-term industrial exploitation of cluster catalysis are not good. The long-term prospects are more attractive; in this uncharted territory, new chemistry and catalysis will surely be found, but the progress is likely to be slow.

Metal clusters provide compelling models of surfaces, and the fields of organometallic chemistry and surface science are destined to intersect broadly. The insights provided by clusters have already proven to be valuable in the advancement of surface science. The metal cluster-metal surface analogy has impressive validity in terms of structure and bonding--but not reactivity or catalytic activity.

Metal clusters are even better models of small metal particles with adsorbates than they are of simple metal surfaces with adsorbates. The clusters are truly intermediate structures between the simplest metal complexes and metal particles, and with some metals (e.g., Os), they almost seem to offer a continuum in the large group of carbonyl clusters, with almost every nuclearity between 3 and 10 or more being represented by well-characterized compounds. This family also offers a wide range of hydrocarbon ligands, including some that bridge metal centers, and it provides tantalizing models of surface organometallic structures yet to be discovered.

Metal clusters anchored to surfaces may already be the best-characterized supported metal catalysts in terms of structure, for the structures are often based on definitive crystal and molecular structures of very close analogues. A few of these materials (e.g., $HOs_3(CO)_{10}$-O-M≤) yield dazzlingly simple and understandable spectra and electron micrographs; we can be confident of the structures of the bound clusters and optimistic that characterization of these simple surface structures will help with the development of the physical characterization techniques for surfaces generally and the logical extension of these techniques to the more complicated supported metal aggregates and crystallites of industrial metal catalysts. The physical characterization techniques of greatest promise today appear to be the workhorse infrared spectroscopy, the still-difficult but rapidly developing EXAFS and solid-state NMR, and high-resolution transmission electron microscopy.

The chemistry of metal clusters on surfaces, especially those on metal oxide surfaces, has arisen and developed strongly in the preceding five years. Although ill-defined mixtures are the typical result of the reactions of organometallic compounds with surfaces, there are now numerous surface organometallic analogues of molecular species, and the reactivities of some of these are beginning to be established. There is an opportunity for rapid advancement and a need for further development of methodology for quantitative determination of reactivity (e.g., techniques such as temperature-programmed decomposition), surface spectroscopy, and synthesis and characterization of exact molecular analogues of surface structures incorporating ligands that mimic the surface.

The application of supported molecular metal clusters as catalysts is fraught with all the difficulties of homogeneous catalysis by metal clusters, and more; most of the difficulties are related to the lack of stability of the clusters, complicated by the presence of the support. The support is a ligand or group of ligands that may contribute to the lack of stability and limit the access of reactive ligands to bond to the clusters and participate in a catalytic cycle.

We see glimmers of hope in the several examples of catalysis by supported clusters, apparently including CO hydrogenation at about 550 K, but we recognize the severe restrictions on preparation of stable cluster catalysts. Effort might fruitfully be directed at some of the metals giving the most stable clusters (Ru, Os, Ir) and at novel supports other than metal oxides, which might offer stabilizing ligands. The reactant ligands are important as well, and the catalytic reactions of stabilizing ligands such as CO are therefore appealing.

The great and immediate promise and attraction of supported metal clusters is the scientific opportunities they offer for investigation of *molecular* catalysis on surfaces. We stand at a frontier in the design of catalysts with specific combinations of metals, ligands, and supports (themselves ligands)--we have visions of determining the effects of supports as we can determine the effects of ligands in homogeneous catalysis. Molecular catalysts on supports can be characterized in the absence of complicating solvents; with the aid of powerful spectroscopic methods such as EXAFS, we will some day isolate intermediates and understand the molecular details of catalytic cycles. The advancement of understanding of surface catalytic processes can receive a boost comparable to that provided by the surface science of single crystals--and the restriction of the high-vacuum apparatus does not pertain.

We foresee development of molecular cluster chemistry and catalysis on surfaces and in cages as a frontier of organometallic chemistry and surface science; the clusters and catalysis are the ties pulling these disciplines together, and the rewards will be a deeper fundamental understanding of catalysis and the technology that will ultimately spring from this understanding.

B. C. Gates
L. Guczi
H. Knözinger

CLUSTER INDEX

The cluster index is organized according to the metal in alphabetical order. For each metal the clusters are listed with respect to increasing nuclearity. Bimetallic clusters are listed under the heading of one metal component only.

CHROMIUM

COBALT

COBALT, bimetallic clusters

GOLD

IRIDIUM

IRON

IRON, bimetallic clusters

OSMIUM, bimetallic clusters

PLATINUM, bimetallic clusters

RHENIUM

RHENIUM, bimetallic clusters

RHODIUM

SUBJECT INDEX